EVOLUTIONARY GENETICS

EVOLUTIONARY GENETICS
Concepts and Case Studies

Edited by
Charles W. Fox
Jason B. Wolf

OXFORD
UNIVERSITY PRESS
2006

GOSHEN COLLEGE LIBRARY
GOSHEN, INDIANA

OXFORD
UNIVERSITY PRESS

Oxford University Press, Inc., publishes works that further
Oxford University's objective of excellence
in research, scholarship, and education.

Oxford New York
Auckland Cape Town Dar es Salaam Hong Kong Karachi
Kuala Lumpur Madrid Melbourne Mexico City Nairobi
New Delhi Shanghai Taipei Toronto

With offices in
Argentina Austria Brazil Chile Czech Republic France Greece
Guatemala Hungary Italy Japan Poland Portugal Singapore
South Korea Switzerland Thailand Turkey Ukraine Vietnam

Copyright © 2006 by Oxford University Press, Inc.

Published by Oxford University Press, Inc.
198 Madison Avenue, New York, New York 10016

www.oup.com

Oxford is a registered trademark of Oxford University Press

All rights reserved. No part of this publication may be reproduced,
stored in a retrieval system, or transmitted, in any form or by any means,
electronic, mechanical, photocopying, recording, or otherwise,
without the prior permission of Oxford University Press.

Library of Congress Cataloging-in-Publication Data
Evolutionary genetics: concepts and case studies/edited by Charles W. Fox, Jason B. Wolf.
p. ; cm.
Includes bibliographical references.
ISBN-13 978-0-19-516817-4; 978-0-19-516818-1 (pbk.)

1. Evolutionary genetics.
[DNLM: 1. Genetics, Population. 2. Evolution. 3. Genotype. 4. Models, Genetic. 5. Variation (Genetics)
QH 455 E928 2005] I. Fox, Charles W. II. Wolf, Jason B.
QH390.E94 2005
572.8'38—dc22 2005011132

9 8 7 6 5 4 3

Printed in the United States of America
on acid-free paper

Preface

Evolutionary genetics is a broad field that has seen particularly rapid growth and expansion in recent years. This diverse field is unified by a set of mirror-image goals: (1) to understand the impact that evolutionary processes have on the patterns of genetic variation within and among populations or species and (2) to understand the consequences of these patterns of genetic variation for various evolutionary processes. Research in evolutionary genetics stretches across a continuum of scale, from studies of DNA sequence evolution (e.g., Chapters 7 and 9) to studies of multivariate phenotypic evolution (e.g., Chapter 20), and across a continuum of time, from ancient events that lead to current species diversity (e.g., Chapter 28) to rapid evolution seen over relatively short time scales in experimental evolution studies (Chapter 31).

A major cause of the recent growth and expansion of evolutionary genetics has been the modern revolution in molecular biology, which has fueled the growth of areas of evolutionary genetics focused on the analysis of sequence data, the genotype–phenotype relationship, and genome evolution. Although many of the questions at the forefront of the field have been around since the early days of evolutionary genetics (e.g., since the Modern Synthesis), the availability of relatively inexpensive high-throughput genetic technology and the resulting large databases of molecular genetic data has led to the emergence of many new areas of study and a sort of revolution in evolutionary genetics. The signature of this revolution is clearly seen in this volume, in which the majority of chapters discuss patterns or processes that occur at the molecular level or have been influenced by the availability of molecular data.

Although we may define evolutionary genetics as a single integrated field, there is a continuum in the degree to which research is evolutionary versus genetic. At one extreme, evolutionary genetics informs molecular geneticists, whose primary interest may be finding and characterizing genes affecting traits, of the consequences that population subdivision and linkage disequilibrium have on their interpretation of associations between loci and trait expression (e.g., Templeton et al. 2005). At the other extreme, evolutionary biologists may use the results of these "gene discovery" studies to identify genes that underlie evolutionary important genetic variation (e.g., Beldade et al. 2002). However, differentiating research into the extremes of these categories is becoming increasingly difficult as evolutionary approaches permeate genetics just as molecular biology permeates evolutionary biology.

The development of this book was initiated late in 2002. It was conceived as a companion to *Evolutionary Ecology: Concepts and Case Studies* (edited by Fox et al. 2001), also published by Oxford University Press. Our primary objective in this book, as in its companion volume, is to provide a collection of readings that will introduce students to concepts and contemporary research programs in evolutionary genetics. Our hope when conceiving this volume was that it might be adopted as a text for graduate courses and seminars, as has been the case for *Evolutionary Ecology.* We thus targeted the level of this book so that it can be used by advanced undergraduates, graduate students, and established researchers in genetics or evolution looking for a concise introduction to evolutionary genetics. Authors were asked to target this audience while writing, and reviewers and the editors focused on making the volume accessible to this audience while reviewing each chapter.

Chapter authors are all leading researchers in their fields and were chosen to provide their particular perspectives on a topic. Chapters thus represent the current stage of evolutionary genetics better than any single-authored textbook could, and the diversity of authors introduces readers to the diversity of ideas, approaches, and opinions that are the nature of science. However, a multi-authored textbook presents special challenges. Authors vary in the level at which they present material and in the amount of background that they expect readers to have. Authors also vary in their writing styles, the way that they organize their chapters and, of course, each has a unique perspective on the overall field. We have attempted to minimize this variation through author guidelines and by aggressively editing and revising chapters. However, some variation among chapters is unavoidable and reflects the variation in styles and approaches common throughout science.

As with any book, especially an edited volume, this book is not comprehensive. To keep the length of the book practical, and the price affordable, we had to impose restrictions on chapter length and the number of references. This allowed us to increase the diversity of subjects covered but at the expense of depth of coverage. Most topics could fill an entire book (and many are indeed the subject of entire books). Chapters are intended to serve as introductions to their topic, focusing on basic concepts rather than becoming comprehensive reviews (the reference limit was intended to minimize the latter). Such a format imposed unavoidable limitations on authors and, as editors, we take responsibility for the necessary omission of missing topics and the lack of many additional references that are perhaps equally appropriate as examples or case studies. Chapters include a "Suggestions for Further Reading" section to guide readers on where to go next for additional coverage of the topic. We hope that readers will be inspired to delve more fully into at least some of the research areas and thus discover the vast literature that we have been unable to include here.

The volume is structured into six parts. Although this might suggest that there are six clearly defined sets of topics, such structuring is somewhat artificial. Evolutionary genetics is a highly integrated field with no clear lines dividing research topics. The structure of the book is simply a convenient way of collecting more related topics together. We start with a collection of chapters presenting many of the principles of evolutionary genetics that serve as the foundation for the rest of the subject (Part I). For this part readers need have only a decent background in genetics, though a background in evolutionary biology will certainly be helpful. Later parts of the book assume an understanding of both general concepts of genetics and the concepts presented in earlier parts. Parts II–IV are ordered hierarchically starting at the basic level of biological complexity, the DNA sequence (Part II), building through development (Part III) to studies of complex phenotypes (quantitative genetics; Part IV) and on to the interactions between individuals and their environment (sexual and social selection; also Part IV). These parts are followed by one on the genetics of species differences and speciation (Part V) that integrates across the hierarchy of complexity to investigate what is often considered the most fundamental problem in evolutionary biology: the origin of species. Lastly we include a part illustrating how the theoretical, conceptual, and empirical approaches developed in previous chapters are applied to specific problems in biology (Part VI). The potential choice of topics here is enormous, but we could choose only a couple of representative examples that we find particularly exciting.

Because we enforced length restrictions on chapters, many important and exciting topics were necessarily left out. Other topics were outside the expertise of the authors or were important topics that did not fit well into the structure of the chapters. We thus include a large number of boxes focusing on specific topics presented largely independently of the main body of the text with which they are associated. With the exception of Box 24.1 (which we use to introduce Part V, Genetics of Speciation), all boxes appear within the pages of the chapters to which they are most relevant. Many were written by the same author as the chapter that they complement; these largely expand on topics mentioned in the main body of the chapter or they present a topic that did not fit well in the main body of the chapter. Other boxes were written by scientists who did not write full chapters; these boxes read more like mini-chapters. Most could indeed have been full chapters but, alas, the realities of publishing prevented us from including every chapter we would want. We also included three boxes on model organisms in biology (in Part VI) since so much of what we know about evolutionary genetics, and biology in general, comes from studies of model organisms. The choice of box topics reflects the views of the editors, the reviewers, and the many chapter authors who suggested topics for boxes.

Lastly, we have compiled a glossary of terms. Initially we asked authors to include footnotes or tables defining the terminology of their field but the large number of submissions made this impractical, so we converted these (at the suggestion of multiple authors) to a glossary at the end of the text. It is by no means a comprehensive glossary of genetics or even evolutionary genetics terms. It is intended to aid the reader by providing definitions for terms that might be considered jargon special to some areas of research, or terms that you know you once learned but may have since forgotten; that is, the terminology not necessarily standard in a working scientist's vocabulary. The glossary entries are largely written by the chapter authors, heavily supplemented (and edited) by the editors; we have thus given the appropriate author credit after each entry. In a few cases we have included multiple entries for a single term because multiple entries were submitted by authors and the difference between those entries was itself informative.

Each chapter and box was reviewed by at least one other contributor to the book and, in most cases, one or more external reviewers. We are truly indebted to all these reviewers for generously donating their time and providing thorough and constructive reviews. Without their help it would not have been possible to produce such a volume given the vast diversity of topics covered and the limits of the editors' expertise. We thus thank the external reviewers, including Hiroshi Akashi, Cerise Allen, Bill Atchley, Scott Carroll, James Crow, Mary Ellen Czesak, Tony Frankino, Oscar Gaggiotti, C. William Kirkpatrick, Larry Leamy, Susan Lindquist, Curt Lively, Manyuan Long, Bryant McAllister, Tami Mendelson, Debra Murray, Joshua Mutic, John Obrycki, Susan Perkins, Massimo Pigliucci, Richard Preziosi, Will Provine, David Queller, Glenn-Peter Saetre, Laura Salter, Douglas Schemske, Hamish Spencer, Marc Tatar, Eric (Rick) Taylor, Lindi Wahl, Gunter Wagner, John Wakeley, Bruce Walsh, Joe Williams, and a few others who asked to remain anonymous. We also thank Lisa Hitchcock, Denise Johnson, and Oriaku Njoku for help proofreading chapters and references.

Finally, and most importantly, we thank the authors for their willingness to invest the substantial amount of time needed to write excellent chapters and boxes. The success of the volume ultimately depends on the quality of the contributions by authors. We are fortunate to have recruited an outstanding group of scientists who dedicated tremendous time and effort to making this project a success. Thank you for being such a wonderful group of people with which to work!

Charles W. Fox
Jason B. Wolf

Contents

Contributors xiii

Part I – Principles of Evolutionary Genetics

1. From Mendel to Molecules: A Brief History of Evolutionary Genetics 3
   Michael R. Dietrich
2. Genetic Variation 14
   Marta L. Wayne and Michael M. Miyamoto
   Box 2.1. Maternal Effects 19
   Timothy A. Mousseau
3. Mutation 32
   David Houle and Alexey Kondrashov
4. Natural Selection 49
   Michael J. Wade
   Box 4.1. Defining and Measuring Fitness 52
   Daphne J. Fairbairn
5. Stochastic Processes in Evolution 65
   John H. Gillespie
   Box 5.1. The Probability of Extinction of an Allele 68
   Box 5.2. Mutational Landscape Model 70
6. Genetics and Evolution in Structured Populations 80
   Charles J. Goodnight
   Box 6.1. Epistasis and the Conversion of Genetic Variance 87
   Jason B. Wolf

Part II – Molecular Evolution

7. Detecting Selection at the Molecular Level 103
Michael W. Nachman

8. Rates of Molecular Evolution 119
Francisco Rodríguez-Trelles, Rosa Tarrío and Francisco J. Ayala
Box 8.1. Timing Evolutionary Events with a Molecular Clock 122
Box 8.2. Testing the Hypothesis of the Molecular Clock 125

9. Weak Selection on Noncoding Gene Features 133
Ying Chen and Wolfgang Stephan

10. Evolution of Eukaryotic Genome Structure 144
Dmitri A. Petrov and Jonathan F. Wendel

11. New Genes, New Functions: Gene Family Evolution and Phylogenetics 157
Joe Thornton

12. Gene Genealogies 173
Noah A. Rosenberg
Box 12.1. Horizontal Inheritance 176

Part III – From Genotype to Phenotype

13. Gene Function and Molecular Evolution 193
Simon C. Lovell
Box 13.1. The Role of Gene Interaction Networks in Evolution 200
Stephen R. Proulx

14. Evolution of Multidomain Proteins 211
László Patthy

15. Evolutionary Developmental Biology 222
David L. Stern
Box 15.1. Hox Genes 224
Box 15.2. Functional Assays in Nonmodel Organisms 229

16. Canalization 235
Mark L. Siegal and Aviv Bergman
Box 16.1. Computational Modeling of the Evolution of Gene Regulatory Networks 243

17. Evolutionary Epigenetics 252
Eva Jablonka and Marion J. Lamb

Part IV – Quantitative Genetics and Selection

18. Evolutionary Quantitative Genetics 267
Derek A. Roff
Box 18.1. Individual Fitness Surfaces and Multivariate Selection 269
Jason B. Wolf

19. Genetic Architecture of Quantitative Variation 288
James M. Cheverud
Box 19.1. Genotypic Values: Additivity, Dominance, and Epistasis 289
Box 19.2. Genic Values and Genetic Variances 290
Box 19.3. How to Perform a QTL Analysis 291
Box 19.4. Evolutionary Morphometrics 294
Christian Peter Klingenberg
Box 19.5. Modularity 304
Jason G. Mezey

20. Evolution of Genetic Variance–Covariance Structure 310
Patrick C. Phillips and Katrina L. McGuigan
Box 20.1. What Is a Covariance? 311
Box 20.2. Pleiotropic Effects 313
Box 20.3. Evolution of the G Matrix 316

21. Genotype–Environment Interactions and Evolution 326
Samuel M. Scheiner

22. Genetics of Sexual Selection 339
Allen J. Moore and Patricia J. Moore

23. Social Selection 350
Steven A. Frank
Box 23.1. Coefficients of Relatedness 352

Part V – Genetics of Speciation

Box. Species Concepts 367
James Mallet

24. The Evolution of Reproductive Isolating Barriers 374
Norman A. Johnson

25. Genetics of Reproductive Isolation and Species Differences in Model Organisms 387
Pawel Michalak and Mohamed A. F. Noor
Box 25.1. The Dobzhansky–Muller Model 392

26. Natural Hybridization 399
Michael L. Arnold and John M. Burke
Box 26.1. Potential Outcomes of Natural Hybridization 400

27. Population Bottlenecks and Founder Effects 414
Lisa Marie Meffert
Box 27.1. Models of the Shifts in Selection Pressures Experienced by Bottlenecked Populations 415

28. Theory of Phylogenetic Estimation 426
Ashley N. Egan and Keith A. Crandall
Box 28.1. Philosophical and Methodological Differences in Phylogenetics 434

Part VI – Evolutionary Genetics in Action

29. Evolutionary Genetics of Host–Parasite Interactions 447
Paula X. Kover
Box 29.1. The Coevolutionary Consequences of Tolerance versus Resistance 448
Box 29.2. *Arabidopsis* as a Model Organism in Evolutionary Genetics 453
Kentaro K. Shimizu and Michael D. Purugganan
Box 29.3. Evolution of Virulence 456

30. The Evolutionary Genetics of Senescence 464
Daniel E. L. Promislow and Anne M. Bronikowski
Box 30.1. Demography of an Age-Structured Population 466
Box 30.2. *Drosophila* as a Model Organism in Evolutionary Biology 471
Jeffrey R. Powell

31. Experimental Evolution 482
Adam K. Chippindale
Box 31.1. *E. coli* as a Model Organism in Evolutionary Genetics 485
Richard E. Lenski

32. Evolutionary Conservation Genetics 502
Richard Frankham

Glossary 513

References 527

Index 575

Contributors

Michael L. Arnold
Department of Genetics
University of Georgia
Athens, Georgia 30602, USA

Francisco J. Ayala
Department of Ecology and Evolution
University of California
Irvine, California 92697, USA

Aviv Bergman
Department of Pathology and Molecular Genetics
Albert Einstein College of Medicine
New York, New York 10461, USA

Anne M. Bronikowski
Department of Ecology
Evolution and Organismal Biology
Iowa State University
Ames, Iowa 50011, USA

John M. Burke
Department of Biological Sciences
Vanderbilt University
Nashville, Tennessee 37235, USA

Ying Chen
Department of Ecology and Evolution
University of Chicago
Chicago, Illinois 60637, USA

James M. Cheverud
Department of Anatomy and Neurobiology
Washington University School of Medicine
St. Louis, Missouri 63110, USA

Adam K. Chippindale
Department of Biology
Queen's University
Kingston, Ontario K7L 3N6, Canada

Keith A. Crandall
Department of Microbiology and Molecular Biology
Brigham Young University
Provo, Utah 84602, USA

Michael R. Dietrich
Department of Biological Sciences
Dartmouth College
Hanover, New Hampshire 03755, USA

Ashley N. Egan
Department of Microbiology and Molecular Biology
Brigham Young University
Provo, Utah 84602, USA

Daphne J. Fairbairn
Department of Biology
University of California
Riverside, California 92521, USA

Steven A. Frank
Department of Ecology and Evolutionary Biology
University of California
Irvine, California 92697, USA

Richard Frankham
Department of Biological Sciences
Macquarie University
NSW 2109, Australia

John H. Gillespie
9849 Martingham Circle
St. Michaels, Maryland 21663, USA

Charles J. Goodnight
Department of Biology
University of Vermont
Burlington, Vermont 054065, USA

David Houle
Department of Biological Science
Florida State University
Tallahassee, Florida 32306, USA

Eva Jablonka
The Cohn Institute for the History and Philosophy of Science and Ideas
Tel Aviv University
Tel Aviv 69789, Israel

Norman A. Johnson
Department of Entomology
Program in Organismic Biology and Evolution
University of Massachusetts
Amherst, Massachusetts 01003, USA

Christian Peter Klingenberg
Faculty of Life Sciences
University of Manchester
Manchester M13 9PT, United Kingdom

Alexey Kondrashov
National Center for Biotechnology Information
National Institutes of Health
Bethesda, Maryland 20894, USA

Paula X. Kover
Faculty of Life Sciences
University of Manchester
Manchester M13 9PT, United Kingdom

Marion J. Lamb
Senior Lecturer (retired)
Birkbeck College
University of London, United Kingdom

Richard E. Lenski
Department of Microbiology and Molecular Genetics
Michigan State University
East Lansing, Michigan 48824, USA

Simon C. Lovell
Faculty of Life Sciences
University of Manchester
Manchester M13 9PT, United Kingdom

James Mallet
Department of Biology
University College London
London NW1 2HE, United Kingdom

Katrina L. McGuigan
Center for Ecology and Evolutionary Biology
University of Oregon
Eugene, Oregon 97405, USA

Lisa M. Meffert
Department of Ecology and Evolutionary Biology
Rice University
Houston, Texas 77251, USA

Jason G. Mezey
Department of Biological Statistics and Computational Biology
Cornell University
Ithaca, New York 14853, USA

Pawel Michalak
Department of Biology
University of Texas
Arlington, Texas 76019-0498, USA

Michael M. Miyamoto
Department of Zoology
University of Florida
Gainesville, Florida 32611, USA

Allen J. Moore
Centre for Ecology and Conservation
University of Exeter in Cornwall
Tremough, Penryn TR10 9EZ,
United Kingdom

Patricia J. Moore
Centre for Ecology and Conservation
University of Exeter in Cornwall
Tremough, Penryn TR10 9EZ,
United Kingdom

Timothy A. Mousseau
Department of Biological Sciences
University of South Carolina
Columbia, South Carolina 29208, USA

Michael W. Nachman
Department of Ecology and Evolutionary Biology
University of Arizona
Tucson, Arizona 85721, USA

Mohamed A. F. Noor
DCMB Group/Biology
Duke University
Durham, North Carolina 27708, USA

László Patthy
Institute of Enzymology
Biological Research Center
Hungarian Academy of Sciences
Budapest H-1518, Hungary

Dmitri A. Petrov
Department of Biological Sciences
Stanford University
Stanford, California 94305, USA

Patrick C. Phillips
Center for Ecology and Evolutionary Biology
University of Oregon
Eugene, Oregon 97405, USA

Jeffrey R. Powell
Department of Ecology and Evolutionary Biology
Yale University
New Haven, Connecticut 06520, USA

Daniel E. L. Promislow
Department of Genetics
The University of Georgia
Athens, Georgia 30602, USA

Stephen Proulx
Department of Ecology, Evolution and Organismal Biology
University of Iowa
Ames, Iowa 50011, USA.

Michael D. Purugganan
Department of Genetics
North Carolina State University
Raleigh, North Carolina 27695, USA

Francisco Rodríguez-Trelles
Fundacion Publica de Medicina Genomica
Hospital Clínico Universitario
Universidad de Santiago de Compostela
15706 Santiago, Spain

Derek A. Roff
Department of Biology
University of California
Riverside, California 92521, USA

Noah A. Rosenberg
Department of Human Genetics and Bioinformatics Program
University of Michigan
Ann Arbor, Michigan 48109-2218, USA

Samuel M. Scheiner
Division of Environmental Biology
National Science Foundation
Arlington, Virginia 22230, USA

Kentaro K. Shimizu
Department of Genetics
Box 7614
North Carolina State University
Raleigh, North Carolina 27695, USA

Mark L. Siegal
Department of Biology
New York University
New York, New York 10003, USA

Wolfgang Stephan
Department of Biology II
University of Munich
Grosshaderner Strasse 2
82152 Planegg-Martinsried, Germany

David L. Stern
Department of Ecology and Evolutionary Biology
Princeton University
Princeton, New Jersey 08544, USA

Rosa Tarrío
Fundacion Publica de Medicina Genomica
Hospital Clínico Universitario
Universidad de Santiago de Compostela
15706 Santiago, Spain

Joseph W. Thornton
Center for Ecology and Evolutionary Biology
University of Oregon
Eugene, Oregon 97403, USA

Michael J. Wade
Department of Biology
Indiana University
Bloomington, Indiana 47405, USA

Marta L. Wayne
Department of Zoology
University of Florida
Gainesville, Florida 32611, USA

Jonathan F. Wendel
Department of Botany
Iowa State University
Ames, Iowa 5001, USA

Jason B. Wolf
Faculty of Life Sciences
University of Manchester
Manchester, M13 9PT, United Kingdom

I

PRINCIPLES OF EVOLUTIONARY GENETICS

1

From Mendel to Molecules: A Brief History of Evolutionary Genetics

MICHAEL R. DIETRICH

Biologists have been grappling with selection ever since Darwin. Historians also face a problem of selection—not natural selection, but the selection of which events to include in their narratives. No historical narrative can be complete in the sense of including every event, actor, and idea. Historians must choose which events they will include and which they will not. Writing a survey of the history of evolutionary genetics in such a short space makes this problem of selection especially acute.

A number of different approaches have been taken to the history of evolutionary genetics. Will Provine has suggested that the history of evolutionary biology is one of persistent controversy (Provine 1989; see also Lewontin 1974). Certainly one could write a history of evolutionary genetics in terms of the disputes between, for instance, the Mendelians and Biometricians, Sewall Wright and R. A. Fisher, saltationists and gradualists, the classical and balance approaches, and neutralists and selectionists (Provine 1986, 1990; Beatty 1987b; Dietrich 1994, 1995, 1998; Smocovitis 1996; Skipper 2002). Such an antagonistic view of evolutionary genetics complements histories emphasizing the great collaborations that have also characterized the history of the subject, such as those between Theodosius Dobzhansky and Sewall Wright, E. B. Ford and R. A. Fisher, or indeed those within any of the many laboratory groups working in the twentieth century (Provine 1986). More institutionally minded historians have emphasized the rise of societies, journals, and funding sources (Smocovitis 1996; Cain 1993). At the same time, others have documented the development of theoretical and experimental tools and techniques, such as the use of chromosomal inversions, electrophoresis, sequence data, population cages, computer simulations, and the vast array of evolutionary models and concepts (Lewontin 1981, 1991; Kohler 1991; Powell 1994; Gayon & Veuille 2001).

In this brief history, I will focus on the major controversies that have marked the history of evolutionary genetics in the twentieth century with special emphasis on the nature of genetic variability and the evolutionary processes acting upon this variability. This approach captures key developments in evolutionary genetics such as the resolution of the conflict between Mendelism and Darwinism and the continuing impact of molecular biology and molecular techniques.

MENDELIANS, DARWINIANS, AND THE ORIGINS OF EVOLUTIONARY GENETICS

The study of evolution and heredity have been intertwined since at least Gregor Mendel's and Charles Darwin's separate efforts to make sense of the origins of varieties and the stability of species. Mendel's experiments with many different species sought to explore the idea that new stable varieties could be created through hybridization (Olby 1979). His famous series of experiments with the garden pea quantified the instability of his hybrid crosses as it documented their hereditary patterns. Darwin's much less quantitative approach to hereditary stability or continuity across generations put much greater emphasis on processes of evolutionary change and the problem of the origin of heritable variation. The differences between Mendel and Darwin were exaggerated after the rediscovery of Mendel's work in 1900 by Carl Correns, Hugo De Vries, and Erich von Tschermak. At this time, Darwinian evolution was criticized as insufficient for the production of new species (Bowler 1983). Evolution was widely acknowledged, but the processes of evolution remained in dispute. Hugo De Vries, for instance, articulated his Mutation Theory as a saltationist alternative to Darwinism during this period. Even Darwin's early defenders expressed concern about Darwin's account of the power of natural selection (Provine 1971).

Darwin acknowledged two forms of variation: continuous or blending variations and "sports" or monstrosities. Although he admitted that his knowledge of variation was insufficient, Darwin thought that continuous variations were the source of heritable variation for natural selection. "Sports" were larger, structural deviations, which Darwin thought were too rare and too harmful to be of evolutionary significance. Fleming Jenkin's criticisms of his views in the *Origin of Species* caused Darwin to take the idea of "sports" or discontinuous variation more seriously. Although "Darwin's bulldog," T. H. Huxley, advocated discontinuous variation, advocacy of this view is often associated with the early Mendelians, Hugo De Vries and William Bateson (Provine 1971; Kim 1994).

Darwin developed his own theory of blending inheritance as a physiological theory called "pangenesis." Like other material theories of heredity that would follow Darwin's in the late nineteenth century, Darwin postulated hereditary particles, pangenes, which corresponded to different body parts and were collected and transmitted via the gametes. While Darwin's cousin, Francis Galton, helped to refute this theory, he supported blending inheritance by developing statistical tools for precisely describing the similarities between characters. Using correlation and regression, Galton reconsidered heredity from a statistical point of view. Because he understood characters to be continuous, Galton believed that their distribution was best described by a normal distribution. The effects of selection were reconsidered in terms of effects on population means and variances. Selection could shift the mean of a population over a number of generations to create a new characteristic population mean. The relationship between parent and offspring was presented in terms of a law of ancestral heredity where a particular character of an offspring can be determined from the diminishing contribution of its ancestors (Provine 1971; Kim 1994). Galton's *Natural Inheritance* (1889) inspired Karl Pearson and W. F. R. Weldon to develop a statistical approach to biology and evolution that they called biometrics. Within the biometrical tradition, Weldon and others applied statistical methods to support gradual Darwinian evolution by natural selection. Weldon himself collected statistical evidence from crab carapaces, which he thought demonstrated the effect of selection in reducing population variability as well as the size of the carapace front. These and other efforts convinced the Biometricians that statistical methods were essential for understanding evolution and heredity.

William Bateson had also been impressed with Galton's work, but was not convinced that statistical methods were the best tools or that either evolution or heredity should be understood as continuous or blending. In 1894, Bateson argued in his book, *Materials for the Study of Variation with Special Regard to Discontinuity in the Origin of Species*, that discontinuous variations were common and saltational evolution of new species was probably the norm. The dispute between Bateson and the Biometricians began with Weldon's hostile review of his book. It was transformed into the Mendelian–Biometrician controversy when Bateson read Mendel's paper in 1900. Bateson translated Mendel's paper into English and immediately began championing it as the key to heredity and evolution. As a result, Weldon and Pearson would debate the significance of Mendel's paper vociferously over the next 10 years.

The dispute between the Mendelians and Biometricians was at once about genetic variation (continuous vs. discontinuous) and evolutionary change (gradual vs. saltational) as well as the appropriateness of statistical methods, and was overlaid with a struggle for authority and position within English biology. During the course of this dispute, the Biometricians and Mendelians drew on extended networks of biologists, and historian Kyung-Man Kim argues that the controversy was resolved by members of this extended network, not by the principal antagonists who remained strongly polarized (Kim 1994). A. D. Darbishire, for instance, set out to refute Mendelism with a set of experiments on albino and waltzing mice. Following Galton, Darbishire reasoned that as the proportion of albino mice forming the parental and grandparental generations increased so should the percentage of albino offspring (Darbishire 1904). Darbishire's evidence in 1904 seemed to support exactly this interpretation until both William Castle and William Bateson wrote devastating critiques reinterpreting Darbishire's results in Mendelian terms (Castle 1905; Kim 1994). Darbishire himself was convinced when he tested his hybrids and realized that some of the mice that produced only albino offspring did so because they were dominant. In this case, statistical analysis of external appearance was not a reliable guide to genetic constitution. Darbishire's defection infuriated Pearson, but this was one of several conversions (Kim 1994).

More biologists joined the Mendelians after Wilhelm Johannsen introduced his pure line approach. Beginning in 1901, Johannsen sought to test whether selection could change the mean of a population's character distribution. Using a continuous distribution of bean size and weight, Johannsen selected for large, medium, and small beans. He discovered that after many generations of selection he could isolate a number of pure lines from the original distribution. Pure lines had stable characters and selection no longer had an effect on their individual means. Selection had made a difference in the original population because it was selecting among different pure lines, not because it was selecting within a pure line. Johannsen's distinction between the distribution of a character (phenotype) and the underlying pure line (genotype) was essential for resolving the Mendelian–Biometrician controversy. As early as 1904, English mathematician G. Udny Yule recognized this as a way to reconcile the biometrical description of phenotypes with Mendelian descriptions of genotypes. This route to reconciliation was reinforced with evidence for multiple factors, which allowed Mendelians to explain a continuous character distribution as the result of the interaction of many genes, each of small effect. By 1910, these developments had begun to significantly depolarize this controversy as many biologist recognized the compatibility of the Mendelian and biometrical approaches (Kim 1994).[1]

[1]Historical interpretations of this controversy have themselves been the subject of controversy concerning the relative roles of evidence and social factors in the course of the dispute. See Kim (1994).

THE DEVELOPMENT OF POPULATION GENETICS

Regardless of the outcome of the Mendelian–Biometrician controversy, the use of statistical methods formalized a population approach to evolution in the early twentieth century. At a time when even the basic language of genetics had yet to be standardized, it is not surprising that different approaches to the mathematical description of evolution would also arise. The rise of mathematical population genetics is usually associated with the work of three founders: Sewall Wright, R. A. Fisher, and J. B. S. Haldane. Their work set the foundations for population genetics, as each attempted to formally reconcile Mendelism and Darwinism (Provine 1971).

Wright was an American biologist trained in genetics at Harvard University by William Castle. His early interest in mammalian genetics led him to create the method of path analysis as a staff scientist at the US Department of Agriculture. By 1921, he had developed his method of path coefficients to describe the effects of inbreeding, assortative mating, and selection. When he joined the faculty of the University of Chicago in 1925, Wright shifted his thoughts from guinea pig colonies and cattle herds to evolving natural populations. By 1931, he had articulated his shifting balance theory of evolution in his now classic paper "Evolution in Mendelian Populations" (Wright 1931; Provine 1986).

R. A. Fisher was an English biologist trained at Cambridge in mathematics. Introduced to Mendelism and Biometry at Cambridge, Fisher sought to reconcile the two by understanding the biometrical properties of Mendelian populations. This approach led him to characterize similarities within Mendelian populations in terms of their variance and the contributions to variance from genetic sources, environmental sources, dominance, and gene interactions. Fisher's approach emphasized natural selection acting in very large natural populations. He set out his general theory in his 1930 book, *The Genetical Theory of Natural Selection* (Provine 1971, 1986).

J. B. S. Haldane was also an English biologist with broad interests. He studied mathematics at Oxford before switching to classics and philosophy. Beginning in 1922, Haldane sought to analyze the mathematical consequences of natural selection. Starting from simple Mendelian models using two alleles at a single locus, Haldane went on to consider selection with self-fertilization, inbreeding, overlapping generations, incomplete dominance, isolation, migration, and fluctuating selection intensities (Provine 1971). Haldane's series of nine papers on selection culminated in his 1932 book, *The Causes of Evolution*. In the appendix to this book, Haldane compares his views to those of Fisher and Wright. While he agrees with elements of both of their views, Haldane differed from Fisher by placing greater emphasis on strong selection of single genes, migration, and epistasis. He sided with Fisher, however, in thinking that Wright put too much emphasis on random genetic drift (Provine 1971; Gillespie, Ch. 5 of this volume).

While Fisher, Wright, and Haldane approached evolution and population genetics from different mathematical perspectives, their disagreements were not about mathematics, but about evolutionary processes and concepts and their representation in different mathematical models. According to Will Provine, Fisher and Wright were engaged in a series of disputes from 1929 until 1962 when Fisher died (Provine 1986, 1992). While they debated many things, the core of their difference lay in their general theories of evolution: Wright's shifting balance theory and Fisher's large population theory. Wright's approach incorporated an array of evolutionary processes and emphasized population subdivision (Goodnight, Ch. 6 of this volume). Fisher argued that natural selection was the dominant process and that large populations were the optimum. These differences were most apparent around the issue of the relative importance of random genetic drift. Although Wright continued to elaborate his views, his early work on the shifting balance theory gave random drift a considerable role in evolution. To counter Wright's view, Fisher and his colleague E. B. Ford studied yearly fluctuations in the gene (allele) frequencies of the moth *Panaxia dominula* from 1941 to 1946. They found that the fluctuations they observed were too great to be accounted for by the action of random genetic drift. Instead, they proposed that the fluctuations were the result of random fluctuations in the strength of natural selection. As this dispute intensified and extended in the 1950s to results on banding patterns in the snail *Cepaea nemoralis*, Wright began to modify his views, limiting the action of random drift to large, but subdivided populations where it could serve as a means for generating novel genotypic combinations (Provine 1986, 1992). The Wright–Fisher debate has resurfaced in recent years with new protagonists (Skipper 2002), but the original debate was especially influential because it occurred just as Neo-Darwinism was being articulated in the evolutionary synthesis (Provine 1992).

THE EVOLUTIONARY SYNTHESIS

The evolutionary synthesis is identified by historians with both the emerging discipline of evolutionary biology and the integration of previously divergent fields such as paleontology, zoology, botany, systematics, and genetics. According to this interpretation, the synthesis refers to a time beginning in the 1930s when a range of arguments were offered to show that different fields relevant to evolution were in fact compatible with each other. These compatibility arguments helped spur on the emergence of evolutionary biology as a field of inquiry—as a new and centrally important discipline (Smocovitis 1996). Compatibility arguments do not necessarily imply that there was widespread agreement on a new synthetic theory of evolution. As Provine and others have argued, there was little agreement about the mechanisms of evolution during the 1930s and 1940s. Instead Provine suggests that we reconsider this period as an evolutionary constriction—"a vast cut-down of the variables considered important to the evolutionary process." According to Provine, "The term 'evolutionary constriction' helps us understand that evolutionists after 1930 might disagree intensely with each other about effective population size, population structure, random genetic drift, levels of heterozygosity, mutation rates, migration rates, etc., but all could agree that these variables were or could be important in evolution in nature, and that purposive forces played no role at all" (Provine 1988).

The foundation for the evolutionary synthesis was communicated in a number of now classic texts: R. A. Fisher's *The Genetical Theory of Natural Selection* (1930), Theodosius Dobzhansky's *Genetics and the Origins of Species* (1937), Julian Huxley's *Evolution: The Modern Synthesis* (1942), Ernst Mayr's *Systematics and the Origin of Species* (1942), G. G. Simpson's *Tempo and Mode in Evolution* (1944), and G. L. Stebbins' *Variation and Evolution in Plants* (1950).

Dobzhansky's work represented the state of the art in animal genetics and population genetics.

Trained in the Soviet Union and influenced by the work of Nicolai Vavilov and Iurii Filipchenko, Dobzhansky began his career studying variability in natural populations of *Coccinellidae* and *Drosophila melanogaster*. To further his understanding of genetics, he received funding from the Rockefeller Foundation to join T. H. Morgan's famous Fly Group in 1927 (Provine 1981). At Columbia and later Cal Tech, Dobzhansky excelled at the business of *Drosophila* genetics. First with A. H. Sturtevant and later in collaboration with Sewall Wright, Dobzhansky turned to evolutionary genetics—taking *Drosophila* genetics from the laboratory to the field. Dobzhansky's 1937 book, *Genetics and the Origin of Species,* articulated a program of research for evolutionary genetics. The theoretical underpinnings of Dobzhansky's program were deliberately borrowed from Wright's shifting balance theory. Unlike Wright's papers, however, Dobzhansky's presentation was non-mathematical and served to widely popularize the shifting balance theory (Provine 1981). *Genetics and the Origin of Species,* thus, translated one of the dominant general theories of evolution into a research program for evolutionary genetics.

Dobzhansky's evolutionary program was challenged in 1940 by Richard Goldschmidt's *The Material Basis of Evolution*. Goldschmidt had been Director of the Kaiser Wilhelm Institute of Biology in Berlin before he was forced to emigrate in 1936. Once in the United States, Goldschmidt challenged the gradualist model of evolution promoted by Dobzhanksy and others. According to Goldschmidt, Dobzhanksy had not demonstrated that his view fit the evidence any better than the view that there were bridgeless gaps between species which could only be crossed by either systemic mutations (large rearrangements of chromosomal structure) or mutations in developmentally important genes. Goldschmidt's developmentally oriented, saltationist alternative immediately inspired a hostile reaction by the Neo-Darwinians. Subsequent editions of Dobzhansky's *Genetics and the Origin of Species* devoted many pages to Goldschmidt's refutation, as did later work by Mayr and Simpson. This negative response to Goldschmidt's views bolsters Provine's interpretation of the synthesis as a constriction. In fact, opposition to Goldschmidt's saltationism became a defining feature of Neo-Darwinism (Dietrich 1995).

Ernst Mayr's *Systematics and the Origin of Species* (1942) responded to Goldschmidt's claims, but was modeled on Dobzhanky's *Genetics and the Origin of Species*. Where Dobzhansky synthesized genetics with evolutionary biology, Mayr added concepts of speciation and species. Trained as an ornithologist in Germany under Hans Stresseman, Mayr was the world's expert on bird systematics. Although developed with avian exemplars, Mayr argued for the generality of his Biological Species Concept (Mallet, Species Concepts box, pp. 367–373 of this volume) and model of geographic speciation. If Dobzhansky was the first to set the intellectual agenda for evolutionary genetics, Mayr broadened that agenda. Moreover, Mayr was absolutely central to the effort to institutionalize and support the development of evolutionary biology as a discipline. Together with G. G. Simpson, who articulated the contributions of paleontology for the synthesis, Mayr, Dobzhansky, and other scientists in the Northeastern United States discussed the similarities and differences in their approaches to evolution in the Committee on Common Problems in Genetics and Paleontology, which met from 1943 to 1945 when the Society for the Study of Evolution was founded (Smocovitis 1996; Cain 1993). Because of World War II, Mayr, Simpson, and Dobzhansky were somewhat isolated from biologists in England (Huxley and Fisher) and evolutionary biologists on the West Coast of the United States (Stebbins). This temporary isolation may be one reason why Dobzhansky, Simpson, and Mayr were so influential in the development of Neo-Darwinism, and why Neo-Darwinism seemed particularly focused on animal systems. The considerable effort of Stebbins and others to bring plants into the synthesis is surely also a result of the interesting differences between plant and animal genetics (Smocovitis 1996).

The architects of the evolutionary synthesis played a central role in the promotion of evolutionary biology and especially evolutionary genetics. Dobzhansky's work on the genetics of natural populations, in particular, was hailed as an exemplar of Neo-Darwinism (Mayr 1944; Stern 1944). Significantly, during the 1940s Dobzhansky's own research program narrowed. From 1938 to 1976, Dobzhansky and his collaborators produced a series of 43 influential papers under the title of "The Genetics of Natural Populations" (GNP) (Lewontin 1981). Early work in the GNP series was often conducted in collaboration with Sewall Wright and sought to explore different aspects of the shifting balance theory using data from characteristic chromosomal inversion of different natural populations.

Because Dobzhansky thought that selection had little effect on inversion frequency, his work with Wright concentrated on breeding structures and the impact of random drift. As early as 1943, however, Dobzhansky's attention begins to shift toward selection favoring heterozygotes. By 1950, the GNP series and Dobzhansky's research program began increasingly to address problems of heterosis and balancing selection (Beatty 1987a). This transition from drift to selection is emblematic of the emerging view in the 1950s that natural selection is the predominant process of evolution. Dubbed the "hardening of the synthesis" by Stephen Jay Gould, the constriction characteristic of the synthesis period had produced a type of pan-selectionism that would dominate evolutionary biology into the 1970s (Gould 1983). Focusing on selection to the exclusion of other processes did not guarantee that consensus. Instead, new controversies emerged concerning the form of selection and the availability of genetic variation.

GENETIC VARIABILITY AND THE CLASSICAL–BALANCE CONTROVERSY

In the 1950 and 1960s, Dobzhansky's research on balanced polymorphisms fueled a major controversy in evolutionary genetics concerning the genetic variability of natural populations, the nature of selection, and the genetic effects of atomic radiation. In 1955 at the meeting of the Cold Spring Harbor Symposium on Quantitative Biology, Dobzhansky articulated two diametrically opposed positions on these issues: the classical position and the balance position. The classical position, according to Dobzhansky, held that "evolutionary changes consist in the main in gradual substitution and eventual fixation of the more favorable, in place of the less favorable, gene alleles and chromosome structures." Most loci, according to the classical position, should be homozygous. Heterozygotes were rare and had four possible sources: (1) deleterious mutations that are eventually eliminated by selection, (2) adaptively neutral mutations, (3) "adaptive polymorphisms maintained by the diversity of the environments which the population inhabits," and (4) rare beneficial mutants which are on their way toward fixation (Dobzhansky 1955). According to Dobzhansky, the main proponent of the classical position was H. J. Muller. The balance position, according to Dobzhansky, held that most loci should be heterozygous. Homozygotes would still occur, but they would not be as advantageous as overdominant heterozygous combinations. In terms of genetic variation, the issue at stake between the classical and balance positions was the relative number and importance of heterozygous superior or overdominant loci. Dobzhansky cast himself as the primary advocate of the balance position.

Muller never agreed with Dobzhansky's characterization of the classical and balance positions, but he had articulated something close to the classical position. An original member of Morgan's Fly Group, Muller was a world leader in genetics having won a Nobel Prize in 1948 for his research on the production of mutations with X-rays. In 1950, he published "Our Load of Mutations," which provided a new way to assess the genetic damage created by mutation. Accepting the premise that the vast majority of mutations are harmful to some degree, Muller argued that in a population of constant size, each mutation leads to one "genetic death"—to one individual that fails to reproduce. The number of deleterious alleles possessed by an individual represented that individual's deviation from a genetic ideal—that person's genetic load. Because he had pioneered much of the early work on the genetic effects of radiation, Muller was adamant about the genetic loads that exposure to radiation could produce. This concern reflected the damaging effects of radiation on genetic material and was motivated by the recent use of atomic weapons in World War II and was heightened by the ongoing Cold War arms race and testing programs. Thus, it was natural that, when Muller discussed factors that would increase genetic loads and put human populations at risk, radiation was prominent (Beatty 1987b).

Muller's radiation fears were exacerbated by a series of ambiguous results from irradiation experiments conducted in the 1950s and 1960s. Bruce Wallace, a student of Dobzhansky's, had been collaborating with J. C. King to study the effects of radiation exposure in *Drosophila*. Setting a control population as the standard, Wallace and King exposed flies to acute and chronic doses of radiation. If Muller was correct, the radiation should induce deleterious mutations and lower the fitnesses of the treated populations relative to the control population. The flies receiving chronic irradiation did indeed have a lower adaptive value, but the acutely irradiated flies had a higher adaptive value. Interpreting this result in light of the balance position,

Wallace and King argued that improvement of the acutely irradiated population "could exist not merely *in spite of* but *because of* the original treatment" (Wallace & King 1951). Wallace and King's results were meant to invite further research, which they did, but they also invited controversy. Wallace himself continued to refine his radiation experiments, while Muller worked with a graduate student, Raphael Falk, to perform similar experiments. None of these experimental efforts were convincing in the end, in part because it was impossible to pin down the exact effects of the irradiation—it was unclear then that irradiation was producing new overdominant loci. Despite efforts to bring the disputants together to work out their differences, by the 1960s the classical–balance controversy had stalemated (Beatty 1987b).[2]

By linking genetic variability to radiation, the stakes in this controversy had been raised beyond those of an intellectual dispute in evolutionary genetics. Both Muller and Dobzhansky saw themselves as struggling for the future of humankind. Hope of some empirical resolution depended on a way of detecting genetic differences more precisely. The tools for addressing this issue had been developing within biochemistry and molecular biology for a number of years. However, the introduction of molecular tools and data into evolutionary genetics would fundamentally alter the classical–balance controversy rather than settle it (Dietrich 1994; Lewontin 1974).

THE ELECTROPHORETIC REVOLUTION

Electrophoresis had been developed in biochemistry as a means for separating molecules by charge and size. In the early 1960s, geneticist Jack L. Hubby began to adapt electrophoresis for use with *Drosophila*. When Richard Lewontin moved to the University of Chicago to collaborate with him in 1964, Hubby's original program of research changed significantly. Lewontin was a student of Dobzhansky's and had been following the classical–balance debate closely. When Lewontin arrived in Chicago, he had a list of criteria for experimentally resolving how much heterozygosity there was per locus in a population. In his words,

> Any technique that is to give the kind of clear information we need must satisfy all of the following criteria: (1) Phenotypic differences caused by allelic substitutions at single loci must be detectable in single individuals. (2) Allelic substitutions at one locus must be distinguishable from substitutions at other loci. (3) A substantial proportion of (ideally, all) allelic substitutions must be distinguishable from each other. (4) Loci studied must be an unbiased sample of the genome with respect to the physiological effects and degree of variation. (Hubby & Lewontin 1966, p. 578)

Hubby and Lewontin's work tried to meet these criteria and provide a reliable measure of the amount of heterozygosity found in *D. pseudoobscura*. Their survey of 18 loci revealed what they understood to be a high degree of polymorphism; the average heterozygosity was 11.5%. Lewontin and Hubby proposed several alternatives to explain this variation. The possibility of neutral alleles was considered, and ruled out because local populations did not have the high levels of homozygosity predicted if drift were prevalent. They also considered the possibility of a large number of overdominant loci, but recognized that so many heterotic loci would carry with them a large segregational load (Lewontin & Hubby 1966). Almost immediately three different groups proposed truncation selection models to address this problem. It looked as if electrophoresis had provided important evidence in favor of the balance position. This sense of resolution was short-lived, however, as the advocacy of neutral molecular evolution, beginning in 1968, redrew the conceptual landscape.

Apart from the classical–balance controversy, electrophoresis had a tremendous impact upon the experimental practice of evolutionary genetics. From 1966 to 1984, the genetic variability of 1111 species was measured using electrophoresis. This "find 'em and grind' em" approach expanded the scope of evolutionary genetics, drew more people to consider the problem of explaining variability, and demonstrated the power of molecular techniques for evolutionary biology (Lewontin 1991). Electrophoresis was only a part of the molecular biology boom going on in the 1960s, however. After James Watson and Francis Crick discovered the double helical structure of DNA in 1953, molecular biologists and biochemists began to address the evolution of DNA, RNA, and proteins, as well as their coding

[2]See the transcript of the Macy Conference at http://hrst.mit.edu/hrs/evolution/public/archives/macyconference/macy.html

properties and interrelations. In the 1960s and 1970s, the new field of molecular evolution would incorporate new data from electrophoresis, immunological assays, hybridization, and sequencing. In doing so it would transform significant parts of evolutionary genetics (Dietrich 1998).

NON-DARWINIAN EVOLUTION AND THE NEUTRALIST–SELECTIONIST CONTROVERSY

Molecular evolutionary genetics developed in the late 1960s with the spread of experimental techniques, such as electrophoresis, and with theoretical developments that embraced these new molecular data. The most significant theoretical or conceptual developments associated with the molecularization of evolutionary genetics were the introduction of the molecular clock and the advocacy of neutral molecular evolution or, as it was called at the time, Non-Darwinian evolution.

In 1965 Emile Zuckerkandl and Linus Pauling articulated what was later referred to as "the most significant result of research in molecular evolution" (Wilson et al. 1977). After comparing the amino acid sequences of proteins from different lineages, Zuckerkandl and Pauling discovered that the differences in amino acid sequence were "approximately proportional in number to evolutionary time" (Zuckerkandl & Pauling 1965). In other words, the rate of amino acid substitution was approximately constant. Zuckerkandl and Pauling christened this constancy the molecular clock (Morgan 1998; Rodríguez-Trelles et al., Ch. 8 of this volume). The value of the molecular clock for systematics was quickly recognized, but the evolutionary mechanisms underlying the clock's constancy were ambiguous until Motoo Kimura, Jack King, and Thomas Jukes made their case for neutral molecular evolution.

Motoo Kimura was a Japanese biologist who had worked with James Crow and Sewall Wright in the United States on mathematical population genetics. As Crow's student, Kimura was familiar with the classical–balance controversy and was sympathetic to the classical position, as was Crow. The possibility of neutral alleles had been frequently mentioned in the course of the classical–balance controversy, but none of the participants seemed to have taken them seriously as an alternative to a system of alleles under some form of selection. Indeed in 1964, Crow and Kimura developed the infinitely many alleles model which, while it presented a model of mutation for neutral alleles, was primarily aimed at demonstrating the high loads produced by more complex models of overdominant alleles. Kimura later shifted his perspective on neutral alleles from a mathematically tractable case to a description of a biological reality. He did so in response to both the high genetic variability observed by Lewontin and Hubby and an array of biochemical evidence for neutral alleles being presented and discussed at the first conferences on molecular evolution, such as the Evolving Genes and Proteins conference in 1965 where Zuckerkandl and Pauling christened the molecular clock. Indeed Kimura's 1968 argument for neutral molecular evolution is based on data about rates of molecular change presented at the Evolving Genes and Proteins conference, including the hemoglobin data presented by Zuckerkandl and Pauling (Dietrich 1994). Kimura's colleague Tomoko Ohta estimated the rate of amino acid change in mammalian hemoglobin, primate hemoglobin, mammalian and avian cytochrome *c*, and triosephosphate dehydrogenase from rabbits and cattle. Kimura then calculated the rate of evolution for a mammalian genome. Kimura's estimate of 1.8 years for the average time taken for one base pair replacement carried with it an intolerable cost of selection. The only way to avoid this high cost or substitutional load was to postulate that most of the observed substitutions were in fact selectively neutral (Kimura 1968).

Kimura's position was strongly reinforced the next year by Jack King and Tom Jukes who strongly advocated the importance of neutral mutations and genetic drift. Jukes was a biochemist by training and an early molecular evolutionist. He had attended the Evolving Genes and Proteins conference and had published a book on the subject entitled *Molecules and Evolution* in 1966. Like many other biochemists interested in evolution, Jukes recognized the existence of neutral substitutions, but to develop his views he needed the help of a population geneticist. Jukes sought out Jack King, a young biologist with training in evolutionary genetics. Together they assembled a broad range of evidence from biochemistry and molecular evolution to directly counter G. G. Simpson's and Emil Smith's claims for panselectionism at the molecular level (Dietrich 1994). Under the intentionally provocative title of Non-Darwinian Evolution, they presented a case for neutral molecular evolution that included Kimura's cost of selection argument as well as arguments based on the significance of synonymous mutations, correlation between the genetic code and the amino acid composition of proteins, higher rates of change at third positions of codons, and overall constancy of the rate of molecular evolution. The response to Kimura, King, and Jukes was immediate and hostile. Bryan Clarke and Rollin Richmond, for instance, offered point by point counterarguments to the evidence presented by King and Jukes, thereby inaugurating the neutralist–selectionist controversy (Clarke 1970; Richmond 1970).

In 1969, Kimura used the constancy of the rate of amino acid substitutions in homologous proteins to argue powerfully for neutral mutations and the importance of random drift in molecular evolution (Kimura 1969b). At the same time, Kimura was also calling on his earlier work on stochastic processes in population genetics (Gillespie, Ch. 5 of this volume) to forge a solid theoretical foundation for the neutral theory. Kimura's diffusion equation method provided the theoretical framework he needed formulate specific models which in turn allowed him to address issues such as the probability and time to fixation of a mutant substitution as well as the rate of mutant substitutions in evolution (Kimura 1970). Working in collaboration with Tomoko Ohta, Kimura also extended the neutral theory to encompass the problem of explaining protein polymorphisms. This was a central concern of population genetics, and Kimura and Ohta were able to show that protein polymorphisms were a phase in mutations' journey to fixation (Kimura & Ohta 1971a).

In 1971 the Sixth Berkeley Symposium on Mathematical Statistics and Probability devoted a session to Darwinian, Neo-Darwinian, and Non-Darwinian evolution. By this time, the debate between the neutralists and selectionists was well under way. Although few tests had been done, there had been quite a bit of talk about the ability of rival hypotheses to explain a wide variety of data and the positions were well articulated. James Crow was charged with giving a review of both sides of the debate to start the conference session. Crow was disposed toward the neutral theory, but was more skeptical than either Kimura or Ohta. As a participant in the classical–balance controversy, Crow had experienced the frustration of trying to find definitive tests for either position; as a result he valued the neutral theory because it offered quantitative predictions that could be tested and seemed to move beyond the classical–balance stalemate (Crow 1972).

At the same symposium, G. L. Stebbins and Richard Lewontin attacked the neutral theory as a testable hypothesis. According to Stebbins and Lewontin, the neutral theory in its simplest form predicts that allele frequencies will vary from population to population, but in *D. pseudoobscura* and *D. willistoni*, widely separate populations show very similar allele frequencies. A migration rate as low as one migrant per generation could account for the similarity. Because assumptions about migration rate could always explain away allele frequency data, Stebbins and Lewontin charged that no observation could contradict the neutral theory's prediction. They even directly appealed to Karl Popper's philosophy of science and labeled the neutral theory "'empirically void' because it has no set of potential falsifiers" (Stebbins & Lewontin 1972). Yet, Stebbins and Lewontin did not reject the idea of neutral mutation and the effects of random drift; instead they claimed that the nature of evolutionary processes was unresolved and encouraged the diverse pursuits of selectionists and neutralists (Stebbins & Lewontin 1972).

Stebbins and Lewontin's concerns about testing the neutral theory would be compounded over the next 10 years. Despite an abundance of data from electrophoretic surveys, using this data to test predictions from the neutral theory was not as straightforward as it had been supposed. Tests proposed by Warren Ewens in 1972 and later refined by Geoff Watterson in 1977 were designed for electrophoretic data, but when applied did not have the statistical power to discriminate between neutrality and selection (Lewontin 1991). The consequence of this and other difficulties with testing the neutral theory was that neutralists put more stock in the molecular clock as evidence in support of neutrality.

In 1971, Tomoko Ohta and Motoo Kimura asserted that the "remarkable constancy of the rate of amino acid substitutions in each protein over a vast period of geologic time constitutes so far the strongest evidence for the theory (Kimura 1968; King and Jukes 1969) that the major cause of molecular evolution is random fixation of selectively neutral or nearly neutral mutations" (Ohta & Kimura 1971). Kimura had shown that for neutral changes the rate of substitution was equivalent to the rate of mutation. Because the rate of mutation was understood to be the result a stochastic process similar to radioactive decay, the rate of substitution could also be understood as constant generated by an underlying stochastic process. The rate of selected substitutions, however, was subject to changes in selection intensity and population size and so could not be expected to be constant over any long period of time.

Whether recognized as a proxy for the neutralist–selectionist debate or not, the molecular clock was the subject of intense debate. For instance, because the molecular clock was a stochastic clock, some variability in its rate was expected. By as early as 1974, however, Walter Fitch and Charles Langley argued that the rate of substitution was not as uniform across different lineages as it ought to be if the neutralist explanation was correct (Langley & Fitch 1974). Morris Goodman and others joined in this line of criticism, adding evidence of slowdowns and speedups from various lineages. In response, Kimura admitted that the rate of molecular evolution was not perfectly uniform, but in his opinion, "emphasizing local fluctuations as evidence against the neutral theory, while neglecting to inquire why the overall rate is intrinsically so regular or constant is picayunish. It is a classic case of 'not seeing the forest for the trees'" (Kimura 1983). Selectionist critics were undeterred. With growing evidence that rate variability was much more pronounced than had been supposed, John Gillespie proposed a selectionist episodic molecular clock that he claimed could explain patterns of substitution better than Kimura's neutralist explanation (Gillespie 1984). To answer Gillespie's claims, neutralists revised their models of substitution to accommodate greater variability. The amount of variability that can be accommodated by the clock concept remains an open question (although see Rodríguez-Trelles et al., Ch. 8 of this volume).

The neutralist–selectionist controversy itself was transformed during the 1980s with the introduction of DNA sequence data. As a graduate student working with Richard Lewontin, Martin Kreitman learned how to sequence DNA in Walter Gilbert's laboratory at Harvard. Kreitman then sequenced ADH genes in *Drosophila* looking for evidence of polymorphisms. Kreitman's detection of polymorphisms in the DNA sequences suggested that there was selection at the ADH locus and that differences between synonymous and non-synonymous sites were significant. Kreitman would develop the analysis of patterns of nucleotide sequence comparisons into the Hudson–Kreitman–Aguadé test and the McDonald–Kreitman test. These statistical tests and others allowed evolutionary geneticists to detect selection at the molecular level (Kreitman 2000). Where earlier tests had been unable to discriminate between neutrality and selection, these tests applied to nucleotide sequence data succeeded.[3]

Accompanying the availability of DNA data was a significant change in attitude toward neutrality. When Kimura proposed the neutral theory in 1968, the dominant attitude of biologists was that natural selection was the only important cause of evolutionary change at any level of organization. This panselectionist attitude informed the early opposition to the possibility of neutral molecular evolution. By the mid-1980s, however, the dominant attitude among evolutionary geneticists using molecular data was that the neutral theory provided the starting place for investigation in the sense of being the accepted null model (Kreitman 2000). Why hypotheses of neutral molecular evolution became accepted as null hypotheses at this time has yet to be investigated by historians, but the rise of neutral null models seems to coincide with increased availability of DNA sequence data, increasing use of molecular clocks in systematics, increasing use of coalescents, and the spread of tests such as the Hudson–Kreitman–Aguadé test.

CONTROVERSY AND THE HISTORY OF EVOLUTIONARY GENETICS

By emphasizing controversy, I have presented one perspective on the history of evolutionary genetics. The controversies of evolutionary genetics highlight the interplay of theory and experiment, the impact of new concepts and results, as well as the power of personality and politics. Controversies, such as those between the Mendelians and Biometricians or Fisher and Wright were often heated and sometimes quite personal. Like all criticism in science, however, controversies also present the possibility of change. The controversies of evolutionary genetics typically began as highly polarized disputes, but the positions in question developed, sometimes radically, sometimes more subtly. These transformations allowed the controversies to depolarize by enabling some participants to disengage, revise their opinions, or change their focus. Whether the future of evolutionary

[3]The history of these tests as well as a discussion of their development and significance by Martin Kreitman and Richard Lewontin are available at http://hrst.mit.edu/hrs/evolution/public/kreitman.html.

genetics is doomed to persistent controversy is hard to say, but controversy has been an unavoidable feature of its past.

SUGGESTIONS FOR FURTHER READING

Provine (1986) provides an excellent overview of the development of evolutionary genetics as it traces the life of Sewall Wright. The earlier debate between the Mendelians and Biometricians is expertly analyzed in Kim (1994). Because it also includes commentaries by other historians of genetics, Kim (1994) provides a useful introduction to the debates among historians, sociologists, and philosophers over scientific controversy. Lewontin et al. (1981) is a collection of Theodosius Dobzhansky's papers in the Genetics of Natural Population series. This very influential set of papers is contextualized by two extensive introductions, one by Provine and the other by Lewontin. The impact of molecular biology on evolutionary genetics and the rise of molecular evolution are examined in Dietrich (1994).

Dietrich MR 1994 The Origins of the Neutral Theory of Molecular Evolution. J. Hist. Biol. 27:21–59.

Kim K 1994 Explaining Scientific Consensus: The Case of Mendelian Genetics. Guilford Press.

Lewontin RC, Moore JA, Provine WB & Wallace B (eds) 1981 Dobzhansky's Genetics of Natural Populations I–XLIII. Columbia Univ. Press.

Provine W 1986 Sewall Wright and Evolutionary Biology. Univ. of Chicago Press.

Acknowledgments I am grateful to James F. Crow, Richard C. Lewontin, William Provine, Robert Skipper, and Michael J. Wade for their thoughtful comments on earlier drafts of this chapter. Any remaining errors are my own.

2

Genetic Variation

MARTA L. WAYNE
MICHAEL M. MIYAMOTO

Genetic variation provides the underpinning of modern biological thought. From evolutionary biologists studying finches in the field, to drug development in the pharmaceutical industry; from the developmental geneticists trying to understand the bodyplan of a mouse, to the researchers investigating the genetic basis for alcoholism; genetic variation gives us a handhold on the phenotype, which is otherwise a complex and slippery construct. Phenotypes are produced by genes, the environment, and the interaction between genes and the environment. There are few phenotypes for which variation occurring in nature is entirely environmental. However, beyond a conviction that organisms must ultimately be the products of their genes, it is very difficult to justify such a statement. This is in part because we still can not describe the complete genotype–phenotype map for any but the simplest traits. Regardless, genetic variation has been found for virtually every trait ever examined, suggesting that genetic variation as a cause of phenotypic variation is likely to be rampant.

It is impossible to study the impact of the environment on a trait if all organisms experience precisely the same environment, that is the environment does not vary at all from one individual to another. Likewise it is impossible to study the role of genes in producing a phenotype without any genetic variation, that is if all individuals are genetically the same. Thus, variation is central, as the differences among individuals serve as markers that allow one to study the genetic and environmental factors responsible for specific traits. The origin of the study of genetics and evolution began with genetic variation: Mendel began his study of sweet peas with the study of "sports" (mutant varieties); Darwin began his study of evolution with the study of heritable pigeon varieties produced in response to artificial selection by pigeon fanciers.

From the perspective of evolutionary biologists, genetic variation is the fundamental requirement for evolution. Evolution is frequently defined quite concisely, particularly in textbooks or PhD qualifying examinations, as a change in allele frequencies over time. Contained in this definition (which is a very narrow one that will be expanded throughout this chapter) is the implicit requirement that a locus that contributes to evolution must not be fixed for one allele, that is that genetic variation must be present for evolution to occur. Such a definition, while precise in some respects, fails to capture several important details. First, what is an allele? What about larger changes in chromosomal evolution, such as genome-wide duplications or gross chromosomal rearrangements—do these not also contribute to evolution? Second, what mechanisms cause the changes in allele frequency, however broadly we may define an allele, and hence cause evolution; and what are the relative contributions of these different mechanisms?

This chapter will concern itself first with the question of what genetic variation consists of: specifically, what is an allele? The definition of an allele is far from static, but rather changes with every increase in our knowledge about genetics and molecular biology. For example, an allele in the broadest sense may be a single nucleotide change or a change in chromosome number, structure, or the distribution of genes throughout the genome. Throughout the chapter, we strive to emphasize a synthesis of functional genetic variation, combining molecular, mechanistic definitions of alleles with their genetical properties. Functional properties of alleles contribute to their roles in evolution. We begin by enumerating types of genetic variation identified at the molecular level, including selective expectations for molecular variation. Next, we link this molecular variation to genetical properties such as dominance and additivity. The origin of genetic variation is also briefly discussed from a functional perspective, as is the inseparable action of selection and drift to create the spectrum of genetic variation that we see. Finally, we consider how a functional, synthetic perspective on genetic variation challenges several classic evolutionary paradigms.

VARIATION AT THE MOLECULAR LEVEL

New molecular variation arises through a spectrum of changes in a genome sequence, encompassing single base substitution through point mutation and genome-wide duplication through polyploidization. Thus, genetic variation constitutes a rich and diverse topic, affording many different ways to hierarchically organize this information. Given the current state of biology, with its emphasis on mechanisms and thereby molecules, we start the definition of an allele at the most reductionist level: the DNA (or RNA) molecule. One of the luxuries of the post-genomic era is that we now can precisely describe far more types of sequence changes at the molecular level, and estimate the relative abundance of such events, at least within the genome of an individual. Molecular alleles are presented from the simplest (single base changes) to the most complex (changes affecting entire genomes, such as genome-wide duplications). A complementary discussion of mutations and their effects is provided by Houle & Kondrashov (Ch. 3 of this volume).

Single Nucleotide Base Changes

The most straightforward type of genetic variation is the single nucleotide base mutation (Figure 2.1).

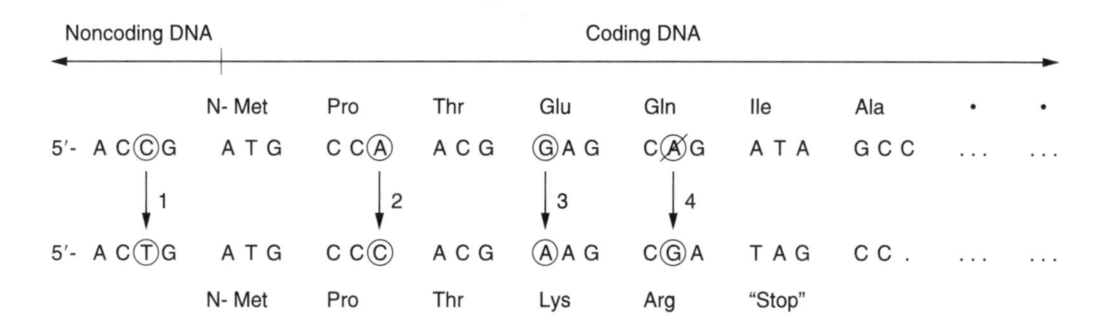

FIGURE 2.1. Four different types of mutations as illustrated with the 5′-end of a hypothetical protein-coding gene. At the top, the original DNA sequence of this gene is shown, along with the amino acid sequence for the amino (N)-terminus of its encoded polypeptide product. In turn, the DNA and polypeptide sequences that result from the four mutations are given at the bottom. To facilitate comparison, the coding regions of the DNA sequences are labeled as such and are presented as base triplets relative to their encoded amino acids. The four mutations are numbered and are defined in the lower left corner of the figure. In the case of mutation 4, the strikethrough highlights a deletion of the marked "A" in the original sequence. In addition to representing a point deletion, this mutation also constitutes a frameshift mutation (i.e., one that alters the downstream reading frame of this gene). In this case, this frameshift mutation results in a new amino acid and premature termination (as indicated by "stop") of the encoded mutant polypeptide.

These include heritable substitutions of one base for another. Substitutions may be broadly classed into transitions (purine to purine, i.e., adenosine to guanine or the reverse; pyrimidine to pyrimidine, i.e., cytosine to thymine or the reverse) or transversions (purine to pyrimidine or vice versa). Relative rates of transitions and transversions are well understood for a wide range of organisms (Graur & Li 2000); in general, transitions are more common than transversions because of (1) rare tautomeric shifts (proton shifts) that result in noncanonical base pairing (e.g., G with T); and (2) the relatively fast rate of mutation of C to T in CpG dinucleotide pairs (where the 5′ C is methylated). Base changes are an extremely common type of mutation, and are caused by errors in genome replication. Although DNA polymerase has a proof-reading function, most base substitution mutations arise from DNA replication (Drake et al. 1998). Thus, spontaneous mutations for base substitutions are inevitable and universal. Base substitutions are also caused by a variety of mutagens, including that favorite mutagen of classical geneticists, ethylmethane sulfonate, or EMS.

Base substitutions may occur in the noncoding sequence that comprises the majority of most organisms' genomes, or in the protein coding sequence (CDS) of the DNA or RNA (Figure 2.1). Genes that occur in the noncoding sequence, which is defined to be sequence that is never transcribed or else is transcribed but not translated (introns, untranslated regions or UTRs), have traditionally been expected to be evolutionarily unimportant and irrelevant to the phenotype of the organism. Recent advances in our understanding of the sources of variation for regulation of protein abundance have challenged this view (see Phenotypes at the Molecular Level: Regulatory Variants, below). In contrast, the evolutionary relevance of base substitutions in the coding sequence is expected to be specific to the context of the base. Some coding sequence substitutions result in amino acid replacements, and these are generally expected to be under stronger selection than those coding sequence changes that do not result in protein changes. Because the genetic code is redundant for many codons at the first and third positions, many first and third position mutations do not result in changes to the amino acid sequence; some, however, are also replacement changes. Base mutations at second positions always lead to amino acid replacements or stop codons. Mutations that change the amino acid sequence are referred to as replacement or non-synonymous changes. In turn, nonsynonymous mutations that result in amino acid replacements versus the incorporation of premature stop codons are known as missense mutations versus nonsense mutations, respectively. Those that do not are referred to as synonymous mutations.

Synonymous coding changes are generally assumed to be under weaker selection than nonsynonymous changes, even weaker than base changes in untranslated regions of genes (Graur & Li 2000). Nevertheless, even these weakly selected changes can leave their marks at the molecular level when population sizes are large and selection is thereby most efficient. For example, in yeast (where population sizes are large), the use of synonymous codons for the same amino acids (e.g., CUA, CUC, CUG, CUU, UUA, and UAA for leucine) is not uniform but is skewed such that their frequencies are correlated with the relative abundances of their corresponding cognate tRNAs (Ikemura 1985). Such codon usage biases (codon bias) are most evident in highly expressed genes (i.e., those most likely under the strongest selection).

Length Changes

Genetic variation at the molecular level may also be caused by variation in sequence length. Sequence length variation is caused by insertions or deletions to the sequence and is more generally referred to as indel variation. Collectively, base substitutions or indels of a single nucleotide that are polymorphic at the population level are known as SNPs (single nucleotide polymorphisms). There are three major mechanistic models for indel variation (transposable elements, unequal crossing over, and DNA slippage), though origin of length variants is not considered to be exclusive to these mechanisms.

Transposable elements (TEs) are genetic units that do not have a fixed place in the genome, but rather can move from one locus to another, sometimes by duplicating themselves and sometimes by excising themselves from the DNA (Petrov and Wendel, Ch. 10 of this volume). TE variation is considered to be a major source of indel variation, as well as a major source of genetic variation in natural populations (Kazazian 2004). The genetic differences between TEs themselves is a fascinating subject which is beyond the scope of this chapter. In brief, the insertion of a TE in a new site, particularly by duplication of the element, can result in a local increase in sequence length. Shorter insertions are more common, and may be due to the imprecise excision of TEs from genes (McDonald 2000). Short deletions may also result from imprecise excision of TEs. Many classical mutations are commonly caused by insertions, including multiple alleles of the *white* locus in *Drosophila melanogaster* (the first visible mutation isolated in this species). Unequal crossing over is another common source of length polymorphism. Unequal crossing over is caused by mispairing of genes during meiosis, and subsequent recombination between the genes. Such crossing over requires that there be extensive similarity between sequences, so it is generally restricted to tandemly duplicated genes, or genes containing tandem repeated sequences (resulting in gain or loss of the area between and/or including the repeats). Unequal crossing over is a major source of indels for longer DNA sequences and chromosomal regions. In turn, for short, tandemly repeated sequences (i.e., microsatellite repeats and homonucleotide runs), indels are primarily the result of strand slippage and mispairing of the template versus replicating strands during DNA replication (DNA slippage). Here loops can form in either the template and/or replicating strands, in such a way that the tandem repeats of the former pair with repeats of the latter which are not their direct counterparts. The end result is that subsequent rounds of replication can then lead to mutant DNA duplexes with increased or decreased numbers of these short tandem repeats.

Gene and Genome Multiplication Events

In addition to single gene duplication events, not infrequently entire genomes have been duplicated one or more times in evolutionary history (autopolyploidy). While individual genes have long been thought to be duplicates of one another (Thornton, Ch. 11 of this volume), the first organism that was proposed to be the product of an ancient genome-wide duplication event is *Saccharomyces cerevisiae*, the common laboratory species of yeast (Wolfe & Shields 1997). Evidence for the autotetraploid state of yeast comes first from many genes with apparently redundant function, that is, no assayable phenotype on knockout; and second and more convincingly, from the discovery of 55 regions of colinear, duplicated genes with whole gene deletions interspersed within these regions.

Polyploidy events are generally more common in plants than in animals, perhaps occurring in up to 70% of angiosperms (Soltis & Soltis 1999; Arnold & Burke, Ch. 26 of this volume). In general, plants are more tolerant of changes in ploidy than animals, perhaps because they are more flexible in development and therefore more tolerant of differences in gene dosage. In particular, plants are commonly allopolyploids (polyploids created from interspecific hybridization), while the more rare polyploid animals are usually autopolyploids, which are frequently parthenogenetic due to meiotic problems. Moreover, plant species are frequently populations of lineages of multiple allopolyploidy events. Genome restructuring, both intra- and intergenomic, is common in both auto- and allopolyploids. Thus, polyploidy is likely a major source of evolutionary novelty in plants (Soltis & Soltis 1999).

Genome-wide multiplications in early vertebrates were emphasized by Spring (1997), who hypothesized that the presence of four homologous copies of certain *hox* genes in vertebrates were derived from invertebrates such as *Drosophila* and primitive vertebrates like *Amphioxus,* and was due to two serial genome-wide duplications resulting in an octoploid genomic state in vertebrates. He also noted that for many genes, both members of the *Hox* cluster and members of gene families, only three copies are extant. He suggested that because interspecific hybridization events are rare in vertebrates (other than certain fish or amphibians), autopolyploidy could be a major source of evolutionary novelties. He further suggested that this novelty was likely to be the outcome of regulatory changes rather than coding sequence changes, such that redundant functions might be rendered tissue-specific. There are now studies in multiple taxa supporting the auto-octoploid hypothesis in chordates, including humans (McLysaght et al. 2002); further, strong evidence exists for a subsequent duplication event in the common ancestor of ray-finned fishes (Taylor et al. 2003), such that they are 16-ploid.

Other Mutations in Chromosome Structure and Number

In the previous two sections, duplication (and its converse, deletion) and polyploidy were emphasized, because of their recognized roles in the origins of biological diversity. As noted above, gene and genome duplications provide the new coding and regulatory sequences for the origins of new protein functions and subfunctionalizations of their ancestral roles. However, a number of other mutations are known that involve changes in chromosome structure and/or number and these are also often of evolutionary and biological significance.

In addition to duplications and deletions, chromosome structure can also change by inversions and translocations. In an inversion, a chromosomal segment becomes flipped by 180° such that it is now oriented in the reverse direction. If the centromere is included in the inverted segment, then the inversion is known as pericentric. Otherwise, it is called a paracentric inversion. Inversions are considered particularly interesting because recombination is suppressed in inversion heterozygotes, allowing for the possibility of ratchet-like mutation accumulation and/or co-adapted gene complexes (Powell 1997).

In turn, translocations refer to the products of crossing over between nonhomologous chromosomes. Such crossing over can involve a unidirectional transfer of chromosomal material from one nonhomologous chromosome to the other or a bidirectional exchange of segments between the two. The two types of translocations are known as nonreciprocal and reciprocal, respectively. Two special types of reciprocal translocations are Robertsonian fissions and fusions. In Robertsonian fissions, a chromosome with a more central centromere (i.e., metacentric or submetacentric) interacts with a minute "donor" chromosome to split the former into two smaller and separate acrocentric chromosomes (those with near-terminal centromeres). In Robertsonian fusions, two acrocentric chromosomes interact such that the two become united into one larger metacentric or submetacentric chromosome. In the process, a minute "donor" chromosome is generated as well.

In contrast to polyploidy, aneuploidy refers to the gain or loss of individual whole chromosomes (rather than to the duplication of the entire genome). Trisomy is the gain of a whole chromosome, whereas monosomy corresponds to its loss. In addition, changes in chromosome number can be linked to Robertsonian fissions and fusions of different nonhomologous chromosomes. Here, as the minute "donor" chromosomes are readily lost, Robertsonian fissions and fusions can lead relatively quickly to a subsequent increase or decrease in chromosome number, respectively. Such a mechanism has been invoked to explain the difference in chromosome number between humans and great apes, with their $2N$ counts of 46 versus 48 chromosomes, respectively (de Pontbriand et al. 2002).

Of these additional sources of change in chromosome structure and/or number, inversions, translocations, and Robertsonian translocations are most important to studies of natural variation. Such changes in chromosome structure and number are frequently present as polymorphisms in natural populations and geographic populations and closely related species are often distinguished by such chromosomal differences.

Epigenetic Changes

Epigenetic changes may be defined as heritable changes in gene expression that are not the result of sequence alterations (Murphy & Jirtle 2003; Jablonka & Lamb, Ch. 17 of this volume). Epigenetic changes can often be reset every generation, in contrast to sequence changes, which are reset according to the site-specific mutation rate of the organism (i.e., rarely). Thus, epigenetic effects are often transient. Methylation is believed to be the primary mechanism of epigenetic effects, but the exact mechanism by which methylation occurs and is reset remains an open question (Vermaak et al. 2003).

Why did epigenetic effects evolve? One intriguing hypothesis is that methylation evolved as a host response to intragenomic conflict, specifically, to silence TEs (McDonald 1999). McDonald points out that mutation rates caused by TEs are far lower in mammals, which have sophisticated genome-wide mechanisms of imprinting, than in *Drosophila*, which does not. However, this hypothesis has been criticized for failing to address directly the role of sexual dimorphism in methylation patterns by genomic imprinting (i.e., why silence TEs in only one parent rather than in both; see below) (Spencer et al. 1999).

One especially interesting example of epigenetics is genomic imprinting, which is defined as a parental-specific expression pattern, or parent-of-origin effect. That is, only the allele from one of the parents is expressed in the offspring rather than biallelic expression. Upward of 70 genes are known to be imprinted in mammals, and probably closer to 200 (Murphy & Jirtle 2003). Imprinting is an exciting area of research for several reasons: it may be an important evolutionary mechanism for intersexual conflict over reproductive investment (Haig 2000), and/or for silencing TEs or enabling genome-wide duplications (see below).

Another interesting argument is that methylation resulting in gene silencing was a necessary condition for the successful maintenance of polyploid genomes (Bird 1995a,b). The idea is that silencing could preserve the appropriate gene dosage

BOX **2.1.** Maternal Effects
Timothy A. Mousseau

Most biologists consider that individual phenotype results from genes inherited from the mother and the father, together with the direct influences of the environment experienced by the developing offspring. However, inherited genes and direct environmental effects are only a few of the many factors underlying the phenotypic variation that is subject to natural selection (Mousseau & Fox 1998). In particular, mothers can profoundly influence the phenotype of their offspring above and beyond the genes they contribute and these maternally effected sources of phenotypic variation can play a major role in trait evolution.

Although maternal effects are defined in a variety of ways depending on the question and application, I will broadly define them as all sources of offspring phenotypic variance due to mothers above and beyond the genes that she herself contributes (Figure 1). As such, most maternal effects are associated with variation in propagule size or quality, parental care, host choice, or mate choice. In genetic terms, maternal effects are usually described as a source of environmental variance among offspring that is mediated by either genetic or environmental influences on the maternal phenotype. See Roff (Ch. 18 in this volume) for genetic methods of quantifying maternal effects.

Sources of Maternal Effects Variation

Maternal Effects on Propagule Size or Composition

In many species there is a positive relationship between maternal size and neonate size and this variation may sometimes influence offspring development and fitness.

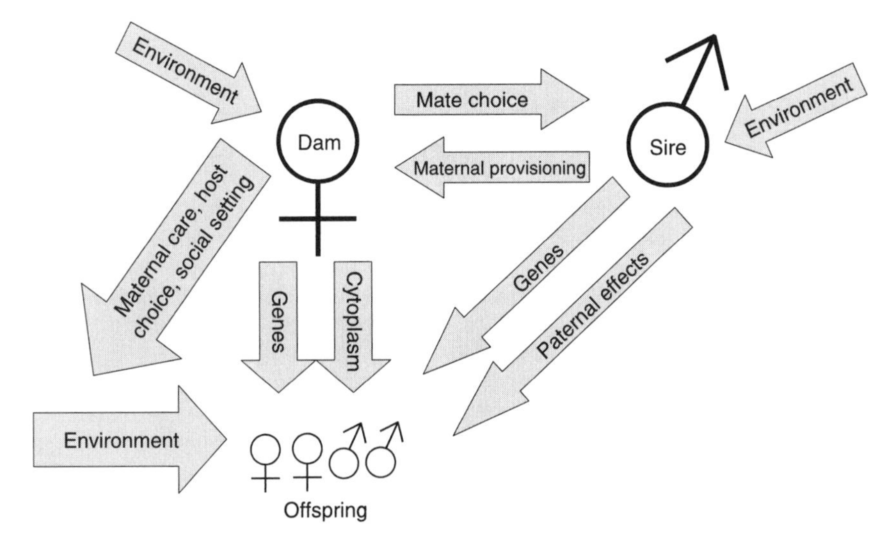

FIGURE 1. A few of the many sources of offspring phenotypic variation that are mediated by mothers. Both mothers and fathers contribute nuclear genes. Mothers also directly influence the amount and quality (i.e., constituents) of cytoplasm allocated to each offspring which can be influenced by her environment, the amount of provisioning (i.e., nuptial gift) given to her by her mate, and her ability to differentially allocate resources among offspring. Mothers may also influence the quality of environment experienced by developing young via maternal care, choice of host, timing of propagule dissemination, and the social setting into which offspring are placed. Female mate choice can influence the quality and quantity of paternal provisioning to both mother and offspring, the quality of paternal care, and the genetic contributions to offspring from fathers. The size of the arrows does not necessarily reflect relative importance.

(continued)

BOX 2.1. *(cont.)*

In many species, maternal diet will influence the number, size, and/or quality of her offspring though the adaptive significance of such effects has rarely been assessed. However, in some cases, mothers are able to adaptively adjust propagule size in response to predicted environmental conditions for developing young. For example, in the seed beetle, *Stator limbatus*, mothers can rapidly change egg size according to the host upon which eggs are laid (Mousseau & Fox 1998).

Although propagule size per se is likely to be important under many conditions, mothers also control the deposition of extranuclear developmental messages in the egg (e.g., hormones, mRNAs, immunofactors). These cytoplasmic factors are often influenced by the environment experienced by the mother (e.g., photoperiod, temperature, resource quality; Mousseau & Fox 1998; Roach & Wulff 1987), and can lead to significant developmental effects including variation in offspring growth rate, diapause or dormancy, wing and color polymorphisms, and offspring behavior (e.g., propensity to disperse). There have been many recent studies with birds suggesting that mothers can modify allocation of hormones (e.g., androgens) that subsequently affect offspring development and behavior.

Recent developments in developmental biology indicate that maternally derived transcription factors play a major role in offspring development and ultimate phenotype. For example, in *Drosophila*, asymmetrically distributed maternal factors initiate a cascade of spatially organized zygotic gene action that provides the blueprint for the larval body at the blastoderm and subsequent stages of development (Akam 1987). In addition, it has been suggested that maternal messages that program terminal differentiation of germ and soma cell lines may have been instrumental in the evolution of multicellular organisms (Buss 1987).

Parental Care and Maternal Effects

There are many examples of the importance of postzygotic maternal effects on offspring fitness. The most obvious include provisioning of developing embryos in mammals; there are even examples of "in utero" care in insects (e.g., roaches, Holbrook & Schal 2004). Post-parturition care is commonly observed and can include lactation in mammals, and provisioning in birds, reptiles, fish, and insects. The importance of such care has obvious consequences and has been well documented for a wide variety of organisms.

Maternal Host Choice and Offspring Fitness

For many species, the most important determinant of offspring survival will be the choice of environment in which offspring are deposited by mothers. This is especially true for parasites and parasitoids which tend to specialize on a limited range of hosts. Females able to discriminate and select high-quality environments for their developing young will have higher inclusive fitness. It seems likely that host preferences have often evolved in response to variation in host quality to developing young. A similar effect is observed in turtles and crocodilians in which nest temperature can influence the gender of offspring (environment sex determination; ESD).

Sexual and Social Influences on Maternal Effects

Maternal condition (and its subsequent effects on egg constituents) can be influenced by nutritive contributions from the sire or helpers in the social group. In many species males will provide females with nuptial gifts prior to, and during, copulation and these nutrients are incorporated in the eggs prior to ovulation. The quality or quantity of these gifts can influence female choice with subsequent effects on offspring from both direct nutrient investment by the male and the indirect influence of his genes on offspring.

BOX **2.1.** *(cont.)*

There is growing evidence that in many species mothers can respond to the genetic or phenotypic quality of fathers by differential (and preferential) allocation of egg constituents to offspring of high-quality mates. Studies of mallard ducks suggest that females mated to high-quality males produce larger eggs but that such investment comes at a cost to later reproduction (Cunningham & Russell 2000). A recent study of crickets has found that females mated to high-quality males produce high-quality sons but low-quality daughters (Fedorka & Mousseau 2004) and that such differential investment is an adaptive response to fitness variation among offspring (high-quality sons have higher lifetime fitness than high-quality daughters).

Maternal effects can also be mediated via social milieu. For example, in Clutton-Brock et al.'s (1984) classic study of red deer, male offspring born to high-ranking mothers have significantly higher lifetime reproductive success than those born to subordinate females and mothers adjust the sex ratio of their offspring to increase their inclusive fitness. Social status and numbers of helpers in a group have also been found to influence offspring development and fitness (e.g., Russell et al. 2003). Conversely, recent studies of birds (e.g., Badyaev et al. 2002) have found that many birds have the ability to adjust the sex ratio (or birth order of different-sexed offspring) in response to ambient environmental conditions or social setting. In these examples, the environment experienced by the mother leads to phenotypic and fitness effects on offspring.

Maternal Effects and Evolutionary Response to Selection

The evolutionary significance of maternal effects stems from both the fitness consequences of transgenerational plastic responses to environmental heterogeneity and the longer term evolutionary responses of adaptive traits to environmental change. Many maternal effects are homologous to the well-studied phenomenon of phenotypic plasticity except that with maternal effects the environmental trigger is experienced by the maternal generation and the phenotypic consequences are expressed by offspring (Mousseau & Fox 1998). In the case where the genes associated with maternal receptivity and offspring response are independent, natural selection acting across generations will favor linkage disequilibrium between "cause and effect" to promote an appropriate response (e.g., Wolf & Brodie 1998). In other cases, linkage between generations may result from pleiotropy. Longer term consequences of maternal effects result from the expectation that maternally effected traits will have higher amounts of additive genetic variation as a result of sex limited expression (Wolf & Brodie 1998), and the fact that total heritability of a given trait will reflect the summation of both maternal and direct (i.e., in the offspring) additive genetic influences and their genetic covariance. If this covariance is negative, then maternal and direct genetic influences may cancel each other, thus deterring evolutionary response to selection (Kirkpatrick & Lande 1989). However, if the covariance is positive, response to selection can be dramatically enhanced. This property of maternally effected traits has long been capitalized on by animal and plant breeders as a means for rapid selection for economically important traits. In recent studies of red squirrels (e.g., McAdam & Boutin 2004), it has been found that a large positive genetic covariance between direct and maternal genetic influences on offspring development can generate responses that are up to 5 times that predicted by simple, single-generation genetic models.

Although it is often difficult to assess the adaptive significance or even measure the fitness consequences of maternal effects, it is apparent that they are displayed by a wide variety of organisms and can influence a great number of traits. Thus, given this diversity, it is always necessary to consider the possible impact of maternal effects on evolutionary response to selection.

in the presence of inappropriate numbers of alleles. Although the idea was originally presented in the context of "preadaptation," one might also argue that polyploidization could have set the stage for selection on a mutation conferring imprinting in a post-polyploid environment.

The evolutionary origins of imprinting remain obscure. In addition to the dosage argument presented above, one particularly interesting argument to explain the evolution of parental-specific inheritance is the intersexual conflict argument articulated by Haig (2000). Haig suggests that it is in the father's best interests to promote the growth of their offspring, even at the expense of the mother's health or future reproductive success. Thus, the prediction is that genes responsible for fetal growth will frequently be imprinted such that the paternal allele is the only one expressed. Likewise, mothers will be selected to imprint their alleles for fetal growth, and/or to remove the father's imprint on those genes for regulating growth, so that they do not sacrifice too many resources in the production of this single offspring.

GENETICAL DESCRIPTIONS OF VARIATION

Genetic variation provides the insights into gene function. This has been recognized by generations of geneticists, who created variation by mutagenesis to be able to study genetic differences if the desired variation was not readily available. Long before the molecular descriptions of mutations were accessible, genetical definitions were invented and remain in use today.

Dominance Variation

Dominance variation is described as the nonadditive relationship among alleles of an individual gene, or, the way that one allele influences the behavior of another at the same locus (Roff, Ch. 18 of this volume). Initial genetical descriptions of variation were confined largely to dominance variation. This was in part historical, due to the rediscovery of Mendel by De Vries and colleagues and thus an expectation that dominance would be pervasive; and in part due to the ease of distinguishing such alleles using methodology such as complementation tests.

The evolution of dominance was a subject vigorously debated by Wright and Fisher in the pages of *American Naturalist* from 1928 to 1934, and is concisely summarized in Charlesworth (1979). The center of the debate was whether dominance evolves as a property of the wild-type allele in response to selection to nullify the effects of being heterozygous when the wild-type allele was paired with new, deleterious mutations (Fisherian view); or instead is an intrinsic property of enzyme function, unrelated to fitness per se (Wrightian view).

Interestingly, Wright's understanding of the phenomenon as an outgrowth of biochemistry and physiology meant that the link between dominance and molecular mechanism was understood relatively shortly after the discovery of the structure of DNA. Muller noted that in general recessive alleles represented loss of function alleles with respect to the phenotype of the organism, while dominant alleles generally represented gain of function alleles. Wright had already described recessivity in terms of enzyme function, so the understanding of null alleles as protein-absent and thus enzymatic function-absent was straightforward and easily associated with fly mutations such as *white* which Muller was familiar with. The wild-type fly eye is red; *white* is a recessive null allele of the protein that, in its functional (dominant) state, allows pigment to be deposited in the eye.

Landmark studies of enzyme kinetics and metabolic flux (Kacser & Burns 1981) have demonstrated that heterozygous phenotypes tend to be close to homozygous dominant phenotypes because, in general, the fractional response in flux to a fractional change in an enzyme will be small (Figure 2.2). The main reason for this is that metabolic pathways tend to be complex and polygenic. Here, this point speaks to a mechanistic understanding, in this case biochemical, of dominance, but later it will also allow for a structural connection to epistasis.

Additive Variation

Additive genetic variation is the most general type of genetic variation in natural populations. Additive variation is simply the case where alleles from multiple loci each contribute independently to the trait of interest, adding to or subtracting from the baseline trait value to create an overall trait mean (Roff, Ch. 18 of this volume). Quantitative traits are those traits which are controlled by many, rather than a few, loci; and which are affected by the environment. In addition, quantitative traits often approximate a continuous distribution (consider human height),

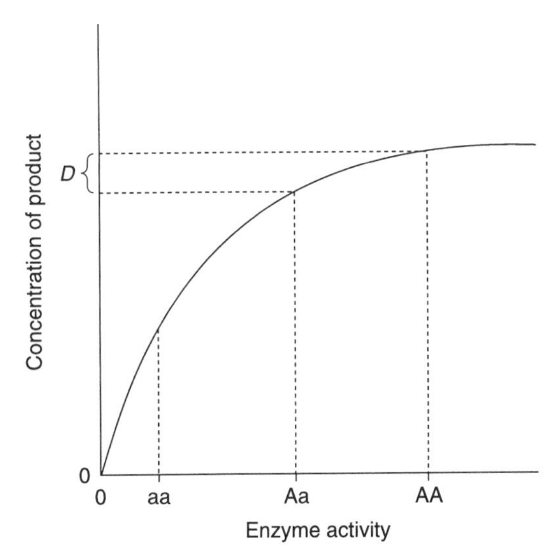

FIGURE 2.2. Relationship between enzyme activity and product concentration illustrating the biochemical basis of dominance as proposed by Kacser and Burns (1981). This figure is modified from Figure 4.5 of Lynch and Walsh (1998). The enzymes of different genes do not usually function alone, but rather together in complex biochemical pathways. Furthermore, the "wild type" version for the enzyme of each gene typically functions well such that *AA* homozygotes lie close to the asymptote or plateau. Taken together, these two observations support a hyperbolic relationship where large decreases in enzyme activity (i.e., from *AA* homozygotes to *Aa* heterozygotes) result in only small reductions in product concentration. Correspondingly, only a small (and quite often negligible) difference in phenotype is observed between *AA* homozygotes and *Aa* heterozygotes, thereby resulting in dominance.

though binary or threshold traits also exist (such as disease status, i.e., affected vs. unaffected). Finally, the environment tends to play a large role in quantitative traits (Roff, Ch. 18 of this volume). Quantitative geneticists from Fisher onward use the infinitesimal model for genetic variation as a null hypothesis (Falconer & Mackay 1996). This model posits that genetic variation is caused by an infinite number of loci with additive alleles of minute effects. Thus, the genome is a composite of plus and minus signs, which sum to the trait value.

Arguably the most important point about additive variation is that it determines at least the short-term response to selection. Because alleles are inherited independently, rather than in combination as intact genotypes, dominance and epistasis are not inherited from the parents. Thus, it is the additive variation that determines response to selection. This is the basis of Fisher's fundamental theorem: the rate of increase in fitness of any organism at any time is equal to its (additive) genetic variance in fitness at that time. It is also the basis of the breeders' equation, stating that the response to selection is a function of the fraction of phenotypic variation that is additive (i.e., heritability, h^2) multiplied by the amount of selection (selection differential).

Although the infinitesimal model is clearly a simplified starting point for the study of quantitative variation, one basic prediction holds: additive genetic variation has been found for almost every trait ever studied, and there do seem to be many loci involved with at least some quantitative traits (Barton & Keightley 2002). The remaining problem then is the very small allelic effects predicted by Fisher. We are interested in the distribution of allelic effects in natural populations. However, because it is extremely difficult to measure effects on an allele-by-allele basis, sources of information about the distribution of allelic effects come primarily from studies of quantitative trait loci (QTL; Cheverud, Ch. 19 of this volume) and mutation accumulation studies (Houle & Kondrashov, Ch. 3 of this volume), which may not accurately represent the actual range of natural variation. One significant exception has been that of Mackay and colleagues, who introgressed natural alleles of a candidate bristle gene locus, *Delta*, into a common background and then examined the effects of the introgressions. They concluded that most of the variation was accounted for by a few mutations of large effect, with the remainder due to a larger number of loci with small effects (Lyman & Mackay 1998).

Most of our information about the distribution of allelic effects comes from QTL mapping and mutation accumulation lines. Considerable effort has gone into describing allelic effects of QTL (Cheverud, Ch. 19 of this volume), but this is problematic owing to the Beavis effect, a tendency to overestimate the effects of individual QTL (Otto & Jones 2000); and overall ascertainment bias due to the inability to detect mutations of very small effect without prohibitively large experiments. However, given these caveats, the data suggest an exponential distribution of QTL effects (Barton & Keightley 2002). Distributions of allelic effects garnered from mutation accumulation experiments using small laboratory populations (because small populations allow for the fixation of deleterious, as well as neutral and beneficial, alleles by drift; and thus estimate total mutation rate more accurately) are also interesting, and likewise suggest that the distribution of effects is a negative exponential (Keightley & Lynch 2003).

Molecular descriptions of additive variation are more complex than for dominance variation, partly because there are so many possible mechanisms and models are less articulated than for dominance or epistatic variation; and partially because additive alleles are less well characterized at a molecular level. Still, in those rare cases where additive quantitative variation has been localized to a nucleotide polymorphism, these polymorphisms tend to not be in coding regions, suggesting that the source of much additive variation may be regulatory (Doebley 1992).

Epistatic Variation

Epistasis is a complex topic, and made more so because the word is used in two senses by biologists, depending on their training. The simplest definition is that used by molecular geneticists, who use epistasis to define any contact between genes or gene products. That is, if proteins interact with one another, this is epistasis *sensu* molecular. Unfortunately, this definition is so simple and broad as to be largely useless. Quantitative geneticists mean something more precise than a mere contact between molecules, defining epistasis to mean an interaction between alleles at different loci that is nonadditive. This is much closer to the epistasis in the sense it is usually encountered in genetics textbooks. There are several kinds of epistasis, depending on how many alleles from two loci are involved in creating phenotypes: two alleles (one from each of two additive loci) is additive × additive epistasis; three alleles (one from an additive locus, the other two from a locus exhibiting dominance) is additive × dominant; and all four (two alleles from each of two loci, each exhibiting dominance) is dominant × dominant. Epistatic variance *sensu* quantitative is notoriously difficult to detect, requiring advanced intercross designs and very large sample sizes (Falconer & Mackay 1996). Epistasis has been well documented, however, at the level of the QTL (Dilda & Mackay 2002). Despite the difficulties in estimation, the central role epistasis plays in evolutionary theory in topics ranging from speciation to the viability of small populations to genetic load makes it a topic of perennial interest.

A mechanistic understanding, in this case biochemical, informs our genetical description of epistasis as it did for dominance (see above). To re-emphasize, fractional changes in an enzyme (i.e., between the heterozygote and dominant homozygote) will result in only small fractional responses in flux, since enzymes function in metabolic pathways that are complex and polygenic (Figure 2.2). Thus, Fisherian "modifiers of dominance" may arise as an outcome of epistasis *sensu* molecular within the extant pathway, rather than from genes functioning uniquely as modifiers. Further evidence of the extent of protein interaction is seen from microarray studies. In particular, a single mutation can theoretically perturb expression in many genes not obviously interacting directly with the mutated gene product; data from yeast support these theoretical conclusions (Bergman & Siegal 2003). These data are clearly an example of flux, but it is unclear whether they are epistasis *sensu* molecular only, or both molecular and quantitative. Interestingly, epistasis *sensu* quantitative has also been demonstrated on a dramatic scale with microarrays by Anholt et al. (2003). Not only were more than 530 genes coregulated in response to one or more new mutations for olfaction (epistasis *sensu* molecular), but in a sample of 21 of these coregulated genes for which mutant stocks were available, 67% of the epistatic interactions at the level of transcription (i.e., *sensu* molecular) were also epistatic by quantitative genetic definitions when crossed to the same new mutations for olfaction (i.e., multiplicative effects on phenotype).

PHENOTYPES AT THE MOLECULAR LEVEL

In the previous section, we described the various changes in DNA sequence and epigenetic effects that comprise genetic variation. How do these changes manifest themselves in the organism? We can consider phenotypes at the molecular level, which are effectively the mechanistic or functional effects of the mutations. The mechanistic effects are essential to the link between genotype and phenotype, and thus the link between genotype and fitness.

Structural Variants

Historically, as discussed in Chapter 1, the attention of evolutionary biologists has focused on structural genetic variation (alterations to the DNA sequence). This is in part because of the importance of the central dogma (DNA to RNA to protein) in our understanding of how phenotypes are constructed (see Intersections Between Structural and Regulatory Genetic Variation, below), and in part because of technological limitations that precluded measuring other types of genetic variation (e.g., regulatory; see below). Structural variants are also the best-understood mutations evolutionarily and mechanistically. Additionally, there is ample evidence that structural genetic variation plays an important role in natural populations (e.g., *Adh* in *Drosophila melanogaster*: Hudson et al. 1987).

Structural mutations are generally assumed to be base changes, specifically, those for amino acid replacements due to single base mutations (i.e., nonsynonymous missense mutations). All amino acid replacements are not equal. Biochemically, the amino acids may be more or less similar to one another. For example, amino acids may be polar or nonpolar, acidic or basic. Predicting the higher-level structure of a protein based on primary structure (amino acid sequence) is a fast-moving field of growing sophistication (Kretsinger et al. 2004). Many models of protein structure are based on physicochemical properties of the amino acids of the sequence, but protein crystallography has often revealed the shortcomings of such approaches (Baker & Sali 2001).

For example, certain amino acids play special roles in protein secondary structure. Two examples are proline, which is recognized as a helix-breaker, and cysteine, which can form disulfide linkages. Even models based on secondary structure, while powerful in combination with biochemical properties of individual amino acids, are less reliable than might be desired. In short, the effect of a given amino acid replacement is highly context-dependent, depending on its position with regard to protein secondary, tertiary, or even quaternary structure. Thus, selection pressures against the same replacement in two different proteins, or even in different locations within the same protein, may be very different (Gaucher et al. 2002).

One classic example of an evolutionarily relevant protein variant is that of the sickle cell allele of hemoglobin in humans. Wild-type homozygotes (Hb^+/Hb^+) are more susceptible to malaria than sickle heterozygotes (Hb^+/Hb^s), while sickle homozygotes are susceptible to anemia under certain conditions (Dickerson 1983). Most sickle alleles appear to be descended from a common ancestor, with a single amino acid replacement at position 6 of human β-hemoglobin, a glutamic acid to valine. Other mutations in β-hemoglobin give rise to different anemias. In contrast, other disease alleles in humans, while largely protein variants, appear to have multiple origins and affect multiple amino acids within a given protein which all cause similar phenotypes (e.g., Tay–Sachs disease: Online Mendelian Inheritance in Man 2000). Our understanding of sickle cell anemia was certainly facilitated by its likely single origin and its single amino acid difference.

Structural alleles need not be single base substitutions to affect protein sequence, however. Intact TEs contain multiple stop codons that can disrupt protein sequence, for example (Mitchell et al. 1991). Short deletions or insertions of coding sequence may be in frame (multiples of three that thus do not disrupt the reading frame of the protein's triplet code), but may still alter protein structure and/or function. Indels that disrupt the reading frame, i.e., frameshift mutations, also can clearly alter a protein to the extent that they are non-neutral (Figure 2.1). However, depending on the position of the indel, or the particular protein, even these are not necessarily under strong selection (e.g., if the indel occurs in a surface loop that is not essential to a protein's function).

Regulatory Variants

In a landmark series of papers, Wilson and colleagues (1977) were among the first to emphasize that evolutionary novelty may be generated by structural (protein) variation, or by regulatory variation. In its purest form, regulatory variation would involve any change in gene dosage in a particular developmental context. Thus, the same protein present earlier, later, or in quantitatively different amounts would be considered a regulatory mutation, in contrast to a different protein product altogether.

How such regulatory variation might arise mechanistically was and remains a complex question. Basically any change affecting any stage in the central dogma, from transcript expression to mRNA half-life to protein processing or export, could be a regulatory mutation in that it would change gene dosage. Investigations at all these levels (the "transcriptome," the "proteome," even the "phenome") do indeed demonstrate heritability for regulatory variation (Oleksiak et al. 2002). In fact the molecular mechanisms that could potentially give rise to regulatory variation are nearly infinite.

The technological ability to detect regulatory variation is recently developed and still evolving, but a theoretical understanding of where regulatory variation comes from in terms of mutational models remains woefully lacking. For example, does a 2-fold change in the phenomenon assayed as "gene expression" come from a single mutation or multiple mutations? What molecular commonalities, if any, underlie 2-fold changes and are they systematically different from 10-fold changes? This situation of a data flood preceding molecular models of DNA sequence or other biochemical changes that give rise to the variation being measured is very different from that in which measurement of genetic variation in coding sequence, and selective expectations for that variation, arose. While DNA sequencing technology was still extremely challenging and data were scarce, the triplet code and its sources of redundancy were understood. Thus, simple null predictions could be generated for DNA sequence variation well in advance of our having the data to evaluate them. For example, changes that affected amino acids were expected to be under stronger selection, and more likely to be adaptive, than synonymous changes. When cloning and sequencing a gene was still enough work for an entire PhD thesis, the simple predictions were developed into the first statistical model of neutrality, the HKA test (Hudson et al. 1987; Nachman, Ch. 7 of this volume).

The opposite situation is true for regulatory data, which continue to accumulate at ever faster rates thanks to developments in microarrays and other technologies for global expression analysis. There are few a priori expectations of selective outcomes of regulatory variation, and very minimal if hard-won models of how such variation arises (Crawford et al. 1999; Phinchongsakuldit et al. 2004). Part of this is because regulatory variation is grossly more complicated and multidimensional than the triplet code. Thus, both experimental and computational studies of promoters, enhancers, and silencers have focused more on their identification and consensus sequences than on quantitative predictions of mutational outcomes and on how variation affects the activity of these *cis* elements (Goodrich et al. 1996). Thus empirically based models for the source of regulatory variation are rare. Instead, approaches that are purely computational, either mathematical or statistical, remain the primary resource for evolutionary geneticists wishing to make sense of the flood of data (Werner 2003).

Interestingly, most of the support for studying regulatory variation by evolutionary biologists arises from molecular quantitative genetics. On the methodologically challenging path to link quantitative phenotypic differences to sequence polymorphisms, workers in systems ranging from plants (Doebley 1992), to flies (Long et al. 1998), to fish (Shapiro et al. 2004) have demonstrated that protein variants are unlikely to explain at least some phenotypic differences, as SNPs map either to introns or to upstream untranslated regions. However, this remains far from a predictive understanding of how molecular regulatory differences are manifested in phenotypic differences. Indeed linking expression and proteomic data to phenotypic outcomes may be regarded as the challenge for the current generation of evolutionary geneticists.

Again due to the lack of empirical models, studies linking observed regulatory mutants to actual selective differences are rare. However, some notable exceptions exist. The work of Stephan and colleagues stands out as a classically elegant approach to the problem. Through an exacting process of mutagenizing an intron, they were able to demonstrate that compensatory mutations preserving secondary structure not only increased mRNA amounts, but increased protein amounts and had an effect on fitness (Chen & Stephan 2003). Similar results were obtained for mutations in the 3′ untranslated region (Baines et al. 2004). Microbial geneticists have also been successful in demonstrating that regulatory variation leads to fitness variation. For example, in yeast adapted to different artificial glucose-limited environments, Botstein and colleagues observed repeatable changes in gene expression in genes related to glycolysis (Ferea et al. 1999).

Intersections between Structural and Regulatory Genetic Variation

Structural and regulatory variation are presented above as distinct phenomena for purposes of presentation. However, this is clearly not the case, as may be demonstrated by considering expression. Gene expression variation is certainly a type of regulatory variation. However, it is at some level controlled by transcription factors, which are proteins, and hence subject to structural variation. Protein variation in transcription factors clearly influences transcript abundance in their target genes. Indeed, expression variation of a given gene may map to *cis*-acting factors in the gene itself, or *trans*-acting factors such as transcription factors (Schadt et al. 2003). Thus the two types of variation are neither binary nor mutually exclusive, but rather are more likely to be selected in concert.

ORIGIN AND MAINTENANCE OF GENETIC VARIATION

Ultimately, mutation is the origin of all genetic variation. Mutations are reassorted between lineages by a variety of mechanisms, but foremost among these are sexual reproduction and recombination. These topics have been the subject of many reviews, recent and classical (Morgan 1932; Otto & Lenormand 2002), and thus are not discussed extensively here. In brief, sexual reproduction, that is meiosis, is the basis of Mendel's first and second rules (segregation and independent assortment). Sex and recombination provide a mechanism to group beneficial mutations together within a single individual (rather than requiring multiple beneficial mutations to arise on a single chromosome), an important evolutionary innovation. Likewise, recombination allows beneficial mutations to fix in a population without dragging along deleterious mutations; similarly, deleterious mutations can be removed from a population without dragging along beneficial mutations. Without recombination (or some other mechanism of genetic exchange), chromosomes and populations would accumulate deleterious mutations and gradually decline in fitness. For example, lack of recombination causes the decay and eventual loss of the heterogametic sex chromosome, including our own Y chromosome (Charlesworth 2003).

Intra-Population Sources

Mutations are the source of within-population variation. Here, we will focus on the distribution of fitness effects of new mutations, emphasizing those caused by transposable elements (TEs) at the end. Molecular descriptions of mutations include the same types of alleles found in the population at large, and have already been presented in this chapter in the section Variation at the Molecular Level.

The distribution of effects of new mutations on fitness has been a hotly debated topic for at least the last 40 years. Most investigators agree that the vast majority of new mutations are deleterious. However, the fraction that are selectively neutral, and the fraction that are beneficial, remain contentious. There remain two major schools of thought that can be grouped into the nearly neutral camp and the selectionist camp. Although these groups may be thought to distinguish themselves largely by theories about the source of evolutionary novelty and the relative importance of the forces directing evolutionary change, these parameters are set by what we expect the frequencies of beneficial mutations to be. The nearly neutralists think that after removing the large class of unconditionally deleterious mutations, the remaining mutations are largely "nearly" neutral (Ohta 1992). Nearly neutral mutations may not actually be selectively neutral, but rather mildly deleterious or mildly beneficial; however, due to the relatively small sizes of the populations they are found in, their frequencies are largely controlled by genetic drift, rather than by natural selection. Thus the majority of fixations between populations and species will not be due to adaptation per se but rather to stochastic forces. In contrast, selectionists believe that although most mutations are unconditionally deleterious, only a tiny fraction of new mutations are selectively neutral; and more mutations are expected to be beneficial (Drake et al. 1998).

The insertion of TEs at random into the coding sequence of genes generally is strongly selected against (Charlesworth & Langley 1986) because this usually disrupts the protein at a gross level, either by truncation because of stop codons in the TE, frameshifts in the host protein, or other damage to the integrity of the protein. However, there are examples of TE insertion being potentially adaptive, for example, in cases where TE insertions upstream of a gene can then be coopted as enhancers or where the TE becomes incorporated into the protein coding sequence of a gene (Brosius & Gould 1992; Jordan et al. 2003). The question of the fraction of mutations that are caused by TEs versus other mechanisms is taxon-specific, with estimates ranging from as much as 50% in flies (Charlesworth & Langley 1986) to less than 1% in humans (Luning & Kazazian 2000).

Extra-Population Sources

Additional genetic variation can be introduced into the gene pool by the incorporation of genes that originated in other populations, species, and/or groups. Here, the transmission of these genes is "horizontal," since these contributions are the result of genetic exchanges across different lineages (rather than the "vertical" passage of traits from ancestors to descendants within the same lineage).

Because of local environmental differences and finite population sizes, different populations will diverge from each other over time due to both selection and drift (Goodnight, Ch. 6 of this volume). The genetic differences among populations that arise as a result of this differential selection and drift can then be introduced into the gene pool of a specific population of the same species by migration. Thus, migration can take new genetic variants that arise in one part of a species' range and distribute these new types to other conspecific populations. In the process, the gene pools of the different populations of the species become homogenized as genetic variation is shared among its different populations.

In an analogous fashion, different species can sometimes hybridize, thereby leading to the introduction of new genetic variation into the gene pool of one (or both) of the parental species (Arnold & Burke, Ch. 26 of this volume). For several closely related species (including many that are widely recognized and accepted taxonomically), interspecific hybridization is of known importance, as either an ongoing process or one in the recent past for the introduction of new genetic variation in one or both of the parental groups (Grant & Grant 2002).

In many cases, interspecific hybridization is restricted to a narrow contact zone between the parental species and/or to the introgression of just one or a few genes from one parent into the gene pool of the other. Because of their reproductive isolating mechanisms, the progeny of such interspecific crosses most often show reduced fitness and thus can be a "sink" for genetic variation (Servedio & Noor 2003). However, hybrid vigor, and evolutionary novelty, can sometimes result through hybridization (Rieseberg et al. 2003). Of perhaps broadest significance (as emphasized before) is the fact that many flowering plant species most likely represent allopolyploids that originated as a consequence of interspecific hybridization (Soltis & Soltis 1999).

Populations of bacteria and single-celled eukaryotes can also undergo sexual conjugation, and thus, migration and interspecific hybridization can apply to them. However, migration and interspecific hybridization in microorganisms are often facilitated by plasmids that promote the efficient spread of new genetic variation within a microbial population and/or species. Furthermore, in addition to migration and interspecific hybridization, microbial populations can also receive new genetic contributions from the outside via mechanisms of horizontal transmission that are largely unique to them. These additional external sources of horizontal transfer include transduction, transformation, and endosymbiosis (Dyall et al. 2004; Syvanen & Kado 2002).

Transduction refers to the horizontal transmission of genes among individuals and populations by a viral agent, whereas transformation corresponds to the capacity of microorganisms to pick up and incorporate external DNA into their own genome. Regarding transformation, this free DNA can represent the degraded fragments of dead organisms in the environment, or the digested products from the endocytotic ingestion of one cell by another. On rare occasions, the incorporation of one or more cells by another may not result in the degradation of the incorporated cell, but rather to a permanent symbiosis, as is widely accepted for the origin of complex eukaryotic cells. Thus, according to this theory of endosymbiosis, eukaryotic cells are chimeric (Margulis 1993; Katz 1998), as is developed further below.

Although horizontal transmission due to migration and interspecific hybridization is common in multicellular organisms, horizontal gene transfer without hybridization is probably rare, with only a few well-documented cases to date (one interesting exception is the acquisition of *P* elements in *Drosophila melanogaster*: Kidwell 1993). In sharp contrast, such horizontal transfer is widely accepted as a major source of new genetic variation in microorganisms. For example, it is estimated that approximately 25% of the genome for *Escherichia coli* (the common intestinal bacterium of humans and other mammals) is the result of lateral gene transfers from other groups, rather than direct vertical inheritance within its own lineage (Lawrence and Ochman 2002).

The three domains of life are the Archaea, Eubacteria, and Eukaryota, with the former two groups traditionally referred to as the Prokaryota. Although several alternative versions exist, a favored hypothesis for the origin of the eukaryotic nuclear genome is that it arose from the fusion of genomes for an archaeal versus eubacterial ancestors (Rivera & Lake 2004). This genome fusion was possible because of the direct endosymbiotic relationship between these two prokaryotic ancestors. Subsequently, a second endosymbiosis between this hybrid lineage and an ancestor of the purple bacteria (i.e., the group that includes *E. coli*) led to the origin of the eukaryotic mitochondrion. Thereafter, a number of genes from the mitochondrial genome were transferred to the nuclear genome. In an analogous way, chloroplasts were acquired from cyanobacteria by some eukaryotic lineages. In turn, numerous horizontal gene transfers were also occurring (as they continue to do) among different microbial groups.

Genetic Drift and Selection

Given that all genetic variation has its ultimate origin in mutation, and that it is reassorted by recombination and sex, what other forces give rise to the variation that we see? The two major contributors are the opposing processes of genetic drift and natural selection, and the interactions between these processes as mediated by population size.

Genetic drift changes allele frequencies by sampling error, a stochastic process (Gillespie, Ch. 5 of this volume). Without regard to its effect on fitness, a new mutation has a probability $1/2N$ of fixation in a diploid organism, where N is the population size. The smaller the population, the more likely that the new mutation will be retained. Consider the following scenario of a very small population ($N = 2$), with one heterozygous (*Aa*) male and a heterozygous female. Given these two heterozygous parents, the probability of their having an *AA*, *Aa*, or *aa* offspring is 1/4, 1/2, and 1/4, respectively. Now assume that these two parents have two offspring who replace them. The probability of the two offspring being *AA* is 1/16 ($1/4 \times 1/4$), which is also the probability that both are *aa*. Thus, the probability of this population becoming fixed for either allele, *A* or *a*, by drift in the next generation is 0.125 (1/16 + 1/16).

In contrast, selection has to do with the intrinsic effects of the allele with respect to fitness (Wade, Ch. 4 of this volume). New mutations with a positive selection coefficient (beneficial) should be fixed, and new mutations with a negative selection coefficient (deleterious) should be lost. However, this ignores drift and population size.

Selection is most powerful in infinitely large populations. For selection to predominate over drift, the selection coefficient for an allele (s) must be $> |1/N_e|$ (i.e., greater than the absolute inverse of the effective population size N_e, where N_e is defined as the size of a theoretical population that is drifting to the same degree as the actual population (Li 1978)). Thus in a population of $N_e = 100{,}000$, selection will predominate over drift even if s for the different alleles is only on the order of 10^{-5}. Conversely, in sufficiently small populations, genetic drift is a more powerful force for fixing mutations than selection, even for mutations with fairly large fitness effects. For example, in a population of $N_e = 20$, drift will predominate over selection even for alleles with $s = 0.02$ (i.e., a 2% advantage of one allele over another). This observation partially explains the interest of conservation geneticists in describing minimum viable population sizes for endangered organisms (Frankham, Ch. 32 of this volume).

Gene and Genome Duplications

The existence of gene duplications begs several questions (Thornton, Ch. 11 of this volume). First, why are duplicated genes preserved if their function is truly redundant? Second, if they are not preserved, how are they eliminated: by random deleterious mutation including silencing, or by silencing only? Third, are duplicated genes recruited for new functions, and if so, how?

The existence of widespread gene duplications as a major source of evolutionary novelty was first proposed by Ohno (1985), who speculated that duplicate copies of many genes would be redundant and therefore one copy or another would be expected to lose function over time due to the random accumulation of deleterious mutations—so long as function was preserved in at least one copy. However, the redundancy of functional copies could also give rise to the evolution of novel functions in the new copy. Duplicate genes exist with little apparent divergence between copies; this and the presence of nulls without phenotype has been taken as a genetic indicator of redundancy. A given gene will be duplicated at a rate of approximately 1% per million years in animals, and the average time to silencing for a duplicated gene in animals is approximately 4 million years (Lynch & Conery 2000).

Redundancy is expected to be more common for developmental genes than for housekeeping genes (Cooke *et al.* 1997). If developmental error (transient, nonheritable failure of transcriptional machinery) occurs at a rate greater than the germline mutation rate for loss of function, selection could maintain functionally redundant genes. Further, redundancy may be at a single stage only, and in fact apparently redundant genes have evolved new functions at other stages.

However, global predictions of elimination of redundant duplicate genes by deleterious mutations ignore their functional properties. By invoking the genetical and molecular properties of proteins, Spring (1997) suggested that for multidomain proteins, random point mutations leading to missense mutations might be highly deleterious because protein–protein interactions would be disrupted (i.e., dominant negative mutations due to intra- and inter-protein interactions between mutated and nomutated alleles; this could be true even for proteins that do not have multiple domains). This would lead to purifying selection maintaining multiple functional copies, and hence a higher persistence time for duplicated genes for such proteins. The persistence of multiple redundant copies would thereby selectively protect against deleterious missense mutations, opening the door for evolving new functions, particularly with respect to new patterns of expression. It is also possible to imagine that gain-of-function missense mutations would potentially be adaptive in this situation, as might mutations that confer tissue-specificity. This possibility was explored in great detail in a series of papers by Lynch and Force, who suggested that in small to medium populations, given a higher frequency of mutations conferring subfunctionality than mutations completely silencing the protein, duplicated genes are more likely to be preserved than lost; while in larger populations, neofunctionalization is more likely than subfunctionalization (summarized in Walsh 2003). However, recent work by Lander and colleagues in yeast indicates that both subfunctionalization and neofunctionalization are probably rare events, at least in this species, as 90% of duplicated genes were lost after the original genome-wide duplication event; and in genes where both members of the pair are retained, accelerated amino acid replacement rates are seen in both copies only 5% of the time (Kellis et al. 2004). However, criteria for accelerated evolution were quite stringent and also restricted to amino acid sequence variation, with no analysis of differential expression via promoter or other noncoding divergence; and there was no opportunity for tissue-specific expression.

In the context of the deleterious dominant negative effect of missense mutations, null mutations that eliminate expression in redundant proteins might be selectively neutral. Indeed, a scenario with many multiple copies of genes with epistatic properties (and hence potential for dominant negative phenotypes) might well select for gene silencing, as speculated by Bird (1995b).

However, this phenomenon is unlikely to explain the large number of redundant genes, if only a limited number of genes would exhibit a "dominant negative" phenotype (Cooke 1998). However, data based on the number of genes in the *Drosophila* genome (Adams et al. 2000) suggest that genetic complexity is indeed a result of gene interactions, rather than numbers of genes per se. This argues for widespread epistasis, which in turn suggests that the number of dominant negative, albeit epistatic, mutants may be greatly underestimated.

Gene duplications potentially could play a major role in speciation under the Dobzhansky–Muller model. In this model, mutations accumulate independently in separated populations that are neutral within their home population but are deleterious in hybrids or hybrid progeny (Johnson, Ch. 24 of this volume). Epistatic interactions between the mutations in both the home populations and the hybrids are implicit in this model. Lynch and Force proposed an epistasis-free model of speciation involving random loss of copies of duplicated genes. If one population lost one copy of the gene and the other population lost the alternative copy, which should occur approximately 50% of the time, then hybrid individuals might produce gametes without any functional copies, thus resulting in hybrid breakdown in the F_2 generation (Lynch & Force 2000).

FUTURE DIRECTIONS

We have considered the links between selection on expression, sequence variation, and epigenetics in the context of single genes, gene duplications, and polyploidy, thus uniting genetics, genomics, and function. We close by highlighting two of the many areas ripe for future research. First, we encourage further studies on the relative frequencies of mutations with respect to their mechanisms of change (e.g., point mutations due to replication errors versus chromosomal rearrangements due to TEs) and their phenotypic effects (e.g., advantageous versus deleterious, nearly neutral, and neutral). These studies will help to establish the primary sources and nature of the genetic variation that underlies all biological diversity and evolution. Second, we call for additional studies of functional sources of genetic variation in gene expression from the genotype to phenotype levels. These investigations will help to determine how genetic variation at the molecular and cellular levels is linked to the differences among individuals, populations, and species. In these ways, a more unified understanding of the origins, maintenance, and evolution of genetic variation will emerge, one that views this diversity in terms of its importance to all other levels in the hierarchy of life.

CONCLUSION

There are a large number of ways one might approach writing an introductory chapter on genetic variation, and at least as many opinions as to which is the most useful. Rather than creating a laundry list of types of genetic variation with no connection between mechanistic and evolutionary properties, we have endeavored to integrate molecular, phenotypic, functional, genetical, and genomic descriptions of genetic variation into what we hope is, as a whole, more useful than the sum of its discrete parts. Throughout we emphasized the continuity between quantitative and population genetics, for quantitative traits are constructed from a series of single genes. We have also emphasized the relationships between sequence and expression polymorphism. Finally, we have considered the links between selection on expression, sequence variation, and epigenetics in the context of single genes, gene duplications, and polyploidy, thus uniting genetics and genomics.

SUGGESTIONS FOR FURTHER READING

Avise (2004) provides a comprehensive overview of the utility of molecular data in comparative studies of populations and their species to the major domains of life. This text reviews the various ways that such data are collected, analyzed, and interpreted, and the concepts that have emerged from these comparative molecular studies. Futuyma (1998) represents the classic text for an introductory course in evolution. It provides broad coverage of the entire field and is written to be easily accessible to anyone in the life sciences. Graur and Li (2000) is the classic introductory text for the study of molecular evolution. Its balanced coverage of empirical data, theory, and statistical methods allows easy access to this field for many different disciplines in biology. Hedrick (2005) is one of several books that provide an excellent introduction to the classic and modern concepts and approaches in population genetics. The focus of these textbooks is on the demographic and evolutionary factors that introduce, maintain, and alter alleles in gene pools. Lewin (2004) is the classic reference for modern molecular genetics and biology. It provides a comprehensive overview of genetic concepts, processes, and approaches at the molecular and cellular levels. Strachan and Read (2004) is the classic reference for the molecular genetics and biology of humans, which remain the ultimate model organism. Despite its title, this textbook also reviews advances in human population genetics and molecular evolution as well as in molecular and cellular biology.

Avise JC 2004 Molecular Markers, Natural History, and Evolution, 2nd ed. Sinauer Assoc.

Futuyma DJ 1998 Evolutionary Biology, 3rd ed. Sinauer Assoc.

Graur D & W-H Li. 2000 Fundamentals of Molecular Evolution. Sinauer Assoc.

Hedrick PW 2005 Genetics of Populations, 3rd ed. Jones and Bartlett Publishers

Lewin B 2004 Genes VIII. Pearson Education

Strachan T & AP Read 2004 Human Molecular Genetics 3. Garland Science

Acknowledgments We thank Ed Braun and Laura Higgins for thoughtful suggestions and comments; and Michele Tennant for invaluable assistance with the references.

3

Mutation

DAVID HOULE
ALEXEY KONDRASHOV

Mutation is the ultimate source of all genetic variation, and genetic variation is absolutely necessary for any sort of evolution to proceed (Wayne & Miyamoto, Ch. 2 of this volume). These two facts should make the study of mutation the foundation of evolutionary genetics. Unfortunately, this is very far from the case. Both the processes of mutation and the pattern of effects of those mutations are relatively little known compared with the properties of the standing genetic variation within populations and among populations and species. However, these properties can only be understood in light of mutation.

Two classes of mutation are particularly important to evolution: those with beneficial and those with deleterious effects on fitness. Evolutionary change depends on the input of beneficial mutations. Unfortunately, such mutations are usually quite rare and hard to study. The vast majority of mutations that affect fitness decrease it. Because these mutations are common, their effects in total can have very large evolutionary consequences. Such disparate phenomena as inbreeding depression, sexual selection (Moore & Moore, Ch. 22 of this volume), recombination and sexual reproduction, and senescence (Promislow & Bronikowski, Ch. 30 of this volume) are all affected by, or even explained by, the commonness of deleterious mutations.

Genetic information is encoded in nucleic acids; genetic variation is created when the sequence of these is altered. In all cellular organisms DNA is the genetic material. This makes the study of mutation at the DNA level both natural and important. The focus at this level is on the rate at which various sorts of alterations occur. It is also necessary to understand the effects that these alterations have on the phenotype of the organism. This latter aspect of mutation is a key to resolving one of the major controversies in evolutionary biology: whether evolution is limited by the supply of genetic and phenotypic variation (Gould & Lewontin 1979), or by natural selection (Charlesworth et al. 1982).

The difficulties of the study of mutation stem largely from one simple fact: individually, mutations are very rare events. This has always made the direct study of mutation, consisting of recording new mutations as they arise, exceptionally tedious. For example, one of the best direct studies of mutation in mammals is that of Russell and Russell (1996), who report examinations of over 1,000,000 mice to find just 46 visible mutations.

The alternative to this tedium is to fit a model to genetic variation within a population or to the variation found among populations or species. These model-based approaches utilize data on contemporary variation, which are relatively easy to gather, to obtain information about mutations that happened over a considerable period of time. Consequently, much of what we know about mutation comes from model-fitting. Such efforts have a very important Achilles heel—if the model is not correct, the results can be very misleading. A model is necessary because the variation studied, while certainly due to mutation, has also been filtered by natural selection and genetic drift. This tension between direct observation and indirect, assumption-based approaches is very common in evolutionary biology.

CONCEPTS

What Is Mutation?

Most, but not all, changes in DNA sequence are mutations. The exception is reciprocal recombination, where both homologous DNA sequences are broken and rejoined at the same point. This does change the DNA sequence, but its reciprocal nature conserves information. Small-scale alterations of one or a few base pairs are mostly caused by errors in DNA replication and repair, and thus are unambiguous mutations. Sequences that emerge after large-scale changes often contain clearly identifiable, rearranged pieces of old sequences, so that recombination as well as errors per se may be involved. Many of these changes that can be viewed as either mutation or nonreciprocal recombination. Such processes will be considered here, because they share the rarity and irregularity of small-scale mutations.

This definitional difficulty carries over to the language of variation. A bit of DNA sequence that has been changed from the copy in its parent is clearly a "mutation," but when two sequences differ, we cannot assume without direct knowledge of their ancestry, that one variant is the "mutation" while the other is "ancestral," "normal," or "wild-type," although in practice a deleterious or low-frequency variant is often referred to as a mutant. This is justified for deleterious variants, which are usually lost due to natural selection, and thus cannot be ancestral; in contrast, a low-frequency variant may, nevertheless, be ancestral. Genetic variants are variously called alleles when referring to alternate forms of a gene or haplotypes when referring to longer DNA sequences. We will adopt the term "variant" as the most general, and least loaded term.

Replication, repair, and recombination of DNA sequences are incredibly complex molecular processes, each involving interactions of dozens of macromolecules. However, we will ignore biophysical and biochemical aspects of these processes, and adopt a simple transmission genetics approach, concerned with inputs, outputs, and rates.

Classification of Mutations

Mutational events may affect anything from 1 base pair to entire genomes. Small-scale mutations that affect only a few nucleotides are categorized in Table 3.1. Small-scale mutation depends on the local sequence context. The most important aspect of context is whether the sequence consists of repeated sequences (for example AGAGAG, known as "periodic"), or a more typical sequence where the progression of base pairs is more or less unpredictable ("complex").

Mutations in short periodic sequences, called micro- and minisatellites, are mostly deletions and insertions, usually of lengths that are multiples of the period length. They often occur with rates

TABLE 3.1. Classification of mutational events involving small numbers of base pairs

Event	l_1[a]	l_2	Example	Description
Nucleotide substitution	1	1	AGC → ATC	Any single base pair change
Transition	1	1	AGC → AAC ACC → ATC	Base pair substitution of a purine (A or G) with another purine, or of a pyrimidine (T or C) with another pyrimidine. More common than transversions
Transversion	1	1	AGC → ATC	Base pair substitution of a purine with a pyrimidine, or a pyrimidine with a purine. Less common than transitions
Deletion	≥ 1	0	AGGC → AC	One or several successive nucleotides are removed
Insertion	0	≥1	AC → AGGGC	Insertions of one or several successive nucleotides
Complex events—	1	>1	AGC → ATTC	Combined indel/substitution very rare. Gene
combine substitution and deletion or insertion	>1	1	AGGC → ATC	conversion (see Table 3.2) can cause complex changes
	2	2	AGGC →ATTC	Simultaneous substitutions do occur at appreciable rates

[a] l_1 is the length of the affected sequence before mutation; l_2 is the length of the sequence after mutation.

that are orders of magnitude higher than those of mutations in complex sequences. Microsatellites are unlikely to have important functions, but their frequent mutations provide abundant data for tracing ancestry. Even within complex sequences, the context can affect mutation rate significantly (Kondrashov & Rogozin 2004). An important case is the elevated mutation rate in mammals of the dinucleotide CpG. CpG pairs consist, in both strands, of a cytosine followed by a guanine, where p refers to the phosphate that joins them. The impact of a local context is so important that, say, insertions into complex sequences versus microsatellites, or transitions within versus outside CpG are often considered as separate types of mutations.

Large-scale mutations are categorized in Table 3.2. With the exception of long deletions, these changes usually involve recombination of pre-existing segments of sequences. The reason for this is that de novo origin of a long DNA sequence is a rather unlikely occurrence. Thus, sequences added by a large-scale mutation are usually copies, often modified, of pre-existing sequences. The probability that such a sequence (which can code for a protein domain, or even an entire protein) would be functional is substantial (Thornton, Ch. 11 of this volume). Creation of homologous sequences by duplication can catalyze further large-scale events, as it creates the opportunity for nonreciprocal recombination. Transposable elements provide a particularly important example, as they can also lead to

TABLE 3.2. Mutational events involving large numbers of base pairs

Event	Where does new sequence come from?	Example	Details
Deletion	—	$A[S_1]G \rightarrow AG$	Removed sequence can be very long. Rates increase if flanking sequences are similar
Tandem duplication	Neighboring sequence	$A[S_1]G \rightarrow A[S_1][S_1]G$	S_1 may be a rather long sequence, sometimes a large proportion of the chromosome
Nontandem duplication LGT	Non-neighboring sequence	$AG \rightarrow A[S_2]G$	S_2 is a sequence of any length from elsewhere in the genome. In practice only detected when S_2 is > 20 bp
Transposable element (TE) insertion	TE	$AG \rightarrow A[T_1]G$	T_1 is DNA derived from a TE. T_1 often consists of a fragment of TE sequence. Rate may be quite high. A common mode of nontandem duplication. Often insertion is accompanied by other changes (i.e., short duplications)
Lateral gene transfer	Other genome	$AG \rightarrow A[S_3]G$	Important in prokaryotes
Inversion		A[GT ... CC]A → A[GG ... AC]A	Rotates a sequence
Gene conversion	Homologous sequence		Sequence of one homolog converted to that of the other
Transposition	Sequence changes location		Usually through breakage and fusion of chromosomes
Chromosome break	None		Usually deleterious, but may increase chromosome number
Chromosome fusion	None		Usually deleterious. May decrease chromosome number
Aneuploidy	Chromosomal duplication or loss		Usually lethal, or very deleterious
Polyploidization	Same genome: autopolyploidy		Fixed more frequently in plants than animals
	Different genome: allopolyploidy		Often caused by hybridization. Fixed more frequently in plants than animals

small duplications through transposition (Petrov & Wendel, Ch. 10 of this volume).

Classification of Phenotypic Effects

Variant DNA sequences may or may not have an effect on the phenotype of their carriers. Those that do may have a cascading series of effects, from the regulation or function of a gene, through the biochemical, developmental, or physiological levels, and ultimately on the readily observable phenotype of the organism. It is useful to distinguish loss-of-function variants, for example a frameshift deletion, from those mutations that leave an altered functional gene, such as an amino acid substitution. Mutations that alter function usually result in quantitative changes in amount or timing of expression. Occasionally such changes may lead to a qualitatively different effects, such as new substrate specificity of an enzyme.

The most important phenotype is fitness, the capacity to produce offspring, which includes survival and reproduction (Fairbairn, Box 4.1 of this volume). The effect on fitness of a variant sequence is an important determinant of its evolutionary fate. However, fitness has surprisingly little effect on the fate of any single variant, as all start out rare where stochastic effects are very strong (Gillespie, Ch. 5 of this volume).

Mutations can be conveniently classified according to their fitness effects into the following categories:

1. Lethal mutations kill the individuals that carry them.
2. Deleterious mutations reduce fitness relative to alternative states, but not to zero. Considerable evidence suggests that such mutations are more common than lethals.
3. Neutral mutations do not affect fitness much, either positively or negatively. These too are likely to be common.
4. Advantageous mutations increase fitness, and therefore will be favored by natural selection. These are probably the rarest type of mutation.

This categorization is context-dependent. An advantageous variant in one environment may be neutral or deleterious in other circumstances (Scheiner, Ch. 21 of this volume). Also, the fitness of a variant might depend on the genotype it finds itself in (epistasis).

In diploid organisms, an important additional consideration is the dominance of a variant. Dominant mutations have their full phenotypic effects when present in heterozygous condition, while recessives only affect the phenotype in homozygous condition. If the mutant heterozygote is intermediate between the two homozygotes, the mutant is partially dominant. If the heterozygote is exactly intermediate, the variant is said to act additively.

The concept of dominance is itself phenotype-specific. For example, Mendel's wrinkled pea allele is recessive when the smoothness of the seed coat is examined, but when the amount of starch in the seed is measured, the heterozygote is intermediate to two homozygotes. When subjected to quantitative analysis, the vast majority of even major variants seem to be neither completely recessive nor completely dominant. Variants with minor phenotypic effects tend to be partially dominant, and are often nearly additive.

How to Study Mutation

Classification of the types of mutations (Tables 3.1 and 3.2) makes the study of mutation sound altogether straightforward. It is important to realize that our ability to study these different classes of mutations varies widely depending on their rarity and the nature of their phenotypic effects. In Table 3.3, we classify the study of mutation according to two criteria suggested in Kondrashov (1998): the time scale over which mutation is studied, and the type of data that is used to detect mutations.

Three different classes of characteristics may be used profitably to detect mutations: DNA sequences, phenotypes, and fitness. Studies of mutation almost invariably cover one of three time frames. First there is the direct study of mutation through comparison of parents and offspring. At a slightly longer time scale, one can set up a mutation-accumulation (MA) experiment. To do so, one maintains the population under conditions that minimize the impact of natural selection on the fate of any mutations that may arise (see Case Studies for examples). Finally, the comparative method infers mutation rate from the rate of divergence between species.

The use of different time frames allows different aspects of mutation to be investigated. The chief reason for these differences is the degree to which we can assume a realistic model for the interaction

TABLE 3.3. Categorization of approaches to the study of mutation, with reviews or examples of successful studies

	Generations separating samples		
	1 Direct	10 to 10^3 Mutation accumulation	$>10^5$ Comparative
DNA	Weber and Wong 1993	Denver et al. 2004; Schug et al. 1997	Nachman and Crowell 2000
Phenotype	Kondrashov 2002	Houle et al. 1996	Lynch 1990
Fitness	Woodruff et al. 1983	Mukai et al. 1972	

between mutation and natural selection. With direct studies, the need for assumptions about natural selection is minimal. Almost all types of mutations may be observed in the offspring. As the time frame of the study lengthens, the necessary assumptions about natural selection become more stringent. Only mutation rates to neutral alleles can simply be inferred over long time periods. As a result, the same data can lead to very different conclusions about the overall mutation rate, depending on the assumptions chosen.

Despite the complications in applying models to divergence data, the essential neutral theory behind such models is easy to grasp. If we consider a population of genetic variants with no impact on fitness, that is neutral variants, whether they are lost from the population or will become fixed (rise to a frequency of 1) depends only on genetic drift, the luck of sampling during reproduction. Lucky variants will become fixed; the vast majority will be lost just by chance. The chance that each particular variant will be fixed in the future is proportional to its frequency right now: rare variants are likely to be lost, common ones likely to be fixed. Now, let us consider the fate of each new neutral variant. If there are N diploid individuals in the population, each new variant starts out at a frequency of $1/2N$, and thus has a chance of $1/2N$ of rising to fixation. On the other hand, with a mutation rate m per gamete, the number of new mutations in each generation is $2Nm$. Multiplying these two together gives the surprisingly simple rate of neutral evolution: $k = 2Nm \times 1/2N = m$. This rate is the divergence from the ancestral sequence; species diverge at twice this rate because variants arise along both branches to the common ancestor.

In reality, variants have a range of effects on fitness from undetectable to lethality. Their ability to persist in the population and so be detected also depends on the effectiveness of natural selection at influencing frequencies; this depends on the size of the population. In a population where N is small, genetic drift (luck) will be a relatively strong force, swamping out small differences in fitness. However, when N is large, even tiny differences in fitness reliably discriminate higher and lower fitness variants. Mutation-accumulation experiments are therefore designed to maximize the impact of drift, either by making N as small as possible, or by equalizing family sizes (Shabalina et al. 1997). Thus, the influence of natural selection is minimal in a direct study, somewhat higher in a mutation-accumulation study, and very large in a comparative study. The result is that the neutral model can be applied to an uncertain and decreasing proportion of variants as the time scale of the study increases. Even at the DNA level, it is difficult to be sure that a particular segment really evolves at the neutral rate. For example, evolutionary biologists have treated pseudogenes, altered sequences derived from functional genes, as neutral (e.g. Nachman and Crowell in Case Studies, below). However, there are at least two possible mechanisms for selection on pseudogenes. First, recombination between pseudogenes and their parent gene is deleterious, so deletions of pseudogenes may be favored by natural selection. Second, the discovery of naturally occurring nonprotein-coding genes (such as micro-RNAs) that can regulate expression of their homologous genes suggests that some apparent pseudogenes may play such a selected role.

A second major disadvantage of comparative studies is that the number of generations that separate species is usually known only very approximately. As the time scale of any comparison becomes longer, these uncertainties become very large. The number of generations in a lineage since the Mesozoic era will hardly be ever known with any confidence at all. As a result, comparative data are usually summarized as a mutation rate per unit time. While this may be useful in some contexts, such as calibrating a molecular clock (e.g. Rodríguez-Trelles et al. Ch. 8 of this volume), it does not tell us what we want to know about mutation rates in organismal terms.

These considerations would seem to make direct studies preferable, were it not for the fact that the longer the time period, the greater the number of mutational events that can be assayed. Because mutations are individually rare, direct studies can only be informative when there is an efficient mechanism for screening enormous numbers of individuals for mutations. Such screening is available for many inherited phenotypes in human societies with advanced health care. At the other end of the biological spectrum, microbial populations can be rapidly screened for the converse sort of mutations that restore function at a defective gene (reviewed in Drake 1991). For other species, the direct data are limited. Furthermore, because the phenotypic impact of most mutations is usually small, we need to be able to infer from the minority of mutations that are observable the properties of the full spectrum of mutations.

Of the three different classes of characteristics that may be used to study mutation, DNA sequences are the most conceptually straightforward. The challenge with the use of sequence data is that care must be taken to account for the possibility of errors in scoring. This has so far limited the use of sequence data in direct or mutation accumulation studies. For example, to detect a sample of base pair mutations, which typically occur at a rate of 10^{-8} per generation, over a 100 generation mutation-accumulation experiment, one needs the ability to sequence many more than 10^6 nucleotides with an error rate well below 10^{-6}. The necessary methods are emerging and are starting to be applied (Denver et al. 2004). Exceptions are provided by sequences with especially high mutation rates, such as microsatellites or mitochondrial DNA.

The other two categories of data (phenotypes and fitness) refer to whole-organism characteristics. Fitness is, in some respects, just another phenotype, but is by definition under strong natural selection. Mutations with large effects on these phenotypes, such as genetic diseases in humans or visible and lethal mutations in *Drosophila*, can be counted. This is the basis for the direct studies in Table 3.3. By connecting such changes to the DNA changes responsible for them, as explained below for human genetic diseases, they can be used to gain very detailed information about mutation rates.

However, mutations with the very largest phenotypic effects used in direct mutation studies are themselves rarely of evolutionary significance, as they usually reduce fitness. Quantitative trait locus and developmental studies of species differences suggest that both detectable-if-you-know-what-to-look-for and small effect variants are the major sort of variation that allows evolution. Their cumulative effects are usually studied in a mutation-accumulation experiment. In most such experiments, an initially inbred genotype is replicated and selection on each replicate minimized by lowering N for each replicate, for example by selfing or brother–sister mating. The rate at which variation in the phenotype accumulates is used to measure the increase in phenotypic variance due to a single generation of mutation, V_M. In addition, a change in the mean of the accumulation lines indicates that mutations are biased in their effects. For example, fitness and its key components of viability, fecundity and mating ability, are maximized by natural selection, suggesting that the mutations that arise will on average decrease fitness. Information on V_M and mutational bias can sometimes be combined to give a very crude estimate of the overall mutation rate of all genes affecting fitness, as in Mutation Accumulation in *Drosophila* in the Case Studies section below.

Mutation Rates

The most detailed picture that we have of mutation rates in eukaryotes is for humans. This is due to several factors. First, human–chimpanzee is the only species pair for which the total number of generations since their divergence is reasonably well known, facilitating comparative studies (see Nachman and Crowell in Case Studies, below). Second, data on Mendelian diseases provide the only large-scale phenotypic screenings for de novo mutations in any eukaryotic species for direct estimates (Kondrashov 2002). For our species, the comparative and direct approaches suggest a very similar mutation rate of about 2×10^{-8} per nucleotide per generation. The fact that the two methods give essentially the same estimate is quite encouraging. Substitutions account for about 95% of this total, with short insertions and deletions accounting for almost all of the remainder. Large-scale mutations are generally rare.

This simple picture conceals substantial complexities based on the sequence that surrounds a particular site. In mammals, substitution rate at CpG sites is elevated by a factor of more than 10. The reason for this is that the C tends to be methylated, which misleads DNA polymerase. This confusion results in a transition from C to T, and the destruction of the CpG context. Oddly, this results in a lower mutation rate of noncoding DNA than in coding DNA. In coding regions, natural selection to maintain a particular sequence can preserve CpG pairs despite this high mutation rate away from them; in noncoding regions mutation pressure rapidly destroys them. Duplicated sequences are another frequent source of mutational hot spots because nonhomologous recombination between the similar sequences leads to large-scale mutations. For example, about half of the mutations causing hemophilia A are caused by recombination of the gene with its nearby pseudogene.

Another complexity in the human data is that it appears that mutation rates are substantially higher in males than females (Drake et al. 1998), contrary to the impression created by the well-known maternal age effect on Down syndrome. Part of this effect is probably due to the fact that the spermatogonia divide continuously throughout a male's lifespan, while oocytes essentially stop dividing before birth. It has been estimated that the number of divisions in the germ line of a 30-year-old human is 31 for a female and 400 for a male; this difference increases with paternal age.

These facts point up the difficulties in generalizing about mutation rates across species. The methylation that gives rise to the CpG bias is not universal, and other biases undoubtedly arise in other groups. While human mutations are overwhelmingly single-base substitutions, mutations in *Drosophila* tend to involve more short insertions and deletions, and more large-scale mutations. The details of gametogenesis and details of life history, such as the average age at reproduction, can have a big effect on mutation rates, even when the cellular details of meiosis and replication remain the same.

With these difficulties in mind, data on mutation rates in some other well-studied DNA-based systems are shown in Figure 3.1. In the nematode worm *Caenorhabditis elegans* mutation rates were obtained by sequencing random nuclear sequences in a mutation-accumulation experiment (Denver et al. 2004). Mutation rates are relatively well known in viruses (phages) and bacteria from direct and mutation-accumulation studies, as it is possible to rapidly screen huge numbers of individuals for novel phenotypes (Drake 1991). With far more investigator effort, substantial direct phenotypic mutation assays have also been performed in several model systems (Schalet 1960; Woodruff et al. 1983; Russell & Russell 1996; Drake et al. 1998). In all these studies, investigators first screen for phenotypic mutations at previously identified loci. These values are then combined with the size of the locus and an estimate of what proportion of all DNA changes will result in a mutant phenotype to arrive at an overall mutation rate. These estimates may be inaccurate; for example, the direct *C. elegans* mutation rate from sequencing is one order of magnitude higher than the estimate that Drake et al. (1998) obtained using data from direct phenotypic assays. Finally, there is a large amount of comparative data from which mutation estimates can be calculated. As an example of this sort of estimate we have used the calculations of Keightley and Eyre-Walker (2000), who calculated rates based on the assumption that synonymous sites evolve at the neutral rate.

The data in Figure 3.1 make it clear that the mutational properties of organisms are extremely different. To help interpret this variation, the genome sizes and number of replication events per generation are shown in Table 3.4. Figure 3.1a shows that mutation rates per base pair per generation vary over three orders of magnitude. Viruses seem to sacrifice accuracy for speed, while multicellular organisms with long generation times accumulate mutations over many cellular replication events. When the numbers of mutations over the whole genome are summed, it is clear that mutation rates on average rise with the genome size of the organism. Drake (1991) called attention to the fact that viruses and unicellular organisms have a fairly constant mutation rate per genome. It is now clear, however, that this relationship does not hold for multicellular organisms (Figure 3.1b; Drake et al. 1998).

A second point concerning Figure 3.1 is that the discrepancies between mutation rates obtained using different approaches can be very substantial. The mutation rates obtained from assuming that third-base-pair positions evolve at the neutral rate are substantially lower than those obtained using direct evidence in *Drosophila* and *Mus*. In humans, there is no such discrepancy, with synonymous rates being within a factor of 2 of the direct estimate. This difference may be caused by the larger population

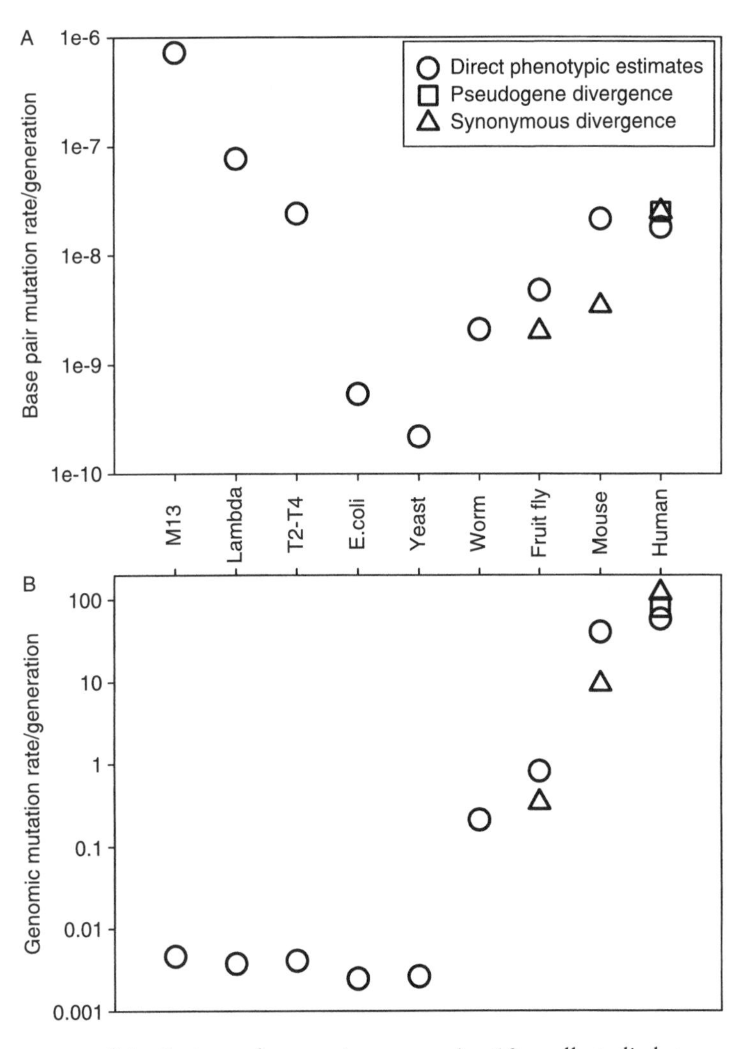

FIGURE 3.1. Estimated mutation rates in 10 well-studied taxa. (A) Mutation rates per base pair per generation. (B) Haploid mutation rates per genome per generation. Circles, direct and mutation accumulation estimates; square, estimate assuming that divergence in pseudogenes is neutral (Nachman & Crowell 2000); triangle, estimates assuming that divergence in synonymous base pairs is neutral (Keightley & Eyre-Walker 2000).

sizes of *Drosophila* and *Mus* relative to that of *Homo*, allowing a larger proportion of human variants with very small deleterious effects to evolve at the neutral rate.

The above estimates do not reflect whole gene duplications, which may be critical for long-term changes in the genome (Petrov & Wendel, Ch. 10 of this volume; Thornton, Ch. 11 of this volume). Genomic data from several species suggest that there is about a 1% chance that a duplicate copy of gene will be fixed per million years (Lynch & Conery 2000). For *Drosophila melanogaster*, this suggests that the fixation rate of duplications is about 3×10^{-10} per gene per generation. This rate may be higher or lower than the actual duplication rate, depending on whether duplications are deleterious (for example due to recombination between duplicates) or advantageous. Bearing in mind that

TABLE 3.4. Genome sizes, generation times and average number of replication events/life cycle for taxa with mutation data in Figure 1.

Taxon	Common name	Genome sizes	Generation time (days)[a]	Rep./gen.[b]
M13 phage		6.4×10^3	0.01	1
Lambda phage		4.9×10^4	0.01	1
T2 and T4 phages		1.7×10^5	0.01	1
Escherichia coli		4.6×10^6	0.01	1
Saccharomyces cerevesiae	Brewer's yeast	1.2×10^7	0.1	1
Caenorhabditis elegans	Nematode worm	1×10^8	4	9
Drosophila melanogaster	Fruit fly	1.7×10^8	12	30
Mus musculus	House mouse	3×10^9	275	43
Homo sapiens	Human	3×10^9	7300	215

[a]These times are minimum generation times under ideal conditions. Such conditions may not be typical of those in nature.
[b]Estimated number of replication events per generation.

a gene duplication will affect many nucleotides, the overall rate of duplications may affect as many or more base pairs as do base pair mutations.

Mutations and Fitness

Mutations that affect fitness are more likely to be deleterious than advantageous. As a result, the rate of evolution of DNA sequences is on average inversely related to their functional importance. Within protein-coding genes, synonymous sites evolve faster than nonsynonymous sites, and homologous exons are much more similar than homologous introns. Direct estimates of the fitness effects of new mutations generally corroborate this by showing a decline in fitness during mutation accumulation experiments (see Mutation Accumulation in *Drosophila* in Case Studies, below). This is hardly surprising: spoiling the product of 3.5 billion of years of evolution is easier than improving it.

However, there are parts of some genes where nonsynonymous substitutions occur faster than synonymous, suggesting that many replacements of amino acids were advantageous. For example, the anitgen binding region of the HLA gene in the major histocompatibility complex in humans and mice has a higher rate of amino acid substitutions than synonymous substitutions. This suggests that positive selection for diversity or at least change in antigen-binding makes advantageous amino acid changes quite frequent (Hughes & Nei 1989).

Another key property of DNA sequences in eukaryotes is that large portions are irrelevant to fitness. Only approximately 2% of the human genome codes for proteins, and only a minority of the remainder evolves slower than the neutral rate, and is therefore functionally important (estimates range from 15% to 3%; Shabalina et al. 2001; Dermitzakis et al. 2002). Mutations affecting the remainder are phenotypically silent and selectively neutral. Quantitatively, the fraction of neutral sequences is smaller in compact genomes of bacteria (where over 80% of sequences code for proteins) and much larger in eukaryotic genomes. From these figures, together with the data on per nucleotide mutation rates referred to above, the 100 mutations expected per human genome per generation can translate into from 4 to 14 deleterious mutations (assuming that 50% of mutations in coding sequences are deleterious).

Direct evidence makes it clear that most mutations that affect fitness have relatively small deleterious effects (see Mutation Accmulation in *Drosophila* in the Case Studies section below). A striking confirmation of this fact is that systematic knockouts of genes in eukaryotes reveal that less than 30% of all genes are essential to viability. For example, the fitness effects of knocking out nearly every gene in the yeast genome have been measured. Under typical laboratory conditions, only 18.7% of the genes are essential, while quantitative decreases in fitness are detectable in another 15% of the knockouts (Giaever et al. 2002). Since any gene that is not capable of affecting fitness will rapidly be destroyed by mutation, the remaining genes must either have effects on fitness that are too small to be detectable, or be advantageous under conditions not found in these experiments. Most spontaneous mutations must have smaller effects on fitness than the whole-gene knockouts used in this experiment.

Unfortunately, our ignorance regarding the mutation rate to beneficial mutations is quite profound. Recent work holds out hope that this ignorance is curable. For example, microbiologists have long been able to exploit the lack of recombination to detect fixation of beneficial mutations. When a novel beneficial variant arises, the genetic background in which it occurs increases along with it. Thus, perturbations of neutral allele frequencies are good markers of fixation. A recent study exploited this fact to estimate that the genomic beneficial mutation rate in a set of *E. coli* populations was 4×10^{-9}/replication (Imhof & Schlotterer 2001), or only one millionth of the total genomic mutation rate shown in Figure 3.1. It is not easy to generalize from this, as presumably the evolutionary history of a population and the constancy of the environment will have an impact on this rate. Comparative data can also be used to estimate beneficial mutation rates if one is prepared to accept a fairly simple set of assumptions about the distribution of effects. For example, fixation events are slightly more likely to involve amino acid substitutions than predicted from the pattern of within-species polymorphism in two *Drosophila* species. Under a simple model this suggests that about 45% of the amino acid substitutions are due to positive selection (Smith & Eyre-Walker 2002).

Mutations and Phenotypes

The importance of mutation for evolution also depends on the precise pattern of effects on the phenotype. For example, the degree to which the evolution of two parts of the body may be decoupled depends on whether and how often mutations that affect the parts in different ways arise (Wagner & Altenberg 1996). While our knowledge of the molecular and fitness effects of mutation are far from comprehensive, we are more ignorant of these important phenotypic properties. We know a bit about the amount of phenotypic variation produced by mutation. The study of the correlated effects of mutations is just beginning (see Mutation Accumulation in *Drosophila* in the Case Studies section below).

The basic challenge is that discrete mutations of large effect, such as lethals, visibles, or human genetic diseases, can be readily observed, but these are irrelevant to long-term evolution as they have extremely large deleterious fitness effects. Mutations with small positive or negative effects on fitness are the ones that we need to understand, and it is precisely these mutations that are most difficult to detect and study. These questions have so far been addressed by relatively crude mutation-accumulation experiments in which the aggregate properties of unknown numbers of mutations are studied.

These studies make it clear that the effects of mutation differ with phenotype (Houle et al. 1996). Figure 3.2 summarizes estimates of the mutational variance, V_M, in seven species expressed as coefficients of variation, CV_M. A CV_M of 1% means that after one generation of mutation, the standard deviation among initially identical lines is 1% of the mean. The traits are classified by their presumed relationship to fitness. Life history traits are measures of viability, fecundity, or mating ability, and are expected to be closely related to fitness. Morphological traits are features such as bristle number or leaf size that are probably under stabilizing selection. Growth traits reflect the size of the organism during growth, which may or may not be closely related to fitness.

Two facts are apparent from Figure 3.2. First, the variation in CV_M is large, ranging from 0.1% to over 4%. This reflects variation both within and among species. Species-level CV_Ms are correlated with generation time and genome size, suggesting that, as for molecular mutation rates, large genomes and/or large numbers of cell divisions increase mutational variance (Lynch et al. 1999). Second, it is clear that morphological traits accumulate variance less fast than life history traits (median CV_M 0.24% vs. 1.47%). At least part of the explanation for this difference seems to be that larger numbers of loci affect life history traits, because they summarize variation in the overall function of the organism (Houle 1998). Thus the concept of mutational target size—the number of base pairs which, when mutated, affect a trait—can help to explain both among- and within-species variation in the impact of mutations on phenotypes.

Why Are Mutation Rates What They Are?

Mutation rates themselves may evolve. Since the 1930s it has been clear that three major factors potentially determine the outcome of this process: the inevitability of some mutation, the costs of making replication as accurate as possible, and the possible advantages of beneficial mutations. Thus, there are three possible sorts of equilibrium mutation rates: the minimum possible, an optimal rate

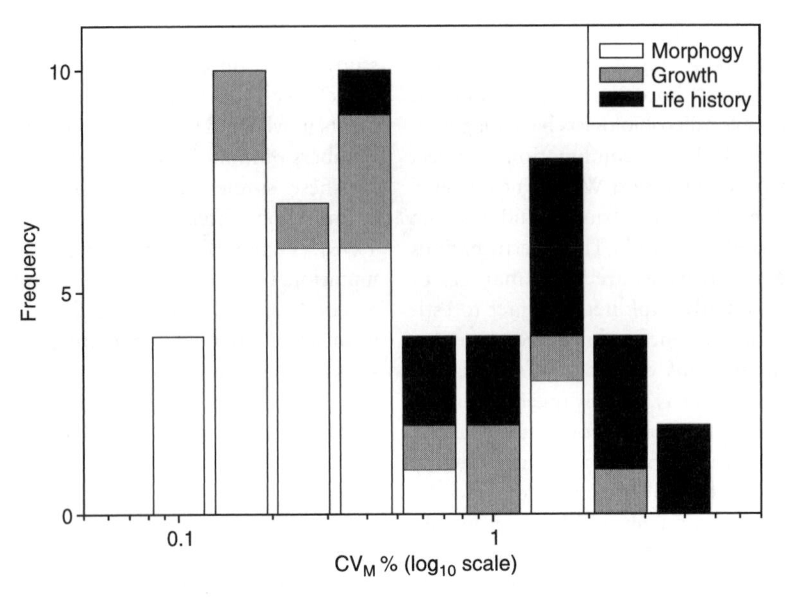

FIGURE 3.2. Mutational coefficients of variation from a review of studies prior to 1995 (Houle et al. 1996). See text for explanation.

determined by the costs of fidelity, and an optimal rate determined by selection for evolvability (Kondrashov 1995).

It is clear that the first two processes must influence mutation rates. Mutation is unavoidable because recovery of destroyed information is generally not possible. For example, when bases on complementary DNA strands do not match, it is difficult to see how a repair mechanism could always determine which of the two bases in the mutant that should be repaired. Similarly, physical considerations show that, if one attempts to reduce the error rate in DNA replication and repair to zero, the cost of these processes, in terms of both time and energy, could be large. Thus, the first hypothesis, that mutation rates are minimal, is not tenable. This is also suggested by the inverse relationship between genome size and mutation rates among microorganisms (Figure 3.1). What is controversial is the effect of selection for evolvability on mutation rate.

The fact that mutation is essential for evolution has made the notion that its rate is tuned to allow evolution attractive to many. Furthermore, there is plenty of evidence for variation in mutation rate within species that could be exploited by natural selection. The fact that most eukaryotes employ meiotic recombination to produce genetic variation shows that variation-generating adaptaions are possible (regardless of why exactly sex is good). As indicated above, the rate of mutation depends on the DNA sequence, and the chromosomal context it is in. There is also substantial evidence for environmental effects on mutation rates. Nutritional deficiencies, the presence of mutagens, and temperature can change both rates and patterns of mutation. All this suggests that the variation necessary to alter mutation rates is abundant.

In a few cases, organisms seem to have evolved portions of their genomes to be susceptible to mutation. For example, a region of the gene specifying host recognition in the *Bordetella* bacterium mutates at a very high rate because the bacterium has harnessed a retrotransposition-like process that targets that region (Doulatov et al. 2004). The attractiveness of the notion that mutation rates are adaptive must be tempered by the evidence that the vast majority of mutations are deleterious, and therefore costly to the individual in whose genome they occur (Johnson 1999). In sexual organisms, this creates a typical conflict between group-level evolution and individual selection over the fate of a variant that increases the mutation rate. The high mutation rate variant can increase in frequency when it causes a beneficial variant; however, this advantage benefits the high mutation variant only as long as it and the beneficial variant remain together in the same genotype. This may not be long at all if the loci involved are unlinked. On the other hand, most copies of the high mutation variant will produce a steady stream of deleterious variants that will reliably decrease the transmission of the high mutation rate variants. From the point of view of a population, more experimentation is ultimately beneficial if it can speed adaptation. However, from the viewpoint of the individual, it would be much better if another individual took the risk of experimentation. It takes special conditions for the advantages of mutation to the population to outweigh its costs to the individual (Sniegowski et al. 2000).

Closely related to the idea that the overall mutation rate is adjusted to promote evolvability is the idea that organisms can increase their mutation rate in relation to their need for variation, so-called adaptive mutation. The converse idea that mutations occur at random with respect to their usefulness has been a cornerstone of evolutionary thinking since the late nineteenth century, so evolutionary biologists reacted with outrage when adaptive mutation was revived again by Cairns et al. (1988). These researchers observed mutation rates that restored growth in nondividing "stationary phase" cells in *E. coli* and other single-celled organisms are higher than the rates of the same mutations when the cells are growing. This basic observation that has now been made in many microorganisms (Foster 2000).

There are three potential explanations for the increase in mutations observed in stationary phase. The increase can be adaptive in two senses: The strong version is that the rate of beneficial mutations can be increased at need, which we can call "directed mutation." The weak version is that the overall rate of mutation may be increased when variation is needed, or "hypermutability." Finally increased mutation may arise because of an unavoidable breakdown in normal repair and replication.

Foster (2000) and others have pinned down the mechanisms which underlie several cases of high mutation in stationary phase. In each case the effects are not confined to genes where mutations might be adaptive, ruling out the directed mutation hypothesis. However, the specific mechanisms by which the mutations arise, including DNA synthesis initiated by recombination and activation of transposable elements, do not necessarily suggest a general breakdown of fidelity. Thus hypermutability is real, but its adaptive signficance is still not clear. The deleterious consequences of increases in mutation are not readily avoided, unless death of the cell is certain in the absence of mutation.

It is much more difficult to imagine adaptive increase of mutation rates in multicellular organisms. In fact, the study of mutation in cancerous cells of multicellular eukaryotes suggests that similar hypermutability may occur in tumors, where it facilitates the evolution of high tumor growth rate and resistance to chemotherapy, resulting in death of the organism. The fact that similar phenomena occur whether or not they can be adaptive favors breakdown in normal replication, recombination, and repair machinery. The isolation of the germline from the soma in most multicellular animals is a powerful argument against the generality of adaptive mutation.

What Limits the Rate of Evolution, Mutation or Selection?

One of the central paradoxes of evolutionary biology is that most of the time, organisms do not evolve at all (reviewed in Gould & Eldredge 1993). There are two sorts of explanation for this stasis (reviewed by Hansen & Houle 2004). First, many believe that this is due to stabilizing selection which is somehow maintained over very long time periods. The weakness in this hypothesis is simply that it is difficult to see why selection should be constant over periods of tens of millions of years. The other alternative is that the kinds of variation necessary for populations to evolve in response to whatever novel selection pressures come up are often not produced (Gould & Lewontin 1979). Such limitations on variation are usually referred to as constraints, but this suggests that the necessary variation is never produced by mutation.

The common argument against the constraint hypothesis is that nearly every trait studied does display genetic variation. This is insufficient to resolve the issue because all aspects of the phenotype will be selected simultaneously. It is not enough to produce variation in each trait, the variation must also be relatively free of entangling effects on other selected traits. The capacity of the genome to produce appropriate sorts of phenotypic variation may determine which of the many pressures that natural selection places on an organism it will be capable of responding to. If mutations tend to affect a limited number of phenotypes, or phenotypes that tend to have similar selection pressures on them, the structure of variation is said to be modular (Mezey, Box 19.5 of this volume; Wagner & Altenberg 1996). Thus not only the rate but also the nature of mutation may itself be shaped by natural selection.

CASE STUDIES

Nachman and Crowell

Nachman and Crowell (2000) studied the rate of mutation in humans using the comparative approach. We are unusually confident of both the time since we diverged from chimpanzee lineage, and of the number of generations that this represents. Nachman and Crowell took advantage of this by studying divergence of DNA sequences that are among the most likely to be neutral and thus evolve at the mutation rate: processed pseudogenes.

A processed pseudogene is a bit of DNA that has been reverse transcribed from a messenger RNA back into DNA and incorporated into the genome. Differences of a pseudogene from the homologous gene, such as frameshift deletions or insertions, make it clear that the pseudogene cannot encode a protein, hence the name. Processed pseudogenes can be recognized because they have had their introns edited out, and often have a poly-A sequence attached. Therefore, pseudogenes are expected to have no function, and consequently to evolve at the neutral rate, although this may not always be so (see above).

A potential complication with the use of pseudogenes is that there may be many pseudogenes derived at different times from the ancestral gene. If an older pseudogene in one species were to be compared with a younger one in the other species then a very misleading picture of the rate of divergence would be obtained. Nachman and Crowell were able to guard against this possibility by a careful choice of methods. They used polymerase chain reaction (PCR) amplification to obtain material directly from genomic DNA. For each pseudogene, they chose one of their PCR primers to lie in the genomic DNA outside the pseudogene itself. Thus, only sequences that possessed the same flanking sequence in both humans and chimpanzees would amplify. This ensures that each pair of compared human and chimpanzee pseudogenes was orthologous, that is, it originated from the pseudogene already present in the last common ancestor.

In total Nachman and Crowell sequenced 18 different pseudogenes in a chimpanzee and two humans. In each individual, 16 kb was sequenced. Overall, they found 199 differences between the human and chimpanzee sequences, for a divergence of 1.2%. These differences consisted of 131 transitions, 52 transversions, and 16 insertion-deletion variants. The insertion-deletion variants were all of 4 nucleotides or shorter. These data are biased against detection of large insertions and deletions as this would tend to preclude recognition of a pseudogene in the first place. CpG contexts accounted for about 25% of all the substitutions, with such sites having a 10-fold higher rate of substitution than non-CpG contexts. The estimated rate of substitution and mutation differed by a factor of 6 among different pseudogenes. These differences were statistically significant, suggesting that the region in which the pseudogenes inserted influenced their mutation rates.

The simple neutral divergence model outlined above assumes that a single copy of DNA is split into two lineages at the time of species divergence. In reality the ancestral species was very likely to already have DNA sequence variation at the time of speciation. This means that some of the differences fixed in each lineage after speciation are actually variants that arose before the time of speciation. To compensate for this an assumption about the effective size, N, of the ancestral species must also be made. When the number of generations since divergence is not much greater than N, which is likely to be the case for humans and chimps, the effect of this adjustment can be substantial.

Thus, there are three unknown factors that still must be taken into account to convert the 1.2% divergence into an estimate of the mutation rate per generation: N before speciation, the time since divergence, and the average generation time since divergence. Nachman and Crowell considered N values up to 10^5, divergence times between 4.5 and 6 million years ago, and generation times of 20 and 25 years. This gives an estimate of the number of generations separating chimps and humans of between 360,000 and 600,000. The range of possibilities suggests mutation rates per base per generation pair between 1.3 and 3.4×10^{-8}. This agrees very well with direct estimates of human mutation rates (Kondrashov 2002), and more recent analyses of a much larger human–chimp data set.

Mutation Accumulation in *Drosophila*

Studying mutations with small effects on the phenotype and fitness is both important for understanding evolution, and difficult experimentally. Much of the data on such mutations come from mutation–accumulation experiments on *Drosophila melanogaster*, due to its 2 week generation time, readily observed morphology and life history, and the genetic tools available.

The most important such tools are balancer chromosomes, so called because they allow the preservation of a sampled chromosome from the disruptive effects of recombination. Balancer chromosomes have three key properties: a multiply inverted gene order, a dominant morphological marker, and a recessive lethal mutation. When an individual is heterozygous for chromosomes with very different gene orders, chromosomes can pair at meiosis, but recombination between them results in duplications and deficiencies in the products, and inviable gametes. Therefore, a fly heterozygous for a balancer chromosome will only give rise to gametes that carry the unrecombined balancer or wild-type chromosomes. As shown in Figure 3.3, this fact can be exploited to capture (or "extract") and replicate single chromosomes from any population, allowing their properties to be studied.

In the mutation-accumulation experiments we want to discuss (Mukai et al. 1972; Houle et al. 1994) a second chromosome balancer was used to extract and replicate a single test chromosome, then to preserve independent copies in heterozygous condition (Figure 3.3). The second chromosome in *D. melanogaster* contains about 40% of the genome. These copies start out genetically identical, but diverge over time as each copy independently accumulates spontaneous mutations. If the heterozygous fitness effects of mutations are small, then they will accumulate at very close to the mutation rate. To detect the effects of mutation, inversion heterozygotes are crossed, and the ratio of test chromosome homozygotes to inversion heterozygotes observed in the offspring, as shown in the last two rows of Figure 3.3.

Mukai utilized this basic design several times to study the effects of mutation on egg-to-adult viability, the probability that an egg survives to become an adult. In his 1972 paper, three test chromosomes were each replicated 50 times to make sublines. Each subline was then subjected to the accumulation process. Every 10 generations, the relative homozygous viability of each subline was measured. From this, Mukai estimated the rate at which viability went

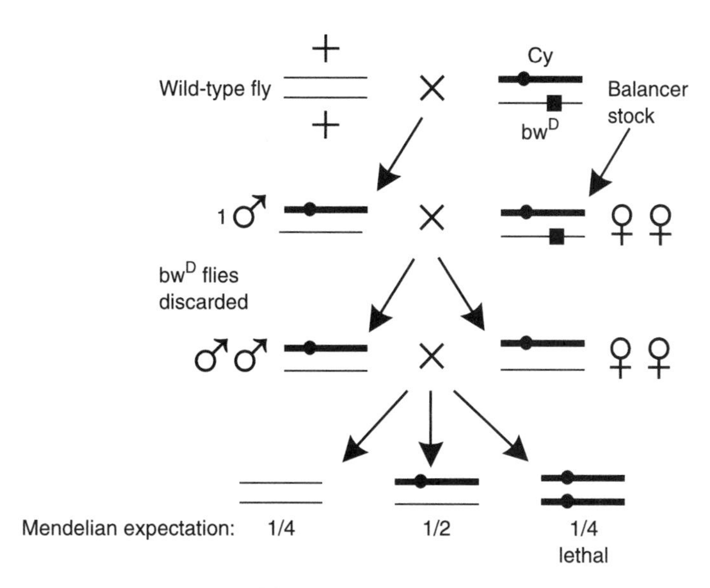

FIGURE 3.3. The use of balancer chromosomes to extract intact chromosomes and to serve as standards for the measurement of viability. The thick line denotes a multiply inverted chromosome (the balancer), while thin lines are chromosomes with the usual gene order. The circle denotes the dominant mutation Curly (*Cy*), which causes the adults to have wings that curl upward. The square denotes the mutation brown-dominant (bw^D), which causes its carriers to have brown rather than red eyes.

GOSHEN COLLEGE LIBRARY
GOSHEN, INDIANA

down over generations (M), and the rate at which the variation among sublines went up (V_M).

The distribution of homozygous viabilities showed a minority of chromosomes acquired lethal recessive mutations, while most chromosomes showed modest declines in fitness, with very few chromosomes having reductions of viability between 50% and 100%. The rate at which chromosomes acquired lethal mutations was 0.006 per chromosome per generation. Even with the lethals excluded, M was negative, as shown in Figure 3.4, as expected due to a preponderance of deleterious mutations, while V_M was positive as expected since each line accumulates its own unique set of individually rare mutations.

These two quantities, M and V_M, together give some indication of the mutation rate and the effects of the nonlethal mutations. To see this, imagine two cases with the same decline in the mean: if the variance among sublines had not increased at all, this would indicate the presence of a very large number of mutations that each had very small effects. Conversely, if the variation among sublines was high, this would indicate that a few mutations with large effects must have occurred. If all mutations had exactly the same effect on homozygous viability this relationship would be mathematically precise: $M = Us$, where U is the total mutation rate on the chromosome to alleles affecting viability by amount s, and $V_M = Us^2$. In reality, s does vary from variant to variant, and this increases V_M, so Mukai et al. could set a lower limit to the mutation rate $U > M^2/V_M$ and an upper limit on the average s, $s < V_M/M$. For this experiment $U > 0.06$ when all the nonlethal chromosomes were considered, and $U > 0.17$ when the few chromosomes with homozygous viabilities near 0 were excluded. Thus the mutation rate to variants with small homozygous effects on viability is at least 10 times greater than the recessive lethal mutation rate. When extrapolated to the whole genome, this suggests a deleterious mutation rate greater than 0.4 per genome per generation.

The high genomic deleterious mutation rate estimates of Mukai et al. (1972) have proved to be quite controversial. Many subsequent studies have undertaken similar estimates in *Drosophila* and other organisms; some broadly support Mukai's results, while others do not (Lynch et al. 1999).

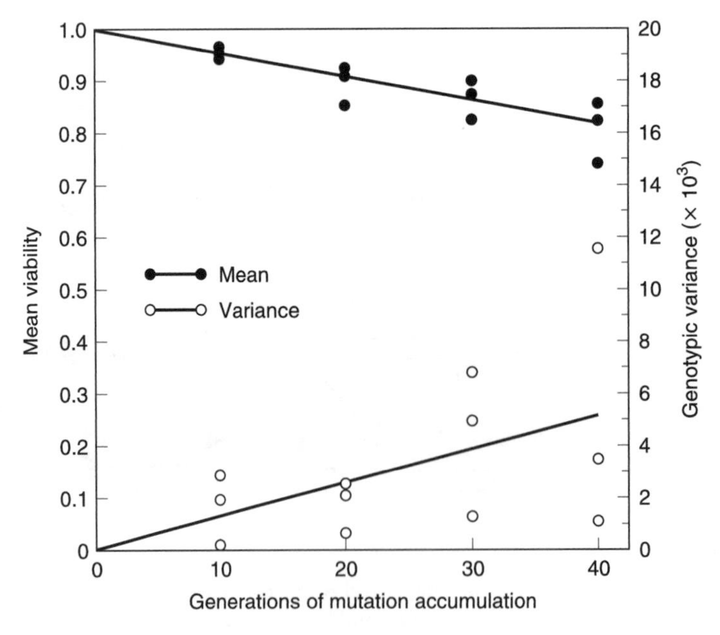

FIGURE 3.4. Changes in the mean viability (left scale, filled circles) and genetic variance (right scale, open circles) in viability over 40 generations of mutation accumulation in *Drosophila melanogaster*. Mukai et al. (1972), reprinted with permission of the Genetics Society of America.

More troubling is that estimates of the genomic mutation rate in *Drosophila*, such as the direct estimate of 0.8 shown in Figure 3.1, are not much larger than this estimate. If the direct estimate is correct, then half of all mutations in *Drosophila* must be deleterious. The direct estimate is itself subject to uncertainty but where the discrepancy lies is not certain.

Two other potential gaps in the Mukai et al. study are less well known. Mukai et al. assumed that balancer heterozygotes provide a standard for viability to compare with that of the heterozygotes. This is almost certainly incorrect due to partially dominant or epistatic effects of the mutations themselves. Second, early development in *Drosophila* is primarily driven by maternal message, so that the homozygous effects of maternal-acting genes are not assayed with their technique.

Houle et al. (1994) used the same technique as Mukai et al. to accumulate mutations, but focused instead on the pleiotropic effects of mutation on life history traits. This work was undertaken largely to test the mutation-accumulation theory for senescence, one of the major theories concerning the evolution of lifespan (Promislow & Bronikowsi, Ch. 30 of this volume). The mutation-accumulation theory requires that many mutations occur that decrease fitness of old individuals, while leaving early-life fitness unimpaired. Natural selection is less effective at removing mutations that act late in the lifespan than those that act earlier, potentially causing a decrease in lifespan. To test for such mutations, Houle et al. studied female fecundity in young flies (5 and 6 days old) and old flies (27 and 28 days old), as well as average male and female lifespan. If the mutation-accumulation hypothesis is correct, there should be variation in late fecundity that does not affect early-life fecundity.

After 44 generations of mutation accumulation, there was significant genetic variation due to mutation for all four traits. The mutational correlation of early and late fecundity was not different from a perfect correlation of 1, but was significantly higher than 0, suggesting that mutations tended to affect early- and late-life fitness components similarly. Consistent with this, the correlations of both fecundities with lifespan were also positive and not different from 1. Mutations that affect one aspect of fitness also seem to impair all aspects of fitness—a fly with poor fecundity early in life will also live less long and have lower fecundity late in life. This substantially undermines the mutation-accumulation hypothesis for the evolution of lifespan.

FUTURE DIRECTIONS

Knowledge of the rates and properties of mutation is critical to understanding evolution. The study of mutation is difficult, and so has been relatively neglected by evolutionary biologists. The result is that our generalizations about mutation rest on somewhat shaky ground. Consequently, we expect that the study of mutation will be among the most productive areas of evolutionary genetics in the future. Here are a few areas where we hope to see great progress over the next 10 years.

1. Improvements in sequencing technology have now made it possible to directly assay mutations in DNA sequences after a mutation-accumulation experiment. Such studies should increase rapidly in number. These data should remove much of our uncertainty about the molecular aspects of mutation, as it will no longer be necessary to find phenotypic evidence for a mutation before molecular analysis.
2. Availability of complete genome sequences for closely related species pairs will tell us what proportion of genomes are subject to natural selection. Combined with direct data on mutation rates, this will reveal what proportion of mutations affect fitness.
3. Automated phenotyping is being pursued in several systems, making it possible to rapidly screen for mutants or increases in variation in many traits simultaneously. This will be necessary to understand the correlation structure of mutational effects.
4. After many years of relative eclipse, the study of advantageous mutations should take its rightful place as the most important aspect of evolutionary biology. New theory and large amounts of molecular data should make it much easier to detect the signature of positive selection, for example in bacterial cultures, or from population survey data.
5. One of the most neglected issues in evolutionary genetics is what limits evolutionary progress (Hansen & Houle 2004), with one of the major possibilities being limits on the phenotypic effects of mutation. Progress in

the previous three areas may make it possible to test this idea.

SUGGESTIONS FOR FURTHER READING

We know of very few general works on the role of mutation in evolution, that is those that cover the range of issues raised in this chapter. Molecular mechanisms of mutation are introduced in Lewin's series of *Genes* books (e.g., 2004). Graur and Li (2000) provide an overview of molecular evolution with a reasonable emphasis on mutation as a driving force. For information on molecular mutation rates the review of Drake et al. (1998) and the recent paper by Denver et al. (2004) introduce the important literature. For phenotypic mutation rates Drake et al. (1998) and Lynch et al. (1999) provide reviews. For the evolution of mutation rates generally, Kondrashov (1995) gives an overview of hypotheses. Foster (2000) and Sniegowski et al. (2000) provide a quick summary of the evidence for and against the adaptive values of environmental responsiveness of mutation rate.

Denver DR, Morris K, Lynch M & WK Thomas 2004 High mutation rate and predominance of insertions in the *Caenorhabditis elegans* nuclear genome. Nature 430:679–682.

Drake JW, Charlesworth B, Charlesworth D & JF Crow 1998 Rates of spontaneous mutation. Genetics 148:1667–1686.

Foster PL 2000 Adaptive mutation: implications for evolution. BioEssays 22:1067–1074.

Graur D & W-H Li 2000 Fundamentals of Molecular Evolution. Sinauer Assoc.

Kondrashov AS 1995 Modifiers of mutation–selection balance: general approach and the evolution of mutation rates. Genet. Res. 66:53–69.

Lewin B 2004 Genes VIII. Oxford Univ. Press.

Lynch M, Blanchard J, Houle D, Kibota T, Schultz S, Vassilieva L & J Willis 1999 Perspective: Spontaneous deleterious mutation. Evolution 53:645–663.

Sniegowski PD, Gerrish PJ, Johnson T & A Shaver 2000 The evolution of mutation rates: separating causes from consequences. BioEssays 22:1057–1066.

4

Natural Selection

MICHAEL J. WADE

Natural selection causes change in the phenotypic composition of a population through the differential birth and death of its members. Natural selection can operate simultaneously at many different levels in the hierarchy of biological organization, from the cellular (gametic selection) to the individual (Darwinian or mass selection), to the population (kin, group, and inter-demic selection) as well as to the species or community level. That is, we can consider populations of gametes, cells, individuals, populations (called "metapopulations"), communities, and species as collections of biological entities whose members persist, reproduce, and die as a result of differences in phenotype. Differences in the direction or strength of natural selection at different levels in the biological hierarchy contribute to the origins of biodiversity.

The study of adaptive evolution among individuals by natural selection in an ecological context has been the focal point of ecological genetics for nearly eight decades (Ford 1975), a field grounded in the interactions among mathematical theory, laboratory genetics, and empirical observations from field populations. However, "natural selection is not evolution" (Fisher 1930) unless there is also *heritable* variation of the right sort (see below) in both fitness and phenotype among the units of selection. Not all natural selection results in a genetic response in a population and, conversely, not all of evolution is caused by natural selection. Thus, the process of natural selection can be formally separated from the evolutionary genetics that govern the response to selection. Nevertheless, much of our understanding of natural selection, including the theoretical foundations of ecological genetics, comes from population genetic theory, in which natural selection acting directly on genes is synonymous with adaptive evolution. In the standard, single-gene population genetic theory introduced in undergraduate and graduate textbooks (sometimes referred to as "bean bag" genetics), natural selection *is* evolution because the models assume a direct and unchanging relationship between genotype and fitness. Nature, however, differs greatly from the depiction of these simplified models because the relationships between phenotype and fitness and between genotype and phenotype are complex and dynamic in natural populations.

Natural selection is often contrasted with random genetic drift, which causes random changes in the genetic composition of finite populations (Gillespie, Ch. 5 of this volume). The magnitude of the allele frequency change caused by random drift is inversely proportional to N, the breeding size of a population (i.e., $1/2N$ in diploids). Although the *average* allele frequency change caused by drift across a group of populations is zero, it can be quite difficult, especially in short-term, observational studies of a particular population, to determine whether or not an observed association between genotypic value and fitness is spurious, owing to the chance vagaries of random drift, or causal, owing to natural selection. Furthermore, natural selection acting on allelic variation at one locus generally affects the strength of random genetic drift at all other loci, linked and unlinked, via the effects of variation in offspring numbers on the effective population size. Thus, it is difficult in natural populations to cleanly separate the effects of natural selection from those of random genetic drift as is often done in theory. Whenever directional natural selection operates in a population, it *increases* the force of random genetic drift. But, the converse is not true. Indeed, just the opposite holds: whenever random genetic drift is acting, genetic variation within a population is reduced, and the unfettered action of evolution in response to natural selection is *decreased*.

In this chapter, I will explore first the definition of natural selection and the properties that a population of cells, individuals, demes, or other entities must possess in order for natural selection to operate. Second, I will introduce the standard population genetic models of natural selection and illustrate the conclusions drawn from them. Next, I will introduce the phenotypic selection approach and show why the strength of natural selection on a particular phenotype equals its covariance with relative fitness, a relationship that has been called "the second fundamental theorem of natural selection" (Robertson 1966; Hill & Robertson 1968; Price 1972; Frank 1995a). This relationship is particularly useful for the study of natural selection acting on phenotypes in natural populations. I will show that this approach to natural selection is not mathematically different from that taken in the population genetic theory discussed previously. Rather, the difference lies in the different ways the genotype–phenotype relationship is conceptualized in the two approaches. The difference is important in a practical sense because phenotypic studies of natural selection in the field tend to use the quantitative genetic approach while studies using sequence or gene-based data draw tend to draw inferences using population genetic theory.

The two approaches can be reconciled by noting that the rate of change of the average phenotype across generations within a single population depends on the additive component of the covariance between fitness and phenotype. When there is a fixed relationship between the phenotype and fitness, the genetic response of a population to natural selection depends on the additive genetic component of the variation of the target phenotype or the "narrow-sense" heritability. The notion of heritability is typically associated with the additive effects of genes, transmitted from parents to offspring, and the additive effect is often considered an inherent property of the gene itself, rather than a statistical construct. Evolutionary genetic theory frequently categorizes genes into groups with large versus small effects or "good" versus "bad" effects with respect to fitness. In the "gene's eye view" of adaptive evolution advocated by some, this focus on the gene is considered a conceptual advance over an "individual"-centered theory of Darwinian adaptation. However, in all but the simplest models, the additive effect of a particular gene is a statistical property, which can change with the abiotic environment in the case of genotype-by-environment interactions, with the social environment in the case of indirect effects, and with the frequencies of other alleles in the case of epistasis. In a genetically subdivided metapopulation, nonadditive components of genetic variation, owing to interactions among genes and between genes and environment, and indirect genetic effects predominate in determining the character of evolution by natural selection. These interactions can cause changes in the sign and the magnitude of the additive effect of a gene from deme to deme. Thus, not only do the local selective forces change from deme to deme within a metapopulation owing to changes in environment, but also the nature of the genetic response to selection changes whenever interactions are present. In a metapopulation, even spatially uniform directional selection among individuals can be genetically diversifying with some types of epistasis (gene–gene interactions) or genotype-by-environment interactions.

Fourth, I will partition the variance in relative fitness into components useful for determining the relative strengths of natural selection and sexual selection, when fitness variation is partitioned between males and females, or the relative strengths of individual and group selection when it is partitioned within and among demes in a metapopulation. The variance in relative fitness represents the "opportunity for selection," and it sets an upper bound on the rate of phenotypic evolution that is possible in a population. The underlying rate of genetic evolution must be smaller than the maximum phenotypic rate because of the complex relationship between genotype and phenotype. When much of the fitness variation among individuals is not heritable, the group mean may be a better predictor of an individual's genotype than is the individual's own phenotype. Finally, I will illustrate how directional natural selection necessarily affects the strength of random genetic drift, by reducing the effective population size, and thus slowing the ultimate rate of adaptive phenotypic evolution.

DARWINIAN NATURAL SELECTION

Darwin's theory of evolution by natural selection rests upon three fundamental features of natural populations and each is essential for his process of adaptive change:

1. There must be phenotypic variation among the members of a population, that is, differences among individuals in morphology, physiology, and behavior.
2. There must be fitness variation among the members of a population, that is, variations in phenotype must cause some individuals with certain trait values to survive longer and/or have higher reproductive rates than individuals with other values of these same phenotypes (Fairbairn, Box 4.1 of this volume).
3. The individual variation in phenotype and fitness must have a heritable component so that offspring tend to resemble their parents.

Whenever these three properties exist for a population at any level in the biological hierarchy, adaptive evolution by natural selection will occur and transform the population over time. The population is phenotypically transformed because the units better able to survive and reproduce leave more offspring than those less able and the offspring tend to resemble their parents. That is, the population changes because the heritable qualities of its membership change over the generations. When the relationship between phenotype and fitness remains stable for long periods of time, a near-perfect fit between organism and environment is possible.

Natural selection occurs in any system whose members have these properties of *replication*, *variation*, and *heredity* (Lewontin 1970; Maynard Smith 1978). When the system consists of cells, the variation among cell lineages in replication and death rates, and the similarity of daughter to mother cells, gives rise to among-cell selection, which determines tissue shape. When such a process operates among cells within the germline or among male gametes competing to fertilize female ova, it can result in gametic selection or meiotic drive, one of the strongest evolutionary forces known. When selection occurs among individuals, among groups, or among species, it is called individual selection (sometimes mass selection), group or inter-demic selection, or species selection, respectively.

Although each of the three properties is essential for adaptive evolution, in evolutionary studies it is very useful to separate the investigation of natural selection, which affects the phenotypic distribution of a population within a generation, from the investigation of heredity, which connects the phenotypic distributions of parents and offspring across generations. That is, it is useful to study natural selection independent of heredity. In genetic studies, the environment is often viewed as a source of unwanted variation that obscures the relationship between phenotype and genotype. In contrast, in studies of natural selection, the environment is the context that gives rise to associations between phenotype and fitness, so that investigations into the causes of natural selection are necessarily ecological. Methodologies also differ between genetics and ecology. Controlled breeding studies are necessary to investigate heredity mechanisms while experimental manipulations of environmental and phenotypic distributions are essential to the analysis of the causes of selection. Clearly, disciplines such as functional genomics will require both approaches.

Darwinian Natural Selection Acting on Single Genes

Darwin's process of evolution by natural selection can be illustrated with the single-gene, two-allele models of selection in a randomly mating population from population genetic theory. There are a number of these models as shown in Table 4.1, with the most general model in the last column. Let the frequency of the A allele at a locus be p and the frequency of the alternative allele, a, be q, where q equals $(1 - p)$. Let G_{AA}, G_{Aa}, and G_{aa} be the frequencies of the three genotypes AA, Aa, and aa, after reproduction and before natural selection in a randomly mating population. These are (p^2), $(2pq)$, and (q^2), respectively. The genotype frequencies are a simple function of the allele frequencies that conform to the expected values of the Hardy–Weinberg equilibrium. Thus, the genetic composition of the population can be described in terms of a single allele frequency, p or $\{G_{AA} + [0.5]G_{Aa}\}$. Natural selection, acting through the differences in genotypic viability, changes the frequencies of these three genotypes and consequently the allele frequency in the population from p before selection to p' after selection. The change in allele frequency, $\Delta p = p' - p$, measures the amount or the rate of genetic evolution in the population in a single generation.

To calculate Δp, we need to consider how natural selection changes the genotype frequencies from G_{AA}, G_{Aa}, and G_{aa}, to G'_{AA}, G'_{Aa}, and G'_{aa}. The average viability fitness, W, in the population equals the sum of the products of genotype frequency times

BOX 4.1. Defining and Measuring Fitness
Daphne J. Fairbairn

The concept of fitness is central to the definition of natural selection and is essential for both theoretical and empirical investigations of Darwinian evolutionary processes. However, although one can extract a general consensus that fitness is some measure of success in contributing descendents to future generations (see Table 1), providing an operational definition of fitness is more elusive. The problem is that that no single parameter accurately predicts fitness (i.e., the genetic contribution to future generations; $W_{(Z)}$ in this chapter) under all demographic scenerios (e.g., Roff 1992; Charlesworth 1994; Benton & Grant 2000; Brommer 2000). In empirical studies, the problem is compounded by the near impossibility of accurately estimating the proposed parameters for real organisms under natural conditions. Here I provide a brief introduction to these issues, but readers are referred to the above citations for more complete treatments of the theoretical issues and to Fairbairn and Reeve (2001) and Wolf (Box 18.1 of this volume) for consideration of the empirical difficulties. For simplicity, I will restrict the discussion to selection in which the "biological entities" are individuals competing within populations, but readers should keep in mind that this can be generalized to other levels of selection.

To envisage the connection between the general definition of fitness and demographic parameters that can be estimated for individuals within natural populations, it is helpful to begin with the concept of lifetime reproductive success,

$$R_0 = \sum_{x=1}^{\infty} l_x m_x,$$

where l_x is the probability of surviving to age x, and m_x is the number of offspring produced at age x. As a population parameter, R_0 is the net reproductive rate for that generation. However, when R_0 is estimated for individuals, l_x reduces to 0 or 1 for each age class, and R_0 becomes simply the number of offspring produced or lifetime reproductive success (*LRS*) for each individual. This can easily be extended to organisms with continuous age distributions by integrating over ages:

$$R_0 = \int_0^{\infty} l_x m_x dx.$$

Lifetime reproductive success is the most common measure of lifetime fitness used in field studies and is particularly appropriate for organisms with synchronized reproduction and relatively constant population sizes. In species with prolonged parental care, *LRS* is often defined to include offspring survival to independence or even to first reproduction (e.g., Clutton-Brock 1988). This has the advantage of incorporating the indirect effects of parental behavior into the estimate of parental fitness. However, because such estimates confound parental and offspring fitness, they cannot be incorporated into standard genetic models of evolutionary response to selection (Wade, Ch. 4 of this volume; Wolf, Box 18.1 of this volume).

If the age schedule of reproduction varies among individuals within a generation, particularly if population size changes rapidly, the age schedule of reproduction becomes an important determinant of fitness. In such cases, r, the instantaneous rate

BOX 4.1. *(cont.)*

of increase, is a more appropriate measure of fitness. This can be estimated by solving the "characteristic equation,"

$$\int_{0}^{\infty} e^{-rx} l_x m_x dx = 1$$

or, for discrete age classes,

$$\sum_{x=1}^{\infty} e^{-rx} l_x m_x = 1$$

(Roff 1992; Brommer 2000). Alternatively, r can be estimated as $\ln(\lambda)$, where λ is the dominant eigenvalue of the Leslie matrix (Caswell 2001).

The importance of the age schedule of reproduction in determining r is illustrated by the relationship, $r \approx \ln(R_0)/\tau$, where τ is the mean generation time. Thus, for a given R_0 or LRS, r varies inversely with generation time: earlier age at first reproduction translates into higher lifetime fitness. This relationship also illustrates why R_0 can be legitimately substituted for r as a measure of fitness if generation time is constant across all individuals being compared. This circumstance is not rare in natural populations. For example, many plants and insects have annual life cycles with a synchronizing season of dormancy. For such species, R_0 or LRS measured to the end of the reproductive season would be an appropriate measure of fitness even if population size were highly variable.

Density- or frequency-dependence poses particular problems for estimates of individual fitness. Theoretical treatments typically use invasibility criteria to determine the strategy or phenotype that maximizes lifetime fitness in these conditions (i.e., the evolutionarily stable strategy [ESS] or evolutionarily unbeatable strategy [EUS]). These models suggest that the appropriateness of different fitness estimators depends critically upon the presence and form of density-dependence but, as a general rule, R_0 tends to perform better than r when population size is regulated by density-dependent processes. In some cases, measures of fitness related to the expected maximum sustainable population size appear to be more appropriate than either r or R_0, but it is not clear how these can be converted to empirical estimates of individual fitness. Charlesworth (1994), Mylius and Diekmann (1995), Brommer (2000), and Benton and Grant (2000) provide more thorough treatments of these issues.

Regardless of which measure of fitness is most appropriate for a particular demographic scenario, it is important to recognize that these are measures of the "absolute fitness." The fitness parameters that enter into population or quantitative genetic models to predict evolutionary trajectories and optimal phenotypes are relative fitnesses. Measures of absolute fitness such as LRS or r are most commonly converted to estimates of relative fitness by dividing each estimate by the population mean fitness (e.g., $w_z = W_z/\bar{W}$; Wade, Ch 4 of this volume; Wolf, Box 18.1 of this volume) although, in single-locus population genetic models, division by the maximum absolute fitness is also common. This conversion from absolute to relative fitness can result in a disconnect between ecological and evolutionary perspectives. Population ecologists study demographic trajectories and are primarily interested in absolute fitness, for example

(continued)

BOX 4.1. *(cont.)*

interpreting $r > 0$ or $R_0 > 1$ as indicating demographic increase. Relative fitness, in contrast, provides no information about demographic trajectories because division by the mean absolute fitness (W) necessarily results in a population mean relative fitness (w) of unity, regardless of the actual demographic trajectory of the population. Relative fitnesses (w_z) therefore indicate only the relationship between the absolute fitness of that individual or genotype and the population mean fitness (W_z). They imply nothing about the actual values of r, R_0, or any other measure of absolute fitness.

While both r and R_0 are used extensively in theoretical analyses of adaptive evolution, the difficulty of estimating either as a function of phenotype in wild populations has limited their application in empirical studies. More commonly, researchers attempt to measure only certain components of fitness. This approach is generally more feasible and appropriate when the aim is to ascertain the adaptive significance of the trait in question. For example, the hypothesis that an exaggerated secondary sexual characteristic in males is favored by sexual selection for that trait could be tested simply by measuring the covariance between the trait value and mating or paternity success (Wade, Ch. 4 of this volume; Wolf, Box 18.1 of this volume). In some cases, the fitness parameter may itself be a component of lifetime fitness (i.e., survival, fecundity or mating success). However, in other cases, surrogate fitness measures can be used on the assumption that they correlate with lifetime fitness. For example, behavioral ecologists may use energy intake rate as a surrogate measure of fitness in studies comparing foraging strategies.

The use of surrogate measures of fitness and studies of only certain components of lifetime fitness has proven very valuable for testing adaptationist hypotheses. However, the dynamics of trait evolution in response to selection (assuming $h^2 > 0$) are ultimately determined by the relationship between trait values and lifetime fitness. Trade-offs between the fitness effects of trait values during different phases of the life history often produce nonlinearities in the relationship between trait values and lifetime fitness that would not be obvious from studies limited to only a subset of life history stages (Preziosi & Fairbairn 2000). Studies that combine analyses of fitness components with estimates of lifetime fitness offer the best hope of both deducing the adaptive significance of the trait in question and understanding the evolutionary dynamics of the trait distribution.

TABLE 1 Definitions of fitness found in the glossaries of evolutionary biology textbooks

Definition	Source
The success of an entity in reproducing; hence, the average contribution of an allele or genotype to the next generation or to succeeding generations	Futuyma 1998
The average number of offspring produced by individuals with a certain genotype relative to the number produced by individuals with other genotypes	Ridley 2004
The extent to which an individual contributes genes to future generations, or an individual's score on a measure of performance that is expected to correlate with genetic contribution to future generations (such as lifetime reproductive success)	Freeman and Herron 2001
1. The ability of an individual or population to leave viable and reproductively effective progeny, relative to the abilities of other individuals or populations. 2. The average contribution of an allele or genotype to the next or subsequent generations, compared with those of other relevant alleles or genotypes	Price 1996
The relative reproductive success of individuals, within a population, in leaving offspring in the next generation. At the genetic level, fitness is measured by the relative success of one genotype (or allele) compared with other genotypes (or alleles)	Kardong 2005

TABLE 4.1. Single-gene, two-allele models of natural selection in a randomly mating population

		Genotypic fitnesses			
Genotype	Frequency before selection	Recessive deleterious allele	Additive effects	Heterozygote advantage	General
AA	p^2	1	$1 + 2s$	$1 + s$	W_{AA}
Aa	$2pq$	1	$1 + s$	1	W_{Aa}
aa	q^2	$1 - s$	1	$1 + t$	W_{aa}

the genotypic fitness or $\{W_{AA}G_{AA} + W_{Aa}G_{Aa} + W_{aa}G_{aa}\}$. The relative fitness of a genotype, say w_{AA}, equals its fitness *relative to* the population mean fitness or the ratio, W_{AA}/W. As a result, G'_{AA} equals $w_{AA}G_{AA}$. Hence, the relative genotypic fitnesses are a useful multiplicative transformation of the set of genotype frequencies before selection to the set after selection. We can now calculate the allele frequency after selection, p' or $\{G'_{AA} + [0.5]G'_{Aa}\}$ as equal to $\{w_{AA}G_{AA} + [0.5]w_{Aa}G_{Aa}\}$. The single-gene, two-allele models of natural selection specify the values of the genotypic frequencies (Table 4.1) to derive the textbook expressions for Δp. For example, a recessive deleterious allele evolves according the genotypic viability fitnesses of column three of Table 4.1 with mean fitness, W, equal to $(1 - sq^2)$. As a result, we find

$$\Delta q = q' - q, \quad (4.1a)$$

$$\Delta q = \{w_{aa}G_{aa} + [0.5]w_{Aa}G_{Aa}\} - \{G_{aa} + [0.5]G_{Aa}\}, \quad (4.1b)$$

$$\Delta q = \{(1 - s)q^2 + [0.5]2pq\}/(1 - sq^2) - q. \quad (4.1c)$$

Remembering that $(q^2 + pq)$ equals q, this reduces to

$$\Delta q = (q - sq^2) - q(1 - sq^2)/W, \quad (4.1d)$$

$$\Delta q = -\, spq^2/W. \quad (4.1e)$$

Natural selection is very slow to remove deleterious recessives from a population because the rate of evolutionary change, Δq, is proportional to q^2 when the $q \sim 0$.

A second example of selection on an "additive" locus can be obtained using the genotypic viability values in the fourth column of Table 4.1. In this case, W equals $(1 + 2ps)$, and we find that

$$\Delta p = p' - p, \quad (4.2a)$$

$$\Delta p = \{w_{AA}G_{AA} + [0.5]w_{Aa}G_{Aa}\} - \{G_{AA} + [0.5]G_{Aa}\}, \quad (4.2b)$$

$$\Delta p = \{(1 + 2s)p^2 + [0.5](1 + s)2pq\}/(1 + 2ps) - p, \quad (4.2c)$$

$$\Delta p = \{p + 2sp^2 + spq\}/(1 + 2ps) - p, \quad (4.2d)$$

$$\Delta p = spq/W. \quad (4.2e)$$

The sign of Δp is determined by the selective value of the *A* allele, *s*, which can be either negative or positive. The rate of evolution of an additively acting allele, such as *A*, is faster than the evolution of a recessive allele (cf. Equation 4.1e) because in the latter, when the allele is rare, most of the copies of the allele occur in heterozygotes which have normal viability. Thus, changing the relationship between genotype and fitness from recessive to additive has a large effect on the rate of evolutionary change as measured by Δp.

Finally, consider the case of balancing selection as depicted in column five of Table 4.1. Proceeding as above, we find that

$$\Delta p = \{(1 + s)p^2 + [0.5](1)2pq\}/(1 + sp^2 + tq^2) - p, \quad (4.3a)$$

$$\Delta p = \{p + sp^2\}/(1 + sp^2 + tq^2) - p, \quad (4.3b)$$

$$\Delta p = \{p + sp^2\}/(1 + sp^2 + tq^2) - p, \quad (4.3c)$$

$$\Delta p = pq(sp - tq)/W. \quad (4.3d)$$

In this case the sign of Δp and the direction of evolution depend upon the quantity $(sp - tq)$, which can be negative, positive or zero. If we set $(sp - tq)$ equal to zero and solve for p^*, we find that p^* equals $t/(s + t)$, which can lie between 0 and 1.

The allele frequency, p^*, is an *evolutionary equilibrium* because there is no change in allele frequency at this value, that is, $\Delta p = 0$. It is also a *stable equilibrium* because, when p lies above the equilibrium value (i.e., $p > t/[s + t]$), the allele frequency change is negative ($\Delta p < 0$), reducing p toward p^*. Conversely, when p lies below the equilibrium value (i.e., $p < t/[s + t]$), the allele frequency change is positive ($\Delta p > 0$), increasing p toward p^*. In this case, natural selection returns a population perturbed away from p^* back to the evolutionary equilibrium!

All these cases use the same general formula but vary in the genotypic viability fitnesses. In each case, the viability fitnesses were considered constant and unchanging throughout the evolutionary trajectory of the population. Indeed, because of the simple relationship between genotype and fitness in the case of random mating, the final expression for rate of evolution, Δp, was formulated in terms of allele frequencies and not genotype frequencies. Although the individual genotypes interact with the environment to determine the viability fitnesses, in these equations (4.1e, 4.2e, and 4.3d), it appears as though the genes determine fitnesses directly.

Darwinian Natural Selection Acting on Phenotypes

Notice that in all the columns of Table 4.1, under the heading of genotypic fitnesses, the genotypes differ from one another. Not only is such fitness variance required for natural selection, but also the strength of selection is *proportional* to the variance in fitness: *the greater the variance in relative fitness, the stronger the force of natural selection.* The variance in relative fitness sets an upper bound on the *rate* of evolution, phenotypic or genetic. Large differences in fitness between individuals mean strong natural selection. The absence of fitness differences between individuals (no fitness variation) means that natural selection and adaptive evolution are not possible.

The upper limit to the rate of evolution by natural selection is set by the variance in relative fitness because, as I show below, it equals the magnitude of the difference between the average breeding parent and the average individual. The breeding parents are the subset of all individuals in the population that contribute offspring and, consequently, genes to the next generation. The mean fitness of this subset minus the mean of the unselected population (i.e., the population including the breeding parents) divided by the mean before selection equals the variance in relative fitness.

Mean, Variance, and Covariance of Phenotype and Fitness

To consider phenotypic selection on individuals, we need the concepts of the mean and variance for both phenotype and fitness as well as the covariance between them. Imagine that each individual in a population is assigned two values: (1) z for its phenotype; and, (2) $W(z)$ for its net viability and fecundity fitness. The phenotypic trait, z, might be a continuous measurement, such as length or body weight, or a discontinuous measurement, such as color or number of vertebrae. It can even be the frequency of an allele within an individual so that, referring to the genotypic models above, an *AA* homozygous individual would have a phenotypic value of 1, an *Aa* heterozygote would have a phenotypic value of 0.5, and an *aa* homozygote would have a value of 0.

To describe the population, we describe the distribution of phenotypes in it. Let $p(z)$ be the frequency of individuals with phenotypic value z. For the phenotypes just assigned to the genotypic model, this means that the genotype frequency, G_{AA}, corresponds to $p(1)$, and similarly G_{Aa} corresponds to $p(0.5)$, and G_{aa} to $p(0)$. The mean phenotype in the population, Z, is defined as the sum of the products of the individual phenotypic values times their frequency or, for a continuous phenotype, we integrate over the distribution to obtain

$$Z = \int zp(z)dz. \qquad (4.4)$$

The mean, Z, describes the state of the population and we want to know how evolution by natural selection changes this state. For this reason, we want to compare the mean phenotype in the population before selection, Z, with the value after selection, Z'. This is typically expressed as the difference, $\Delta Z = Z' - Z$. For the genotypic model, this definition means that the mean phenotype equals $\{(1)(G_{AA}) + (0.5)(G_{Aa}) + (0)(G_{aa})\}$ or p_A, the frequency of the A allele. Thus, the change in the mean phenotype for the single-gene, two-allele model is Δp. The population genetic theory introduced above and the phenotypic selection theory are united by the same equation.

By this same definition, the mean fitness of the population *before selection* is

$$W = \int W(z)p(z)dz. \quad (4.5)$$

For the genotypic model, this definition means that the mean fitness in the population equals $\{(W_{AA})(G_{AA}) + (W_{Aa})(G_{Aa}) + (W_{aa})(G_{aa})\}$ or W, just as defined above.

After selection has acted, the fraction of individuals with phenotypic value z is no longer $p(z)$ but rather $p'(z)$. If we define the relative fitness of an individual with phenotypic value z as the ratio $W(z)/W$, we can again use the multiplicative transformation to convert the distribution before selection to the distribution after selection:

$$p'(z) = (W[z]/W)p(z). \quad (4.6)$$

The mean phenotype *after* selection thus equals the mean of the new, post-selection phenotypic distribution or

$$Z' = \int zp'(z)dz, \quad (4.7a)$$

$$Z' = \int zw(z)p(z)dz. \quad (4.7b)$$

Natural selection acting through the relative fitness differences among individuals in the population has changed the mean from Z to Z'. We describe this change as

$$\Delta Z = Z' - Z, \quad (4.8a)$$

$$\Delta Z = \int zw(z)p(z)dz - Z. \quad (4.8b)$$

Noting that both $\int p(z)dz$ and $\int w(z)p(z)dz$ equal 1 and that Z' is the mean of the product of relative fitness and phenotypic value, $w(z)p(z)$, we can recognize the last expression as a *covariance*, which is defined as the mean of a product minus the product of the means. Thus, we can rewrite the last expression as

$$\Delta Z = \mathrm{Cov}(z,w[z]). \quad (4.8c)$$

Revisiting our several earlier expressions for Δp, we find that they are also covariances defined as the mean of the product of phenotype and relative fitness, $\{(1)(w_{AA})(G_{AA}) + (0.5)(w_{Aa})(G_{Aa}) + (0)(w_{aa})(G_{aa})\}$, minus the product of the mean allele frequency within individuals, $p = \{(1)(G_{AA}) + (0.5)(G_{Aa}) + (0)(G_{aa})\}$, and the mean relative fitness, $1 = \{(w_{AA})(G_{AA}) + (w_{Aa})(G_{Aa}) + (w_{aa})(G_{aa})\}$.

The change in mean phenotype caused by natural selection *is* the covariance between phenotype and relative fitness. This change caused by natural selection must be transmitted across generations, from surviving breeding parents to their offspring. In the single-gene, two-allele models with random mating, allele frequencies do not change from parent to offspring in the absence of evolutionary forces such as mutation or gametic selection. Thus, in the population genetic models there is no further change in allele frequency so the change in allele frequency from newborns to newborns across generations is equal to the change within a generation or Δp.

This is not so in the phenotypic selection models. The offspring mean at the start of the next generation lies between the selected and unselected means of the previous generation, that is, between Z' and Z, respectively. This happens because h^2, the heritability of the parental phenotypes, is always less than or equal to 1, and only the fraction of phenotypic change, $h^2\Delta Z$, is transmitted across generations to the offspring at the start of the next generation. Heritability is always less than or equal to 1 because z, the phenotypic value of an individual, depends not only upon its genes but also upon the environmental factors that it experiences during development. If all individuals were genetically identical, they still might vary because of different experiences of the environment during development prior to measuring z. We could perform selection on these environmentally caused phenotypic differences and cause a change in mean phenotype, ΔZ. However, the change could not be transmitted across generations because all individuals, despite variations in z, are genetically identical.

The Variance in Relative Fitness and the Strength of Selection

The mean fitness, W', of the parents contributing to the next generation is defined by

$$W' = \int W(z)p'(z)dz = \int [W(z)w(z)]p(z)dz. \quad (4.9)$$

The difference in mean fitness, $\Delta W = W' - W$, between breeding parents and the unselected population is obtained by subtracting Equation. 4.5 from Equation. 4.9 to give

$$\Delta W = (W' - W) = \int(W[z]w[z])p(z)dz - \int W(z)p(z)dz. \quad (4.10)$$

Again recognizing that both $\int p(z)dz$ and $\int w(z)p(z)dz$ equal 1, we can write the relative change in mean fitness as the ratio, $\Delta W/W$, or

$$\Delta W/W = \int w^2(z)p(z)dz - \{\int w(z)p(z)dz\}^2, \quad (4.11a)$$

$$\Delta W/W = V_w, \quad (4.11b)$$

where V_w is the *variance* in relative fitness of the population *before selection*, defined as the mean of the square minus the square of the mean. (The variance can also be viewed as the covariance of something with itself; in this case, the covariance of relative fitness with relative fitness.) It follows that, the greater the variance in relative fitness, the stronger is natural selection because natural selection *is* the variance in relative fitness.

In a constant environment, offspring mean fitness must be less than or equal to that of the parents because fitness is determined by the relationship between phenotype and environment. The offspring mean phenotype is always less than or equal to that of the parents for two reasons. First, the heritability of a phenotype is a fraction less than or equal to 1. Thus, the offspring mean phenotype at the start of the next generation lies between the selected (Z') and unselected (Z) means of the previous generation. Second, the environmental factors that determine the relationship between phenotype and fitness may change from one generation to the next. Since fitness characterizes the relationship between organism and environment and is not a property of either one, a change in either phenotype or environment affects the offspring mean fitness. For this reason, Crow (1958, 1962) defined I, the "opportunity for selection," as

$$I = V_W/W^2 = V_w. \quad (4.12)$$

He argued that it set an upper bound on the rate of evolutionary change in the mean of *all* phenotypes.

The Covariance between Quadratic Phenotypes and Relative Fitness

With strong directional selection or with disruptive or stabilizing selection, natural selection also affects the phenotypic variance, $V(z)$, which is defined as

$$V(z) = \int z^2p(z)dz - \{\int zp(z)dz\}^2. \quad (4.13)$$

The variance after selection, $V'(z)$, is similarly defined over the distribution $p'(z)$. Thus, the change in the variance, ΔV, resulting from selection is given by

$$\Delta V(z) = V'(z) - V(z) = \int z^2p'(z)dz - \{\int zp'(z)dz\}^2 - \int z^2p(z)dz + \{\int zp(z)dz\}^2, \quad (4.14a)$$

$$\Delta V(z) = \int z^2w(z)p(z)dz - \{\int zw(z)p(z)dz\}^2 - z^2p(z)dz + \{\int zp(z)dz\}^2,$$

$$\Delta V(z) = \int z^2w(z)p(z)dz - \{\int z^2p(z)dz\}\{\int w(z)p(z)dz\} - \{\int w(z)p(z)dz\}^2 + \{\int p(z)dz\}^2,$$

$$\Delta V(z) = \mathrm{Cov}(z^2, w[z]) - \{Z + \Delta Z\}^2 + \{Z\}^2,$$

$$\Delta V(z) = \mathrm{Cov}(z^2, w[z]) - 2Z\Delta Z - \{\Delta Z\}^2,$$

$$\Delta V(z) = \mathrm{Cov}(z^2, w[z]) - 2Z\mathrm{Cov}(z,w[z]) - \mathrm{Cov}^2(z,w[z]),$$

$$\Delta V(z) = \mathrm{Cov}([z - Z]^2, w[z]) - \mathrm{Cov}^2(z,w[z]). \quad (4.14b)$$

With stabilizing selection, individuals with extreme phenotypes far from the mean (i.e., large values of $[z - Z]^2$) have the lowest fitness so that $\mathrm{Cov}([z - Z]^2, w[z])$ is negative, reducing the phenotypic variance. Conversely, whenever individuals with extreme phenotypes have the highest fitnesses (i.e., disruptive selection), the phenotypic variance is increased by selection because $\mathrm{Cov}([z - Z]^2, w[z])$ is positive. Even if the $\mathrm{Cov}([z - Z]^2, w[z])$ is zero, changes in the mean in either direction, owing to directional selection, will *reduce* the phenotypic variance because the term, $\mathrm{Cov}^2(z,w[z]) = \{\Delta Z\}^2$, is always positive. This reduction in the phenotypic variance as a result of directional selection is consonant with Darwin's pre-genetic discussion of natural selection's role in keeping characters "true and constant."

The single-gene, two-allele model of balancing selection introduced above can also be viewed from this same covariance perspective. In this model, the necessary conditions for an intermediate equilibrium value of the allele frequency, p^*, are identical to those for a nonzero equilibrium value of $V^*(z)$ where, at equilibrium, $\Delta V(z) = 0$. The variance in this model is defined as in Equation. 4.13 or, in words, as the difference between the mean of the squared values, that is, $\{(1)^2(G_{AA}) + (0.5)^2(G_{Aa}) + (0)^2(G_{aa})\}$, and the square of the mean value, that is, $(p)^2$. This gives us $V(z)$ before selection equal to $(pq/2)$. Following the exposition above, we could derive an expression for $\Delta V(z)$.

When the distribution of phenotypic values in a population is significantly skewed, the skew can be represented as $\mathrm{Cov}([z - Z], [z - Z]^2)$ or, more simply, $\mathrm{Cov}(z, [z - Z]^2)$. In this case, selection acting strictly on z will result in indirect selection on the variance in phenotype, $V(z)$. This is one reason why multivariate normality is so critical to the study of phenotypic selection in natural populations as a means of identifying agents of natural selection. The transformation of the observed phenotypic distribution to multivariate normality ensures that all covariances between linear and quadratic terms are zero and facilitates analysis. Whether or not nature engages in the same kinds of transformations is open to question (Wade & Kalisz 1989, 1990).

When one assumes that the phenotypic variance remains constant across generations despite directional selection, then one is implicitly making assumptions about the higher moments of the phenotypic distribution or the mating system and the underlying genetics. Unlike the linear multiplicative changes in the mean, Z, resulting from heredity, $h^2{}_z$, mating and transmission may increase or decrease the phenotypic variance of the offspring relative to the parents. For example, with *stabilizing selection* on additively determined phenotypes, the extreme individuals purged from the population will tend to be multilocus homozygotes and the surviving parents will have an excess of multilocus heterozygotes. Segregation from these heterozygous parents will reconstitute the extreme individuals leading to a dynamic balance between selection and segregation in their opposing effects on the phenotypic variance. The mating system affects this balance by influencing the frequency of heterozygotes. On the other hand, with *disruptive selection* the surviving parents will consist of an excess of multilocus homozygotes and random mating among the survivors will reduce the variance of the offspring phenotype distribution. The potential increases or decreases in the phenotypic variance and covariance that attend transmission are more difficult to account for than the changes in the mean.

This treatment can be extended to describe the effects of selection on the covariance between traits, $\mathrm{Cov}(x, y)$, simply by noting that the variance is the special case of the covariance of a variable with itself. Thus, Equation. 4.14b becomes

$$\Delta\,\mathrm{Cov}(x, y) = \mathrm{Cov}([x - X][y - Y], w[x,y]) - \mathrm{Cov}(x,w[x,y])\mathrm{Cov}(y,w[x,y]). \quad (4.15)$$

As above, this derivation does not require continuously distributed phenotypes. For example, the linkage disequilibrium between alleles at two bi-allelic loci, A and B, can be expressed as a covariance. As above, let M_{ij} be the frequency of allele A_1 within individuals of genotype ij and N_{ij} the frequency of allele B_1 within these same individuals so that p_{A1} is the average value of M_{ij} over the genotype frequency distribution ($M_{.j} = \Sigma_i\Sigma_j\ G_{ij}M_{ij}$) and, similarly, p_{B1} is the average value of N_{ij} ($N_{i.} = \Sigma_i\Sigma_j\ G_{ij}N_{ij}$). Furthermore, let x_{ij} be the product of M_{ij} and N_{ij}; so that, $x_{ij} = (M_{ij})(N_{ij})$. Also define d_1, d_2, d_3, and d_4 as the frequencies of the gametes A_1B_1, A_1B_2, A_2B_1, and A_2B_2, respectively, and note that $(d_1 + d_2)$ is p_{A1} and $(d_1 + d_3)$ is p_{B1}. Also note that, at the gametic level, the product of M and N is 1.0 for d_1 but zero for all other gametes, so that X_{gamete} is d_1. The gametic linkage disequilibrium, D_{gametic}, or the gametic covariance (Hartl 1980, p. 110), is defined as

$$D_{\text{gametic}} = d_1 - p_{A1}\,p_{B1}, \quad (4.16a)$$

$$D_{\text{gametic}} = d_1 - (d_1 + d_2)(d_1 + d_3),$$

$$D_{\text{gametic}} = d_1 - (d_1 - d_1d_4 + d_2d_3) = d_1d_4 - d_2d_3, \quad (4.16b)$$

$$D_{\text{gametic}} = X_{\text{gametic}} - (M_{..})(N_{..}), \quad (4.16c)$$

$$D_{\text{gametic}} = \mathrm{Cov}_{\text{gametic}}(M_{ij}, N_{ij}). \quad (4.16d)$$

The genotypic covariance between M_{ij} and N_{ij} is half the gametic covariance when the gametes are randomly united, permitting the genotypic frequencies, G_{ij}, to be expressed as products of the d_k as follows:

$$D_{genotypic} = Cov_{genotypic}(M_{ij}, N_{ij}), \quad (4.17a)$$

$$= \Sigma_i \Sigma_j G_{ij} x_{ij} - (\Sigma_i \Sigma_j G_{ij} M_{ij})(\Sigma_i \Sigma_j G_{ij} N_i), \quad (4.17b)$$

$$= (d_1^2 + d_1 d_2 + d_1 d_3 + [d_1 d_4 + d_2 d_3]/2) - (d_1 + d_2)(d_1 + d_3), \quad (4.17c)$$

$$D_{genotypic} = (d_1 d_4 - d_2 d_3)/2, \quad (4.17d)$$

$$D_{genotypic} = D_{gametic}/2. \quad (4.17e)$$

Selection causes a within-generation change in the genotypic covariance which can be represented as $\Delta Cov(M_{ij}, N_{ij})$, that is, $Cov'(M_{ij}, N_{ij}) - Cov(M_{ij}, N_{ij})$. This difference equals

$$\Delta Cov(M_{ij}, N_{ij}) = \Sigma_i \Sigma_j G'_{ij} M_{ij} N_{ij} - (M'_{ij})(N'_{ij}) - Cov(M_{ij}, N_{ij}). \quad (4.18a)$$

Substituting $[M.. + Cov(M_{ij}, w_{ij})]$ and $[N.. + Cov(N_{ij}, w_{ij})]$ for M'_{ij} and N'_{ij}, respectively, and simplifying gives

$$\Delta Cov(M_{ij}, N_{ij}) = Cov([M_{ij} - M..][N_{ij} - N..], w_{ij}) - Cov(M_{ij}, w_{ij}) Cov(N_{ij}, w_{ij}). \quad (4.18b)$$

Clearly, the formulation of Equation. 4.15 does not depend on a continuous distribution.

Reproduction and transmission across generations introduces an additional change in the covariance across loci as a result of recombination. Let $Cov''(M_{ij}, N_{ij})$ be the value of the genotypic covariance after selection and transmission (recombination between loci A and B at rate c), that is, at the start of the next generation. With random mating, the change in the genetic covariance, $\Delta Cov'(M_{ij}, N_{ij})$, that is, $Cov''(M_{ij}, N_{ij}) - Cov'(M_{ij}, N_{ij})$, caused by recombination equals $-cD'/2$. The effect of transmission on the genotypic covariance is always the *opposite* of the sign of D. Hence, it is always toward zero or no association. The evolutionary implication is that maintenance of a permanent, nonzero disequilibrium requires a balance between the effects of selection and transmission.

Partitioning the Variance in Fitness

Variance in relative fitness is essential for the action of natural selection as we saw above and this variance arises from multiple causes, environmental or genetic, as well as chance (see above). The variance in relative fitness sets an upper limit on the rate of all evolutionary change and the phenotypic covariances among different traits further limit the degree to which selection can be focused on a single trait without causing deleterious consequences for other traits. For these reasons, the partitioning of the total variance in relative fitness into different components is useful for determining where and to what extent the force of natural selection is focused on a particular trait, life history stage, or level of selection. Below, I partition the variance in relative fitness into components of natural and sexual selection, individual and group selection, and selfing and outcrossing, to illustrate the range and utility of the approach.

Sexual Selection and the Sex Difference in Variance in Fitness

The opportunities for selection in males, I_{males}, and in females, $I_{females}$, are defined as sex-specific ratios of fitness variance to the square of mean fitness, which is the variance in relative fitness. The sex-specific variances in relative fitness are related to one another through the sex ratio and mean fitness. Reproductive competition among the members of one sex can increase the variance in fitness of that sex so that selection in one sex can be stronger than selection in the other (Shuster & Wade 2003). To see this, let p_i be the fraction of males that have i mates. The *average* number of mates per male, $\Sigma_j jp_j$, is the sex ratio, R, expressed as the ratio of the number of mating females over the number of mating males. If $W_{females}$ and $V_{females}$ are the mean and variance of offspring across females, then the variance in male reproductive fitness, V_{males}, is

$$V_{males} = \Sigma p_j(jV_{females}) + \Sigma p_j(jW_{females} - RW_{females})^2, \quad (4.19a)$$

$$V_{males} = RV_{females} + W_{females}{}^2 V_{males}. \quad (4.19b)$$

Clearly, the variance in male reproductive fitness exceeds that of females whenever there is variation among males in the numbers of mates.

Since mean male fitness, W_{male}, equals $RW_{females}$, dividing Equation 4.19b by the square of mean male fitness gives

$$I_{males} = (1/R)(I_{females}) + I_{mates}, \quad (4.20)$$

where I_{mates} is the ratio V_{mates}/R^2, or the opportunity for selection arising from the variance among males in mate numbers. When R is 1, the sex difference in the opportunity for selection, $I_{males} - I_{females}$, equals I_{mates}. Thus, I_{mates} is a standardized measure of the strength of sexual selection (Wade 1979; Wade & Arnold 1980; Wade 1995; Shuster & Wade 2003).

Selfing, Outcrossing, and the Variance in Relative Fitness

In a population of N hermaphrodites, let the ith individual have fitness W_i, consisting of W_{Si} selfed offspring and $(1/2)(W_{Oi})$ outcrossed offspring. The factor of (1/2) discounts the outcrossed offspring in proportion to the genetic contribution of the ith individual. The mean fitness of this population, W, equals $\{W_S + (1/2)W_O\}$, where W_S and W_O are the mean numbers of selfed progeny and outcrossed progeny per individual, respectively.

The variance in fitness, V_W, is defined as

$$V_W = \Sigma_i \{W_{Si} + (1/2)W_{Oi} - [W_S + (1/2)W_O]\}^2/N \quad (4.21a)$$

$$= \Sigma_i \{[W_{Si} - W_S] + (1/2)[W_{Oi} - W_O]\}^2/N$$

$$= \Sigma_i \{[W_{Si} - W_S]^2 + (1/4)[W_{Oi} - W_O]^2 + (2)(1/2)[W_{Si} - W_S][W_{Oi} - W_O]\}^2/N,$$

$$V_W = V_S + (1/4)V_O + \mathrm{Cov}(W_S, W_O). \quad (4.21b)$$

To obtain the variance in relative fitness or I, the opportunity for selection, we divide V_W by the square of the mean fitness, W:

$$I_{Total} = V_W/(W)^2 = \{V_S + (1/4)V_O + \mathrm{Cov}(W_S, W_O)\}/(W)^2. \quad (4.22)$$

Let $V_S/(W_S)^2$ or I_S be the opportunity for selection on selfing, I_O or $V_O/(W_O)^2$ the opportunity for selection on outcrossing, and $Co\text{-}I_{SO}$ or $\mathrm{Cov}(W_S, W_O)/(W_I W_O)$ the opportunity for selection on the *trade-off* between selfing and outcrossing.

With these definitions, we can rewrite I_{Total} from Equation 4.22 as

$$I_{Total} = (p_S)^2(I_S) + (p_O)^2(I_O) + 2p_S p_O\, Co\text{-}I_{SO} \quad (4.23)$$

were p_I is (Ws/W), the average fraction of individual fitness derived from selfing, and p_O equals $([1/2]W_O/W)$, the average fraction of individual fitness derived from outcrossing.

When the mean number of selfed progeny equals the mean number of outcrossed progeny ($W_S = W_O$) and both are equally variable ($V_S = V_O$) and uncorrelated ($\mathrm{Cov}[W_S, W_O] = 0$), then p_S equals (2/3) and p_O equals (1/3) and Equation 4.18 becomes

$$I_{Total} = (2/3)^2(I_S) + (1/3)^2(I_O) + 2(2/3)(1/3)\, Co\text{-}I_{SO} \quad (4.24a)$$

$$I_{Total} = (0.44)(I_S) + (0.11)(I_O). \quad (4.24b)$$

Because the mean and variance of selfed and outcrossed offspring were assumed to be equal, I_S equals I_O. Thus, the contribution of selfed progeny to the total opportunity for selection is *four* times that of outcrossed progeny. In this circumstance, phenotypes that improved selfing would evolve much more rapidly than those that improved outcrossing by a similar amount and the selection pressure to preserve selfing or vegetative growth as a means of reproduction is stronger than that to preserve outcrossing. In this way, the differential contributions of selfed and outcrossed offspring to the opportunity for selection in hermaphroditic species share a similarity with the differential opportunity for selection on males and females in species with separate sexes.

More generally, if the mean number of selfed offspring equaled k times that of outbred offspring, then p_S equals $(2k/[2k + 1])$ and p_O equals $(1/[2k + 1])$ and Equation 4.23 becomes

$$I_{Total} = (2k/[2k + 1])^2(I_S) + (1/[2k + 1])^2(I_O) + 4(k/[2k + 1]^2)Co\text{-}I_{SO}. \quad (4.25)$$

The coefficients of I_S and I_O are equal at 0.25 when k is 0.50. This corresponds to the case where the viability of inbred offspring is uniformly equal to half that of outbred offspring. However, at this point, the coefficient of $Co\text{-}I_{SO}$ is 0.50, which is *twice* that of selection on either selfed or outcrossed offspring. As a result, when selection is equally strong on traits producing selfed offspring as it is on traits producing outcrossed offspring (see intersection of the lines in Figure 4.1), the fitness trade-off between selfing and outcrossing becomes the most predominant factor in mating system evolution.

Group and Kin Selection and the Variance in Fitness among Groups

Group selection is a controversial topic because it is difficult to establish that groups have the necessary properties of replication, variation, and heredity. The processes of group formation, dispersion, colonization, fission, and local extinction, together with environmental variation, determine the amount of phenotypic variation, among groups and the extent to which it is heritable (Wade 1996). When groups fission or disperse along lines of kinship, they are more likely to have the requisite heritable variation than groups formed in other ways.

The variance in relative fitness can be partitioned into within-and among-group components in a manner similar to that above for the separate sexes. For simplicity of derivation, let each of the M groups contain N individuals, that is, the same fraction, (1/M), of the total population, (MN). If the lifetime fitness of the ith individual in the jth group is W_{ij}, and mean fitness, $W_{..}$, equals $\Sigma_j \Sigma_i (W_{ij})/MN$, then the variance in fitness can be partitioned into two components. Beginning with the definition of the variance in fitness, we have:

$$V_W = \{\Sigma_j \Sigma_i (W_{ij} - W_{..})^2\}/MN, \quad (4.26a)$$

$$V_W = \{\Sigma_j \Sigma_i (W_{ij} - W_{.j} + W_{.j} - W_{..})^2\}/MN, \quad (4.26b)$$

by adding and subtracting the jth group mean fitness, $W_{.j}$ (= $\Sigma_i W_{ij}/N$). Recognizing that both sums, $\Sigma_j (W_{.j} - W_{..})$ and $\Sigma_j \Sigma_i (W_{ij} - W_{j.})$, are zero, the square can now be separated into two components:

$$V_W = \{\Sigma_j \Sigma_i (W_{ij} - W_{.j})^2\}/MN + \{\Sigma_j \Sigma_i (W_{.j} - W_{.j})^2\}/MN, \quad (4.26c)$$

$$V_W = \{\Sigma_j V_j\}/M + \{\Sigma_j (W_{.j} - W_{.j})^2\}/M, \quad (4.26d)$$

which reduces to

$$V_W = V_{\text{Within Groups}} + V_{\text{Among Groups}}. \quad (4.26e)$$

Dividing Equation. 4.26 by the square of the mean fitness, $W_{..}^2$, partitions the total opportunity for selection into within- and among-group components:

$$I_{\text{Total}} = I_{\text{Within Groups}} + I_{\text{Among Groups}}. \quad (4.27)$$

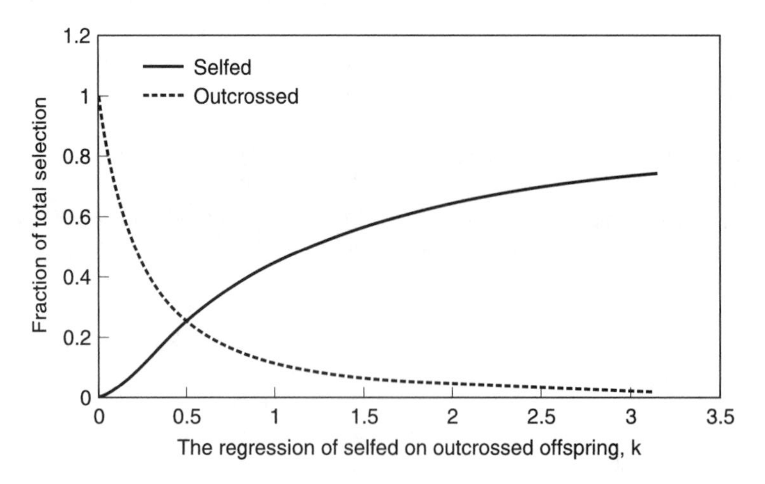

FIGURE 4.1. The effect on the variance in relative fitness of different relationships between the numbers of selfed (W_S) and outcrossed offspring (W_O) where k is a scalar constant such that W_S equals kW_O. (See text for further explanation.)

Variance in Fitness and Random Genetic Drift

Natural selection affects the strength of random genetic drift experienced by many genes through the relationship between the variance in offspring numbers the effective population size. For a single haploid locus, in a closed population of hermaphrodites and without mutation, the general relationship between N, the actual population size, and N_e, the effective population size, can be derived as follows. Let k_i be the number of gametes that the ith parent at generation t contributes as offspring to the next generation, $t + 1$, where $\Sigma_i k_i$ is N_{t+1}. Thus, k_i is W_i. The number of ways that a pair of genes can be drawn at random without replacement from among these N_{t+1} gametes equals $(\Sigma_i k_i)(\Sigma_i k_i - 1)/2$. To be identical by descent, a gene pair must come from the same parent, which can happen in any of $(\Sigma_i k_i[k_i - 1])/2$ different ways. Thus, the fraction of ways that a pair of genes can be i.b.d. is

$$(\Sigma_i k_i[k_i - 1])/([\Sigma_i k_i][\Sigma_i k_i - 1]). \quad (4.28)$$

Noting that $(\Sigma_i[k_i]^2)$ equals $N_t(V_W + W^2)$ and $(\Sigma_i k_i)$ equals $N_t W$, where W is the mean fitness of the parents, Equation 4.28 can be rewritten as

$$(V_W - [W][W - 1])/([W][NW - 1]), \quad (4.29a)$$

$$(I + 1 - [1/W])/(N - [1/W]). \quad (4.29b)$$

For a population of constant size, W is 1, and, by equating Equation 4.29b with $(1/N_e)$ from standard theory, we then derive the effective population size with natural selection as

$$N_e = (N - 1)/I. \quad (4.30)$$

Thus, the greater the opportunity for natural selection, I, the smaller the effective population size, N_e, and, thereby, the stronger the force of random genetic drift. This means that current natural selection acting on one trait will diminish necessarily the future opportunity for a response to selection at a later time on another, independent trait. Thus, hereditary variation is not simply an independent "multiplier" of natural selection but is necessarily affected by it.

With sexual selection, the opportunity for selection on males, I_{males}, tends to exceed that of females, $I_{females}$, (see Equation 4.20 above), so the effective size of paternally inherited, haploid genes, like those that are Y-linked, will tend to be lower than that of maternally inherited, haploid genes, like those of the mitochondrial genome (Wade & Shuster 2004). As a result the effective number of males, $N_{e,males} = (N_{males} - 1)/I_{males}$, will tend to be smaller than the effective number of females, $N_{e,females} = (N_{females} - 1)/I_{females}$. This in turn will influence expected diversity of paternally inherited, haploid genes, like those that are Y-linked, relative to maternally inherited, haploid genes, like those of the mitochondrial genome (Wade & Shuster 2004). Because maternally inherited cytoplasmic and paternally inherited Y-linked traits evolve independently of one another (unless there is inbreeding; Wade unpublished.) and are governed by distinct effective numbers, the strength of selection on one does not affect the strength of selection on the other. This independence allows the strength of sexual selection to be estimated from patterns of maternally inherited and paternally inherited molecular diversity.

For a haploid gene with the standard assumption that the migration rate is much greater than the mutation rate, the equilibrium probability of identity by descent, or the among-population variation, is

$$F \sim 1/(2N_e m + 1). \quad (4.31)$$

Substituting Equation 4.29 for N_e and rearranging, gives

$$F \sim I/(2[N - 1][m] + I), \quad (4.32)$$

which can be applied to maternally inherited mitochondrial haplotypes, F_{mt}, or paternally inherited Y-linked haplotypes by adding the appropriate sex-specific subscripts.

When the sex ratio is even ($N_{females} = N_{males}$) and the sex-specific migration rates are equal ($m_{male} = km_{female}$; where k, a constant of proportionality, is set to 1.0), then the ratio of the difference, $\{(1/F_{mt}) - (1/F_Y)\}$, to the sum, $\{(1/F_Y) + (1/F_{mt}) - 2\}$, gives

$$\{(1/F_{mt}) - (1/F_Y)\}/\{(1/F_Y) + (1/F_{mt}) - 2\} = (I_{mates})/(I_{male} + I_{female}). \quad (4.33)$$

Thus, a function of maternally inherited and paternally inherited diversity can be used to estimate the fraction of total selection represented by sexual selection. (Note that, when $k > 1$, the migration rate of females exceeds that of males, and Equation 4.33 is an *underestimate* of the proportion of sexual selection. The comparable nonequilibrium solution is similar (see Wade & Shuster 2004).)

SUMMARY AND CONCLUSION

The basic logic of evolution by natural selection was discovered and discussed in detail by Darwin, but the quantitative investigation of his theory required additional theoretical developments. Key developments were Fisher's (1930) insight that the phenotypic process of natural selection could be formally separated from the problem of inheritance and Crow's insight (1958) that the variance in relative fitness or the opportunity for selection sets an upper bound on the rate of adaptive evolution. Robertson (1966) and Price (1970, 1972) made the important discovery that natural selection *is* the covariance between phenotypic value and relative fitness, a relationship which greatly facilitates the study of phenotypic selection in natural populations (e.g., Shuster & Wade 2003).

In the sections above, I have tried to show the conceptual utility and the empirical potential of these fundamental contributions to the theory of evolution by natural selection. They can be used to formally connect quantitative genetic theory with population genetic theory and thereby illustrate the similarity and differences in these two theoretical approaches to evolutionary genetics. The opportunity for selection is particularly useful since it can be applied to so many different aspects of natural selection. I illustrated its utility in partitioning selection between males and females, between selfed and outcrossed offspring, and between different levels of selection. These partionings reveal an interesting similarity between selection acting on males and females in species with separate sexes and the action of selection on selfed and outcrossed offspring in hermaphroditic species.

The opportunity for selection is also useful for illustrating the necessary interaction between natural selection and random genetic drift. The variance in offspring numbers with directional selection tends to be greater than random and, thus, to reduce the effective population size and enhance drift. This effect can be employed in the interpretation of patterns of molecular diversity of maternally and paternally inherited genes to estimate the proportional strength of sexual selection.

SUGGESTIONS FOR FURTHER READING

To further explore these concepts, I recommend as an excellent and broader introduction to evolutionary genetic theory, the chapter by Kirkpatrick (1996) entitled "Genes and adaptation: a pocket guide to theory." For a more nuanced and detailed introduction to the topics covered in this chapter, the first nine chapters of the book *Introduction to Quantitative Genetics* by Falconer and Mackay (1996) are recommended; this book should be required reading for all graduate students interested in evolutionary genetics. I also highly recommend Hedrick's (2005) *Genetics of Populations*, especially chapters 2 and 3.

Falconer DS & TFC Mackay 1996 Introduction to Quantitative Genetics. Addison Wesley.

Hedrick PW 2005 Genetics of Populations. Jones and Bartlett.

Kirkpatrick M 1996 Genes and adaptation: a pocket guide to theory. pp. 125–146 in MR Rose & G Lauder, eds. Evolutionary Biology and Adaptation. Sinauer Assoc.

5

Stochastic Processes in Evolution

JOHN H. GILLESPIE

Pity the poor mutation: its most likely fate is extinction even if it is more fit than any other allele at its locus. When lost, these mutations play no significant role in evolution and evolutionists have no record of their short-lived existence. On the other hand, when an advantageous mutation is lucky enough to fix in the population, there are a number of consequences that excite evolutionists of various stripes. Darwin focused on the idea that the species is better adapted to its environment as a result of the substitution. A population geneticist might note that there is reduced variation around the site of the fixed mutation; a molecular evolutionist might use the substitution as a basis for a new phylogeny; a functional evolutionist might wonder why this particular substitution rather than another one that would seem to be even more fit; a statistician might use the timing of the substitution as part of a description of the temporal sequence of evolution. At the core of all of this interest is a random event that includes the time at which the substitution occurred, the functional properties of the substituted allele, and the neighboring pieces of the genome that were carried along with the allele as it swept though the population.

In this chapter, we will be particularly interested in the random times at which alleles enter the population and in their subsequent dynamics as well as those of linked portions of the genome. These are arguably the most important stochastic processes in evolution. The source of the randomness is somewhat mysterious, but obviously includes demographic stochasticity—the apparent randomness in the numbers of offspring among individuals—and the random nature of mutations, including both their probabilities of occurrence and the particular alterations in their DNA sequence. The source must also include random temporal and spatial variation in fitness, mutation and migration rates as well as changes induced by events in neighboring regions of the genome. When epistatic interactions from other evolving loci are added to the mix, the stochastic landscape for alleles becomes unimaginably complex. Our confidence that we do understand some aspects of this landscape comes from an examination of each source of randomness in isolation followed by a judgment call on its relative importance. In doing this, there is a major schism between the dynamics of very rare alleles—new mutations being the prime example—and more common alleles. The first two sections in this chapter are concerned with rare alleles, which are said to form a boundary process. The next three sections deal with common alleles, which are alleles that leave the boundary region to begin a calmer sojourn.

The randomness of evolution is nearly impossible to investigate through observation or experimentation. Most of our understanding comes from mathematical and computer modeling. This is precisely what we shall use in this chapter. Beyond basic algebra, there are only three critical prerequisites for this chapter: an understanding of the Poisson and geometric distributions and of the Taylor series expansion of a function. As a special case of the latter, we will use the geometric series in a couple of places.

CONCEPTS

The Fate of a New Mutation

When a new mutation appears in a population it is likely to disappear within very few generations even if it carries a substantial fitness advantage. This simple statement adumbrates much of what is to follow. But first we must see that it is true.

A simple warm-up problem is as follows: What is the probability that a new neutral mutation in a population of N haploid individuals will be lost in a single generation? Assume for simplicity that there will be exactly N individuals in the next generation as well. We could imagine that the next generation is obtained by simply reaching into the current generation and drawing out N gametes at random with replacement. (In this context, "with replacement" captures the idea that individuals are sending random numbers of offspring into the next generation.) As the frequency of the new mutation is $1/N$, the probability that it will fail to cross into the next generation in a single draw is the probability that some other allele is chosen, $1 - 1/N$. As the draws are independent, the probability that the mutation is never drawn is

$$\left(1 - \frac{1}{N}\right)^N \approx e^{-1} = 0.3678.$$

(The convergence of the expression on the left to $1/e$ for large N may be found in a basic calculus book that discusses the definition of e. You can prove it yourself by writing out the binomial expansion of $(1 - 1/N)^N$, letting $N \rightarrow \infty$ and noting that the resulting power series is the Taylor series expansion of e^{-1}.) From this we conclude that the probability that a new neutral mutation is lost in its first generation of life is about one third.

Our calculation has one major surprise: The population size does not appear in the final answer even though it was an integral part of the underlying model. Given this, it should be possible to restate the problem in such a way that N plays no role in the model or the calculation. The simplest approach is to assume that each individual has, on average, one offspring, and that the number of offspring is Poisson-distributed:

$$\text{Prob}\{i \text{ offspring}\} = \frac{e^{-1}}{i!}.$$

A mean of 1 is chosen because we are modeling a stable population whose mean size is not changing. The probability that our new mutation has no offspring is e^{-1} just as we saw with the previous calculation. If you jumped to the conclusion that this implies there was a hidden assumption in the previous approach that the number of offspring is Poisson-distributed, you are correct!

The second approach allows an immediate generalization to the case where the new mutation has a slight selective advantage accrued through a larger mean number of offspring, $\mu = 1 + s$, where s is small and positive. Now the probability of no offspring is

$$e^{-(1+s)} = e^{-1}e^{-s} \approx e^{-1}(1 - s),$$

where the final approximation uses the first two terms of the Taylor series expansion of e^{-s}. The surprising aspect of this result is that mutations with a considerable selective advantage still have a significant probability of being lost in the first generation. For example, the probability of loss in a single generation of an allele with a 1% selective advantage ($s = 0.01$) is $0.99/e = 0.3642$, which is very close to the probability for a neutral allele ($s = 0$). Most evolutionists would consider 1% selection to be very strong selection. The reason there is always a substantial probability that a new mutation is lost in a single generation is traceable to the assumption that the mean number of offspring per individual is close to 1. The probabilities of two, three, four, or more offspring must be balanced by the probability of zero offspring to keep the mean near 1. It should be obvious that the larger the variance in offspring number, the higher will be the probability of zero offspring.

I have a strong preference for the second approach to the problem. The dynamics of rare alleles, including new mutations, are essentially independent of the size of the population. The contrary view is hard to defend: Why should the immediate fate of a new mutation be different in a population of 10,000 versus 10 million individuals? When alleles do become common, their dynamics are influenced by the population size.

We turn now to the fate of new mutations for more than the first generation. Figure 5.1 graphs the probability of ultimate extinction for neutral alleles and alleles with 2% and 10% selective advantages over 100 generations. As is evident, the probability

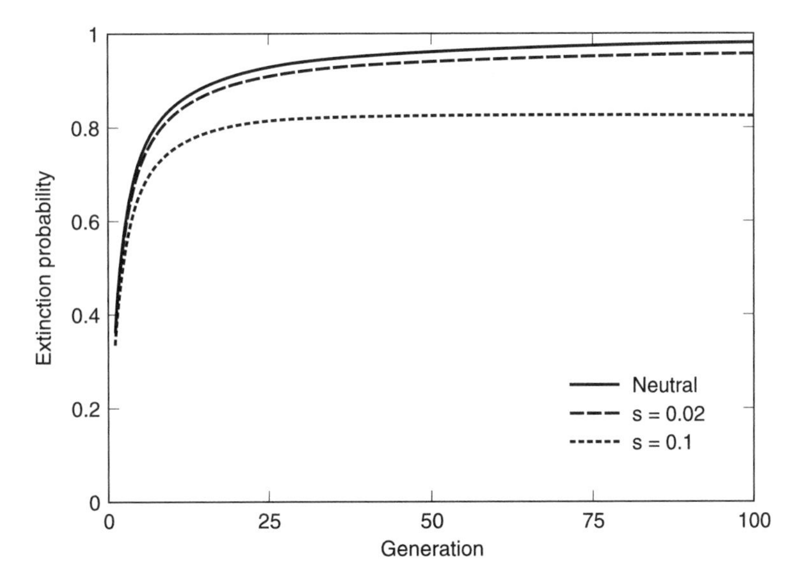

FIGURE 5.1. The probability that a new mutation will have been lost by a particular generation.

of extinction within the first few generations is remarkably similar among all three mutations. In particular, the probability of loss in the first generation is close to one third, as we already know. After a few generations, the extinction probabilities of advantageous mutations appears to asymptote while that for the neutral allele continues on toward 1. The fixation probability of the mutation with a 10% selective advantage is particularly striking as, even though this would be considered very strong selection, 80% of the time this wonderful mutation is lost.

As these extinction probabilities are among the most important quantities that we have in evolution, it is important to know how they were obtained. The curves in Figure 5.1 come from the theory of branching processes, which is beyond the scope of this chapter. Feller (1968) remains one of the clearest expositions of these processes and is highly recommended. A derivation of the probability of ultimate extinction, the asymptotic values in Figure 5.1, may be found in Box 5.1. There it is shown that the probability of ultimate survival of a new mutation with a selective advantage of s is approximately $2s/\sigma^2$, where σ^2 is the variance in offspring number. If the offspring distribution is Poisson, then the survival probability is approximately $2s$. Thus, an allele with a 1% advantage (huge) has a 2% chance of ultimate survival (not so huge).

There are a few other facts to be gleaned from Figure 5.1. The first is that most of the action appears to occur within the first few generations. Thus, we could say that the time scale for the stochastic dynamics of newly arising mutations in the first phase of their existence is on the order of a few generations. This is much faster than most other evolutionary processes. For example, in a population of a million individuals this is about a million times faster than genetic drift. Should these new mutations become common, the time scale of their subsequent dynamics slows down rather dramatically. For example, the time scale of directional selection is proportional to the reciprocal of the selection coefficient, $1/s$. For 1% selection, hundreds of generation are required for selection to change allele frequencies appreciably. It is not too far wrong to picture a substitution as beginning with a few hectic generations in which the allele frequency is bouncing around at random followed by a relatively smooth (almost deterministic) ride to fixation in roughly $1/s$ generations.

Boundary and Origination Processes

Populations have a steady input of new mutations that bounce around at random before facing almost certain extinction. These rare alleles form a

BOX 5.1. The Probability of Extinction of an Allele
John H. Gillespie

The probability of ultimate extinction of a new mutation may be calculated in many different ways. Here we will use an approach that is particularly streamlined. Let p_i be the probability that a particular allele has i offspring and let r be the probability of ultimate extinction of this allele when it begins its sojourn with a single copy. The probability of ultimate extinction may be expanded as follows:

$$r = \sum_{i=0}^{\infty} p_i r^i. \tag{1}$$

Each term in the sum is the probability that the original mutant has i offspring, p_i, times the probability that all i of them ultimately go extinct, r^i. There is an implicit assumption that the dynamics of the i offspring are independent, which makes their joint probability of extinction the product of their individual extinction probabilities. Such an assumption seems plausible when you picture a few of these mutations bouncing around in a population of, say, a million individuals. There is no general solution to Equation 1, although we can obtain a useful approximation under the assumption that the mean number of offspring is close to 1. Toward this end, note that the sum can be viewed as a function of r, $f(r)$, with the following useful properties:

- $f(1) = \sum p_i = 1$,
- $f'(1) = \sum p_i i r^{i-1} \mid r = 1 = \sum i p_i = \mu$,
- $f''(1) = \sum p_i i(i-1) r^{i-2} \mid r = 1 = \sigma^2 + \mu^2 - \mu$,

where μ and σ^2 are the mean and variance in the number of offspring from a single individual.

If the mean number of offspring is close to 1, then the extinction probability will be close to 1 as well. Write $r = 1 - \delta$, where the survival probability δ is small and positive. With this notation, a truncated Taylor series expansion of f,

$$f(1-\delta) \approx f(1) - f'(1) - f'(1)\delta + \frac{1}{2} f''(1)\delta^2,$$

and the properties of f listed in the previous paragraph it is easy to obtain an approximate solution to Eqution 1:

$$\delta \approx \frac{2(\mu - 1)}{\sigma^2 + \mu^2 - \mu}.$$

Set $\mu = 1 + s$, assume that s is small and positive and that $\sigma^2 >> s$ to obtain

$$\delta \approx \frac{2s}{\sigma^2}.$$

Thus, alleles with smaller variance in offspring number have a greater chance of survival than those with larger variances. If the offspring distribution is Poisson, then $\sigma^2 = 1 + s$ and the survival probability is approximately $2s$, which is Fisher's classic result.

process that is like a boiling caldron out of which bubbles the occasional—very lucky—favorable mutation. We will call the caldron the boundary process and the times at which alleles jump out the origination process. Once out of the caldron, the fate of alleles is impacted by a number of random forces that will be discussed in subsequent sections.

The traditional way to describe the origination process is to consider the repeated introductions of the same advantageous mutation. To be concrete, imagine that the population is fixed for the A_1 allele and that this allele mutates to A_2 at a rate v. In a haploid population of N individuals, the number of A_2 mutations introduced into the population each generation will be Poisson-distributed with mean Nv:

$$\text{Probability of } i \text{ new } A_2 \text{ mutations} = \frac{e^{-Nv}(Nv)^i}{i!}.$$

(The Poisson is appropriate because there are a lot of DNA replications, N, with a small probability of mutation during any one of them, v, and because the mutation events in separate replications are independent.) If the number of new mutations in a single generation is i (Poisson-distributed with mean Nv), then the probability that none of these mutations survives, given i, is approximately

$$(1-2s)^i.$$

The mean of this probability with respect to the Poisson distribution for the number of new mutations is

$$\sum_{i=0}^{\infty}(1-2s)^i \frac{e^{-Nv}(Nv)^i}{i!} = e^{-2Nvs} \approx 1-2Nvs,$$

where the final approximation is valid when, as will be assumed throughout, $2Nvs \ll 1$. Thus, the probability that a new mutation will enter the population in a particular generation and ultimately fix is $2Nvs$. We could call it a success when a new mutation arises that ultimately fixes in a particular generation and a failure when this does not occur. This language shows that the time until the appearance of the first mutation that ultimately fixes has a geometric distribution. That is, the probability that the origination time is i generations is

$$2Nvs(1-2Nvs)^{i-1}.$$

As the mean of a geometric random variable is the reciprocal of the probability of success, the mean time until the first substitution occurs is

$$\bar{t} = \frac{1}{2Nvs}. \quad (5.1)$$

From this we infer that the waiting time for a mutation to jump out of the caldron is shorter in larger populations and at loci with higher mutation rates and selection coefficients.

What about the next substitution? Traditionally, population geneticists assume that the mean time until this origination is also $1/2Nvs$, and so on for all originations. If so, then the rate of substitution is just the reciprocal of the mean time between originations,

$$\rho = 2Nvs. \quad (5.2)$$

This approach has one conspicuous peculiarity: Why should the selection coefficients of all originating mutations be the same? If a substitution increases the level of adaptation of a species, would not subsequent advantageous mutations have smaller selection coefficients? Explicit models of adaptive evolution from Fisher's (1958) original geometric model up through Orr's (2002) recent work all have the property that a sequence of substituting alleles has decreasing selection coefficients. (See Box 5.2 for a simulation of one of Orr's models.) In fact, they all have the property that evolution stagnates after a few substitutions either because the supply of advantageous mutations has been exhausted or because the selection coefficients become so small that originations fail to appear in an evolutionarily reasonable span of time. Under these models, evolution produces a small burst of substitutions and then . . . nothing. An intriguing aspect of these models is that the number of substitutions in a burst is insensitive to the mutation rate, population size, and selection coefficients. For example, in some of the models the mean depends on the logarithm of the population size; in others it is independent of population size (Gillespie 2002). Box 5.2 shows how to study some properties of a burst using computer simulations.

For continuous evolution we need only add one new element: a changing environment. If the environment changes, a new burst of substitutions can occur. We expect a similar burst of substitutions to follow each change in the environment. If λ is the

BOX 5.2. Mutational Landscape Model
John H. Gillespie

Alan Orr (e.g., Orr 2002) has been working on some fascinating properties of a simple model of adaptive evolution called the mutational landscape model. Under this model, we imagine that some allele with fitness w_0 is currently fixed in the population. Suppose that there are n alleles one mutational step away from the fixed allele and that the fitnesses of these n alleles are assigned at random and independently from the same probability distribution. (A normal distribution works fine here.) The neighboring alleles are labeled such that A_1 is the most fit allele, A_2 is the next most fit, and so forth. If selection is sufficiently strong, then only alleles that are more fit than the fixed alleles can themselves become fixed. The probability that the ith allele fixes is

$$\frac{s_i}{s_1 + s_2 + \ldots + s_m}, \quad i = 1,2,\ldots,m$$

if there happen to be m alleles that are more fit than the fixed allele. In this expression $s_i = w_i - w_0$ is just the selection coefficient of the ith allele. Thus, the probability that a particular allele fixes is proportional to its selection coefficient, as seems obvious. A simulation of this model begins with $n + 1$ alleles with random fitnesses and the fixed allele being the jth most fit allele. j should be viewed as a parameter of the model. Figure 1 shows the results of 10 replicate evolutions of the model with n = 10,000, j = 20, and fitnesses drawn from a normal distribution with standard deviation 0.01, Notice that even though there are 19 alleles that are more fit than the originally fixed alleles, evolution usually (7 out of 10 times) stops after one or two substitutions. While, on average, the biggest jump in fitness occurs with the first substitution, the replicates show a great deal of scatter in the fitnesses of fixed alleles.

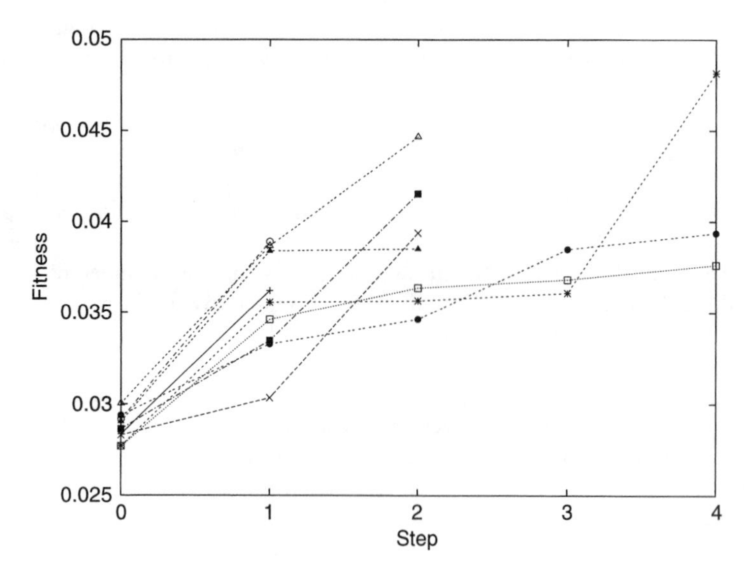

FIGURE 1. Ten trajectories from the mutational landscape model.

BOX 5.2. *(cont.)*

The python code for this simulation is below. Readers are encouraged to play around with the model and find its properties.

Python code for mutational landscape simulation

```
from random import *
from math import *
random_fitness = lambda: gauss(0.0,0.01)

def fitness_array(w, n, rng):
    "Returns a list of those alleles more fit than w"
    better_alleles = [w]
    for i in xrange(n):
        r = rng()
        if r > w: better_alleles.append(r)
    better_alleles.sort()
    return better_alleles

def choose_allele(fitnesses):
    """Returns the index of the fixed allele
    The first allele in fitnesses is the standard."""
    selection_coefs = [f - fitnesses[0] for f in fitnesses]
    sum_sc = sum(selection_coefs)
    probabilities = [s / sum_sc for s in selection_coefs]
    running_sum = 0.0
    r = random()
    i = 0
    for p in probabilities:
        running_sum += p
        if running_sum > r: return i
        i += 1

def evolution(n, start, rng):
    initial_alleles = [rng() for i in xrange(n+1)]
    initial_alleles.sort()
    alleles = initial_alleles[-start:]
    fixed_alleles = [(0, alleles[0])]
    while len(alleles) > 1:
        fixed_allele = (choose_allele(alleles))
        fixed_alleles.append((fixed_allele, alleles[fixed_allele]))
        alleles = fitness_array(fixed_alleles[-1][1], n, rng)
    return fixed_alleles
```

rate of change per generation in those aspects of the environment that affect our locus, then the rate of substitution per generation is just the rate of change of the environment, λ, times the mean number of fixations following a change,

$$\rho = \lambda \mu_x. \tag{5.3}$$

We have arrived at a rate of substitution that is essentially independent of the population size, the mutation rate, and the strength of selection and about as different from our previous rate, $\rho = 2Nvs$, as we could imagine. There is no consensus in population genetics about which rate is better or, in fact, whether either is worthy of our attention.

In this section we have given a description of the stochastic dynamics of originations of advantageous mutations. This is an important stochastic process not only because of its consequences for adaptation, but because of the hitchhiking of genome segments that invariably accompanies the substitution of a new mutation. In the next three sections we will look at the stochastic forces that impact mutations once they leave the caldron and begin their sojourn in the realm of common alleles.

Genetic Drift

The origins of much of mathematical population genetics, including the stochastic theory, can be found in Fisher's monumental 1922 paper. In this paper, Fisher considers two models that incorporate randomness. The first appears in a treatment of the survival probability of "individual genes" as described previously. Fisher went on to treat "Factors not acted on by selection," and, in so doing, introduced genetic drift to population genetics:

> If p be the proportion of any gene, and q its allelomorph in a dimorphic factor, then in n individuals of any generation we have $2np$ genes scattered at random . . . Further, if a second generation of n individuals be now formed at random, the standard departure of p from its previous value will be
>
> $$\sigma_p = \sqrt{\frac{pq}{2n}}, \ldots$$

He went on to conclude that the rate of loss of genetic variation is $1/4n$. (Later, Sewall Wright corrected this result to $1/2n$.) The loss of genetic variation due to this "Hagedoorn effect" he thought was insignificant:

> As few specific groups contain less than 10,000 individuals between whom interbreeding takes place, the period required for the action of the Hagedoorn effect, in the entire absence of mutation, is immense.

Curiously, Fisher does not motivate or attempt to justify his assertion that the "standard departure of p' is $\sqrt{pq/2n}$. (Standard departure is an old name for standard deviation.) Today, we typically model genetic drift with binomial sampling and from this obtain a standard deviation of $\sqrt{pq/2n}$ for a diploid population. The reason for using binomial sampling is not at all obvious or even plausible. In this section will we (1) illustrate the problem with binomial sampling, (2) show that Fisher's standard deviation is reasonable without binomial sampling, and (3) discuss some of the properties of genetic drift.

The motivation for mathematical models of genetic drift is usually somewhat contrived. The bag-of-marbles metaphor that is often used in teaching genetic drift is a wonderful device for learning the consequences of binomial sampling, but the biological underpinnings are not at all clear. To see this, imagine a haploid population with two alleles, A_1 and A_2, with frequencies p and $q = 1 - p$. If there are N individuals in the population, then Np of these will be A_1 and Nq will be A_2. To form the next generation, we imagine that the number of offspring for each individual is chosen at random from the same probability distribution. Let the number of offspring from the ith A_1 be X_i and from the ith A_2 be Y_i. The allele frequency in the next generation will be

$$p' = \frac{\text{Number of } A_1 \text{ offspring}}{\text{Total number of offspring}} = \frac{X}{X+Y},$$

where

$$X = X_1 + X_2 + \ldots + X_{Np}$$
$$Y = Y_1 + Y_2 + \ldots + Y_{Nq}.$$

The bag-of-marbles metaphor suggests that the distribution of the number of A_1 alleles in the next generation, i, is the binomial distribution

$$\text{Prob}\{X = i\} = \binom{N}{i} p^i q^{N-i}.$$

But why? In our model, the number of A_1 individuals in the next generation is called X and its distribution can be that of any positive integer-valued random variable. For example, X could have a geometric, negative binomial, Poisson, or some other distribution. And why should the total number of individuals, $X + Y$, be N? In fact, a binomial distribution is only appropriate when two conditions are met: The total number of offspring is set at some fixed number and the distributions of X and Y are Poisson. Biologically, we are in trouble because studies of offspring distributions in natural populations seldom, if ever, find Poisson distributions. In general, the variance in the number of offspring is larger than the mean. The idea of fixing the total number of offspring at N before forming the next generation is strange as well. Population sizes of most species fluctuate significantly over relatively small numbers of generations. In our model, that fluctuation is embodied by the distribution of $X + Y$, which cannot be set at a fixed number N without disrupting the biological appeal of the model.

Does this mean that genetic drift is a flawed concept? Not at all; it only means that binomial sampling is not a good mirror of the demography of natural populations. Fortunately, much of theoretical population genetics does not deal directly with the binomial distribution but uses large N approximations such as diffusion models. For these, the binomial distribution is only used to obtain the variance of p', which can be obtained directly without any reference to a binomial distribution. To see this, first rewrite p' as

$$p' = \frac{E\{X\}}{E\{X\}+E\{Y\}}\left(\frac{1+\dfrac{X-E\{X\}}{E\{X\}}}{1+\dfrac{X-E\{X\}+Y-E\{Y\}}{E\{X\}+E\{Y\}}}\right), \tag{5.4}$$

where the notation $E\{X\}$ refers to the expected value (or mean) of X. An application of the delta method of statistics will allow us to obtain useful approximations of p'. Toward this end, define

$$\delta_x = \frac{X-E\{X\}}{E\{X\}},$$

and

$$\delta_{x+y} = \frac{X-E\{X\}+Y-E\{Y\}}{E\{X\}+E\{Y\}}$$

(The δs are small when N is large because of the law of large numbers.) Note that $E\{\delta_x\} = E\{\delta_{x+y}\} = 0$. Substitute the δs into Equations 54 and ignore powers of δs greater than 1 to obtain

$$p' \sim \frac{E\{X\}}{E\{X\}+E\{Y\}}\left(1+\delta_x-\delta_{x+y}\right).$$

Now let us specialize and assume that the mean number of offspring for both alleles is 1 and that the variance in offspring per individual is σ^2. Thus, $E\{X\} = N_p$, $E\{Y\} = N_q$, Var $\{\delta_x\} = \sigma^2 / Np$, Var $\{\delta_{x+y}\} = s^2/N$, Cov $\{\delta_x, \delta_{x+y}\} = 2\sigma^2/N$, and Var $p' = pq\sigma^2/N$ The square root, $\sigma\sqrt{pq/N}$, gets us close the Fisher's standard deviation, the only difference being the factor σ. As $\sigma^2 = 1$ when the offspring distribution is Poisson, Fisher's model appears to be one with a hidden assumption that the offspring distribution is Poisson. (Convince yourself that $E\{p'\} = p$, in agreement with the fact that genetic drift does not change the mean allele frequency of the population.)

Let us step back from the algebra and see what we have accomplished. The model is a simple model of a haploid population with two alleles in which the source of randomness is the (random) number of offspring produced by each individual. For this model, the variance in the change in the allele frequency (due to the random number of offspring per individual) is the familiar $pq\sigma^2/N$, although we are not invoking binomial sampling. Populations with larger variances in offspring number will have a larger variance in allele frequency change, as we would expect. In this context, the effective size of the population is called the variance effective size and is $N_e = N/\sigma^2$. For example, a population of $N = 1000$ individuals with $\sigma^2 = 1$ has the same effective size as one with $N = 2000$ and $\sigma^2 = 2$. We are happy with this as both have the same variance in p'

Many properties of genetic drift depend only on the mean and variance of p'. The most important of these is the decay in the heterozygosity of the population. The heterozygosity of a two-allele population is defined as $H = 2pq$. In random mating diploid species, H is the frequency of heterozygotes. In other species, such as haploids or inbreeding plants, H is used as a measure of genetic variation in the populations without any implications about the frequency of heterozygotes. Our job is to discover the effects of genetic drift on H. In particular, given a population in which the frequency of the A_1 allele is p (hence, $H = 2pq$), we will find the expected value of H in the next generation, $E\{H'\} = E\{2p'(1-p')\}$:

$$\begin{aligned} E\{H'\} &= 2\left(E\{p'\}-E\{p'^2\}\right) \\ &= 2\left(E\{p'\}-\text{Var}\{p'\}-E\{p'\}^2\right) \\ &= 2E\{p'\}(1-E\{p'\})\left(1-\frac{\text{Var}\{p'\}}{E\{p'\}(1-E\{p'\})}\right). \end{aligned}$$

The expression for $E\{H'\}$ may be simplified for genetic drift by recalling that drift does not change mean allele frequencies, $E\{p'\} = p$. Thus

$$E\{H'\} = H\left(1 - \frac{\text{Var}\{p'\}}{pq}\right), \qquad (5.5)$$

which, using our result that Var $\{p'\} = pq\sigma^2/N$, becomes the familiar

$$E\{H'\} = H\left(1 - \frac{\sigma^2}{N}\right).$$

The expected change in H in one generation is given by

$$\Delta H = -\frac{\sigma^2}{N}H.$$

From this we conclude that the time scale of genetic drift in a haploid population is on the order of the population size. It takes N generations to do anything interesting. This is why Fisher said "the period required for the action of the Hagedoorn effect, in the entire absence of mutation, is immense."

At this point our calculations show that the expected heterozygosity decreases under genetic drift. But why? Must we simply say that this is how the calculations came out, or can we gain more intuition by poking around a bit? In fact, we can with the aid of Figure 5.2, which plots H as a function of p. Imagine a population in which the value of p is represented by the middle vertical line. In forming the next generation, random changes will cause p to increase or decrease by a small amount, labeled Δp in the figure. Notice that a small change up in p will not increase H to the same extent that a small change down in p of the same magnitude will decrease H. If we assume that changes up and down in p are equally likely (only approximately true), then it is clear that the expected value of H will decrease. Significantly, this argument works just as well if the initial value of p is to the right of the mode at $p = 1/2$. Thus, the reason that H decreases on average is because H is a concave function of p. The variance in p' appears in the formula because H is a quadratic function of p.

Our calculations show that genetic drift tends to remove genetic variation from populations and does so very slowly. As evolutionary forces go, drift is a very weak force. To claim that it is the most important stochastic force in evolution would be tantamount to asserting that other stochastic forces have no significant impact on regions of the genome for time periods on the order of N generations. For large populations, this seems unlikely.

Genetic Draft

When the substitution of a new favored mutation occurs, a portion of the genome that is closely linked to it is carried along for the ride in a process known as hitchhiking. This can lead to the fixation of alleles that are very closely linked to the advantageous

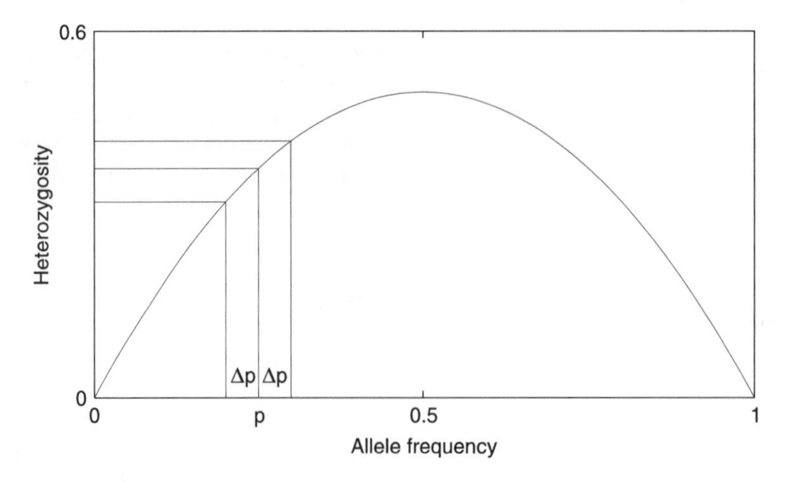

FIGURE **5.2.** The relationship between changes in the allele frequency 'p' and the consequent changes in the heterozygosity, H.

mutation and to a change in frequency of ones that are more loosely linked. As the timing of a selective sweep is random and as the alleles that happen to be linked to the advantageous mutation are there by chance, a selective sweep causes a significant random perturbation of a small bit of the genome. The stochastic process induced by these random sweeps is called genetic draft because of some surprising similarities to genetic drift.

The essence of genetic draft is best illustrated in a simple model where a two-allele neutral locus is tightly linked to another locus that is experiencing selective sweeps at a rate ρ. There are two sources of randomness here: the first is the set of times at which the sweeps occur; the second is due to the chance event of whether the A_1 or the A_2 allele is initially linked to a favorable mutation at the selected locus.

Consider a particular selective sweep that begins when a new advantageous mutation (that ultimately fixes) enters the population. The advantageous mutation will initially be linked to either the A_1 or the A_2 allele at the neutral locus. If the frequency of the A_1 allele is p, then the probability that the new mutation is initially linked to the A_1 allele is also p. When the fixation is complete, the allele that was initially linked to the advantageous mutation will be fixed along with it. We could summarize the two possible outcomes as follows

$$p' = \begin{cases} 1 \text{ with probability } p \\ 0 \text{ with probability } q. \end{cases}$$

As p' is a random variable, it has a mean and a variance, which are

$$E\{p'\} = p$$
$$\mathrm{Var}\{p'\} = pq.$$

This random event shares two properties with genetic drift: the mean allele frequency does not change and the variance in the allele frequency is proportional to pq.

Selective sweeps occur at a rate ρ, which implies that the probability a sweep begins in a particular generation is also ρ. If a sweep is not initiated, then p will not change. Thus, the change in a given generation may be written

$$p' = \begin{cases} 1 \text{ with probability } \rho p \\ 0 \text{ with probability } \rho q \\ p \text{ with probability } 1-\rho. \end{cases}$$

The moments for p' per generation are

$$E\{p'\} = p$$
$$\mathrm{Var}\{p'\} = \rho pq.$$

As far as first- and second-order moments go, this model of genetic draft is the same as a model of genetic drift with $\rho = \sigma^2/N$. In fact, as the calculation for the rate of decay of the expected heterozygosity used only the variance of p', we have immediately that

$$\Delta H = -\rho H$$

Two technical details were glossed over in this development of genetic draft. The first concerns a time scale argument. We implicitly assumed that the time required to complete a selective sweep is very short relative to the time between sweeps. This allows us to view sweeps as instantaneous events occurring within a single generation. The second concerns the distribution of the time between sweeps, which was implicitly assumed to be exponentially distributed. Thus, the times of selective sweeps form a Poisson process; equivalently, the origination process is a Poisson process.

Our model of genetic draft thus far does not allow for recombination between the selected locus and the neutral locus. When there is recombination, a hitchhiking neutral allele might not fix as in the no-recombination case described above. Rather, recombination leads to an uncoupling of the hitchhiking allele and the advantageous mutation. The full dynamics of this two-locus model are quite complicated and have yet to receive a complete mathematical treatment. To get around this inadequacy, we can invent a model, called the pseudohitchhiking model (Gillespie 2000), that captures the main effects of hitchhiking with recombination. The model is based on the idea that when a new advantageous mutation enters the population, it is initially linked to only a single copy of one of the neutral alleles. That one copy will increase in frequency until it is released by recombination. Meanwhile, the frequencies of all the other alleles are reduced by the same factor because they are not hitchhiking. If the final frequency of the hitchhiking copy is y, then we can write our model as

$$p' = \begin{cases} p(1-y) + y & \text{with probability } \rho p \\ p(1-y) & \text{with probability } \rho q \\ p & \text{with probability } 1-\rho \end{cases}$$

Once again, the mean of p' is p, but now the variance is

$$\mathrm{Var}\{p'\} = \rho y^2 pq.$$

The value of y is determined by the strength of selection at the selected locus, the rate of recombination between the neutral and the selected locus, and the boundary process during the initial phase of the selective sweep. Because of these stochastic elements, y is actually a random variable. As a consequence, the variance in p' is more properly written as $\mathrm{Var}\{p'\} = \rho E\{y^2\}pq$. Finally, let us add genetic drift to the model, to obtain

$$\mathrm{Var}\{p'\} = pq\left(\rho E\{y^2\} + \frac{1}{2N}\right),$$

where N is the effective size of the diploid population.

When the two stochastic effects are combined, we have a new variance effective size of the population, which is simply one half the reciprocal of the coefficient of pq,

$$N_e = \frac{N}{1 + 2N\rho E\{y^2\}}.$$

This result is quite remarkable. Notice that if we keep ρ fixed and increase N, the effective size of the population increases, but eventually asymptotes at $1/2\rho E\{y^2\}$ as illustrated in Figure 5.3. Thus the effective size of a population under the influence of both genetic drift and genetic draft may be only weakly dependent on the actual population size. At the limit, it can be independent of population size. As a consequence, levels of genetic variation may be insensitive to population size as well. For example, the expected number of mutations that differ between a pair of randomly drawn DNA sequences in a neutral population is $4N_e\mu$ (Houle and Kondrashov, Ch. 3 of this volume), or, with drift and draft,

$$\frac{4Nu}{1 + 2N\rho E\{y^2\}}.$$

In these formulae, ρ is treated as a constant independent of population size. Eq. 5.3 is from a model that has this property. On the other hand, population geneticists often use $\rho = 4Nvs$ (the diploid equivalent of Eq. 5.2), in which case the variance effective size becomes

$$N_e = \frac{N}{1 + 8N^2 vsE\{y^2\}}.$$

Under this model the effective size will eventually decrease with increasing population size as illustrated in Figure 5.3. The inflection of the curve occurs where genetic draft begins to dominate genetic drift as the main stochastic force. This occurs when

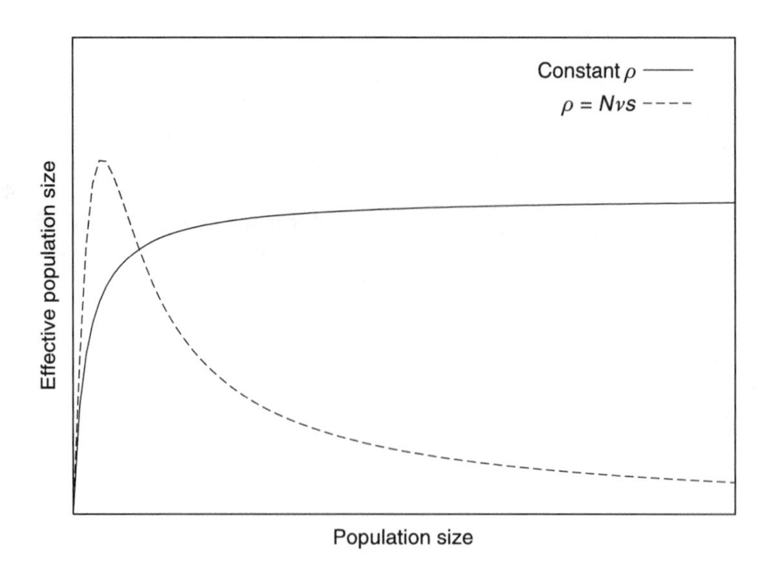

FIGURE 5.3. The relationship between the population size and the effective population size under genetic draft.

$\rho E\{y^2\} > 1/2N$. As a consequence, the amount of genetic variation at the neutral locus will also eventually decrease with increasing population size. Anyone familiar with genetic drift will recoil from the idea that neutral variation can decrease with increasing population size. The reason for this is that the rate of hitchhiking is increasing with increasing population size ($\rho = 4N\nu s$) and, as a consequence, neutral variation is reduced. The nature of the functional relationship between ρ and N in evolution is not at all clear. Although we frequently use $\rho = 4N\nu s$ in discussions of adaptive evolution, this may well be a poor choice as the underlying model is so at variance with our usual notions of adaptive evolution. It is clear that we cannot understand the relationship between neutral variation and population size without knowing more about ρ.

No matter what the functional relationship between ρ and N, our results on genetic draft suggest that levels of genetic variation will not necessarily show a strong dependency on population size. This is a good thing as one conspicuous conundrum in population genetics has been that our drift-based theories predict a strong dependency of levels of variation on population size but observations generally show a weak dependency at best. This is often called Lewontin's paradox, as he was the first to clearly state the problem (Lewontin 1974). Genetic draft provides an explanation for Lewontin's paradox: genetic drift is not important in natural populations.

If genetic draft is to be considered an important force acting on common alleles, we should be able produce some numbers to support this idea. We could ask: In regions of normal recombination in *Drosophila,* could genetic draft account for observed levels of polymorphism at silent sites if they were neutral and if genetic drift were not operating? Consider a typical block of 10,000 bases on an autosome. Assuming that the recombination rate between a neighboring pair of nucleotides is 10^{-8}, then the recombination rate between any pair of nucleotides within this block is less than about $r = 0.0001$. A sweep anywhere in this block with a selection coefficient of about $s = 0.001$, hence $r/s < 0.1$, will cause the entire block to be nearly homozygous. If we assume that such substitutions occur at a rate $\rho = 10^{-7}$ within the block, then, by Eq. 5.6 with $N = 8$, the heterozygosity should be about 0.02, which is close to the observed value in *Drosophila simulans*. If we assume that the rate of substitution per silent site is about 10^{-9}, then all we require is that about 1 out of every 100 substitutions is strongly selected.

Random Environments

The last stochastic process to be discussed is selection in a random environment. Our goal is not to give an extensive treatment of this rich area, but rather to give a few simple results that can be used to compare the role of this source of randomness with those discussed earlier. All our work will focus on a two-allele, additive diploid model in which the fitnesses of the three genotypes are:

$$\frac{A_1A_1}{1+U_t} \quad \frac{A_1A_2}{1+(U_t+V_t)/2} \quad \frac{A_2A_2}{1+V_t}.$$

Assume that the selection coefficients U_t and V_t change at random through time and that their values in a particular generation are independent of all values in previous generations. That is, assume that U_t and V_t are not autocorrelated. If p is the frequency of the A_1 allele, then the change in its frequency in a single generation is

$$\Delta p = \frac{1}{2}\frac{p_tq_t(U_t - V_t)}{1+p_tU_t + q_tV_t}.$$

A lot can be learned by an examination of the mean and variance in Δp. Toward this end, assume that the values of U_t and V_t are very small, which allows the denominator of Δp to be expanded as a geometric series

$$\Delta p \sim \frac{1}{2}pq\left[U - V - pU^2 - (1-2p)UV + qV^2\right],$$

where all subscript ts have been suppressed for clarity and powers of U and V greater than 2 have been ignored. Our final simplifying assumption is that the means of the selection coefficients are zero, $E\{U_t\} = E\{V_t\} = 0$, and their variances are equal, $\mathrm{Var}\{U_t\} = \mathrm{Var}\{V_t\} = \sigma_e^2$. With these moment assumptions, the mean change becomes

$$E\{\Delta p\} = \sigma_e^2 pq(1/2 - p),$$

and the variance in the change is

$$\mathrm{var}\{p'\} = \frac{1}{2}\sigma_e^2 p^2 q^2.$$

A totally unexpected property of this calculation is that the mean change in p suggests that there is balancing selection. Note that when $p < 1/2$, $E\{\Delta p\} > 0$ and when $p > 1/2$, $E\{\Delta p\} < 0$. This indicates that selection pushes p toward one half, on average. Why is this unexpected? The underlying additive model is a model of directional selection, which, when fitnesses are constant, leads to the fixation of the favored allele. $E\{\Delta p\}$ suggests that random changes in the fitnesses of a model that does not maintain polymorphism will turn it into a model of balancing selection that does maintain polymorphism. More work needs to be done to prove that this is, in fact, a model of balancing selection. The reason is that while $E\{\Delta p\}$ shows a mean push toward one half, $\text{Var}\{\Delta p\}$ indicates that there is an opposing dispersive force pushing alleles away from the interior. More mathematics is needed to show that the centripetal force wins out over the centrifugal force and thus that polymorphism is maintained. A complete treatment of this problem may be found in Gillespie (1991).

To find the effects of random environments on the heterozygosity, we could simply plug our expressions for $E\{p'\} = p + E\{\Delta p\}$ and $\text{Var}\{p'\}$ into Eq. 5.5. Try it; you will hate it! Unlike the cases of genetic drift and draft, the expected value of H' does not depend only on H. As a consequence, we cannot write down a simple expression that gives us the heterozygosity at some arbitrary generation in the future. For drift and draft, $\text{Var}\{p'\}$ exactly cancels the factor pq in the final term in Eq. 5.5. Heterozygosity is a population statistic that is convenient only when drift and/or draft are the main stochastic forces. This convenience disappears in random environment models. In fact, there is no known statistic that is useful for following genetic variation in models with random environments.

Random environment models have many technical aspects, such as the problems with $E\{H'\}$, that make them difficult to analyze. As a result, they have been largely ignored in population genetics. This is unfortunate as it is clear that environments do change and that adaptive evolution is driven by these changes.

SUMMING UP

One of the most important problems in evolutionary genetics is assessing the relative contribution of the various evolutionary forces to forming the genetic structure of populations. As an aid to thinking about this problem, consider the following table, which summarizes the processes discussed in this chapter. The rate is a crude measure of the rate of change of the process; the reciprocal of this rate is the time scale over which the process unfolds.

Process	*Rate*
Boundary process	σ^2
Genetic drift	$1/N$
Genetic draft	$\rho E\{y^2\}$
Origination process	ρ
Random environments	σ^2_e
Deterministic selection	s

Among the stochastic processes, there can be no denying the importance of the boundary process and its progeny, the origination process. Within the caldron that we call the boundary, many evolutionary forces are at work. These can never be studied as any observation will necessitate sampling the entire population. However, we can say with some confidence that there are two important sources of randomness for very rare alleles: variation in the numbers of offspring and random changes in fitnesses. Other sources of randomness operate on time scales that are generally irrelevant to the short-term survival probabilities of such alleles.

A small fraction of all mutations cross out of the boundary at random times that make up the origination process. The model giving rise to $\rho = 4Nvs$ carries with it the view that the determining stochastic elements are demographic stochasticity (the boundary dynamics) and the random input of mutations. While popular, this view seems at variance with the notion that evolution occurs in response to changes in the environment. A better view has random changes in the environment being the most important stochastic element determining the times at which originations occur. If the environment does change at a particular generation, then the dynamics of the boundary region will determine the lag until mutations cross into commonness. Unlike the boundary process, we can actually make observations relevant to the origination process. We can estimate ρ from molecular evolution studies and we can even learn a limited amount about the time between originations. For example, the variance in this time is often greater than it would be if the origination process were a Poisson process.

Once an allele crosses into the interior, we do not know which process dominates. Typically, one uses genetic drift as a benchmark by asking if a particular rate is greater than $1/N$. For example, in a population of, say, 10^7 individuals, selection in a random environment will dominate drift if $\sigma^2_e >> 10^{-7}$. ($\sigma^2_e << 10^{-7}$ the alleles are called effectively neutral.) The great neutralist–selectionist debate of the 1970s was over the relative magnitude of σ^2_e or s and $1/N$. In the genetic draft section, we argued that draft may well dominate drift ($N\rho E\{y^2\} >> 1$), but we do not know if either of these dominates selection for common alleles. Moreover, there is no clear experimental program that will help decide this issue for the majority of alleles found segregating in natural populations. The main argument against genetic drift as the dominant stochastic force for common alleles is the insensitivity of most measures of molecular variation to population size. This argument carries with it the assumption that most of the observed molecular variation is effectively neutral. If this assumption is questioned, then it becomes much more difficult to disentangle the stochastic forces.

There is reason to be hopeful that progress will be made in the next few years in our understanding of stochastic processes in evolution. I foresee at least three intertwining threads. The first of these is observations on a genomic scale. Whole-genome samples will tell us about the importance of hitchhiking in reducing genetic variation and about the haplotype structure of populations, which is greatly affected by stochastic processes acting on common alleles. The second thread involves functional studies on strongly selected molecular polymorphisms and substitutions that can be measured and understood in an ecological context. Although the selection coefficients of most alleles will be too small to be measured, strongly selected loci, although just the tip of the iceberg, should give us some sense of the hulk lying below. Finally, computers have become fast enough that they can be used to study realistic stochastic processes on a scale unthinkable only a few years ago. Here I must make an observation that population geneticists have been reluctant to leave the comfort of a very few models that have been around for almost 80 years. Much of our intuition is based on models with just a few parameters ($1/N$, u, r, m, and s) of similar magnitude at a locus with just two alleles. This situation has been made worse in recent years with the ascension of coalescent theory, which takes an even more restrictive view of evolution. While this was necessary 20 or so years ago, fast computers make all models accessible to investigation. I hope that students reading this chapter will be motivated to let their biological intuition guide them toward realistic models of genetic evolution and then to learn to program computers to study those models. Only when a rich variety of stochastic models are on our plates can we be sure that we are tasting the relevant ones.

SUGGESTIONS FOR FURTHER READING

Two classic papers on stochastic processes in evolution are Fisher (1922) and Wright (1955). Both are worth reading! Maynard Smith and Haigh (1974), Kaplan et al. (1989) and Gillespie (2001) give good accounts of hitchhiking and its interaction with population size. For selection in a random environment see Felsenstein (1976) and Gillespie (1991). For a modern treatment of the Wright–Fisher model and all its stochastic ramifications, there is no better source than Ewens (2004).

Ewens WJ 2004 Mathematical Population Genetics. I. Theoretical Introduction. Springer.

Felsenstein J 1976 The theoretical population genetics of variable selection and migration. Annu. Rev. Genet. 10:253–280.

Fisher RA 1922 On the dominance ratio. Proc. R. Soc. Edinb. 42:321–341.

Gillespie JH 1991 The Causes of Molecular Evolution. Oxford Univ. Press.

Gillespie JH 2001 Is the population size of a species relevant to its evolution? Evolution 55:2161–2169.

Kaplan NL, Hudson RR & CH Langley 1989 The hitchhiking effect revisited. Genetics 123:887–899.

Maynard Smith J & J Haigh 1974 The hitchhiking effect of a favorable gene. Genet. Res. 23:23–35.

Wright S 1955 Classification of the factors of evolution. Cold Spring Harbor Symp. Quant. Biol. 20:16–24D.

6

Genetics and Evolution in Structured Populations

CHARLES J. GOODNIGHT

Structured populations, or metapopulations, are sets of small populations tied together by migration, extinction, and recolonization. Because of the importance of population structure, the genetics of metapopulations cannot be modeled as simple extensions of standard single population systems. Genetic evolution, typically defined as changes in allele frequency (Dobzhansky 1937; see Futuyma 1998 for a more complete discussion), is normally considered to occur as a result of exactly four forces: selection, mutation, migration, and genetic drift. Within a single population only the first two, selection and mutation, are the primary forces of interest. Migration necessarily involves more than a single population, and becomes the cohesive force that distinguishes a metapopulation from a set of independently evolving populations. Genetic drift also takes on special meaning in a metapopulation. Whereas in a single population the consequences of small population size are limited to the effects on variation within demes, in metapopulations the effects of small deme size on the variation among demes also becomes an important area of study. Although generally considered as a force within single populations, selection also needs to be considered specifically within the context of metapopulations since selection, especially when it is coupled with genetic drift, may have very different consequences in a metapopulation compared with the well-understood effects it has in unstructured populations.

Many of the models that will be discussed in this chapter are based on a "linear additive" model. These models, which can be directly traced to Fisher's (1930) pioneering work on this subject, assume that each locus acts independently of other loci, and although dominance is allowed in these models, generally the effects of drift on dominance tend to be ignored. Furthermore these models frequently assume random mating and random interactions within populations. They do not model the "real world" in the sense that it is obviously true that enzymes work in pathways, and inevitably interact, indicating that the genes determining these enzymes must also interact. Similarly, it is rare that there is true random mating and interactions within populations. For these reasons it is tempting to dismiss linear additive models as overly simplistic. The reality is, however, that population and quantitative genetic models that ignore genetic complications work exceedingly well within a population. Indeed, they work so well that there is ample reason to consider such models to be fully sufficient for describing evolution, and especially evolution by natural selection (Coyne et al. 1997; Barton & Turelli 2004). However, this apparent adequacy is deceptive, as they are really only adequate under the assumptions; that Fisher originally used in developing the models. Thus, the linear additive model is a reasonable approximation of genetic evolution only when applied to a single large population with approximately random mating. The further the conditions deviate from Fisher's ideal the less adequate these models become. Metapopulations represent an extreme deviation from Fisher's original assumptions; thus it becomes important to discuss how these basic models need to be modified to begin to adequately understand evolution in metapopulations.

The population genetics of metapopulations is a broad topic; however, I will restrict myself primarily to two topics: genetic drift and changes in allele frequency due to small population size, and selection in metapopulations. In the final section I will consider both the consequences of individual selection acting within individual subpopulations, or demes, and the effects of higher level selection acting simultaneously within and among demes.

CONCEPTS

Assortative Mating and Wright's Inbreeding Coefficient

Consider a population with a single locus of interest with two alleles, A_1 and A_2. The frequency of the A_1 allele is p, and the frequency of the A_2 allele is $q = (1-p)$. If the population is infinitely large, and there is no selection, mutation, or assortative mating, the genotype frequencies will follow Hardy–Weinberg–Castle proportions. A convenient metric for exploring such a system is the "gene number," or number of A_1 alleles in the genotype (Table 6.1). It is the among-genotype variance in gene number that is of interest when examining the effects of genetic drift:

$$\sigma^2_{HW} = 2pq. \tag{6.1}$$

It is worth noting that the importance of the Hardy–Weinberg–Castle equilibrium is that it represents the situation in which no evolution is occurring. Indeed, the four forces of evolutionary change—migration, mutation, selection, and genetic drift—are precisely the converse of the conditions that must be met (along with the assumption of random mating) for the Hardy–Weinberg–Castle equilibrium to hold.

Next, consider a situation in which a proportion f of the population assortatively mates, whereas the remaining proportion, $(1 - f)$, mates randomly (Table 6.1, line 3). In this case the variance will become

$$\sigma^2 = (1 + f)\sigma^2_{HW}. \tag{6.2}$$

Finally, consider a population that has complete assortative mating (Table 6.1, line 4). That is, individuals only mate with individuals that have the same genotype. In this case the heterozygous class will eventually disappear, and the variance will be twice that in the random mating population:

$$\sigma^2 = 4pq = 2\sigma^2_{HW}. \tag{6.3}$$

Thus, partial assortative mating causes an increase in the population variance that is proportional to the degree of assortative mating.

Wright's inbreeding coefficient, f (Wright 1931), used in the equation above is defined as the correlation among gametes that combine to form an individual. In this example f is the correlation among gametes that occurs because individuals choose mates that match their genotype. This is an example of assortative mating based on identity by state. That is, the alleles that combine to form an individual that generates a correlation between gametes are chemically the same. For example, two A_1 alleles are identical by state, whereas an A_1 and an A_2 allele are not identical by state.

Inbreeding Due to Small Population Size

In Wright's original conception of the inbreeding coefficient he was primarily focused on the increase in the correlation among gametes that occurs as a result of small population size. In this situation it is identity by descent that generates the correlation between the gametes that combine to form an individual. Two alleles are identical by descent (IBD) if they are derived from the same allele at some point in the past. Note that identity by descent implies identity by state; however, the converse is not true.

TABLE 6.1. Genotype frequencies for a system with random mating, complete assortative mating, and partial assortative mating

Genotype	A_1A_1	A_1A_2	A_2A_2
"Gene number"	2	1	0
Random mating	p^2	$2pq$	q^2
Partial assortative mating	$fp+(1-f)p^2$	$(1-f)2pq$	$fq+(1-f)q^2$
Assortative mating	p	0	q

Inbreeding is similar to assortative mating in that the frequency of homozygotes increases and the frequency of heterozygotes decreases. However, they are different in that assortative mating affects only those loci influencing the trait under discrimination and other closely linked loci, whereas inbreeding affects all loci.

Inbreeding is also different in that with small population size the variance within demes declines; however, in a metapopulation with multiple demes the variance among demes will increase. As a result, the total variance in a metapopulation among individuals (ignoring population structure) increases to the $(1 + f)\ \sigma^2_{HW}$ found for assortative mating.

To see this it is easiest to use coancestries, or intraclass correlations (Cockerham 1954; Goodnight 1987; Tachida & Cockerham 1989). Intraclass correlations are average correlations between elements drawn from the same group. For single-locus genetical systems this would be the average correlation between the alleles from the same group, where the "group" may be an individual or a deme. For example, the intraclass correlation of an individual with itself would be the average correlation between the alleles drawn from the same individual:

$$\Theta_{T,A} = \frac{1+f}{2}. \tag{6.4}$$

The subscript "T" can be thought of as referring to "total," as this is the total correlation of an individual with the metapopulation, and the subscript "A" identifies that this coancestry measures the correlation due to additive effects.

Using a similar logic the coancestry between alleles drawn from the same deme is found by sampling one allele from the deme and comparing it with other alleles in the same deme. Typically in this comparison the sample pool of available alleles is considered to be infinite so that the probability of sampling the same allele twice (correlation = 1) is zero. Under the assumption of random mating the probability that two alleles drawn at random from a population are identical by descent is exactly equal to f. Thus:

$$\Theta_{D,A} = f. \tag{6.5}$$

The subscript D refers to the coancestry of two individuals sampled from the same deme. Similarly, the subscript B refers to the coancestry of two individuals sampled from different demers.

The variance within demes can be found as 2 times the difference in coancestries times the genetic variance in the original population. The factor of 2 comes in because there are two effects. (Since we are modeling a diploid there are two alleles, each affecting the phenotype.) In other words the variance among individuals within demes due to additive effects is equal to the variance among individuals minus the variance among demes:

$$\begin{aligned}\sigma^2_{\text{within(A)}} &= 2\left(\Theta_{T,A} - \Theta_{D,A}\right)\sigma^2_{HW} \\ &= 2\left(\frac{1+f}{2} - f\right)\sigma^2_{HW} \\ &= (1-f)\sigma^2_{HW}.\end{aligned} \tag{6.6}$$

To find the variance among demes recognize that in a metapopulation with an infinite number of demes and no migration the coancestry between two individuals from different demes will be zero. Thus:

$$\Theta_{B,A} = 0$$

and

$$\begin{aligned}\sigma^2_{\text{Between(A)}} &= 2\left(\Theta_{D,A} - \Theta_{B,A}\right)\sigma^2_{HW} \\ &= 2f\sigma^2_{HW.}\end{aligned} \tag{6.7}$$

Adding the variance within and among populations recovers the result for assortative mating, $\sigma^2_{\text{total(A)}} = (1+f)\sigma^2_{HW}$.

This is a somewhat novel way to derive a well-known result (e.g., Hedrick 2005). There are several advantages of using the coancestry approach presented here. First, this approach makes explicit the relationship of population structure to analysis of variance (ANOVA). That is, a "metapopulation" can be viewed as a nested hierarchy in which variances can be partitioned using ANOVA. Second, this approach provides a general method for calculating variance components that can potentially be expanded to cover any form of genetical effect or interaction (Tachida & Cockerham 1989).

A Two-Locus Example: Additive by Additive Epistasis

As an example, consider an interaction between alleles at different loci, additive-by-additive (A × A) epistasis. Adding this second locus considerably complicates the problem, requiring several additional descent measures; however, the methods outlined for a single locus can be applied here as well. Consider an interaction involving two loci, the *A* locus and the *B* locus. The two alleles at a locus are identical by descent with probability f. Because of the possibility of linkage the probability that two two-locus allele pairs (each pair consisting of one *A* allele and one *B* allele) are simultaneously identical by descent requires a different measure, f_4, which in most circumstances is approximately equal to f^2 (Whitlock et al. 1993). The subscript "4" identifies that this is a measure involving four alleles (two at each locus), and that all four alleles are in the same deme. Finally, f_4 refers specifically to four alleles (two two-locus allele pairs) derived from two parental gametes. For other coancestries descent measures for three gametes (γ_4), and four gametes (δ_4) are needed. To extend these models to metapopulations with migration between demes, interdeme measures comparing two two-locus allele pairs in different demes are required. These will have subscripts identifying the distribution of alleles among demes. For example $\theta_{2,2}$ identifies the probability of identity by descent for a pair of two-locus allele pairs on two parental gametes in which each allele pair is split between two demes (Goodnight 2000b).

The two-locus coancestry of an individual with itself is the average correlation between two-locus allele pairs drawn with replacement from the same individual. There are three possible outcomes for these draws: (1) both alleles the same, (2) one allele the same, the second being the same with probability f, and (3) neither being the same, but both related with probability f_4. Thus:

$$\Theta_{\mathrm{T,AA}} = \frac{1 + 2f + f_4}{4} \approx \left(\frac{1+f}{2}\right)^2. \qquad (6.8)$$

Using similar reasoning the two-locus coancestry for two individuals in the same deme becomes

$$\Theta_{\mathrm{D},AA} = \frac{f_4 + 2\gamma_4 + \delta_4}{4} \approx f^2. \qquad (6.9)$$

As with the single-locus measure, in a metapopulation with a large number of demes $\Theta_{\mathrm{B},AA} = 0$.

Using these measures the variance within demes due to two-locus additive-by-additive epistasis is

$$\begin{aligned}\sigma^2_{\mathrm{within}(AA)} &= 4\left(\Theta_{\mathrm{T},AA} - \Theta_{\mathrm{D},AA}\right)\sigma^2_{AA}\\ &= 4\left(\frac{1+2f+f_4}{4} - \frac{f_4 + 2\gamma_4 + \delta_4}{4}\right)\sigma^2_{AA}\\ &\approx (1-f)^2\sigma^2_{AA}\end{aligned}$$

and the variance among demes is

$$\begin{aligned}\sigma^2_{\mathrm{between}(AA)} &= 4\left(\Theta_{\mathrm{D},AA} - \Theta_{\mathrm{B},AA}\right)\sigma^2_{AA}\\ &= 4\left(\frac{f_4 + 2\gamma_4 + \delta_4}{4} - 0\right)\sigma^2_{AA}\\ &\approx 4f^2\sigma^2_{AA}\end{aligned}$$

where σ^2_{AA} is the additive by additive epistatic variance when the population is in two-locus Hardy–Weinberg–Castle equilibrium. This approach can, in principle, be used for nearly any genetic interaction, such as interactions involving three or more loci (Tachida & Cockerham 1989).

Recursion Equations for Inbreeding Coefficients

To use descent measures it is necessary to develop recursion equations that describe how they change as a function of population size. It is instructive to examine the recursion equation for f, the single-locus probability of identity by descent for the alleles that combine to form an individual, as a simple model of the logic behind deriving recursions for descent measures in general.

In a finite population with random mating an individual will be the result of selfing with probability $1/N$ and the result of mating between two parents with probability $(1 - 1/N)$. For a selfed individual, with probability 1/2 both alleles will be derived from the same parental allele and will, by definition, be identical by descent and with probability 1/2 they will be derived from different alleles, and will be identical by descent; and with probability f. If the individual is the result of outcrossing, the alleles that combine to form it will be identical by descent with probability f. Using this logic, in the next generation f will be:

$$f_{t+1} = \frac{1}{N}\left(\frac{1+f_t}{2}\right) + \left(1 - \frac{1}{N}\right)f_t.$$

While somewhat different from the way this equation is traditionally given, this formulation better expresses the biological justification for the equation. The "traditional" form of this equation (Hedrick 2005), $(f_{t+1} = (1/2N) + (1-(1/2N))f_t)$, works well for a single locus with random mating where the genes can be considered separately. However, this logic fails with any degree of complexity such as partial selfing, which calls for a separate descent measure for selfed versus outcrossed progeny. This formulation is also necessary when deriving two-locus coancestries in which the probability of identity by descent for the two-locus allele pairs will be affected by the recombination rate, r (Goodnight 1987).

Selfing, F_{IT}, F_{IS}, and F_{ST}

Sewall Wright (Wright 1969; Crow & Kimura 1970) recognized that there were two ways an individual could become inbred. One way that this could occur is if a population went through a period of small population size. With small population size the increase in the probability of randomly chosen alleles in different individuals in the same deme being indentical by descent (IBD) is expected to increase based on the formulae given above. Wright referred to the probability of randomly chosen pairs of alleles being IBD as F_{ST}, where S refers to subpopulation and T refers to the total metapopulation. Alternatively, an individual can become inbred if it is the result of the mating of relatives, that is, the result of "pedigree inbreeding." For example, in an outbred population the product of a full-sib mating has a 0.5 chance of having alleles that are identical by descent. This pedigree inbreeding is inbreeding due to mating structure within the population. Wright referred to the probability of the alleles combining to form an individual being IBD relative to the subpopulation as F_{IS}, where I refers to the individual and S refers to the subpopulation.

The difficulty arises when an individual is inbred both due to being a member of a small population (F_{ST}) and as a result of pedigree inbreeding (F_{IS}). To resolve this it is easiest to work with the probability that alleles are *not* IBD. The probability that the two alleles combining to form an individual are IBD relative to the total metapopulation is F_{IT}, where I refers to the individual and T refers to the total metapopulation. Thus, the probability that two alleles combining to form an individual are not IBD due either to small population size or to pedigree inbreeding is $(1 - F_{IT})$. The probability that a pair of alleles in an individual are not IBD is equal to the probability they are not IBD due to small population size times the probability that they are not IBD due to pedigree inbreeding. Thus:

$$(1 - F_{IT}) = (1 - F_{IS})(1 - F_{ST}).$$

Algebraic manipulation can then be used to recapture more usable forms of this equation, such as

$$F_{IT} = F_{ST} + F_{IS}(1 - F_{ST}).$$

Another example of how these hierarchical levels of inbreeding coefficients come into play is to return to the recursion equation for the inbreeding coefficient. With nonrandom mating the probability of IBD depends on the type of crossing that occurs. That is, the probability that the two alleles that combine to form an individual are IBD is different from the probability that two alleles chosen at random are IBD. As an example consider a population with selfing at the level of S. Then the probability that the two alleles that combine to form an individual are IBD is F_{IT}, and the recursion equation becomes

$$F_{IT(t+1)} = S\left(\frac{1+F_{IT(t)}}{2}\right) + (1-S)\left(\frac{1}{N}\left(\frac{1+F_{IT(t)}}{2}\right) + \left(1-\frac{1}{N}\right)F_{ST(t)}\right).$$

This expression emphasizes that with any deviation from random mating it is necessary to distinguish the product of selfed offspring from outcrossed offspring.

Genetic Variance Components within Demes: A Markov Chain Approach

There are two aspects of partitioning variance components in a subdivided population to consider. First, the distribution of variance components within finite demes is important. In a system with no gene interaction the variance within demes is entirely due to "additive variance." However, if there are gene interactions such as dominance or epistasis the within-deme variance must be divided into components due to these different effects. This is important because individual selection within demes can act only on the additive genetic variance; thus at least this component must be clearly separated from the other sources of variance. Second, the partitioning of variance within and among demes must also be considered. The additive case was examined above; however, each form of genetic effect has its own characteristic contribution to the variance among demes. In addition, there are two aspects of the variance among demes that must be considered: the variance in deme means, and the among-demes variance in genetic effects. This second effect, the variance in genetic effects, is not generally recognized, since it does not occur in the additive systems that are typically modeled.

Consider a system with two loci, each with two alleles. In this case there are nine possible genotypes. These nine genotypes can be viewed as elements in a weighted analysis of variance, where the weightings are the genotype frequencies. Since there are nine "treatments" there will be eight degrees of freedom, and thus eight independent genetic effects (Goodnight 2000a,b).

An important point about these orthogonal genetic effects is that any two-locus two-allele genetic effect at any allele frequency can be divided into these eight genetical effects. This partitioning can be done by regression (Goodnight 2000a,b).

To examine the effects of drift it is necessary to determine the distribution of allele frequencies resulting due to small population size. One way to do this is to use Markov chain models. Single-locus Markov chains are the easiest to model, and are what will be developed here. To date, two-locus Markov chains have been modeled by combining two single-locus matrices. The reason for this is that a transition matrix that accounts for all possible degrees of linkage disequilibrium quickly becomes unreasonably large. Combining two single-locus Markov chain models is a reasonable approximation, provided that linkage disequilibrium can be assumed to be small.

Where comparable, coancestry models and Markov chain models provide identical results (Goodnight 2000a). The models differ in several respects, however. Coancestry models are analytical models that provide the expected outcome of genetic drift, in terms of the expected change in variance components within and among demes. These models are very powerful in that they are independent of the details of the distribution of genetic effects, or even how many loci are contributing to the effects. However, this power comes at the cost of providing no information on the distribution of genetic effects among demes. In contrast, Markov chain models are stochastic models that are quite sensitive to the assumptions about the distribution of genetic effects, number of alleles per locus, and number of loci. However, they provide information on the among-deme distribution of allele frequencies, and as a consequence the distribution of genetic effects among populations.

Markov chain models assume that the deme population size is a constant, N. Assuming there are two alleles, A_1 and A_2, at the A locus (the locus of interest) there will be $2N$ alleles at a locus. Any deme in the metapopulation can be placed in one of $2N + 1$ categories based on the number of A_1 alleles. Thus, the distribution of demes with different allele frequencies can be described by a vector:

$$\mathbf{X} = \begin{bmatrix} P(A_1 = 0, A_2 = 2N) \\ P(A_1 = 1, A_2 = 2N - 1) \\ \bullet \\ \bullet \\ \bullet \\ P(A_1 = 2N - 1, A_2 = 1) \\ P(A_1 = 2N, A_2 = 0) \end{bmatrix}.$$

The $\mathbf{X}$ vector will change with time because of the finite population size. Assuming that mating is random and there is no selection this change can described by a transition matrix, $\mathbf{T}$. The elements of the matrix are the binomial probabilities that a population of type i will give rise to a population of type j, where i and j are the number of A_1 alleles in the population current and subsequent generation respectively. Thus the elements of the transition matrix become

$$\mathbf{T}_{ij} = \binom{2N}{j}\left(\frac{i}{2N}\right)^{j}\left(1 - \frac{i}{2N}\right)^{2N-j}.$$

Using this transition matrix the distribution of population types can be found using linear algebra:

$$\mathbf{X}_{t+1} = \mathbf{X}_t\mathbf{T}$$

This model will give the distribution of allele frequencies for any time during the course of genetic drift in a metapopulation. A number of interesting observations can be made concerning the distribution of allele frequencies using the Markov chain model. Perhaps most important is that the two fixation classes ($A_1 = 0$, $A_1 = 2N$) are absorbing boundaries. As a result genetic drift will ultimately lead to the fixation of one of the two alleles, with the frequency of the two classes of population (fixed for A_1 or fixed for A_2) being equal to the allele frequencies in the original population.

More interesting, however, is the potential for using this model to examine the variance within and among populations. For this it is most interesting to move to a two-locus Markov chain. Consistent with current methods the transition matrix is made by combining transition matrices for two independent binomial sampling events:

$$\mathrm{T}'_{i'j'} = \mathrm{T}_{ij}\mathrm{T}_{kl}$$

where T_{ij}, etc., represent the ijth element of the transition matrix. T is the transition matrix for a single-locus Markov chain, and T′ is the transition matrix for a two-locus Markov chain. The elements of the two-locus transition matrix are related to the single-locus transition matrix in that $i' = ((i - 1)(2N + 1) + k)$ and $j' = ((j - 1)(2N + 1) + l)$.

This model will provide the distribution of the two-locus joint allele frequencies that result from genetic drift. If we have two-locus genotypic values we can use these allele frequency data together with the genotypic values and the regression approach described above to calculate the variance components expected as a result of genetic drift. Figure 6.1 shows the additive genetic variance due to each of the genetical as a function of inbreeding coefficient (see Goodnight 2000a). Notice that in the early stages of inbreeding much of the gene interaction effects are expressed as additive variance. In the later stages of inbreeding the additive genetic variance is lost due to an overall loss in genetic variation. In general there is a conversion of the variance due to more complex forms of gene interaction (digenic epistasis) toward variance due to simpler forms of gene interaction (additive and dominance variance) as inbreeding proceeds.

Genetic Variance Components among Demes in a Metapopulation

The result that there is a conversion of genetic interactions into additive genetic variance raises an important point. In a randomly mating population the additive genetic variance is equal to variance in breeding values, or twice the variance in average effects. This suggests that there are only two ways to increase the additive genetic variance. One is to increase the effective number of alleles (defined as $1/f$; Crow & Kimura 1970). The second is to increase the range of values of the average effects of alleles. Drift necessarily decreases the effective number of alleles; thus an increase in the additive genetic variance must be due to a spreading of the average effects of alleles. This begs the question, however, whether this spreading of average effects of alleles is a simple change of scale, or whether it involves a reordering of the average effects.

To answer this question define the local average effect of an allele to be the mean deviation from the metapopulation mean of an individual containing

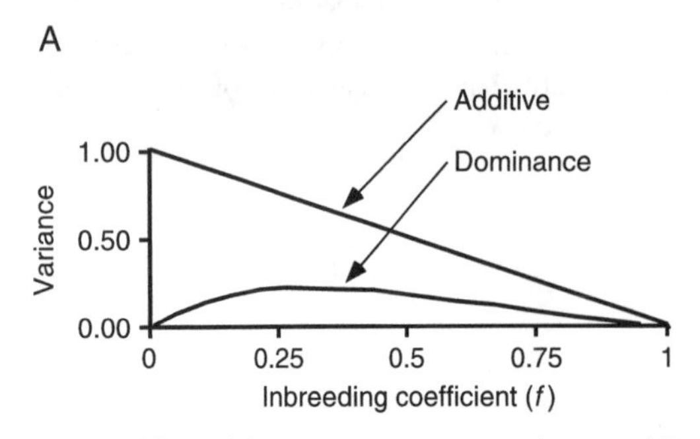

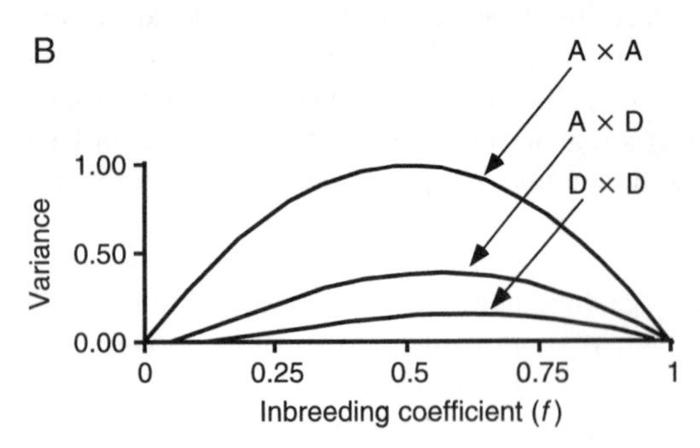

FIGURE 6.1. Additive genetic variance due to additive effects, dominance (A), additive by additive epistasis, additive by dominance (and dominance by additive) epistasis, and dominance by dominance epistasis (B) as a function of inbreeding coefficient. The additive genetic variance is a statistical property of a population. As inbreeding proceeds the genetic interactions are expressed as additive variance.

BOX 6.1. Epistasis and the Conversion of Genetic Variance
Jason B. Wolf

Epistasis, in the evolutionary genetic sense, occurs when the phenotypic effects of genotypes at one locus are dependent upon the genotypes present at other loci (see Phillips 1998; Wolf et al. 2000; Cheverud, Ch. 19 of this volume). This relationship is, in principle, symmetrical and it is often simplest to think of epistasis as an interaction between the genotypes or alleles at these different loci. Essentially, all of the impacts that epistasis has on evolutionary processes result from the fact that epistasis makes the effects of alleles at a locus dependent upon the genetic background provided by the rest of the genome. Here I provide a brief introduction to how this context-dependent nature of genetic effects can lead to the evolution of genetic variances, where the contributions of loci to genetic variance components shift due to changes in allele frequencies.

Physiological epistasis resulting from interactions between loci can contribute to the additive and epistatic genetic variance components (denoted V_A and V_I respectively), and in some cases may also contribute to the dominance component. Their relative contributions to the different components of variance shift with changes in allele frequencies in a process that has been called "conversion." The term conversion is used to reflect the way that the contributions of loci are, in a sense, "converted" from one component to another as allele frequencies change, without producing major changes in the total genetic variance (see Lopez-Fanjul et al. 2000; Meffert, Ch. 27 of this volume). Nearly all studies of the conversion processes have focused on the evolution of V_A because it is V_A that determines the evolutionary response to selection (i.e., the change in allele frequencies and the mean phenotype). However, the conversion process can involve changes in the contribution of loci to all three genetic variance components. Most studies have also focused on the effects of drift events at population bottlenecks on the contribution of loci to variance components (Cheverud & Routman 1996; Goodnight 1988), although presumably all processes that affect allele frequencies could have similar effects.

For simplicity I will illustrate the concept of conversion using the special case of additive-by-additive epistasis, but the general points apply equally well to other forms of epistasis (see Goodnight, Chapter 6 of this volume, for more on other forms of epistasis). Additive-by-additive epistasis occurs when the additive effect of alleles at one locus depend on the genotype present at another locus and vice versa.Table 1 shows the genotypic values associated with the nine two-locus diploid genotypic combinations for additive-by-additive epistasis. In additive-by-additive epistasis, the effects of alleles at each of the two loci reverse across the two homozygous backgrounds of the other locus, and both alleles at both loci have no effect when on a heterozygous background. For example, the A_1 allele has a positive additive effect when on the B_1B_1 background, since it "adds" to the genotypic value (i.e., the genotypic value goes from

TABLE 1. An example of additive-by-additive epistasis values in the table are the genotypic values for the nine genotypes in a two-locus system with two alleles per locus.

	A_1A_1	A_1A_2	A_2A_2
B_1B_1	1	0	−1
B_1B_2	0	0	0
B_2B_2	−1	0	1

(continued)

BOX 6.1. *(cont.)*

−1 to 0 when one A_1 allele is added to a genotype and from 0 to +1 when a second A_1 allele is added) and a negative effect when on the B_2B_2 background. In this case, changes in allele frequencies at the *B* locus would be expected to lead to a change in the overall average additive effect of the A_1 allele (i.e., the overall effect of the A_1 allele averaged over the *B* locus backgrounds) since it would shift the average genetic background provided by the *B* locus. Such a change in the genetic background provided by the *B* locus would lead to an evolutionary change in V_A contributed by the *A* locus since it would lead to a change in the overall additive effects of the *A* locus alleles.

The values of V_A, V_I and V_G (the last denoting the total genetic variance) as a function of allele frequencies at two loci are illustrated in Figure 1. In Figure 1A you can see that, at intermediate allele frequencies (i.e., when none of the alleles is near fixation, with frequencies ranging from about 0.2 to 0.8), V_G is relatively constant (ranging from 0.25 to 0.34 in this example). However, within this same range, we can see that V_A and V_I are much more variable. For example, when allele frequencies are at 0.5 at the two loci, all of the genetic variance is epistatic (in this example $V_I = 0.25$ and $V_A = 0$), but the value of V_A goes up rapidly as you move away from a frequency of 0.5 at either locus (e.g., when both loci have one allele at a frequency of 0.8, $V_I = 0.10$, and $V_A = 0.23$). Since the total variance is the sum of the additive and epistatic variances, we can view changes in allele frequencies as converting variation from one component to the other without having a major effect on the total genetic variance. This shows why the concept of conversion is logical—genetic variance does not vary much within a range of intermediate allele frequencies, but rather, it is converted between components. Epistatic variance is greatest at intermediate allele frequencies because there is a lot of variation in genetic background provided by each locus. This makes the effects of alleles at each locus very variable and variation is, therefore, epistatic. However, when frequencies move toward extreme values at one locus, the variation in the genetic background provided by that locus goes away and the other locus starts to appear mostly additive and contributes primarily to V_A.

Consider a simple case where the B_2 allele is very rare while the two alleles at the *A* locus are at equal frequency (0.5). In this case, almost the entire population lies within the first row (B_1B_1) of Table 1. In this case, the *A* locus is almost completely additive since there is no variation in the genetic background provided by the *B* locus

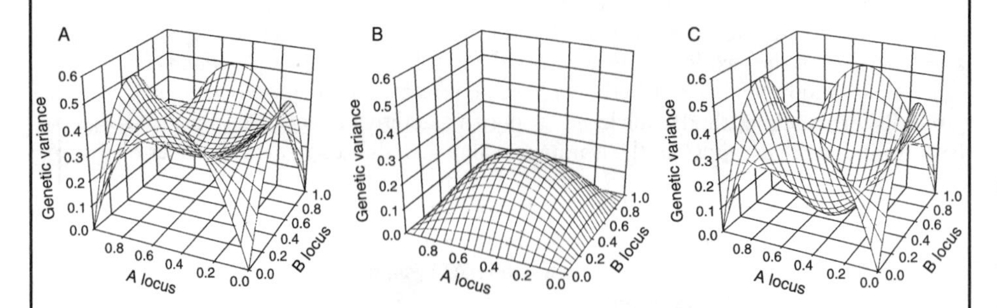

FIGURE 1 Components of variance for a two-locus two-allele system with additive-by-additive epistasis. (A) the total genetic variance as a function of allele frequencies at two loci (labeled locus *A* and *B*). (B) The epistatic component of variance. (C) The additive component of variance. Note that the scales for the variances are the same in all three figures, so that B and C can be added together to produce A.

BOX 6.1. *(cont.)*

(note that the effect of the A_1 allele is positive and adds 1 to the genotypic value). The B locus does not contribute to genetic variances since nearly all individuals have the same B locus genotype (B_1B_1). Now, if the B_2 allele were to reach an allele frequency of 0.5, then the average value associated with each of the three genotypes at the A locus would be zero and its contribution to the additive genetic variance would be zero. If we assume Hardy–Weinberg proportions of genotypes, we would expect the A_1A_1 genotype to have a value of +1 one quarter of the time (on the B_1B_1 background), 0 one half of the time (on the B_1B_2 background) and −1 one quarter of the time (on the B_2B_2 background), giving it an average genotypic value of 0. For the same reason, the A_1A_2 and A_2A_2 genotypes would have average genotypic values of 0. Thus, the A locus would go from being nearly entirely additive when the B_2 allele was very rare to having no overall additive effect (since the average genotypic values of all three genotypes are equal) when the B_2 allele reaches a frequency of 0.5. Although neither locus would contribute to the additive variance, there would still be a large amount of genetic variation contributed by the pair of loci, but it would all be epistatic due to the interaction of the two loci (see Figure 1).

To understand the effects of drift on genetic variances we can take the previous example and reverse the direction of allele frequency change. That is, we would start with a population with an allele frequency at the B locus of 0.5, where there is no additive variation, and move toward the B_2 allele becoming very rare, where the A locus contributes additive variation and almost no epistatic variation. This process captures the scenario examined by Cheverud and Routman (1996) in their model of the conversion process. They assumed that populations start with allele frequencies of 0.5 at a pair of epistatic loci and that drift events at population bottlenecks move the populations away from the frequency of 0.5. From Figure 1 we can easily see what should happen—because most populations end up with some allele frequency other than 0.5 (since drift randomly alters allele frequencies, very few will end up with allele frequencies of exactly 0.5 after a bottleneck) they necessarily have a V_A that is greater than the starting V_A, which was 0. Therefore, the average V_A after a population bottleneck is greater than before, and the bottleneck resulted in the conversion of V_I to V_A. Note, however, if you started at some other value, where there was already V_A and less epistatic variance, that population bottlenecks could, on average, reduce V_A (e.g., when an allele at one locus is rare while the two alleles at the other locus are at equal frequency), increase V_A (when one of the two alleles at each locus is very rare—i.e., in the corners of Figure 1A, B, and C) or have no effect on V_A. Therefore, the actual influence of bottlenecks on genetic variances will be strongly dependent upon initial allele frequencies (see Meffert, Ch. 27 of this volume).

Acknowledgments Thanks go to Lisa Meffert and Joshua Mutic for constructive comments on the earlier draft of this box.

the allele with the rest of the genome having come at random from the deme in question. Whereas an allele has only a single average effect, it has a different local average effect in every deme. It is useful to divide the local average effect into its components (Wade & Goodnight 1998). Thus, the local average effect of the ith allele in the jth deme, α_{ij}, can be divided into an allelic component, $\alpha_{i\bullet}$, a deme component, $\alpha_{\bullet j}$, and an allele by deme interaction α_{i*j}:

$$\alpha_{ij} = \alpha_{i\bullet} + \alpha_{\bullet j} + \alpha_{i*j}.$$

This partitioning is done using a least squares analysis of variance, thus the components will be orthogonal and the variances can be written as

$$\mathrm{Var}(\alpha_{ij}) = \mathrm{Var}(\alpha_{i\bullet}) + \mathrm{Var}(\alpha_{\bullet j}) + \mathrm{Var}(\alpha_{i*j}).$$

In the absence of effects such as inbreeding depression that directionally alter the local average effects, 2*Var($\alpha_{i\bullet}$) is the additive genetic variance in the ancestral outbred population. In the absence of gene interaction Var($\alpha_{\bullet j}$) equals the among-deme variance in deme means, although in the presence of gene interaction this will not be true. It is the Var(α_{i*j}) that is most interesting. In Figure 6.2 Var(α_{i*j}) is plotted as a function of inbreeding coefficient for each of the forms of gene interaction. For additive by dominance epistasis the variance in the allele by deme interaction is plotted separately for the additive and dominance locus. Importantly, in the absence of gene interaction (i.e., only additive effects) Var(α_{i*j}) is equal to zero. The nonzero allele by deme interaction results from the increase in additive genetic variance associated with drift in systems with gene interaction. What this means is that the spreading of average effects within demes is associated with the shuffling of their local average effects.

Var(α_{i*j}) is the differentiation of demes for local average effects. For example, if Var(α_{i*j}) is zero, as it is in the additive case, the relative difference between a pair of alleles will be a constant. On the other hand, if Var(α_{i*j}) is nonzero, as it is when there is gene interaction, then the relative difference in local average affect between a pair of alleles will not be constant, and alleles may even change relative rankings. This differentiation of average effects is very different from the differentiation of deme means that is normally considered "population differentiation." Figure 6.3 is a graph of the differentiation of population means due to the different forms of pure gene interaction. Notice that the additive effects generate variation in deme means but have no effect on the

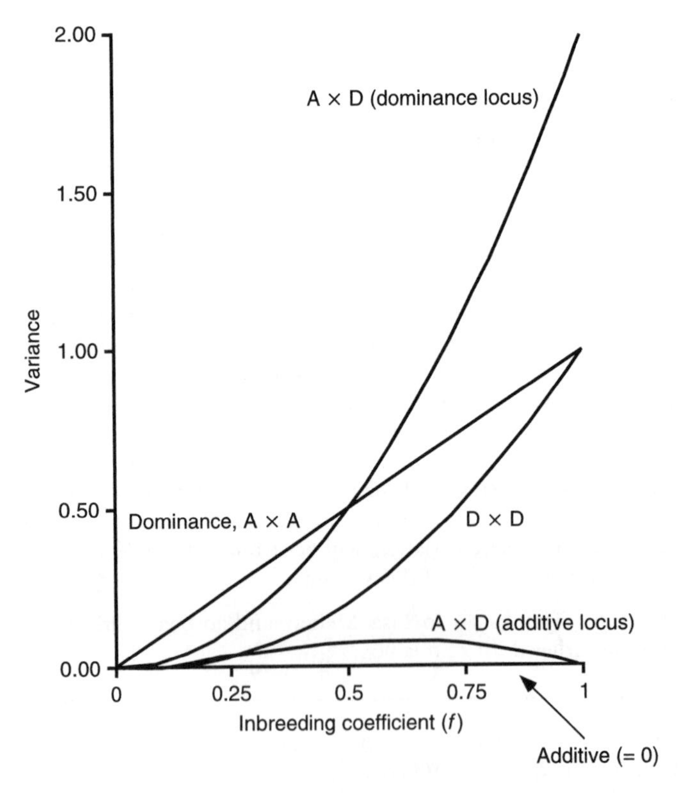

FIGURE 6.2. The variance in the allele by deme interaction, Var(α_{i*j}), as a function of inbreeding coefficient. In an additive system the variance in the allele by deme interaction is zero, indicating that the alleles maintain their relative ranking in all demes. When there is gene interaction Var(α_{i*j}), will be nonzero, indicating that alleles are changing their relative ranking.

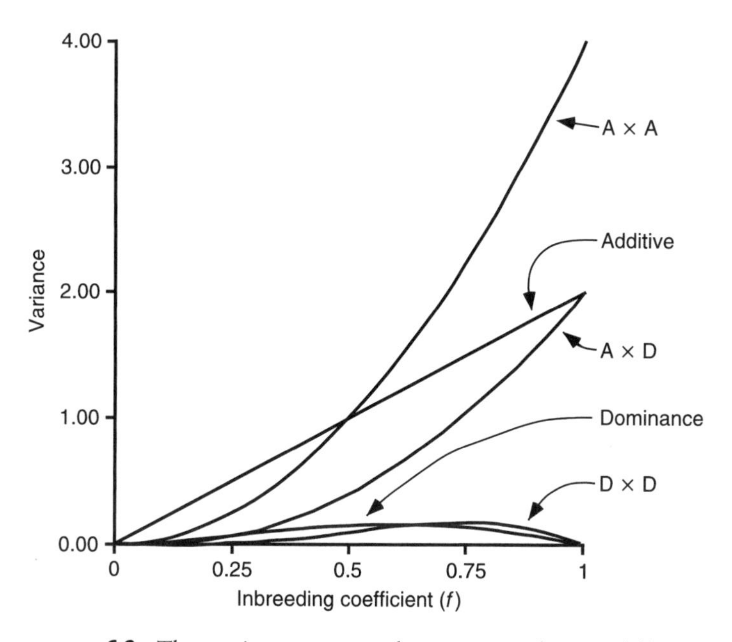

FIGURE 6.3. The variance among deme means due to different genetic effects as a function of inbreeding coefficient. Note that the variance among deme means is not related to the variance in the allele by deme interaction (Figure 6.2).

differentiation of average effects. Conversely, dominance and dominance-related epistatic interactions cause little differentiation for deme means, but strongly influence differentiation for average effects.

Differentiation for average effects is the form of population differentiation that is most closely allied with processes such as speciation. This provides an explanation as to why it has been so difficult to determine how much differentiation of population means is necessary for speciation to occur. Because these two forms of population differentiation are uncorrelated, differentiation of population means has no influence on the process of speciation.

Epistasis and Speciation

The biological species definition Mayr (1963) primarily concerns the barriers to gene flow (Mallet, Species Concepts box, pp. 367–373 of this volume). One of the primary barriers to gene flow will be genetic incompatibilities. These "postmating isolating mechanisms" (Dobzhansky 1937; Mayr 1963) require some form of gene interaction (Muller 1939; Orr 1995) to generate the genetic incompatibilities. Many forms of dominance and epistasis can potentially generate the necessary incompatibilities for speciation to occur. For many forms, however, it may be difficult or impossible to maintain the necessary genetic variation in the original population. As a simple example, consider underdominance ($A_1A_1 = 1$, $A_1A_2 = -1$, $A_2A_2 = 1$). This can easily lead to speciation as different populations become fixed for the A_1A_1 genotype or the A_2A_2 genotype. The difficulty is, however, that it is unlikely that both alleles would be segregating in the original population prior to the isolation that is typically needed for speciation to occur. It is for this reason Muller (1939; Orr 1995) proposed that mutations arising after isolation may be responsible for speciation. Muller suggested that negative between-population gene interactions would arise as a by-product of new mutations and evolutionary processes occurring within the two isolated populations.

Epistasis provides a mechanism for maintaining variation within a population in a neutral or near-neutral state, while simultaneously providing a mechanism for rapid divergence between subdivided populations. The mechanism by which this could occur can be illustrated using dominance by dominance epistasis. Consider a pair of loci with an interaction that is the inverse of the dominance by dominance interaction, and a component of overdominance:

	A_1A_1	A_1A_2	A_2A_2
B_1B_1	1	−1	1
B_1B_2	−1	2	−1
B_2B_2	1	−1	1

At a allele frequency of 0.5 at both loci there will be no marginal effects on fitness for either locus, and therefore no selective changes in allele frequency. Furthermore, deviations from an allele frequency of 0.5 will be opposed by selection since there is mild overdominance at both loci. Thus, in a large population at equilibrium there is selection favoring the double heterozygote, and the population should remain stable at an allele frequency of approximately 0.5 for both loci. However, following population subdivision genetic drift or slight changes in the selective environment in the two populations (putting one of the loci under directional selection in one of the populations) could easily drive the populations to any of the four "corners" ($A_1A_1B_1B_1$, $A_1A_1B_2B_2$, $A_2A_2B_1B_1$, $A_2A_2B_2B_2$). Notice that if either of the alleles at either locus is lost or becomes rare, either by selection or by drift, the interaction becomes an underdominant one in which fixation at the other locus is assured. Thus, this interaction, which is stable in a large population, becomes unstable leading to population divergence when populations are small and subject to genetic drift, or the selective environment changes adding directional selection to either of the loci. Goodnight (2000b) gives a similar example, using additive by dominance epistasis.

Multilevel Selection

The variance among groups is particularly interesting if this contributes to differences in fitness. Group selection occurs when population differentiation contributes to fitness differences. Indeed Wade (1977) defines group selection as the differential survival and reproduction of groups. In a similar definition Goodnight et al. (1992) define group selection as occurring when the fitness of an individual is a function of group membership. Common wisdom has it that group selection is of little importance in evolution (e.g., Maynard Smith 1964, 1976; Harrison & Hastings 1996). However, experimental work clearly shows that group selection is far more powerful than traditional theory would predict (see Goodnight & Stevens 1997 for a review).

The reasoning that group selection is normally considered less effective than individual selection is based on models that consider only additive effects (Wade 1978; Goodnight & Stevens 1997). In the absence of gene interaction differentiation among populations is limited both because only one form of population differentiation occurs (differentiation of population means) and because the amount of population differentiation in an additive system is limited.

Goodnight and Stevens (1997) argued strongly that a major cause of the higher than expected response to group selection was due to genetically based interactions among individuals. Evidence for this assertion comes from several sources. Supportive of this thesis is the study of Goodnight (1990a,b), who examined the response to selection acting on two species communities of *Tribolium castaneum* and *Tribolium confusum* flour beetles. In this study he had eight community selection treatments: selection for high and low values for each of four traits, population size, and migration rate each species. He found that two species communities of these beetles responded to community selection in a manner that was qualitatively similar to previous studies of group selection (e.g., Wade 1977; Craig 1982; Goodnight 1985).

In itself this response to selection is interesting; however, more interestingly, in the second paper (Goodnight 1990b) reported on experiments in which the communities that had responded to community selection were disrupted in three different treatments. In the first treatment the intact community was assayed. In the second treatment the communities were broken up, and assayed as single-species populations. In the third treatment the communities were broken up into single-species populations; however, the ecological, but not the genetic, structure of the communities were restored by adding a naïve competitor of the opposite species. The results of these assays was that all the traits showed a response to selection in the intact communities but none showed any differences in the single-species populations. In the reconstructed communities (with one selected strain and a naïve competitor) migration rate in both species showed a response to selection, but the effect of selection on population size completely disappeared. These results show that the response to community selection for population size resulted from a genetic change in both species, and that the change involved an interspecies interaction that was genetically very similar to within-individual epistasis. This result is perhaps not surprising given *Tribolium* biology. Park (1948, 1954) showed that population size in *Tribolium* is primarily determined by cannibalism rates, and that cannibalism rates depend both on the tendency of a strain to be cannibalistic, and the timing of development (since only certain stages are vulnerable to cannibalism).

Because of these nonadditive interactions experimental studies of group selection have shown significant and often surprisingly large responses to group selection or surprising large degrees of population differentiation even when populations were not highly structured. For example, Wade (1982a) found significant levels of population differentiation even with migration rates as high as 25% (4 individuals per generation), and Wade and Goodnight (1991) showed that differential migration provided sufficiently strong group selection to lead to a significant response even though the migration itself led to a reduction in the inbreeding coefficient.

Importantly these genetically based interactions among individuals are not part of traditional theory on group selection. These experimental studies show that group selection is capable of acting on components of genetic variance that are not available to individual selection, and rarely incorporated into current thinking.

A second reason why group selection has normally been considered to be a minor force in evolution is that originally group selection was considered to occur only through the differential extinction of discrete groups (e.g., Maynard Smith 1964, 1976). The metapopulation model of semi-isolated discrete demes with random mating within demes is the most theoretically tractable and conceptually easiest way to conceive of group selection. However, real world populations often are not divided into easily identified discrete groups. Rather they may form a "viscous" population, which is a population in which neighbors mate with each other, and offspring settle near to each other and their parents. In these viscous populations the fitness of an individual may be dependent on the phenotype of its neighbors. In such a situation there is the potential for selection occurring at several levels of organization at the same time. Such a situation has been referred to as "multilevel selection" (Goodnight & Stevens 1997). Selection among overlapping neighborhoods is not mathematically different from what has been traditionally described as "group selection" and has the potential to be quite common (Goodnight et al. 1992).

Contextual Analysis and Multilevel Selection

One approach to measuring multilevel selection that has been suggested is contextual analysis (Heisler & Damuth 1987, Goodnight et al. 1992, Stevens et al. 1995). This is a multiple regression approach that simultaneously measures selection both on individual traits, and on "contextual" or group level traits. Although introduced as a statistical method for measuring multilevel selection in natural populations (see the case studies at the end of this chapter), it can also be used for theoretical studies as well.

Goodnight et al. (1992) examined contextual analysis by theoretically considering individuals that expressed a trait, Z. These individuals were assembled into groups, with each group having a group mean of the trait, $\bar{Z}$. They assigned fitnesses based on models of classic group selection, hard selection, and soft selection, and contextual analysis was used to interpret these systems. This provided a test as to whether contextual analysis correctly identified the levels of selection when the levels were known. In the group selection model individual fitnesses were considered to be solely a function of the group mean of the trait; the individual traits did not affect fitness except to the extent that they affected the group mean of the trait. Satisfyingly, contextual analysis identified this as pure group selection with no individual selection acting. In the hard selection model fitnesses were assigned based only on the individual trait, without regard to the group mean of the trait. In accordance with intuition, contextual analysis identified this as a system with only individual selection acting. In the soft selection model fitnesses were assigned based on the individual trait relative to the value of other individuals in the same deme. That is, each deme produced the same number of individuals after selection, and the individuals were chosen based on selection acting independently within each deme. Intuitively it is apparent that the fitness of an individual is affected by the group in which it is found, and indeed, contextual analysis revealed that group and individual selection were acting in opposition to each other in the soft selection case. These studies show that contextual analysis does correctly identify the level at which selection is acting in a theoretical setting (Goodnight et al.

1992), and can be used to analyze multilevel selection in natural populations, even in situations where there are no clearly differentiated groups.

Contextual analysis can also be used to examine the relationship between kin selection and group selection (Goodnight et al. 1992). The heart of kin selection is Hamilton's rule (Hamilton 1964a,b):

$$\frac{\text{Cost}}{\text{Benefit}} < \text{Relatedness}.$$

The contextual analysis equivalent of this is

$$\frac{\left(r_{wz} - r_{w\bar{z}}r_{z\bar{z}}\right)^2}{\left(r_{w\bar{z}} - r_{wz}r_{z\bar{z}}\right)^2} < r_{z\bar{z}}^2$$

where r_{wz} and $r_{w\bar{z}}$ are the correlation between relative fitness and the individual and group level trait respectively and $r_{z\bar{z}}^2$ is the fraction of variance among groups. The squared correlation between fitness and the individual trait with the effects of the group trait removed, $(r_{wz} - r_{w\bar{z}}r_{z\bar{z}})^2$, is a measure of the strength of individual selection, and $(r_{w\bar{z}} - r_{wz}r_{z\bar{z}})^2$ is a measure of the strength of group selection. Thus in Hamilton's equation "cost" can be interpreted as the strength of individual selection, "benefit" can be interpreted as the strength of group selection, and "relatedness" can be interpreted as the fraction of the variance that is among groups.

Interestingly this points out some ways in which Hamilton's rule should be reinterpreted. First, contextual analysis shows that Hamilton's rule can be divided into group and individual selection, yet contextual analysis applies only to within-generation selection. What is missing is the heritability of the group and individual trait. Simple models of kin selection consider only an additive trait and the group mean of that additive trait. However, in contextual analysis the individual trait and the group traits are considered to be different traits that may have different heritabilities. Given the demonstrated importance of interactions among individuals in the response to group selection in many cases it will be unlikely that the heritabilities of the group and individual traits will be identical. If Hamilton's rule is to be used to predict between-generation changes, questions of heritability will have to be addressed. Second, Hamilton considered that the only cause of variance among groups was relatedness, and that the relationship between the fraction of variance among groups and relatedness was linear. Neither of these assumptions needs be true. Many factors not having to do with relatedness can contribute to between-group differences. In human populations cultural differences are important and need not be influenced by relatedness (e.g., Boyd & Richerson 1985; Cavalli-Svorza & Feldman 1973). In noncultural organisms shared symbionts or parasites could contribute to group differences whether or not individuals are related. Finally, given the importance of interactions among individuals in population differentiation it would not be surprising to find that relatedness had a nonlinear effect on variance among groups. None of these factors invalidates Hamilton's rule; rather they call into question a simplistic interpretation of this equation.

Wright's Shifting Balance Theory

The distinctive roles of nonadditive genetic effects, genetically based interactions among individuals, and multilevel selection in metapopulations is brought together in Wright's shifting balance thoery (WSBT). Wright originally described his process in 1931 (Wright 1931, 1977), before most of the theoretical advances that can be applied to it were developed. In light of these advances, Wright's original formulation may seem naïve (Coyne et al. 1997, 2000). Nevertheless, when used as a general outline for describing evolution in structured populations WSBT remains compelling (Wade & Goodnight 1998; Goodnight & Wade 2000).

From his experimental work Wright derived a series of generalizations about populations (Wright 1977). The most important of these generalizations are (1) that traits are polygenic, (2) that there are large networks of epistatic interactions (principle of universal epistasis), and (3) that every gene affects many traits (principle of universal pleiotropy). From these generalizations Wright concluded that there were multiple ways an organism could achieve a high fitness, which he described as an "adaptive topography." Although Wright never formalized this model, he envisioned a topography in which a set of axes describe aspects of the genotype of an organism, and a single "vertical" axis describes fitness. Particularly favorable genotypic values conferring high fitness were considered to be "adaptive peaks" that were separated by regions of low fitness or "adaptive valleys." This model was never put in formal mathematical terms (likely an impossibility in 1930), thus Wright was not explicit about what the genotypic axis described. This remains a major source of controversy surrounding this model.

Wright was particularly interested in the development of novel adaptations, which he considered to be crossing from one adaptive peak to another. In his conception the evolution of a novel adaptation involved a peak shift, which could not occur as a result of simple directional selection. He reasoned that peak shifts could occur through a three-phase process. The phases he identified were (1) a phase of random drift, (2) a phase of mass (individual) selection, and (3) a phase of interdeme (group) selection. He reasoned that in order to cross from one peak to another a population would have to descend into a fitness valley before climbing the new peak. This decline in fitness can only occur by stochastic forces such as genetic drift. He further reasoned that it would be extremely unlikely that a single population would randomly locate a new adaptive peak; however, the probability of locating a new peak would be greatly enhanced for a set of demes tied together by a low level of migration, that is, in a metapopulation. Thus, he reasoned that phase 1 of random drift was a process in which a metapopulation explored an adaptive topography. When a deme came under the influence of a local adaptive peak Wright reasoned that it would enter phase 2, the phase of mass selection. During this phase simple individual selection would drive it to near the local adaptive peak. The result of this would be that different demes would end up climbing different local adaptive peaks, with some peaks higher (conferring higher fitness) than others. During phase 3, the phase of interdeme selection, Wright suggested that populations on the different adaptive peaks "compete" through differential migration, with those demes on the highest fitness peaks sending out more migrants and those on lower peaks tending to receive migrants. He reasoned that this differential migration would lead to the metapopulation coalescing on the highest adaptive peak.

In the context of traditional population genetics WSBT seems like an implausible mechanism for adaptive evolution (Coyne et al. 1997, 2000); however, in light of the unique features of the genetics of metapopulations described in this chapter it becomes more plausible. Most of our understanding of the genetics of metapopulations has arisen since Wright described his model, and as a consequence are not explicitly incorporated in the model. For this reason Wade and Goodnight (1998) revisited WSBT with an eye toward modifying this conceptual model to bring it in line with some of these recent developments in evolutionary biology. They argue that phase 1, the phase of random drift, takes on special significance when there are epistatic gene interactions. In particular, genetic drift can lead to random changes in local average effects, and potentially lead a population down a new evolutionary trajectory. Phase 2, the phase of mass selection, becomes not just a refinement of adaptations, but actually a diversifying force as selection acts on different gene complexes, even if it is acting in a uniform manner at the phenotypic level. Thus selection would act to magnify the small differences generated by genetic drift. Finally, phase 3, the phase of interdeme selection, takes on far greater significance in light of the demonstrated effectiveness of group selection, even when there are moderate levels of gene flow between demes.

The importance of Wright's shifting balance process in evolution remains an open question. On the basis of traditional models it is unlikely to be an important mechanism (Coyne et al. 1997, 2000). However, on the basis of models incorporating gene interactions and interactions among individuals it may be an important evolutionary mechanism (Wade & Goodnight 1998; Goodnight & Wade 2000). An evaluation of the relative importance of WSBT awaits more sophisticated models and experiments on the genetics of complex systems.

CASE STUDIES

Teosinte and Maize

When there is gene interaction the total variance within demes can be partitioned into variances due to individual genetic effects. Of particular interest is the additive genetic variance, which is the proportion of the total genetic variance that can contribute to a lasting response to individual selection. The partitioning of genetic variance components is a statistical property of the allele frequencies within a population; thus, as allele frequencies change as a result of genetic drift the additive genetic variance will change as well.

The most convenient way to perform this partitioning is to use multiple regression. Using an appropriate regression model any system in which genotypic values can be assigned and in which the genotype frequencies are known can be partitioned into appropriate genetic variance components.

As an example, consider the two-locus interaction found in a teosinte × maize (corn) cross described by (Doebley et al. 1995). Teosinte is the progenitor of maize, thus, this cross is of particular interest to both corn breeders and those interested in the origins of modern maize. In this study Doebley et al. (1995) identified two quantitative trait loci (QTL), BV302 and UMC107 (the names refer to molecular markers associated with the QTL). These two loci affected a number of traits, including PEDS, the percent of cupules lacking a spikelet. In maize kernels the cupule lacks a spikelet, whereas in teosinte kernels the cupule has a spikelet. Thus, this trait can be interpreted as the percent of maize-like kernels in the ear. The genotypic values for this cross are:

	U_TU_T	U_TU_M	U_MU_M
B_TB_T	0	0	0.8
B_TB_M	0.3	0.2	1.3
B_MB_M	1.1	0.4	7.3

Here the *U* locus is the UMC107 locus, and *B* is the BV302 locus. T refers to the allele derived from teosinte, and M refers to the allele derived from maize. The cross was done into a teosinte background, which accounts for the low percent of maize-like kernels even with both alleles being of maize origin.

To perform the regression the genotypic values listed above are used as dependent variables, and the eight genetical effects described by Goodnight (2000a,b) are used as independent variables (Table 6.2). The regression is performed using the genotype frequency as a weighting factor, and using type 1 sums of squares. Because type 1 sums of squares are used, the order in which the independent variables is added must be in the order in which they are listed in Table 6.2 (from right to left). The values for the independent effects have been adjusted so that the effect variance is 1 (see Goodnight 2000a,b for a description of how this is done).

Notice that there are nine "observations" (genotypes) and eight independent variables. This means that there are no degrees of freedom left over for error. This regression provides a means of partitioning variances; however, unlike traditional uses of regression it provides no statistical tests. When the regression is done using sequential or type 1 sums of squares the sum of squares for each independent variable is equal to the contribution of that variance component to the overall genetic variance. As genotype frequencies change, the contribution to the different variance components will change as well. The type 1 sums of squares track this change. Type one sums of squares are sensitive to unbalance. In standard statistical tests this would be considered a shortcoming; however, in partitioning genetic variances the unbalance is driven by changes in allele frequency, allowing the type 1 sums of squares to correctly track the conversion of nonadditive to additive genetic effects.

The regression of PEDS on the different genetic effects is shown in Table 6.3 for the allele frequencies of 0.5 for both loci (as listed in Table 6.2), and for an allele frequency of 0.25 for the teosinte alleles at both loci. Note the shift in variance components.

TABLE 6.2. The values used in the regression in the teosinte × maize example.

			Independent variables							
Genotype	Freq.	Dep.Var.	Add A	Add B	Dom A	Dom B	A×A	A×D	D×A	D×D
$U_TU_TB_TB_T$	0.0625	0	1.414	1.414	−1	−1	2	1.414	1.414	−1
$U_TU_TB_TB_M$	0.125	0.3	1.414	0	−1	1	0	−1.414	0	1
$U_TU_TB_MB_M$	0.0625	1.1	1.414	−1.414	−1	−1	−2	1.414	−1.414	−1
$U_TU_MB_TB_T$	0.125	0	0	1.414	1	−1	0	0	−1.414	1
$U_TU_MB_TB_M$	0.25	0.2	0	0	1	1	0	0	0	−1
$U_TU_MB_MB_M$	0.125	0.4	0	−1.414	1	−1	0	0	1.414	1
$U_MU_MB_TB_T$	0.0625	0.8	−1.414	1.414	−1	−1	−2	−1.414	1.414	−1
$U_MU_MB_TB_T$	0.125	1.3	−1.414	0	−1	1	0	1.414	0	1
$U_MU_MB_MB_M$	0.0625	7.3	−1.414	−1.414	−1	−1	2	−1.414	−1.414	−1

Genotype is the nine possible two-locus two-allele genotypes for UMC107 (U) and BV302 (B). Frequency (Freq.) is the expected frequencies for the genotypes at a gene frequency for both loci of 0.5. This serves as the weighting factor in the regression. The dependent variable (Dep. Var.) is the genotypic values measured in the experiment. The independent variables (Add A, etc.) is the orthogonal genetical effects from Table 6.1 adjusted so that the (weighted) variance due to the effect is 1.

TABLE 6.3. The results of the regression partitioning the total variance into the genetical components listed in Table 6.1. The regression is done at a gene frequency of 0.5 ($p_T = 0.5$) and 0.25 ($p_T = 0.25$) of the teosinte allele at both loci.

Source	Variance ($p_T = 0.5$)	Variance ($p_T = 0.25$)
Additive A	0.551	2.772
Additive B	0.633	3.721
Dominance A	0.141	0.343
Dominance B	0.456	0.802
A × A	0.456	1.440
A × D	0.361	0.317
D × A	0.195	0.211

Figure 6.4 is a plot of the variance components for the BV302 and UMC107 interaction in the teosinte × maize cross plotted against the inbreeding coefficient. In this cross (with an allele frequency of 0.5 at both loci) 59.6% of the genetic variation is nonadditive, and 39.2% is epistatic. There is a substantial conversion of nonadditive to additive genetic variance, the variance in the population mean is much larger than would be expected under the additive model, and the Var(α_{i*j}) is nonzero for

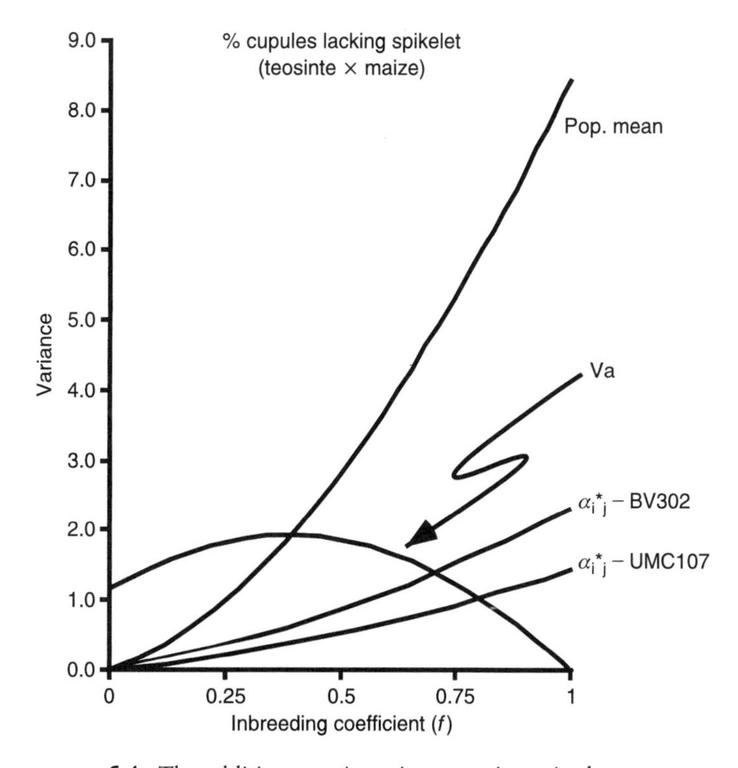

FIGURE 6.4. The additive genetic variance, variance in deme means, and variance in the allele by deme interaction for the BV302 and UMC107 QTL alleles of the teosinte × maize cross described by Doebley et al. (1995). Note that the additive genetic variance increases with inbreeding, the variance among demes is greater than expected for an additive model, and the variance in the allele by deme interaction for both loci is nonzero, all consistent with this being an epistatically interacting pair of loci.

both loci. The average effects of the BV302 allele are plotted in Figure 6.5 as a function of the frequency of the UMC107 and BV302 alleles. These are average effects, rather than local average effects therefore the surface for the maize BV302 allele is a straight line at 0 when the frequency of the BV302 teosinte allele is 0 (front edge top surface), and similarly the BV302 teosinte allele has an average effect of 0 when it is fixed (back edge bottom surface). The back corner of Figure 6.5 is a "teosinte-like" genotype, whereas the front corner is a "maize-like" genotype. In a teosinte-like background the difference between the maize and teosinte allele is very small, that is, the two alleles are nearly neutral with respect to the extent to which they make the plant look like corn or teosinte. On the other hand, in a maize background (front corner) the difference is large with the teosinte allele strongly converting the corn phenotype back into a teosinte phenotype. This suggests that the BV302 maize allele may have always been present in the teosinte population; however, its effects would have been small. As the teosinte was selected to become more maize-like the effects of this allele would have been magnified,

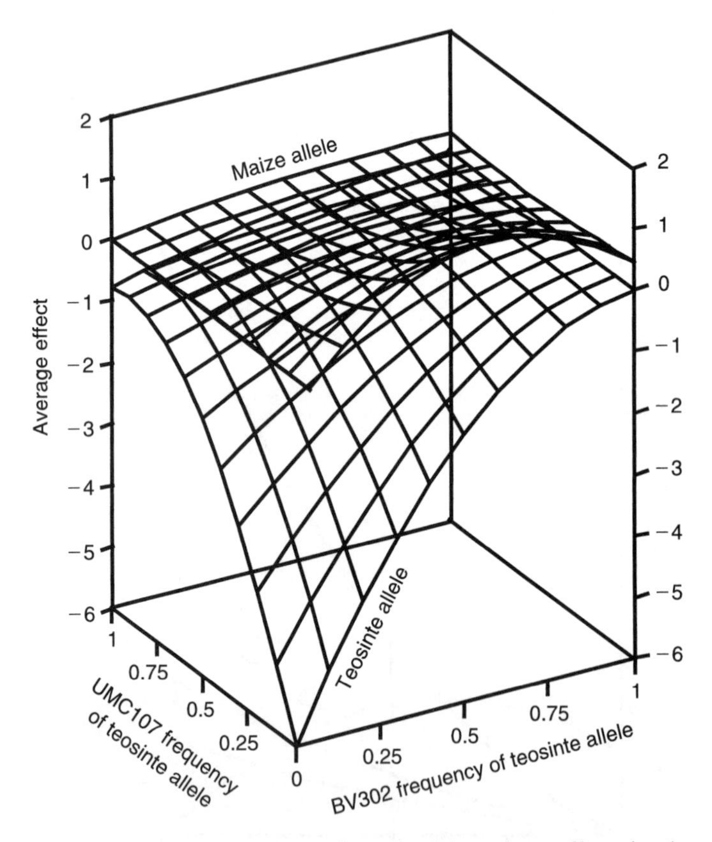

FIGURE 6.5. A three-dimensional graph of the average effects for the two BV302 alleles. The horizontal axis represent populations with different frequencies of the teosinte alleles at the two loci, the vertical axis represents the average effects. The "back" corner represents a teosinte type background (both loci fixed for the teosinte allele) and the "front" corner represents a "maize" type background (both loci fixed for the maize allele). Note that the alleles at the BV302 locus are nearly neutral in a teosinte background, but are strongly differentiated in a maize background. Also note that it is average effects rather than local average effects that are plotted. I need to change that fact!

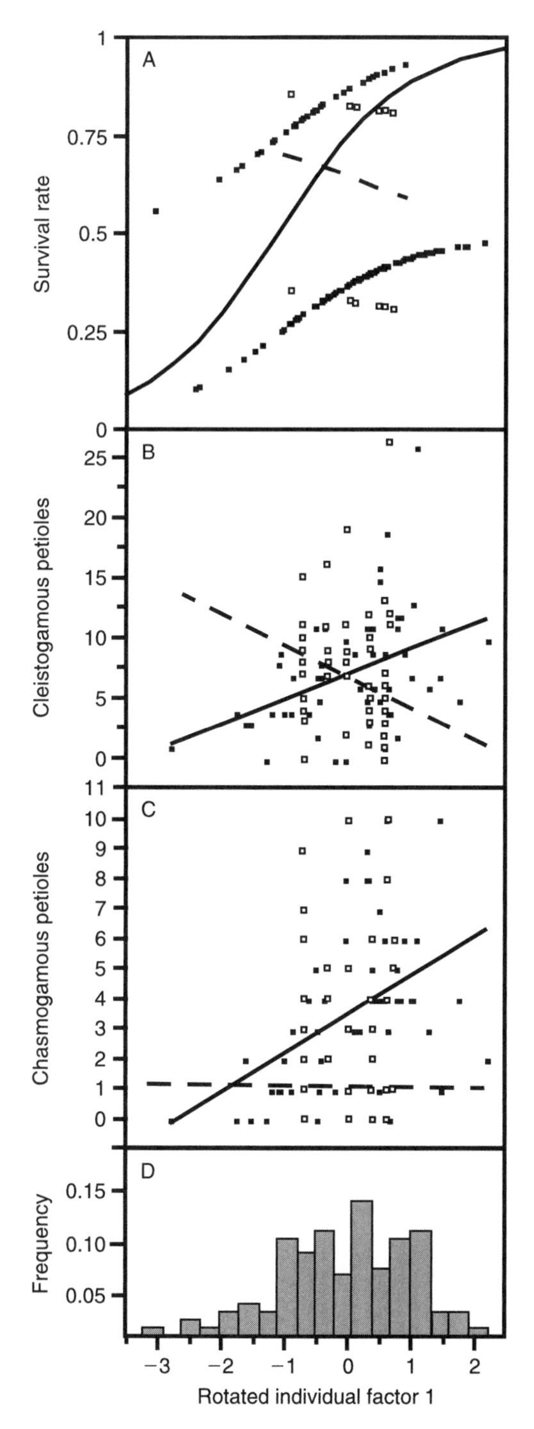

FIGURE 6.6. The correlation of the first rotated factor (RF1, related to "size") with (A) probability of survival rate to first reproduction, (B) number of cleistogamous petioles, (C) number of chasmogamous petioles, in 1990. The unbroken line is the log linear regression of the fitness component on the individual trait (RF1). The dashed line is the regression of the fitness component on the group mean of the trait. Filled squares are individual values and open squares are group values. The frequency distribution of the RF1 for all individuals in the study is shown in (D). A significant regression indicates selection is acting; thus, individual selection favors large values of RF1 (large size) for all three traits and group selection favors small values of RF1 for both survival rate and cleistogamous petioles.

and it could have contributed to the domestication of corn. This illustrates how gene interaction can provide a pool of hidden genetic variation that can be released by selection or drift.

Contextual Analysis in *Impatiens capensis*

Contextual analysis (Heisler & Damuth 1987; Goodnight et al. 1992; Stevens et al. 1995) has been suggested as a means of analyzing multilevel selection in natural populations. This method is identical to the regression analysis of selection introduced by Arnold and Wade (1984a, b) for analyzing individual selection, except that group-level traits are incorporated into the analysis. As with the analysis of individual selection, this is an analysis of within-generation selection. This method measures only the forces affecting within-generation change, and does not provide insight as to whether the observed selection will lead to evolutionary change, that is whether the traits under selection are heritable. As with standard individual selection analysis, a fitness trait is measured on each individual. For example, Stevens et al. (1995) used survivorship, chasmogamous (open pollinated) and cleistogamous (self-pollinated) seed set in *Impatiens capensis* as measures of fitness. Also, as with individual selection analysis, a number of individual-level traits were measured. Stevens et al. (1995) measured a number of traits including plant height, size of the first leaf after the cotyledons, and number of leaves at a date near maturity. However, in addition to the individual traits a number of contextual traits, traits measured on the group, or in this case local neighborhood, were also measured. Only one such trait was included in this analysis: number of plants within a 25 cm radius of the individual. Finally, a number of group mean traits were included, including group mean plant height, and group mean number of leaves per plant. Typically a multiple regression is done using a fitness measure as the dependent variable, and the individual, contextual, and group mean traits as independent variables. A significant partial regression on a contextual trait or on a group mean trait indicates that multilevel selection is occurring. However, Stevens et al. (1995) decided that a path analysis (effectively a factor analysis) would be more appropriate, thus their analysis differed from a simple regression approach. Three underlying individual-level factors and their group means were identified and used in the regression. The results (Figure 6.6) show that primarily individual selection is acting on chasmogamous seed production, but that cleistogamous seed production is being affected by both group and individual selection acting in opposition. Larger plants produce more seeds, but plants in groups with smaller mean height produce more seeds. It is a common observation in plant population biology, known as the constant yield law (Harper 1977), that the yield in seed production per unit area is constant and independent of plant density over a wide range of planting densities. Contextual analysis identifies this as an interaction between group and individual selection acting in opposition.

SUGGESTIONS FOR FURTHER READING

The recent book *Ecology, Genetics, and Evolution of Metapopulations* (Hanski & Gaggiotti 2004) gives an excellent overview of many aspects of metapopulations ranging from ecological considerations to the implications for speciation. Of particular interest are those chapters addressing evolutionary aspects of metapopulations. These include chapters on selection and drift (Whitlock), coalescent theory (Wakeley), metapopulation quantitative genetics (Goodnight), life history evolution (Ronce and Olivieri), selection in metapopulations (Wade), and speciation (Gavrilets). Current controversies concerning Wright's shifting balance theory are covered well in a series of papers that appeared in *Evolution* (Coyne et al. 1997, 2000; Wade & Goodnight 1998; Goodnight & Wade 2000). These papers take very different views on the nature of evolution in structured populations, and illustrate why this is currently an active area of research and debate.

Coyne JA, Barton NH & M Turelli 1997 Perspective: A critique of Sewall Wright's shifting balance theory of evolution. Evolution 51:643–671.

Coyne JA, Barton NH & M Turelli 2000 Is Wright's shifting balance process important in evolution? Evolution 54:306–317.

Goodnight CJ & MJ Wade 2000 The ongoing synthesis: a reply to Coyne, Barton and Turelli. Evolution 54:317–324.

Hanski I & O Gaggiotti (eds) 2004 Ecology, Genetics, and Evolution of Metapopulations. Elsevier/Academic Press.

Wade MJ & CJ Goodnight 1998 The theories of Fisher and Wright: when nature does many small experiments. Evolution 54:1537–1553.

II

MOLECULAR EVOLUTION

7

Detecting Selection at the Molecular Level

MICHAEL W. NACHMAN

Selection acts on the phenotype but can leave its signature at the molecular level. For example, if a new mutation arises at a locus and confers a fitness advantage, it may spread quickly to all individuals in a population and thereby eliminate all preexisting variation at the locus. Because selection is a *deterministic* process, patterns of DNA variation caused by selection can often be distinguished from patterns caused by other processes, such as genetic drift or migration. This basic idea has led to a proliferation of theoretical and empirical studies aimed at detecting the effects of selection at the molecular level.

This approach for studying selection is very different from more direct approaches that focus on observations of phenotypic change over several generations (e.g., Grant & Grant 2002), and it has both advantages and disadvantages. One advantage is that selection coefficients that are small but biologically meaningful may be impossible to measure directly but will still leave an imprint in patterns of DNA sequence variation. Another advantage is that this approach gives us a picture of selection over evolutionary, as opposed to ecological, time scales. In other words, we can detect selection that has happened many generations ago. This approach also allows us to ask questions about selection, even without knowledge of the phenotype. For example, we can ask how much selection has occurred in the recent history of a species even if we are unaware of the agent of selection. Finally, this approach holds promise that we might discover genes associated with the evolution of novel traits. One of the chief disadvantages is that when we do find a signature of selection, it is very difficult to make the link to the phenotype, much less to the environment. The literature contains many examples of genes or genomic regions that have been clearly influenced by selection, but where the specific polymorphisms under selection and their functional consequences are unknown.

The neutral theory of molecular evolution (Kimura 1983) serves as the null hypothesis for most statistical tests in this field, and so I first describe it briefly below. Next, I describe three simple models of selection and the patterns of DNA sequence variation expected under each. Many statistical tests for detecting selection at the molecular level have been developed in the last two decades. I have grouped these tests into five basic kinds. I describe them and present a few key applications of each kind of test to real data. The genomics revolution has opened up the possibility of looking for selection at virtually every gene in a genome; I discuss the promise and difficulties of these genome-wide studies. Finally, I present a few particularly compelling case studies of selection at the molecular level.

CONCEPTS

The Neutral Theory of Molecular Evolution

Proposed by both Kimura (1968) and King and Jukes (1969) and later developed in great detail by Kimura (1983), the neutral theory states that most mutations are deleterious, but of the remaining ones, a negligible proportion are advantageous and the vast majority are neutral with respect to fitness. The fate of these neutral mutations is governed by random genetic drift. According to this theory, it is these neutral mutations that we see as polymorphisms within species or as fixed differences between species. Deleterious mutations, although far more numerous, are eliminated quickly by selection and thus contribute neither to polymorphism nor to divergence. The neutral theory accounts for many observations in population genetics and molecular evolution, and it also has tremendous heuristic value. Because it is simple, mathematically tractable, and makes several straightforward predictions, it serves as the null model in many statistical tests. These tests are usually based on one of three mutational models. The infinite alleles model (Kimura & Crow 1964) assumes that each new mutation creates a new allele in a population. The infinite sites model (Kimura 1969a) is typically used to model DNA sequence evolution and assumes that each new mutation occurs at a site that has not previously mutated. The stepwise mutation model (Ohta & Kimura 1973) was originally developed for allozyme data but is now frequently used to model microsatellite loci; it assumes that mutations arise in steps, and that alleles can only mutate to neighboring states.

One key prediction of the neutral theory concerns the amount of variation expected within and between species. Under a neutral model, the amount of genetic variation in a population represents a balance between the input of new mutations and their loss or fixation due to genetic drift. For example, under the infinite sites model, at mutation–drift equilibrium the expected heterozygosity is $4N_e\mu$, where N_e is the effective population size and μ is the neutral mutation rate (Kimura 1969a). The rate of evolution along a lineage, v (i.e., the substitution rate or the fixation rate) is equal to the number of new mutations entering a population each generation, $2N\mu$, times the probability of fixation for a new mutation, $1/2N$. Therefore, the rate of evolution (v) is simply equal to the mutation rate (μ) and does not depend on the population size (Kimura 1983). This simple but important result implies that the amount of divergence between orthologous copies of a gene will depend only on the mutation rate and the amount of time separating the copies. Thus, the neutral theory predicts that the ratio of variation within a species to divergence between species will be the same for different genes, since both depend on the mutation rate (Figure 7.1A).

A second important prediction concerns the distribution of allele frequencies at mutation–drift equilibrium. At steady state, the expected heterozygosity can be specified for any mutational model, but new alleles are constantly entering the population due to mutation and are leaving due to stochastic fixation or loss. Nonetheless, there is an expected distribution of allele frequencies that can be described based on the observed heterozygosity and the sample size. This "neutral distribution" consists of many low-frequency alleles and an increasingly smaller number of increasingly higher frequency alleles (Figure 7.1B).

Both these predictions serve as the basis for some of the statistical tests described below. There are many other predictions that follow from the neutral theory that have also been developed into statistical tests, and some of these are described in detail elsewhere (e.g., Kreitman 2000; Yang & Bielawski 2000; Luikart et al. 2003).

Models of Selection

At the molecular level, selection can act to fix alleles (positive, directional selection), eliminate alleles (negative or purifying selection), or maintain two or more alleles in a population (which I will refer to as "balancing selection" to include heterosis, spatially or temporally varying selection, and any other selection that maintains variation). These different forms of selection will affect the shape of the genealogy of alleles in a population: positive and purifying selection will produce shallower genealogies and balancing selection will produce deeper genealogies (Figure 7.2). Positive and purifying selection will reduce genetic variation since shallower genealogies have less time over which mutations can arise, and balancing selection will increase genetic variation since deeper genealogies have more time over which mutations can arise. The distribution of allele frequencies is also skewed by these different genealogies: in general, positive or purifying selection can produce an excess of low-frequency alleles while balancing selection can produce an excess of intermediate-frequency alleles. Finally, positive selection will lead to increased rates of evolution at the sites under selection.

An important concept for studies of selection at the molecular level is that linked sites will have correlated evolutionary histories. Thus, selection can affect patterns of DNA sequence variation at genes near to those that are the target of selection. This suggests that we might be able to find evidence of selection in the genome even when the targets

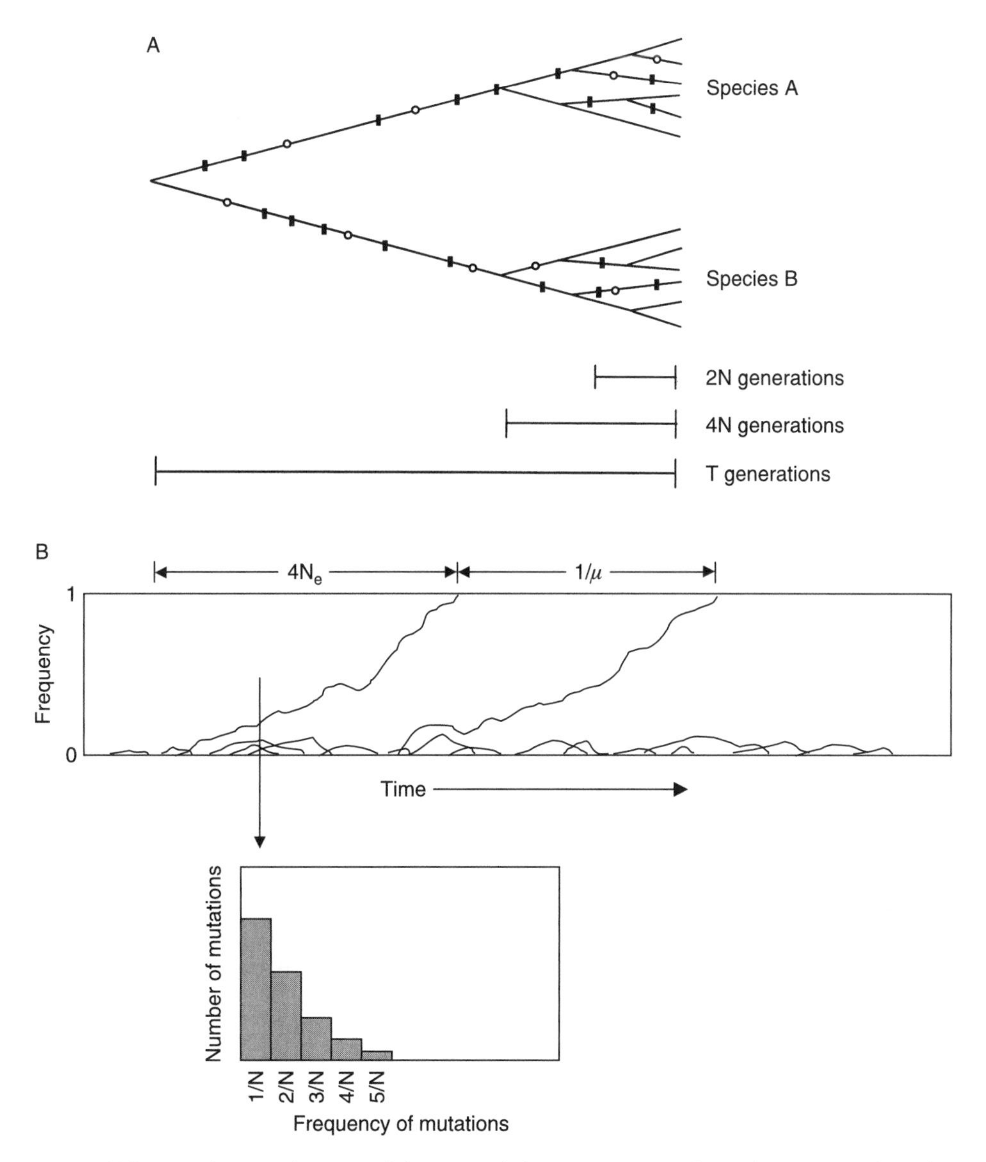

FIGURE 7.1. Two key predictions of the neutral theory. (A) Hypothetical gene genealogy for two species under the neutral model. The average time separating one allele from each species is T generations, and therefore the average sequence divergence between each species is $2\mu T$. The average coalescence time for two randomly chosen alleles within a species is $2N$ generations, and therefore the average sequence divergence between two randomly chosen alleles within a species is $4N\mu$. The average coalescence time for all alleles within a species is $4N$ generations. These expectations all have large variances. The mutation rate, μ, may vary among loci. For example, the open circles represent mutations at a locus with a low mutation rate, and the ratio of polymorphism within species A to fixed differences between A and B is 2:5 for this locus. The filled rectangles represent mutations at a locus with a high mutation rate, and the ratio of polymorphism within species A to fixed differences between A and B is 4:10 for this locus. Under the neutral theory, the ratio of polymorphism to divergence is expected to be the same for different genes (HKA test) or different classes of sites within a gene (MK test). (B) Hypothetical distribution of allele frequencies under the neutral model. Most new neutral mutations are lost due to drift. A small fraction are fixed by drift, and this takes $4N$ generations, on average. The mutation rate, μ, per gamete per generation is equal to the substitution rate, v, per lineage per generation. Thus, the time between successive fixations is $1/\mu$. At mutation–drift equilibrium, there are many low-frequency polymorphisms and an increasingly smaller number of higher frequency polymorphisms.

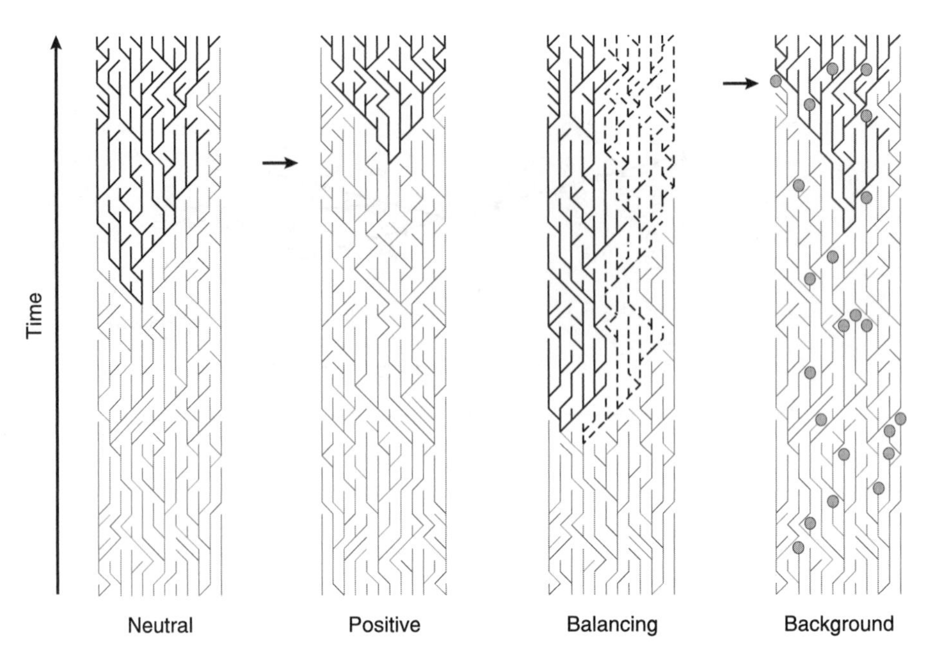

FIGURE 7.2. Hypothetical gene genealogies for a population of 12 haploid individuals under different models of selection (from Bamshad & Wooding 2003). A neutral model is given first, followed by a model of positive directional selection (which results in a shortened coalescence time for all alleles in a sample, thus reducing heterozygosity; the arrow indicates the onset of positive selection), a model of balancing selection (which results in a lengthened coalescence time for all alleles in a sample, thus increasing heterozygosity), and a model of background selection (which results in a shortened coalescence time due to the pruning of branches by deleterious mutations, shown with circles). The arrow in the last panel indicates a branch that has been pruned by background selection, relative to the neutral model in the first panel. Reprinted from Bamshad & Wooding (2003) with the permission of Nature Publishing Group.

are unknown. Conversely, when we do find evidence of selection, the target of selection may still lie at a considerable genomic distance.

Genetic hitchhiking refers to the adaptive fixation of an advantageous mutant and the associated fixation of linked, neutral variants (Maynard-Smith & Haigh 1974). In the aftermath of a selective sweep, variation at the gene under selection and at linked sites will be reduced or eliminated (Figure 7.3). The strength of this hitchhiking effect (i.e., the amount of reduction) will depend on the strength of selection and the rate of recombination in the region (Kaplan et al. 1989). In regions of low recombination, there will be little opportunity for linked neutral sites to become decoupled from the selected site during the typically short sojourn time of an adaptive fixation. In regions of high recombination, linked sites may become decoupled from selected sites. Thus, if genetic hitchhiking is common, we might expect to see a general correlation between levels of neutral nucleotide diversity and recombination rate for different genomic regions. First demonstrated in *Drosophila melanogaster* (Begun & Aquadro 1992), this pattern has now been seen in many organisms, including humans (see Case Studies). One trivial explanation for this pattern might be that recombination is mutagenic; i.e., that high nucleotide diversity results from a greater input of new mutations in some genomic regions. A simple test of this idea is provided by comparing recombination rate with interspecific divergence for different regions of the genome. If recombination is mutagenic, then regions of high recombination should show higher rates of evolution between species. This pattern is not seen in *Drosophila* (Begun & Aquadro 1992), but a weak positive correlation

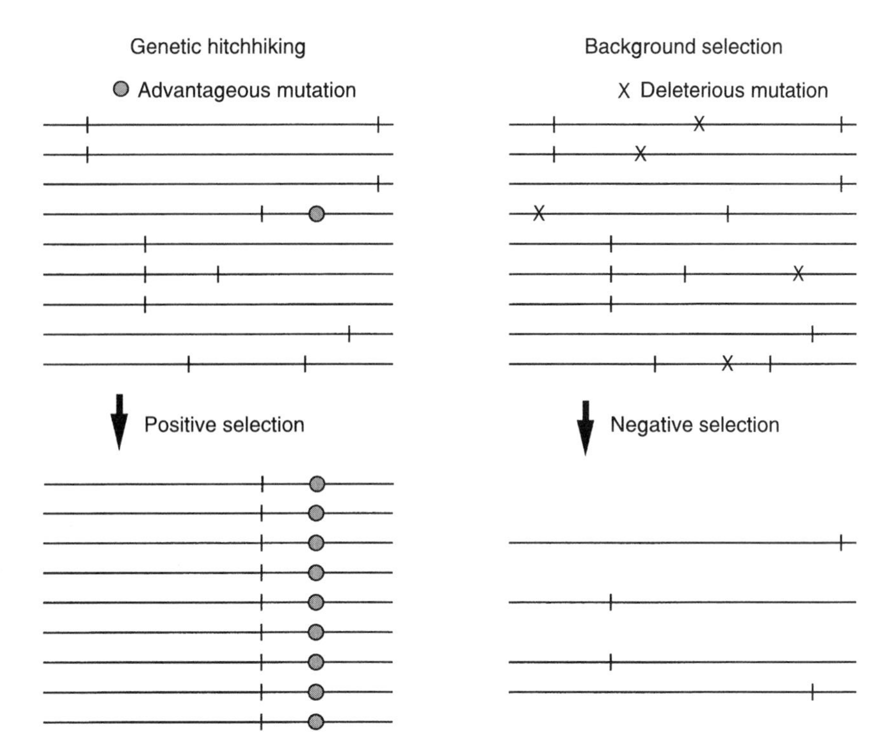

FIGURE 7.3. Schematic model of genetic hitchhiking and background selection without recombination (from Nachman 2001). Horizontal lines depict haplotypes in a population, and vertical marks depict neutral mutations. Under genetic hitchhiking, an advantageous mutation arises and is fixed by positive selection, dragging linked neutral variants with it. In the aftermath of a complete selective sweep without recombination, all individuals possess the same haplotype. If recombination occurs during the selective sweep (not shown), some variation may remain in the population. Under background selection, deleterious mutations arise and are eliminated by selection, eliminating linked neutral variants with them. In the presence of recombination (not shown), neutral variants may escape elimination. Formally, background selection is equivalent to a reduction in the effective population size by a fraction f_o, the equilibrium frequency of chromosomes free of deleterious mutations.

between recombination rate and interspecific divergence is seen in mammals (Hellman et al. 2003). Although positive selection will lead to increased rates of evolution for sites under selection, it does not increase the rate of evolution at linked sites (Birky & Walsh 1988). Thus, the association between recombination and divergence cannot be explained by genetic hitchhiking and instead suggests that recombination is associated, perhaps causally, with higher mutation rates in some species.

Another model of selection at linked sites involves purifying selection. Background selection (Charlesworth et al. 1993) is the removal of deleterious mutations by selection and the associated removal of linked neutral variants (Figure 7.3). Like genetic hitchhiking, this process can reduce genetic variation, particularly in regions of low recombination. Background selection thus represents an alternative hypothesis that might explain the correlation seen between nucleotide diversity and recombination rate (Begun & Aquadro 1992).

Statistical Tests of the Neutral Model

In the standard neutral model, genetic variation is assumed to be selectively neutral, and populations are assumed to be panmictic (randomly mating), constant in size, and at equilibrium with respect to mutation and drift. Point mutations in DNA sequences are typically modeled using the infinite sites model (Kimura 1969a) in which each new mutation occurs at a site that has not previously mutated. Deviations from these assumptions may lead to rejection of the null hypothesis. For example, many statistical tests of neutrality will return a significant result if there are changes in population size, even in the absence of selection. Many such "population-level" processes can lead to rejection of the null hypothesis, and thus an important goal of molecular studies of selection is to distinguish the effects of demographic processes (e.g., migration, nonrandom mating, population growth, population bottlenecks) from selection. One way to do this is to compare patterns of variation at multiple loci; in general, selection will affect only specific genes, while population processes will affect all loci in the genome.

Another important consideration is that different statistical tests differ tremendously in their power to detect particular non-neutral events. Some tests seem to be effective in detecting more recent selection while others seem to more effective in detecting older selection. Below, I describe different statistical tests, and where such information exists, I describe the power and utility of the tests for detecting particular non-neutral patterns.

I have grouped statistical tests of neutrality into five general categories. This classification is based mostly on the kinds of data that are used, but it is somewhat arbitrary and other classifications are possible.

Tests Based on the Distribution of Allele Frequencies

The first test of this kind (Watterson 1978) was based on the infinite alleles model of mutation (Kimura & Crow 1964) rather than the infinite sites model, and it predated population-level DNA sequence data by several years. Watterson's test compares the observed number of alleles and the observed heterozygosity in a sample. An excess of heterozygosity, given the observed number of alleles at a locus, is consistent with balancing selection or a population contraction, while a deficit of heterozygosity is consistent with positive directional selection, a population expansion, or the presence of weakly deleterious alleles in the sample.

Tajima (1989) introduced a test that is conceptually similar but is based on the infinite sites model; it compares the observed number of polymorphic sites corrected for sample size (θ_W) with the observed nucleotide heterozygosity (θ_π) in a sample. Formally, both θ_W and θ_π are estimators of the neutral mutation parameter, $4N_e\mu$. At equilibrium $\theta_W = \theta_\pi = 4N_e\mu$. The test statistic, D, is constructed as the difference between these two estimators $[D = (\theta_\pi - \theta_W)/\mathrm{Var}(D)]$ and therefore takes on positive values when $\theta_\pi > \theta_W$ and negative values when $\theta_\pi < \theta_W$. Since low-frequency mutations contribute proportionally more to θ_W than to θ_π negative values of D occur when there is an excess of low-frequency polymorphisms and positive values of D occur when there is an excess of intermediate-frequency polymorphisms. An excess of low-frequency polymorphisms is consistent with positive directional selection, since in the aftermath of a selective sweep variation is eliminated and new mutations arise at low frequencies. It is also consistent with the presence of mildly deleterious alleles (which remain at low frequency because of selection against them) or with a recent population expansion (since rare alleles are preserved in expanding populations). Conversely, an excess of intermediate-frequency polymorphisms may be indicative of balancing selection or a population contraction (since rare alleles are lost during population contractions). Recently, several authors have also shown that sampling from subdivided populations may create a skew in the distribution of allele frequencies (e.g., Hammer et al. 2003). Tajima (1989) showed that for small insertion-deletion polymorphisms (indels) in *D. melanogaster*, the value of D was not significantly different from 0, while for large indels, D was significantly negative, leading him to conclude that large indels are more harmful than short indels.

Many other tests based on the distribution of allele frequencies have subsequently been developed. These different tests focus on different aspects of the data. For example, Fu and Li's (1993) test focuses on the number of singletons in a sample, and subsequent tests by Fu (1997) utilize other aspects of the data. Fay and Wu's (2000) test focuses on the number of derived, high-frequency mutations. For most of these tests, the significance is assessed through coalescent simulations to generate a distribution for the test statistic under the null hypothesis. Since population-level processes can lead to deviations, a key problem is choosing the appropriate null model. For example, in humans, most loci show an excess of rare alleles. Since this pattern is seen at many loci, it is probably a consequence of population expansion in the history of our species. Detecting a signature of positive selection against this backdrop of a genome-wide skew toward rare alleles is difficult. One approach is to simulate a null distribution with a model that includes the appropriate changes in population size. Recently, this approach was used by Wooding et al. (2004) to conclude that the values of Tajima's D seen at the PTC locus in humans are indicative of selection acting at this locus. This gene is polymorphic in most human populations and is involved in the ability to taste bitter substances.

Several researchers have conducted power analyses of these tests under different models of selection or changes in population size (e.g., Simonsen et al. 1995; Fu 1997). These studies generally show that fairly large samples are needed to achieve reasonable power, and that even with these samples, the ability to detect selection (or changes in population size) is restricted to fairly short windows of time following the event, but not necessarily immediately after the event. For example, following a complete selective sweep, variation is entirely eliminated and these tests consequently have no power. Once new mutations arise, the tests have some power to detect a skew in the allele frequency spectrum, but power is again lost once a neutral distribution is reached ($< 4N$ generations).

These tests have been used to detect both positive directional selection and balancing selection. In most cases, demographic explanations can be ruled out only when it is shown that the distribution of allele frequencies at other loci is distinct.

Tests Based on Intraspecific Polymorphism and Interspecific Divergence

A second class of tests utilizes data from variation within and between species. These tests are based on the basic idea that the ratio of polymorphism to divergence is expected to be the same for different genes under neutrality, even if the actual amount of variation is different among genes. For example, histones are extremely conserved proteins and therefore have a low neutral mutation rate (i.e., the fraction of sites at which changes do not affect function is very small). Most intergenic, noncoding DNA, on the other hand, is generally not conserved and therefore has a higher neutral mutation rate (i.e., changes at most sites do not affect function). Histone coding sequences are therefore expected to show less variation than intergenic noncoding sequences, both within and between species. However, the ratio of polymorphism to divergence should be the same for both loci provided that they are evolving neutrally (i.e., all mutations are either deleterious or neutral).

The first formulation of this basic idea into a statistical test was by Hudson et al. (1987), and this has become known as the HKA test. The test incorporates any number of loci and utilizes polymorphism data from one or two species and divergence data from interspecific comparisons. The test calculates expected values of polymorphism and divergence using least-squares estimators, and compares these with observed values to generate a test statistic that is approximately χ^2 distributed. Loci are assumed to be independent (i.e., unlinked), but the test is conservative if they are linked. By comparing multiple loci, the HKA test is able to disentangle locus-specific selection from population-level effects. For example, if a population has undergone a severe contraction in one species, genetic variation will be reduced at all loci, and the test will not be significant. If selection has reduced variation at just one locus, however, the ratio of polymorphism to divergence will be lower at that locus, and the test may be significant. In principle, rejection of the null model might be due to either elevated or reduced polymorphism or elevated or reduced divergence. However, as described above, selection at linked sites will affect levels of polymorphism (Maynard-Smith & Haigh 1974) but not levels of divergence (Birky & Walsh 1988), so unless the sites surveyed are themselves the target of selection, deviations in polymorphism levels are more likely to be the cause of any non-neutral pattern. This raises an important point: significant rejection of the null model in an HKA test may be due to selection at the sites surveyed, or it may be due to selection at linked sites, potentially at a considerable distance from the surveyed loci. For example, in *D. melanogaster* nucleotide variability is reduced over a considerable distance near the tip of the X chromosome (Begun & Aquadro 1991) and along the fourth chromosome (Jensen et al. 2002), presumably as a result of either positive or negative selection, but the targets of selection are unknown. If two loci which are both experiencing similar selection are compared in an HKA framework, the null model may not be rejected.

A related test compares the number nonsynonymous and synonymous mutations within and between species (McDonald & Kreitman 1991) and has become known as the MK test. The ratio of polymorphism to divergence for these two kinds of mutations is expected to be the same under neutrality, following the same logic as for the HKA test. Deviations can be tested using a simple 2×2 test of independence, such as a χ^2 test or a Fisher's Exact Test. In principle, deviations could be due to either an excess or deficiency of counts in any of the four cells (synonymous polymorphisms, synonymous divergence, nonsynonymous polymorphisms, nonsynonymous divergence), although it is typically assumed that selection on amino acids is more likely than selection on silent mutations. Thus, in practice, it is often assumed that silent changes are neutral and that deviations reflect selection on nonsynonymous mutations. This assumption may not be valid in situations where codon bias due to selection occurs (Chen & Stephan, Ch. 9 of this volume).

There are some important and often unappreciated differences between the HKA and MK tests. The HKA test compares two loci that are freely recombining with respect to each other while the MK test compares two classes of sites that are interspersed. The loci being compared in an HKA test have independent evolutionary histories, and may have shorter or longer gene genealogies (and therefore more or less polymorphism) due to chance. The two classes of sites in an MK test are completely interspersed and are therefore expected to share the same gene genealogy. The HKA test therefore has two sources of variance (the sampling variance and the evolutionary variance) while the MK test has only one (the sampling variance). This difference is reflected in the statistical treatment of each test: the MK test can be performed with any conventional test of independence while the HKA test utilizes a framework that incorporates evolutionary variance. A second consequence of the interspersion of sites in the MK but not the HKA test is that selection at linked sites must be considered in the latter but not the former. In other words, a significant rejection of the MK test implies that the sites surveyed are themselves the target of selection. The tests are thus complementary: the HKA test can be used to detect selection at a distance but it typically does not provide direct information about the genes under selection, while the MK test will not detect selection at a distance but it will provide information about specific genes under selection. The MK test has been used to provide evidence of strong positive selection at several genes in *Drosophila* (e.g., McDonald & Kreitman 1991; Eanes et al. 1993) and other organisms (Ford 2002).

Tests Based on Population Differentiation

Cavalli-Sforza (1966) first suggested that differences among loci in measures of population differentiation, such as F_{ST}, could be used to infer the action of selection, and this was later developed into a statistical test by Lewontin and Krakauer (1973). The idea behind this test is that gene flow among populations will generate some average value of differentiation for most loci, against which outliers can be identified. In principle, these outliers could lie in either direction: local adaptation might produce unusually high levels of differentiation at some loci, while balancing selection acting similarly in different populations might produce below-average levels of differentiation at some loci. The test developed by Lewontin and Krakauer (1973) was criticized by Nei and Maruyama (1975) and Robertson (1975) who argued that the expected variance in F_{ST} for the test was not valid under many models of population structure. However, the fundamental logic of comparing levels of population differentiation among loci has served as the basis for several newer tests. For example, Beaumont and Nichols (1996) and Beaumont and Balding (2004) have used coalescent simulations to develop a null distribution of F_{ST} values, conditioned on observed heterozygosity under various models of population structure. Stephan et al. (1998) also used coalescent simulations to generate expected distributions of F_{ST} under neutral and selection models.

Another approach for detecting selection based on comparisons among populations is to look at levels of heterozygosity for different loci in different populations. Schlötterer et al. (1997) and Schlötterer (2002) pointed out that in comparisons of heterozygosity at multiple loci from two or more populations, it is possible to disentangle the effects of selection from the effects of demography. For example, if two populations differ in size, the expected level of variation in the smaller population will be lower for all loci. However, individual loci that have been under selection may have even less variation than this genome-wide difference.

Thus by looking at locus and population combinations, it is possible to identify genomic regions that have been under selection.

Tests Based on Linkage Disequilibrium

Linkage disequilibrium (LD) is the nonrandom association of alleles (nucleotides) at different genes (sites). Positive, directional selection will decrease genetic variation, but can also lead to an increase in LD, especially if the selected allele has not yet been fixed. This idea has motivated several tests that look for an excess of LD (Kelly 1997; Toomajian & Kreitman 2002; Sabeti et al. 2002). In general, these tests are likely to be most sensitive at identifying recently selected alleles, since complete selective sweeps will eliminate variation. In humans, it appears that selection can create LD over considerable genomic distances (e.g., Saunders et al. 2002).

Tests Based on Fixation Rates and Patterns

All the tests described above require data on genetic variation within species. A different class of tests is typically based on patterns of molecular evolution between species, and these tests can therefore detect selection that has occurred in the more distant past. These tests are based on the partitioning of DNA sequences into sites at which mutations will change the amino acid (nonsynonymous sites) and sites at which mutations will not change an amino acid (synonymous sites). For a given gene, the numbers of synonymous and nonsynonymous sites are counted. Sequences from two species are then aligned, and the observed numbers of synonymous and nonsynonymous mutations are counted. If no selection is operating, the observed numbers of synonymous and nonsynonymous mutations should occur in proportion to the numbers of each kind of site. Expressed on a per site basis, the ratio of nonsynonymous substitutions per nonsynonymous site (K_A or d_N) to synonymous substitutions per synonymous site (K_S or d_S) should be 1. Under purifying selection K_A/K_S (or d_N/d_S) will be less than 1. If $K_A/K_S > 1$, this suggests that positive selection has driven the fixation of amino acids. This test can also be applied to variation within a species, but unless the number of observed differences is large, the test has little power. For many genes, K_A/K_S in interspecific comparisons is on the order of 0.1–0.2, reflecting the fact that most genes are under considerable selective constraint. In general, selection has to be quite strong to drive K_A/K_S above 1. Thus, this test has little power to detect weak selection. For example, genes with high K_A/K_S values that are still less than 1 (e.g., 0.8) might reflect weak positive selection or relaxed constraint. Some of the first methods were proposed by Miyata and Yasunaga (1980) and Li et al. (1985). The first applications of these methods to demonstrate positive selection were by Hill and Hastie (1987) who studied serine protease inhibitors, and Hughes and Nei (1988) who showed that $d_N/d_S > 1$ for the antigen recognition sites of the major histocompatibility complex class I loci in both humans and mice.

Subsequent studies have modified this basic idea to incorporate a maximum likelihood framework for estimating d_N/d_S and for hypothesis testing. These newer models have two main advantages: they incorporate lineage-specific effects and codon-specific effects. For example, models developed by Yang and Nielsen (2002) allow one to test the hypothesis that all sites have a d_N/d_S ratio that is drawn from a single distribution against the alternative that some sites have $d_N/d_S < 1$ while other sites have $d_N/d_S > 1$. This is biologically sensible since it is unlikely that positive selection will affect all codons within a gene. Even in genes subject to positive selection, many codons are probably still subject to selective constraints. There are many variations of these models, and they are summarized along with empirical examples of positive selection from the literature by Yang and Bielawski (2000). A major result of molecular evolutionary studies of the last two decades is that many of the genes that are under positive selection are involved in either immunity or reproduction (Ford 2002). This is probably a consequence of the fact that both classes of genes are involved in co-evolutionary processes in which selection pressures are constantly changing.

Genomics and Selection

The recent completion of whole-genome sequences from many organisms is making it possible to conduct tests for selection on a genome-wide scale. In principle, with enough markers at sufficient density, one might be able to "scan the genome" to identify many or most of the regions that have recently been under selection. This approach has been used with tests based on linkage disequilibrium

(Huttley et al. 1999), the distribution of allele frequencies (Payseur et al. 2002), patterns of population differentiation (Akey et al. 2002), relative levels of heterozygosity (Kayser et al. 2003; Storz et al. 2004) and patterns of interspecific evolution (Clark et al. 2003), among others (reviewed in Luikart et al. 2003). These studies often include thousands of loci and therefore thousands of tests. This presents a formidable statistical challenge. With 1000 tests, for example, we expect 50 to be significant at the 0.05 level under the null model. Separating these false positives from a true signature of selection is difficult. At present there is no clear solution, but two approaches have been used. One is to use a very conservative P value, thus minimizing the likelihood that low probability test results will be due to chance. A second is to simply identify outliers in the observed distribution and to treat these outliers as candidates for genes under selection. Some authors have looked for confirmatory evidence that a particular locus may be under selection by looking at adjacent markers (e.g., Kayser et al. 2003), although this neglects the fact that linked sites will have correlated evolutionary histories even under the null model. One approach for confirming that a particular locus is a truly under selection is to use different tests that rely on independent data. For example, if a locus shows significantly elevated K_A/K_S between species and significantly reduced variation within species in an HKA test, we have greater confidence that selection has operated (Karn & Nachman 1999). Ultimately, however, a statement about selection is a hypothesis of function. The best evidence that a gene is currently under selection comes from functional studies in which allelic variants are shown to be associated with functional differences that affect fitness.

CASE STUDIES

There are many good examples of selection at the molecular level. Here I restrict discussion to recent studies that incorporate DNA sequence data. Thus, I do not provide examples from the earlier allozyme literature, although these are some of the best examples linking molecular variation to fitness differences in a known ecological context (see Ford 2002). I also restrict discussion to examples that rely heavily or exclusively on population-level data, thus excluding the many interesting molecular evolutionary studies that document selection over longer evolutionary time scales between species (reviewed in Yang & Bielawski 2000; Ford 2002).

Reduced Nucleotide Variation in Low-Recombination Regions

One very general result to come out of studies of selection at the molecular level is that genomic regions with reduced rates of recombination show reduced levels of neutral variation. The was first demonstrated in an HKA framework by Begun and Aquadro (1991) for the tip of the X chromosome and by Berry et al. (1991) for the fourth chromosome in *D. melanogaster*. Begun and Aquadro (1992) later provided evidence of a general genome-wide correlation between levels of genetic variation and local rates of recombination (Figure 7.4A), a result that was initially interpreted as evidence of recurrent selective sweeps until Charlesworth et al. (1993) presented a model of background selection as an alternative explanation. The correlation between recombination and nucleotide heterozygosity now appears to be quite general, having been demonstrated in humans (Nachman et al. 1998; Nachman 2001) and many other organisms. Background selection and genetic hitchhiking are not mutually exclusive, and both are likely operating to some degree (Kim & Stephan 2000). Distinguishing between these models is difficult and has been the subject of considerable recent work. These models make different predictions concerning the relative levels of genetic variation on the X chromosome and autosomes (Begun & Whitley 2000), the distribution of allele frequencies (Jensen et al. 2002), patterns of population differentiation (Stephan et al. 1998), and levels of variation at markers with different mutation rates (Payseur & Nachman 2000).

The observation of reduced variation in low-recombination regions is instructive in several respects. First, it seems likely that some form of selection at linked sites is responsible for the pattern in most cases (but see Hellmann et al. 2003), although it is unclear whether the pattern is caused predominantly by positive or negative selection (Kreitman 2000). The biological implications are obviously very different for these two models of selection, and thus resolving the relative contributions of each is important for understanding the underlying biology. Second, it seems that we are detecting the effects of selection at linked sites, but the actual targets of selection (i.e., the sites affecting function, either beneficially or detrimentally) are unknown and may lie at a considerable genomic distance from the sites that have been surveyed. Thus, although there is now a fairly large empirical and theoretical literature on genetic hitchhiking and background selection, and despite the fact that recombination and heterozygosity appear to be correlated in many organisms, the full evolutionary significance of this pattern remains elusive.

Despite the fact that the relative contributions of positive and negative selection to the correlation between heterozygosity and recombination are unknown, this correlation still implies that the dynamics of variation at many or most sites in the genome may be governed by selection, even if these sites are not functionally important themselves (Ford 2002). This suggests that some parameters estimated in neutral models (such as N_e) from many genes may be substantial underestimates of true values.

Adh in *Drosophila melanogaster*

In contrast with the example above, an example where the target of selection has been clearly

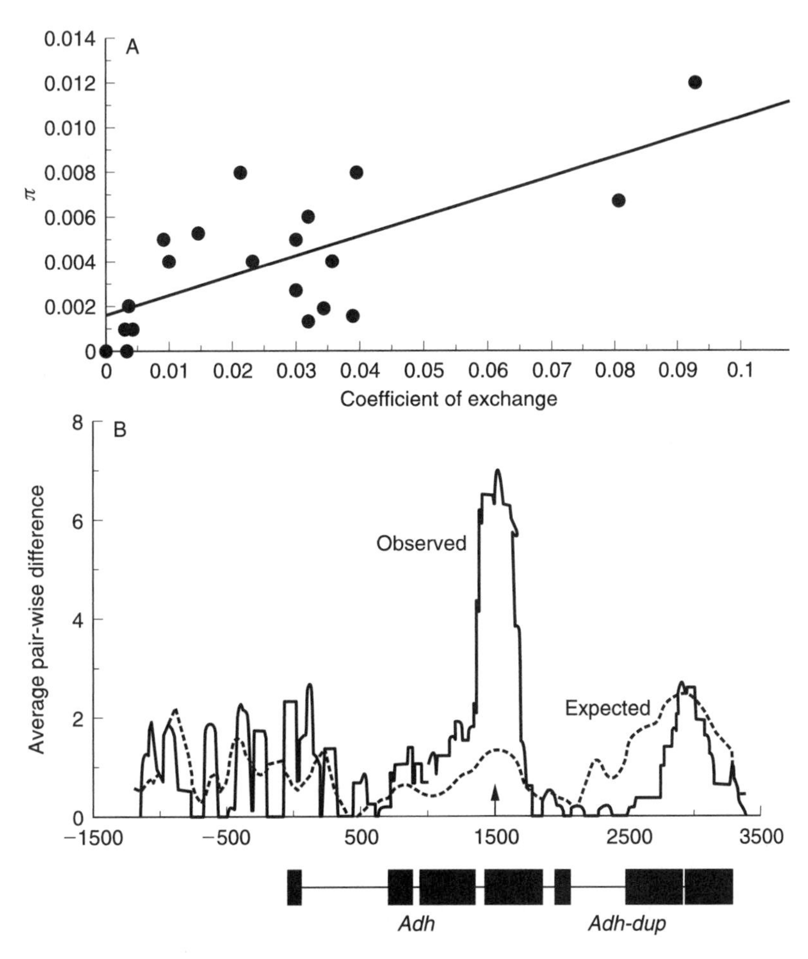

FIGURE 7.4. Three examples of selection at the molecular level in *Drosophila*. (A) Nucleotide variability is positively correlated with recombination rate in *D. melanogaster.* Begun and Aquadro (1992), reprinted with the permission, © 1992 National Academy of Sciences, USA. (B) Sliding window analysis of *Adh* in *D. melanogaster*, showing a peak of polymorphism surrounding the target of selection. Kreitman and Hudson (1991), reprinted with permission of the Genetics Society of America.

(continued)

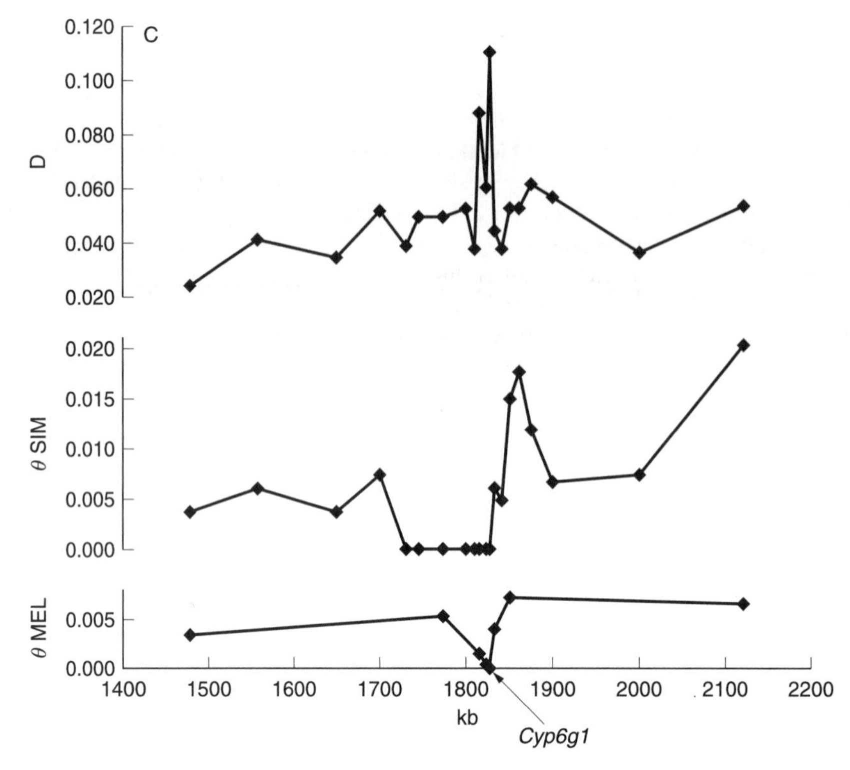

FIGURE 7.4 *(cont.)* (C) Patterns of polymorphism (θ) and divergence (D) at *Cyp6g1* and neighboring loci in *D. melanogaster* (MEL) and *D. simulans* (SIM). Schlenke and Begun (2004), reprinted with the permission of Nature Publishing Group. *Cyp6g1* shows reduced polymorphism in both species, but increased divergence, relative to neighboring genes. See text for discussion of these three examples.

identified is *Adh* in *D. melanogaster*. The HKA test was first applied to this locus to show that the ratio of polymorphism to divergence is higher for the coding region of the gene compared with the 5′ regulatory region, an observation that was interpreted as evidence for balancing selection (Hudson et al. 1987). Kreitman and Hudson (1991) later used a sliding-window approach to show that the peak of polymorphism was centered on the known fast/slow allozyme polymorphism in the third exon (Figure 7.4B). The hypothesis that this locus is under some form of balancing selection is further strengthened by the observation of clinal variation with latitude along the eastern coast of North America (Berry & Kreitman 1993). Interestingly, the MK test comparing nonsynonymous and synonymous substitutions within and between species of *Drosophila* reveals an excess of nonsynonymous mutations between species, suggesting that positive directional selection has fixed amino acid substitutions over evolutionary time. Thus, this locus serves as a good example of the potential complexity of selective forces. We see evidence for balancing selection in the recent history of *D. melanogaster* (based on patterns of variation within and between populations) as well as evidence for positive directional selection deeper in the history of *D. melanogaster* (based on patterns of nonsynonymous substitution between species).

Cyp6g1 in *D. simulans* and *D. melanogaster*

A remarkable case of parallel evolution at the molecular level is provided by studies of *Cyp6g1* in both *D. simulans* and *D. melanogaster*. Schlenke and Begun (2004) studied patterns of intraspecific variation and interspecific divergence at this locus and at neighboring loci. Their study was initially motivated by interest in another gene in the same region of the *D. simulans* genome. Schlenke and Begun sequenced eight inbred *D. simulans* lines from a single population from California. They surveyed 28 primarily noncoding regions, each about 900 bp in length, spanning over 3 Mb of the genome. Surprisingly, they discovered a region of approximately 100 kb that was completely devoid of polymorphism but that did not show reduced divergence between *D. simulans* and *D. melanogaster*. This suggests that the low level of polymorphism in *D. simulans* is not caused by a lower mutation rate. Their observation is extremely unlikely under a neutral model, but is consistent with recent, strong selection somewhere within this 100 kb region. Schlenke and Begun also surveyed a subset of the 28 loci in *D. melanogaster*. One locus, *Cyp6g1*, showed reduced variation in both species, but elevated levels of divergence between species, consistent with strong positive selection in the recent history of both species (Figure 7.4C).

Previous work by Daborn et al. (2002) showed that the insertion of an *Accord* transposable element upstream of *Cyp6g1* is polymorphic in *D. melanogaster* and is associated with higher expression of this gene. Moreover, the *Accord* insertion is associated with insecticide resistance in *D. melanogaster*. Remarkably, Schlenke and Begun found that the insertion of a different transposable element (*Doc*) was polymorphic in *D. simulans*, and was also inserted upstream of *Cyp6g1*. Just as in *D. melanogaster*, the insertion of the transposable element in *D. simulans* was associated with higher *Cyp6g1* expression. Finally, the *Doc* insertion in *D. simulans* seems to be associated with slightly stronger resistance to DDT insecticide. While the link to DDT resistance for *Cyp6g1* in *D. melanogaster* is strong (Daborn et al. 2002; Schlenke & Begun 2004), the link between DDT resistance and *Cyp6g1* expression in *D. simulans* is weaker, and the story appears more complicated. For example, Schlenke and Begun found that selection associated with the *Doc* insertion upstream of *Cyp6g1* in *D. simulans* was found in a California population, with no evidence of selection in a population from Zimbabwe, despite the fact that DDT is still used in Zimbabwe but has been banned from California for more than 30 years. They suggest that selection at *Cyp6g1* in *D. simulans* may have been caused by some other insecticide or contaminant (or even a natural toxin), with the *Doc* insertion conferring some cross-resistance to DDT.

Several lessons come from these studies. First, in both species it seems that adaptation resulted from the insertion of a transposable element upstream of a gene in a way that altered expression. In a remarkable example of parallel evolution, the same gene is involved in both species, although changes in expression were caused by the insertion of a different transposable element in each case. In *D. simulans*, this genomic region was studied out of interest in a completely different gene, and the discovery of adaptation at *Cyp6g1* was fortuitous. This represents one of the best examples of a scan for selection based on patterns of DNA sequence variation where the actual target of selection seems to have been identified and where functional consequences have been demonstrated. It is important to recognize, however, that the agent of selection is still unknown in *D. simulans*. This underscores the difficulty of making the complete link from genotype to phenotype to environment.

From Phenotype to Genotype: *Tb1* in Maize and Coat Color in Mice

Two unrelated examples provide case studies of a different approach, in which investigators started with a phenotype of interest and then used mapping and statistical analyses of DNA sequence variation to identify the causative genes and to better understand the nature of selection. In both cases, the agent of selection was known in advance, and thus the link between phenotype and environment was clear.

Doebley and colleagues (Doebley et al. 1997; Wang et al. 1999) have been studying the genetic basis of morphological evolution by studying the genes underlying the domestication of maize from teosinte, its wild ancestor. One major difference between these plants is that teosinte typically has long branches with tassels at the ends while maize has short branches with ears at the ends. Through mapping and cloning, Doebley et al. (1997) identified *teosinte branched1* (*Tb1*) as the gene that largely controls this difference. Wang et al. (1999) then sequenced the *Tb1* gene and the upstream nontranscribed 5′ region of this gene in a sample that included both maize and teosinte. In the transcribed region of the gene, they found that maize had 39% of the heterozygosity seen within teosinte, but in the 5′ region, maize had only 3% of the heterozygosity seen within teosinte (Figure 7.5A). Wang et al. (1999) performed HKA tests comparing

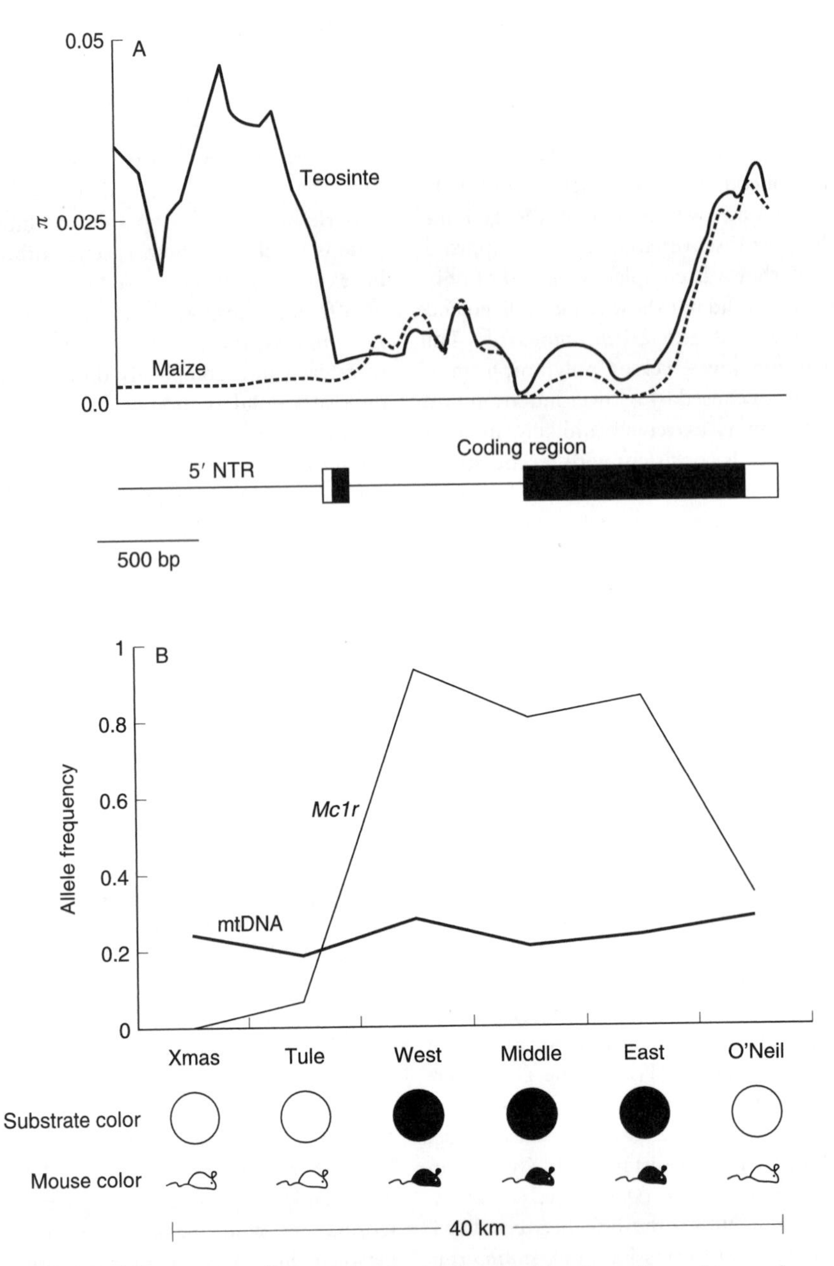

FIGURE 7.5. (A) Sliding window analysis of polymorphism in maize and in teosinte at the *Tb1* gene and at the 5′ regulatory region of *Tb1*. Maize and teosinte show similar levels of variability throughout the transcribed region, but maize shows drastically reduced polymorphism in the 5′ region, suggesting that an adaptive fixation occurred in the regulatory region of *Tb1* during the domestication of maize. Wang et al. (1999), reprinted with the permission of Nature Publishing Group. (B) Change in allele frequency at *Mc1r*, a gene underlying coat color in pocket mice, and for a presumably neutral locus across five populations (from Hoekstra et al. 2004). Three populations are on light-colored rocks and three populations are on dark lava, as indicated. Mitochondrial DNA(mtDNA) haplotypes show similar frequencies across these populations, while *Mc1r* alleles show dramatic changes in frequency, consistent with selection on this gene.

the untranscribed 5′ region with other, presumably neutral genes, and showed that the ratio of polymorphism to divergence at the 5′ region of *Tb1* was significantly reduced. This suggests that a beneficial mutation was fixed in the 5′ regulatory region of *Tb1* during maize domestication. This is also consistent with the observation that the maize *Tb1* allele is expressed at a higher level than the teosinte *Tb1* allele (Doebley et al. 1997). Remarkably, an HKA test between the 5′ region of *Tb1* and the transcribed region of *Tb1* was significant, showing that the effects of selection were limited to a narrow genomic region (about 2 kb). Wang et al. were able to use this observation, together with estimates of the recombination rate, to provide an estimate of the strength of selection on this gene during maize domestication ($0.04 < s < 0.08$) and the time over which selection drove this allele to fixation (315–1023 years). Finally, a phylogenetic analysis of maize and teosinte alleles in the 5′ region helped pinpoint the specific teosinte ancestor for maize, and suggested that maize domestication occurred in the Balsas river valley of southwestern Mexico. This example shows how statistical analysis of DNA sequences helped provide insight into the nature of selection for a trait whose genetic basis had been previously identified. It also provides a cautionary note. Wang et al. (1999) failed to find a fixed difference between teosinte and maize in the 5′ regulatory region of *Tb1*, suggesting either that the functional site lies upstream of the sequenced region, or that the genetic basis is more complex and involves interactions with other sites elsewhere in the genome.

Nachman and colleagues (Nachman et al. 2003; Hoekstra et al. 2004) have been studying the genetic basis of adaptive color differences in rock pocket mice (*Chaetodipus intermedius*). These mice inhabit rocky areas of the Sonoran desert. They are typically light-colored and live on light-colored rocks. However, in several different areas, melanic mice are found on dark rocks. The close match between the color of the mice and the color of the rocks on which they live is presumably an adaptation to avoid predation from owls and other vertebrate predators. Thus, this is a situation where the ecological importance of alternative phenotypes is reasonably well understood. The light color is ancestral and is geographically widespread, while the dark color is derived and is found on several different, geographically isolated lava flows. Nachman et al. (2003) used association studies with markers in candidate genes to identify the genetic basis of these adaptive differences. In one population in Arizona, allelic variation at the melanocortin-1-receptor gene (*Mc1r*) was found to be perfectly associated with color variation in the mice. One allele of *Mc1r* (the *D* allele) was distinguished by four amino acids from other *Mc1r* alleles and was only found in dark animals. Patterns of DNA sequence variation at *Mc1r* revealed that the D alleles were nearly devoid of genetic variation, suggesting that a selective sweep had recently driven this allele to its current frequency. This hypothesis of selection was supported by in vitro functional studies which showed that the *D* allele encodes a receptor with hyperactive function relative to the other *Mc1r* alleles. Increased activity of this receptor is known to be associated with dark color in laboratory mice. Thus the *Mc1r* gene appears to be responsible for adaptive melanism in one population of pocket mice in Arizona. Surprisingly, Nachman et al. (2003) found that similar melanic phenotypes in mice from populations in New Mexico were not caused by changes at *Mc1r*. Thus, population genetic studies revealed that adaptive dark color has evolved independently in this species through changes at different genes. In the Arizona population, Hoekstra et al. (2004) compared patterns of population differentiation at *Mc1r* with patterns of population differentiation at a presumably neutral mitochondrial DNA locus (Figure 7.5B) to estimate the strength of selection on the *Mc1r* D alleles ($s > 0.05$). This example thus represents a situation where the link from genotype to phenotype to environment is reasonably clear. Even in this situation, however, population genetic analyses cannot tell us which of the four amino acid mutations that comprise the *Mc1r* D allele is responsible for the phenotypic differences in color. In principle, this might be solved by introducing each mutation separately into the gene, and testing the function of these different receptors in vitro.

FUTURE DIRECTIONS

While we have made tremendous progress detecting selection at the molecular level in the last two decades, there are at least three clear directions for future work. First, additional studies that make the link between genotype and phenotype in an explicit ecological context are needed. Despite the extensive evidence for selection in different regions of the genome and in different species, there are still relatively few examples where the functional significance of allelic variants is well understood in a particular environmental setting. Functional studies that dissect the biochemical consequences of genetic variation, combined with phenotypic studies of the fitness effects of functional differences, will help us to fully understand the evolutionary significance of genetic variation. Such studies will also help make the connection between ecological and evolutionary timescales.

A second area of needed research concerns selection on regulatory regions. Adaptation may often result from changes in gene regulation rather than changes in gene structure, yet most studies of selection at the molecular level have focused on coding regions of genes. One difficulty is that in many cases the specific regulatory elements of genes are unknown. Empirical studies to identify regulatory elements and theoretical studies to develop appropriate statistical methods for analyzing regulatory sequences will provide real insight into the genetics of adaptation.

A third area in which future work is likely to be especially rewarding are studies that take full advantage of complete genome sequences. We now have the potential to interrogate every gene in the genome to ask questions about selection. For example, it will soon be possible to perform MK tests on every gene in the *D. melanogaster* or human genome. In principle, such data should allow us to ask questions about the relative frequency of different kinds of selection, about the kinds of genes that are under selection, and about the amount of selection in the recent history of a species. New theoretical and statistical work will be needed to overcome the inherent problems associated with multiple tests, but these studies have the potential for the first time to give us a detailed genome-wide view of the genetics of adaptation.

SUGGESTIONS FOR FURTHER READING

A good review of basic concepts is provided by Kreitman and Akashi (1995). Ford (2002) provides a comprehensive review of specific examples where selection has been detected at the molecular level. There are two recent reviews that focus specifically on humans: Bamshad and Wooding (2003) and Vallender and Lahn (2004). Luikart et al. (2003) provide a nice overview of studies that use genome-wide data to detect selection.

Bamshad M & SP Wooding 2003 Signatures of natural selection in the human genome. Nat. Rev. Genet. 4:99–111.

Ford MJ 2002 Applications of selective neutrality tests to molecular ecology. Mol. Ecol. 11:1245–1262.

Kreitman M & H Akashi 1995 Molecular evidence for natural selection. Annu. Rev. Ecol. Syst. 26:403–422.

Luikart G, England PR, Tallmon D, Jordan S & P Taberlet 2003 The power and promise of population genomics: from genotyping to genome typing. Nat. Rev. Genet. 4:981–994.

Vallender EJ & BT Lahn 2004 Positive selection on the human genome. Hum. Mol. Genet. 13:R245–R254.

Acknowledgments I thank David Begun for discussion and Jeff Good for comments on this chapter.

8

Rates of Molecular Evolution

FRANCISCO RODRÍGUEZ-TRELLES
ROSA TARRÍO
FRANCISCO J. AYALA

Changes in DNA that give rise to new sequences are called mutations. Mutational changes are the primary source of variation and novelty in evolution. Mutations are caused by the exchange of one nucleotide for another, by insertion or deletion of one or more nucleotides ("indels"), and by reorganization of the genetic material. Mutations arise constantly, irrespective of their effect on the fitness of the organism that carries them. Yet, mutations do not occur at random with regard to genomic location, nor do all types of mutations occur in equal frequencies. A new mutant DNA sequence arising in an individual may be lost or spread and even become fixed in a species by genetic drift (i.e., random sampling errors on allele frequencies that occur as genes are transmitted from one generation to the next in finite populations), and/or natural selection. The process whereby a mutant replaces alternative sequence variants, or alleles, in a species is called "substitution"; substitutions that change the encoded amino acids are named "replacements."

CONCEPTS

Number of Substitutions

The process of nucleotide substitution can take up to thousands or millions of years to complete. To detect evolutionary changes in a DNA sequence, it is necessary to compare two or multiple homologous sequences, which are sequences that have descended from a common ancestral sequence. Homologous sequences from different species often differ in length, as a result of insertions and deletions that might have occurred in either of the lineages since their divergence from a common ancestor. To compare homologous sequences it is first necessary to identify the location of the indels by means of sequence alignment. Sequence alignment methods work by opening gaps (i.e., stretches of one or more null bases usually denoted by hyphens) in the sequences so as to maximize the similarity between paired sequences at each site, while minimizing the number and extension of the gaps invoked. A penalty needs to be imposed for each gap, so as to avoid the trivial solution of obtaining a perfect match between two sequences by the expeditious procedure of introducing enough gaps in each sequence so that each nucleotide associates with an identical nucleotide in the alternative sequence.

The observed number of nucleotide substitutions is the most basic and commonly used variable in molecular evolution. It embodies two parameters of critical import: the total number of nucleotide substitutions, and the pace at which they accumulate. The number of nucleotide substitutions is critical to disentangling the *pattern* of molecular evolution (i.e., the branching history of lineage diversification) because it reflects the degree of relatedness between the sequences. The *processes* underlying this pattern are concealed in the second parameter, the rate of molecular evolution.

Reliability of the number of substitutions hinges upon the premise of positional homology (i.e., a claim that aligned nucleotides at a given site descend from a common ancestral nucleotide), which highlights the importance of accurate alignment for the study of molecular evolution. Exact methods that guarantee the discovery of all optimal alignments for more than a few sequences demand enormous computational time. Instead, heuristic (i.e., approximate), less computationally demanding methods, are used (e.g., Jeanmougin et al. 1998). Different assumptions with respect to the penalties associated with gaps, degree of similarity between character states, phylogenetic relationships among sequences, and so forth, can produce quite different alignments; in addition, alignment efficiency can greatly vary across different parts of an alignment, being lowest in regions with the highest level of divergence. Because amino acid sequences always diverge more slowly than their encoding DNA sequences, alignment of protein-encoding nucleotide sequences is preceded by the alignment of their protein products as a guide. Several criteria have been devised to deal with regions of uncertain positional homology (e.g., Castresana & Moreira 1999; Löytynoja & Milinkovitch 2001).

When the degree of divergence is substantial, the observed number of sequence differences in an alignment may underestimate the number of evolutionary substitutions that actually happened. This is because the number of possible states is only four for each nucleotide site and 20 for each amino acid site, so that a site that is identical in two sequences may have experienced numerous evolutionary substitutions but has eventually converged to the same state. To circumvent this multiple-substitutions or "multiple-hits" problem, it is necessary to make assumptions regarding the probability that each nucleotide is replaced by each alternative nucleotide. For DNA sequences these assumptions are mathematically expressed as a 4×4 matrix or **Q** matrix, in which each element Q_{ij} represents the probability of change from base i to base j. Models of substitution have been developed that allow explicit accommodation of complex substitution patterns, such as unequal frequencies of the four nucleotides, unequal frequencies of the different substitution types (for example, transition:transversion biases), and unequal substitution frequencies across sites. Substitution models have also been developed for the description of the substitution process at the codon (61×61 **Q** matrix) and amino acid (20×20) levels (see Yang 1997). The models apply to (1) homogeneous stationary substitution processes, in which nucleotide (or codon or amino acid) frequencies and substitution probabilities between nucleotide states are assumed to have remained the same across lineages and through evolutionary time, as well as to (2) nonhomogeneous nonstationary substitution processes.

Parameter values of the substitution models can be efficiently inferred by means of maximum likelihood methods. Given an arbitrary substitution model and a tree (i.e., a hypothesis about the phylogenetic relationships among the species), the statistical method of maximum likelihood chooses amongst competing sets of parameter values of the substitution model by selecting the one that makes the observed data (i.e., the sequences being compared) the most likely evolutionary outcome. Obtained quantities are the maximum likelihood estimates of the parameters. Maximum likelihood methods have the advantage over pair-wise distance methods that they compare all the sequences simultaneously taking into account previous knowledge about their phylogenetic relationships. To avoid incurring potential circularities, assumed trees should ideally come from data sources that are independent of the sequence information used to estimate the parameter values. Maximum likelihood methods are also superior to parsimony methods, which do not account for multiple hits (such as two successive substitutions at the same site, e.g., $A \rightarrow G \rightarrow C$).

Choice of a reasonably realistic substitution model is critical to obtain accurate estimates of the true number of nucleotide (or amino acid) substitutions. Simplistic models, which neglect relevant features of the data, do not correct properly for multiple substitutions and often lead to highly biased conclusions (see Tarrío et al. 2001). At the other extreme, overparameterized models yield estimates with inflated variances and are unduly expensive computationally. Statistically objective criteria, such as maximum likelihood ratio or Akaike information statistics (see Kishino & Hasegawa 1990; Posada & Crandall 1998), provide a rationale for choosing between competing representations of the substitution process. These methods assume that the phylogenetic relationships among the species are known. The likelihood ratio test is for the comparison of two models, of which one is a more general version of the other. The constrained model, which represents the H_0 or null hypothesis, is said to be nested within the more general model (H_1). The test yields significant results when the more general model provides a better description of the evolution of the sequences than the constrained model. The Akaike information criterion (AIC) does not assume nesting of the models, and penalizes models that have more parameters than necessary. Unlike the above methods, Bayesian methods produce results that are not conditional on the correctness of the assumed tree topology. Bayesian inference is currently becoming a computationally more feasible approach than maximum likelihood for examining complex evolutionary models (Shoemaker et al. 1999; Huelsenbeck et al. 2001; Aris-Brosou & Yang 2003).

The number of substitutions between two homologous sequences is determined by the product of the rate of substitution (i.e., the number of substitutions per unit time) and the length of time the sequences have been diverging. Knowing the divergence dates allows calculating the absolute rates at which the observed numbers of substitutions accumulated in different sequences. Yet the age of the sequences is usually unknown, so that an observed number of substitutions can be large either because it reflects a long period of evolutionary time or because the rate of substitution has been high. Molecular evolution studies have to rely on relative rates, i.e., comparisons of substitution numbers between sequences that started to diverge at the same time, regardless of the date. In general, it is impossible to tease rate and time apart, unless one is willing to assume that the function linking substitution number and time is known. If that were indeed the case, the number of substitutions would become a universal tool, not only for exploring the mechanisms and processes of evolution, but also for reconstructing phylogenetic relationships among species and timing the evolutionary events, in an analogous way to the dating of geological times by measuring the decay of radioactive elements.

Rate of Molecular Evolution

In their pioneering work, Emile Zuckerkandl and Linus Pauling (1965) compared protein sequences from hemoglobins derived from several divergent vertebrates. They plotted the number of amino acid replacements between the protein sequences against the species age inferred from fossil evidence. Their study unveiled a striking constant rate of accumulation of amino acid changes over evolutionary time. Later, similar observations were reported for several other proteins and RNA molecules in various taxa. This constant rate was unexpected because of the variable rates typically observed in morphological evolution among species and through evolutionary time. The constancy of molecular evolutionary rates led Zuckerkandl and Pauling to the formulation of the hypothesis of the molecular clock of evolution (Ayala 1986). This hypothesis asserts that the number of amino acid replacements (or nucleotide substitutions) between species increases "linearly" with the time elapsed since they diverged from their last common ancestor. The time of remote events, as well as the branching relationships among lineages leading to contemporary species for which fossil information was lacking, could be determined on the basis of amino acid (or nucleotide) differences (Box 8.1). A notable feature of the hypothesis of the molecular evolutionary clock is multiplicity: every one of the thousands of proteins or genes of an organism is an independent clock, each ticking at a different rate but all measuring the same historical events.

A theoretical foundation for the clock was provided by Motoo Kimura, who developed a "neutral theory of molecular evolution," endowed with great mathematical simplicity. Notably, the theory states that a diploid species of N_e individuals (where N_e is the effective population size, or the fraction of individuals that produce progeny) with a neutral mutation rate per year, μ_y, produces mutations at a rate $2N_e{\cdot}\mu_y$. The probability that a neutral mutation eventually becomes fixed is equal to its frequency in the species at the time it arises, that is, $1/2N_e$. Thus, the rate of substitution is $k_y = 2N_e{\cdot}\mu_y{\cdot}(1/2N_e) = \mu_y$. For a protein with a proportion f_0 of sites that are selectively neutral the neutrality theory predicts that the rate of molecular evolution in species, k_y, is simply equal to the rate of production of neutral mutation in individuals, $\mu_y f_0$.

Adaptively neutral mutations are those whose absolute selective values are smaller than the reciprocal of the effective population size. If so, natural selection will not play a significant role in determining the ultimate fate of the alleles (elimination or fixation). The remaining ($\mu_y(1 - f_0)$) mutations, including deletereous and favorable changes, would not contribute to estimates of the rate of molecular evolution because deleterious mutations would be rapidly removed or kept at very low frequencies by purifying selection, and favorable, positively selected changes would be very rare and would not have a measurable effect on the overall evolutionary rates of nucleotide and amino acid substitution. If the underlying mutation rate is

BOX 8.1. Timing Evolutionary Events with a Molecular Clock

The basic idea underlying molecular dating procedures is to convert numbers of substitutions between sequences into time, on the basis of a reference substitution rate (i.e., the number of substitutions expected per unit time). Reference rates used for extrapolation are calibrated using evolutionary events for which dates are known from the fossil record. The simplest approach is as follows. Suppose three orthologous proteins are related as in Figure 1. Let us assume that the average number of amino acid replacements per site between A and B is $K_{AB} = 1$, and that C differs from either A or B by $K_{AC} = K_{BC} = 10$. Also, it is known from the fossil record that A and B split from a common ancestor (t_C) 100 Myr ago. Thus, the absolute rate of molecular evolution between A and B would be $r_{A+B} = K_{AB} / [\ 2\ (100 \times 10^6)] = 5 \times 10^{-9}$ replacements per site per year. If we assume that r_{A+B} is equal to the rate between C and AB, the estimated divergence time between C and AB would be $t_T = [(K_{AC} + K_{BC}) /4] / 5 \times 10^{-9} = 1000$ Myr. Critical steps of this procedure are the accuracy with which the number of substitutions is known, and the calibration of the reference rate. Recent methodological developments, including maximum likelihood and Bayesian approaches, allow relaxing the rate constancy assumption for extrapolating from the reference rate to the estimated rates of divergence (see text).

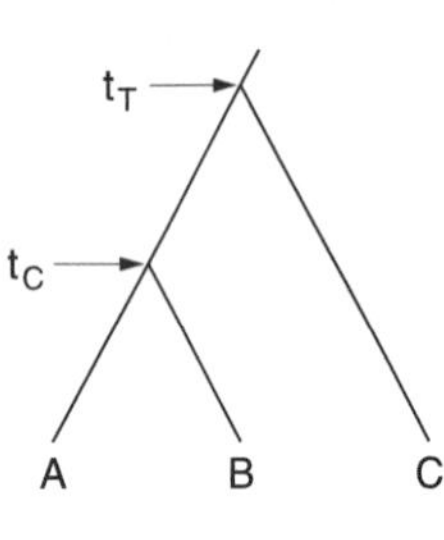

FIGURE 1

uniform and the function of the protein does not change over time, so that the proportion of sites at which mutations are neutral remains the same across lineages and evolutionary time, then the rate of molecular evolution should be constant, independent of factors such as generation time, living conditions, and the population size. That the rate of neutral evolution is independent of population size fluctuations might be intuitively understood considering that the expected number of neutral mutations in a growing population increases approximately as the reciprocal of the average time required for a neutral mutation to reach fixation (specifically, $4N_e$ generations in diploid populations; Kimura & Ohta 1969). In large populations, there occur more neutral mutations, but it takes longer to reach fixation, so that the two effects cancel out. According to the neutrality theory, f_0 may vary among proteins, which would account for different molecules (different proteins or different genes) having different rates of substitution; that is, the greater the proportion of neutral sites (or the lower the degree of functional constraint), the faster the rate of molecular evolution.

The neutrality theory differed from previous accounts that had assumed that most amino acid changes would be either favorable or deleterious. The neutrality model advanced definite quantitative predictions regarding the amount and patterns of genetic variation within species and between different species, in this way providing a null model to test for the occurrence of natural selection (as a departure from the expectations of neutral evolution). The neutrality theory established a clear quantitative connection between functional constraint and the rate and pattern of molecular evolution. Functionally important amino acid sites would not be those that evolved between species, but rather those that remained constant through evolutionary time. This link between sequence conservation and function provided the ground for the development of comparative sequence approaches as powerful means for addressing positional homology inferences.

It may be worth recalling that the hypothesis of the molecular clock was elicited by the early observation of proteins that happened to evolve at approximately constant rates per unit of chronological time in the lineages under consideration. Mutations, however, occur on a per-generation, rather than in a per-chronological-time basis. Consequently, species with shorter generation times should produce greater numbers of neutral mutations per year, provided the number of germline cell divisions per gamete is constant across taxa, and hence should exhibit faster rates of molecular evolution than species with longer generations. Kimura accommodated the apparent absence of a generation-time effect on the rate of protein evolution by assuming that neutral mutations occur at a constant rate per year throughout the different evolutionary lineages, regardless of the length of their generation time (Kimura 1968; Kimura & Ohta 1971b). The apparent constancy of the rate of protein evolution was embraced by Kimura as strong evidence supporting his neutrality theory of evolution.

Using data based on the DNA hybridization method, Tomoko Ohta (1972) soon found that nonprotein-coding DNA diverged proportionally (as the inverse) to the number of generations, rather than proportionally to chronological time. Because noncoding DNA had been characterized, par excellence, as neutral, Ohta's finding posed a serious problem to the neutrality theory, which prompted her to develop the "nearly neutral" theory of molecular evolution.

The nearly neutral theory represents an important modification of the neutrality theory. It is predicated on the empirical fact that generation times and population sizes tend to be inversely related. As originally conceived, the nearly neutral theory proposes that there is a continuum of allelic effects, from deleterious through slightly deleterious to neutral mutations, rather than only two categories of mutants, deleterious and neutral, as proposed by the neutral theory (Ohta & Kimura 1971). Slightly deleterious mutants become effectively neutral when their selective disadvantage drops below $1/N_e$, the reciprocal of the population size. Organisms with small effective population sizes, and therefore large proportions of effectively neutral mutations, tend to have long generation times, and hence low substitution rates, and vice versa. Specifically, if the distribution of selection coefficients against mutations follows an exponential distribution, the effects of generation time and effective population size might exactly cancel each other out. The cancellation between generation time and population size was later shown to ensue from other (nonexponential) distributions of fitness effects as well (Ohta 1992). If the counteracting effects of generation time and effective population size occur for all lineages through evolutionary time, it follows that constant neutral mutation rate per generation will result in constant neutral rate of replacement per year. On the other hand, the divergence of the noncoding fraction of the DNA would manifest a generation time effect because mutations in this fraction would be strictly neutral.

The molecular clock is not deterministic, but stochastic. This means that, although the underlying rate of substitution is constant, the duration of time between consecutive substitutions ("ticks") is not uniform, like in a metronome, but fluctuates randomly, such as in radioactive decay, reflecting the fact that the genetic drift of random mutations is in itself a random sampling process. The inherent irregularity of the substitution process imposes substantial uncertainty on molecular dating because, even under a perfect clock, it is not possible to know precisely the time taken to produce the observed number of substitutions. The simplest neutral model predicts that the number of substitutions per unit time has the properties of a Poisson distribution, in which the variance of the number of substitutions among lineages is expected to be equal to the mean of the number of substitutions. If all species diverge at the same time and accumulate substitutions at the same rate since their separation, the variance-to-mean ratio of the number of substitutions among lineages should not be significantly different from one. The "index of dispersion," measuring the deviation of this ratio from the expected value of 1, is a way to test whether observations fit the theory (Kimura 1983; Gillespie 1989). Empirical results involving numerous genes and taxa have shown that the index of dispersion is often significantly greater than 1. In the case of synonymous substitutions, this outcome has been explained as being the result of lineage effects, such as effective population size, generation time, metabolic rate, and other lineage-specific factors that affect all nucleotide sites equally (see below). Lineage effects, however, are insufficient to account for the dispersion of the rate of amino acid replacement. The observed inflated variance of the rate of protein evolution invalidates the extrapolation of the earliest studies based on a few proteins, and has been referred to as the "overdispersed molecular clock" (Gillespie 1989). The claim of overdispersion has been contested on the ground that statistics based on the dispersion index are overly sensitive to uncertainties of the phylogenies assumed in order to perform the tests (e.g., Goldman 1994).

The controversy about the magnitude of the dispersion index has raised two additional issues. One is the concern about the appropriateness of the Poisson model as a descriptor of the process of amino acid replacement, and how this could affect the validity of the neutrality theory. The other issue is the applicability of the hypothesis of the molecular clock. The assumption that rejection of the Poisson model led to refutation of the neutrality theory prompted an extension of the nearly-neutral theory of molecular evolution (Ohta 1972, 1973, 1995).

In the extended version of the nearly neutral theory, the equilibrium substitution rate that results from the counteracting effects of generation time and effective population size is allowed to vary from one to another lineage through evolutionary time (Ohta 1973, 1995). In small populations, subjected to larger stochastic fluctuations in allele frequencies, genetic drift would overcome natural selection for nearly neutral mutations. Lineages that experience pronounced reductions in effective population sizes would undergo a greater rate of fixation of slightly deleterous alleles, thereby exhibiting a greater rate of molecular evolution. In this way, the nearly-neutral theory can account for the excess variability in substitution rates inferred from overdispersion patterns. From a theoretical, as well as operational, perspective, the nearly neutral theory and other supplementary hypotheses have the discomforting feature of involving additional parameters, often not easy to estimate. It is of great epistemological significance that the original proposal of the neutrality theory is (i) highly predictive and, therefore, (ii) eminently testable. The two properties, really two sides of the same coin, become diluted in modified versions of the theory (Ayala 1997).

Recent studies have shown that allowing base substitutions to occur according to more complex stochastic processes than the Poisson process produces an increase in the variability of substitution rates (Zheng 2001). The rate of substitution can theoretically be both constant and overdispersed with respect to the Poisson expectation. It has, thus, been argued that rejection of the simple Poisson process has no bearing on the validity of the neutrality theory (Takahata 1987; Zheng 2001; Bastolla et al. 2003).

The issue of the molecular clock (predicated on the constancy of the rate of substitution) is more rigorously approached with relative-rate tests and likelihood-ratio tests (Box 8.2) than with tests based on the index of dispersion. Use of these procedures with more realistic models of substitution than the Poisson model has shown that the molecular clock is contravened for most genes and species groups, except for sequences from closely related species.

Decomposing the Rate of Molecular Evolution

The number of substitutions is impacted by three different effects: locus, lineage, and locus × lineage interactions. The causes underlying locus and lineage effects operate consistently in the same direction through evolutionary time, giving rise to definite rate-substitution patterns. However, locus × lineage interactions often change unpredictably, causing the number of substitutions to vary erratically from one locus to another and across evolutionary lineages.

Locus Effects

Early in the study of molecular evolution, it was realized that rates of substitution vary across different nucleotide positions (e.g., average substitution rates at the first (r_1), second (r_2), and third (r_3) codon positions are almost always in the order $r_2 < r_1 < r_3$; Figure 8.1), different genes (e.g., fibrinopeptides evolve faster than histones), different genomic regions (typically, protein-coding regions evolve slower than "junk" DNA regions), or different genomes within an organismal lineage (for example, mitochondrial genomes evolve faster than nuclear genomes in mammals). The phenomenon of different sites exhibiting different rates of change is globally referred to as among-site rate variation. The ubiquity of among-site rate variation means that molecular evolutionary rates cannot be extrapolated from one locus to another. Among-site rate variation has nothing to do with the concept of overdispersion, such as was defined by Gillespie (1989), which refers to the variation in the rate of evolution not attributable to gene or lineage effects.

The rate of substitution has been recently partitioned into gene, lineage, and gene-by-lineage interaction effects by means of analysis of variance (ANOVA) for mammalian nuclear and mitochondrial genes (Smith & Eyre-Walker 2003). Gene effects account for most of the variability of the rate of nonsynonymous substitutions (replacements) in both nuclear (63.7%) and mitochondrial (44.3%) genes, but are relatively unimportant for synonymous substitutions. Presumably, a large fraction of the locus effect on the mammalian rate of nonsynonymous substitution is attributable to among-site rate variation within genes. Although mutation

BOX 8.2. Testing the Hypothesis of the Molecular Clock

Methods of molecular dating that rely on the premise of the molecular clock require that the sequence data conform to the rate constancy assumption. A number of statistical tests have been devised to evaluate the molecular clock assumption. The simplest method is the relative-rate test, which is based on comparison of pair-wise sequence distance estimates. Suppose that we have three species (or sequences), A, B, and C, as in Figure 1. If species A evolved at the same rate as species B, then the distance (i.e., expected number of substitutions) from ancestral node 0 to the present should be the same for species A and B; that is $d_{0A} = d_{0B}$. The ancestor 0 may be unknown, but this handicap is circumvented using a third species, called "outgroup," known to have branched off before the split A–B (species C in Figure 1). The null hypothesis (H_0) then becomes $d_{AC} = d_{BC}$, which can be tested using the statistics $d = (d_{AC} - d_{BC})$ by means of a Z-test, using the variances of the estimated pair-wise sequence distances in order to obtain the standard error of the test statistic. Relative-rate tests have been extended to comparisons between two groups of species. The relative-rate method does not test whether the outgroup species has a different rate from the two ingroup species. The method fails when the third species is not a true outgroup, and it should not be used with remotely related outgroup species, because the large amount of evolution between C and 0 will overwhelm true rate differences between A and B.

An increasingly common method to evaluate the molecular clock hypothesis is the likelihood-ratio test. Unlike the relative-rate test, the likelihood-ratio test is applicable to any number of species. Provided that the phylogenetic relationships of the species are known, the likelihood-ratio gives a criterion to decide whether a model of evolution (H_1, or alternative hypothesis) that allows lineages to evolve at different rates, provides a better description of the observed data (i.e., likelihood score L_1) than a special case of it (H_0) that constrains lineages to evolve at the same rate (i.e., likelihood score L_0). For nested hypotheses, the likelihood-ratio statistic, with the form $2\Lambda = 2(L_1 - L_0)$, is approximately distributed as a chi-square random variable with $n - 2$ degrees of freedom, n being the number of species. A 2Λ value greater than the critical value indicates that there is more rate variation than expected if all lineages evolved at equal rates.

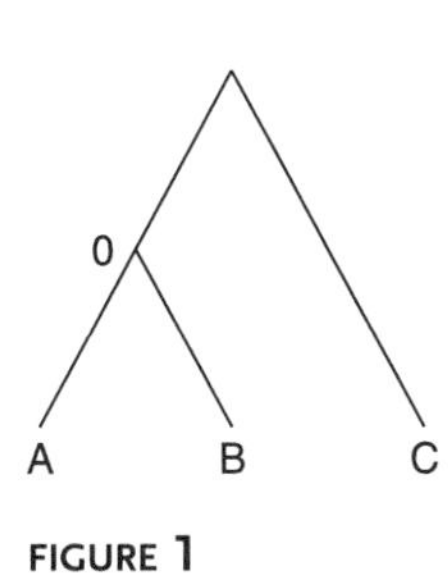

FIGURE 1

Caution needs to be exercised in interpreting the results of molecular clock tests. Failure to reject the clock hypothesis does not necessarily mean that lineages evolve at a constant rate. For instance, molecular clock tests such as those just described cannot detect variations of rate within a lineage. In addition, molecular clock tests generally have low statistical power, which decreases with the length of the sequences and the number of variable sites. In most typical data sets, relative rate tests fail to detect rate variation between lineages even when the rate of one lineage is 4 times the rate of the other (see text).

rates may vary from one site of a gene to another, nonsynonymous site-specific substitution rates are more likely the result of different selective constraints at different sites determined by the functional requirements of each gene or protein.

Among-site rate variation is a critical feature that must be considered when modeling the process of nucleotide substitution. Heterogeneity in substitution rates among sites can be accommodated in substitution models using the discrete gamma approximation (Yang 1996a). This approach is based on early observations that the distribution of sites, grouped according to their number of substitutions, fits a negative binomial function (Uzzell & Corbin 1971). The shape of the gamma distribution is summarized by a single parameter denoted as α. The value of α is inversely related to the extent of among-site rate variation. When $\alpha = 1$ the distribution of rates across sites is L-shaped, meaning that most sites exhibit low substitution rates or do not change at all, while a few sites evolve very fast. When $\alpha > 1$ the distribution of rates is bell-shaped, meaning that most sites evolve at intermediate rate, while only a few sites change either very slowly or very fast (Figure 8.1). In the absence of positive selection, α can be interpreted as a measure of functional constraint (Zhang & Gu 1998), so that highly constrained, slowly evolving genes, such as histone genes, yield lower α values, whereas unconstrained, fast-evolving genes, such as pseudogenes, yield greater α values.

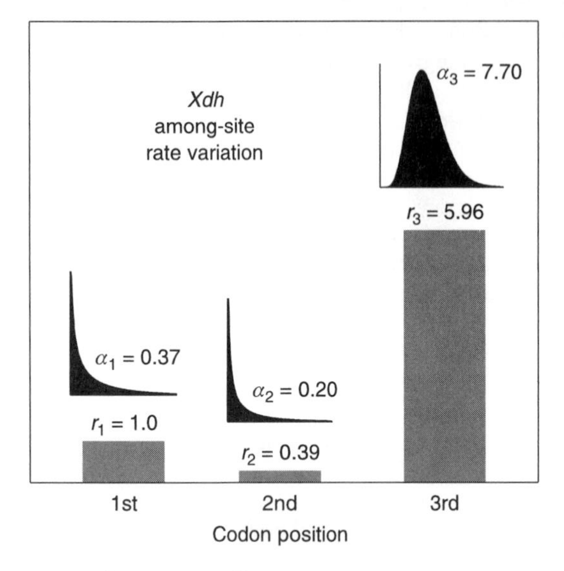

FIGURE **8.1.** Locus effects. Diagrammatic representation of the rate variation among nucleotide sites of the *xanthine dehydrogenase* locus (*Xdh*) for several *Drosophila* species. Among-site rate variation is partitioned into two main components: (i) a fixed component, or the average effect of the codon position (i.e., r_1, r_2, and r_3, for first, second, and third codon positions, respectively); and (ii) a random component, or the rate variation among sites of the same codon position (i.e., α_1, α_2, and α_3). The random component is modeled using the discrete gamma-distribution approach of Yang (1996), assuming eight categories of rates. Second codon positions exhibit both the slowest rate of evolution (r_2 = 0.39 substitutions per site, versus r_1 = 1.00, and r_3 = 5.96) and the highest among-site rate variation (α_2 = 0.20, versus α_1 = 0.37, and α_3 = 7.70), reflecting that they are subjected to the greatest functional constraints, which change widely from one site to another. Histogram bars are scaled proportionally to evolutionary rates of the average first, second, and third codon positions. Gamma distributions above the bars represent the rate variation among sites pertaining to a given codon position. Modified from Rodriguez-Trelles et al. (1999).

Assuming that substitution rates are constant among sites can lead to erroneous phylogenetic relationships and their timing when they are not. For instance, ignoring among-site rate variation causes underestimation of the actual number of substitutions at fast-evolving sites; the bias is more pronounced for greater evolutionary distances, because more multiple substitutions are expected when the sequences have been diverging for a longer time. This causes molecular clock extrapolations to overestimate dates that are younger than the reference time, and to underestimate dates that are older than the reference time.

Choosing the right gene is a principle of economics in molecular systematics. Characters are phylogenetically informative when their rate of evolution matches the relevant divergence times. Widespread occurrence of variation in the average substitution rates among genes has proven very useful for reconstructions of different parts of the tree of life, because genes (or proteins) can be chosen for inferring phylogenetic relationships at the appropriate level of phylogenetic divergence. For instance, fast-evolving genes, such as animal mitochondrial genes, are useful for resolving recent separations (say below the order category for mammals), but are inappropriate for disentangling branching patterns at deeper divergence times, which are better addressed using ribosomal or other slowly evolving nuclear genes (Graybeal 1994).

Lineage Effects

Widespread application of relative rate tests and likelihood ratio tests has revealed that different lineages evolve at different rates of change (Box 8.2). For example, substitution rates in murid rodents (rats, mice, and hamsters) are, on average, 2 to 3 times faster than the rates in hominid primates (apes and humans) (Li et al. 1996); within the *Sophophora* subgenus of the genus *Drosophila*, the species of the *saltans* and *willistoni* groups accumulated A and T nucleotide bases at a significantly faster rate than taxa of the *melanogaster* and *obscura* species groups (Figure 8.2) (Rodriguez-Trelles et al. 1999, 2000; Bergman et al. 2002). In contrast, the variation in nonsynonymous substitution rates is

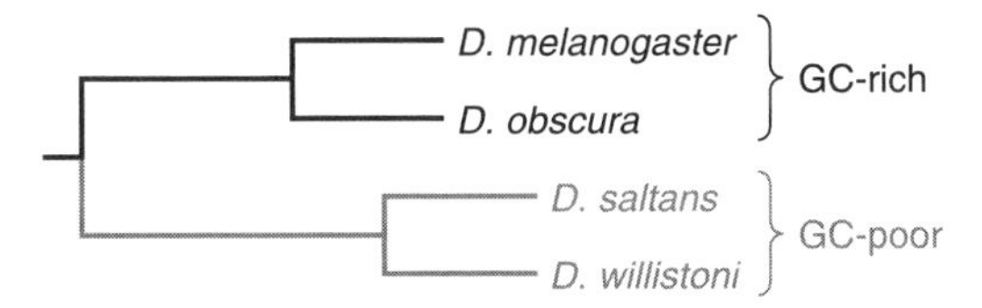

FIGURE **8.2.** Lineage effects. Differences in the rates of amino acid replacement among several lineages of the *Sophophora* subgenus of the genus *Drosophila*. The species of the *D. saltans* and *D. willistoni* groups (gray) evolve faster than the *D. melanogaster* and *D. obscura* species groups (black). Differences in the rate of amino acid replacement between the species groups are caused by a shift toward increased AT content of the genome, a process started in the last common ancestral lineage of the *saltans* and *willistoni* groups, after its split from the lineage that gave rise to the *melanogaster* and *obscura* groups (Rodriguez-Trelles et al. 1999, 2000; Bergman et al. 2002).

dominated by gene effects. Lineage effects are the main factor accounting for the variation in synonymous substitution rates in mammalian nuclear (36.5%) and mitochondrial (50.3%) genes (Smith & Eyre-Walker 2003).

Substitution rate differences among lineages have been explained on several not mutually exclusive grounds:

1. Generation-time differences, because lineages with shorter generation times replicate their germline DNA more frequently and, thus, accumulate more copy-error mutations per unit time than species with longer generations; in addition, shorter generation times accelerate the rate of substitution per unit time, because the time to fix new mutations is shortened. This hypothesis accounts for the observation that various genes evolve faster in rodents than in cows or primates, which have longer generation times. In this respect, the rate of substitution might better be measured as a per generation rather than a per unit time rate, in particular for the majority of animals, which have separate germline and somatic cells. The relationship between generation time and replacement rate might be obscured if the number of germline cell divisions per gamete changes from one taxon to another.
2. Differences in population size, because the nearly-neutral theory predicts that the rate of fixation of slightly deleterious mutations should increase with decreasing population size. It has been suggested that population size and generation time are often inversely related (there are more mice than elephants), which thus might partially cancel out, and yield approximate constant rates over time (see above). But this inverse relationship can hardly be very general, because organisms with short generation times may range several orders of magnitude in population size. Some *Proechimys* spiny rats and *Spalax* moles have narrow ecological niches and much smaller population sizes than the cosmopolitan house mice and common rats; the population size of the widespread *D. willistoni* is surely more than six orders of magnitude greater than that of *D. insularis*, which is confined to two small Caribbean islands. Population sizes may, however, have been very different in the past.
3. Species-characteristic differences in polymerases or other biological properties that affect the fidelity of DNA replication, and hence the incidence of mutations, neutral or not. For example, murid rodents have less efficient excision-repair mechanisms than humans (Hart & Setlow 1974), which is consistent with the higher rate of molecular evolution in murid rodents than in hominoids. Animal mitochondria are the site of oxidative phosphorylation, which might explain why mitochondrial DNA evolves faster than nuclear DNA, if enhanced metabolic rate is associated with greater concentrations of mutagenic oxygen radicals that are by-products of cellular respiration. Because the mutation rate is affected by the properties of the DNA replication-repair system, it will be affected by changes in this system caused by mutations that alter its efficiency. If variation in mutation rates among individuals is heritable and associated with fitness, then different species might evolve different optimum mutation rates in response to natural selection, that is, different balances between the costs of enhanced accuracy and repair, and the costs of deleterious mutations. The observation that different lineages often exhibit different characteristic substitution rates has caused many to abandon the concept of a universal molecular clock in favor of the existence of

local molecular clocks. Related species with similar life histories, generation times, metabolic activities, DNA replication-repair systems, and so forth, have similar rates of evolution that are often very different from those of distantly related species with very different biological features. Assuming a global molecular clock when there are substitution rate differences among lineages may lead to spurious date estimates.

Molecular dating studies have addressed lineage effects in different ways. In some cases analyses have been exclusively circumscribed to molecules that pass some global molecular clock criterion, such as the relative rate test (e.g., Kumar & Hedges 1998). Other studies have opted for excluding the taxa that evolve significantly faster or slower than the average rate as determined, for instance, with the linearized tree method (Takezaki et al. 1995). These or any other sorting approaches can potentially bias date estimates. For example, common global molecular clock tests have low statistical power (as pointed out earlier, the relative-rate test may miss rate variation between lineages in cases where the rate of a lineage is as much as 4 times the rate of another; Bromham et al. 2000), which decreases with the length and variability of the sequences. Using the relative-rate criterion will favor data sets in which short and/or slowly evolving sequences are overrepresented.

If the substitution process is intrinsically overdispersed, then it is inappropriate to circumscribe the analyses to loci that fit the global molecular clock criterion. Alternatively, lineage effects can be addressed by relaxing the global clock assumption (Sanderson 1997; Rambaut & Broham 1998; Thorne et al. 1998; Huelsenbeck et al. 2000; Yoder & Yang 2000; Aris-Brosou & Yang 2002). In this way, predictions of time are made by calibrating the rate of evolution separately for each gene in each taxonomic group. One approach is to construct local molecular clock models, where independent evolutionary rates are assigned to some lineages while all the other lineages are assumed to evolve at the same rate. The fit of a local clock model can be evaluated by comparison with that of its corresponding simpler global clock model using a likelihood ratio test. This approximation is appropriate when the lineages that evolve at different rates can be identified beforehand. However, if the hypothesis about rates is derived from the same sequence data used for testing the hypothesis, the method becomes too liberal, with the null model being rejected more frequently than expected on the basis of statistical significance level. In general, date estimates are sensitive to the clock assumption, even when the data fit the global clock assumption by the likelihood ratio test.

Locus × Lineage Interaction Effects

Locus × lineage interaction effects are the variation ascribable to locus-specific changes in the rate of substitution from one lineage to another and through evolutionary time. Locus × lineage interactions are equivalent to Gillespie's (1989) definition of overdispersion. Unlike locus effects, which are consistent across different lineages, and lineage effects, which affect all loci the same, locus × lineage interactions are erratic and unpredictable.

Locus × lineage interactions have been demonstrated in comparisons of distantly related lineages. For instance, they have been shown in a comparative investigation of three well-known proteins—glycerol-3-phosphate dehydrogenase (GPDH), superoxide dismutase (SOD), and xanthine dehydrogenase (XDH)—across representatives of the three multicellular eukaryote kingdoms: animals, fungi, and plants (Rodriguez-Trelles et al. 2001). That study disclosed that: (i) The three proteins evolve erratically through time and across lineages. Thus, for example, the rate of evolution of GPDH increases as less related species of dipterans are compared: $\sim 2 \times 10^{-10}$ replacements per site per year between *Drosophila* species, 2.5 times greater (5×10^{-10}) between species of different genera, and more than 10 times greater (23×10^{-10}) between species of different families. (ii) Moreover the patterns of acceleration and deceleration are erratic; they differ from locus to locus, so that one locus may evolve faster in one lineage than another, whereas the opposite may be the case for another locus. For instance, the rates between drosophilid genera and between fungi are 5 and 40.0 for GPDH, 34.9 and 24.9 for SOD, and 31.7 and 13.7 for XDH (Figure 8.3; Rodriguez-Trelles et al. 2001). The occurrence of locus × lineage interactions has also been demonstrated in comparisons involving less distantly related species. For instance, the growth hormone gene has gone through a very rapid burst of replacements in ruminant artiodactyls and primates, but not in other mammalian lineages

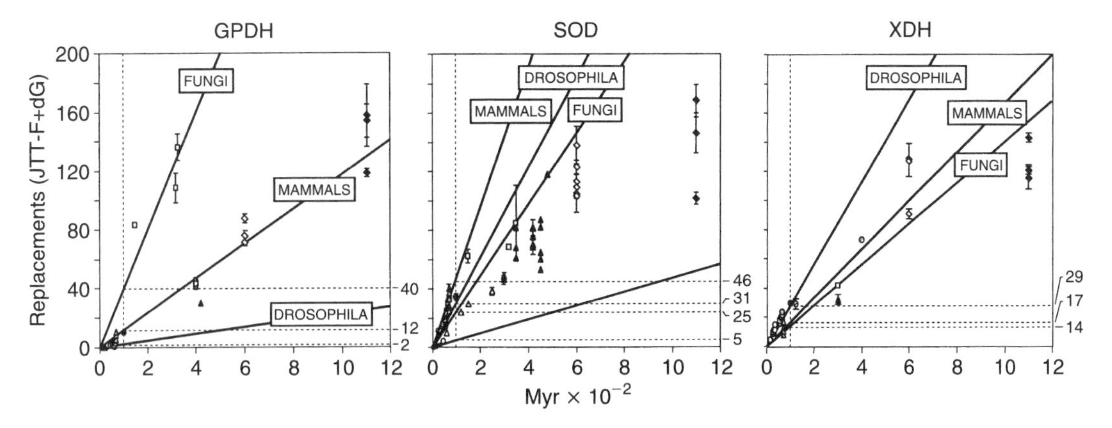

FIGURE 8.3. Locus × lineage interaction effects. Rates of glycerol-3-phosphate dehydrogenase (GPDH), superoxide dismutase (SOD), and xanthine dehydrogenase (XDH) amino acid replacement for different lineages and evolutionary time-scales. Time (abscissa) is in million years (Myr) × 10^{-2}. The numbers on the right are rates of replacement × 10^{-10} per site per year. These rates correspond to comparisons between drosophilid subgenera, mammal orders, or fungi. Other points in the figure are for other comparisons, such as between kingdoms (1100 Myr) or animal phyla (600 Myr). Modified from Rodriguez-Trelles et al. (2001).

(Wallis 1994). In mammals, locus × lineage interactions account for 8.5% and 13.5% of the variability of the amino acid replacement rate of nuclear and mitochondrial genes, respectively (Smith & Eyre-Walker 2003). Yet, mammals represent species fairly closely related, so that they are all classified within the class Mammalia, and it seems likely that the relative importance of locus × lineage interactions will increase as the taxa compared are more different, such as species from different classes or different phyla.

Several models have been conceived to explain locus × lineage interactions, including neutralist (e.g., Takahata 1987; Bastolla et al. 2003) and selectionist accounts (Gillespie 1989, 1991; Cutler 2000). Most neutrality models are motivated by the notion first expressed in the covarion model of evolution (Fitch 1971a) that the rate of a site might be modified by a substitution occurring at a distinct site of the molecule, with which it thus covaries. For instance, the "fluctuating neutral space" model (Takahata 1987) assumes that a neutral mutation at one site changes the proportion of neutral mutations at other sites, so expanding (or reducing) the neutral space, or the degree of selective constraint of the protein. A mechanistic model of this idea, called the "structurally constrained neutrality model" (SCN model; Bastolla et al. 2003), invokes effects of the three-dimensional structure of the proteins. The SCN model hinges upon the premise that the higher evolutionary conservation often observed for protein fold (i.e., three-dimensional configuration) compared with protein sequence is due to the structural requirement of thermodynamic stability. The model assumes that there are only two kinds of sites: neutral and highly deleterious. Many different protein sequences can yield the same fold, even though sequences differ in the number of structural neighbors they can mutate to without altering protein activity. Structural neighbors tend to have similar proportions of neutral sites (i.e., neutral spaces) and, therefore, they also exhibit similar rates of neutral substitution. As the three-dimensional structure of the protein evolves, this autocorrelation between the substitution rates of neighboring sequences will increase the variability of the neutral substitution rate. The SCN model can afford low to moderate levels of variation, say overdispersion values (sensu Gillespie 1989) no larger than 4, but is unable to account for large locus × lineage interaction effects, such as those evidenced by Gillespie (1989) or Rodriguez-Trelles et al. (2001) (Bastolla et al. 2003). The extreme substitution rate differences reported in those and other studies might be better accounted for by bursts, or episodes, of adaptive evolution (Gillespie 1989; reviewed in Cutler 2000).

Organisms are continually adapting to the physical and biotic environments, which change endlessly in patterns that are unpredictably and differently significant from gene to gene and from species to species. Demonstration of adaptation at the molecular level has, however, proved remarkably difficult. Several (complementary) tests of positive selection have been devised in the context of population genetics (Kreitman & Akashi 1995). A frequently used method exploits the prediction that, if observed replacements are neutral, the ratio of replacement to synonymous fixed differences between species should be the same as the ratio of replacement to synonymous polymorphisms within species (McDonald & Kreitman 1991). If a large fraction of amino acid replacements are driven by positive selection, the replacement to synonymous ratio of divergence should be inflated above that of polymorphism. Application of this and other population genetics tests of molecular adaptation to individual loci has allowed identification of a number of cases where the observed patterns of genetic variation depart from the strict neutral expectation (Kreitman & Akashi 1995; Wayne & Simonsen 1998). Yet, in most instances, population genetics tests lack power to determine the precise source (or sources) for the departures, among the likes of varying population size, fluctuating environments, or different forms of selection (Yang & Bielawski 2000).

A different approach to account for locus × lineage interactions is based on comparison of silent (i.e., synonymous; dS) and amino acid changing (i.e., nonsynonymous; dN) substitution rates in protein-coding sequences. The ratio $\omega = dN/dS$ measures the difference between the two rates. If a replacement mutation is neutral, it will be fixed at the same rate as a synonymous mutation, with $\omega = 1$. If the amino acid change is deleterious, then purifying selection will reduce its fixation rate, thus $\omega < 1$. Only when the amino acid change confers adaptive advantage will it be fixed at a higher rate than a synonymous mutation, with $\omega > 1$. Therefore an ω value significantly greater than 1 is interpreted as unequivocal evidence for adaptive evolution driven by positive Darwinian selection. Because silent and replacement sites are interspersed in the same segment of DNA, the effects of factors such as phylogeny and effective population size will be shared, providing a natural control (Yang & Bielawski 2000). This approach has been implemented in the framework of maximum likelihood, and its power has been greatly improved with new developments that allow investigation of lineage-specific episodes of molecular adaptation at the level of single amino acid sites (Yang & Swanson 2002). Increasing application of this method is revealing many instances of molecular adaptation in diverse systems, from viruses to humans (Yang & Bielawski 2000). Yet the results of this approach have been put into question by simulation studies showing that it tends to reject the neutral model more often than specified by the significance level of the test (Suzuki & Nei 2002; Zhang 2004).

FUTURE DIRECTIONS

Knowledge of the rates of molecular evolution has so far been bounded by a representation of the genome that is conceptually atomizing and biased. For the most part, this representation has consisted of largely invariant triplet-code-structured sequences and the more variable noncodifying sequences anchored to them (introns, and 3′ and 5′ flanking regions). To a large extent, this partial representation came about because of powerful PCR methods, which only return evolutionarily conserved targets with no information of their genomic location. Lack of this contextual knowledge makes it difficult deciding orthology versus paralogy in dealing with gene families, or just judging homology when sequences are overly variable, as is often the case with less-constrained nonprotein-coding sequences. In this "bean bag" representation of the genome, functional elements have to be approached without regard to most of their connections.

Wholesale sequencing of genomes has changed it all. The possibility of matching species DNAs over their full length allows the beam of comparative analysis to be shone into almost every corner of the genome. It furnishes positional coordinates to assure homology when sequence similarity alone does not suffice. Throughput genomics generates vast amounts of data, impossible to handle without networked computers. Rapidly developing bioinformatic tools are helping to identify, catalogue, and integrate mounting numbers of known and newly discovered functional components. These include elements related to gene activity, but also other constituents, such as matrix scaffold attachment regions, that perform structural or still-to-be-deciphered roles. Functional assay as a generalized approach to unveil sequence performance is overly expensive. Sequence conservation is the hint most practical to be exploited. Consequently, rates of substitution are at the core of these advances.

Duplication is the most important mechanism for the origin of new genes. Yet, little is known about the rates and patterns of duplication, the evolution of gene location, and the importance of natural selection for the acquisition of new functions. In many instances the acquisition of new functions might be preceded by the evolution of the regulatory apparatus governing gene expression. In particular, *cis*-regulatory sequences—clusters of binding sites for transcription factors—are thought to hold the key for understanding the evolution of phenotypic differences between species (reviewed in Wray et al. 2003; Rodriguez-Trelles et al. 2003). This is because of their complex organization into independent modules such that, unlike coding sequences in which mutations affect protein function every time the protein is expressed, mutations in *cis*-regulatory sequences may have minor or no pleiotropic effects. Empirically, the pattern of *cis*-regulatory sequence evolution has qualitatively been described by conserved blocks of DNA separated by unalignable gaps (Bergman & Kreitman 2001). There can be large substitution rate differences between promoters, although some are largely conserved. For example, comparative analysis of the *Dlx5/Dlx6* intergenic region across zebrafish and mammals unveiled a highly conserved segment shown by functional analysis to be the site of cross-regulatory interactions between *Dxl* genes in the embryonic forebrain. This means that *cis*-regulatory regions can be conserved for time periods as long as 5×10^8 years (Müller et al. 2002). But frequently cis-regulatory regions diverge much faster, and quickly lose any trace of homology detectable by comparative analysis.

There is not a linear relationship between functional differentiation and the amount of *cis*-regulatory sequence evolution. Sequence divergence with retention of regulatory function has been shown for the enhancer-driving *even skipped* (*eve*) expression in closely related species of *Drosophila* (Ludwig et al. 2000). High sequence turnover of *cis*-regulatory regions calls for comparisons of closely related species. But in these cases it may be difficult to distinguish functional from passive conservation due to close relatedness. An analogous problem can arise at any phylogenetic depth if there are nonfunctional mutational cold-spots. In general, caution should be exercised in interpreting sequence conservation as a signature of selection. The evolution of *cis*-regulatory sequences, and that of noncoding regions in general, is only beginning to be placed in a quantitative analytical framework.

The neutral theory can only be addressed with a genomic estimate of the proportion of amino acid replacements in protein evolution that is driven by positive selection. Polymorphism data from an increasing number of genes examined under improved detection methods place this number to be 25–45% in *Drosophila* and humans (Bierne & Eyre-Walker 2004; Smith & Eyre-Walker 2002; Fay et al. 2002). These percentages are considerably higher than those proposed by the neutral theory, although they are preliminary estimates. Even if adaptive molecular evolution turns out to be more frequent than assumed under the strict neutrality theory, there are problems with the hypothesis that extreme locus × lineage interaction effects are caused by episodes of positive Darwinian selection. Specifically, for fluctuating selection to generate large overdispersion of the rate of protein evolution, it would be required that the environment fluctuate at a rate approximately similar to the rate of replacement (Gillespie 1993; Cutler 2000). A typical vertebrate protein, some 500 amino acid residues long, experiences a replacement every 2 Myr. The physical and biotic environment of most species apparently changes much more rapidly. For example, during the last glaciation, just 10,000 years ago, a large proportion of the world was covered by ice, which may have impacted many physiological processes and, therefore, the proteins involved.

The generality of locus × lineage interaction effects implies that rates of molecular evolution are not predictable. All that would remain of the molecular clock is that evolution is a time-dependent process and, thus, the longer the time elapsed, the larger the number of changes. With respect to the timing of evolutionary events it seems unlikely that the problem of the unpredictability of rates of molecular evolution can be circumvented with local molecular clock methods, precisely because (as discussed above in connection with lineage effects) this approach is valid only if the lineages that might exhibit different evolutionary rates are specified a priori. Bayesian methods offer a more promising alternative. These methods rely on the assumption that rates of evolution change throughout the phylogeny according to some random process, so that the methods do not require prior specification of which lineage might have evolved differently from the others and when. A Bayesian model of episodic molecular evolution has recently been used (Aris-Brosou & Yang 2003) to time the origin of the Metazoa. The resulting estimates support a late Precambrian explosive diversification of the animal phyla, in accord with the fossil record but in contrast to the considerably older dates often obtained by molecular clock studies (see Rodriguez-Trelles et al. 2002). In addition to these methodological advances, efforts at determining the molecular evolutionary time scale must combine data for several genes, as many as possible. Accumulation of data over lineages and genes might lead to more reliable dates, as a consequence of the "law of large numbers." People take on the average longer to travel between Los Angeles and New York than between Los Angeles and San Francisco, even though some may drive, others fly, some stop along the way, and so on; the average time of very many travelers may likely be approximately proportional to the distance traveled.

SUGGESTIONS FOR FURTHER READING

Bierne and Eyre-Walker (2004) provide a critical review of the various ways in which population geneticists are using rates of molecular evolution to understand the importance of positive Darwinian selection in the evolution of the protein-coding fraction of the genome. Bromhan and Penny (2003) review the concept of the molecular clock, emphasizing its utility as a null model for hypothesis testing. Cutler's (2000) historical overview of the index of dispersion as an approach to test the neutral theory provides a critical evaluation of different models proposed to account for observed overdispersed rates of molecular evolution. Huelsenbeck and Crandall's (1997) paper is an excellent, understandable introduction to maximum likelihood as a framework for learning about the patterns and processes of molecular evolution through hypothesis testing. Yang (1996b) is a classical contribution illustrating how alternative representations of the rate of molecular evolution can impact evolutionary inferences.

Bierne N & A Eyre-Walker 2004 The genomic rate of adaptive amino acid substitution in *Drosophila*. Mol. Biol. Evol. 21:1350–1360.

Bromham L & D Penny 2003 The modern molecular clock. Nat. Rev. Genet. 4:216–224.

Cutler DJ 2000 Understanding the overdispersed molecular clock. Genetics 154:1403–1417.

Huelsenbeck JP & KA Crandall 1997 Phylogeny estimation and hypothesis testing using maximum likelihood. Annu. Rev. Ecol. Syst. 28:437–466.

Yang Z 1996 Among-site rate variation and its impact on phylogenetic analyses. Trends Ecol. Evol. 11:367–372.

9

Weak Selection on Noncoding Gene Features

YING CHEN
WOLFGANG STEPHAN

The neutral theory of molecular evolution proposed that most of the variation observed at the molecular level is due to neutral mutations and random genetic drift, rather than positive Darwinian selection (Kimura 1983). However, as more protein sequence data became available, it was clear that the strictly neutral theory fails to explain two major observations, namely, the generation-time effect and the narrow range of heterozygosities. It was observed that the rate of protein evolution is roughly constant per year, rather than per generation as expected by the strictly neutral theory. Furthermore, the estimates of effective population sizes (N_e) based on heterozygosity data tend to fall into a relatively narrow range for diverse groups of organisms which were thought to have very different population sizes (Ohta & Gillespie 1996). These findings were later explained by a modified version of the neutral theory, the nearly neutral theory of Tomoko Ohta. Ohta proposed that deleterious mutations with selection coefficients (s) around $1/N_e$ might be quite common among amino acid substitutions.

There is a clear distinction between strong selection and weak selection acting on slightly deleterious mutations: strong selection is defined by $|N_es| >> 1$, suggesting that selection changes allele frequencies essentially in a deterministic way; in contrast, weak selection posits $|N_es| \sim 1$, such that random genetic drift becomes important. The nearly neutral theory differs from the strictly neutral theory by predicting a negative correlation between the evolutionary rate and the species population size: the larger the population size, the less likely that mutations are effectively neutral and can be fixed in the population, and the slower the evolutionary rate. In addition to explaining patterns of amino acid substitutions, Ohta's nearly neutral theory has since also been invoked to describe weak selection on synonymous sites leading to the unequal usage of codons within a given codon family, a phenomenon known as codon bias (Grantham et al. 1980).

After the initial focus on nonrandom synonymous codon usage, other aspects of a gene have been suggested to be under weak selection as well. For example, even though the highly conserved elements involved in splicing (e.g., splice sites) are usually maintained by strong selection, other features of introns (in particular, intron length; e.g., Carvalho & Clark 1999) are likely to be under weak selective constraints. Similarly, regulatory sequences in the 5′ and 3′ regions of genes (with the exception of core promoter regions such as the TATA box or the CAAT box) are often considered as evolving nearly neutrally (Ludwig et al. 2000). Another aspect of a gene that appears to be under weak selection is the precursor messenger RNA (pre-mRNA) or messenger RNA (mRNA) secondary structure. As Stephan and Kirby (1993) have shown, evolution in the pairing regions of RNA secondary structures may not be much slower than in unpaired regions due to the frequent occurrence of correlated nucleotide substitutions ("covariations") maintaining the base pairing.

Here, we describe the functional importance of these features of genes in relation to the strength and form of selection operating on them. We will not focus on the coding function of genes, but rather emphasize the gene features that do not alter protein coding. In this sense, we refer to codon usage, regulatory sequences, pre-mRNA and mRNA secondary structures, and intron length as noncoding gene features.

CONCEPTS: EFFECTS OF WEAK SELECTION ON NONCODING GENE FEATURES

With the exception of strongly conserved regulatory elements, noncoding regions (i.e., 5′ and 3′ untranslated regions, introns, and intergenic regions) and synonymous sites within coding regions are conventionally regarded as neutral, since mutations at these sites do not lead to changes in amino acids (Kimura 1983). However, even though the selective constraints associated with encoding a functional protein are expected to be the strongest, other weak selective forces that do not affect the protein-coding capacities can also play an important role in the evolution of DNA sequences, such as selection for optimal synonymous codons, and selection maintaining or avoiding mRNA secondary structures. Furthermore, mutations in regulatory regions, with the exception of some very well conserved motifs, could be subjected to weak selection. Gene and intron lengths may also be weakly selected, due to either their specific gene functions and regulations, or their local recombination environments. To demonstrate these findings, we discuss theoretical and empirical advances in the following areas: codon usage bias, compensatory evolution in RNA secondary structure, and evolution of regulatory sequences and intron length.

Synonymous Codon Usage

The genetic codon table (Figure 9.1) is redundant: 20 amino acids are represented by 61 codons, with 18 of them represented by more than one codon. However, synonymous codons that encode the same amino acid are not used with equal frequencies (Grantham et al. 1980). For example, GGC, a member of the 4-fold degenerate glycine codon family, is the optimal codon in *Drosophila melanogaster.* It is used 43% of the time, instead of the expected 25% if all synonymous codons for glycine are used with equal frequency. The nonrandom nature of codon usage is found in both prokaryotic and eukaryotic genomes.

What evolutionary forces are responsible for codon bias? Two alternative models have been proposed: the translational selection model and the mutational bias model.

The translational selection model suggests that natural selection plays a major role in the codon bias patterns observed. Many empirical results, mainly in verifying the following three predictions

		Second Position of Codon					
		T	C	A	G		
First Position	T	TTT Phe [F]	TCT Ser [S]	TAT Tyr [Y]	TGT Cys [C]	T	Third Position
		TTC Phe [F]	TCC Ser [S]	TAC Tyr [Y]	TGC Cys [C]	C	
		TTA Leu [L]	TCA Ser [S]	TAA Stop	TGA Stop	A	
		TTG Leu [L]	TCG Ser [S]	TAG Stop	TGG Trp [W]	G	
	C	CTT Leu [L]	CCT Pro [P]	CAT His [H]	CGT Arg [R]	T	
		CTC Leu [L]	CCC Pro [P]	CAC His [H]	CGC Arg [R]	C	
		CTA Leu [L]	CCA Pro [P]	CAA Gln [Q]	CGA Arg [R]	A	
		CTG Leu [L]	CCG Pro [P]	CAG Gln [Q]	CGG Arg [R]	G	
	A	ATT Ile [I]	ACT Thr [T]	AAT Asn [N]	AGT Ser [S]	T	
		ATC Ile [I]	ACC Thr [T]	AAC Asn [N]	AGC Ser [S]	C	
		ATA Ile [I]	ACA Thr [T]	AAA Lys [K]	AGA Arg [R]	A	
		ATG Met [M]	ACG Thr [T]	AAG Lys [K]	AGG Arg [R]	G	
	G	GTT Val [V]	GCT Ala [A]	GAT Asp [D]	GGT Gly [G]	T	
		GTC Val [V]	GCC Ala [A]	GAC Asp [D]	GGC Gly [G]	C	
		GTA Val [V]	GCA Ala [A]	GAA Glu [E]	GGA Gly [G]	A	
		GTG Val [V]	GCG Ala [A]	GAG Glu [E]	GGG Gly [G]	G	

FIGURE 9.1. The standard codon table.

of the translational selection hypothesis, support it. First, if there is selective pressure on synonymous codons to improve the translation efficiency and/or accuracy, then highly expressed genes should be under stronger selection than weakly expressed genes and thus have higher codon bias. This has indeed been found in various organisms including *Escherichia coli, Saccharomyces cerevisiae, D. melanogaster, Caenorhabditis elegans,* and *Arabidopsis thaliana.* Second, the translational selection model predicts that codon usage is positively correlated with transfer RNA (tRNA) abundance, and this has been verified in *E. coli, S. cerevisiae, D. melanogaster,* and *C. elegans.* Third, if mutations from optimal codons to nonoptimal codons carry a selective disadvantage, then synonymous substitutions should be biased toward major codons. It has been found that synonymous substitution rates correlate negatively with codon bias in *E. coli, D. melanogaster,* and *C. elegans.* However, the last point should be interpreted with caution, as the synonymous substitution rate was found to be uncorrelated with codon bias when a codon-based maximum likelihood method was used to estimate the synonymous substitution rate (Dunn et al. 2001; but see also Bierne & Eyre-Walker 2003). Codon bias has also been found to be positively associated with functional constraint at the protein level in *Drosophila,* indicating that there is a selective pressure on synonymous codons for translational accuracy (Akashi 1994, 2001).

However, among mammals, codon bias seems to be shaped more by mutational patterns than selection (Eyre-Walker 1991). The fitness differences between synonymous codons are expected to be very small (Kimura 1983; Akashi 1995, 2001). If selection coefficients for optimal codons in various organisms are of the same order, then it is expected that in species with smaller effective population sizes, such as humans, selection is less effective, and mutation will have a bigger role in shaping codon usage patterns; while in species with large effective population sizes, such as *E. coli,* selection is the major determinant (Shields et al. 1988). It is also possible that effects of selection at synonymous sites in mammals are masked by other factors such as isochores and CpG islands (Bernardi et al. 1985).

Therefore, it is of great interest to determine what selective pressure is required to maintain the codon bias patterns in various organisms. Several attempts have been made to estimate the selection coefficients of changes from nonoptimal to optimal codons using population genetics theory. Kimura first treated the problem of nonrandom usage of synonymous codons with a stabilizing selection model (Kimura 1981). He assumed the relative availability of isoaccepting tRNA species is the major factor determining codon bias, and the highest fitness for a gene is achieved when the relative frequencies of synonymous codons in the mRNA exactly match those of the isoaccepting tRNAs in the cell. Thus, in this stabilizing selection scheme it is advantageous to use some nonoptimal codons. This model also treated the nucleotides in a gene as independent (unlinked). The results show that even a small selective difference among synonymous codons can produce the strong codon bias observed in nature. However, this model fails to explain the observed positive correlation between codon bias and gene expression.

An alternative scheme is based on the selection–mutation–drift model: selection favors optimal codons to increase translational efficiency and accuracy, and decrease proofreading costs, while mutation and genetic drift allows the persistence of nonoptimal codons (Li 1987; Bulmer 1991).

Li (1987) provided a population genetics study of codon bias using both a single-site and a multi-site model. In the single-site model, each nucleotide site in a gene is treated as independent. In the multi-site model, all nucleotide sites in a gene are treated as completely linked and various selection schemes are considered. His results show that a strong codon bias can be produced even when the selective difference between synonymous codons is quite small: selection is effective when $|N_e s| = 2$ in a haploid population, or $|N_e s| = 1$ for a diploid population. His results also suggest that the interactions between synonymous codons under weak selection are synergistic rather than additive; that is, instead of a fixed s for each nonoptimal codon, the selective disadvantage of having one nonoptimal codon is negligibly small but the disadvantage increases with the number of nonoptimal codons in a gene. This is because in the additive model, selection would either be too small to overcome the effect of random drift or prolonged bottleneck, or so large that the mutational load on the population becomes unrealistic. However, for the synergistic model, the effect of adding one more nonoptimal codon is minimal until the number of nonoptimal codons in a gene exceeds a critical value, and then selection against nonoptimal codons becomes effective.

Bulmer (1991) also presented a population genetic model of synonymous codon usage based on selection–mutation–drift equilibrium. Using a biochemical model, he determined the magnitude of the selective forces acting on codon usage in unicellular organisms (such as *E. coli* and yeast) in which growth rate and rate of protein synthesis may be equated with fitness. If the speed of translation is the dominant selective force in *E. coli,* the selective disadvantage of a nonoptimal codon in a protein is about 1%. This yields $|2N_es|$ of 2–4 for ribosomal protein genes, and 1–3 for aa-tRNA synthetase genes. However, the selection–mutation–drift model predicts $|2N_es|$ to be about 10^5 for ribosomal protein genes and 10^4 for aa-tRNA synthetase genes, which are several orders of magnitude larger than the values estimated from *E. coli* growth rates. The discrepancy of the estimated $|2N_es|$ values between the population genetic model and the biochemical model could be due to inadequate estimation of the selection coefficient using the biochemical model, counterbalancing selection pressures from maintaining DNA or mRNA secondary structure, or the failure of the population genetic model to take into account the genetic structure of clonal organisms and the periodical selective sweeps of favorable alleles.

Selection on synonymous codons has also been investigated by comparing patterns of polymorphism and divergence at synonymous sites in *Drosophila* (Akashi 1995). Assuming that all synonymous sites evolve independently and the population is at equilibrium, it is estimated that $|N_es|$ for synonymous mutations in *D. simulans* is between 1.3 and 3.6, and less than 1 for *D. melanogaster* (due to a 3- to 6-fold smaller effective population size). Using a different comparative method, McVean and Vieira (2001) confirmed that the selection coefficient for synonymous codon usage in *D. melanogaster* is not significantly different from 0.

Compensatory Evolution in RNA Secondary Structure

Since mRNA undergoes transcription, splicing, transportation and translation, and can be regulated at many steps, it is not surprising that regulatory elements in mRNA are found not only at the DNA sequence level, but also in higher-order RNA structures. Secondary structures of mRNAs can play significant roles in gene regulation, intron splicing, and RNA localization. One of the best-characterized examples is the secondary structure of the *Drosophila bicoid (bcd)* mRNA. *bcd* is an essential gene for early embryonic pattern formation and is maternally transcribed and localized to the anterior pole of the embryo. The *cis*-acting sequences necessary for mRNA localization fall within a large (~700 nucleotides), phylogenetically conserved secondary structure in the 3′ UTR (Macdonald 1990). Without the conserved secondary structure and proper localization, the embryos fail to develop correctly.

Inferring RNA Secondary Structures

In contrast to ribosomal and transfer RNAs that—for the most part—have canonical secondary structures, the secondary structure of each pre-mRNA and mRNA must be individually determined. Several approaches are available for inferring RNA secondary structures. One of them is thermodynamic prediction of RNA secondary structure by free energy minimization (Zuker et al. 1991). This method generates reliable results for small RNAs. However, it becomes rather inaccurate when applied to large RNAs, where the length is 1000 nucleotides or more. At this time, DNA sequence comparison is still the most incisive method for inferring secondary structures of large RNAs. The inference of an RNA secondary structure from DNA sequence comparison is usually based on the Woese–Noller criterion (Noller and Woese 1981); that is, a putative helix of a RNA secondary structure is considered "proven" if two or more covariations, caused by independently occurring base substitutions in the complementary sequences of a putative helix, are detected in sequence comparisons. However, this is only a heuristic criterion, which does not account for divergence levels and the number of species in the comparison.

A more rigorous statistical method to detect RNA secondary structures from aligned sequence data was proposed by Muse (1995). His approach relies on a likelihood ratio test (LRT) to identify potential pairing regions showing constraints for base pairing interactions. The advantage of this approach is that it does not rely on the *ad hoc* rules used in the comparative method. The pattern of nucleotide substitution at paired sites is compared with that of unpaired sites and a pairing parameter, λ, is estimated. The relative evolutionary conservation of each predicted pairing is quantified by calculating a LRT statistic. A drawback to the LRT approach has been the requirement to specify the coordinates of potential pairing stems before application of the test. Thus, so far, this method has primarily been used to test structures previously predicted by phylogenetic-comparative analysis (Kirby et al. 1995; Muse 1995; Parsch et al. 1997).

A method that extends Muse's likelihood approach by predicting secondary structures without a priori knowledge of the location of putative pairing regions was developed by Parsch et al. (2000). The first step in this procedure is to generate a complete list of potential pairing regions meeting specified stability and conservation criteria based on an alignment of homologous RNA-encoding sequences. For each helix, the constraint for base pairing is then estimated by Muse's LRT statistic. In the third and last step, compatible helices are combined to form a secondary structure model of the entire RNA molecule that maximizes the LRT value of the total structure.

The phylogenetic comparison method has been used successfully to predict secondary structures of messenger, ribosomal, and transfer RNA. However, a major requirement of this method is the availability of multiple sequences from moderately divergent species.

Modeling Compensatory Evolution in RNA Secondary Structures

Kimura (1985) initially proposed the compensatory neutral evolution model to study the interaction between two amino acids in a protein. Compensatory mutations, by definition, are a pair of mutations at different loci (or nucleotide sites) that are individually deleterious but neutral in appropriate combinations. Similar to a folded protein with interacting amino acid residues, the double-stranded stems in RNA secondary structures are also expected to undergo compensatory evolution, since they are composed of complementary base pairs. If a particular stem structure plays a functional role in RNA processing or stability, then selection is expected to preserve its secondary structure, rather than the primary DNA sequence. Thus, individual mutations within stems are expected to be deleterious if they break up the pairing of an intact secondary structure; however, fitness can be restored, when a second, compensatory mutation occurs at the appropriate position on the opposite strand of the stem and reestablishes the pairing.

Kimura's model provides the theoretical basis of the substitution process in stems of RNA secondary structures. The model assumes two linked sites, with two variant nucleotides at site 1 and two at site 2. Furthermore, the model assumes that a compensatory change occurs at one base pair at a time. A change from an AU pair to a CG pair may go through a CU or an AC intermediate. Because the mutations occurring in the first step destabilize the stem, it is assumed that they are selected against; that is, the fitness of the intermediate haplotypes AC and GU is lower than that of the combinations AU and GG (which, to a first approximation, may be set to 1). Linkage is characterized by the rate of recombination between the two nucleotides in the initial base pair. Kimura's model allows only unidirectional mutations. A more general model (Figure 9.2), with individual mutation rate parameters and different selection coefficients for the deleterious intermediate states, was described in Innan and Stephan (2001). Under strong selection ($|N_e s| >> 1$), the expected transition time from one base pair to another depends on the recombination rate between the two nucleotide substitutions (Kimura 1985). Recombination reduces the rate of compensatory evolution by breaking up favorable combinations of double mutations. However, mismatches, including not only GU wobble pairs but also other noncanonical pairs, have frequently been observed in phylogenetic analysis of RNA secondary structures, indicating that selection against deleterious intermediate states may be relatively weak (Rousset et al. 1991; Parsch et al. 2000). For this reason, Innan and Stephan (2001) introduced reversible mutations into the compensatory evolution model and assumed a broad range of selection coefficients. They found that under weak selection ($|N_e s| \sim 1$), the rate of compensatory evolution is independent of the recombination rate. Furthermore, they observed that the rate of molecular evolution in stem regions of some well-characterized pre-mRNA and mRNA secondary structures (including those of the *bcd* 3′ UTR mentioned above) is only about five times slower than in the adjacent unpaired regions of the molecule, which is consistent with $|N_e s| \sim 1$. Their model also predicts that linkage disequilibria due to compensatory fitness interactions should be rare in natural populations, unless selection against the intermediates is very weak. Thus, it is difficult to

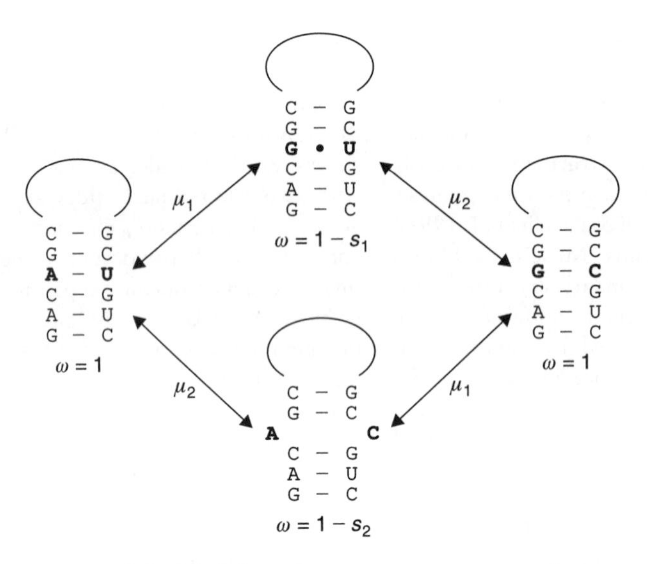

FIGURE **9.2.** Model of compensatory evolution. Bi-directional mutations are considered. The fitness values (for a genic selection scheme) are represented underneath each secondary structure (s_1, $s_2 > 0$). The base pair that undergoes a compensatory change is indicated in bold.

detect epistatic selection maintaining RNA secondary structures by analyzing patterns of intraspecific DNA sequence variation.

Regulatory Elements

Eukaryotic genes are modulated by many noncoding *cis*-regulatory elements. Some regulatory elements are identified by interspecific sequence comparisons as being evolutionarily conserved. They are likely under strong purifying selection for their essential function. Other regulatory regions show considerable levels of interspecific divergence and intraspecific polymorphisms. One such example is the stripe 2 enhancer element in the 5′ flanking region of the *even-skipped (eve)* gene in *Drosophila* (Ludwig 2002). The *eve* stripe 2 element regulates the correct segmental expression of *eve* mRNA, and is one of the best-characterized eukaryotic enhancers. However, a relatively high level of variation, both intra- and interspecific, has been found in the 671 bp enhancer region in both *D. melanogaster* and *D. simulans*. Statistical tests show that the level of nucleotide variation and divergence is compatible with neutrality. Therefore, even though the *eve* enhancer region is functionally constrained, the patterns of molecular evolution appear to be neutral. These two seemingly contradictory observations might be reconciled by the multiplicity and redundancy of transcription binding sites within the enhancer: if a large number of redundant binding sites are involved in enhancer function, each mutation will have a quite small selection coefficient and is controlled by genetic drift rather than selection, so that the rate of substitution within the region can be high even though the enhancer is functionally constrained (Ludwig et al. 2000; Ludwig 2002).

Intron Length

The mRNA precursors of most genes in higher eukaryotes contain introns that need to be removed by nuclear pre-mRNA splicing. Intron lengths vary widely among different genes and among homologous genes of different species. A few splicing signals and regulatory elements, such as the 5′ and 3′ splice sites, the branch point, and the polypyrimidine tract in large introns, are either essential or important to splicing; however, most of the intronic regions seem to lack functional constraints and evolve neutrally. Even though introns are removed before translation and usually do not have any effects on the protein sequences, the cumulative time and energetic cost incurred by DNA replication and transcription of introns in the genome may not be negligible. Thus, although most nucleotide sites in introns evolve neutrally, the length of introns could be under selection, especially in highly expressed genes or genes with large introns (Castillo-Davis et al. 2002). Yet, the majority of introns have persisted over long periods of time, and the level of intron length variation within and between species is high, indicating that intron length cannot be a trait under strong selection.

Several studies have investigated whether intron length is under weak selection. The first evidence of selection acting on intron length came from Carvalho and Clark (1999). They investigated the relationship between intron length and recombination rate of 1817 *D. melanogaster* introns, and found a significant association. In regions of low recombination, where selection is less effective due to the Hill–Robertson effect, longer introns occur preferentially. This is consistent with the notion that longer introns have a higher cost of transcription and are thus more deleterious. In addition to *D. melanogaster*, the negative correlation between intron length and recombination rate is also found in humans, and may exist for all eukaryotic genomes (Comeron & Kreitman 2000).

In addition to selection, insertion-deletion mutation could also influence intron length. An excess of deletions relative to insertions has been observed in many organisms (Petrov et al. 1996; Comeron & Kreitman 2000). This led Comeron and Kreitman (2000) to argue that if both mutation and selection work in the same direction (i.e., reducing intron length), introns would rapidly collapse to a minimal size required for splicing. However, since the distributions of intron size are rather broad in most species studied so far, the authors propose that, if the deletion bias is true, selection must favor longer introns in some instances, such as introns acting as enhancers of within-gene recombination (or reducers of Hill–Robertson interference). In this way, selection could offset the directional mutation pressure to shorter introns, thus creating a broad equilibrium distribution of intron length.

In contrast to *Drosophila* and humans, data from *Caenorhabditis elegans* show a significantly positive correlation between intron length and recombination rate (Duret 2001). This is likely due to the differences in the rate of transposable element (TE) insertions. In *C. elegans*, TE insertions, which are usually much longer than the small indels identified in Comeron and Kreitman's (2000) study, are preferentially located in regions of high recombination. Thus, the correlation between intron length and recombination may only reflect variation in the pattern of TE insertion rather than the strength of selection (Duret 2001).

The effect of selection on intron length has also been questioned by Lynch (2002). He proposed a model of intron evolution with only random genetic drift and mutation as the relevant evolutionary forces. His model ignores the selective disadvantage of introns due to reduced transcriptional efficiency, but rather focuses on the elevated mutation rate to nonfunctional alleles in intron-containing genes, the logic being that intron-containing genes have more essential sites, such as the 5′ and 3′ splice sites, that can not be mutated without loss of proper function. The reduction of N_e in regions of low recombination promotes the fixation of introns. Lynch claims that the association between intron length and recombination rate is only a "passive" response to differences in N_e, with no need to invoke selection. However, it is unclear how his weak mutation-pressure hypothesis differs in essence from Carvalho and Clark's (1999) weak selection-pressure hypothesis. Furthermore, it is doubtful whether the effective population size of clonal organisms is greater than 10^{10}, as Lynch claimed, so that the very weak mutation pressure prevents the establishment of functional introns in these organisms.

EXPERIMENTAL CASE STUDIES

In this section we will discuss some experimental studies of weak selection that have been done in multicellular organisms. In particular, we focus on functional analyses of codon bias and mRNA secondary structures, using *Drosophila* as a model organism. These studies are largely (though not always) consistent with the results of the theoretical and statistical analyses presented above.

Synonymous Codon Usage

A large amount of experimental work has shown that by optimizing the coding region, or even only the 5′ end of the coding region, toward the preferred codons of the host cell, the level of gene expression and heterologous protein production can be increased significantly in a variety of hosts, such as *E. coli,* yeast, plants, and mammalian cells (e.g., Andersson & Kurland 1990; Kim et al. 1997).

However, few attempts have been made to examine in vivo the relationship between codon bias and level of gene expression in a multicellular organism, such as *Drosophila.* The common perception has been that the effects of weak selection on synonymous codons are too small to be measured in the laboratory, and that such effects may only be important if accumulated genome-wide and over evolutionary time scales.

To demonstrate experimentally the importance of codon bias in a multicellular organism and, in particular, to show the importance of codon bias in enhancing translation rate and fidelity, Carlini and Stephan (2003) explicitly introduced nonoptimal codons into the *D. melanogaster Adh* gene and analyzed their effects in vivo. Since laboratory measurements are usually not sensitive enough to observe very small fitness differences, in order to maximize the probability of observing the effects of synonymous codon changes, leucine codons of the *Adh* gene were targeted. This is because *Adh* is highly expressed and among the top 2% of the highly biased genes in *D. melanogaster* (Duret & Mouchiroud 1999). Furthermore, leucine is one of the most biased codon families in *D. melanogaster* (Li 1987; McVean & Vieira 2001).

Single nucleotide substitutions were made to change the leucine codons from optimal (CTG or CTC) to nonoptimal (CTA) ones. *Adh* transgenes carrying one (1 Leu), six (6 Leu), and ten (10 Leu) mutations were introduced into flies by germline transformation. The effects of these synonymous codon changes were then assayed by comparing the ADH activity of the transformant stocks with that of the wild-type transformant stocks. Since the synonymous codon changes do not affect the protein sequence, any difference in ADH activity is completely due to the differences in ADH protein level.

The authors found a measurable decrease in ADH activity with the introduction of the nonoptimal codons. The average ADH activities of the wild-type *Wa-F* control and 1 Leu, 6 Leu, and 10 Leu lines were 98.8, 88.9, 80.3, and 75.0, respectively (Table 9.1). A linear regression of activity (relative to the wild-type) on the number of unpreferred synonymous mutations was performed, and a 2.13% decrease in ADH activity per unpreferred mutation was found. Differences in ADH activity among the four genotypes were tested using a nested analysis of variance were highly significant ($P < 0.01$). Post-hoc tests were performed to test for significant differences in pair-wise comparisons, and the ADH activity of the wild-type control line was significantly greater than that of 6 Leu ($P < 0.05$) or 10 Leu ($P < 0.01$). These results are consistent with the model of translation selection discussed in synonymous codon usage, p. 134. The phenotypic effects of these nonoptimal codons were further examined through ethanol tolerance essays in adult transformant flies. It was demonstrated that not only do synonymous substitutions affect the abundance of the ADH protein, the decrease in ADH abundance also affects the fitness of adult flies by a reduction of ethanol tolerance under laboratory conditions (Carlini 2004).

The reduction in ADH activity in the transformant lines was then used to estimate the fitness effects of the synonymous substitutions. First, the authors applied the saturation theory of molecular evolution (Hartl et al. 1985) to the data. The saturation theory transforms the differences in enzymatic

TABLE 9.1. ADH activities of control and mutant transgenic fly lines

Genotype	No. of independent insertion lines assayed	Average ADH activity (± SD)[a]
Wa-F (control)	$n = 10$ lines	98.75 ± 12.79
1 Leu	$n = 9$ lines	88.92 ± 6.24
6 Leu	$n = 16$ lines	80.33 ± 10.82
10 Leu	$n = 15$ lines	75.01 ± 22.35

From Carlini and Stephan (2003).
[a]ADH activity is expressed in standard units (micromole NAD^+ reduced per minute per milligram of total protein × 100).

activity to fitness effects. Using the relationship between ADH activity and fitness, the selection coefficient of each nonoptimal synonymous change was estimated to be 4.0×10^{-5}, and with the standard estimate of $N_e = 10^6$ for *D. melanogaster*, $|N_e s|$ becomes rather large. Second, following an extension of Bulmer's (1991) selection–mutation–drift model, they also estimated $|N_e s| > 1$, perhaps even much greater than 1.

These results differ considerably from previous theoretical estimations of $N_e s$ for synonymous codon change (Akashi 1995; McVean & Vieira 2001). However, these estimates of the selection coefficient of optimal to nonoptimal codon change might be an overestimation for an average codon due to the following reasons: (i) Carlini and Stephan (2003) used a very highly expressed gene and one of the most biased codons. The CTA codon they introduced in the transformant lines is highly unpreferred; in fact, the frequency of CTA codons in *Adh* in all species of the *melanogaster* subgroup is zero (Nakamura et al. 2000). Meanwhile, estimates of *s* from sequence comparisons are based on analyses of the average across all codon families and genes of various expression levels (Akashi 1995). (ii) All synonymous codons are not under the same selective pressure (McVean & Vieira 2001), and the equilibrium assumption of codon bias may be violated in *D. melanogaster* (Akashi 1995). It is therefore conceivable that selection at leucine codons in the highly expressed *Adh* gene is an order of magnitude stronger (i.e., 10^{-5}) than the genome-wide average of 10^{-6}. For these reasons, more studies are needed to estimate the selection coefficients of synonymous codon changes both experimentally and theoretically before firm conclusions about the magnitude of the strength of selection in codon bias can be drawn.

mRNA Secondary Structure

Evidence for the maintenance of pre-mRNA and mRNA secondary structures has been found in both introns and UTRs (Stephan and Kirby 1993; Kirby et al. 1995). Here, we will discuss two experimental results of the predicted secondary structures in the *Adh* pre-mRNA and mRNA. The main goal of these studies was to test empirically Kimura's (1985) model of compensatory evolution (see Compensatory Evolution in RNA Secondary Structure, p. 136). Furthermore, since the predicted structures are close to regions of the *Adh* gene with regulatory importance (i.e., translation initiation sites and intron branch points), their roles in gene expression were examined.

Hairpin Structure in an Adh *Intron*

A hairpin structure possibly involved in intron processing has been predicted in intron 1 of the *Adh* gene in diverse *Drosophila* species (Kirby et al. 1995). Chen and Stephan (2003) evaluated the function of the putative hairpin structure by systematically eliminating either side of the stem. The effects of these mutations and the compensatory double mutation on intron splicing efficiency and ADH protein production were assayed in *D. melanogaster* Schneider L2 cells and germline transformed adult flies. Mutations that disrupt the putative hairpin structure immediately upstream of the intron branch point were found to cause a reduction in both splicing efficiency (6.1%) and ADH protein production (15%). In contrast, the compensatory double mutant that restores the putative hairpin structure was found to be indistinguishable from the wild-type in both splicing efficiency and ADH level. The results are consistent with Kimura's compensatory evolution model: mutations that disrupt the conserved hairpin structure are expected to be deleterious and deviate from the wild-type in splicing efficiency and ADH protein level; on the other hand, the compensatory mutation that restores the hairpin structure and is expected to be neutral, has a splicing efficiency and ADH protein level not significantly different from the wild-type.

The hairpin structure is likely to be under weak selection. First, extensive quantitative variation in ADH protein abundance and activity has been observed in natural populations of *D. melanogaster*, much larger than the range of differences found in the transformant lines (discussed in Chen & Stephan 2003). This suggests that the small differences in ADH abundance observed have only small fitness effects. Second, theoretical models of compensatory evolution have predicted that when selection against deleterious intermediate states is strong, the rate of compensatory evolution is very low. However, the observation that evolution in stems of RNA secondary structures is generally only 5 times slower than in neighboring unpaired regions (Innan & Stephan 2001) suggests that selection against deleterious intermediates is weak. This is consistent with the facts that mismatches in stems are found in interspecific sequence comparisons of the *Adh* intron 1 within the *melanogaster* subgroup, and covariations occur within the subgenus *Sophophora* (Kirby et al. 1995).

It was also observed by mutational analysis that a more stable secondary structure (with a longer stem) in this intron decreases both splicing efficiency (10%) and ADH protein production (21%). In most species groups, a length of around 6 bp for the stem of the hairpin structure appears to be optimal; in the *obscura* group, however, the stems are 2 to 3 bp longer; and in the Hawaiian species, the stem is even longer (Stephan & Kirby 1993). One possible explanation for this observation is that the reduction in splicing efficiency is caused by a more stable secondary structure of the entire intron (Chen & Stephan 2003). This may indicate that secondary structures in introns (even if they have a function such as in gene regulation) must not exceed a certain level of thermodynamic stability. Since an important determinant of the stability of RNA secondary structures is the GC content of the primary sequence, this result may explain why introns are relatively AT-rich in most species (further discussed in Chen & Stephan 2003).

Long-Range Compensatory Interactions in the Adh *Gene*

Long-range interactions between the 5′ and 3′ ends of mRNA have been suggested to be important in the initiation of translation and the control of mRNA stability (reviewed in Parsch et al. 1997). Stephan and Kirby (1993) used the comparative phylogenetic method to predict mRNA secondary structures in the *Drosophila Adh* gene and found preliminary evidence for long-range RNA–RNA interactions between the protein-coding region and the 3′ UTR. Parsch et al. (1997) extended this study by a mutation analysis of the hypothesized long-range interaction. They identified two putative pairing regions between exon 2 and the 3′ UTR in *D. melanogaster.* A mutation was made at a silent third codon position in the putative pairing region of exon 2 (very close to the translation initiation site). The mutant transformant lines showed a significant 15% reduction in ADH activity compared with the wild-type transformants. A compensatory mutation in the 3′ UTR restored the ADH activity to wild-type level. However, transformant lines containing only the point mutation in the 3′ UTR did not show significant differences in ADH activity from the wild-type. This result is inconsistent with predictions of Kimura's classical compensatory evolution model where both intermediate states should be deleterious.

A possible explanation for this observation is that the mutation made in the 3′ UTR is very close to a highly conserved region of the *Adh* 3′ UTR. Deletion of this conserved region resulted in a 2-fold increase in ADH activity due to an underlying 2-fold increase in mRNA level. Baines et al. (2004) investigated this problem by making a series of single and compensatory double mutations in a background transformant line with the conserved 3′ UTR region deleted. Their results fit the compensatory evolution model in that both intermediate states differ significantly from the background control line while the compensatory double mutant does not. The authors showed that the conserved region plays a dual role in both long-range pairing and the negative regulation of *Adh* expression, so that in the previous experiment (Parsch et al. 1997) the reduction in ADH activity due to a disruption of long-range pairing may have been masked by a decrease in the negative regulation of mRNA. Thus, the results of this series of experiments (i) confirm Kimura's model of compensatory evolution, and (ii) demonstrate the importance of RNA–RNA long-range interaction in the regulation of gene expression.

FUTURE DIRECTIONS

Most of the empirical studies of weak selection on gene structures have been done using a single gene, usually the highly expressed *Adh* gene in *D. melanogaster,* as a model system and a representative of the entire genome. However, due to the nature of weak selection, some aspects of it can not be elucidated by just looking at individual genes; rather, the patterns of selection are only obvious when genome-wide data are examined. With a greater accessibility to genomic sequence data, it should now become possible to further study the patterns of weak selection, to differentiate the effects of weak and strong selection, and to systematically analyze the interference between linked, strongly and weakly selected sites.

In addition, it is important to measure experimentally the fitness effects of weak selection. For this reason, it is necessary to increase the sensitivity of the available experimental methods.

Furthermore, the type of interactions between weakly selected sites, whether synergistic or additive, needs to be clarified. It is possible that in some cases only the cumulative effects of weak selection on a genome-wide scale can be detected. Selection coefficients for individual genes would then have to be inferred.

SUGGESTIONS FOR FURTHER READING

Akashi (2001) presents an excellent review on studies using large-scale genome sequences and gene expression data to understand the mechanisms of molecular evolution, including both synonymous codon usage bias and protein evolution. Chen et al. (1999) extend Kimura's classic compensatory evolution model to describe epistatic interactions within gene regions, with a particular focus on the pairing regions of mRNA secondary structures. They provide a comprehensive overview of the concept, model, and preliminary experimental results on compensatory evolution in mRNA secondary structures. Ohta and Gillespie (1996) present a concise history of four decades of work devoted to the development of the neutral and nearly neutral theories of molecular evolution that gives an excellent perspective to anyone entering the field of molecular evolution.

Akashi H 2001 Gene expression and molecular evolution. Cur. Opin. Genet. Dev. 11:660–666.

Chen Y, Carlini DB, Baines JF, Parsch J, Braverman JM, et al. 1999 RNA secondary structure and compensatory evolution. Genes Genet. Syst. 74:271–286.

Ohta T & JH Gillespie 1996 Development of neutral and nearly neutral theories. Theor. Pop. Biol. 49:128–142.

Acknowledgments We thank our colleagues J. Baines, J. Braverman, D. Carlini, H. Innan, D. Kirby, S. Muse, J. Parsch, and S. Tanda who contributed to this long-term project. Our research has been supported by NIH and the University of Munich.

10

Evolution of Eukaryotic Genome Structure

DMITRI PETROV
JONATHAN F. WENDEL

The discipline of molecular evolution has until recent years focused on understanding patterns and processes of evolutionary change at the level of individual nucleotides, amino acids, and genes. Emerging from the fields of population genetics and evolutionary theory, and enabled by the advent of molecular biology, the nascent discipline of molecular evolution spawned a seemingly endless series of novel insights into the molecular nature of genetic variation and the fundamental forces that are responsible for molecular evolutionary change. Thus, within the span of a couple of decades, our understanding of the evolution of genes, mutational biases, natural selection, and genetic drift was dramatically enhanced. Notwithstanding these many advances, evolutionary processes operating at scales above that of individual genes were largely inaccessible until the last decade or so. These processes operate at the level of the whole genome, shaping assemblages of interacting and coevolving genes, influencing the size and structure of the genome, and, ultimately, affecting its function. Although beyond the scope of traditional molecular evolutionary studies, the exploration of genome-wide processes has in recent years been advanced by an explosion in genome sequencing efforts and in other relevant technologies.

As the first eukaryotic genomes were sequenced, it became clear that many features of eukaryotic genomes are more evolutionarily labile than previously thought. It has long been recognized that amounts of nongenic DNA can vary by two or even three orders of magnitude (Gregory 2001a; Mirsky & Ris 1951), even among similar organisms (e.g., within insects or plants), but insights into the precise nature of these differences were lacking. The rapidly accumulating genome sequencing and survey data revealed in exquisite detail for the first time the nature of the size variation, showing how it largely reflects differences in the type and copy numbers of a very large spectrum of transposable elements and other repeated sequences. The new data also revealed a number of surprises, including the fact that gene complements may vary among closely related species and sometimes even among populations of individual species. Moreover, the rates of duplications of genes, chromosome segments, and even whole genomes, have also proven to be extremely high. For example, humans and chimpanzees may differ by as many as 700 gene duplications that have arisen independently in the two lineages since their split approximately 6 million years ago (Mya) (Frazer et al. 2003). Perhaps even more surprising, as many as 1–2% of genes in *Arabidopsis thaliana* show presence/absence polymorphism between two randomly chosen strains (Borevitz et al. 2003). These and many similar examples indicate that processes of molecular evolution, and hence of phenotypic evolution and adaptation, may be driven as much by wholesale additions and removal of genetic material as by the more slowly accumulating changes in the amino acid and nucleotide sequences of phylogenetically conserved proteins.

In this chapter we focus on advances in our understanding of eukaryotic genome structure and evolution made possible by the data explosion of the last several years. Due to space limitations we have restricted ourselves to the evolution of nuclear genomes. In addition we do not discuss many important advances in understanding of karyotypic evolution in general and genome rearrangements in particular. Given the extraordinary variation among organisms with respect to DNA content, we thought it natural to structure our discussion on genome evolution around the central theme of genome size variation. We highlight some of the major advances emerging from ongoing genome studies and consider implications regarding function and adaptation.

CONCEPTS

The Data Explosion

To appreciate the scale and scope of the recent exponential increase in sequence availability one need only visit one of the major web-accessible gateways to this information. Particularly useful sites are provided by the National Center for Biotechnology Information (www.ncbi.nlm.nih.gov), The Institute for Genomic Research (www.tigr.org), and the European Bioinformatics Institute (www.ebi.ac.uk/genomes), which contain links to ongoing as well as "finished" genome sequencing projects for bacteria, viruses, organelles, archaea, and eukaryotes. At present the complete sequences of over 1000 viruses and 100 prokaryotes are available, as are finished or nearly finished sequences of one or more members of all three clades of higher eukaryotes, representing fungi (yeast), animals (human, mouse, rat, pufferfish, mosquito, fruit fly, nematode, *Tetrahymena*, several parasites), and plants (*Arabidopsis*, rice). By the time you read this chapter, the number of completely or almost completely sequenced eukaryotic genomes will undoubtedly have increased much further.

One caveat for the genome sequencing of higher eukaryotes, and for the dozens of additional species with ongoing genome sequencing projects (preceding web sites; see also www.genomesonline.org), is that at present only the genomes of *Saccharomyces cerevisiae* and *Caenorhabditis elegans* are thought to be completely sequenced, whereas in the other genomes listed, one or more heterochromatic regions remain to be sequenced. In addition, it is important to note that the sampling of the genomes is biased toward model organisms with small genomes.

One illustration of the rapid expansion in sequence information is offered by a table at the TIGR web site (www.tigr.org/tdb/contig_list.shtml), which shows the current and past record-holders for the longest contiguous DNA sequence known, which now stands at 87,410,661 nucleotides (human chromosome 14) but was only 315,339 nucleotides (yeast, chromosome 3) scarcely a decade ago. This nearly 300-fold increase in the span of a decade is mirrored in all aspects of data deposition in national and international databases, and is emblematic of the exponential growth in data relevant to understanding processes of genome evolution.

Evolution of Genome Size

It has been known for over 50 years (Mirsky & Ris 1951) that eukaryotic genome sizes vary by many orders of magnitude in a fashion that defies simple explanations based on perceived differences in organismal complexity. Because DNA is the physical substrate of genes, it is natural to think that the more complex organisms would require more genes, and thus that they would also carry more DNA. However, the very first observations (Mirsky & Ris 1951) showed that many apparently simple organisms (such as amoebas) could have over a thousand times more DNA than presumably more complex multicellular organisms. This lack of correspondence between phenotypic complexity and genome size has been termed the C-value paradox (Thomas 1971), where C-value stands for "chromosome complement" and refers to a haploid genome size. The C-value paradox is little better understood today than it was 50 years ago, although genomic sequence data have defined it more clearly and have clarified some aspects of genome size variation.

Figure 10.1 shows the range of genome size variation in different eukaryotic groups. Overall genome size varies by at least 200,000-fold (Gregory 2001a), with order-of-magnitude differences within groups of closely related organisms being common. Examples include protozoans, which exhibit a 5800-fold range in genome sizes (with *Amoeba dubia* having the largest genome known to date), fish (350-fold), algae (5000-fold), and angiosperms (more than 1000-fold) (Cavalier-Smith 1985; John & Miklos 1988). Among terrestrial arthropods, haploid genome size varies by over 100-fold (Gregory 2001a), from 90 Mbp in *Mayetiola destructor* (Hessian fly) (Ma et al. 1992) to 16,700 Mbp in some acridid grasshoppers (Westerman et al. 1987).

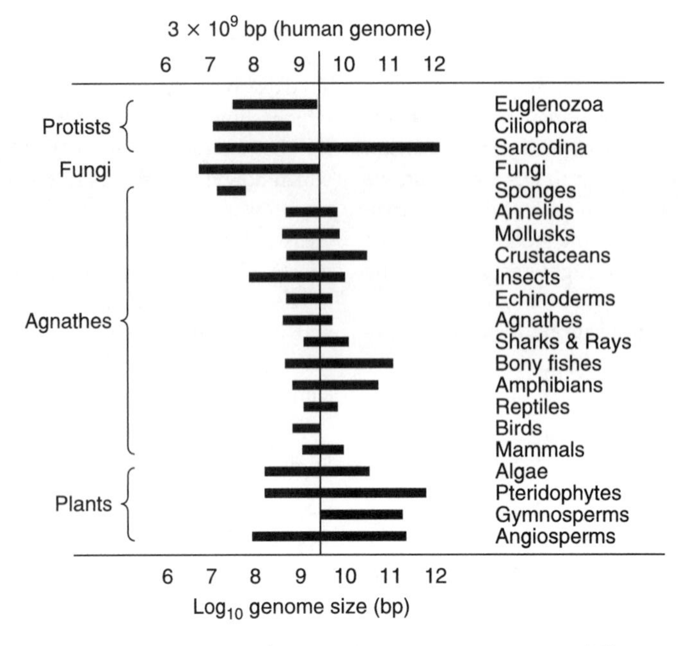

FIGURE 10.1. Range of genome size variation in different eukaryotic groups on a log scale. The vertical line corresponds to the size of the human genome as a reference.

The genome sequencing projects also conclusively confirmed that differences in genome size are not due to variation in the number of genes (Table 10.1). For example, the 2 insects in the table, mosquito and fruit fly, vary 2-fold in genome size yet have practically identical numbers of genes. Similarly, the mouse genome is 20 times as large as that in the flowering plant *Arabidopsis thaliana*, yet the numbers of genes in these two genomes are rather similar. Overall, across eukaryotes the number of genes varies by approximately 10-fold, in sharp contrast to a greater than 200,000-fold variation in total genome size.

Given that variation in gene content fails to account for differences in genome sizes, it becomes self-evident that DNA content variation arises

TABLE 10.1. Sizes of some sequenced eukaryotic genomes, illustrating the lack of correspondence between gene numbers and organismal complexity

Species	Genome size (Mb)[a]	No. of genes[a]
Saccharomyces cerevisiae	12	6,200
Caenorhabditis elegans	100	21,200
Arabidopsis thaliana	125	25,500
Drosophila melanogaster	137	13,700
Ciona intestinalis	162	15,500
Anopheles gambiae	278	13,700
Fugu rubripes	365	32,000
Oryza sativa	466	45,000
Mus musculus	2500	30,000
Homo sapiens	2900	35,000

[a]Data on genome sizes and approximate gene numbers may be found through links at the *National Center for Biotechnology Information* (www.ncbi.nlm.nih.gov), *The Institute for Genomic Research* (www.tigr.org), and from "genomes online" (www.genomesonline.org). See also *Science Online* supplementary data from (Lynch & Conery 2003). Gene number estimates for some organisms (e.g., *O. sativa*) vary widely, so approximate means are shown.

almost exclusively from variation in the amount of noncoding, nongenic DNA. In a certain sense this solves the C-value paradox, but also generates a host of new questions. What is the origin of the nongenic DNA? Which evolutionary forces are responsible for the extraordinary variation in its amount? How much diversity exists in the spectrum of sequence types and classes that comprise this noncoding fraction? Does the nongenic DNA have a function?

There is broad consensus that genomes have much more DNA than demanded by the informational needs of the organism. However, some theories propose that this extra DNA nevertheless has a function because in some cases DNA abundance, rather than its informational content, could have a direct and significant effect on phenotype (Bennett 1971). For instance, it has been suggested that a larger genome size might be adaptive in some organisms because it directly or indirectly increases nuclear volumes (Cavalier-Smith 1978), helps to buffer fluctuations in the concentration of regulatory proteins (Vinogradov 1998), or protects coding DNA from mutation (Hsu 1975). These proposals are buttressed by correlations between physiological, ecological and developmental characters, and genome size, underscoring the potential functional importance of genome size per se (for review see Cavalier-Smith 1985; Gregory & Hebert 1999; Knight et al. 2005).

Not surprisingly, there are also hypotheses about genome size that suggest that the nongenic DNA is useless, maladaptive, "junk" DNA carried passively in the chromosomes (Ohno 1972). More recent versions of the "junk DNA" hypothesis propose that the "junk" DNA is composed mainly of transposable elements (TEs) that accumulate through transposition ("selfish" DNA hypothesis) (Doolittle & Sapienza 1980; Orgel & Crick 1980) or more passively as a consequence of the permissive selective environment provided by small population sizes in higher eukaryotes relative to prokaryotes (Lynch & Conery 2003).

A related but distinct question concerns the evolutionary forces that generate genome size variation. Generally, theories that propose that large genomes have a functional consequence due to size alone also suggest that large genome sizes have been fashioned through directional, positive selection. Junk DNA proposals typically postulate that genome size variation is generated through mutational processes (pseudogene formation, particularly following gene duplication, and TE insertion) and genetic drift overwhelming deleterious effects of "junk" DNA accumulation (Lynch & Conery 2003).

On the other hand, during the process of pseudogene formation, or perhaps more commonly much later, pseudogenes and TE insertions fixed by genetic drift as junk DNA might be "colonized by functions" (Zuckerkandl 2002). In a sense, once "junk" exists in a genome, it becomes available as a substrate for acquisition of novel functions. Alternatively, a large genome fashioned by positive selection may become full of nonfunctional "junk" DNA once the selective environment changes. Noncoding DNA useful at one time may become extraneous later, but nonetheless may persist in the genome for extensive periods by "inertia," thereby raising the possibility that it once again may acquire a global or local relevance to gene or genome function.

Setting aside for the moment the questions about functionality of noncoding DNA, it is important to ask what is known about the processes that guide its evolution. In particular, we would like to understand better which molecular mechanisms generate DNA length mutations, as well as the extent to which selective and neutral forces (such as biased gene conversion) promote or retard fixation of these length mutations within populations (Petrov 2001).

Main Molecular Mechanisms that Add DNA to Genomes

Many molecular mechanisms (Figure 10.2) are capable of generating length mutations. Most dramatic is wholesale genome duplication, which in a single event doubles the entire genome (Wendel 2000). Another major force is the spread of TEs, which are present in all eukaryotic genomes and have the potential to repeatedly add large amounts of DNA (from ~0.5 to ~10 kb in each transposition event) during bursts of transposon activity. In some cases, these bursts may be stimulated by hybridization, generating hundreds of de novo insertions in a single generation (Petrov et al. 1995). Both of these apply upward pressure on genome size.

Genome Duplication and TE Expansions in Maize and Cotton: Case Studies

These two prominent processes, polyploidy and proliferation of TEs, are illustrated well by examples from cotton (Figure 10.3) and maize (Figure 10.4), respectively. In cotton, two species groups that had

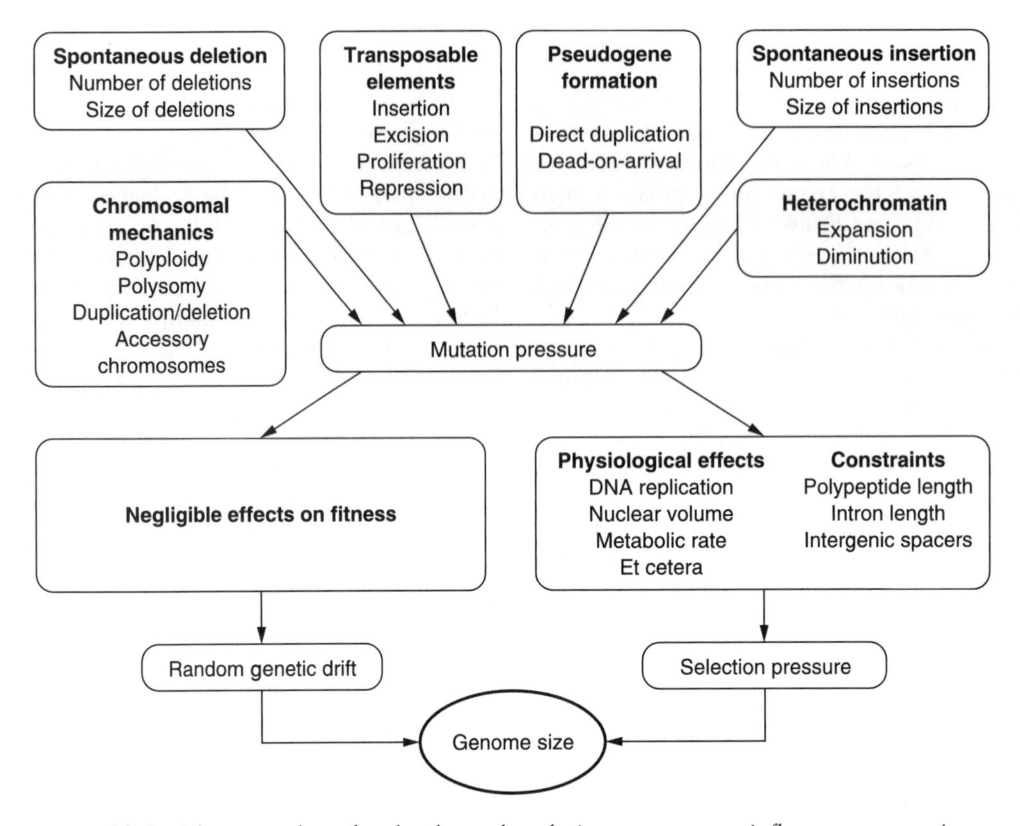

FIGURE **10.2.** A large number of molecular and evolutionary processes influence genome size.

been evolving independently in different hemispheres for perhaps 5–10 million years reunited following a chance trans-oceanic dispersal event approximately 1–2 Mya. This led to hybridization and chromosome doubling among the highly differentiated genomes, generating a new allopolyploid (i.e., coming from the union of the genomes from two different species) lineage that subsequently radiated into modern lineages that include the commercially important cultivated cotton (*Gossypium hirsutum*). This mode of speciation in plants is very common (Wendel 2000). Following polyploidization, most lineages lose much of the duplicated genomic material over evolutionary time only to undergo a later round of allopolyploid (i.e., union of two genomes) or autopolyploid (i.e., duplication of the genome in one species) formation. This repeated, episodic, cyclical process of genome doubling followed by deletion and divergence is evident in all sequenced plant genomes, most clearly by their multiple, nested sets of duplicated genes and chromosome segments (Vision et al. 2000; Wendel 2000).

This latter pattern is also apparent in the modern maize genome. In maize a polyploidization event approximately 20 Mya (Gaut & Doebley 1997) left numerous vestiges of duplicated chromosome segments despite extensive gene loss that followed this genome doubling (Ilic et al. 2003). In addition, maize exemplifies the genomic effects of proliferation of TEs (Figure 10.4), which accumulate preferentially between genes, often inserting into sites near to previous insertions, thus generating intergenic graveyards of dead retrotransposable elements inserted into one another from sequential insertions over millions of years. This process generates a highly clumped gene distribution, with dense genic islands embedded in the sea of TE insertions (Fu et al. 2001). These insertions are also expected to increase genome size, and in fact they seem to have done so to a remarkable degree. Careful dating of TE insertions in several genomic regions of maize suggested that a major increase in TE activity and/or fixation in the last 5 million years increased the maize genome by at least 2-fold (Ilic et al. 2003; SanMiguel et al. 1996).

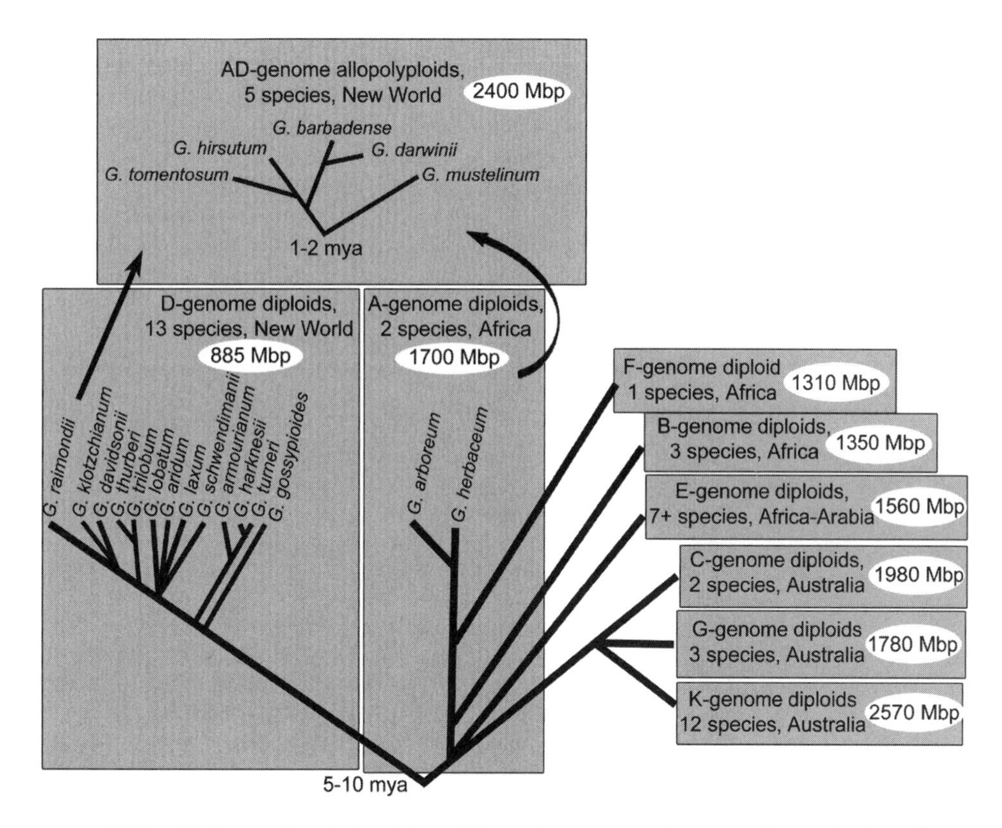

FIGURE **10.3.** Phylogenetic relationships of the major groups of cotton (Gossypium), as determined from multiple molecular data sets (Wendel & Cronn 2003). Following rapid diversification of the genus early in its history, different diploid (all $n = 13$) "genome groups" (designated by single uppercase letters) radiated globally in arid to semi-arid habitats, where they underwent extraordinary morphological, ecological, and genome size evolution (2C DNA contents shown in circles; note the 3.5-fold variation, even among diploids with the same chromosome number). Allopolyploid cottons, which include the commercially important *G. hirsutum* and *G. barbadense,* originated following a trans-oceanic dispersal of an A-genome progenitor to the New World. Sequence data suggest that the A-genome and D-genome groups diverged from a common ancestor 5 to 15 million years ago, and that the two diverged diploid genomes became reunited via polyploidization in a common nucleus and an A-genome cytoplasm.

Genome Duplication and TE Expansions in Eukaryotic Evolution

Whole-genome duplications have undoubtedly played an important role in genome evolution of eukaryotes. We already mentioned that genome duplication in plants is extremely common. Genome duplication in other major eukaryotic groups has also been inferred from the distribution of duplicated genes and from the observation of duplicated syntenic chromosome blocks; examples include ancient genome duplications in the yeast *Saccharomyces cerevisiae* (Wolfe & Shields 1997) and in vertebrates (Abi-Rached et al. 2002; Ceoighe 2003; Friedman & Hughes 2001). More recent polyploidization events have also been found in a diverse set of organisms, including insects (Stenberg et al. 2003), frogs (Ptacek et al. 1994), and rodents (Gallardo et al. 2003).

Although polyploidy has been widespread in eukaryotic evolution, genome duplication has not been particularly frequent in the nonplant lineages studied. For example, in yeast a single genome duplication approximately 100 Mya is detectable (Wolfe & Shields 1997), implying that genome

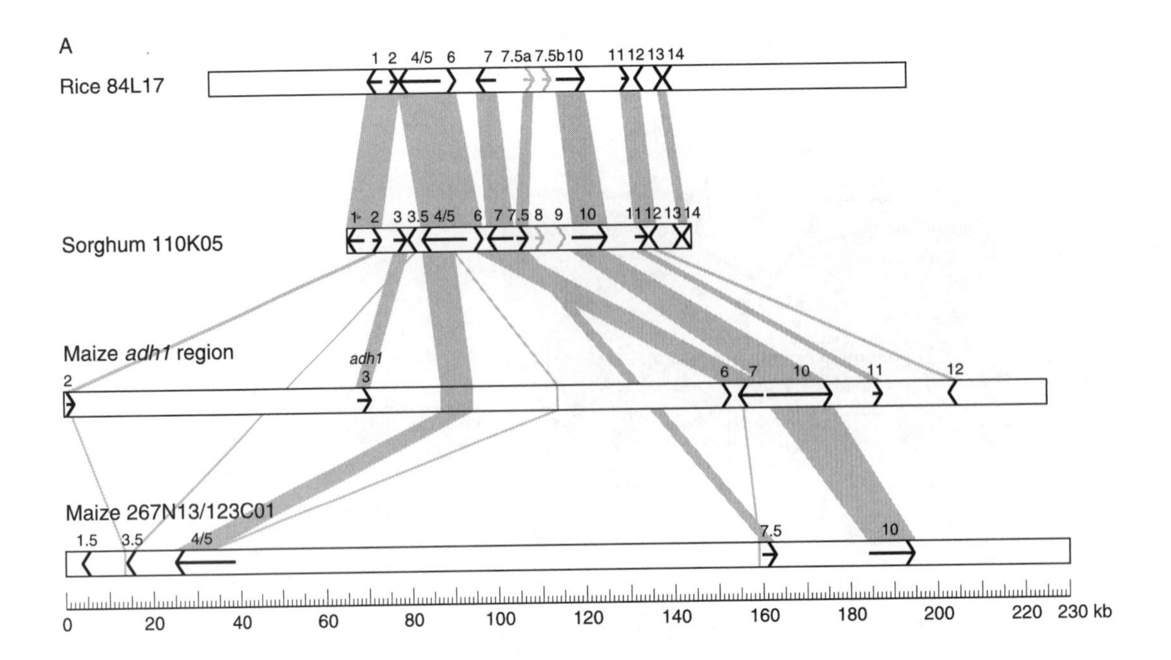

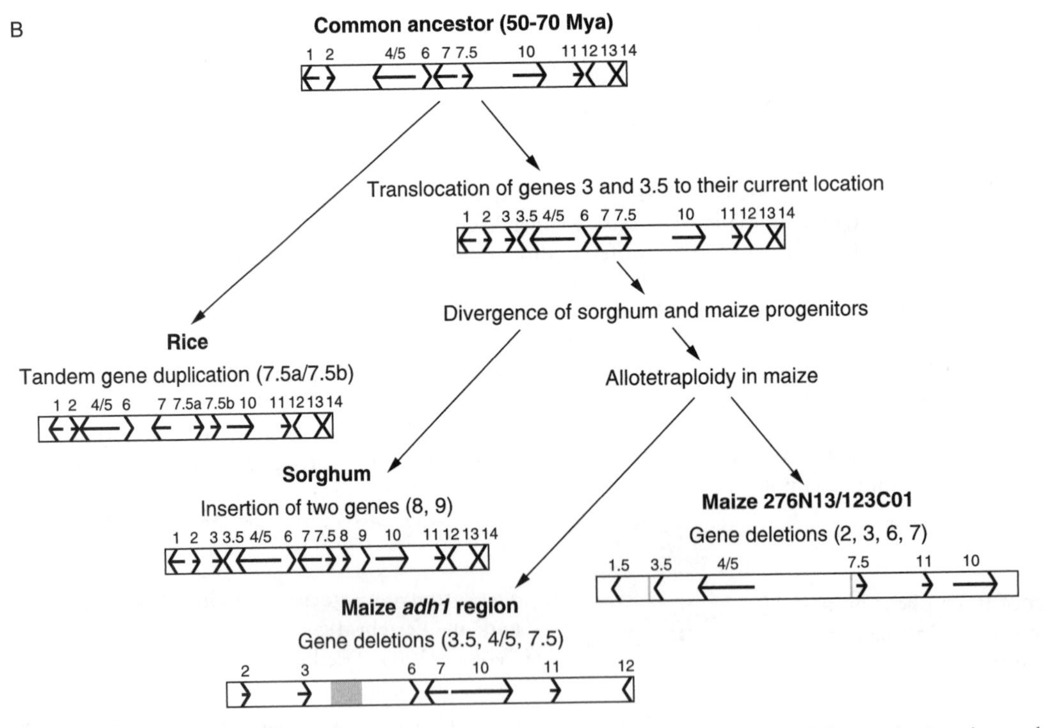

FIGURE 10.4. Transposable element (TE) activity, genome expansion, and gene deletion during the evolutionary history of maize. (A) Homologous regions of the rice, sorghum, and maize genomes, the latter being represented twice because of an ancient genome doubling (polyploidy) event. Numbers denote genes, which are shown as shaded connections when conserved among species. Note the greater distances between the genes in maize than in the other two genomes, reflecting relatively recent TE proliferation. Note also the many missing genes (e.g., 2, 3, 6, 7) in the two duplicated regions, indicating loss of duplicated genes following genome doubling. (B) An evolutionary scenario for the history of this genomic region, which depicts the interplay of gene duplication, gene loss, and genome duplication during organismal divergence.

duplications do not take place much more frequently than once every 100 million years. The rate of genome duplication in most other nonplant eukaryotic lineages studied is unlikely to be much higher, as recent duplications in such lineages are unknown. On the other hand this rate is sufficiently high that any particular nucleotide has fairly similar probabilities of changing by point substitution, being included in small-scale segmental duplication (1–200 kb), or being duplicated by polyploidy (Lynch et al. 2001).

TE expansions also appear to be of great importance in all known genomes. The story of the maize genome is probably one of the most dramatic and best-understood examples of genome expansion due to TE activity. But many other genomes tell us similar stories of episodic, but extremely powerful TE expansions. For example, the human genome shows evidence of an invasion approximately 60 Mya by nonautonomous, retrotransposable *Alu* elements. These elements then proceeded to build up to greater than 1 million copies through periodic expansions, such that *Alu* elements now constitute upward of 10% of the human genome (Lander et al. 2001). Other TEs have also gone through wave-like expansions in the human genome (Lander et al. 2001).

The expansions of TEs are not limited to moderately large genomes such as are found in mammals (Figure 10.1). Even in the compact *Drosophila melanogaster* genome (~170 Mbp), in which fixations of TEs are rare, there is evidence of a massive fixation of thousands of nonautonomous, retrotransposable *DNAREP1* elements approximately 5–10 Mya (Kapitonov & Jurka 2003). It is also likely that yet to be sequenced, heterochromatic regions of the *D. melanogaster* genome will tell us about expansions of other TEs as well. Even in euchromatic regions of *Drosophila*, however, fixations of TEs have been documented (Petrov et al. 2003).

The foregoing discussion of polyploidy and episodic bursts of TE activity begs the question as to whether these processes alone could account for the majority of variation in eukaryotic genome size. Because both mechanisms lead to genome expansion rather than contraction, the consequence would be that genomes exclusively ratchet upward in size over evolutionary time (Bennetzen & Kellogg 1997). This seems unlikely given that many disparate lineages have genomes that are very small, which, by this theory, must have remained very small since the dawn of time. Somehow these small genomes must have avoided TE expansions and polyploidization events for billions of years. Neither seems likely and, as pointed out above, even plants with very small genomes, such as *Arabidopsis*, have experienced several rounds of genome doubling followed by massive loss of redundant DNA (Vision et al. 2000). Moreover, phylogenetic analyses of specific lineages appear to show both increases and decreases in genome size (Wendel et al. 2002b).

Growth and Decline of Eukaryotic Genomes: Variation in the Rate of Deletion and Insertions

The evidence for genome size contraction indicates that there must exist processes that counteract unlimited genome size growth. Indeed many molecular mechanisms that generate deletions are known. These include DNA loss during the repair of double-stranded breaks (Kirik et al. 2000; Orel & Puchta 2003), homologous and illegitimate recombination (Devos et al. 2002), replication slippage, and likely others.

The known deletional mechanisms are capable of reducing genome size across all levels of organization. However, they are not expected to be very quick-acting unless aided by natural selection. This is because the potentially fast removal of DNA by large deletions should be attenuated by the higher probability of such large deletions to remove vital, genic parts of genome (Petrov 2002b; Ptak & Petrov 2002). Such deleterious large deletions will be quickly removed by natural selection and will have no effect on the genome size. Smaller deletions should have an easier time removing nonessential DNA and shrinking the genome without affecting genes. However such small deletions remove DNA rather slowly. In the fastest cases, such as in *Drosophila* (Petrov 2002a), small deletions are expected to remove half of all "junk" DNA in approximately 10–15 million years.

In principle, DNA loss through various deletional processes could counterbalance genome size growth through polyploidization and TE expansions. Indeed, for relatively small indels, deletions predominate in all known cases (Petrov 2002b). Although polyploidization and TE expansion are powerful forces, they are episodic, resulting in sudden, massive changes in genome size. DNA loss through deletions, on the other hand, may be a slower but more persistent process. In the long term they may be equally powerful.

Variation in the Rate of Deletion and Insertions

As noted earlier, pseudogenes are nonfunctional copies of functional genes, and consequently they evolve without the constraints imposed by the requirements of translation into functioning proteins. When we know the sequence of the functional antecedents of pseudogenes, it becomes possible to infer the number and types of indels and nucleotide substitutions that a pseudogene has accumulated since its inception. By comparing rates of different types of substitutions it becomes possible to assess the relative rates of different types of mutational events.

The use of pseudogenes is limited by their rarity in many organisms. Fortunately, practically all genomes contain sequences that by their very nature can be assumed to evolve in a pseudogene-like fashion. For example, insertions of mitochondrial DNA into animal nuclear genomes (the so-called numts) are inevitably noncoding due to the difference in genetic codes between nuclear and mitochondrial genomes (Bensasson et al. 2001). Non-LTR (long terminal repeat) retrotransposable elements also generate 5′-truncated, pseudogene-like copies (termed "dead-on-arrival," or DOA elements) as a natural by-product of their transpositional life cycle (Petrov et al. 1996).

Both numts and DOA non-LTR elements have been used successfully to estimate rates and patterns of nucleotide substitutions, insertions, and deletions in a number of organisms (Petrov 2002b). The first such estimate in two *Drosophila* species groups (*D. melanogaster* subgroup and *D. virilis* group) (Petrov & Hartl 1998; Petrov et al. 1996) indicated that the rates, and more importantly the sizes, of small deletions are much greater in *Drosophila* than in mammals. In fact, in *Drosophila* a nucleotide has about a 3 times higher probability of being removed by a small deletion than of being substituted. In mammals, by comparison, the reverse is true—the probability of change is about 7 times higher than the probability of being lost. This suggested that one reason for the compactness of the *Drosophila* genome is that truly "junk" DNA has a short persistence time in *Drosophila* relative to mammals.

Thus, species may differ in the degree to which their genomes exhibit "housekeeping" skills. In this sense *Drosophila* has little temporal tolerance for messiness and as a result ends up with a very "tidy" genome (Petrov et al. 1996; Singh & Petrov 2004), consisting mostly of functional sequences. The human genome, on the other hand, is more like a messy attic; almost full-length copies of TEs inserted prior to 250 MYA in the human lineage are still present in large numbers (Lander et al. 2001). Although the more ancient insertions may be unrecognizable, they are probably still taking up genomic space.

In principle, a bias in the fixation of small deletions and insertions may arise from mutational biases or from natural selection. One possibility is that natural selection for smaller/larger genome sizes biases the pattern of fixation for small indels. This does not seem to be the case however (Petrov & Hartl 2000). Briefly, natural selection at the level of genome size is expected to affect indels in a predictable, length-dependent manner. The observations do not fit these expectations. For example, if small deletions were advantageous to a significant degree in compact genomes, then much longer insertions of TEs should be strongly deleterious. Occasional fixations of TEs in *Drosophila* make this possibility unlikely.

Direct mechanistic studies also show that mutational patterns of deletions can vary in different organisms. In plants a beautiful study by Kirik and coauthors (2000) showed that during repair of DNA breaks in *Arabidopsis*, which has one of the smallest plant genomes known (*ca.* 125 Mbp), deletions were larger and more frequent than in *Nicotiana*, which has a 20-fold larger genome. Moreover, in *Nicotiana*, 40% of the deletions were accompanied by insertions. In a parallel study (Orel & Puchta 2003), it was recently shown that free ends of DNA molecules were more stable in *Nicotiana* than in *Arabidopsis*, either because of lower exonucleolytic activity or better protection of DNA ends. In either case, this difference in degree of exonucleolytic degradation of DNA breaks may contribute to interspecific differences in genome size (see Case Study below).

This suggestion of a relationship between genome size and rates of small deletion is supported also by the measurements of indel (insertion and deletion) rates in several diverse eukaryotes (reviewed in Petrov 2002b; although see Denver et al. 2004). The highest rates of DNA loss through small indels have been observed in organisms with fairly compact genomes (such as *Drosophila* or *C. elegans*) and the lowest rates are observed in organisms with gigantic genomes (such as *Podisma* grasshoppers or *Laupala* crickets). Using the six data points for which sufficient data exist (Figure 10.5), there is an overall statistically significant ($P < 0.001$) power regression of the form:

(Genome size) $\propto$ (Rate of DNA loss through small indels per bp per bp substitution)$^{-1.3}$. (10.1)

The detection of this relationship based on such limited data is unexpected given the multitude of other forces that affect genome sizes at scales beyond 400 bp (Figure 10.2). These results are intriguing but at best are preliminary as they are based on small amounts of data for a few diverse lineages. They deserve additional experimental attention in the future, however.

Fast Local Increases and Slow Global Decreases?

Phylogenetically broad comparisons of eukaryotic genomes show that larger genomes tend to be larger at all levels of organization. Larger genomes tend to have not only more TEs but more insertions of organellar DNA, more pseudogenes, and longer introns (Deutsch & Long 1999; Vinogradov 1999).

Over smaller time scales the pattern is much less predictable. The correlation of genome and intron sizes is particularly illustrative. There is a clear positive correlation between genome size and intron size across all eukaryotes. On the other hand, in the cases of recent and sharp changes in genome size such correlations are not observed. For example, several species of cotton that differ sharply in genome size show no difference in the length of orthologous introns (Wendel et al. 2002a). Similarly, in the case of the 6-fold growth of maize genome, we find that orthologous introns in Rice (genome size ~ 400 Mbp) and maize (genome size ~2500 Mbp) exhibit no difference in size (Petrov et al. unpublished data).

One explanation for the different patterns observed at short and long time scales is that the powerful forces capable of changing genome size quickly, such as polyploidization and TE expansions, do not appear to significantly modulate intron sizes. Polyploidization can only change the total number of introns, and many TEs (such as particularly active retrotransposable elements) in many cases either tend to avoid inserting into introns or tend to be strongly deleterious whenever they do so. In contrast, the slower acting but persistent small insertion and deletion events are capable of modulating genome size at all levels of organization (as long as they are functionally free to vary in size). However, small indels generate sufficient impacts only over longer

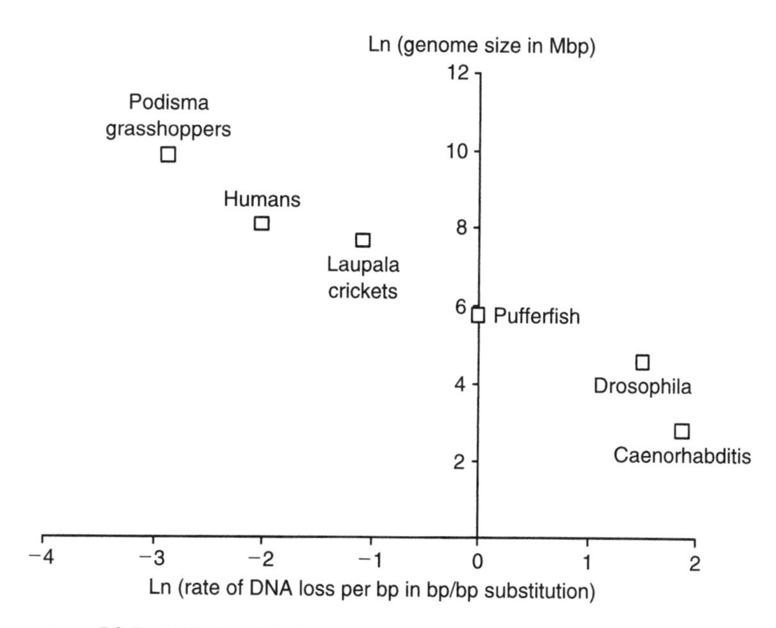

FIGURE 10.5. Relationship between genome size and rate of loss of DNA through small deletions. Negative relationship (depicted on a ln/ln scale) between the rate of DNA loss through small deletions (per nucleotide per base pair substitution) and the genome size.

periods of time. In this regard it is important to note that just as with the relationship between intron size and genome size, the relationship between rates of DNA loss through small indels (Figure 10.5) and genome size is only evident over very long periods of time measured in hundreds of million years.

It is possible that rates of DNA loss through small deletions contribute to an overall baseline genome size that may be saltationally impacted through powerful, quick-acting mechanisms such as episodic TE bursts and polyploidization. Large deletions may also be important in large genomes, where extended areas devoid of genes are common (Devos et al. 2002; Fu et al. 2001; Walbot & Petrov 2001). Such large deletions may help return the genome size back to the original value after powerful increases brought about by TEs and genome duplications.

Functional Significance and Macroevolution of Genome Size

So far we have focused on molecular mechanisms of genome size change and have not directly addressed the phenotypic, ecological, and evolutionary consequences of such change. A large number of correlations between genome size and phenotypic traits of obvious functional significance suggest that genome size may directly affect or coevolve with various phenotypic traits (Gregory 2001b).

Unfortunately we still are largely ignorant about the causal forces that bring about correlations between genome size and life history, phenotypic, or ecological traits. Genome size changes may directly impact phenotype or physiology or may coevolve with those traits in an indirect way. In either case we do not know whether a great deal of genome size change is relatively neutral even if it is directly causative of phenotypic modification. For example, selection at the organismal level may not be strong enough in many cases to stem the growth of TEs or slow down DNA attrition through mutational biases even if genome size directly affects various life history, physiological, or phenotypic characters. It is of course also possible that genome size changes may be driven largely by positive natural selection (Cavalier-Smith 1978). These issues are largely unresolved, but the present rapid growth of genomic data offers some hope of additional insight into the relative importance of natural selection at the organismal level, mutational biases, and TEs in genome evolution.

One interesting wrinkle in this debate is the possibility that natural selection is indeed too weak to significantly alter genome size within populations, but that lineages with large genomes pay a cost at a macroevolutionary level. One prima facie reason for this conjecture is the observation that genome size distribution is often biased toward small genomes (Figure 10.6). One way to explain this bias is to suggest that increases in genome size are rare and that only a few lineages experienced them. But as we already discussed in plants, TEs are ubiquitous and polyploidy is exceedingly common (Wendel 2000). Both these processes operate rapidly on evolutionary time scales. Certainly it appears that there has been enough time and ample means for all plant genomes to become large. But by and large they have not, suggesting that some evolutionary process favors small genomes.

Recently, several studies suggested that lineages with large genomes may pay an ecological cost, in that they are excluded from extreme habitats (Knight & Ackerly 2002) and are less able to invade new habitats (Bennett et al. 1998; Knight & Ackerly 2002). In addition, it has been shown that plant lineages with larger genomes seem to be less species-rich over evolutionary time scales and are more likely to be endangered at present (Knight et al. 2005; Vinogradov 2003). Interestingly, the reduction in diversity and exclusion from extreme habitats is only evident for the lineages with extremely large genomes (Knight et al. 2005). These results, to the extent that they apply generally across eukaryotes, may explain the skew in the distribution of genome size. Lineages with larger genomes may experience higher rates of extinction and/or lower rates of speciation, thereby pruning large genomes from the evolutionary tree.

CASE STUDY

Species-Specific Differences in DNA Break Repair Contribute to the Evolution of Genome Size Differences

Most of our understanding of the evolution of genome size has become apparent through comparative observations of natural patterns of variation. In some cases, though, key insights have emerged from direct experimental evidence. A particularly compelling example concerns experiments in two

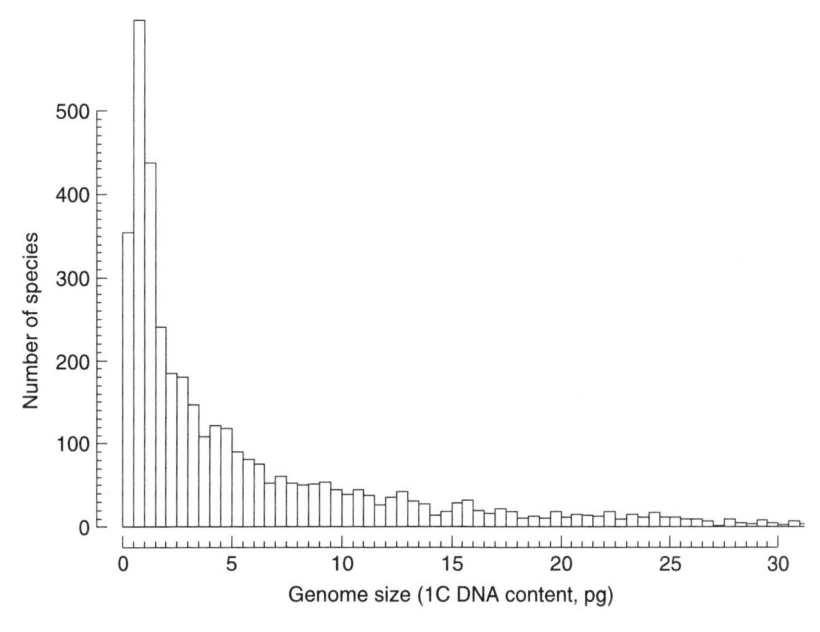

FIGURE 10.6. Histogram of genome size in 3493 angiosperm species from the Kew DNA C-values database (www.rbgkew.org.uk/cval). Note that this histogram is cut-off at a 1C DNA content of 30 pg. The largest genome size in the Kew database is 127.4 pg.

flowering plants, *Arabidopsis* and *Nicotiana* (tobacco), whose genomes differ approximately 20-fold in size (125 vs. 2500 Mb). To explore molecular genetic mechanisms that underlie this difference, Kirik et al. (2000) introduced double-stranded breaks into the two plant genomes using a transgenic system that included a restriction site for a rarely cutting restriction enzyme (I-*SceI*) and a selectable marker. Following induction of double-stranded DNA breaks by transient expression of the restriction enzyme I-*SceI*, cells that underwent break repair and hence reunited DNA ends were evaluated by polymerase chain reaction (PCR) amplification using primer binding sites on the introduced vector that flanked the novel restriction site. Forty amplification products from each of the two species were cloned and sequenced, so that the recombination junctions at the site of DNA breakage could be studied. Interestingly, in *Nicotiana*, filler sequences were observed in about 40% of the cases, ranging in size from 19 to 121 bases. In *Arabidopsis*, however, the plant with the smaller genome, insertions were never observed. For the remaining 60% of the tobacco clones and 100% of the *Arabidopsis* clones, deletions of DNA sequences flanking the breakage site were observed, these ranging in size from about 0.2 to 2.3 kb. The mean deletion size was significantly higher in *Arabidopsis* (920 bp) than tobacco (1341 bp).

Why might deletions be larger in one species than another? To explore this question the same laboratory introduced linearized plasmid molecules into *Arabidopsis* and *Nicotiana* using particle bombardment transformation (Orel & Puchta 2003). Total leaf DNA was isolated 2 days after transformation and the fate of the introduced plasmids was studied using Southern hybridizations and plasmid DNA as probes. The analyses showed that free ends of DNA molecules were more stable in *Nicotiana* than in *Arabidopsis*, either because of lower exonucleolytic activity or better protection of free DNA ends.

The foregoing studies demonstrate that DNA break repair is consistently associated with relatively large deletions in one plant species (*Arabidopsis*) but both insertions and smaller deletions in a second plant species (tobacco), and that these differences between the species reflect variation in the stability of free DNA ends. To the extent that these experimentally introduced DNA breaks and repair events mimic natural mechanisms, they are directly relevant to our understanding of genome size evolution. Extended over an evolutionary time scale, the implication is that species-specific differences in the efficiency and manner of DNA break repair may contribute significantly to genome size variation.

CONCLUSIONS

Eukaryotic genomes are extraordinarily variable. Most conspicuously, eukaryotic genomes vary in size by over 200,000-fold, with practically all this variation due to nongenic DNA. The variability of genome size underscores the power and persistence of many molecular mechanisms that constantly reshape, reorder, and modify genomic landscapes.

Although we have made tremendous progress toward understanding the patterns and processes of genome evolution, many key questions are still unresolved. Probably the most obscure, yet clearly key, question is the degree to which evolution of genome organization is driven by mutational imbalances or natural selection. Clearly, natural selection strongly affects the evolution of many genomic components, most obviously the functional genes. What we do not yet understand is the manner in which selection sees higher levels of genomic organization. Are particular genome structures or overall sizes molded by selection into their present state, or are these features merely the outcomes of selection, genetic drift, and mutational imbalances acting on individual genome components? Putting it differently: To what extent is the overall genome visible to selection?

Notwithstanding the ultimate resolution of this question, it is clear that genome evolution is shaped by a diversity of mutational processes and mechanisms, which, even if selectively near neutral individually, may have important immediate and long-term functional consequences. We are a long way from understanding the interplay between evolutionary forces and functional consequences. An exciting prospect, however, if not an outright promise, is that the present genomic era will soon lead us to a new and much deeper level of understanding in this regard. A true understanding of biology awaits this unraveling of the mysteries of genome function and evolution.

SUGGESTIONS FOR FURTHER READING

For additional reading we would suggest three additional sources. *Genetica* (2002, v. 115) and *Annals of Botany* (2005, v. 95) came out with special issues focused on the evolution of genome size and structure. In addition, in 2004 Academic Press released an edited volume *The Evolution of the Genome* which covers many of the topics discussed here in much greater detail.

11

New Genes, New Functions: Gene Family Evolution and Phylogenetics

JOE THORNTON

Since Darwin, the central project of evolutionary biology has been to explain the origin of biodiversity—to determine how novel species and their characteristics have evolved. With the advent of genome sequencing, an entirely new level of biodiversity has been revealed: the proliferation and diversification of genes within genomes. Discovery of this molecular biodiversity allows evolutionary biology's fundamental question to be asked at a new level: by what mechanisms and dynamics have new genes with novel functions evolved?

The vast majority of genes in sequenced genomes are organized hierarchically into gene families and superfamilies (e.g., Lander et al. 2001). This hierarchy of genes, like the nested organization of living organisms, has been produced primarily by a process of lineage splitting and divergence—in this case, gene duplication followed by independent trajectories of sequence substitution. The formal analogy between speciation and gene duplication means that the same phylogenetic techniques used to reconstruct the tree of life from gene sequence data can also be used to infer the evolutionary history of gene families. A reliable gene family phylogeny provides a scaffold on which patterns of variation in sequence, structure, and function can be interpreted to reconstruct the evolutionary events and dynamics that generated gene family diversity.

The purpose of this chapter is to discuss major concepts and techniques for studying the evolution of gene families. Specifically, this chapter covers how to: (1) interpret gene family trees, (2) use trees to identify and date gene duplications and losses, (3) determine the mechanisms by which a gene family diversified, (4) test hypotheses about how gene functions evolve after duplications, (5) determine the functions of ancestral genes, and (6) link the evolution of function to specific changes in gene sequence and structure. In the last section, I will highlight some of our work on the steroid hormone receptors, a family of transcriptional activators that mediate the cellular effects of steroid hormones.

CONCEPTS

Reading Gene Family Trees: Dating Gene Duplications

Phylogenetic methods infer the hierarchical evolutionary relationships among genes based on patterns of shared or similar nucleotide or amino acid states. There are two major categories of methods for inferring phylogenies (e.g., DeSalle et al. 2002; Felsenstein 2004). Maximum parsimony, a nonparametric technique, seeks the tree that explains shared character states, to the greatest degree possible, as due to inheritance from common ancestors. Parametric methods such as maximum likelihood (ML) and Bayesian Markov Chain Monte Carlo (BMCMC) techniques use explicit probabilistic models of the evolutionary process to find the tree with the highest likelihood of generating the sequence data actually observed. Both sets of methods are theoretically sound, widely applicable, and have their own strengths and weaknesses. A third class, distance-based methods, constructs a tree from the pairwise similarity scores for each pair of sequences;

these methods are generally less robust, powerful, and theoretically justifiable than parsimony, likelihood, and Bayesian analysis. They are used more often by molecular biologists seeking a quick method than by evolutionists or phylogeneticists seeking a rigorous one (see discussion in Thornton & DeSalle 2000a).

Once a gene family tree has been inferred, it provides a framework for reconstructing the evolution of the members' sequences, structures, and functions. The first step is to identify gene duplications and their dates. A tree of gene family members from a single species tracks relatedness due to gene duplication. Each node is a gene duplication event; closely related genes are descended from recent duplications, whereas more distantly related genes trace their last common ancestor to a more ancient duplication. When a gene family tree includes members from multiple species, nodes can represent either gene duplications or speciation events. On a tree such as this, orthologs and paralogs will be interleaved on the phylogeny in a complex pattern that hierarchically reflects the order in which gene duplication and speciation events occurred (Fitch 1970).

Consider the example shown in Figure 11.1. In a gene family tree, each branch that descendes from a node representing a gene duplication contains a replica of part of the taxonomic tree—specifically, the portion produced by speciation events that occurred after the gene duplication. If all of the gene duplication events happened before any of the taxa on the tree diverged from each other, then each branch leading to a group of orthologs will contain the entire species tree. If some duplications occurred before and some after the speciation events, then subtrees of various sizes, one for each paralog, will be arranged in nested fashion in the gene tree.

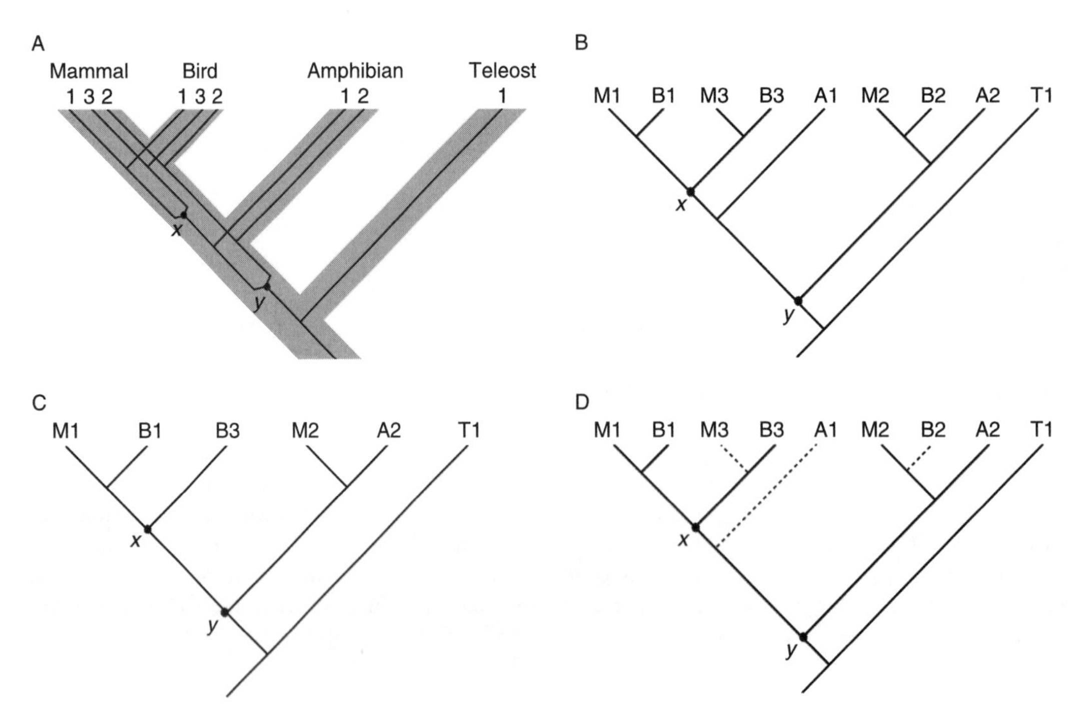

FIGURE 11.1. Hypothetical gene family phylogeny. (A) Gene duplications and speciation events generate a tree of related genes within the genomes of species as they evolve. Thick gray lines show the phylogeny of four major taxa; thin black lines show genes within these taxa's genomes. Filled circles mark gene duplications; unmarked nodes are speciation events. (B) Gene family phylogeny generated by the process in (A). Gene duplications are marked with filled circles. This tree is simply the gene tree in (A) "unfolded." (C) Gene family tree generated by the process in (A), but genes M3, A1, and B2 have been lost or not sequenced. It is still necessary to postulate gene duplications *x* and *y* to reconcile the gene tree with the species tree. (D) Fully reconciled tree from panel (C), with lost or unrecovered genes added back to the phylogeny represented by dotted lines.

Gene duplication events can be located on a phylogeny by comparing the gene family tree with the phylogeny of the species included in it. A gene duplication is obviously required to explain the existence of multiple gene family members within a species, but we need a more explicit rule to determine exactly when during evolutionary history specific gene duplications occurred. A node always has two branches that descend from it. Working backward from the tips of the phylogeny, we evaluate each node by two criteria: (1) Is any single species represented by genes in *both* the clades that descend from the node? If yes, that node is a gene duplication. If the answer to question 1 is no, then: (2) Is the branching order of species compatible with the taxonomic tree? If yes, that node is a speciation event; if no, it is a gene duplication.

The location of a duplication event; on the gene tree gives a lower bound of its age, because the duplication must have occurred prior to the divergence of all the lineages represented within that clade on the gene family tree. An upper bound on the age of a gene duplication can be inferred from speciation events that occurred after the gene duplication on the tree. For example, Figure 11.1B suggests that the duplication labeled *x*—the event that created the paralogous genes 1 and 3—occurred prior to the mammalian–bird split and after the amphibian–amniote divergence; duplication *y* happened after the divergence of tetrapods from teleosts but before the amphibian–amniote divergence. Combined with a priori information on speciation times from the fossil record or molecular clocks, this kind of reasoning allows the chronological date of a gene duplication to be estimated within a restricted window of evolutionary time.

Inferring Gene Loss

Tree-based dating of gene duplications can be used even when some sequences are not included in an analysis because they have been lost from genomes or have not been sequenced. Indeed, the method also allows gene loss events to be specifically located on a phylogeny. This approach uses a "reconciled tree" (Page 1993a), which resolves differences between the species tree and the gene tree by determining gene duplications based on the rules above and then including hypothetical branches for any missing sequences. Thus, the gene tree in Figure 11.1C requires gene duplications *x* and *y* to explain the phylogenetic relationships among the known genes in the family, because the clades descending from these nodes have branching patterns incompatible with the species tree. This inference of gene duplication is possible despite the fact that no species contains all three paralogs in its genome, as would be required to infer gene duplications without a tree.

Once the duplication events have been located, lost or missing genes are added to the tree using a simple rule: every gene duplication should produce two genes in all the taxa that descend from the ancestral species in which the duplication occurred; any exceptions are due to gene loss (or failure to recover a gene). Therefore each of the two clades that descend from a gene duplication node must contain one or more genes from all taxa descended from the taxonomic ancestor in which the gene duplication occurred. If it does not, then a branch that leads to the species from which the gene is missing must be added, and a gene loss postulated along that branch (Thornton & DeSalle 2000a).

This approach assumes that both the gene and species trees are known with confidence; if either tree is weakly supported or erroneous, constructing the reconciled tree can be problematic. Suppose that for some family, the optimal phylogeny based on the sequence data requires seven gene duplications and three losses, but an alternate tree that is only slightly suboptimal for the sequences requires only four gene duplications and no losses. Which is the better reconstruction? Answering this question requires judgment about the relative probabilities of "extra" gene duplications/losses versus "extra" sequence substitutions, where "extra" means more than required under the most parsimonious or maximum likelihood scenario for each type of change considered in isolation. In a parsimony framework, the criteria for inferring the trees and reconciling them can be combined, by calculating the total number of sequence changes plus gene duplications and losses required for each tree. Given a weight that expresses the "cost" of a gene duplications or losses relative to a sequence substitution, the most parsimonious phylogeny overall is the one that minimizes the weighted sum of substitutions and losses/duplications (Goodman et al. 1979; Thornton 2001). An alternative approach is to use a test of confidence in phylogenies inferred from gene sequences to assemble a set of credible phylogenies and then to choose the tree from that set that minimizes gene duplications/losses. This approach has been used in a parsimony framework (Martin 2000), and it could be easily adapted to likelihood or Bayesian phylogenetic methods, as well.

Evolution of New Genes: Identifying Mechanisms

There are many mechanisms that can create new genes (reviewed in Long et al. 2003). For example, entirely new kinds of genes can be created by exon shuffling, which brings together domains of genes from different families to produce a new chimeric gene (Gilbert 1978). New genes can also be produced when adjacent genes fuse into a single open reading frame, or when a gene fuses with an adjacent untranslated region to create a larger gene.

But the diversity of genes within genomes is due primarily to the expansion of existing gene families by frequent gene duplication. One analysis of duplicate genes in eight fully sequenced eukaryotic genomes estimated that on average about one duplication has occurred per gene every 100 million years (Lynch & Conery 2000). This figure implies that virtually all genes have been duplicated several times over the course of eukaryotic evolution. This estimate could be inflated for a variety of reasons (Long & Thornton 2001; Zhang et al. 2001), but the point that gene duplications are frequent is supported by a comparison of the human and mouse genomes, which found that since the divergence of these species there have been about 17 duplications per megabase of aligned genomic sequence—or more than 40,000 duplication events in total (Kent et al. 2003).

There are several mechanistic categories of gene duplication: tandem duplication and transposition may result in copying of a single gene, while duplication of part or all of a chromosome, or global duplication of an entire genome, produces copies of many genes simultaneously. Each of these processes leaves unique traces in the genome. The mechanisms by which a gene of interest evolved can therefore be inferred by interpreting sequences, their phylogenies, and their genomic map positions in a phylogenetic context. The patterns predicted for each mechanism are discussed below and are illustrated in Figure 11.2.

Tandem Duplication

The signature of tandem duplication (Figure 11.2B) is two sister paralogs that are tightly linked on a chromosome. It can be caused by replication slippage or unequal homologous recombination. Studies of the genomes of *Caenorhabditis elegans* and *Drosophila melanogaster* genomes found that about 70% of identified duplicate genes in these species are due to tandem duplication (Rubin et al. 2000). Some 30% of all duplications in human and mouse are in tandem (Kent et al. 2003).

Transposition

The signature of transposition—the reintegration of a DNA or RNA copy of a gene into a new location in the genome (Figure 11.2C)—is unlinked sister paralogs that are not part of larger blocks of multiple duplicated genes. Transposition can occur through DNA- or RNA-mediated mechanisms, each of which leaves a distinct signature. RNA-mediated retrotransposition results in a duplicate gene that has no introns and, in many cases, a poly-A tract in the 3′-untranslated region, as well. Thousands of pseudogenes that carry these traces of retrotransposition have been detected in the human genome (Lander et al. 2001). Many of these retrotranscripts no longer function, but some encode functioning proteins (Lahn & Page 1999). DNA-transposed genes contain the same introns as the original copy and contain no poly-A tract. The conserved terminal sequences that mediate the movement of transposable elements can often be found flanking genes duplicated by transposition (Hughes et al. 2003).

Large-Scale Duplication

Another mechanism for producing paralogs that are dispersed in the genome is duplication of DNA segments that contain many genes—parts of chromosomes, whole chromosomes, or entire genomes (Figure 11.2D)—due to errors in meiosis or hybridization of different species leading to polyploidy. Large-scale duplication simultaneously diversifies all the gene families that have members on the duplicated segment(s). As a result, closely related paralogs are scattered in the genome, but each paralog is linked to members of other gene families duplicated in the same event. Whole-genome duplication is clearly an important process that has been documented in plants, fungi, and animals, usually by showing many examples of copied blocks of linked genes (Kellis et al. 2004; Simillion et al. 2002; Van de Peer et al. 2003). This mode of inference is seldom as simple

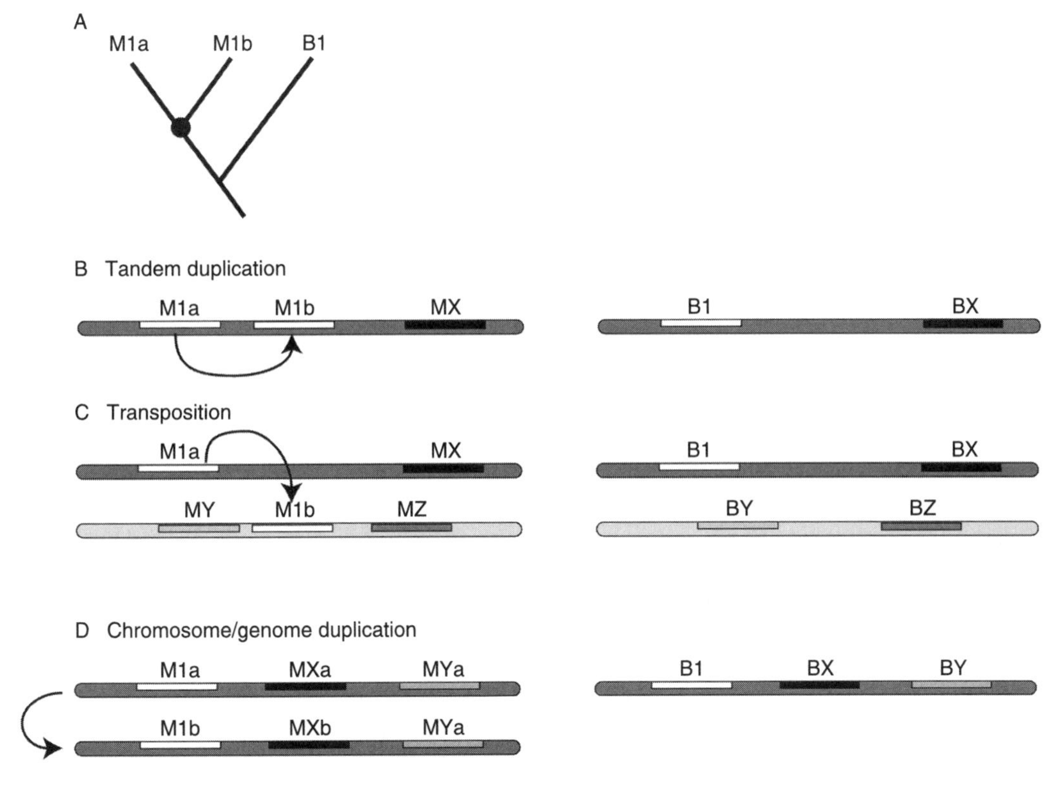

FIGURE 11.2. Mechanisms of gene duplication and associated gene mapping patterns. (A) A simple gene family phylogeny for all other panels in this figure. Gene duplication is marked with a filled circle. (B) Tandem duplication produces duplicate genes M1a and M1b arranged in tandem on the same chromosome. The chromosomes in species M (which has the duplication) are shown at the left; chromosomes in species B (which diverged from M before the duplication) are at the right. Genes X, Y, and Z are members of other families. Arrows show hypothetical duplication events. (C) Transposition produces duplicates that are dispersed in the genome, often to other chromosomes. (D) Whole chromosome or genome duplication produces duplicated clusters of genes on different chromosomes.

as it sounds: some segments copied in genome duplications subsequently undergo additional local duplications, losses that delete duplicate genes, inversions that change gene order, and chromosomal rearrangements that move genes out of the duplicated block. As a result, the expected pattern—fully intact blocks of duplicated genes in identical order—is not always present.

Domain Shuffling

New gene family members (or entirely new families) can be created by combining parts of sequences among more ancient members of the same family or of different families. Most proteins are composed of domains: discrete structural units that have specific and often autonomous functions. Domain shuffling occurs when a domain from one gene is added to or exchanged with one or more domains of another gene, by transposition or nonhomologous recombination. There are a large number of protein domains that are shared among multiple gene families, indicating the importance of domain shuffling in the evolution of new proteins (Patthy 1999a). The traces left by domain shuffling can be recovered by sequence and phylogenetic analysis. When a new gene is created by shuffling the domains of genes from different families, the new gene will have distinct blocks of high sequence similarity to two or more unrelated gene families. When new genes are created by shuffling domains among members of the same family, phylogenetic analysis will show that the shuffled domains of the chimeric gene are most closely related to different paralogs in the family. Separate analysis of the two domains should indicate strong support for incongruent phylogenies. Several statistical tests are available for determining the significance of incongruence between parts of sequences (Farris et al. 1995; Huelsenbeck & Bull 1996; Thornton & DeSalle 2000b).

Evolutionary Dynamics after Gene Duplication: Three Models

How do the functions of genes evolve after duplication? Once a new gene is created, it evolves under the influence of mutation, selection, and drift. Three major models have been proposed to explain what happens to genes subsequent to duplication.

- *The "more-of-the-same" model.* Duplication of a gene per se may confer an immediate fitness advantage if the ability to increase expression of the gene is beneficial (see Ohno 1970). In this case, both copies will be subject to purifying selection and will be maintained in the genome with their ancestral functions.
- *The classic model* was proposed by Ohno (1970). According to this model, one member of a duplicated pair is constrained by purifying selection to carry out the ancestral function, but the other is redundant and is released from selection. The redundant copy then drifts neutrally, eventually meeting one of two fates. If it accumulates mutations that knock out its expression, it becomes a pseudogene, a fate called *nonfunctionalization*; pseudogenes continue to degenerate in sequence, or they may be deleted entirely from the genome. Alternatively—and more rarely—the drifting duplicate may undergo one or more substitutions that by chance confer on it a novel function. If that function is beneficial, the gene's sequence will be optimized by positive selection for the new function, after which it will be constrained by purifying selection. This process has been called *neofunctionalization*.
- *The duplication, degeneration, and complementation (DDC) model* was proposed by Force et al. (1996). Gain-of-function mutations are so much rarer than deleterious ones that virtually all duplicate genes should be expected to become pseudogenes under the classic model. Examination of real genomes shows, however, that a surprisingly large proportion of duplicate genes are retained. The DDC model provides a way for the largely destructive power of mutation to lead to the stabilization rather than loss of duplicated genes. The premise of this model is that many genes have multiple, separable functions: they are expressed at different times or in different tissues under the control of independent regulatory elements. Alternatively, some genes have multiple functions at the biochemical level; for example, crystallins serve both as transparent proteins of the eye lens and as enzymes in other cell types (Piatigorsky & Wistow 1991). After duplication, mutation may knock out one of the functions of copy 1—for example by promoter mutations that eliminate expression of that copy—without fitness consequences, because the gene's function is covered by copy 2. From that point on, copy 2 will be constrained to complement that mutation and is therefore subject to purifying selection. If mutation then knocks out a different function of copy 2, copy 1 will be constrained to fulfill that function. The sequences of both genes are then maintained by purifying selection, because both copies are necessary to carry out all the required functions of the ancestral gene. According to the DDC model, the fact that most mutations are deleterious makes such complementary loss-of-function mutations much more likely than gain-of-function mutations.

Evolutionary Dynamics after Gene Duplication: Testing the Models

How any specific gene evolves after duplication depends on the function of the gene and the population genetic context (Walsh 2003). The post-duplication evolutionary processes associated with each of the three models do leave distinct traces in the sequences, functions, and evolutionary rates of duplicate genes, which can be evaluated on a phylogeny to test hypotheses about the dynamics of any gene duplication that occurred in the past (Figure 11.3). As discussed below, the available data make clear that no single theory explains the

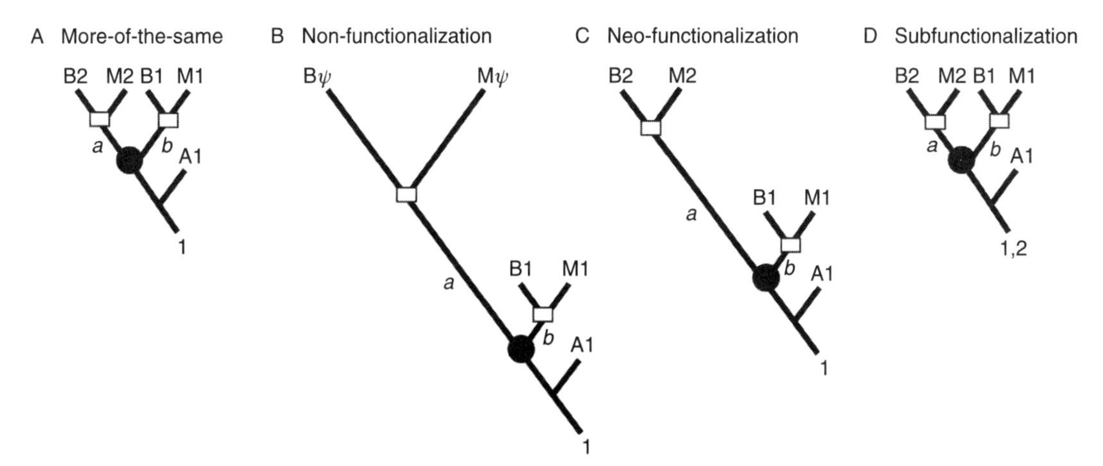

FIGURE 11.3. Rate patterns produced by different evolutionary dynamics after gene duplication. Numbers indicate gene functions. Filled circle, gene duplication. The branches marked *a* and *b* represent equivalent periods of time, beginning at the gene duplication and ending at the speciation event that split taxa M and B. The ratio of these branch lengths is therefore equal to the ratio of the rate of sequence substitution in the two genes during the post-duplication interval. (A) In more-of-the-same selection, having duplicate genes that conserve the ancestral function 1 is advantageous, so both copies are constrained by purifying selection. (B) In nonfunctionalization, one copy is constrained by selection to carry out the ancestral function but the other copy is redundant and diverges at an accelerated rate; it ultimately loses function and becomes a pseudogene (ψ). (C) In neofunctionalization, the redundant copy drifts as in nonfunctionalization, but it amasses substitutions that confer on it a new function 2, after which it is subject to stabilizing selection and a slower rate. (D) In subfunctionalization, the ancestor has multiple functions 1 and 2 which are partitioned among the duplicates. Coding sequences of both duplicates remain constrained by selection, so rates do not increase in either copy.

evolution of eukaryotic genomes: selection for more-of-the-same, the classic model's nonfunctionalization and neofunctionalization, and the DDC model's subfunctionalization have all played important roles in gene family evolution.

More-of-the-Same Selection

Conservation of both copies after duplication due to purifying selection leaves a simple pattern: both copies are constrained, evolving at equally slow rates, and both copies maintain the ancestral function (Figure 11.3A). Selection-for-duplication has been important for a number of gene families in which increased dose is beneficial. Selection for dose on tandem duplicates is usually associated with concerted evolution—the homogenization of paralogs within a genome, typically due to gene conversion—which results in very low sequence divergence among family members within a species (Dover et al. 1993). This model has been corroborated for several protein synthesis genes, including ribosomal RNAs and elongation-related proteins, which are arranged in tandem arrays and have been subject to concerted evolution (Liao 1999). More-of-the-same selection appears to have played little part, however, in the evolution of families of transcription factors and other regulatory proteins, for which dosage increase is of little selective value and no evidence of concerted evolution is present (Semple & Wolfe 1999).

Nonfunctionalization

Nonfunctionalization leaves an obvious pattern: one duplicate is constrained, and the other becomes a pseudogene that degenerates over time and loses expression altogether. Pseudogene sequences are recognizable because their sequences are riddled with stop codons and frameshifts; pseudogenes may also be missing from the genome entirely. Because a pseudogene is redundant and has been released from selection, the rate of sequence divergence in the lineage leading from a gene duplication to a pseudogene is much higher than in the conserved copy (Figure 11.3B).

Nonfunctionalization is clearly a frequent outcome of gene duplication. The analysis of eight eukaryotic genomes estimated that the average half-life of a duplicated gene is 4 million years, which means that the vast majority of duplicates become pseudogenes (Lynch & Conery 2000). Although this estimate has been called into question (Long & Thornton 2001; Zhang et al. 2001), the qualitative conclusion that most duplicates become pseudogenes relatively rapidly unless they are stabilized by selection has been corroborated by numerous other studies. For example, about 17,000 noncoding regions of the human genome contain pseudogenes that have recognizable similarity to functional gene family members (Lander et al. 2001). In the duplicated genome of the yeast *Saccharomyces cerevisiae*, 88% of the duplicates have been subsequently lost as deleted or unrecognizable pseudogenes (Kellis et al. 2004).

Neofunctionalization

In neofunctionalization, the coding sequences of both copies are maintained in the genome; one of the duplicates has the conserved functions of the unduplicated ancestor, but the other copy has novel functions. According to this model, one copy was constrained after duplication while the redundant copy experienced a transient release from constraints until a new function was established. Neofunctionalization therefore predicts an increase in the rate of amino acid replacements on the branch that descends from the duplication and leads to the neofunctionalized copy (Figure 11.3C). After this burst of rapid evolution, the rate of evolution is reduced again as the new function is conserved by purifying selection.

When neofunctionalization occurs, one of the branches descending from a gene duplication will be significantly longer than the other. A branch length is expressed as the estimated number of substitutions that occurred between two nodes; because it reflects both the amount of time elapsed and the rate of substitution, the rate usually cannot be inferred directly from the length. Gene family trees provide a special case, however, that simplifies inferences about relative rates. As Figure 11.3 shows, the amount of time that elapses between a gene duplication and a subsequent speciation event (or the present) is always identical for the two paralogs created in the duplication. As a result, the ratio of the two branch lengths reflects only the relative rates of sequence change during this period of time. Because the duplicated genes existed within the same ancestral species, lineage effects that can affect the divergence rate—such as the mutation rate, generation time, and population size—affect the two copies equally. As a result, differences in branch length reflect only differences in selection pressure, making significant branch length inequality after gene duplication a useful indicator of neofunctionalization.

Branch lengths can differ due to chance alone even if the rates of each gene are approximately equal. To determine whether the deviation from equal rates is statistically significant, a likelihood ratio test can be used. In this case, the hypothesis is that the probabilities of substitution along the branches descending from a gene duplication are unequal; the null hypothesis is that the two branches are of equal length. The likelihood of the null hypothesis can be calculated under a model that constrains the two branch lengths to be the same, whereas the alternative hypothesis allows them to be optimized independently. If the increase in likelihood with unequal branch lengths is statistically significant, then the null model can be rejected, corroborating the hypothesis of neofunctionalization.

An alternative way to test the hypothesis of unequal rates after duplication is to use codon models of evolution that allow the ratio of the rate of nonsynonymous substitutions (K_a) to the rate of synonymous substitutions (K_s) to be calculated in each lineage after gene duplication. (e.g., Yang 2002) This ratio reflects the strength of selection: a ratio close to 1 indicates neutral evolution, a ratio close to 0 suggests strong purifying selection, and a ratio greater than 1 reflects positive selection that fixes amino acid substitutions at a rate greater than caused by mutation and drift alone. The neofunctionalization hypothesis predicts that after duplication the neofunctionalized copy will have a K_a/K_s greater than that for the constrained copy. A likelihood ratio test can be used to determine whether a model that allows the K_a/K_s ratio to vary between the post-duplication lineages provides a significantly better fit to the data than the null model in which the K_a/K_s ratio does not change after duplication. One study, for example, found evidence of asymmetric K_a/K_s ratios after duplication, consistent with neofunctionalization, for 20–30% of gene duplicates in fully sequenced eukaryotic genomes (Conant & Wagner 2003).

Subfunctionalization

The DDC model makes several predictions about the distribution of rates and functions after duplication. First, if subfunctionalization reflects the partitioning of ancestral expression domains among duplicates, then extant duplicate genes should have largely nonoverlapping expression domains, and the union of these domains should equal the expression domains in which the unduplicated copy is expressed in outgroup taxa. A large number of duplicate genes appear to have been subfunctionalized based on this kind of data (Prince & Pickett 2002) Subfunctionalization of expression domains also implies that in most cases divergence rates will not change substantially in either copy after duplication, because the coding sequences of both genes remain constrained to carry out the ancestral biochemical function (Figure 11.3D). If the likelihood ratio test described above does not reject the null hypothesis of equal rates—and the duplicated genes have different functions—the sequence data are consistent with subfunctionalization.

One source of ambiguity in this kind of inference is that expression domains are not the only kind of function that can be partitioned. The ancestral coding sequence's biochemical functions, for example the ability to interact with different ligands or substrates, could also be partitioned after duplication, leading to a release from constraints for different portions of the coding sequence in each copy. If the function maintained in one duplicate requires more sites to be conserved than the function maintained in the other copy, then unequal rates will occur even under subfunctionalization. This makes the DDC model difficult to reject based on branch length analysis, unless the hypothesis is restricted to the degeneration and complementation of expression domains and excludes other kinds of function.

In summary, each of the major models to explain the fate of duplicate genes seems to be valid for some genes. The classic model explains clearly why so many duplicate genes become pseudogenes and are ultimately lost from the genome. The DDC and more-of-the-same models explain why so many genes that do not have novel functions have not been lost. But what about the thousands of genes that do have unique functions: how did these new functions evolve in the first place? Even the subfunctionalization hypothesis ultimately leads us back to neofunctionalization at an earlier time in history, if we want to explain the origin of the multiple functions the ancestral gene had before it was duplicated.

Evolution of New Gene Functions

The ultimate goal of gene family studies is to identify how mutation, drift, and selection have led to the evolution of new and well-optimized functions. For any specific gene family, this question can be formulated in terms of the evolution of structure–function relations: how have evolutionary dynamics at the sequence level generated the structural changes that have conferred on new proteins their unique functions?

The first task in understanding the evolution of structure–function relations is to determine the evolutionary changes that occurred in structure and in function. A phylogenetic approach can be used to identify when specific functions emerged during the evolution of a gene family. To this end, functions must first be coded as characters. For example, the ability to catalyze a reaction can be coded as a binary present/absent character, or the capacity to bind a variety of potential ligands can be coded as either several binary characters or a single multi-state character (e.g., Thornton 2001). The evolution of each character can then be traced on the phylogeny of the family using the parsimony algorithm, which places changes on specific branches so as to minimize the total number of changes in each character (Fitch 1971b). When crystal structures are available for proteins in a family, the evolution of three-dimensional structure can be reconstructed in the same way.

Ultimately, we would like to identify the specific sequence substitutions that generated novel functions. There are two phylogenetic approaches to this problem, derived from the parsimony or likelihood frameworks. The parsimony-based approach identifies phylogenetically diagnostic residues that are reliable markers of membership in a functionally conserved and evolutionarily related class of proteins (Thornton & Kelley 1998; Sarkar et al. 2002). Such residues are defined as the product of changes that occurred on the same branch on which a novel function of interest also emerged and are conserved thereafter in all genes that have the function of interest. The set of phylogenetically diagnostic sites is expected to include those that provide a mechanistic basis for the function. Not all diagnostic amino acid substitutions necessarily contribute to the new functions, but structural and biochemical information can be used to identify the most plausible candidates. Site-directed mutagenesis can then be used to test the functional importance of these changes (e.g., Dean & Golding 1997).

The likelihood approach identifies functionally important sites by finding sequence positions whose evolutionary divergence rates changed on the branch on which the new function emerged (Gu 2003). These methods are based on the principle that if the function of a protein changes, the evolutionary rates of specific sites necessary for the new function (or no longer necessary for the old one) will also change. In contrast, sites that contribute to conserved aspects of function—or are of little functional importance—are expected to have rates that remain unchanged. To identify sites that have undergone rate-shifts associated with the emergence of new functions, the K_a/K_s ratio can be evaluated using codon-based likelihood methods that allow the ratio to vary among sites. Sites with ratios >1 are interpreted to have experienced adaptive evolution for new functions. This model can be further elaborated to allow the K_a/K_s value to be independently inferred on different branches of the tree; a changed K_a/K_s for a codon on the post-duplication branch on which a new function emerged provides evidence that it contributed to adaptive functional divergence. This method has been used to identify positively selected sites in a large number of gene families (Bielawski & Yang 2003)

For anciently duplicated genes, synonymous sites are often saturated with multiple substitutions, so this approach cannot be used. In these cases, protein sequences can be examined for evidence of rate shifts: functionally important sites are identified by finding sequence positions whose rate of amino acid replacement differs among clades of genes that have different functions (Gaucher et al. 2002).

There are concerns about the accuracy of the likelihood approach to this problem. Simulation studies have shown that the maximum likelihood codon approach has a high rate of type I error—false positive inferences that a site has $Ka/Ks > 1$—on simulated sequences that in fact contain no positively selected sites (Suzuki & Nei 2002). An alternative approach, which uses the maximum parsimony algorithm to reconstruct synonymous and nonsynonymous changes across the tree (Suzuki & Gojobori 1999), was much less prone to error. Further, if the evolutionary model is incorrect (if, for example, the evolutionary process is heterogeneous), then the inference of rates will be incorrect as well (Kolaczkowski & Thornton 2004).

A final—and crucial—limitation of the statistical approach to evolutionary novelty is that its most basic assumption may be incorrect: sites that change rates after gene duplication may not necessarily be responsible for functional divergence. One study found that site-specific rate changes do not appear to be any more common after duplication than they are after speciation events that are not associated with functional shifts (Gribaldo et al. 2003). This study found that the character-based approach to identifying diagnostic residues is considerably more reliable than the rate-shift method.

Recreating the Past: Empirical Reconstruction of Gene Family Evolution

All the methods discussed so far in this chapter involve inferring the ancient process of evolution from the pattern of sequences and functions and extant genes, using statistical or character-tracing methods. The hypotheses generated by these inferential methods can be tested empirically using the relatively new technique of ancestral gene resurrection (reviewed in Thornton 2004). This strategy involves inferring the sequences of ancestral genes using maximum likelihood, synthesizing them using DNA manipulation techniques, expressing their protein products in cultured cells, and using molecular techniques to characterize their functions. In this way, scenarios about the functions of ancient evolutionary intermediates can be evaluated using the experimental power of molecular biology. Recent advances in techniques for phylogenetic reconstruction and gene synthesis now make it possible to reconstruct genes hundreds of millions of years old.

Ancestral sequences at any node on a tree can be reconstructed using likelihood-based methods (e.g., Yang et al. 1995; reviewed in Thornton 2004). A parsimony algorithm is also available, but it has lower power and accuracy than maximum likelihood (Zhang & Nei 1997). Given a phylogeny and the amino acid sequences of extant gene products, this method can calculate the probability of all possible states for each site in a sequence at any node on the tree. At each sequence position, the state with the highest probability is the maximum likelihood reconstruction (MLR), and the sequence of these states is the MLR of the ancestral sequence. The MLR sequence is a best estimate rather than a direct discovery of the actual sequence of an ancient gene, but if the extant sequences are reasonably conserved and few sites have undergone parallel changes and reversals, the per-site confidence can be quite high.

Once the sequence of the ancestor is known, a cDNA that codes for the ancestral protein can be assembled from overlapping oligos using polymerase chain reaction (PCR) annealing or enzymatic ligation. The cDNA is then cloned into an expression vector and transfected into cultured mammalian or bacterial cells, where the protein is produced in large quantities. The functions of the protein can then be characterized experimentally using whatever methods are appropriate to the gene family, such as assays of reporter gene activation, enymatic activity, or ligand-binding affinity. This approach has been used in about a dozen studies thus far to resolve the functions of ancestral genes up to a billion years old and test hypotheses about the functions of genes that existed in the deep evolutionary past (see review in Thornton 2004).

CASE STUDY: EVOLUTION OF STEROID HORMONE RECEPTORS

Steroid Hormones and their Receptors

We have used many of the techniques discussed in this chaper to understand the evolution of the steroid hormone receptor gene family. Our goal is to determine how these proteins evolved their diverse and specific functions as hormone-regulated transcription factors. Steroid hormones, which are produced by the gonads or adrenal/interrenal glands of vertebrates, control many aspects of development, reproduction, and physiology. Most of these effects are mediated by intracellular proteins called steroid receptors (SRs). Humans and other amniotes have six SRs: two for estrogens (ERα and ERβ) and one each for androgens (AR), progestins (PR), glucocorticoids (GR), and mineralocorticoids (MR). Each receptor has a specific relationship with one hormone (or a small group of structurally similar hormones). The hormone binds tightly to the receptor's ligand-binding domain (LBD), triggering a change in shape that allows the receptor to bind tightly to specific response elements in the promoters of target genes and activate their transcription (Figure 11.4A). Steroid receptors' well-studied and specific functions make them an ideal case study in gene family diversification.

The evolution of the steroid receptors also exemplifies an important evolutionary puzzle. The classic Darwinian view of evolution argues that complex structures and systems are the result of a gradual process of elaboration and optimization under the influence of selection. This model is certainly correct in some cases; for example, metazoan eyes evolved gradually (and repeatedly) from a primitive light-sensing organ, presumably due to selection for improved perception of the environment (Futuyma 1998). It is not obvious how this model can explain the evolution of tightly integrated systems, in which the function of each part depends on its interaction with other parts. What, for example, is the selection pressure that drives the evolution of a new hormone if there is not already a receptor to transduce its signal? Conversely, what is the function of a new receptor if there is not already a hormone for it to receive? In such systems—which are very common in molecular biology—selection to drive the evolution of or maintain some new element x would seem to require the prior existence of all the other parts, and explaining those in turn is difficult unless x already existed. By reconstructing the detailed history of steroid hormones and their receptors, we hope to provide a Darwinian answer to this important problem.

Steroid Receptor Phylogeny

To understand steroid receptor evolution, we first reconstructed the phylogeny of the family from a large database of broadly sampled steroid receptor protein sequences, togther with those of several closely related proteins in the same superfamily. Some of these were retrieved from public databases, and others were generated in our own laboratory from previously unstudied taxa that occupy critical positions in the animal tree of life. We analyzed these sequences using both maximum parsimony and BMCMC analysis. The optimal trees using the two methods are virtually identical, and most of the nodes on the tree are strongly supported (Thornton 2001; Thornton et al. 2003).

The phylogeny (Figure 11.4B) requires the minimum number of gene duplications and losses to explain the distribution of known receptor sequences.

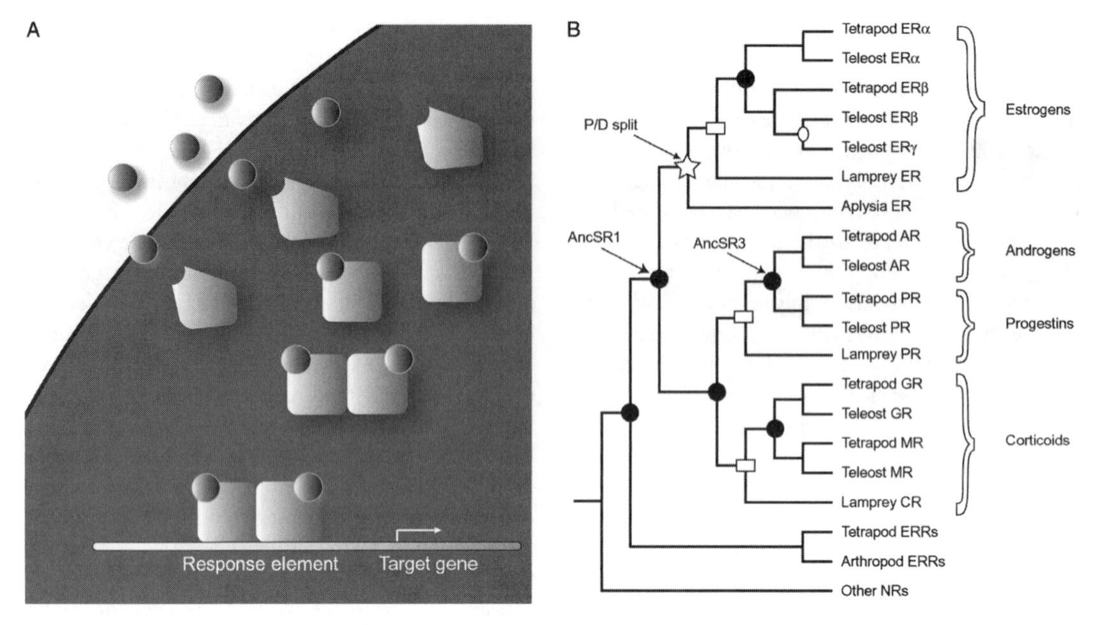

FIGURE 11.4. Phylogeny and mechanism of action for steroid receptors (SR). (A) Simplified mechanism of hormone-activated transcription. Steroid hormones (dark balls) are hydrophobic, so they cross cell membranes by diffusion. One hormone molecule binds tightly to its receptor, conferring a conformational change that allows the receptor to dimerize, enter the nucleus, and bind to specific response elements in the promoters of target genes. The hormone-bound SR attracts coactivator proteins that increase expression of the target gene. (B) Steroid receptor gene tree. A reduced version of a tree inferred by both parsimony and Bayesian analyses of 74 steroid and related receptors is shown. Filled circles, gene duplications that generated the diversity of SRs in all vertebrates. Star, the ancient split of protostomes from deuterostomes. Open squares, the speciation event that split jawed from jawless fish. Open circle marks a duplication specific to the teleosts. The major ligands for each group of receptors are shown at the right.

There are three receptors in the jawless sea lamprey, each of which appears in the tree as the unduplicated sister gene to a pair of SRs that duplicated in the jawed vertebrates: the two estrogen receptors ERα and ERβ, the gonadal steroid receptors AR and PR, and the adrenal steroid receptors GR and MR. The duplications that generated these pairs therefore occurred after the speciation event that separated jawed from jawless vertebrates but before the split of ray-finned fish from tetrapods—between 400 and 450 million years ago based on the fossil record. The duplication of all three receptors in this window is consistent with the hypothesis that two genome duplications occurred deep in the vertebrate lineage (Furlong & Holland 2002; but see Hughes & Friedman 2003). Corroborating this possibility, mapping data from the human genome suggests the AR, PR, GR, and MR are the result of two large-scale duplication events, because there are a large number of other gene families that have duplicate genes on the same four chromosomes as these four closely related SRs (Thornton 2001).

The phylogeny also allowed us to radically push back the known age of the superfamily. It had long been thought that SRs are restricted to vertebrates, because they had only been detected in vertebrates and were lacking from the complete genomes of fruitflies and nematodes. Using PCR, we isolated a full-length SR cDNA from the mollusk *Aplysia californica*, and the phylogeny strongly supports the conclusion that it is an ortholog of the vertebrate ERs (Figure 11.4B; Thornton et al. 2003). The ERs therefore originated by gene duplication before the protostome–deuterostome divergence—at least 600 million years ago—implying that SRs are much more ancient and widespread than previously assumed. The tree shows that the ancient ER gene was then lost during the evolution of ecdysozoans, and, unless there have been additional, independent losses, ERs are expected in other non-ecdysozoan protostomes, such as annelids and platyhelminths.

Neofunctionalization and Subfunctionalization

We next sought to understand how SR function diversified. First, we determined the roles of neo- and subfunctionalization in the evolution of new receptors by examining rate shifts after duplication. We are particularly interested in the function of the ancestral receptor—AncSR1 in Figure 11.4, the single primordial receptor in the common ancestor of all bilaterally symmetrical animals, from which all modern-day receptors evolved by gene duplication. The DDC model predicts that the ancestor was a multifunctional receptor with affinity for numerous ligands and response elements; the specific affinities of modern-day receptors would be due to partitioning of the ancient functions among the ancestral protein's descendants. The classic model predicts that the ancestral receptor already had specific affinity for one type of hormone and response element, and the diverse functions of the modern-day receptors appeared later by neofunctionalization.

To address this question, we first looked at evolutionary rates after duplication of the ancestral receptor AncSR1 as revealed by branch length analysis (Figure 11.5A). After this duplication, the divergence rate in the branch leading to the estrogen receptors was just a fraction of the rate in the branch leading to the receptors for other hormones such as androgens, progestins, and corticoids. A likelihood ratio test indicated that the hypothesis of different rates after duplication is >2500 times more likely than the null hypothesis of equal rates, a difference significant at the $P < 0.0001$ level (Thornton 2001 and unpublished data). This result suggests that the ancestral gene was far more similar in sequence—and presumably in function as well—to the extant estrogen receptors, which were subject to stabilizing selection after duplication. There was an acceleration of the evolutionary rate on the branch leading to the other clade of receptors (AR, PR, GR, and MR); on the same branch, affinity for nonestrogenic steroids first evolved as a derived novelty. Later on the tree, rates were also accelerated on the post-duplication branch leading to the androgen receptor, suggesting that the physiological role of testosterone acting through steroid receptors is likely to be a relatively recent evolutionary novelty unique to jawed vertebrates (Thornton 2001; Figure 11.5A). These findings suggest that the specific interactions that characterize steroid receptors emerged by neofunctionalization—by modification after gene duplication of ancestral receptors affinity for other structurally similar but slightly different hormones.

Neofunctionalization does not tell the entire story of receptor evolution, however. Subfunctionalization seems to have played a role in two more recent duplication events—of the ERα gene in teleost fish and of the ERβ gene in the clawed frog *Xenopus laevis*—both of which seem to be due to ancient whole-genome duplications in these lineages (Hawkins et al. 2000; Wu et al. 2003). In each case, the evolutionary rates of the two duplicate genes are not significantly different from each other, and the key biochemical function associated with the unduplicated ancestral ERs—specifically, the ability to bind estrogens—is conserved in both duplicated receptors. The duplicates are expressed in nonoverlapping domains of brain, reproductive tract, and other tissues, and these domains are subsets of the expression domains found in other "outgroup" species that do not have the duplication. These findings suggest that the expression domains of each ancestral ER have been subfunctionalized independently after duplication in these lineages. This dynamic is likely to explain why a third estrogen receptor gene without apparent new functions has persisted in the genomes of both these taxa for so long.

Resurrecting the Ancestral Receptor

To experimentally test the hypothesis that the ancestral steroid receptor (AncSR1) functioned as an estrogen receptor, we inferred its sequence, resurrected the gene, and characterized its activity in the laboratory (Thornton 2001; Thornton et al. 2003). Only the DNA- and ligand-binding domains (DBD and LBD) are conserved enough to allow ancestral reconstruction, but these are the domains that confer the specific functions in which we are interested. Further, the domains function largely autonomously, so their characteristics can be inferred independently. We inferred the ancestral sequences of these domains by maximum likelihood using the sequence data set and phylogeny described above. A close analysis of the ancestral sequence in light of structure–function relationships indicated that the ancestor probably functioned like an estrogen receptor, for

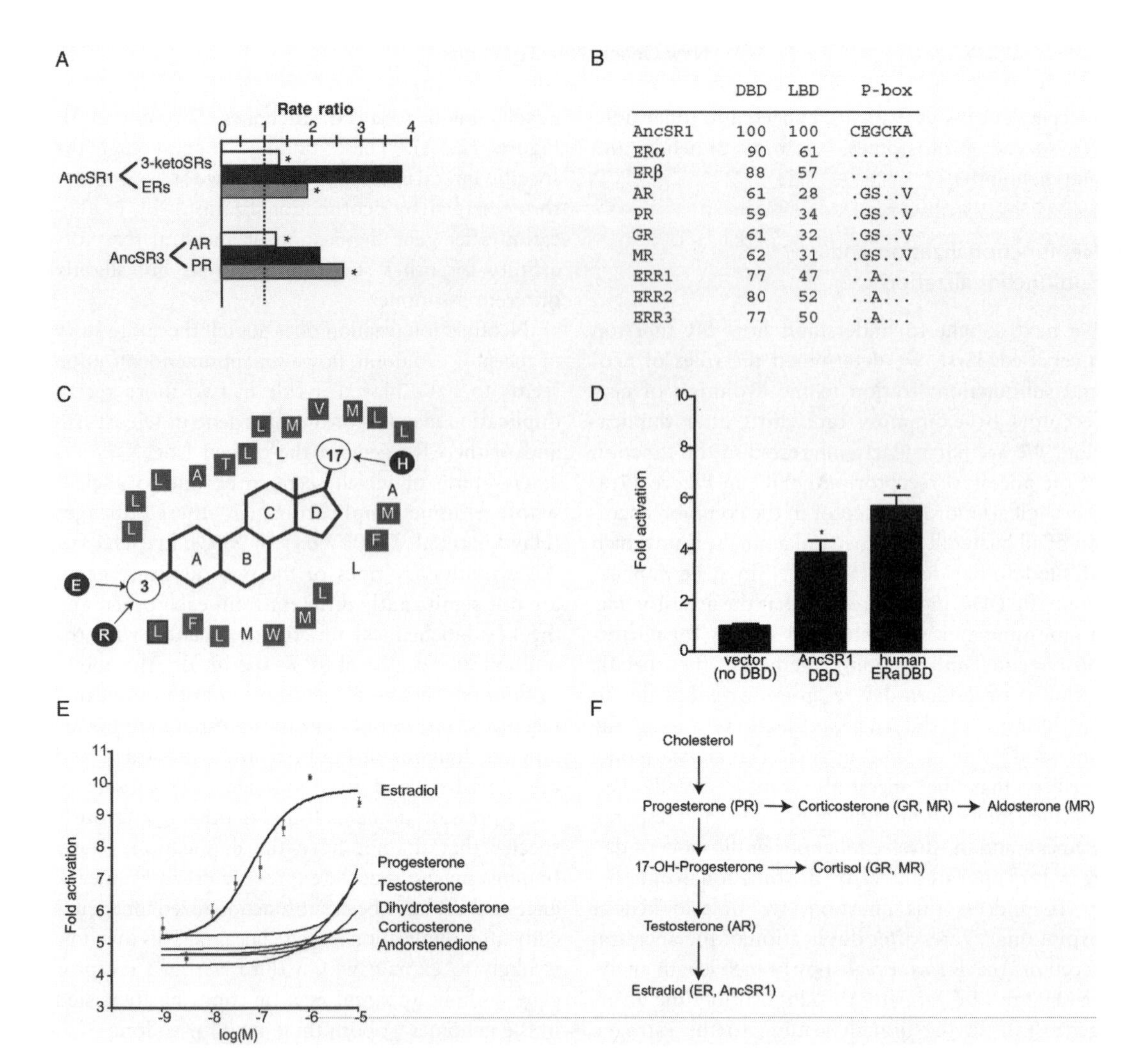

FIGURE 11.5. The ancestral steroid hormone receptor functioned as an estrogen receptor (ER). (A) Analysis of rates after steroid receptor duplications. The graph shows the ratio of the rate of amino acid replacement in the upper branch to that in the lower branch using neighbor-joining (white), parsimony (black), and maximum likelihood (gray). Asterisks, ratio > 1.0, $P < 0.01$. Upper series, the rate in the lineage leading from the ancestral steroid receptor (AncSR1 in Figure 11.4) to the ERs was substantially slower by all methods than that in the branch leading to the other receptors. Lower series, the rate in the lineage leading from AncSR3 (Figure 11.4) to the androgen receptor was also elevated compared with that leading to the progesterone receptor (lower cluster). (B) Sequence similarity of the AncSR1 to human steroid and related receptors. Percent amino acid identity of AncSR1 is shown for the DNA-binding domain (DBD) and the ligand-binding domain (LBD). The sequence in the P-box, a region of the DBD that confers specificity for response elements, is also shown. Dots indicate residues identical to AncSR1. (C) The ligand-binding pocket of AncSR1 is almost identical to that of the human estrogen receptor. Residues lining the pocket of AncSR1 are shown with the steroid ligand, based on crystallography of several extant receptors. Residues shown as white on black background are identical to those in the human ERα. Circled residues interact with substituent groups of the steroid at the 3- and 17-positions and confer specificity for the various steroid hormones. (D) AncSR1 activates transcription from estrogen response elements (EREs). Activation of an ERE-driven luciferase gene is shown for the DBDs of the resurrected ancestral receptor and the human ERα. Asterisks, significantly different from control, $P < 0.01$. (E) AncSR1 activates transcription in response to estrogen but not other hormones. Activation of a luciferase reporter by the ligand-binding domain of the resurrected ancestral receptor is shown over a range of concentrations of various steroid hormones. (F) Simplified synthesis pathway for steroid hormone synthesis. The receptors that bind each steroid are shown. Some arrows represent more than one enzymatic transformation.

two reasons. First, it was far more similar to the extant ERs (90% in the DBD and 85% in the residues that line the ligand-binding pocket of the LBD) than to the other steroid receptors (at most 62% and 44%, respectively). Second, we looked at specific residues that are known from crystallographic and mutagenesis studies to discriminate between estrogens and the other steroid hormones and between estrogen response elements and the elements recognized by the other receptors. Every one of these critical sites in the ancestral protein contained the ER-specific residues; none of the states were identical to those in the other SRs (Figure. 11.5B and C; Thornton 2001).

We then tested these hypotheses experimentally by resurrecting the ancestral SR's functional domains. We synthesized cDNA sequences that code for the ancestral DBD and LBD and cloned them into vectors for high-level expression in mammalian cells. We transfected the ancestral receptor constructs into CHO-K1 cells, an easily cultured mammalian cell line that does not express its own SR genes. Using a reporter transcription assay, we showed that the ancestral DBD activated transcription of genes flanked by estrogen response elements, to which other SRs do not bind, almost as effectively as modern-day estrogen receptors do (Figure 11.5D; Thornton et al. 2003) The LBD activated transcription in a dose-dependent manner in the presence of low levels of estrogens but was insensitive to other steroid hormones (Figure 11.5E; Thornton et al. 2003). Together, these data indicated that the ancestral SR functioned as an estrogen receptor, and the functions of the other members of the SR family are therefore derived (Thornton et al. 2003).

Evolution of Hormone–Receptor Interactions

Our finding that estrogen was the first of the gonadal/adrenal steroids to act as a ligand for an intracellular receptor is surprising, because estrogen is the terminal hormone in a pathway that uses progesterone and testosterone as intermediates (Figure 11.5E). The last hormone to be produced was therefore the first one for which a receptor evolved. The entire steroid synthesis pathway therefore must have existed at a time when only a single hormone–receptor pair had evolved. As receptors diversified, they moved backward up the pathway, establishing relationships with new hormonal ligands.

This surprising result does, however, suggest a Darwinian answer to the riddle with which we began: how can new hormone–receptor pairs evolve and persist if each depends on the prior existence of the other for its function? Our results indicate that progesterone and testosterone—before they were hormones—must have been present as intermediates in the production of the hormone estrogen, and therefore subject to purifying selection. Duplicated receptors evolved increased affinity for these steroids, turning what had been biochemical steppingstones into bona fide hormones. This kind of recruitment of old molecules for new functions by slightly modified gene duplicates may be a major theme in molecular evolution and may explain in large part how the complex molecular systems critical for cell function, development, and homeostasis evolved.

THE FUTURE OF GENE FAMILY EVOLUTION

With the explosive growth of available sequence data and computer power, gene family studies are likely to grow to ever-larger scales. Phylogenies will be better resolved, and the evolutionary process inferred with greater precision, as sampling of family members among genomes improves. This process will move most quickly if genome sequencing projects focus not only on biomedically and economically important species but also seek to fill major gaps in the taxonomic tree of life.

As gene family studies become broader, they should also grow deeper and more focused on the mechanisms and dynamics by which new gene functions evolve. To this end, ancestral gene reconstruction is likely to become more widely used to test hypotheses generated by interpreting sequence data. Site-directed mutagenesis and experimental evolution systems also offer great potential for testing hypotheses about the dynamic and mechanistic basis for the evolution of gene functions.

As the mechanisms by which gene families diversified emerge, we may begin to glimpse the evolution of the larger molecular, cellular, and physiological systems that gene family members constitute. For example, analyses of the phylogeny and evolution of families of enzymes that catalyze major biochemical pathways, of ligands and their receptors, of signal transduction pathways and transcription factor targets, should gradually build up an understanding of the evolution of many of the critical elements of eukaryotic physiology and development. From this foundation, we may begin to illuminate how the biological processes that make life possible were assembled, elaborated, and optimized over deep evolutionary time.

SUGGESTIONS FOR FURTHER READING

Long et al. (2003) have written an excellent review on the evolutionary mechanisms and dynamics by which new genes evolve. Prince and Pickett (2002) have reviewed models for the evolution of gene function after duplication, with an emphasis on subfunctionalization and the DDC model. Bielawski & Yang (2003) provide a concise review of statistical methods for detecting the traces of positive selection on gene family members and for determining sequence sites that have been subject to selective pressures. My review (Thornton 2004) of resurrecting ancestral genes covers the history, methods, applications, and future directions of this strategy.

Bielawski JP & Z Yang 2003 Maximum likelihood methods for detecting adaptive evolution after gene duplication. J Struct Funct Genomics 3:201–12.

Long M, Betran E, Thornton K & W Wang 2003 The origin of new genes: glimpses from the young and old. Nat Rev Genet 4:865–75.

Prince VE, & FB Pickett 2002 Splitting pairs: the diverging fates of duplicated genes. Nat Rev Genet 3:827–37.

Thornton JW 2004 Resurrecting ancient genes: experimental analysis of extinct molecules. Nat Rev Genet 5:366–75.

12

Gene Genealogies

NOAH A. ROSENBERG

Genetic variation at a locus among extant individuals can be viewed as the result of mutations on a scaffold of genetic relationships—a *gene genealogy*. Because patterns of genetic variation contain much information about phenomena such as hybridization, migration, species divergence, and changes in population size, an understanding of gene genealogies is helpful for the application of genetic variation to inference about evolutionary processes. As we will see, gene genealogies, which underlie numerous statistical methods for population genetic analysis, are useful in diverse areas of genetics and evolutionary biology, ranging from phylogenetics to genetic mapping.

The basic nature of the inheritance of genetic material is familiar: copies of corresponding stretches of the genome in different individuals are passed through a series of generations from some piece of DNA in a common ancestor of the individuals. The mutations that occur in transmission leave a pattern of similarities and differences in extant individuals that, albeit imperfectly, records the genealogical history in their DNA sequences. All the processes that affect this history—for example, the size of the population to which the individuals belong, which influences the length of time to the common ancestor—affect the outcome in the DNA sequences, the data available to us today. Thus, to learn about how the population has evolved, we need to know how evolutionary processes affect genealogies, and in turn, how genealogies affect genetic data.

In this chapter, I introduce gene genealogies, which describe relationships among copies of a locus in different individuals, through a discussion of their link to pedigrees, the structures that describe relationships among the individuals themselves. Two initial questions that might be asked about gene genealogies are:

1. What schemes can be used to categorize gene genealogies, and what are the categories?
2. What attributes do we expect gene genealogies to have in specific evolutionary scenarios?

After considering these issues—classification of genealogies and properties of random genealogies—I discuss a variety of examples that illustrate the use of gene genealogies for interpreting patterns of genetic variation.

CONCEPTS

Pedigrees and Gene Genealogies

For haploid organisms, relationships of individuals and those of their genomes are equivalent: when a cell divides, the genomes of the offspring descend directly from the parental genome (but see Box 12.1). For diploids, however, the way in which genomes pass from parents to offspring is more complex. To understand the relationships between diploid genomes, rules that characterize the transmission process of genomes from parents to offspring—Mendel's laws of inheritance—can be used.

Consider an individual, and choose one of its parents. The law of segregation states that for any (autosomal) locus in the genome, (1) the individual has a copy of the locus from the chosen parent, and (2) with probability ½ this copy is inherited from the parent's maternal copy, and with probability ½ it is inherited from the parent's paternal copy. For two loci, the law of independent assortment states that whether the copy inherited at the first locus derives from the chosen parent's maternal or paternal copy does not depend on which grandparent produced the copy at the second locus. Genetic linkage between some pairs of loci produces exceptions to this rule; in these cases, however, modifications can be made to accommodate dependence between loci.

Suppose we are given a set of individuals *S*, whose biological relationships are represented by a pedigree (Figure 12.1A). Consider a locus randomly chosen from the genomes of the individuals. If we use the law of segregation to trace copies of the locus through the pedigree, starting with the set *S*, it is likely that we will eventually reach a single copy from which all copies in *S* descend (Figure 12.1C).[1] All individuals in the figure are biologically ancestral to the individuals in *S*, that is, ancestors in terms of the pedigree. However, only a small fraction of the individuals in the pedigree, by being in lines of descent to *S* from the most recent common ancestor of the copies of the locus in *S*, are *genetically* ancestral at the locus. These genetic ancestors are the only individuals that affect the genotypic state at the locus for individuals in *S*. When we restrict our attention to these ancestors, we obtain the *gene genealogy* for the individuals at the locus.

Using the law of independent assortment, the grandparent from whom the copy from the chosen parent descends at one locus is independent of the one from whom the corresponding copy descends at a second locus. Applying this rule as we trace through a given pedigree, gene genealogies of two unlinked loci are independent. Because most diploid genomes have many independent loci and, thus, many independent gene genealogies, for any set of individuals, many paths are followed by at least one locus. Consequently, a pedigree of a set of individuals can be viewed as describing their "average" gene genealogy: proceeding through a pedigree, each path has the same probability. On average, all paths of a given length (that is, of a fixed number of generations) are taken by equal numbers of loci.

Examples considered by Wollenberg and Avise (1998), Derrida et al. (2000), and Rohde et al. (2004) make the relationship between pedigrees and gene genealogies apparent. The time until all humans share a common ancestor along the male or female line, that is, the time until the genetic ancestor for all human Y chromosomes or mitochondrial genomes, has been estimated at tens to hundreds of thousands of years. However, the most recent common ancestor (MRCA) in terms of the pedigree—the most recent individual to be part of the pedigree of all living humans—might have been surprisingly more recent, perhaps only 2000–7000 years ago (Rohde et al. 2004). In other words, across all loci in the genome, the common ancestor for the gene genealogy whose MRCA is smallest may have lived in historical times.[2]

Terminology

This chapter uses the following definitions, which are generally standard, except where noted. The tips of gene genealogies represent *sampled lineages* (Figure 12.2). In general, each line that connects a descendant to an ancestor is a *lineage*. Nodes, which represent the joining of lineages in common ancestors as time proceeds backward from the present, are *coalescences* or *coalescence events*. Lengths of time that separate coalescences from each other or from sampled lineages are *branch lengths*. A branch that separates two coalescences is *internal*; one that separates a sampled lineage from a coalescence is *external*. A coalescence at which two external branches join is a *cherry*. The *time to the most recent common ancestor* (T_{MRCA}) for a set of sampled lineages is the length of time from the present until the lineages first reach a common ancestor, their *most recent common ancestor* (MRCA). The T_{MRCA} for a genealogy is often called the *coalescence time*, although *coalescence times* can

[1]The exception in which a single copy is not necessarily reached is if life originated multiple times and the copies trace back to more than one of the original genomes.

[2]Technically, there is no guarantee that any living person contains DNA descended from the pedigree MRCA studied by Rohde et al. (2004), as such segments of DNA may have disappeared over time through recombination. However, if the genome had infinitely many possible points at which recombination could occur, and if recombination only happened at each point at most once in evolutionary history, the pedigree MRCA would be the MRCA of the gene genealogy whose MRCA is smallest across all loci.

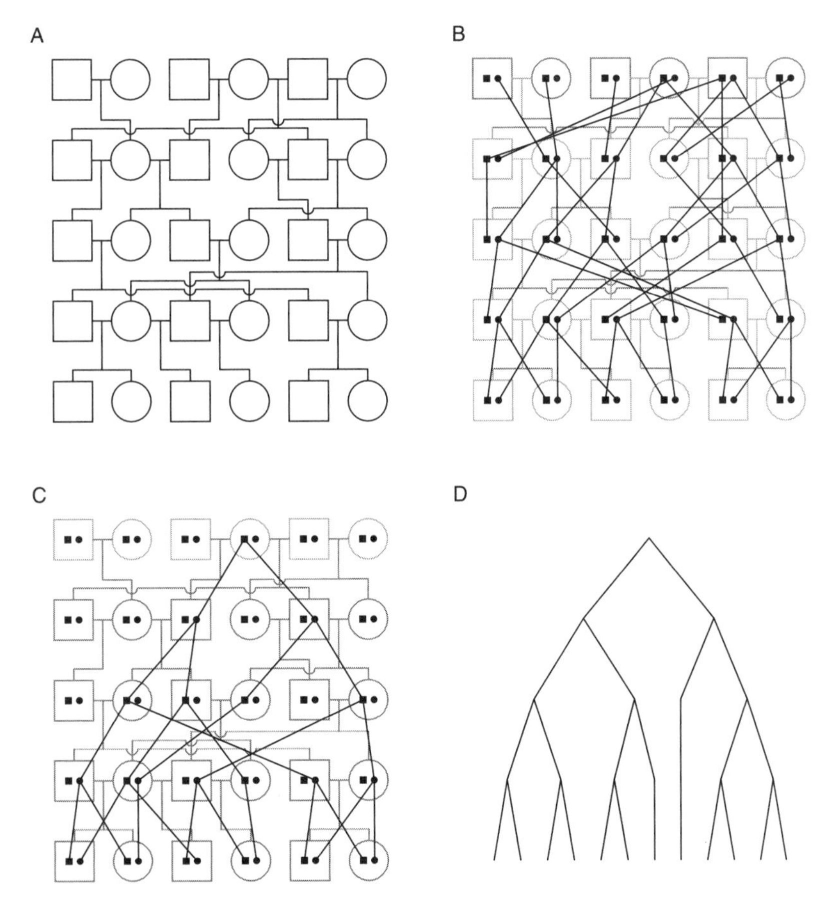

FIGURE 12.1. Pedigrees and gene genealogies. (A) The pedigree for a set of six individuals from the current generation. Empty squares and circles represent males and females, respectively. (B) Application of the law of segregation to a randomly chosen locus, conditional on the pedigree. This diagram shows the transmission paths of a particular locus through the pedigree. Shaded squares and circles respectively represent paternal and maternal copies of a locus. (C) Genealogy of the copies of the locus present in the most recent generation of individuals, showing only the transmissions that contribute to the current generation. (D) Abstracted genealogy obtained by rearranging the order of the copies. Time proceeds downward (in this and subsequent figures).

also refer to lengths of time between successive coalescences. The *root* node represents the MRCA for all sampled lineages in a genealogy; the two branches connected to the root are *basal*.

For a set of sampled lineages, a *locus* is a unit of DNA, ranging in size from a single base pair to a whole chromosome, in which no recombination has occurred in the genetic ancestors of the lineages since the time of their MRCA. In scenarios in which lineages derive from multiple populations, it often does not matter whether the populations are from the same species. Thus, except where otherwise specified, *species* is used to refer to the population of individuals that belong to a "species," and is sometimes interchangeable with *population*.

A *genealogy* or *gene genealogy* for n sampled lineages is a tree specified by the sequence of coalescences that reduce the n lineages to a MRCA, along with the *coalescence times* that separate these events. Two genealogies are *identical* if and only if

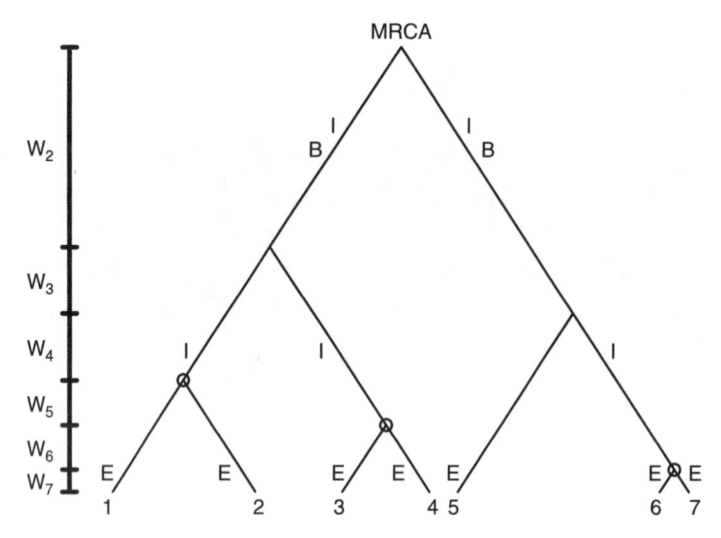

FIGURE 12.2. A genealogy for seven sampled lineages. Cherries are marked with open circles. Each lineage is separated from the root by exactly three branches, except for lineage 5, which is separated by only two branches. The two subgenealogies that coalesce at the root have four and three lineages (lineages 1, 2, 3, and 4 in the subgenealogy on the left, and lineages 5, 6, and 7 in the one on the right). W_n is the length of time during which the sampled lineages have exactly n ancestors. T_{MRCA} for the lineages is represented by the height of the genealogy, $T_n = \sum_{n=2}^{7} W_n$. The total length of all branches in the genealogy is $L_n = \sum_{n=2}^{7} nW_n$. This length is the sum of E_n, the sum of the lengths of the external branches (marked E), and I_n, the sum of the lengths of the internal branches (marked I). Basal branches are marked B.

they have the same sequence of coalescence events and the same coalescence times. A *subgenealogy* containing k of the n lineages includes the MRCA of these k lineages together with all parts of the genealogy that descend from this MRCA. Although it is possible to consider genealogies in which coalescences involve more than two lineages, it is assumed in this chapter that exactly two lineages join in each coalescence.

The major features of a genealogy can be captured in quantities that summarize its shape and size (Table 12.1). These quantities fall into three categories: (1) those that depend only on which lineages participate in coalescences, without regard to when coalescences occur; (2) those that depend only on the coalescence times, without regard to which lineages participate in coalescences; (3) those that depend both on the lineages involved in coalescences and on the coalescence times.

Classification of Genealogies

We frequently have occasion to compare two or more genealogies. For example, to search for signatures of events with genome-wide effects, such as population splits, we can compare genealogies for different loci in the same set of individuals. To determine whether a particular sample is suitably representative of a population, we can compare genealogies for the same locus in several samples.

We may be interested in whether or not two genealogies are identical; because identity of genealogies is rare, however, the equivalence or nonequivalence of attributes of the shapes of two genealogies, such as their *labeled topologies*, is more often of interest. Thus, it is useful to consider various ways in which shapes of genealogies can be classified; for convenience, each of several classification schemes is denoted here by a different letter.

TABLE 12.1. Attributes of a genealogy with n lineages ($n \geq 2$).

Symbol	Description of attribute	Expected value over random genealogies generated by the coalescent distribution[a]	References
Attributes that depend only on the coalescence sequence of the genealogy			
C_n	Number of cherries ($n \geq 3$)	$n/3$	McKenzie and Steel 2000
$X_{n,1}$	Number of branches that separate a randomly chosen lineage from the root	$2\sum_{i=2}^{n} 1/i$	Steel and McKenzie 2001
$X_{n,2}$	Number of branches that separate the two randomly chosen lineages from the root	$\frac{2}{n-1}\left(n-1-2\sum_{i=2}^{n} 1/i\right)^{b}$	Steel and McKenzie 2001
Y_n	Number of branches that separate two randomly chosen lineages from each other	$\frac{4n+4}{n-4}\left(\sum_{i=2}^{n} 1/i\right) - 4$	Steel and McKenzie 2001
$\{l, n-l\}$	Numbers of lineages in the two subgenealogies that coalesce at the root	[c]	Tajima 1983, Slowinski and Guyer 1989
Attributes that depend only on the coalescence times of the genealogy			
$W_{n,k}$	Length of time it takes for n lineages to coalesce to k lineages ($1 \leq k \leq n$)[d]	$2(n-k)/(nk)$	Tajima 1983
T_n	Time to the most recent common ancestor	$2(n-1)/n$	Tajima 1983, Tavaré et al. 1997
L_n	Total length of time in all branches	$2\sum_{i=1}^{n-1} 1/i$	Hudson 1990, Tavaré et al. 1997
Attributes that depend on both the coalescence sequence and the coalescence times of the genealogy			
E_n	Total length of time in external branches	2	Fu and Li 1993, Durrett 2002
I_n	Total length of time in internal branches	$2\sum_{i=1}^{n-1} 1/i$	Fu and Li 1993, Durrett 2002
P_n	Average coalescence time for two randomly chosen lineages	1	Tajima 1983, Durrett 2002
B_n	Average length of time in a basal branch	$2\left(1/n + \sum_{n-2}^{n-1} 1/i^2\right)$	Uyenoyama 1997

[a]For genealogies generated by the coalescent distribution, coalescence sequences follow the Yule distribution. Therefore, the expected values for attributes that depend only on the coalescence sequence do not utilize the exponential distribution of coalescence times under the coalescent. For attributes that depend on the coalescence times, times are measured in units of N generations, where N is the population size.
[b]For each k, with $1 \leq k \leq n$, we can consider $X_{n,k}$, the number of branches that separate the most recent common ancestor (MRCA) of k randomly chosen lineages from the root. Under the coalescent, $P[X_{n,k}=0]=(k-1)(n+1)/[(k+1)(n-1)]$ (Saunders et al. 1984; Steel and McKenzie, 2001).
[c]For each l, with $1 \leq l \leq [n/2]$, $\{l, n-l\}$ has probability $2/(n-1)$, with the exception that if n is even, $\{n/2, n/2\}$ has probability $1/(n-1)$.
[d]The special case $W_{n,n-1}$ is often abbreviated to W_n.

Labeled Histories and Labeled Topologies

The *labeled history* of a genealogy is its sequence of coalescence events (Figure 12.3). Two genealogies of n lineages have the same *labeled history*, or are *H-equivalent*, if they have the same coalescences in the same temporal order. The number of possible labeled histories for genealogies of n lineages is $H_n = n!(n-1)!/2^{n-1}$ (Steel & McKenzie 2001). Each genealogy of n lineages has one of H_n possible labeled histories, and each labeled history is the labeled history of some genealogy.

The genealogies in Figure 12.3A and 12.3B have the same coalescence events, but in different

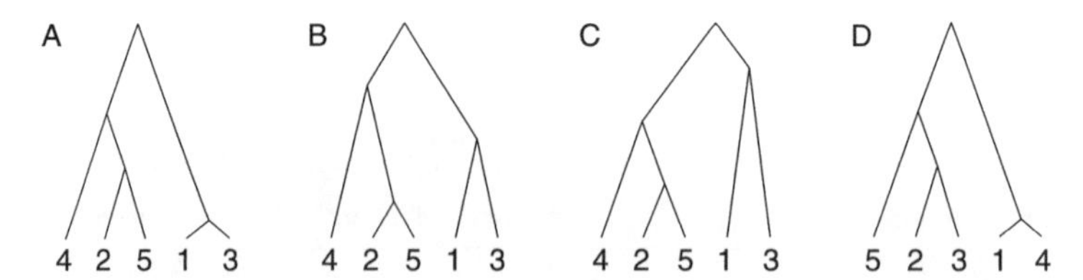

FIGURE 12.3. Labeled histories and labeled topologies for example genealogies. The sequences of coalescence events are: (A) (1,3), (2,5), ((2,5),4), ((1,3),((2,5),4)); (B) (2,5), (1,3), ((2,5),4), ((1,3),((2,5),4)); (C) (2,5), ((2,5),4), (1,3), ((1,3),((2,5),4)); (D) (1,4), (2,3), ((2,3),5), ((1,4),((2,3),5)). The genealogies in (A), (B), and (C) have the same coalescence events and therefore have the same labeled topology, ((1,3),((2,5),4)). Because the order of the coalescences differs for (A), (B), and (C), however, these genealogies have different labeled histories. Although its coalescence times equal those of (A), the genealogy in (D) has different coalescences from those of (A), (B), and (C); thus, it differs from the other genealogies both in labeled history and in labeled topology.

sequences; therefore, they have different labeled histories. However, there is a sense in which these two genealogies are equivalent. The *labeled topology* of a genealogy is its unordered list of coalescence events.[3] Two genealogies of n lineages have the same *labeled topology*, or are *T-equivalent*, if they have the same coalescences but not necessarily in the same order. The number of possible labeled topologies for genealogies of n lineages is $I_n = (2n-3)!/[2^{n-2}(n-2)!]$ (Table 3.1 in Felsenstein 2003). Each genealogy of n lineages has one of I_n possible labeled topologies, and each labeled topology is the labeled topology of some genealogy.

Monophyly, Paraphyly, and Polyphyly

For genealogies whose sampled lineages derive from two species (or populations), (A, B), we may be interested in how the lineages from the two species are interleaved in the genealogy. For each species, the sampled lineages from that species have a *monophyly status*: they are either *monophyletic*—that is, they comprise all the sampled descendants of their MRCA—or they are *not monophyletic*. Lack of monophyly requires that lineages of the other species be descendants of this MRCA.

[3]It is also possible to consider the *unlabeled* topology (Felsenstein 2003, p. 29) and *unlabeled* history (Tajima 1983, appendix 1) of a genealogy.

A genealogy of lineages from two species can be classified into one of four categories (Figure 12.4):

- C1. *Monophyly of A and B*, or *reciprocal monophyly*. The lineages of each species are separately monophyletic.
- C2. *Paraphyly of B with respect to A*. The lineages of species A are monophyletic, and the lineages of species B are not monophyletic.
- C3. *Paraphyly of A with respect to B*. The lineages of species B are monophyletic, and the lineages of species A are not monophyletic.
- C4. *Polyphyly of A and B*. Neither the lineages of species A nor the lineages of species B are monophyletic.

Two genealogies of lineages from two species will be said to have the same *phyletic status* here if they classify into the same one of these four categories.

Suppose now that sampled lineages derive from m species ($m \geq 2$). For each species, the lineages of that species are either monophyletic or not monophyletic. The ordered list of m monophyly statuses for the species is the *M-type* of the genealogy. Two genealogies of lineages from two or more species are *M-equivalent* if and only if they have the same M-type. Each genealogy of lineages from m species has one of 2^m possible M-types.

For each pair of species, the phyletic status of the lineages from the two species can potentially be either C1, C2, C3, or C4. The ordered list of (mC_2)

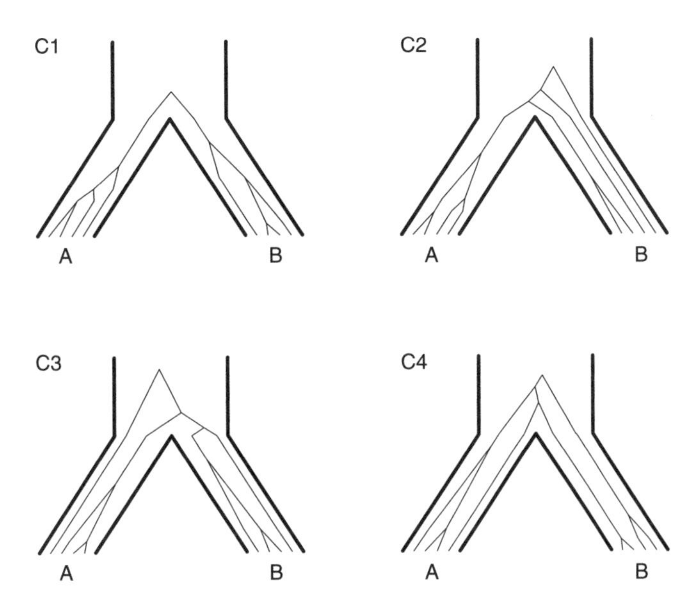

FIGURE 12.4. The four phyletic statuses possible for a genealogy of lineages sampled from two species. Thick lines represent the divergence of an ancestral species into two descendant species, *A* and *B*. Thin lines represent the genealogy of the sampled lineages from the two species. (C1) monophyly of *A* and *B*. (C2) paraphyly of *B* with respect to *A*. (C3) paraphyly of *A* with respect to *B*. (C4) polyphyly of *A* and *B*.

phyletic statuses for the m species is the *P-type* of the genealogy. Two genealogies of lineages from two or more species are *P-equivalent* if and only if they have the same P-type. Note that for $m = 2$, P-equivalence has the same meaning as M-equivalence. For $m > 2$, however, each M-type is the M-type of some genealogy but many of the $4^{({}_mC_2)} = 2^{m(m-1)}$ possible P-types cannot be the P-type of any genealogy. For example, no genealogy for three species—A, B, and C—can have pairs (A, B) and (A, C) in category C2 while (B, C) is in C1.

Collapsed Genealogies

For $m \geq 2$, the phylogeny of m species—the genealogy of the species—has one of H_m possible labeled histories, and one of I_m labeled topologies. To ease comparison between gene genealogies and species phylogenies, it is convenient to classify genealogies of lineages from m species with the same classes as those used for the species phylogeny itself.

The collapsing algorithm in Rosenberg (2002) gives a procedure for mapping a genealogy of n lineages from m species ($n \geq m$) onto the set of H_m labeled histories or to the set of I_m labeled topologies. This algorithm maps a gene genealogy from many species onto a *collapsed genealogy* obtained by considering only the most recent interspecific coalescence for each species (Figure 12.5). Taking into account the order of these coalescences, the genealogy is mapped to its *collapsed labeled history* or *C-type*. Considering the coalescences but ignoring their order, the genealogy is mapped to its *collapsed labeled topology* or *D-type*. Two genealogies of lineages from two or more species are *C-equivalent* if and only if they have the same collapsed labeled histories, and *D-equivalent* if and only if they have the same collapsed labeled topologies. For $m = 3$, because each labeled topology is consistent with only one labeled history, *D*-equivalence has the same meaning as *C*-equivalence. Each of the H_m labeled histories for m lineages can be the collapsed labeled history for some genealogy of lineages from m species; similarly, each of the I_m labeled topologies for m lineages can be the collapsed labeled topology for some genealogy.

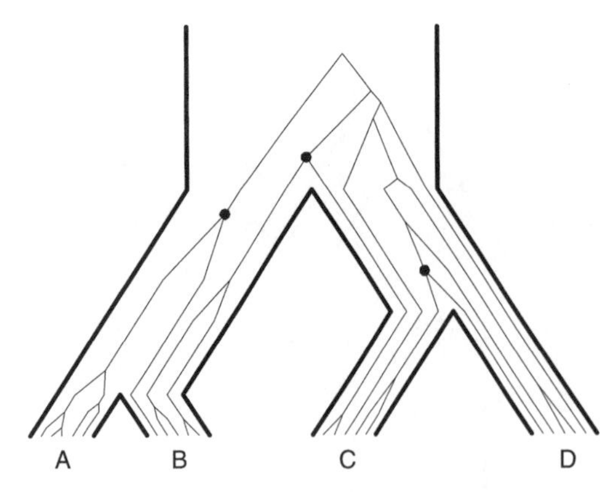

FIGURE 12.5. Classification of a genealogy of lineages from four species. Filled circles indicate interspecific coalescence events used in determining the collapsed genealogy. Note that coalescences of lineages from two species occur prior to species divergence. The genealogy can be classified as follows. M-type: *A*, monophyletic; *B*, not monophyletic; *C*, not monophyletic; *D*, not monophyletic. P-type: (*A*,*B*), C2; (*A*,*C*), C2; (*A*,*D*), C1; (*B*,*C*), C4; (*B*,*D*), C3; (*C*,*D*), C4. C-type: the collapsed labeled history of the genealogy is the sequence (*C*,*D*), (*A*,*B*), ((*A*,*B*),(*C*,*D*)). D-type: the collapsed labeled topology of the genealogy is ((*A*,*B*),(*C*,*D*)). The collapsed labeled topology of the genealogy and the labeled topology of the species phylogeny match, but the collapsed labeled history of the genealogy and the labeled history of the species phylogeny differ.

Random Genealogies

For a given collection of assumptions about the evolutionary process in a set of species—a model—it is of interest to know the probability distribution for a random genealogy, or the genealogy of a random sample of lineages. Such a model can be used to predict patterns of genetic variation for a randomly chosen locus under a specific set of conditions. Although we would like to make predictions under any model, much can be learned using a relatively simple model with one population.

The Coalescent Distribution

Consider a random sample of n lineages from a haploid population of constant size N, with $N >> n$. In each of a series of discrete generations, every lineage chooses a random parent from the previous generation. Under these assumptions, the same as those of the frequently used *Wright–Fisher model* (Ewens 2004), the probability distribution of the genealogy of n random lineages is closely approximated by the *coalescent distribution*, variously termed the *coalescent*, *n-coalescent*, *neutral* or *standard coalescent*, or *Kingman's coalescent* (Kingman 1982; Hudson 1983; Tajima 1983; Nordborg 2001).

Recall that a genealogy consists of two components: its sequence of coalescence events and its set of coalescence times. Under the coalescent, the coalescence times have exponential distributions, so that the time until n lineages reduce to n–1 has exponential distribution with mean $2/[n(n-1)]$ units of N generations. The sequence of coalescence events has a uniform distribution over the set of labeled histories: at any point in time, each pair of lineages has the same probability of being the next pair to experience a coalescence. This uniform distribution, the *Yule distribution* (Aldous 2001), assigns probability $1/H_n$ to each labeled history. Note that under the coalescent, the probability distribution of the labeled topology of a random genealogy is not uniform: the probability that a random genealogy has labeled topology t equals $(2^{n-1}/n!)\prod_{i=3}^{n}(i-1)^{-d_i(t)}$, where $d_i(t)$ is the number of coalescences in the labeled topology from which exactly i sampled lineages descend (Brown 1994; Steel & McKenzie 2001). Table 12.1 lists additional properties of genealogies under the coalescent.

The utility of the coalescent derives from the fact that it describes the distribution of the genealogy of n lineages in diverse evolutionary models besides the Wright–Fisher model, such as scenarios with age structure, horizontal DNA transfer (Box 12.1), or separate sexes (Möhle 2000; Nordborg & Krone 2002). In each of these models, a parameter termed the *coalescence effective size* (Sjödin et al. 2005), or N_e, is required to transform the model into one for which the coalescent applies. In other words, for a given model, if it has a coalescence effective size, the probability distribution of a random genealogy under the model is obtained from the coalescent, substituting N_e for N. One useful case for which the coalescent distribution applies is that of diploidy: a diploid constant-sized population with $N/2$ males and $N/2$ females has coalescence effective size $2N$ (Nordborg 2001).

Many models, however, including some that include time-varying population size, do not have coalescence effective sizes. That is, for every value of N, the distributions of genealogies under these models differ from the coalescent distribution for

BOX 12.1. Horizontal Inheritance

Individuals of some organisms can inherit DNA from individuals other than their parents. This is particularly true for certain haploids, which can replace DNA that they "vertically" inherit from parents with DNA "horizontally" inherited from other individuals of the same species, individuals of other species, or the surrounding environment (Bushman 2002). Such organisms have two types of coalescence, vertical and horizontal.

Because of horizontal inheritance, genealogies in many haploid species might not follow the pattern of bifurcation of genomes expected for haploids. With horizontal transfer, haploid genealogies contain many of the complexities seen in gene genealogies of diploids. Just as recombination enables different parts of the genomes of diploids to have distinct genealogies, horizontal DNA transfer leads to differing genealogies for different parts of a haploid genome (Figure 1). Analogously, as migration in diploids can lead different multi-population genealogies to have different collapsed labeled topologies, horizontal inheritance among individuals from different species can cause such discordances in haploid genealogies.

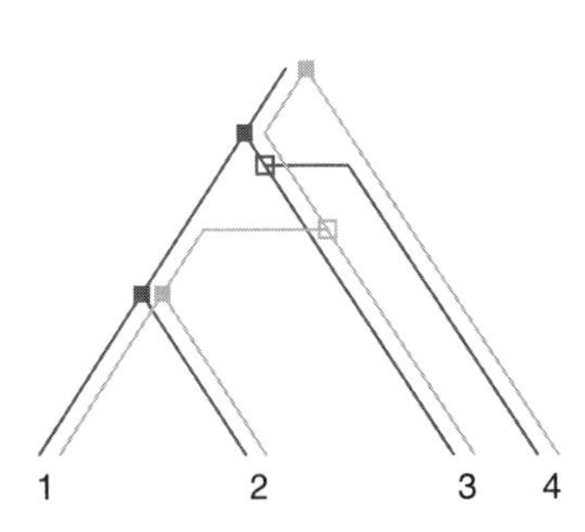

FIGURE 1. Genealogies for two loci. Each locus, one with a darkly drawn and the other with a lightly drawn genealogy, is sampled in a set of four individuals. Vertical and horizontal coalescences are marked with filled and empty squares, respectively. Because horizontal transfer of DNA from one individual to another usually involves pieces of DNA that are small relative to the genome size, it is unlikely that two loci, unless separated by a short distance, would experience horizontal coalescence in the same individual. Because of horizontal inheritance, the two genealogies have different labeled topologies: ((1,2),(3,4)) and (((1,2),3),4) for the dark and light genealogies, respectively.

Recall that in diploids, discordance of collapsed labeled topologies does not require migration among populations. Similarly, in haploids, such discordance can arise even if no horizontal transfers occur between individuals of different species. In other words, discordance of collapsed labeled topologies for genealogies for several regions of a genome can result from horizontal transfer *between* species or *within* species (Figure 2). At the same time, however, horizontal transfers between or within species need not necessarily lead to discordance.

In bacterial studies, it is of interest to identify which genes have and have not been transferred across species, and for those that have been transferred, to identify the donor species (Eisen 2000; Koonin 2003). Because any shape for a haploid genealogy can be produced by many different combinations of horizontal transfers within and between species, it is important to quantitatively evaluate the relative support for different scenarios. Such an endeavor might be advanced by connecting horizontal transfer models to the coalescent.

A Horizontal Transfer Model

Consider a random sample of n individuals from a haploid population of constant size N in a closed environment, with $N >> n$. Suppose that the individuals have independently and identically distributed lifespans that follow exponential distributions with mean 1 generation. When an individual dies, another individual randomly chosen from the population duplicates to replace it. These are the basic assumptions of the *Moran model*, a frequently used neutral model in population genetics (Ewens 2004).

(continued)

BOX **12.1.** *(cont.)*

Looking backward in time from the sample of n individuals, the waiting time until one of the individuals arose from its parent is exponentially distributed with mean $1/n$ generations. The probability that this origin is a (vertical) coalescence is the probability that the parent is ancestral to the other $n-1$ sampled individuals, or $(n-1)/(N-1)$. Using basic properties of exponential random variables, the time until a vertical coalescence is exponentially distributed with mean $(N-1)/[n(n-1)]$ generations. Genealogies in this model follow the coalescent distribution with coalescence effective size $(N-1)/2$.

Now suppose that for each individual, the waiting time until its DNA at a locus of interest is replaced by DNA horizontally transferred from another individual in the population is exponentially distributed with mean $1/\lambda$ generations. Such transfers could potentially occur by *conjugation*, *transduction*, or *transformation*, procedures in which DNA is transferred between cells via plasmids, viruses, or the extracellular environment, respectively (Bushman 2002).

Assuming that horizontal transfers in different individuals are independent, the waiting time (backward in time) until one of the lineages experiences a horizontal

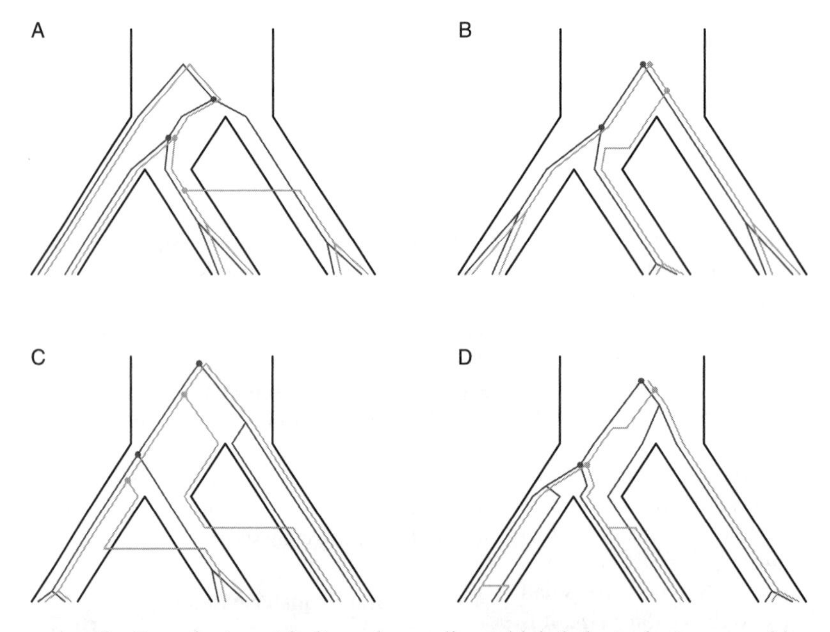

FIGURE 2. Genealogies with discordant collapsed labeled topologies caused by horizontal transfer among individuals of (A) different species, and (B) the same species; genealogies with concordant collapsed labeled topologies despite horizontal transfer among individuals of (C) different species, and (D) the same species. The thick lines represent the species phylogeny; filled circles depict interspecific coalescences used in determining collapsed genealogies. A "species" is interpreted to be a group of individuals each of which has the property that most of its genome coalesces with that of other individuals in the group, horizontally or vertically, more recently than it does with individuals not in the group. *Speciodendric* loci are those loci whose collapsed labeled topologies match the labeled topology of the species phylogeny.

BOX 12.1. *(cont.)*

transfer event (as the recipient of DNA) is exponentially distributed with mean $1/(n\lambda)$ generations. If the individual that donates DNA during this transfer is an ancestor to one of the other $n-1$ sampled lineages, an event that has probability $(n-1)/(N-1)$, horizontal coalescence occurs. If this donor is not an ancestor to the $n-1$ lineages, no coalescence takes place. As before, using the properties of exponential random variables, the time until a horizontal coalescence is exponentially distributed with mean $(N-1)/[\lambda n(n-1)]$ generations.

Considering the vertical and horizontal processes simultaneously, the time until a coalescence of either type has exponential distribution with mean $(N-1)/[(1+\lambda)n(n-1)]$ generations. This distribution has the same form as in models that only include vertical coalescence. In other words, the waiting times in this model follow the coalescent distribution with coalescence effective size $(N-1)/[2(1+\lambda)]$.

Implications of the Model

In comparison with a model that includes vertical coalescence only, the horizontal transfer model has shorter waiting times until coalescence, so that lineages find a most recent common ancestor more rapidly. This is sensible, as horizontal inheritance enables genes to diffuse rapidly through a population. The amount by which horizontal transfer speeds up coalescence depends on λ, which measures the mean number of horizontal transfers experienced by a random individual at the locus of interest during a lifetime of average length. If λ is very small—that is, if most cells die before experiencing any transfers—the presence of horizontal transfer has little effect on genealogies, and most coalescences are vertical.

The horizontal transfer model has a coalescence effective size, so that the coalescent distribution applies to its genealogies. Thus, in the same way used for models without horizontal transfer, it can potentially be generalized to allow multiple genes, populations, or species. This could enable methods originally designed for such problems as the estimation of migration rates (Beerli & Felsenstein 2001; Nielsen & Wakeley 2001) to be applied to estimation of horizontal transfer rates within and among species, and to probabilistic determination of the sources of observed apparent transfers.

population size N. Despite the lack of a coalescence effective size, the labeled history of the genealogy under such models can still have the Yule distribution. For example, although changes in population size affect coalescence times, they do not alter the fact that all pairs of lineages are equally likely to coalesce.

Several strategies are available for determining the properties of models whose genealogies do not follow the coalescent distribution. It is sometimes possible to directly calculate or at least approximate the distributions of random genealogies. Alternatively, it may be possible to obtain the distributions from modified versions of the coalescent. However, the most general strategy for studying genealogies under complex models is simulation from sampled lineages back in time to their MRCA (Hudson 1990). In fact, because backward simulations can often be performed rapidly, they are useful even when the coalescent distribution does apply. Their efficiency results from the fact that simulation from a small sample backward in time to a MRCA requires that only a small number of random variables be generated. The forward approach, which entails simulation of whole populations for a long enough period of time to erase the effects of initial conditions, followed by extraction of genealogies of random sets of lineages, expends considerable effort simulating lineages that are not ancestral to samples.

The coalescent distribution of genealogies is often taken as a "null" distribution, as it represents the behavior of a population under simple assumptions. To understand the impact of complex phenomena on genealogies, distributions of genealogies under various models can be compared with the coalescent qualitatively or quantitatively, using properties such as T_n or L_n from Table 12.1 (Donnelly 1996; Uyenoyama 1997). For example, it is often noted that genealogies from exponentially growing populations are more "star-like" than are those from constant-sized populations (Slatkin & Hudson 1991). In quantitative terms, this observation reflects the fact that random genealogies under exponential growth have elevated values of ratios such as P_n/T_n and $L_n/(nT_n)$ (Rosenberg & Hirsh 2003).

Population Structure

In models with subdivision of populations, by geography or by other variables, the coalescence sequence of a random genealogy does not follow the Yule distribution, as pairs of lineages from the same group are more likely to coalesce than are pairs from different groups. The distribution of the labeled history or labeled topology of a random genealogy may be of less interest, however, than such distributions as that of the M-type or the collapsed labeled topology. Under a given model, these distributions, only applicable for multiple populations (or species), can help in articulating the predictions that the model makes about the processes it considers.

Two Populations

For two populations, the probability distribution of the phyletic status of a random genealogy may be of interest. Consider the *island model*: two haploid populations of size N with a fraction m of the lineages in each population switching populations each generation. With samples of size 2 from each population, for small Nm, the probabilities of scenarios C1, C2, C3, and C4 (Figure 12.4) approximately equal $1 - 14Nm/3$, $5Nm/3$, $5Nm/3$, and $4Nm/3$, respectively (Takahata & Slatkin 1990). From these values, it is observed that as the migration rate decreases to 0, the probability of reciprocal monophyly increases to 1.

The distribution of phyletic status can also be obtained (for any sample sizes) in the *two-population divergence model*, in which an ancestral population splits instantaneously into two descendant populations each of size N (Rosenberg 2003), or (for small sample sizes) in a divergence model that allows descendant populations to be subdivided after divergence (Wakeley 2000). In these cases, it is observed that at divergence, polyphyly is the most likely phyletic status, and as time progresses, reciprocal monophyly becomes most likely. In the two-population divergence model, reciprocal monophyly has probability 0.99 by $6N$ generations after divergence.

Although much is known about random genealogies under the island model (Takahata & Slatkin 1990; Nath & Griffiths 1993), the two-population divergence model (Takahata & Nei 1985; Rosenberg 2003), and other two-population models (Wakeley 2000; Teshima & Tajima 2002), the distributions of attributes of genealogies (Table 12.1) are more difficult to compute with two populations than with one. However, as in one-population models, backward simulation has proven useful for exploring these distributions in two-population scenarios (Hudson 1990; Rosenberg & Feldman 2002).

Three or More Populations

The probability distributions of C- or D-types for random genealogies, which are trivial for one or two populations, become interesting with three or more populations. Perhaps the most useful of these distributions is that of the collapsed labeled topology of a random genealogy. Suppose three populations descend from an ancestral population that split into two groups, one of which subsequently bifurcated again. Suppose also that the time between the bifurcations is t generations and that the population size between bifurcation events is constant at N haploid individuals. If one lineage is sampled from each population, the probability that the (collapsed) labeled topology of a random genealogy is the same as the labeled topology of the population phylogeny is $1 - (2/3)e^{-t/N}$ (Pamilo & Nei 1988). Each of the other two possible collapsed labeled topologies has probability $(1/3)e^{-t/N}$, so that as t increases to infinity, the probability of concordance of the labeled topologies of the gene genealogy and the phylogeny nears 1. A similar calculation for arbitrary sample sizes shows that the probability of topological concordance increases more quickly with t if larger samples are used (Rosenberg 2002).

As is true for the two-population case, probability distributions of complex aspects of genealogies in multi-population models remain elusive, except by simulation. However, some progress has been made in various scenarios (Pamilo & Nei 1988; Wakeley 1998; Wilkinson-Herbots 1998; Degnan & Salter 2005).

CASE STUDIES

Uses of Genealogies

The usefulness of gene genealogies arises from the fact that genetic variation can be viewed as the result of mutations occurring along the branches of genealogies (Figure 12.6). Thus, patterns of genetic variation are affected by the attributes of the genealogies on which mutations have occurred. However, these genealogies are generally unknown. To address this issue, one of two main strategies can be adopted (Rosenberg & Nordborg 2002; Hey & Machado 2003): first, the genealogy can be estimated from the data, and the analysis based on the estimated genealogy. Alternatively, the coalescent and its extensions can be used to sample genealogies from a set of random genealogies consistent with the data, and the analysis averaged over these genealogies. The former approach has the limitation that basing the analysis on the estimated genealogy ignores uncertainty in the estimate. The latter approach, while statistically rigorous, can potentially require intensive computations, so that it can only be applied approximately.

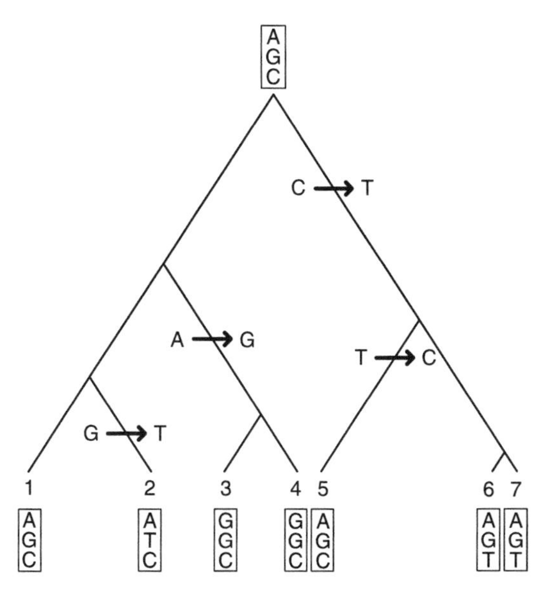

FIGURE 12.6. Mutations on genealogies. Suppose the genealogy in Figure 12.2 describes a locus with three nucleotides, and that the most recent common ancestor (MRCA) has genotype AGC. Mutations on the genealogy lead to the genotypes shown for the sampled lineages. Because mutations can be viewed as random events that occur along the branches of a genealogy, random samples of genotypes can be obtained by placing random mutations on simulated random genealogies. Mutations may obscure the underlying relationships of the lineages: although lineages 1 and 2 are closely related, they differ in genotype; at the same time, 1 and 5 are distantly related but are identical in genotype.

The fact that genealogies underlie patterns of variation has been useful for developing interpretations of particular observations in genetic data. Allowing for mutations, the coalescent model has been used to make various predictions about the distribution of allele frequencies expected across sites in a set of DNA sequences (Tajima 1989; Fu & Li 1993). For example, the comparatively "star-like" nature of genealogies in populations undergoing expansions in size, compared with those from constant-sized populations, is reflected in an excess number of mutations along external branches. The D and D^* statistics of Fu and Li (1993), which are computed from DNA sequences sampled from a population, compare numbers of mutations along internal and external branches. Negative values of these statistics, reflecting an excess of external mutations, indicate that growth in size may have been important in the history of the population.

A need to use gene genealogies arises in many contexts in diverse organisms (Avise 2000; Donnelly & Tavaré 1997; Li & Fu 1999; Knowles & Maddison 2002; Slatkin & Veuille 2002). Several examples are discussed below.

Molecular Phylogenetics

The inference of species genealogies (or phylogenies) from the distribution across species of a genetic character typically relies on the premise that if one lineage is sampled per species, then the genealogy for the character is identical to that of the species. If species are distantly related, this premise generally holds for the coalescence sequence of the gene genealogy, although the coalescence times of the gene genealogy are often considerably larger than those of the species genealogy (Figure 12.5). In this case, the problem of phylogenetic inference is to recover an underlying genealogy that has been obscured by the stochastic occurrence of mutations along its branches (Figure 12.6).

As we have seen, however, especially for closely related species, this basic premise may fail to hold. First, the lineages of one or more of the species may not be monophyletic, so that the choice of lineage affects the shape of the genealogy. Second, the gene genealogy often may have a different labeled topology from that of the species genealogy, so that the choice of locus affects the shape of the genealogy. When these scenarios have nontrivial probabilities, careful consideration of gene genealogies is important to phylogenetic inference. Generally, the solutions to the nonmonophyly and discordance problems involve use of many lineages per species and many independent genealogies, respectively.

A study by Wilson et al. (2003) addresses the problem of nonmonophyly of lineages for a set of 13 human populations. Assuming that the evolution of the populations followed a bifurcating tree, Wilson et al. aimed to estimate the genealogy of the populations. They genotyped 121 individuals for seven linked markers on the Y chromosome. They scanned the space of genealogies of 13 populations, for each population genealogy using the coalescent distribution to simulate gene genealogies of 121 lineages. Their numerical procedure, a Bayesian Markov chain Monte Carlo approach, guaranteed that the possible population genealogies and gene genealogies were visited during the scanning process with frequencies roughly proportional to their likelihoods. Of the population genealogies visited by their population growth model, 91% included a monophyletic grouping of the three African populations. Such a grouping has a probability of only 1/132 for random labeled histories sampled from the Yule distribution. Thus, the analysis was quite confident in the monophyly of these populations.

Discordance between gene and species genealogies is considered in a study of a human, a gorilla, and a chimpanzee. Chen and Li (2001) used genetic data in a study of the classic "trichotomy problem," that of deciding which pair of species, among humans, gorillas, and chimpanzees, has the closest relationship. The divergence of the three species occurred during a short enough period of time that genealogies vary by locus. Unlike in the case of separate groups within the human population, however, the splits among these species occurred long enough ago that nonmonophyly is unlikely for genealogies representing only one of them; thus, attention can be restricted to one lineage per species. Of the gene genealogies estimated by Chen and Li—one for each of 53 noncoding regions—the majority (31/53) showed that the human and chimpanzee had the most similar DNA sequences, favoring a grouping of humans and chimpanzees. By computing a multinomial likelihood to measure the weight of the evidence, Chen and Li concluded that their data provide very strong support for the human–chimpanzee grouping.

Demographic History

Gene genealogies are frequently applied to the reconstruction of population histories from DNA sequences. The inference of population and species phylogenies is one example of this kind of application. A second is the quantitative estimation of parameters of population history, such as times of divergence or migration rates.

Morrell et al. (2003) sequenced nine loci in 25 individuals representing three populations of wild barley: two low-elevation groups from east and west of the Zagros mountains in southwest Asia, and one group from the mountainous region itself. They were interested in the amount of migration among the three populations. Using a procedure that searches the space of possible migration rates and gene genealogies, sampling regions of this space in proportion to their likelihoods of explaining the data, they estimated that approximately one or two migrants move from each population to each of the other two populations in every generation. Morrell et al. suggest that this observation could be a consequence of dispersal via seeds embedded in the fur of migratory animals, or of deliberate dispersal by ancient hunter-gatherer peoples.

Selected Genes and Speciation Genes

One of the aims of genome-wide studies is to identify loci that have been strongly affected by natural selection. Demographic phenomena, such as admixture and migration, affect individuals, and are reflected in patterns of genetic variability across whole genomes. Natural selection, however, is localized to particular regions of the genome. Thus, selected loci can potentially be identified through their deviations from genome-wide averages. One way in which such deviations can be identified is through unusual properties of gene genealogies.

Using 10–20 individuals per species and a popular genealogical estimation method—the neighbor-joining algorithm—Machado and Hey (2003) inferred the genealogies for 16 regions in the genomes of three *Drosophila* species. Genealogies for regions on chromosomes X and 2 came closer to achieving monophyletic concordance—in which lineages from each species were monophyletic and the collapsed labeled topology matched the labeled topology of the species phylogeny—than did genealogies for regions on other chromosomes. Interestingly, laboratory studies have assigned to chromosomes X and 2 the highest densities of hybrid-sterility genes in the genome. Machado and Hey suggest a view in which genotypes on chromosomes X and 2 diverged earlier in speciation than did those of other chromosomes, as it was possible to produce hybrids with differing genotypes on other chromosomes long after hybrids with incompatible types on chromosomes X and 2 were no longer viable.

Experimental Design

Experimental studies of genetic variation require choices about sample sizes, numbers of markers, and statistical methods. Random genealogies can assist in deciding how to optimize studies to obtain maximal information about quantities of interest with minimal effort.

Pluzhnikov and Donnelly (1996) considered various ways of estimating the population mutation parameter Θ, which measures the level of genetic diversity in a set of DNA sequences. Because longer branches in genealogies provide more opportunities for mutations to occur, the information that a data set contains about mutation parameters increases with the branch lengths of underlying genealogies. To improve the precision in an estimate of Θ obtained from a set of DNA sequences, data can be added either by sampling new individuals for the same sequenced region or by increasing the length of the region. Because individual DNA sequences are correlated in that they result from the same genealogies, the addition of individuals provides new information about Θ only if the new individuals represent parts of genealogies that have not yet been sampled. Lengthening the sequence provides additional loci at which recombination could have occurred. Because recombination causes neighboring loci to have different (though correlated) genealogies, additional sequence provides new information if recombination did indeed occur. Pluzhnikov and Donnelly used random genealogies to derive expressions for the variance of estimates of Θ as a function of sample size and sequence length. They determined what allocation of resources to sample size and sequence length led to the smallest variance in the estimate of Θ. For various values of Θ and recombination rates, they found that samples of fairly small size (~3–10) were optimal, with most of the effort devoted to increasing the lengths of sequences from these individuals. Their optimal schemes can be used for future studies that aim to estimate Θ.

A related use of gene genealogies for experimental design is in evaluating statistical methods. Ramos-Onsins and Rozas (2002) were interested in identifying tests useful for detecting population growth. Using extensions of the coalescent for population growth models, they simulated genealogies on which they simulated mutations in order to obtain simulated data sets of DNA sequences. For each simulated data set, they applied 17 tests, observing that their own R_2 test and Fu's F_S test most frequently rejected the null hypothesis that the sequences were drawn from a constant population size model when indeed they were sampled from a growing population. Thus, investigators who wish to detect growth may be more successful if one of these two tests rather than one of the other 15 methods studied is used.

Genetics of Complex Traits

Many traits, including various human diseases, result from the interactions of multiple genetic factors. By searching for alleles that are found more frequently among individuals who have a trait than among those who do not, a genome can be narrowed to a small set of alleles that can be more directly tested for possible effects on the trait. These alleles must have originated as mutations in ancestors of the extant individuals who possess them. Thus, considering the genealogies on which these mutations occurred can help to make predictions about properties of trait loci; these predictions, in turn, can be used to design streamlined strategies to map the loci.

Using a random genealogy model, Pritchard (2001) studied the fraction of the individuals with a disease who possess the disease-susceptibility allele of highest frequency. In the model, mutations could occur from "normal" to "susceptibility" alleles and vice versa. Susceptibility alleles conferred elevated disease risks and selective disadvantages to their possessors. For various assumptions about mutation rates, selection coefficients, and human demographic history, random genealogies were simulated backward to a MRCA, which was assumed to be a normal allele. For each mutation on the genealogy that changed a normal to a susceptibility allele, the number of descendants of that mutation in a sample was tabulated. The mutation rate from normal to susceptibility alleles was observed to be the most important determinant of the fraction of diseased individuals who possessed the most frequent allele. Except at very small values of this rate, only a small fraction of the diseased individuals descended from the highest-frequency mutation. Pritchard concluded that mapping strategies in humans will be most effective if they account for the possibility that disease-susceptibility genes might have many low-frequency mutations, each of which is found in only a small proportion of diseased individuals.

FUTURE DIRECTIONS

The use of gene genealogies has led to new ways of conceptualizing genetic variation. By viewing genetic variation as the result of mutations on branches of genealogies, it becomes possible to reason about the signatures of evolutionary phenomena in data by thinking about how these phenomena affect genealogies. The coalescent enables quantification of the resulting intuitions, and new insights about evolutionary processes continue to follow from the incorporation of new phenomena into genealogical models. Statistical approaches based on gene genealogies continue to find new applications, of which the examples above give only a short introduction.

By considering many possible random genealogies that *could* underlie the pattern of variation at a locus, and by treating independent loci as replicates of the evolutionary process, methods based on genealogies can enable estimation of population history parameters and measurement of the uncertainty in the estimates. Because many uses of gene genealogies cannot yet be incorporated in methods that both quantify uncertainty in estimates and evaluate relative support for alternative models (Knowles & Maddison 2002), however, a major challenge is to develop methods applicable to the complex scenarios that are typically of interest. This endeavor requires computational improvements: while the simulation of random genealogies and data sets can usually be performed quickly, simulation of random genealogies from the conditional distribution of the genealogy given a specific data set is generally slow (Stephens 2001). Use of approximate numerical techniques may lead to greater computational tractability (Hudson 2001; Beaumont et al. 2002). Such tools will be especially useful for forthcoming genome-wide data on genetic variation.

Computational infeasibility is a particular problem in regions with large amounts of recombination. Such regions produce a sequence of correlated genealogies, which can be simulated using an adaptation of the coalescent (Nordborg 2001); however, most existing statistical tools apply only to individual regions with little or no recombination, or to unlinked collections of several such regions. Construction of computationally desirable models of genealogies that are not based on the coalescent may help to deal with this problem (Li & Stephens 2003). Indeed, the development of models of gene genealogies and the statistical methods to which they give rise offers many new challenges for the genomic era.

SUGGESTIONS FOR FURTHER READING

The well-known reviews of Hudson (1990) and Maddison (1997) cover gene genealogies and the coalescent, and the relationship of gene genealogies to species phylogenies, respectively. The rich and thorough survey by Nordborg (2001) and the excellent book of Hein et al. (2005) provide more recent treatments. Material on gene genealogies is also expertly embedded in the context of theoretical population genetics by Ewens (2004) and in the context of phylogenetics by Felsenstein (2003).

Ewens WJ 2004 Mathematical Population Genetics. I. Theoretical Introduction, 2nd ed. Springer.

Felsenstein J 2003 Inferring Phylogenies. Sinauer Assoc.

Hein J, Schierup MH & C Wiuf 2005 Gene Genealogies, Variation and Evolution. Oxford Univ. Press.

Hudson RR 1990 Gene genealogies and the coalescent process. Oxford Surv. Evol. Biol. 7:1–44.

Maddison WP 1997 Gene trees in species trees. Syst. Biol. 46:523–536.

Nordborg M 2001 Coalescent theory. pp 179-212 in Balding DJ, Bishop M & C Cannings, eds. Handbook of Statistical Genetics. Wiley.

Acknowledgments Finanical support was provided by an NSF Postdoctoral Fellowship in Biological Informatics and by a Burroughs Wellcome Fund Career Award in the Biomedical Sciences. I thank Steve Finkel, Peter Morrell, Mark Tanaka, John Wakeley, Jeff Wall, and Jason Wolf for extensive comments on a draft of this chapter.

III

FROM GENOTYPE TO PHENOTYPE

13

Gene Function and Molecular Evolution

SIMON C. LOVELL

Francis Crick (1916–2004), along with James Watson, is best known for predicting the structure of DNA, probably the greatest success in the history of molecular modeling. He also published important work on virus structure where he and Watson argue that the only way a stretch of DNA can encapsulate itself with protein is if multiple repeating subunits are used, he published a paper on helix packing in proteins where he suggests the reader should knock nails into broom-handles to visualize the point being made (Crick 1953; Figure 13.1), he suggested that "selfish DNA" may exist and be "the ultimate parasite," he suggested that there must be "adaptors" in protein synthesis (now known to be transfer RNAs) and, arguably most importantly, suggested the "sequence hypothesis" and the central dogma of molecular biology.

The sequence hypothesis states that sequence information flows from DNA to RNA to proteins. The central dogma states that once that sequence information is in the form of protein it cannot escape, meaning that protein sequence cannot be back-translated into DNA or RNA sequence. It was termed a "dogma" in 1958 because at the time there was very little evidence for it. However, Crick thought it so important that if it were proved not to be correct it would undermine molecular biology (Crick 1970). In the almost 50 years that have passed since the original proposal, the central dogma has not been disproved, although the sequence hypothesis has proved to be incomplete, but not incorrect. The generally accepted meaning of "central dogma" has changed in the molecular biology community to encompass both the sequence hypothesis and Crick's original definition of the term. I will use the current generally accepted definition, although Crick's original strict definition should not be forgotten.

The current molecular biologists' view is shown in Figure 13.2A. It describes the flow of linear sequence information in the cell. This description can be extended, especially if we take a more functional view (Figure 13.2B). The flow of linear sequence information from DNA to protein is still essential for the function of molecules (the function of nonmessenger RNAs excepted), but is by no means sufficient. The protein is not the end of the line. Proteins are usually made from only 20 amino acid types and 5 atom types. It is only the accurate positioning of the amino acid residues in three dimensions that gives rise to the great diversity of functions that proteins can perform. The information that determines the three-dimensional structure is encoded in the sequence, although exactly how is still one of the great unsolved problems of biology. Nevertheless, the attainment of structure is an essential prerequisite of the production of molecular function.

Structure is still not enough. A folded protein in splendid isolation will do nothing. To be functional it needs to bind something. To my knowledge this is universally true: an enzyme needs a substrate, a receptor needs a ligand, and an antibody needs an antigen. It is only with this binding that the lowest form of function—molecular function—is achieved. More complex functions, such as the function of cells, requires the formation of metabolic networks and protein–protein interaction networks and this is the subject of systems biology. Interactions between cells to form tissues and organisms is developmental biology, but is still mediated almost entirely

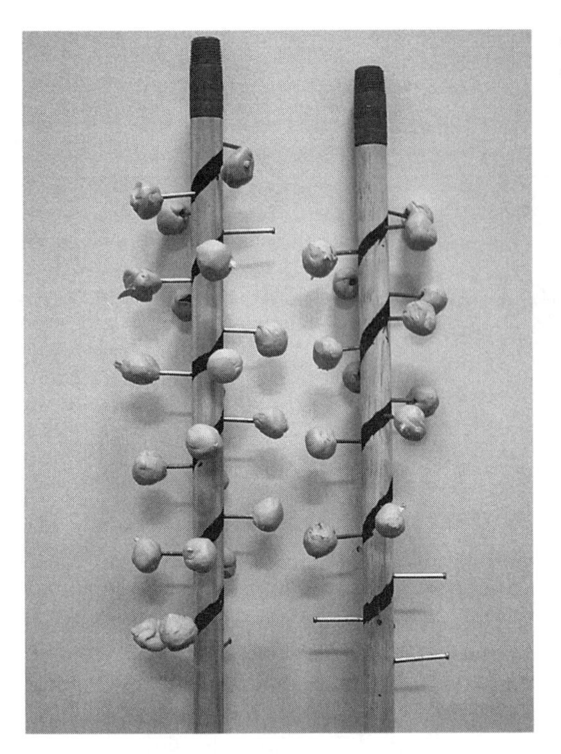

FIGURE 13.1. Models of α-helices made from broom handles, nails and modeling clay by the author, following instructions published by Francis Crick.

though proteins. These proteins' interactions are still determined by their structure, which is determined by their sequence, which is determined by RNA and DNA sequences.

Information can also flow backward, although not in the strict "linear sequence" sense that Crick discussed. An evolutionary view (Figure 13.2C) shows us that evolutionary constraints work in the opposite direction. Selection does not act directly on DNA. Rather, it acts on phenotypes which confer relative survival advantage. Since these phenotypes are determined by proteins and their interactions, selection acts indirectly on proteins. Specifically it selects proteins that have specific appropriate functions. Because these functions are determined by structure, evolution selects some protein structures and eliminates others. This leads to the selection or elimination of certain protein sequences only because sequences determine structure. In turn certain RNA and DNA sequences are selected.

Certain parts of a sequence contribute more to a protein's structure or function and hence to the phenotype. These regions have stronger evolutionary constraints. It is the subject of the remainder of this chapter how these arise, how they can be identified, and how this information can be used.

CONCEPTS

Genes Make RNA Makes Proteins and Proteins Form Three-Dimensional Structures

Evolution is dependent on variation and selection. Where there is variation but no difference in selective advantage, evolution is neutral. Selection arises if a mutation at a site in the genome changes fitness; this selection leads to departure from neutrality. The mutated sequence can have an effect on the organism's fitness in a number of ways, depending on its function.

Noncoding DNA

Noncoding DNA can have an effect on fitness, and so be selected for in one of two ways: directly, through an effect on its own stability, or indirectly, through an effect on RNA and protein production.

Sequences that give rise to structural features of chromosomes such a centromeres and telomeres must be maintained, at least in terms of their general characteristics if not their specific sequence. Mutations in noncoding DNA can also affect fitness by having an indirect effect on the production of proteins. Messenger RNA (mRNA) synthesis is dependent on promoters, transcription-factor binding sites, and, in eukaryotes, enhancer sequences. mRNAs are also processed, that is, their introns are removed. Processing depends on the sequence of the intron, especially at the intron boundary. In eukaryotes many genes can be alternately spliced, resulting in a single gene producing more than one protein. The resulting mRNA must also be translated; this is also dependent upon the sequence of the mRNA immediately preceding the open reading frame. All of these processes require the correct DNA sequence, and so mutations in these regions will potentially lead to strong selection.

Functional RNA

Thousands of genes in the human genome code for RNAs that are not transcribed into proteins (Lander et al. 2001). These mostly produce transfer

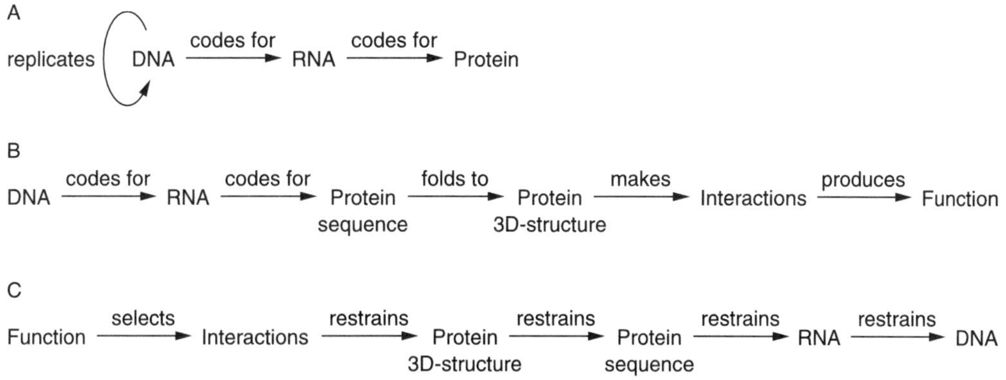

A

replicates DNA —codes for→ RNA —codes for→ Protein

B

DNA —codes for→ RNA —codes for→ Protein sequence —folds to→ Protein 3D-structure —makes→ Interactions —produces→ Function

C

Function —selects→ Interactions —restrains→ Protein 3D-structure —restrains→ Protein sequence —restrains→ RNA —restrains→ DNA

FIGURE **13.2.** Various views of the central dogma. (A) The usual "molecular biology" view, as described by Crick. (B) The "functional" or "systems biology" view. (C) The evolutionist's view. Note that Crick specifically described the flow of linear sequence information in the cell; both (B) and (C) require generalization to "information" in the broad sense.

RNAs, ribosomal RNAs, small nuclear RNAs, and small nucleolar RNAs, although there is a wide range of other classes. As with proteins, the function depends on the three-dimensional structure and this is determined by the sequence of bases in the molecule. This must, presumably, lead to selection and limits on substitution in a similar way to that which will be discussed for proteins, although this is much less widely studied for RNA. Even their number is hard to determine since they do not have open reading frames and are often small, so cannot easily be detected by gene-finding software.

mRNAs, which constitute the majority, may be directly selected for stability and lack of formation of secondary structure, but the majority of their sequence selection is through the agency of their ability to form proteins.

Proteins and their Structure: There Are Many Ways to Skin a Cat...

That genes make proteins affects their evolution. Sixty-four codons and 20 amino acids mean there is an approximate 3-fold redundancy in the genetic code. The result is that some mutations apparently will be synonymous, that is, they will not change the amino acid coded for, and, it would appear, be selectively neutral. In fact, synonymous mutations in coding regions and mutations outside coding regions can be non-neutral, as discussed by Nachman (Ch. 7 of this volume) and by Chen and Stephan (Ch. 9 of this volume). The selection is, however, weaker than for nonsynonymous mutations in protein-coding regions.

If there is a mutation to a different amino acid type, it may still be neutral, favorable, or deleterious. Kimura (1968) and King and Jukes (1969) suggested that although the majority of mutations are likely to be deleterious, producing purifying selection, there will be more neutral mutations than favorable ones. This is known as the "neutralist" point of view. In contrast to this is the "selectionist" view. The selectionists agree with the neutralists that there are three types of mutations (favorable, deleterious, and neutral), and agree that the majority will be deleterious, but disagree over the relative proportion of neutral to advantageous mutations, arguing that neutral mutations will be rare and favorable mutations will be relatively common. The relative merits of this argument are outside the scope of this chapter, but an excellent overview can be found in Page and Holmes (1998). The effects of neutral evolution are discussed in detail in Rodriguez-Trelles et al. (Ch. 8 of this volume).

Even nonsynonymous mutations (those that change the amino acid type) can be nearly neutral (Ohta 1973), although this depends on their role in the protein's structure. An amino acid can be substituted for another that can carry out the same structural role with very little evolutionary cost; that is, with little or no reduction in fitness. The result is that sequence is less conserved than structure, and therefore there are many sequences that can make any given structure.

...Except for the Tricky Bits Around the Ears

The majority of the genome in higher organisms does not code for proteins. Therefore, the majority of mutations will be in noncoding regions. These will be subject to weak, but nonzero, selection (Chen and Stephan, Ch. 9 of this volume). Amino-acid-changing mutations in coding regions have the potential to be under strong selection.

Exactly whether a mutation in a gene is deleterious, advantageous, or neutral depends critically on the protein structure. Any given amino acid will have a number of characteristics, be they related to the physical chemistry, stereochemistry, or hydrogen bonding ability. Any subset of these can be important at a given location in the protein structure. For example, the predominant role of an aspartate on the surface of a protein will probably be to hydrogen bond with solvent as an aid to solubility. Its exact conformation is unlikely to be important and it will probably be flexible. Conversely, an aspartate in the core of a protein is likely to be making specific side chain to main chain hydrogen bonds. Its conformation will be very important, and all buried hydrogen bond donors and acceptors must be satisfied to avoid destabilizing the structure of the protein.

These different structural roles will give rise to very different degrees of conservation, and different patterns of mutation. The aspartate on the protein's surface will be highly mutable, and amongst the most poorly conserved residues in the protein. It is likely that in homologous proteins it will be mutated to a range of other polar residues (Glu, Gln, Asn, Thr, Ser) any of which can make the same protein–solvent interactions. In contrast, the aspartate in the core of the protein will be one of the most conserved residues. Depending on the details of the hydrogen-bonds it makes there is chance it could be replaced by an asparagine, but even this is relatively unlikely.

The probability of mutation to every amino acid type can be calculated. Overington et al. (1992) used a database of structurally aligned proteins and counted the probabilities of mutation from each residue type to every other residue type. Crucially, they calculated not just the substitution frequencies, but first assigned various local environments for each amino acid. The environments used were those shown to change the substitution pattern to a significant degree: secondary structure, solvent accessibility, and a number of hydrogen bonding classes. The resulting environment-specific substitution tables have been used for fold recognition, comparative modeling, and structure verification.

Taking a structural view of mutations explains otherwise-puzzling observations. For example, aspartate and glutamate substitute for each other relatively rarely; similarly substitutions of asparagine to/from glutamine are also rare. Each of these pairs differs from each other by only a single CH_2 group (Figure 13.3), and so it may be expected that they will easily exchanged. In fact they differ greatly in their structural roles. They have markedly different conformations (Lovell et al. 1999, 2000), with the Asp and Asn being highly constrained by the main chain and likely to make local side chain to main chain hydrogen bonds. In contrast Glu and Gln do not have extensive local main chain interactions. They have a larger number of conformations open to them, and are much more likely to be on the surface, flexible, and interacting with solvent.

Proteins Work by Binding Things

The evolutionary constraints on a protein's sequence depend, then, on the specific roles an amino acid has in the structure. There can be a multitude of roles for an amino acid, and these can be decidedly non-obvious (Blundell & Wood 1975). A protein usually has to fold into a three-dimensional

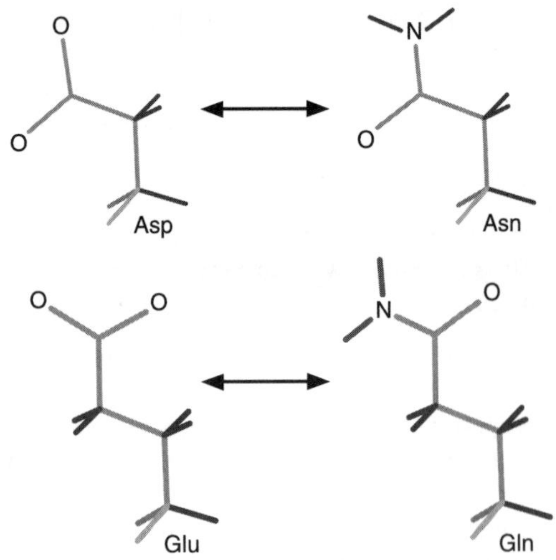

FIGURE 13.3. Asparagine (Asp), aspartate (Asn), glutamine (Glu), and glutamate (Gln). Arrows show common substitutions.

structure (although exceptions are discussed below). This structure has to be both stable and soluble. This leads to the type of evolutionary constraint discussed above. The structure has to be unique, meaning that there must not be alternative low-energy conformations that the protein can fold or refold into: if such competing structures exist, disease states, such as prion disease, can arise (Dobson 2002). Proteins must also be protease-resistant for the most part, although in some cases (for example insulin) must be activated by protease cleavage at a specific site. Many proteins form obligate oligomers, and so the correct binding interface must be produced. Because proteins work by binding other molecules, other transient binding sites must be produced, for example a substrate-binding or receptor-binding site. Enzymes must have additional catalytic residues.

The role of an amino acid in one of these "functional sites" will be markedly different from the role of an amino acid in any other part of the protein. Moreover, it is likely that this role is much more narrowly defined for a functional site residue than for any other: both specific chemistry and specific positioning of atoms will be required. In the majority of cases, therefore, it is unlikely that a different amino acid type could perform the same role, and so the residue will be conserved. In the rare cases where another amino acid type can perform the same role, the substitutions allowed will be highly constrained, and so the substitution pattern will be unusual.

If these unusual substitution patterns, and hence unusual constraints, can be identified, they can be used for identifying functional sites in proteins. Probably the most widely method is "evolutionary trace" (Lichtarge et al. 1996), which identifies highly conserved residues. With this technique a multiple sequence alignment is partitioned into a phylogenetic tree. Residues that are conserved in all branches of the tree, or in specific subbranches, are identified and mapped on the protein's three-dimensional structure, if available. Residues conserved in all branches are likely to be important for binding or catalysis, whereas those conserved in subbranches are likely to determine specificity of binding.

These evolutionary constraints can, however, derive not just from function but from any of the other roles an amino acid may have. They will be present at any position where mutation is not neutral. Neutralists and selectionists agree that this is the majority of sites in a molecule. It is possible to identify evolutionary constraints in a general manner by back-correcting for the expected substitution pattern from an environment-specific substitution table (Chelliah et al. 2004).

This approach has been implemented in the program CRESCENDO. Using the known structure of a protein it is possible to determine the local environment of each amino acid. A combination of 64 environments are used describing all combinations of hydrogen bonding, solvent accessibility, and main chain conformation (Overington et al. 1992). Based on these environments it is possible to predict the probability of substitution to every other amino acid type for each position in the sequence. Because this is done using an environment-specific substitution table derived from the whole database, this is the average substitution pattern for amino acids of that type, in that structural environment. Thus it encodes the evolutionary constraint placed on an amino acid by the structure. If this set of substitutions is compared with the observed substitutions, which can be determined by collecting and aligning homologous sequences, then amino acids under unusual evolutionary constraint will have unusual substitution patterns. These are amino acids under evolutionary constraint from sources other than the structure. We have found that these extra constraints are overwhelmingly likely to be from function. Thus we can identify active sites, regulatory sites, and protein–protein binding interfaces.

The failure to distinguish functional from structural constraints can produce problems. The fungal pepstatin-insensitive carboxyl peptidases and homologs have been classed as members of the aspartic proteinases (A4 family). Huang et al. (2000) mutated conserved acidic residues in order to determine which were catalytic. They identified an Asp and a Glu residue which were both conserved and led to loss of activity when mutated: this seemingly confirmed the classification as an aspartate was required for activity.

Subsequent solution of the crystal structure of a member of the family (Fujinaga et al. 2004) showed that in fact the enzymes had a novel fold, and so were unrelated to the aspartic proteinases. The conserved Glu is indeed catalytic (along with a previously unidentified Gln), but the conserved Asp is some distance from the active site. It is, however, buried, charged, and making four hydrogen bonds (Figure 13.4): a classic example of a residue conserved for structural reasons. It is conserved because there is no other amino acid that can

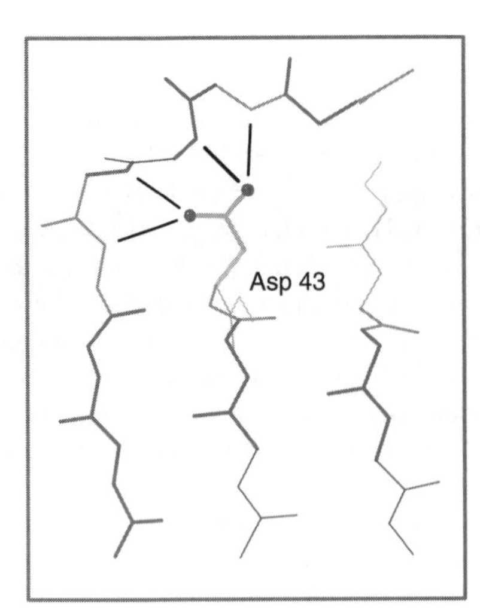

FIGURE 13.4. Buried aspartate (Asp 43) from a pepstatin-insensitive carboxyl peptidase from *Scytalidium lignicolum* (pdb code 1s2k). The aspartate is completely conserved in all homologous enzymes, and, before the structure was solved was thought to be catalytic. It is, however, conserved for structural reasons: it is buried in the core of the protein (it is surrounded on all sides by other parts of the protein) and makes four side chain to main chain hydrogen bonds (black lines).

perform the same role. It is at the end of a secondary-structural element, as these conserved residues often are. It has four hydrogen bonds, so must have two oxygen atoms, and must take a precisely defined conformation to make these bonds. This is a "tricky bit" of the protein structure, which can be made only one way. It is no surprise that mutation of this residue affects activity, but this is almost certainly because it disrupts the protein structure.

Correlated Mutations in Protein Structures

Models of sequence evolution typically assume that substitutions at different sites are independent. In fact this is a simplification. Adjacent amino acids are likely to be in the same type of secondary structure, and have correlated solvent accessibilities, which affect the substitution pattern. As surface loops in proteins accept insertions more readily than buried secondary structures, and adjacent residues are likely to be in similar environments to their neighbors, the rate of acceptance of insertions is also likely to be correlated. Sites distant in sequence can also have correlated substitutions. This can happen when the protein's structure brings them together in three-dimensional space or, alternatively can be due to common functional constraints.

Covariation can be identified directly from the sequence alignments by looking for "mutual information" between columns in a multiple sequence alignment. This approach is particularly interesting when the information is correlated with the protein structure. Wollenberg and Atchley (2000) found that in the basic helix–loop–helix (bHLH) transcription factors, residues on the same face of a helix, which would have similar structural environments, had correlated substitution patterns. Additionally, substitutions of residues in contact with each other, which of course will have similar structural environments, are correlated and often more so than would be predicted from their shared environments alone.

Atchley et al. (2000) have also made efforts to distinguish between the various sources of correlation between sites, with a model that includes the sources of correlation above but also those due to shared evolution history and those due to chance. This suggests that in the bHLH family, residues within a given structural or functional role not only have a characteristic substitution pattern, but also have a characteristic covariation of this substitution pattern.

Nonconserved Functional Sites and Nonconserved Functions

It is the rule that binding sites, active sites, and other functional sites in proteins are unusually well conserved. As with most rules, however, there are exceptions. Moreover, these exceptions are surprisingly common.

In a study of 31 enzyme superfamilies (i.e., proteins that clearly share a common ancestor but which may be very distantly related to each other), Todd et al. (2002) found that 15 of them contained at least one example whereby there was significant active site variation. These superfamily members have the same fold and the same molecular function, but have come up with two or more solutions to the "catalytic conundrum" of how to perform a reaction. In some cases the active site has moved to a completely different part of the protein structure, but in other cases there are mutations and/or structural rearrangements that allow different residues to perform catalytically equivalent roles. In many cases the catalytically equivalent residues are of the same type but from different parts of the protein; in some cases the atoms of the catalytically equivalent residues occupy equivalent positions in space (Figure 13.5).

This may seem at odds with the argument that residues in functional sites are under strong evolutionary constraint, although in reality it is not. Mutations are selected based on their effect on function. If a mutation is functionally advantageous or neutral, it will be accepted. In these examples it seems that a mutation in a functional site is compensated for by the use of a different amino acid from a different part of the protein. Crucially, the compensating amino acid can place equivalent atoms in equivalent positions in the structure. Alternatively the entire active site can "hop" to a different position in the protein over evolutionary time, preserving the same relative spatial position of the side chains. Because the three-dimensional positioning of functional atoms is conserved, even when the residues themselves are not, function is unaffected. It may be said that these are substitutions in sequence but are not substitutions (or at least, are more minor substitutions) in structure and function.

There is clear evidence that enzymes with new activities are recruited from existing enzymes. In seminal work, Horowitz (1945) argued that metabolic pathways evolved backward. This argument attempts to explain the puzzle of the origin of metabolic pathways: if an organism has only a precursor in its environment, and requires a product that can only be synthesized in many steps, how can a metabolic pathway arise? An organism that has only half of the pathway has no selective advantage over one lacking the whole thing.

Horowitz proposed the following solution. Imagine a population of organisms that live in an environment rich in nutrients. These organisms survive by consuming substance A. After a number of generations the organisms reproduce, and substance A will be used up in the environment. Any organism that can synthesize substance A from something else, let us say substance B, will have a selective advantage. We have a one-step pathway: B→A. In time substance B will in turn be depleted and synthesis of B from something else (substance C) will give selective advantage to the organism.

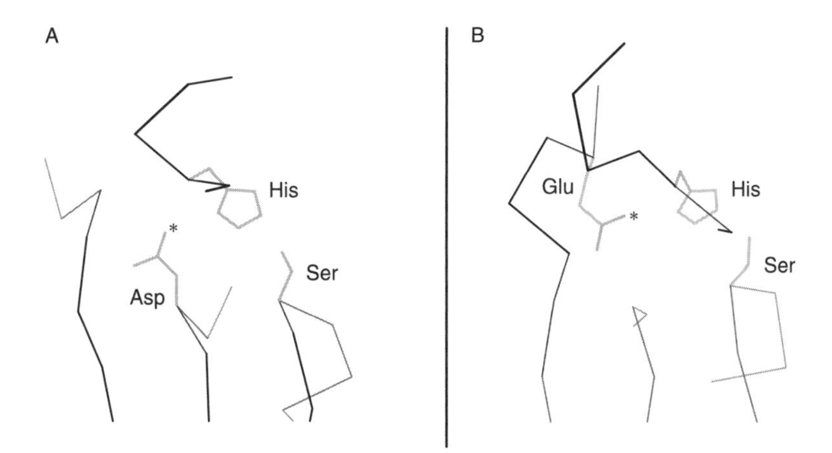

FIGURE 13.5. An example of homologous enzymes where a catalytic residue has "moved" as the sequences have diverged. (A) The catalytic triad from human pancreatic lipase (pdb code 1lpb). (B) The catalytic triad from *Geotrichum candidum* lipase (pdb code 1thg). The histidine and serine are in equivalent positions. In contrast the acidic group has both changed residue type (from aspartate to glutamate), and changed position in the sequence (from β-strand number 6 to β-strand number 7). However, the oxygen atom that takes part in the catalytic charge-relay system (marked with an asterisk) has almost exactly the same position in three-dimensional space, despite these radical changes at the sequence level. The example is taken from Todd et al. (2002).

Already we have the beginnings of a metabolic pathway: C → B → A.

In later work, this was argument was extended (Horowitz 1965). As enzymes catalyze both the forward and reverse reactions they must bind both substrate and product. Therefore, the binding site can be partly conserved and only the reaction chemistry changes. This is an elegant argument, but appears not to be generally true.

The change of function subsequent to gene duplication has been widely studied from a phylogenetic point of view, and is discussed in Thornton (Ch. 11 of this volume).

Gerlt and Babbitt (1998) have shown that homologous enzymes can catalyze different reactions. They have found that in the examples where this is shown to be the case, it is the chemistry that is conserved, rather than the specificity. This would suggest that nature finds the evolution of a different chemical reaction more difficult than evolving a different specificity. Interestingly, when catalytic antibodies are produced in the laboratory, the opposite approach is taken. Typically a protein (antibody) that binds the required substrate is produced and then the catalytic residues are designed and engineered in.

Protein–Protein Interaction Surfaces

Proteins frequently function by binding other proteins. Protein complexes can be divided into two major classes: obligate and transient. Obligate protein complexes are formed immediately after folding and last until the protein is degraded. They are often, although not always, homo-oligomers. Transient complexes are more interesting. The constituents will spend much of the time in a noncomplexed state, and then form the complex in response to some signal to perform a function. They are usually hetero-complexes. Examples are signaling complexes, hormone binding receptors, and antibodies binding antigens. The result of this multitude of protein–protein interactions is a higher level of order: the network (Box 13.1).

Proteins are larger than the nonprotein "small molecules" that are the usual substrates of enzymes. This is always true, although a circular statement of course, as "small molecules" are so called because of their relatively small size. Nevertheless, in a protein–protein complex, two relatively large molecules of approximately equal size bind to each other.

BOX 13.1. The Role of Gene Interaction Networks in Evolution
Stephen R. Proulx

No gene exists in a vacuum. Each functional genetic element in a genome has an effect, at the organismal level, that results from a suite of interactions ranging from extremely general to the highly specific. For example, any gene that functions by producing an RNA or protein product must, at a minimum, successfully promote binding of RNA polymerase, thereby interacting with the gene encoding polymerase. At the other extreme, protein interactions can be highly specific, such as the polymerization of α and β globins to form hemoglobin. Likewise, regulatory genetic elements have organismal effects solely through their interaction with the signaling molecules that they transduce and the genes whose transcription they induce. These interactions can occur at a host of levels of organismal structure:

- between genes and proteins (transcription factors) within a cell,
- directly between proteins within a cell,
- between enzymes and substrates within a cell,
- between gene products that move among cells and genes or gene products in other cells,
- between tissues that interact physiologically,
- between organs that interact physically,
- between behavioral modules.

BOX **13.1.** *(cont.)*

These interactions act in aggregate to produce phenotypes and traits whose basis can be defined as an interaction network. In a general sense, these interactions can always be thought of as genetic networks to the degree that the phenotype is genetically determined. Because all organismal features must interact at some level, at least to the extent that they affect fitness, it is difficult to restrict our consideration of a network to a single level. In practice, however, most studies do this by utilizing data sets that contain only one type of interaction. Two such interaction networks are gene regulatory networks and protein–protein interaction networks.

Gene Regulatory Networks

In gene regulatory networks, the transcription of genes within the network is controlled by transcription factors produced by other members of the network (Davidson et al. 2003b; Siegal & Bergman, Ch. 16 of this volume). Just as no gene exists in a vacuum, a regulatory network must have some external input, at least to determine the initial state of the system. Regulatory networks are thought of as being dynamic, with the rate at which each gene is transcribed changing through time and influencing the rate of transcription of other genes. These networks can form feedback loops and have the ability to form and hold patterns, cyclically fluctuate, or go through predictable phases. For these networks to have some observable effect on phenotype, they must also have some output. This is typically through the transcriptional control of downstream genes that directly alter cell structure, function, or physiology. While gene regulatory networks are ubiquitous, they are most often studied in the context of development.

Protein–Protein Interaction Networks

In a protein-protein interaction network, sets of proteins physically interact with each other to form networks (Uetz et al. 2000). These networks are usually inferred from assays that detect binding between pairs of proteins. By testing for pair-wise interactions between all proteins in a library, the connections in the network can be found. The phenotypic output of such networks is, however, less clear than that of gene regulatory networks. The method simply determines which proteins can interact, but does not reveal anything about how the proteins function or how they interact with other components of the cell.

The Network Approach

The network approach can be powerful because it allows large amounts of data to be boiled down to more manageable statistical descriptions of both whole networks and individual network elements. The real hope for the biological application of network theory is that the statistical descriptions of networks that mathematicians have already developed are related to evolutionary processes. Each node in a network can be characterized by the number of connected nodes (node degree), the number of shortest pathways it lies on (betweenness), and the number of small loops formed around it (clustering[1]), among other features (see Figure 1; Newman 2003). One of the most general descriptors of whole network structure is the degree distribution. The frequency distribution of node degree contains information about how the network

[1]A feature of nodes in interaction networks that describes the density of local interactions. Clustering is usually defined by the number of closed loops of length 3 that a node falls on, divided by the total number of possible closed loops of length 3.

(continued)

BOX 13.1. *(cont.)*

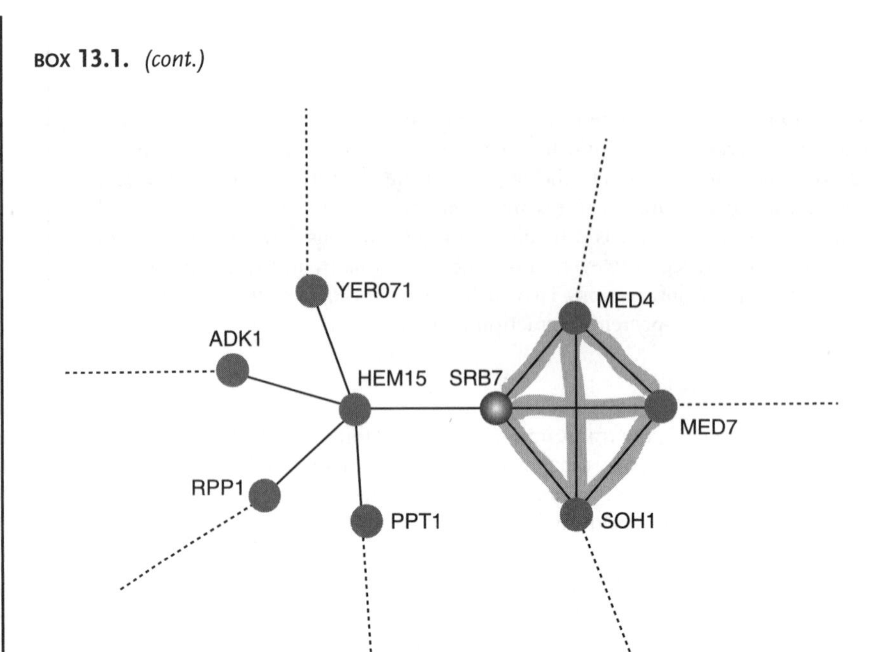

FIGURE 1. A small portion of the protein–protein interaction network obtained from http://mips.gsf.de/proj/yeast/CYGD/interaction/ and centered around the protein SRB7 (involved in transcription). Dashed lines depict connections outside the pictured network and represent multiple connections. SRB7 has degree 4 because it interacts with 4 other proteins. The clustering coefficient is measured by the number of closed triangles in the network, which are highlighted. There is a high degree of clustering on the right of the graph, but not on the left.

could have been created and how removal of nodes will affect the network's performance. Many networks, biological and otherwise, have scale-free degree distributions, where the frequency of nodes of degree k is proportional to a power series (i.e., $p(k) \propto g^{-k}$) (Newman 2003). Scale-free networks are characterized by many nodes with few connections but some nodes with very many connections. Such networks differ from random networks that are created by adding a fixed number of connections between pairs of randomly chosen nodes and have many nodes of intermediate degree (Figure 2). A fundamental question in the evolution of gene networks asks how scale-free networks have evolved and what the biological effects they have. A critical debate centers around whether non-adaptive processes are likely to produce scale-free networks (Wagner 2003), or whether network-wide features are under direct selection. Scale-free networks have been shown to be difficult to break up when nodes are removed at random, a process similar to the mutational knockout of genes within a network (Albert et al. 2000). If robustness of the network, to either mutational or environmental change, is under strong selection (Siegal & Bergman, Ch. 16 of this volume), then a scale-free architecture could emerge as a result of selection.

While the degree distribution is probably one of the easiest network properties to calculate, it is far from the only measure of a genetic network. In fact, from an evolutionary perspective, the most important feature of a genetic network is the way that variation in individual network elements translates into variation at the phenotypic level.

BOX 13.1. *(cont.)*

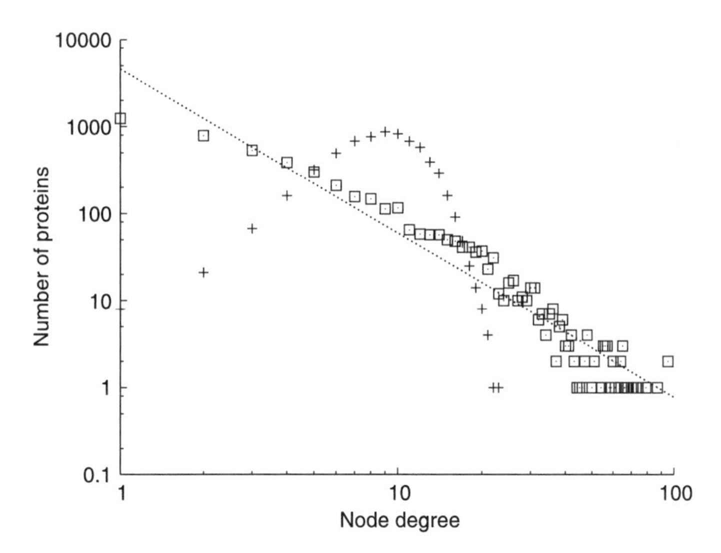

FIGURE 2. A log-log plot of the degree distribution for the yeast protein–protein interaction network. The data were obtained from the entire yeast database available at http://dip.doe-mbi.ucla.edu/(9/05/2004 update). The boxes represent data from the yeast database after log transformation after Wagner (2001). The log-transformed data are fit with a line given by the power law relationship: $p(k) \propto k^{-188}$. A network constructed of 6500 proteins with the number of connections observed in the actual network produces a Poisson degree distribution, shown by +. The randomly connected network produces data that are considerably different from the observed data, suggesting that some other process in responsible for the observed pattern.

Understanding this relationship is at the heart of such important questions as the genetic basis of complex traits and the microevolution of development. In some cases, the node distribution may be related to features of evolutionary importance. For example, Promislow (2004) has shown that the degree of yeast proteins known to be involved in aging is higher than expected based on other yeast proteins. He hypothesizes that more connected proteins are more pleiotropic, that is to say that connectedness is related to an important feature of theoretical models of aging. These results support the antagonistic pleiotropy theory of aging because it predicts that the genes involved in aging are genes that have pleiotropic effects.

The mapping from genetic variance at individual genes within a network onto phenotypic variation is important for several reasons. It determines the pathways in phenotype space that are open to evolution and the speed of phenotypic change. In fact, the **G** matrix of quantitative genetics is an instance of this mapping, under specific simplifying assumptions about the way that genes and alleles interact (Phillips & McGuigan, Ch. 20 of this volume). Epistasis is an emergent feature of this mapping, suggesting that understanding the forces that shape genetic networks will allow us to

(continued)

BOX 13.1. *(cont.)*

make predictions about patterns of epistasis. The epistatic relationships in turn determine how selection on further modification of the genetic network can act. In particular, the amount of variance in fitness due to the current mapping determines the opportunity for canalization through both allelic substitutions and network restructuring (Siegal and Bergman, Ch. 16 of this volume; Proulx & Phillips 2005). Evolutionary biology is currently making a transition from a field where small experiments and small data sets are common to a field where overwhelming amounts of information become rapidly available. There is still a major gap in our ability to relate these new data to the existing theory of evolutionary processes and great opportunities remain for seminal contributions.

The interaction interfaces are large and flat. This is in contrast to enzyme active sites, which tend to be in clefts or pockets in the protein.

The result is that protein–protein interaction surfaces contain 3–4 times as many residues involved in binding than do small molecule binding pockets. It may be expected, therefore, that the contribution of each residue to the binding is one third to one quarter as large as the small molecule case, and, in consequence, that the evolutionary constraint is also smaller.

This is something of an oversimplification: not all parts of a protein–protein interaction surface contribute equally to the binding energy. In a landmark study, Clackson and Wells (1995) mutated residues in interaction surfaces of a hormone and receptor, and measured the effect on binding. They found that there was apparently a large range of energy contributions: mutation of some residues barely affected binding whereas other seemed to contribute a large amount of energy to the binding. These large contributors clustered together in what they termed a "hot spot." Remarkably, the hot spot for binding on the hormone matched up almost exactly with the hot spot for binding on the receptor. The interacting surfaces have a central hydrophobic core (the hot spot) that contributes most of the binding energy, surrounded by a ring of hydrophilic residues. Bogan and Thorn (1998) reviewed a number of similar experiments on 22 different systems and found that many of these conclusions are general.

Residues that have a large contribution to binding energy would be expected to be under stronger evolutionary constraint than those with a smaller contribution. This has been examined by Hu et al. (2000). They found conserved protein–protein binding surfaces, but the degree of conservation was only slightly above that in the rest of the protein. They showed that the conserved residues were of the same type as those in binding hot spots, but did not investigate whether they were the same residues.

In our recent work on functional sites (Chelliah et al. 2004) we were surprised to find that the most difficult aspect of the work was producing the benchmark set. It seems intuitively obvious that some parts of proteins are more important for function than others. The active site residues in an enzyme, which are completely conserved and have a direct role in catalysis are surely more "functional" than those easily substituted ones on the enzyme's surface. Some residues are unequivocally functional. Once you try to determine objectively which residues are in a set labeled "functional" and which are not, it becomes extremely difficult. Catalytic residues of enzymes are normally defined as those that directly take part in the reaction mechanism. These are widely reported in the literature. Substrate binding residues are often understood and reported less well. Residues that hold together an active site or stabilize the active conformation may also be thought of as functional, but are rarely reported as so in the literature. An example of this is the "fireman's grip" threonine from the aspartic proteinases that is completely conserved: if it is mutated catalysis is abolished. Yet it is neither catalytic nor involved in substrate binding. As we have seen, the contribution to binding energy in protein–protein interactions varies, so it is difficult to describe some residues as "binding residues" and others not, just on the basis of whether or not they are in the binding interface.

Rather I propose a *definition* of functionality of a residue that is based on degree of evolutionary constraint. Of course, it is necessary to correct for the degree to which a residue is required for a given structural environment. If environment-specific substitution tables are the correct way to do this, and if the formulation used by CRESCENDO is the correct one for the back-correction, then we have a method for calculating the degree of evolutionary constraint and, therefore, the degree of functionality. Of course other researchers have other definitions, but philosophically they are essentially the same.

What Is Function?

Selection depends on the effect of a mutation on function. Function, however, is decidedly slippery. We can ask, for example: What is the function of the glycolytic enzyme pyruvate kinase? Is it to convert phosphoenolpyruvate (PEP) to pyruvate? To produce ATP? Is it to provide energy to a muscle cell? To allow muscle contraction? To provide the locomotive power for a mating ritual, and therefore reproduction? Given that pyruvate kinase deficiency causes anemia, its function may be thought of as anemia prevention. Alternatively under anaerobic conditions pyruvate is converted to acetaldehyde and ethanol. Is the function of pyruvate kinase to make beer?

Pyruvate kinase has all of these functions and more. Typically genes and their products are described at least in terms of their molecular function (conversion of PEP to pyruvate) and cellular function (metabolism), and organism-level function (energy production). That they are not described in terms of, for example, their functions in society is more a reflection of the training of biologists rather than their lack of such characteristics (anyone who has attended a party or hospital on a Saturday night will appreciate the profound affect ethanol production has on some societies).

The effect of a gene on several different traits is known as "pleiotropy." This is related to, but distinct from, the "hierarchy of function" described here. Typical examples of pleiotropy are the action of a gene product on several different target proteins, or the action of a protein in a range of tissues. This is a "horizontal" effect: a gene product carries out a single function in multiple places, the multiple effects arising from the multiplicity of sites of action. In contrast (and in addition) I argue that there is also a "vertical" component to the view of function: a gene product can carry out a single function in a single place but there is still a multiplicity of functions. This multiplicity derives only from our point of view.

At any level of this hierarchy, function can further be subdivided: functions may be specific, general, or conditional. The specific function of pyruvate kinase (or, rather, one of them) is the conversion of PEP to pyruvate. This function is specific to pyruvate kinase and is possessed by few, if any, other proteins. One general function it has is maintenance of the osmotic potential in the cell, and therefore the Donnan equilibrium. This is a function that arises due to the general character of the protein, and is common to all proteins. A conditional function will be utilized only under a specific set of conditions, and indeed may never be utilized. A typical example of a conditional function would be that of an antibody to an antigen the organism never meets during its lifetime.

Conditional functions lead to problems designing experiments to identify function. Naively, a gene may be deleted and the phenotype of the organism observed. An experiment such as this would conclude that the spare tyre on my car has no function, as I have not needed to use it for the last 10 years, and its removal would lead to no apparent change to the car's "phenotype." This is clearly not a valid conclusion; it merely tells us about the function under the conditions observed. It is impossible to observe an organism under all conditions the organism will meet in its life, or, more importantly, all conditions its ancestors have ever met during their evolutionary history. Moreover, biological systems are highly redundant. Blockage of one metabolic or signaling pathway often leads to induction of an otherwise silent alternative. The body requires blood glucose levels to be maintained, as glucose is the sole source of energy that can be used by the brain. In the absence of dietary glucose, glycolysis is switched off and gluconeogenesis is switched on, producing the required glucose. It is possible that this will never happen in a well-fed laboratory rat, and so disruption of the genes involved in gluconeogenesis would produce no apparent phenotype. Incidentally, the regulation of this switch is (of course) another function of pyruvate kinase.

Phenotype depends on a combination of genotype and environment. This is just as true for molecular evolution as it is for any other level of function; all that varies is the definition of "phenotype" and "environment." The phenotype is the molecular function of the molecule in question, the environment the cellular environment, made up of the molecule in question, its binding partners, and the cellular milieu of solvent, small molecules and macromolecules. Sometimes, as the case of conditional functions described above, the genotype can interact with the environment in such a way that the phenotype is silent. For organisms such as viruses, molecular function and the cellular environment is almost all there is. Consequently there are very few steps between genomes and the top level of function, and so loci are under almost direct selection. For most other organisms, there are many more levels of environment: the environment of the tissue, organ, organism, and population. It is the interaction of a gene and its products with each level of hierarchy of environment that gives rise to the hierarchy of function.

The functional annotation of genes and the genome is currently one of the major efforts in bioinformatics. The gene ontology (GO) system (The Gene Ontology Consortium 2000) and other related schemes do make an effort to include functions at the molecular, cellular, and biological process levels and to capture multiple functions at each of these levels. However, there are an indefinitely large number of these functions. Many of the functions for pyruvate kinase mentioned above are listed by GO, but many are not. Nor are any of the three or four more I have thought of while writing this (for example, heat production in exothermic organisms). It should be realized that this is just a single enzyme produced by a single gene with a single enzyme commission number. Surely this must be the simplest case for functional annotation.

Shrager (2003) suggests that the concept of a static assignment of function to biological systems leads to all of this confusion. In fact we will probably always be able to think of more and more functions for pyruvate kinase given enough time. Shrager suggests that function is meaningless outside of a context, and, more specifically, outside of a process. This can be rephrased as "the function of the genotype is meaningless outside the environment." It is clear that within the process of glycolysis the function of pyruvate kinase is phophorylation and control. Within the process of flight from a predator its function is providing chemical energy that can be converted to kinetic energy. Shrager proposes, therefore, that functions are not pre-assigned at all, but rather are assigned dynamically as the process is considered. This may be correct, but the world seems not to be listening. Genomes are annotated according to old-fashioned, static, specific functions, mainly because they are so useful.

Evolutionary Origins of Genes and Folds

Evolution is usually studied comparatively. In *Systema Naturae*, Linnaeus used comparison of features of modern organisms to produce the famous hierarchical taxonomy. This taxonomy implies evolutionary relationships. Our understanding is greatly enhanced, however, by the inclusion of information from the fossil record, allowing us to see primitive, intermediate, and transitional forms in their appropriate historical order. Molecular biology adds a wealth of complementary information. The single greatest confirmation of Darwinian evolution is the correspondence between sequence-based phylogenies and Linnean classification.

Unfortunately, sequence analysis has few fossils. With the exception of a few preserved samples we can only analyze sequences from modern organisms and use these to imply more primitive forms. Maximum likelihood methods can be used to predict the most likely sequence of a common ancestor (Thornton, Ch. 11 of this volume), but moving further back in evolutionary history to identify the origins of genes and folds is more difficult.

Both genes and proteins come to us fully formed. Modern genes have transcription initiation sites, promoters, translation start sites, introns with appropriate splice signals, stop codons, polyadenylation signals, and transcription stop signals. Similarly, proteins tend to be folded, stable, and soluble with a unique structure (although important exceptions are discussed below). Their origins remain a mystery, though.

Where Do Genes Come From?

There is good evidence that genes derive from gene and genome duplication. Humans probably have two whole-genome duplications in their history, although there is stronger evidence for one than the other (McLysaght et al. 2002). Obviously whole-genome duplication doubles the number of genes, but it is followed by large-scale gene loss. This is probably due to the risk of recombination if there are several identical sequences within the nucleus, which can lead to meiotic errors. Thus the yeast strains that have undergone genome duplication now have approximately the same number of genes as those that have not. Nevertheless, paralogs exist within yeast that derive from this event.

Genes can also be produced by duplications of smaller regions of the genome. This leads to whole clusters of genes with the same order on the chromosome (conserved synteny). This can be seen, for example, in both the human (Lander et al. 2001) and mouse (Waterston et al. 2002) genomes. Individual genes may also be duplicated, probably due to unequal crossing-over during replication. Horizontal transfer from bacteria and viruses has also been identified. Alternatively exons can be shuffled to give rise to new functions.

The mechanism of derivation of new genes and new functions from existing genes is widely studied and relatively well understood. Moreover, the characteristics of their evolution once duplicated are well studied. Both duplication and subsequent evolution are discussed by Thornton (Ch. 11 of this volume).

The question of origins remains. If our only known mechanism of gene production requires the pre-existence of ancestor genes, where did the first one come from? Logically, there must be a method for the production of coding DNA from noncoding sequence. Initial analysis of the various vertebrate genomes failed to find a single example of where this had happened. There is now a documented case where *Alu* repeat DNA has become coding, producing an entire exon (Singer et al. 2004). It has happened recently enough that the origins can be identified from comparative genomics. It would seem, however, that this not a common event, unless such new genes have such extreme positive selection that their origins are obscured, which does not seem likely. How genes originally arose seems, therefore, unclear.

Where Do Protein Folds Come From?

The "fold" of a protein or a protein domain refers to the arrangement in three dimensions of the secondary structural elements, and their connectivity. The three proteins in Figure 13.6 have different folds. G-protein B (Figure 13.6A) is clearly different from the other two as it is a mixed α-helical/β-sheet fold. However, Figure 13.6B and C also represent

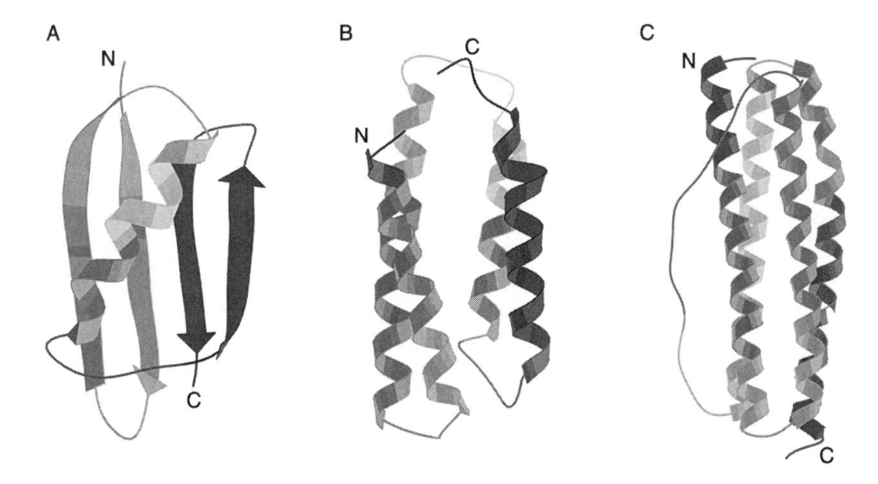

FIGURE 13.6. Three proteins with different folds. (A) G-protein B (pdb code 1igb) is clearly different from the other two, since it contains both α-helices (coils) and β-sheets (arrows). (B) Hemerythrin (pdb code 1hmd) and (C) Ferritin (pdb code 1eum) also have different folds, however, because although they are both four-helical bundles, they are connected in different ways. The N^- and C^- termini are labeled, and the structures oriented so that the N^- termini are at the front left. The differing positions of the C^- termini show that they have different connectivity. Also note that ferritin (C) has a long connection between two helices (green) which is absent in hemerythrin: it is this that allows the different connectivity.

different folds, despite both being four-helical bundles, as the helices are connected differently. This can be seen by the relative positions of the N- and C-termini. They are oriented so that the N-termini are top left. This results in the C-termini having different relative positions (top right front and bottom right back, respectively).

We can study the relationships between folds, and their evolution. We can see how they recombine to give rise to new functions (Patthy, Ch. 14 of this volume). However, uncertainties remain. It is not clear how folds originated and whether this happened once, multiple times, or whether it still happens. Moreover, it is not clear (a) whether protein domains with the same fold are homologous, nor (b) whether protein domains with different folds are not homologous.

Homology can be positively identified. Since there are many sequences that can produce a given fold and large-scale sequence convergence is overwhelmingly unlikely, sequence similarity can be taken as evidence of divergent evolution. Sequences can, however, diverge to such an extent that similarity is undetectable with current methods. When this happens, homology can be implied from structural or functional similarity. It is possible that genes can evolve further so that these signals are very weak or even lost completely. The question then arises: if two proteins share only fold similarity, are they homologous or analogous?

The origins and relationships of the TIM barrel fold is the most widely studied and argued about. It is the most common fold known: 10% of enzymes are TIM barrels and the overwhelming majority of TIM barrels are enzymes. Does this commonness represent a single origin and subsequent divergence into a huge range of enzymic functions, or, conversely, does it represent multiple origins? Attempts to answer this question have centered on looking for fainter and fainter evolutionary signals.

Lesk et al. (1989) argued for analogy based on differences in side chain packing in the central β-barrel. Farber and Petsko (1990) argued for homology amongst the 17 TIM barrel enzymes then known, partly on the basis of structural similarity and partly on the basis of colocalization of the active sites. They argued that the active site could equally well be at either end of the barrel, but in every known case it was found at the C-terminal end. The chances of this happening at random were $1{:}2^n$ where n is the number of examples. At this time $n = 17$ and the random chance 1:1072. There is no known physical reason for the active site always to be at the same end of the barrel. Of course, this does not mean that there is not one to be discovered. In fact, common binding sites in common positions in folds presumed to be analogous have been found in a wide range of proteins (Russell et al. 1998). If these are, indeed, structural analogs, then they must have their common position due to "engineering" considerations, that is, physical constraints.

Nagano et al. (2002) analyzed 889 TIM barrel structures in 76 families and 21 superfamilies. They found that 17 of the 21 superfamilies have hints of common ancestry, 10 of which were strong links. The evolutionary links identified were sequence relationships, functional relationships, and common structural motifs. The trend, at least for the TIM barrels, is that the development of more sensitive techniques identifies more evolutionary relationships. Whether eventually all TIM barrels will be shown to be homologs remains to be seen.

This does not answer the original question though, which was that if proteins share *only* fold similarity, are they homologs or analogs? It is often assumed that they are analogs because there is no evidence that they are not (Russell et al. 1997). Certainly analogs can be made artificially by directed evolution (Krishna & Grishin 2004). We do not know, however, how new folds arise, whether they still arise, or how difficult this process is. There is good evidence, at least in a limited set of cases, that proteins can evolve into different folds (Grishin 2001). It is not clear how commonly this happens, but it is remarkable that it happens at all, given the requirement for close packing in protein cores that must have evolved rapidly in these proteins after the change of fold.

Lupas et al. (2001) have speculated about the origins of folds being from small segments of proteins, probably composed of a handful of secondary structural elements. They suggest that these protein segments were expressed and then associated. Subsequently, rearrangements of the DNA could have brought these protein segments together, to allow them to be expressed as full-sized proteins. Mechanisms of this type are clearly easier for repetitive folds, such as those made from repeated α-helical hairpins, because they can be made by tandem duplication of the original gene segments: all that would be required is that the expressed protein segments self-associate. For nonrepetitive folds two different protein segments would have to associate with each other, and the genes would have to fuse. These two simultaneous events seem much less likely.

If the mechanism Lupas et al. propose is correct, then there are two major conclusions. First, this is a process that is no longer occurring, because we see no evidence of these small segments today—rather, we see whole domains. Second, protein domains are polyphyletic, that is, do not have a single evolutionary origin. In Patthy (Ch. 14 of this volume) the combination of many domains from disparate origins to make whole proteins is discussed. The Lupas mechanism suggests that, similarly, domains themselves may be formed by the accretion of smaller fragments of distinct origin.

There must be mechanisms both for noncoding DNA to become coding, and for protein folds to arise: once there were no genes and no folds, now they exist, so they must have been produced at least once. The extreme view would be, therefore, that each event happened *only* once, and all other genes and folds are divergently evolved from this initial event. This would seem not to be the case, as there are suggestions of several different types of protogenes with evidence of divergence from them, but no evidence of they themselves having a common origin (Lupas et al. 2001). It would seem likely, therefore, that there have been conversions of noncoding DNA to coding DNA multiple times. The exact number of times is, however, impossible to determine.

The opposite extreme view is even less tenable: it is not the case that every gene and fold observed today arose spontaneously as there are clear evolutionary relationships between them. When there are no signs of evolutionary relationships we are unable to tell whether genes and folds are homologs and have diverged so far that we cannot identify the relationship, or whether they are analogs.

FUTURE DIRECTIONS

Structural Conservation

I have discussed above how sequences diverge and how this is affected by protein structure. A much less widely studied problem is how protein structures diverge and how this is affected by changes in sequence. Chothia and Lesk (1986) found that as amino acids were substituted in the core of a protein, the secondary structural elements change position to accommodate the different volumes of substituted side chains. They gave a function to describe the amount of structural change for a given sequence divergence. However, the type and direction of change could not be predicted.

There have been attempts to predict the detail of the structural changes caused by mutations (Reddy et al. 1999), and they lead to modest improvements in accuracy of predicted structures. Nevertheless, we do not have nearly the same degree of understanding of structural evolution as we do of sequence evolution. This is, perhaps, because sequence changes are discrete, and easily quantifiable. Structural changes, in contrast, are continuous. Moreover, methods for determining protein structures lead to degrees of errors in the coordinates often of the same order of magnitude as the structural divergence. Finally, proteins are inherently mobile, and this mobility varies in different parts of the protein structure in ways not well understood. Nevertheless, we will not properly understand how molecular evolution works until we fully understand how sequence affects structure.

Unstructured Proteins

I began by discussing how sequence gives rise to protein structure and function, and how structure and function constrain sequence evolution. It is becoming increasingly clear that there are an important class of proteins that have no defined structure. Instead they function in the unfolded state (Dyson & Wright 2002). These "natively unfolded" proteins may be whole open reading frames, or insertions into otherwise-folded proteins. They are probably much more common than previously appreciated because they are difficult to study using our current techniques for studying protein structure: they do not crystallize and they give poorly dispersed nuclear magnetic resonance spectra.

One of their characteristics is that they tend to have low-complexity sequences, that is, containing fewer than usual amino acids, often with long runs of a single amino acid type. Low-complexity protein sequences are likely to be coded for by low-complexity DNA sequences, and low-complexity DNA sequences tend to be genetically unstable. They can lead to slippage and/or unequal crossing over, and therefore can expand in the genome. I have previously speculated about their origins and the likelihood that they may be nonfunctional (Lovell 2003).

Not only is the discovery of these proteins challenging our traditional views of sequence/structure relationships, they may be important evolutionarily. They clearly evolve rapidly, are mobile, and there is at least one case of an Alu repeat sequence becoming coding (Singer et al. 2004). Whether they are a major driving force in molecular evolution remains to be seen.

Genomic View

In the 1960s and 1970s determination of each new protein structure was a major event requiring many years work, followed by a flurry of papers analyzing the results. Genome sequencing is in a similar state now to protein crystallography then. Fortunately the rate of sequencing is ever-increasing, genome assembly is getting easier as it can be done by homology to already-known genomes, and computational tools (and computers) are developing so that we can make sense of the data. I imagine we will have to wait much less than 30 years to see daily depositions of whole genome sequences into public databases. Perhaps there will even be efforts to sequence all, or at least representative, genomes from all known species ("genomic-omics"?).

In the way that we can build hierarchical databases of genes, sequences, and protein structures and understand their relationships and evolution by comparing them, we are now beginning to do the same with whole genomes. Comparative genomics allows us to know the positions of genes on the genome and allows us to understand gene and genome duplication events and lateral gene transfer. This in turn allows us an unprecedented insight into the origins and history of genes and hence their evolutionary relationships.

Importantly, having a complete catalogue of sequences for an organism not only allows us to tell what does exist in terms of molecules and therefore molecular evolution, it also allows us to say categorically that certain sequences, molecules, or events are absent. Never have we had more unexploited data that can give us a greater understanding of molecular evolution.

SUGGESTIONS FOR FURTHER READING

Page and Holmes (1998) is a very good book on molecular evolution from a phylogenetic perspective. The sections on models of molecular evolution are particularly good, with a very readable discussion on the "neutralist/selectionist" debate. The enduring contributions of Francis Crick to molecular biology, as well as the human aspects of the history of his discoveries, can be found in his autobiography, *What Mad Pursuit.*

Crick FH 1988 What Mad Pursuit: A Personal View of Scientific Discovery. Basic Books.
Page RDM & EC Holmes 1998 Molecular Evolution: A Phylogenetic Approach. Blackwell Scientific.

Acknowledgments Many of the ideas in this chapter come directly from discussions with Tom Blundell, and I thank him for stimulating my interest in this area and providing me with a great deal of information. The "functional" view of the central dogma came to me from Andy Brass, but he claims not to have invented it; I would be interested to hear its original source. The "specific, general and conditional" categorization of function was suggested by the anonymous reviewer of a previously published manuscript. I thank Steven Morrissey for inspiration.

14

Evolution of Multidomain Proteins

LÁSZLÓ PATTHY

Although there is evidence that folded proteins occur frequently among randomly generated amino acid sequences (Davidson & Sauer 1994) it seems much easier to remodel replicas of old protein folds than to invent new folds from scratch. Accordingly, it is generally accepted that creation of the first folded proteins was a rate-limiting step in the appearance of protein-based life and that most extant proteins have arisen from a limited number of ancestral protein folds by divergence. In fact, it has been suggested that the majority of proteins belong to less than 1000 types of protein folds (Brenner et al. 1997). Although more recent analyses suggest that the total number of distinct folds is higher (Coulson & Moult 2002; Koonin et al. 2002), the fact remains that the majority of proteins arose from a surprisingly small number of ancestral protein folds.

There are two major pathways for generating independently evolving replicas of the same protein-coding gene: through speciation and through gene duplication. Following speciation, replicas of the same gene in the diverging species (orthologous genes) usually continue to fulfill the same biological function. Although they start to accumulate mutations independently, acceptance of mutations follows the same pattern: the pattern dictated by the conserved structure and conserved function of orthologous proteins. In the case of orthologous genes with well-established functions, nonsynonymous substitutions are predominantly deleterious whereas synonymous substitutions are predominantly neutral; therefore the rate of synonymous substitutions (K_S) significantly exceeds the rate of nonsynonymous mutations (K_A): usually $K_A/K_S << 1$. Since in most cases the functional and structural constraints of orthologous proteins are essentially the same in different species, the ratio of nonsynonymous to synonymous substitutions is usually a constant value. A major deviation from this K_A/K_S constancy may be observed if orthologous proteins of diverging species adapt to different functional or structural requirements: for example, when the biological activity, specificity, or stability of the protein is altered to adapt to environments. During such adaptations there is positive selection for advantageous amino-acid-changing mutations that make the protein better adjusted to the new needs: there will be an increase in the rate of substitutions at nonsynonymous sites.

Gene duplication is the other major pathway that may lead to independently evolving replicas of a given protein-coding gene. The evolutionary significance of gene duplication is that it gives rise to a redundant duplicate of a gene within a genome and this duplicate may acquire divergent mutations and eventually may emerge as a new gene with a new function (Ohno 1970). Genes derived by a duplication event are said to be paralogous; in contrast orthologous genes are derived by speciation events. Studies in the last decades have clearly shown that gene duplication is the predominant and most important mechanism by which new genes (with new functions) can arise.

The initial fate of duplicated genes is determined by the functional consequences of having extra copies of the same gene and increased amounts of the same protein. Duplication of a gene is disadvantageous if increased amounts of the gene product upset some fine regulatory balance and causes confusion within the cell. In such cases the most likely scenario is that one of the copies is inactivated by deleterious mutations and becomes a pseudogene.

Immediate advantage will be gained through gene duplication if a greater supply of the given gene product increases the fitness of the organism. In such cases the most likely scenario is that both copies will be retained. For example, if an organism is exposed to a toxic environment, there may be selective advantage of overproducing those proteins that decrease toxicity: it has been shown that duplication of genes involved in insecticide resistance occurs at a high rate in insects exposed to insecticides (Devonshire & Field 1991). Similarly, positive selection for immediately advantageous gene duplications has been observed in protozoans exposed to drugs (Ouellette et al. 1991).

The majority of gene duplications, however, are neutral, their final fate being determined by natural selection. In the long run such duplicated genes are unlikely to be fixed unless they acquire a novel and useful function. Accordingly, one of the typical routes of duplicated protein-coding genes is that one of the copies becomes a pseudogene. The greatest evolutionary significance of initially neutral gene duplications, however, lies in the fact that they may lead to the emergence of genes with novel or significantly modified functions. This may happen if one of the duplicated genes retains its original function while the other accumulates molecular changes adapting it to perform a different task from their common ancestor. In many cases both duplicates acquire functions that are different from that of their common ancestor: they may specialize in different subfunctions of their ancestor.

Although neutral duplications (with fully redundant functions) are expected to be evolutionarily unstable and may be lost, they can survive long enough to have a chance to acquire a useful novel function that would make them advantageous. Many types of mutations can endow duplicated protein-coding genes with advantageous novel functions. Most frequently, new functions appear stepwise as a result of several advantageous point mutations: continuous modification of the original function may eventually appear as a novel function. This scenario may be illustrated by the pancreatic proteases trypsin, chymotrypsin, and elastase that are involved in the digestion of proteins present in foodstuffs. Despite the common origin and close structural similarity of their protease domains they differ in their cleavage specificity: elastase cleaves proteins in the vicinity of amino acids with small nonpolar side chains, chymotrypsin cleaves at bulky hydrophobic residues, whereas trypsin cleaves only at arginyl or lysyl residues of proteins. Structure–function studies on these proteases have revealed that substitution of several key residues had to occur during step-by-step divergence of trypsin, elastase, and chymotrypsin from a common ancestor.

New functions of duplicated genes may arise as a result of more dramatic changes, such as when the diverging paralogs experience fusion with entire protein domains. Unlike point mutations, which usually can lead to a change of function only in a stepwise manner, domain duplication, domain shuffling, or domain fusion mutations may frequently lead to a major and immediate change of function. Recent studies have confirmed that protein evolution by assembly from distinct domains (domain shuffling, domain accretion) has played an important role in the evolution of large multidomain proteins and contributed significantly to organismal evolution.

This chapter will focus primarily on the unique features and evolutionary significance of multidomain proteins and the evolutionary mechanisms responsible for their formation.

CONCEPTS

Structural, Functional, and Evolutionary Characteristics of Multidomain Proteins

Analyses of protein structure databases have revealed that the average size of a protein domain of known crystal structure is about 175 residues; proteins that are larger than 200–300 residues usually consist of multiple protein folds (Gerstein 1997). The individual structural domains of such multidomain proteins are defined as compact folds that are relatively independent inasmuch as the interactions within one domain are more significant than those with other domains. Individual domains of multidomain proteins usually fold independently of the other domains.

Some multidomain proteins contain multiple copies of a single type of structural domain, indicating that internal duplication of a gene segment encoding a domain has given rise to such proteins. The duplicated domains may diverge in function

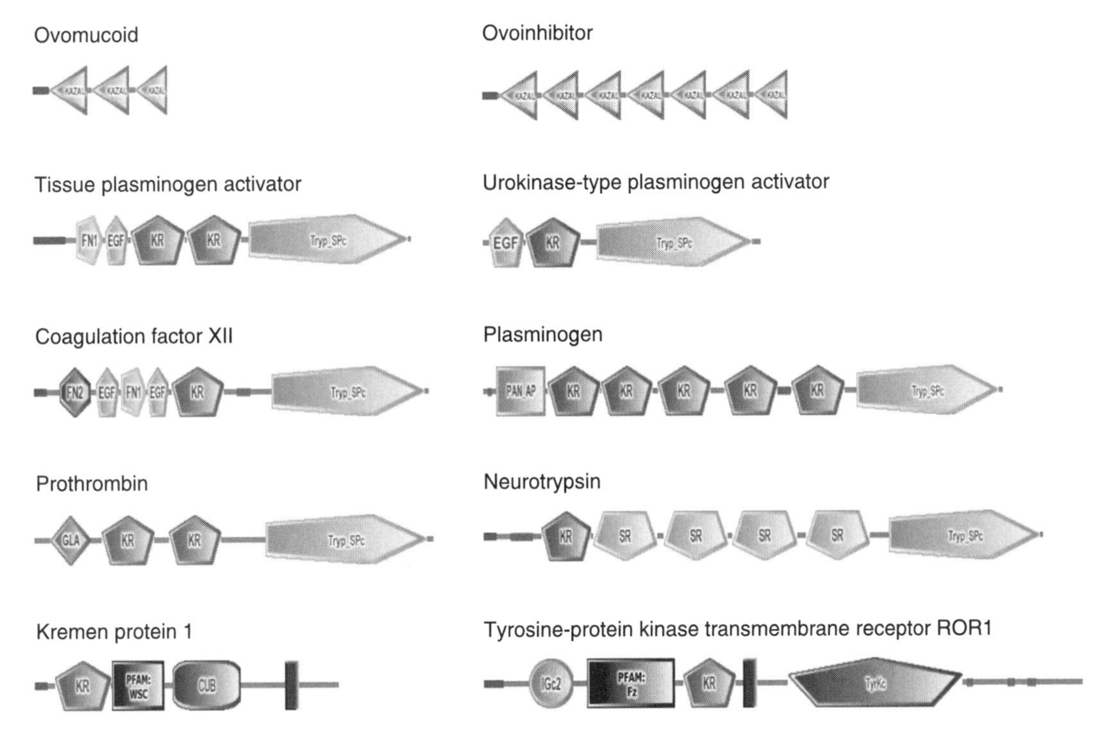

FIGURE 14.1. Domain architecture of some multidomain proteins. The figure shows SMART representations of the proteins (Letunic et al. 2004, *Nucleic Acids Research* D142-4). The domains are designated as follows: CUB: domain first found in C1r, C1s, uEGF, and bone morphogenetic protein; EGF: epidermal growth factor-like domain; FN1: fibronectin type 1 domain; FN2: fibronectin type 2 domain; Fz: frizzled domain; GLA: domain containing Gla (gamma-carboxyglutamate) residues; IGc2: immunoglobulin C-2 type; KAZAL: Kazal type serine protease inhibitors; KR: Kringle domain; PAN_AP: divergent subfamily of APPLE domains; SR: scavenger receptor Cys-rich; Tryp_SPc: trypsin-like serine protease; TyrKc: tyrosine kinase, catalytic domain; WSC: present in yeast cell wall integrity and stress response component proteins. The vertical bars represent transmembrane domains.

just as paralogous proteins diverge. For example, egg white ovomucoid and ovoinhibitor consist of tandem, homologous domains; the individual domains are capable of binding one molecule of trypsin or other serine proteinases (Figure 14.1).

Many multidomain proteins contain different types of domains (i.e., domains that are not homologous to each other). The genes of such multidomain proteins were created by joining two or more gene segments which encode different protein domains. As a typical example, we may refer to the protease tissue plasminogen activator, in which a finger domain, an epidermal growth factor related domain and two kringle domains are joined to a trypsin-like serine protease domain (Figure 14.1). Such multidomain proteins, consisting of multiple domains of independent evolutionary origin, are frequently referred to as mosaic proteins. There are domain types that frequently occur in different architectural contexts in multidomain proteins. For example, the Kringle domain has been used to build a wide variety of multidomain serine proteases such as urokinase, tissue plasminogen activator, plasminogen, coagulation factor XII, prothrombin, neurotrypsin, and transmembrane proteins such as kremen protein 1, ROR1 receptor tyrosine kinase (Figure 14.1). Such versatile building blocks are usually referred to as mobile protein modules.

There are several reasons why formation of multidomain proteins from pre-existing domains may have major evolutionary significance.

1. This is the fastest way to create novel, structurally viable, larger proteins.
2. Acquisition of a new domain with an established function (e.g., binding specificity) can

bring about an immediate change in the function of the recipient.

3. In multidomain proteins a large number of functions (different binding activities, catalytic activities) may coexist, making such proteins indispensable constituents of regulatory or structural networks where multiple interactions (e.g., protein–protein, protein–ligand, protein–DNA interactions) are essential. For example, the domains that constitute multidomain proteins of the intracellular and extracellular signaling pathways (cf. receptor tyrosine kinases) mediate multiple interactions with other compents of the signaling pathways. As another example, we may refer to the different modules of regulatory proteases (cf. proteases of the blood coagulation and fibrinolytic cascades, such as those shown in Figure 14.1) that recognize substrates, bind inhibitors, cofactors, phospholipid membranes etc. and through these interactions control the activation and activity of these regulatory enzymes. Similarly, the coexistence of different domains with different binding specificities is also essential for the biological function of multidomain proteins of the extracellular matrix (cf. fibronectin, proteoglycans, laminins, modular collagens): the multiple, specific interactions among these constituents are indispensable for the proper architecture of the extracellular matrix.
4. As a corollary of their involvement in multiple interactions, formation of novel multidomain proteins is likely to contribute significantly to the evolution of increased organismal complexity since the latter reflects the complexity of interactions among the genes, proteins, cells, tissues, and organs (Patthy 2003). In the next sections I will discuss evidence supporting the increased importance of multidomain proteins in higher organisms and their unique role in the evolution of organismic complextity.

Frequency of Multidomain Proteins in Different Groups of Organisms

Some recent studies on the distribution of protein folds in the three kingdoms of life have analyzed the relative frequency of multidomain proteins in complete proteomes. Wolf et al. (1999) have counted the number of different folds in each protein of the proteomes of Archaea, Eubacteria, and Eukarya and the average fraction of the proteins with each given number of domains was calculated. It has been concluded from these analyses that distributions of single-domain, two-domain, three-domain, etc., proteins in Archaea, Eubacteria, and Eukarya show a very good fit to a model where each next class (e.g., two-domain proteins vs. single-domain proteins, three-domain proteins vs. two-domain proteins) contains approximately 7 times fewer entries then the previous one, a distribution that is compatible with the evolution of multidomain proteins by random combination of domains. The observation that—in this respect—there was little difference between the proteomes of eukaryotes and prokaryotes apparently contradicted the notion that evolution of complex eukaryotes favored (and benefited from) the formation of multidomain proteins since they contributed to their increased organismic complexity.

A more recent analysis, however, provided evidence that there may be a connection between the propensity of protein domains to form multidomain architectures and organismic complexity. Koonin et al. (2002) have shown that—although in all proteomes the domain distribution is compatible with a random recombination model of the evolution of multidomain architectures—the likelihood of domain joining increases in the order: Archaea < bacteria < eukaryotes and there is a significant excess of larger multidomain proteins in eukaryotes.

Evolutionary Origin of the Multidomain Protein Repertoire of Higher Eukaryotes

A major group of multidomain proteins of higher eukaryotes consists of intracellular proteins involved in cytoplasmic or nuclear signaling processes (Bork et al. 1997) that have been assembled from a limited repertoire of intracellular signaling domains (Table 14.1). Comparison of the genomes of single-celled eukaryotes, plants, and metazoans have reavealed that the multidomain proteins involved in their intracellular signaling pathways are strikingly similar (Copley et al. 1999; Aravind & Subramanian 1999). Significantly, components fulfilling basic eukaryotic functions are conserved in all three groups, consistent with the notion that the key intracellular signaling pathways have originated prior to the divergence of the major eukaryotic lineages (Aravind & Subramanian 1999). The fact that many

TABLE 14.1. Intracellular modules found most frequently in multidomain proteins

Module	SMART[a]	Pfam[a]
WD40 repeats	WD40	WD40
Spectrin repeats	SPEC	spectrin
Ankyrin repeats	ANK	ank
EF hand repeats	Efh	efhand
Armadillo/beta-catenin-like repeats	ARM	Armadillo_seq
Domain present in PSD 95, Dlg, ZO-1/2	PDZ	PDZ
Src homology 3 domains	SH3	SH3
Pleckstrin homology domains	PH	PH
Protein kinase C conserved region 2	C2	C2
Calmodulin-binding motif	IQ	IQ
Domain present in Lin-11, Isl-1, Mec-3	LIM	LIM
Src homology 2 domains	SH2	SH2
Calponin homology domains	CH	CH
Domain with 2 conserved Trp (W) residues	WW	WW
Protein kinase C conserved region 1	C1	DAG_PE_bind

The modules are listed in the order of decreasing frequency in nonredundant domain databases.
[a]The abbreviations correspond to those used by the SMART database (http://smart.embl-heidelberg.de) and the Pfam database (http://www.sanger.ac.uk/Software/Pfam/).

components of the intracellular signaling pathways have also been identified in prokaryotes (Bakal & Davies 2000) suggests an even earlier time for the formation of the multidomain proteins involved in these processes. The similarity of the intracellular communication systems of plants, fungi, and animals may be illustrated by the fact that there is no major difference in the relative abundance of the intracellular signaling domains of plants, fungi, and animals (Figure 14.2).

Another major group of multidomain proteins of higher eukaryotes consists of extracellular proteins such as the multidomain proteases of the complement, blood coagulation, and fibrinolytic cascades of mammals and nonprotease constituents regulating these cascades. This group also contains

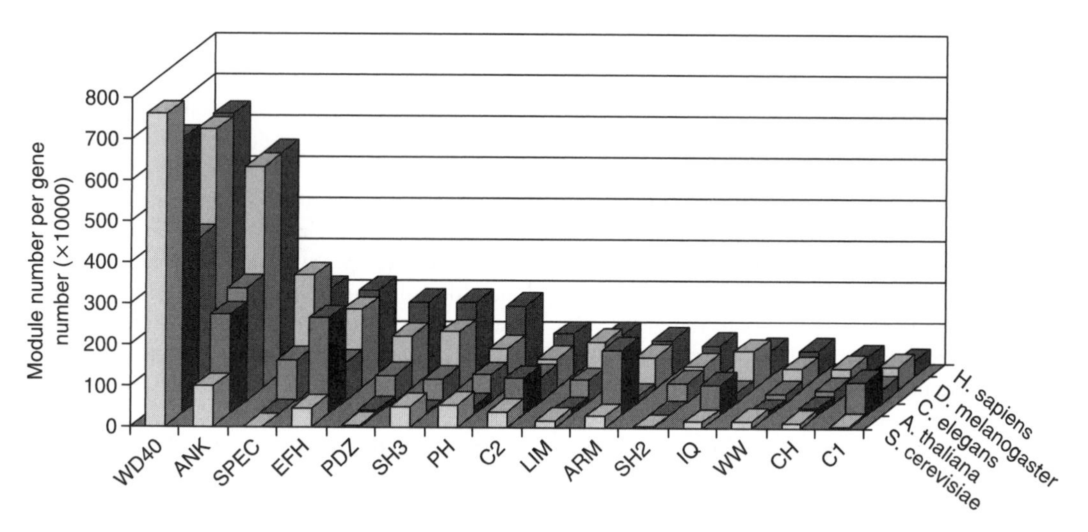

FIGURE 14.2. Frequency of intracellular modules in proteins of some model organisms. The abbreviations of modules correspond to those shown in Table 14.1. The values on the ordinate were calculated by dividing the total number of a module type in a genome with the number of the genes present in that genome. Note that the most common intracellular module families are present in fungi, plants, and animals, although they have expanded in multicellular eukaryotes relative to the single-celled eukaryote *S. cerevisiae*.

TABLE 14.2. Extracellular modules found most frequently in multidomain proteins

Module	SMART[a]	Pfam[a]
Fibronectin type III module	FN3	fn3
Epidermal growth factor (EGF) module	EGF	EGF
Immunoglobulin module	IG	ig
Complement B-type module (sushi module)	CCP	sushi
Thrombospondin type I module	TSP1	tsp_1
LDL receptor type A module	LDLa	ldl_recept_a
EGF-like module of laminin	EGF_Lam	laminin_EGF
Complement C1r module (CUB module)	CUB	CUB
Scavenger receptor module	SR	SRCR
C-type lectin module	CLECT	lectin_C
A-type module of von Willebrand factor	VWA	vwa
C-type module of von Willebrand factor	VWC	vwc
Kunitz-type trypsin inhibitor module	KU	Kunitz_BPTI
D-type module of von Willebrand factor	VWD	vwd
Kringle module	KR	kringle
Factor V/VIII type C module	FA58C	F5_F8_type_C
MAM module	MAM	MAM
Link protein module	LINK	Xlink
Finger module (fibronectin type I module)	FN1	fn1
Thyroglobulin module	TY	thyroglobulin_1
Fibronectin type II module	FN2	fn2
Frizzled module	FRI	Fz
Whey protein module (WAP module)	WAP	wap
Calcium-binding module (gla module)	GLA	gla
SEA module	SEA	SEA
NTR module, Complement C3/C4/C5 module	C345C	NTR
Olfactomedin domain	OLF	OLF
LCCL domain	LCCL	LCCL

The modules are listed in order of decreasing frequency in nonredundant domain databases.
[a]The abbreviations correspond to those used by the SMART database (http://smart.embl-heidelberg.de) and the Pfam database (http://www.sanger.ac.uk/Software/Pfam/).

the various structural proteins of the extracellular matrix as well as multidomain proteases involved in the formation or remodeling of the extracellular matrix. The domain types constituting these extracellular modular proteins (cf. Table 14.2) are clearly distinct from those present in intracellular proteins: they are usually rich in disulfide bonds, reflecting the adaptation of these domains to the extracytoplasmic space. A few module types (e.g., immunoglobulin module, fibronectin type III module) may also occur in some intracellular proteins. For example, the large intracellular muscle proteins twitchin, titin, projectin, and skelemin are entirely composed of multiple copies of immunoglobulin and fibronectin type III modules.

A third group of multidomain proteins comprises transmembrane proteins such as receptor tyrosine kinases, receptor tyrosine phosphatases, and cytokine receptors. The extracellular parts of such multidomain proteins usually consist of domains typical of extracellular multidomain proteins (e.g., those listed in Table 14.2); their intracellular parts may consist of domains characteristic of intracellular multidomain proteins. For example, the receptor tyrosine kinases possess an intracellular tyrosine kinase domain, whereas their extracellular, ligand-binding regions may consist of disulfide-bonded extracellular domains such as the kringle, frizzled and immunoglobulin domains of ROR1 tyrosine kinase (Figure 14.1).

Studies on the evolutionary distribution of extracellular domains frequently used in the assembly of the extracellular proteins (or extracellular parts of transmembrane proteins) have shown that they are present in all major groups of Metazoa but are practically missing from the genomes of plants, yeast, and bacteria, suggesting that they arose and/or were expanded only in the metazoan lineage

(Patthy 1994, 1995). Comparison of the complete genomes of the yeast *Saccharomyces cerevisiae* and the worm *Caenorhabditis elegans* has confirmed that there are many more extracellular domain types in *C. elegans* than in *S. cerevisiae* and, even in the case of those module types that are present both in yeast and Metazoa, the families are greatly expanded in *C. elegans* and human (Copley et al. 1999). Statistical analyses (Gerstein 1997) have also shown that the frequency of these small extracellular domains increases as one moves from simple metazoans to more complex organisms. Similarly, analysis of the proteome of the plant *Arabidopsis thaliana* has confirmed that the module types widely used in the assembly of extracellular multidomain proteins of metazoans are rare or missing in plants (Aravind & Subramanian 1999).

Comparative analyses of the genomes of the fly *Drosophila melanogaster*, the worm *C. elegans*, and the yeast *S. cerevisiae* (Rubin et al. 2000) have shown that the multidomain proteins of the fly and worm are far more complex than those of yeast. These observations clearly suggest that the organismal complexity of metazoans "does not principally depend on the generation of new genes, but on new combinations of protein domains or novel interactions" (Jasny 2000). This conclusion has been confirmed and extended by comparative genomic studies of the human genome: the frequency of multidomain proteins was found to increase as one moves from simple eukaryotes to more complex organisms (Li et al. 2001; Lander et al. 2001). It has been concluded that human genes are more complex than worm or fly genes primarily because "vertebrates appear to have arranged pre-existing components into a richer collection of domain architectures" (Lander et al. 2001). Similarly, comparison of the human genome with the fly and worm genomes has led Venter et al. (2001) to conclude that one of the most important differences between these genomes is that the human genome contains greater numbers of multidomain proteins with multiple functions and domain architectures.

The most important conclusion from the comparison of genomes of different eukaryotes is that the increase in organismic complexity from yeast to vertebrates is paralleled by the generation of new extracellular multidomain proteins with an ever richer collection of domain architectures. This expansion of domain architectures may also be illustrated by the fact that the relative abundance of the extracellular domains is significantly increased in metazoans—especially vertebrates—relative to fungi or plants (Figures 14.3 and 14.4).

The importance of the generation of novel extracellular multidomain proteins for metazoan

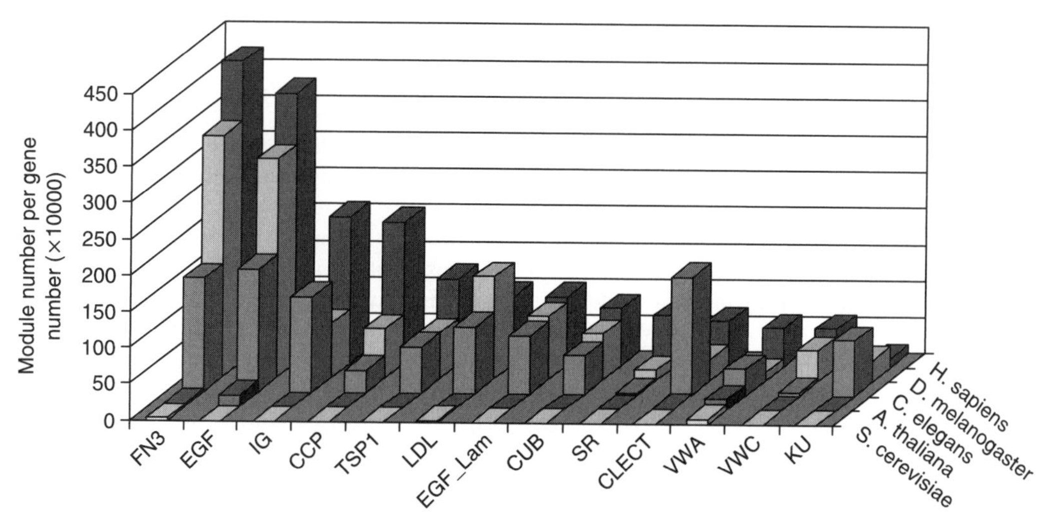

FIGURE 14.3. Frequency of extracellular modules in proteins of some model organisms. The abbreviations of modules correspond to those shown in Table 14.2. The values on the ordinate were calculated by dividing the total number of a module type in a genome with the number of the genes present in that genome. Note that the most common module-types of Metazoa are virtually absent from the genomes of plants (represented by *A. thaliana*) and fungi (represented by *S. cerevisiae*).

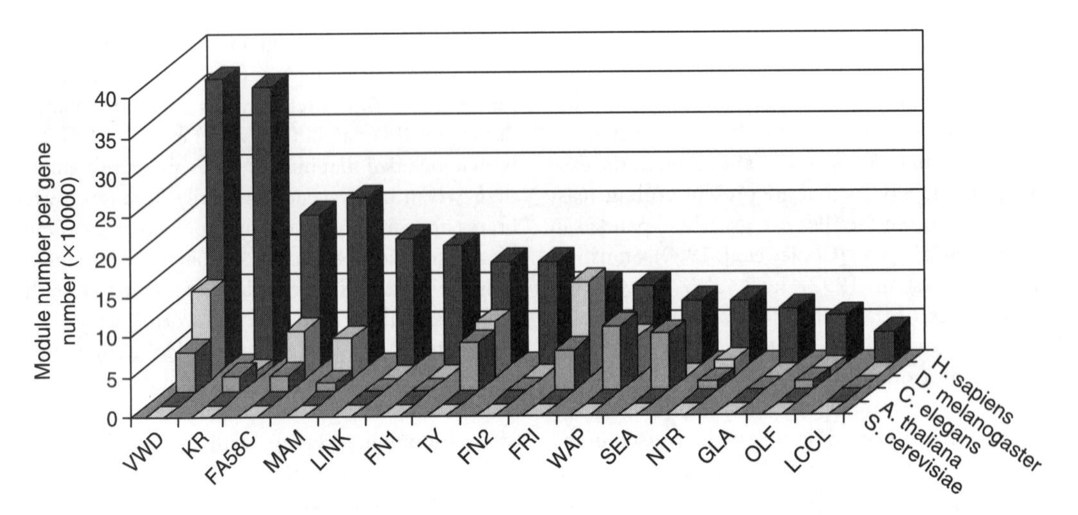

FIGURE 14.4. Frequency of extracellular modules in proteins of some model organisms. The abbreviations of modules correspond to those shown in Table 14.2. The values on the ordinate were calculated by dividing the total number of a module type in a genome with the number of the genes present in that genome. Note that some module families that are rare in invertebrates (represented here by *C. elegans* and *D. melanogaster*) have greatly expanded in vertebrates (represented by *H. sapiens*) to build novel multidomain proteins.

evolution may be best appreciated if we correlate the appearance of novel multidomain proteins with major biological innovations.

Appearance of Novel Extracellular Multidomain Proteins Correlates with the Appearance of Novel Biological Functions in Metazoa

A distinguishing feature of multicellular organisms is that they need mechanisms for the cohesion and communication among different cells. Accordingly, animals have evolved novel, multidomain, extracellular or transmembrane molecules that mediate cell–cell communication and adhesion of cells to other cells or to an extracellular matrix.

Recent analyses on the proteomes of the worm (Hutter et al. 2000) and the fly (Hynes & Zhao 2000) have shown that the most important multidomain proteins characteristic of basement membranes of vertebrates (laminins, type IV collagen) are also present in worm and fly. Since conserved type IV collagens are also present in hydra and sponges, and laminins are known to be present in hydra, it appears that these fundamental consitutents of the basement membranes were formed very early in the evolution of multicellular animals. The multidomain membrane proteins, integrins, that are crucial for the linkage of the extracellular matrix to the cytoskeleton are also very ancient: they are highly conserved in organisms ranging from sponges to mammals.

There are many types of transmembrane proteins involved in cell adhesion that are present in the worm, fly and vertebrates (e.g., cadherins, members of the NCAM family), suggesting that they evolved at an early stage of metazoan evolution. Similarly, the fact that large intracellular muscle proteins (titin, twitchin, projectin) consisting of immunoglobulin and fibronectin type III domains exist in the worm, fly, and vertebrates suggests that these genes must have been formed before the separation of the arthropod, nematode, and chordate lineages. The receptor tyrosine kinases and receptor tyrosine phosphatases so critical for intercellular communications also appear to have emerged very early in metazoan evolution (Ono et al. 1999; Plowman et al. 1999; Suga et al. 1999; Miyata & Suga 2001).

However, many multidomain proteins known from vertebrates appear to be absent from lower organisms, suggesting that they may have been formed only in the chordate lineage. For example, vertebrates use numerous types of large, multidomain collagens to build their endoskeletons (bone, cartilage), to build the tendons that connect bones, and to construct connective tissue that provides structure for vertebrate tissues. Most multidomain collagens typical of vertebrates are missing from invertebrates; it thus appears that these multidomain collagens have evolved in the vertebrate lineage and their formation must have been essential for chordate evolution. Several other constituents of the endoskeleton/extracellular matrix have also been formed during vertebrate evolution: the cartilage-aggregating proteoglycan aggrecan, cartilage link protein, cartilage matrix protein, and fibronectin are missing from invertebrates. Orthologs of well-known vertebrate extracellular matrix proteins such as fibronectin, tenascins, and von Willebrand factor are also absent from the fly and worm genomes.

There are many novel multidomain proteins associated with hemostasis and other plasma effector systems of mammals. Apparently, most multidomain components of the mammalian plasma effector systems arose relatively recently: orthologs of the protease and nonprotease components of the blood coagulation, fibrinolyis, kinin, and complement pathways are missing from the *D. melanogaster* and *C. elegans* genomes. In addition to von Willebrand factor already mentioned, these include prothrombin, blood coagulation factors V, VII, VIII, IX, X, XI, XII, XIIIb, plasminogen, urokinase, tissue plasminogen activator, plasma prekallikrein, and complement factors B, C1q, C1r, C1s, C2, C3, C4, C5, C6, C7, C8α, C8β, C9, H, I. Thrombomodulin, an endothelial cell membrane protein of the anticoagulant pathway, and C1qR, the receptor for complement C1q that mediates phagocytosis, are also absent from fly and worm.

Most components of the coagulation cascade of mammals, however, already have orthologs in fish, suggesting that the blood coagulation network evolved before the divergence of tetrapods and teleosts over 430 million years ago (Davidson et al. 2003). In a recent study the blood coagulation and fibrinolytic cascades of the puffer fish *Fugu rubripes* have been reconstructed by identifying orthologs of genes of mammalian blood clotting factors in its genome (Jiang & Doolittle 2003). This analysis has shown that fishes have practically the complete set of multidomain blood coagulation proteases (prothrombin, factors VII, IX, X, protein C), coagulation factors (protein S, factors V and VIII, thrombomodulin), and constitutents of the fibrinolytic cascade (plasminogen, urokinase, tissue plasminogen activator). It is noteworthy that these fish proteins have the same domain organization as their mammalian orthologs, indicating that the assembly process was completed before the divergence of tetrapods and teleosts. A similar analysis of the genome of the urochordate *Ciona intestinalis* did not turn up genuine orthologs for the blood coagulation/fibrinolyis factors, suggesting that the blood coagulation network evolved 450–550 million years ago, following the divergence of urochordates and vertebrates but before the divergence of tetrapods and teleosts. These findings thus significantly narrow the evolutionary window for the development of the vertebrate coagulation and fibrinolytic cascades.

It thus appears that creation of novel multidomain proteins played a very significant role in vertebrate evolution: many unique aspects of vertebrate biology (endoskeleton, plasma effector systems, etc.) rely on novel multidomain proteins with novel architectures.

Evolutionary Mechanisms Responsible for the Formation of Multidomain Proteins

As mentioned above, the distribution of multidomain proteins in all three kingdoms of life appears to fit a model in which multidomain proteins form by random combination of domains; however, the likelihood of domain-joining was found to be much higher in eukaryotes than in prokaryotes (Koonin et al. 2002).

An obvious factor that might affect the likelihood of domain-joining is the presence or absence of introns at the domain boundaries. Although there is clear evidence that multidomain proteins may be constructed without the assistance of introns, for example in bacteria (de Chateau & Bjorck 1996), it has long been suspected that the existence of introns may dramatically increase the chances of recombination to shuffle the exonic sequences (Gilbert 1978).

Studies on the complete sequences of the genomes of numerous Eubacteria, Archaea, and Eukarya have confirmed that whereas introns are rare or absent from eubacterial and archaeal genomes, they are usually present in eukaryotes. It thus seems very likely that the greater propensity of domains to form multidomain architectures in eukaryotes than in prokaryotes may be—at least in part—due to the presence of introns in eukaryotes and their absence from prokaryotes.

Within Eukarya the genomes of multicellular organisms are significantly more intron-rich than those of single-cell eukaryotes. It is therefore to be

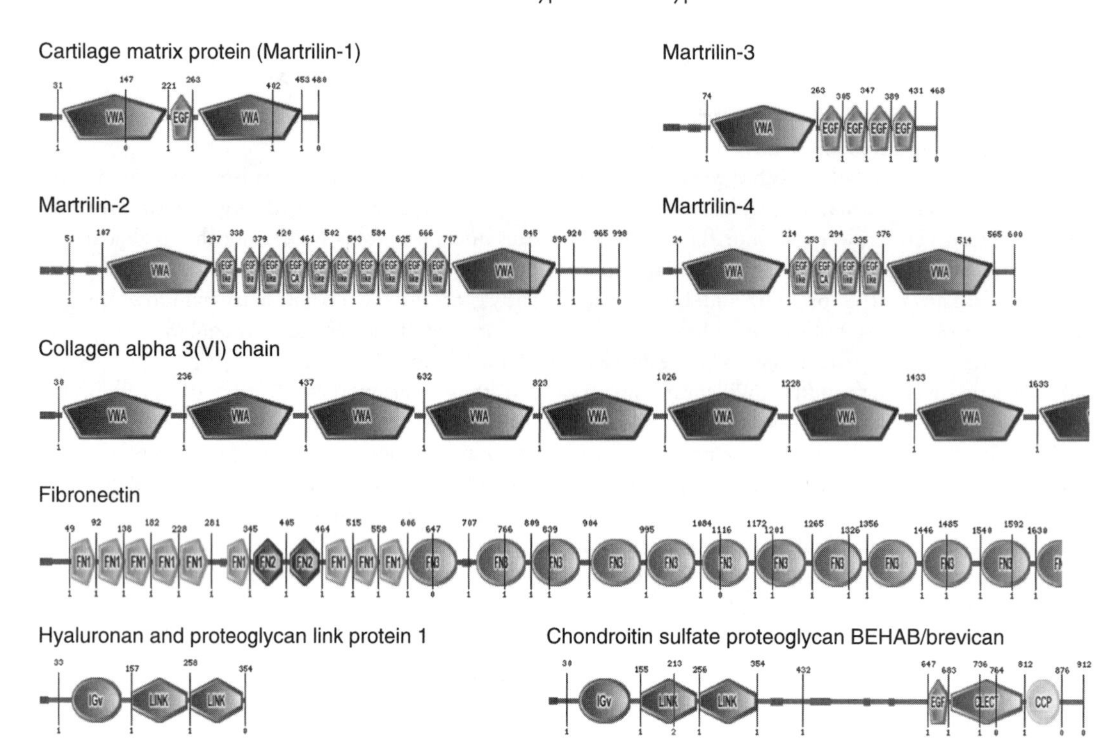

FIGURE 14.5. Correlation of the domain organization of some multidomain proteins assembled from class 1-1 modules with the exon–intron structures of their genes. The figure shows SMART representations (Letunic et al. 2004, *Nucleic Acids Research* D142-4) with included intron positions mapped onto a ClustalW alignment of orthologous genes. The numbers refer to the position and phase class of introns. The class 1-1 domains are designated as follows [synonyms are given in brackets]: CCP [domain abundant in complement control proteins; SUSHI repeat: short complement-like repeat (SCR)]; CLECT [C-type lectin (CTL) or carbohydrate-recognition domain (CRD)]; EGF [epidermal growth factor-like domain]; FN1 [fibronectin type 1 domain]; FN2 [fibronectin type 2 domain]; FN3 [fibronectin type 3 domain]; IGv [immunoglobulin V-type]; LINK [link (hyaluronan-binding)]; VWA [von Willebrand factor (vWF) type A domain]. Note that phase 1 introns are found at the boundaries of the constituent class 1-1 modules.

expected that as a consequence of the evolution of more intron-rich genomes, exon shuffling might become increasingly important in the metazoan lineage (Patthy 1999a). The first proof for exon shuffling has indeed come from vertebrates: from studies on the proteases of blood coagulation and fibrinolytic systems characteristic of this group of organisms (Bányai et al. 1983; Ny et al. 1984; Patthy 1985). Subsequent studies have also shown that this evolutionary mechanism has been widely used in the creation of a variety of multidomain proteins of animals, including the nonprotease constituents of the plasma effector cascades, most constituents of the extracellular matrix, extracellular parts of transmembrane proteins and receptor proteins, and the intracellular muscle proteins discussed above.

As a signature of exon shuffling, the exon–intron organization of genes produced by this mechanism shows a clear correlation with the mosaic structure of multidomain proteins (Figure 14.5). Furthermore, since only, "symmetrical modules" (i.e., modules flanked by introns of the same phase) are suitable for domain duplication and domain insertion by exon shuffling (Patthy 1987), in the genes of proteins created by exon shuffling inter-module introns show a characteristic intron-phase distibution (Patthy 1987, 1991, 1994). Since the vast majority of the modules used in the assembly of extracellular multidomain proteins (e.g., those listed in Table 14.1) are class 1-1 modules (i.e., both introns flanking the module are phase 1), the inter-module introns split the reading frame in phase 1 (Figure 14.5).

In contrast to the overwhelming evidence for exon shuffling in the case of the Metazoa-specific extracellular and transmembrane multidomain proteins assembled from "extracellular" domains (Table 14.1), in the case of most intracellular multidomain proteins there is no clear evidence for a role of exon shuffling. Since the majority of multidomain proteins involved in intracellular signaling were formed prior to the divergence of eukaryotes (see above), probably in organisms with intron-poor genomes, it seems likely that these multidomain proteins were formed through recombination in exons.

FUTURE DIRECTIONS

An implicit assumption of the random recombination model of multidomain proteins is that all protein domains have a chance to be used for the construction of multidomain proteins. Nevertheless, it appears that different domain types are not equally represented in multidomain proteins of the various organisms. Whereas some domain types—usually called mobile protein modules—have been found frequently in multidomain proteins with different domain architectures, some protein domains exist only in a limited number of domain combinations, while others occur only as stand-alone proteins.

There may be several explanations for such differences. It is possible that the apparent exclusion of some domain types from this game just reflects our ignorance: the domain family is improperly defined (e.g., the homology of a stand-alone domain with domains of multidomain proteins has not yet been recognized). It is also possible that the present status of a "stand-alone domain" may change when new multidomain proteins containing the given domain type are discovered (e.g., as a result of more genome analyses).

However, the domain-usage bias may turn out to be real. This may be due to the greater propensity of certain domain types to participate in recombination events (e.g., domains flanked by introns) or the greater chance of survival of certain domain types within multidomain proteins (e.g., domains with exceptional folding efficiency, domain types of higher functional versatility). Systematic analyses on the structure, function, and genomic features of domains will be needed to clarify the relative contributions of such factors to the domain-usage bias of multidomain proteins.

15

Evolutionary Developmental Biology

DAVID L. STERN

Evolutionary developmental biology (EDB, often shortened to "evo-devo") includes a broad array of studies, stretching from phylogenetic analysis of developmental genes to comparative studies of embryos and gene expression patterns to microevolutionary studies of the genetic basis for morphological change. This is a partial list and, in general, studies that attempt to combine evolutionary and developmental studies have been placed under the heading of EDB. While the field is broad and rich, I attempt in this chapter to highlight a few important conceptual findings in EDB. This brief summary will, of necessity, omit many exciting avenues of research that fall under the umbrella of EDB. Nonetheless, I believe there is a coherence to EDB studies that is reflected in the broad conceptual strokes I outline in this chapter. For example, the major finding of EDB has been the discovery that many of the genes controlling animal development are conserved across wide phylogenetic expanses. This finding is, in many ways, the unifying theme of EDB. From it, we draw the two major avenues of EDB research to date: first, how are conserved genes used in divergent taxa during development and second, since so many developmental genetic mechanisms appear to be strongly conserved, how do differences between species arise and generate the dramatic morphological differences between divergent taxa?

As an aid in understanding the research addressing these questions it is useful to recognize that there are two major methodological approaches that have been used. While some may consider a division based on methodologies to be an artificial distinction, I believe it allows us to rank the evidence in support of particular hypotheses and thus to think clearly about the meaning of particular results. I call the two major approaches Comparative EDB and Genetical EDB.

Comparative EDB involves the analysis of expression patterns and gene function of developmental genes in a comparative framework. As in any other comparative analysis, these studies are best performed in a phylogenetic context to allow identification of the polarity of events and the identification of independent evolutionary transitions. Such studies are not, of course, limited to the macroevolutionary level, although Comparative EDB is the predominant mode of analysis at the macroevolutionary level. Note that I do not limit Comparative EDB to the analysis of gene expression patterns. Comparative EDB can include the comparison of anything related to development, including gene functions. Such comparative data can provide evidence for correlations between gene expression patterns or gene functions and morphological change. This approach does not provide proof that a particular gene has caused a particular morphological transition. For example, even if a particular gene is expressed in a novel domain and this expression can cause the observed morphological change, the real evolutionary change could lie in a gene upstream. In addition, there is always the possibility that a correlation between an expression pattern and morphology reflects a chance correlation and that the gene, in fact, has nothing to do with the patterning of a particular morphological characteristic. As I discuss in more detail below, this type of chance correlation is unlikely to be relevant to most work in EDB since additional information on gene function often provides compelling evidence for a gene's role in developmental patterning. One final issue that is not often recognized is that Comparative EDB is, for a variety of reasons, a biased approach to examining the genetic basis for morphological evolution. The bias arises at several levels. First, the choice of a particular gene for study often reflects knowledge of the function of the gene in one or several model systems. The choice of genes is thus limited by our knowledge of the genetic architecture guiding development in these model systems. Second, Comparative EDB has involved a bias toward detecting and reporting changes in gene expression patterns, since these are much easier to assay than changes in post-transcriptional modifications or changes in protein function. There is, therefore, some possibility that we have overestimated the importance of changes in *cis*-regulatory regions in morphological evolution because such changes are easier to detect.

Genetical EDB, in contrast, utilizes methods that test for evolved gene functions by testing candidate DNA regions within a novel genomic context. I have used the term Genetical because all such approaches include some component of genetic recombination (or a proxy such as transgenesis) in the method. (Note that Comparative and Genetical approaches are often utilized in the same studies.) As I will discuss in detail, there are two major advantages of Genetical EDB. First, such studies can provide proof of the role of a particular piece of DNA in an evolved difference. Second, such studies can provide, in principle, unbiased genome-wide surveys for relevant evolutionary changes. I believe that such studies provide the most important data for constructing a theory of EDB and it is likely that this mode of Genetical EDB will become an increasingly important feature of the field.

SOME HISTORY

While consideration of development was important in the thinking of evolutionary biologists of the nineteenth century, the New Synthesis of the early twentieth century left little room for developmental biology. The precise roles of genes in development was not seen to be crucial to constructing a working synthesis of Mendelian genetics and Darwinian natural selection. While there were various detractors to this dominant view throughout the mid-twentieth century, our current conception of the central issues in development and evolution can be traced to intellectual developments beginning in the 1970s (Bonner 1982; Gould 1977; Raff & Kaufman 1983). Notably, however, it is unlikely that EDB would have seen the explosive growth it is now enjoying if it had not been for the efforts of a small cohort of developmental biologists in the 1980s and early 1990s. At that time, some developmental biologists felt that they had gained sufficient understanding of the molecular regulation of embryology to begin turning their attention to comparative questions. In addition, the discovery in 1984 that the Hox genes (Box 15.1) were conserved between invertebrates and vertebrates provided a strong impetus to explore the roles of such conserved patterning genes in divergent taxa. This phase of EDB (represented almost entirely by Comparative EDB) has been remarkably productive (Carroll et al. 2001; Wilkins 2002). In this sense, contemporary EDB has been constructed largely by developmental biologists who have rarely spent much time considering issues in contemporary evolutionary biology (such as those in the other chapters of this book). Thus, in the eyes of many evolutionary biologists, the field of EDB often seems to stand apart from contemporary evolutionary biology and the relevance of the recent flush of new data has not yet been fully appreciated by many evolutionary biologists.

The remainder of this chapter is organized as follows. First, I review some of the major questions and central assumptions from Comparative EDB and describe several biases that may arise in Comparative EDB. Second, I explore the major approaches used in Genetical EDB and illustrate these approaches with a variety of case studies. In a third section, I review some of the major lessons from both Comparative and Genetical EDB.

COMPARATIVE EDB

The main goals of Comparative EDB can be divided, broadly, into two issues: the testing of hypotheses of homology and the search for the genetic and developmental changes that are related to evolutionary changes in development and morphology. Here I review some advances in these two avenues of research.

Testing Homology

The recognition that superficially dissimilar organs, such as a bird wing, a human arm, and a whale fin, may all have evolved from a common ancestral

BOX 15.1. Hox Genes

If the modern era of evolutionary developmental biology can be thought to have "started" at any particular time, it started with the discovery of the widespread conservation of the Hox genes. The Hox genes are a family of transcription factors found in all animals that are involved in patterning the anterior–posterior axis of animals. The Hox genes are typically clustered in a complex, with the 5′ to 3′ order of the genes conserved amongst most animals. Remarkably, the 5′ to 3′ position of Hox genes within a complex correlates with the position at which they function along the anterior–posterior axis; genes located relatively 5′ within a cluster act in relatively anterior positions. The term Hox derives from the fact that the first such genes were found to contain a conserved sequence of amino acids called a homeobox, because mutations in these genes cause homeotic transformations. William Bateson defined the term homeosis in 1894 to refer to variants found in nature where one organ had apparently been transformed into the likeness of another organ (Bateson 1894). In a typically prescient mode, Bateson felt that these variants provided important clues to understanding the evolution of development and morphology.

The correlation of Hox gene position along the chromosome and Hox gene function along the anterior–posterior axis of the fruit fly was first described in a landmark, and Nobel-prize winning, paper by Ed Lewis (1978) Using a model of arthropod evolution that is now thought to be incorrect, Lewis hypothesized that these genes arose as tandem duplications during the evolution of insects from annelids and that each duplicate allowed the specification of an additional domain along the anterior–posterior axis. Lewis' paper inspired molecular biologists to clone these genes and test his bold model of arthropod evolution. Three remarkable findings arose soon after the cloning of the *Drosophila* Hox genes. First, it was found that all of the homeotic genes of the *Drosophila antennapedia* and *bithorax* complexes contained homeodomains (McGinnis et al. 1984b). Second, through Southern hybridization of one homeodomain to the whole genome of the fruit fly, it was determined that the fly had many genes containing homeodomains. Third, through Southern hybridization of the homeodomain to genomic DNA from a diversity of organisms, a so-called zoo-blot, it was discovered that all animals contained many of these homeodomain-containing genes (McGinnis et al. 1984a). In the subsequent 20 years, Hox gene orthologs and complexes have been identified in a wide diversity of animals and it is now clear that the ancestral metazoan contained at least several Hox genes and that the diversity of Hox genes in extant metazoans has evolved by the tandem duplication of genes within complexes, as Lewis predicted, and the wholesale duplication of entire complexes, sometimes followed by gene loss (Holland 1999). The fundamental insight provided by the observed conservation of structure and function of the Hox genes was that animal diversity was likely to have resulted primarily from changes in mechanisms that had remained largely conserved at the level of protein structure and function. Therefore, morphological evolution is likely to have proceeded primarily through changes in the regulation of conserved genes. While similar predictions had been made based on more limited data and from first principles (Britten & Davidson 1969; Jacob 1977; King & Wilson 1975), the data from the Hox genes profoundly altered our view of morphological evolution and ignited interest in evolutionary problems amongst many developmental biologists.

organ was central to Darwin's argument for common descent and the identification of such homologies played a strong part in nineteenth-century evolutionary biology. While there have always continued to be biologists with strong interests in identifying homologies, this avenue of research has not been a central concern of contemporary evolutionary biology or evolutionary genetics. EDB has provided a new kind of data with which to explore hypotheses of homology: data on the underlying developmental mechanisms, which have provided a deeper understanding of the patterns of morphological evolution and reignited interest in identifying homologies. Several books provide excellent reviews of many case studies (Carroll et al. 2001; Davidson 2001; Wilkins 2002). Here, I review one fairly straightforward study in which EDB data has provided a striking test of a classical hypothesis of homology.

Insect wings develop in a dorsolateral position on the thorax. The presumptive ancestors of insects, the Crustacea, do not possess any obvious limb-like organs in a similar body position and two hypotheses have been forwarded to explain the origin of insect wings. The first hypothesis is that insect wings evolved de novo in their current position as an outgrowth from the dorsal thorax. Alternatively, the wing might have evolved from a dorsal branch of the multibranched appendages of the crustacean limb. The latter hypothesis has two advantages. First, it does not require the invention of an entirely new organ and second some crustaceans use this dorsal branch for respiration and osmoregulation, which is similar to one model for the origin of insect wings from gill-like structures. To test between these alternative hypotheses, Averof and Cohen (1997) posited that if wings evolved from a dorsal branch of appendages in Crustacea, wings and the dorsal branches of crustacean limbs may share expression of certain genes involved in specifying this domain. In particular, they examined the expression of two genes that play important roles in specifying the wing fate in *Drosophila melanogaster*: *pdm* (also called *nubbin*) and *apterous*. They observed that both genes are expressed specifically in the dorsal branch of the multibranched limbs of two crustacean species, supporting the hypothesis that insect wings are homologous to the dorsal branch of multibranched appendages of crustaceans.

It is worth examining this case in slightly more detail. This study provides strong evidence in support of the homology of insect wings and crustacean limbs not simply because of the observed correlation of expression patterns. The real power of this approach derives from the detailed understanding of gene function gained by studies in model systems. *Pdm/nubbin* and *apterous* are key genes involved in specifying wing identity. If either *apterous* or *pdm/nubbin* functions are removed from a developing fly, then the fly develops without functional wings. In this case, as with many genes originally identified in model systems, the gene names provide clues to gene functions. Flies carrying some alleles of *apterous* are, in fact, apterous, while flies carrying some alleles of *nubbin* have wings that are reduced in size and misshapen. Thus, *apterous* and *nubbin* must be involved in regulating many genes, either directly or indirectly, that are involved in specifying the development of wings. That is, it is unlikely that the expression patterns of these genes would evolve quickly without causing the degeneration of the wing. This functional understanding adds considerable weight to the argument that wings and the dorsal branch of crustacean appendages are homologous, but it does not provide proof.

Searching for Molecular Correlates of Morphological Macroevolution

The second major avenue of research in comparative EDB has involved the search for the molecular correlates of morphological differences. Although such efforts have focused primarily on distantly related taxa, this approach can be used at any phylogenetic level. This approach again flows primarily from our knowledge of gene function in model systems. The primary assumption is that developmental genes have retained their function (for example, specifying eyes or limbs) in distantly related taxa and that changes in expression patterns of these genes are therefore related to changes in morphological features. I use the word "related" advisedly. Comparative EDB is not able, on its own, to determine whether evolution of a focal gene has been the cause of the morphological transition, but the claim made for these examples is usually weaker: that changes in gene expression are related to changes in morphology. I first provide three case studies that illustrate the power of this approach and then review several case studies that illustrate some of its biases.

We return to the subject of insect wings and this time explore the evolution of wing morphology within the insects. As is well known, the Diptera have one pair of "normal" wings and one pair of highly reduced and heavily modified hindwings called halteres. Classical genetic work in *D. melanogaster* indicated, rather remarkably, that the differences between the forewing and the haltere are controlled, ultimately, by the action of a single gene, *Ultrabithorax* (*Ubx*). Certain mutations of *Ubx* are capable of transforming the haltere into an organ that is morphologically indistinguishable from a forewing (Lewis 1978). This led to the simple hypothesis that the haltere evolved from the hindwing of an ancestral insect that had two pairs of normal-looking wings by the adoption of expression of the *Ubx* gene in the hindwings. Warren et al. (1994) tested this hypothesis by examining the distribution of *Ubx* protein in the hindwing of a butterfly. They found that cells in the hindwing of the butterfly express *Ubx*, similar to the expression of *Ubx* in the haltere of *Drosophila*. In *Drosophila*, *Ubx* controls the morphology of the haltere by regulating a large set of downstream genes (Weatherbee et al. 1998). Therefore, a reasonable hypothesis is that *Ubx* also generates the differences between the hindwing and the forewing of the butterfly by regulating a set of downstream genes. There is no evidence that *Ubx* itself has evolved in this case. Instead, the differences between a haltere and butterfly hindwing suggest that many of the downstream linkages, the particular set of genes that are regulated by *Ubx*, have evolved.

In a second contrasting example, differences in *Ubx* expression were observed to be correlated with morphological transitions, suggesting that changes in the regulation of *Ubx* itself *are* related to some morphological differences. Averof and Patel (1997) examined the distribution of *Ubx* (strictly speaking, they used an antibody that recognized both the *Ubx* and *abd-A* proteins) in a variety of crustaceans. These investigators focused on a key difference between these taxa, which is the segmental boundary marking the transition from the larger swimming appendages in the posterior and the smaller, specialized feeding appendages called maxillipeds found on more anterior segments. They compared species that exhibit 0, 1, 2, or 3 pairs of maxillipeds and found that the boundary of *Ubx/abd-A* expression was precisely correlated with the shift from swimming appendages to maxillipeds in every species. Given the known role of *Ubx* in specifying segment identities in *Drosophila*, a likely hypothesis is that *Ubx* is involved in specifying the morphology of the swimming appendages and that evolutionary changes in the expression domain of *Ubx* are related to differences in the boundary between swimming appendages and maxillipeds. This parallels the role of *Ubx* in specifying halteres in flies and hindwings in butterflies.

A final example illustrates the use of Comparative EDB at a lower taxonomic level, between species of a single genus. This study first focused on a description of the function of the *bric-a-brac* (*bab*) gene, a transcription factor, in defining sexually dimorphic pigmentation in the model organism *D. melanogaster* and went on to examine the correlation between expression patterns of this gene and the distribution of sexually dimorphic pigmentation in a range of species from the genus *Drosophila* (Kopp et al. 2000). Kopp et al. found that *bab* represses the development of dark pigmentation. They also showed that *bab* expression is regulated by genes of the homeotic and sex-determination pathways, causing differential regulation of *bab* in the abdomen of males and females, leading to the different pigmentation patterns in male and female *Drosophila*. They then went on to demonstrate that expression of *bab* varies between different species of *Drosophila* in patterns that are strongly correlated with the distribution of sexually dimorphic pigmentation in different species. This study provides strong comparative evidence that changes in the regulation of *bab* (either by changes in the *cis*-regulatory region of *bab* or by changes in regulatory genes upstream of *bab*) are responsible for the evolved pigmentation patterns.

Bias Toward Detecting Changes in Expression Patterns

In almost all studies of Comparative EDB one of two kinds of experiments are performed. The temporal and spatial distribution of either the messenger RNA (mRNA) or protein product of a developmental gene is examined. Study of the mRNA distribution is usually accomplished by cloning a small part of the ortholog of a known developmental gene for in situ hybridization to the mRNA. Study of protein distribution is usually performed by immunohistochemistry with an antibody that cross-reacts with the homolog across different taxa or with an antibody that was raised to bind specifically to the product of the gene cloned from a new species. Only rarely have investigators performed both experiments in the same study, although this is becoming increasingly common. One example will illustrate why it is important to perform both experiments.

We return to the specification of maxillipeds in crustaceans. Abzhanov and Kaufman (1999) showed that the morphological transition between normal appendages and maxillipeds is recapitulated in the first thoracic segment during the ontogeny of the terrestrial isopod, *Porcellio scaber*. Remarkably, the Hox gene *Scr* is transcribed in this limb throughout ontogeny, but *Scr* protein is observed only temporally coincident with the morphological transformation. Given the known function of *Scr* in *D. melanogaster*, it seems likely that *Scr* is playing an instructive role in this morphological transformation. Therefore, post-transcriptional regulation of *Scr* is crucial to defining the specific limb morphology expressed throughout ontogeny. This example illustrates how a study of both mRNA and protein distribution can provide important clues to the real molecular changes regulating developmental evolution.

There are additional limitations to the study of gene function evolution when assays are limited to in situ hybridization to mRNA and immunohistochemistry with antibodies. For example, transcription factors are often found in active and inactive states depending on whether they have been secondarily modified, for example by phosphorylation or by partial cleavage. In addition, neither approach provides an accurate estimate of quantitative levels of mRNA or protein. In addition, the distribution of mRNA and protein is rarely analyzed at a detailed cellular level, where small quantitative differences in protein level may cause morphological evolution. For example, in a study of the role of *Ubx* in patterning trichomes on the legs of *Drosophila* species, I performed genetic experiments that strongly suggested changes in the distribution or quantitative level of *Ubx* have caused a slight shift in the distribution of leg trichomes between the closely related species *D. melanogaster* and *D. simulans* (Stern 1998). However, direct examination revealed fairly similar patterns of the distribution of *Ubx* protein in these two species. It is likely that current methods of assaying protein distribution do not provide sufficient resolution of quantitative protein levels to provide visual confirmation of the genetic experiments. If many aspects of microevolutionary changes in morphology are due to slight changes in protein distribution or level, technological limitations may thwart our ability to study these phenomena. This example also illustrates the importance of Genetical EDB; the expression assays in the absence of genetic experiments would have suggested that *Ubx* was not involved in trichome pattern evolution.

Finally, the amino acid sequences of proteins themselves evolve, of course, and the developmental consequences of such protein evolution are rarely explored. Recently, two studies have demonstrated that protein evolution of *Ubx* is likely to have been important in the evolution of the arthropods. These studies examined how the function of *Ubx* genes isolated from a crustacean (Ronshaugen et al. 2002) and an onychophoran (Galant & Carroll 2002) differed from that of the *D. melanogaster Ubx* protein. Both studies focused on the role of *Ubx* in repressing limbs in the abdomen of insects. This is a well-characterized function of *Ubx* in *D. melanogaster* and the repression of abdominal limbs was a key step in the evolution of insects from crustaceans. Both studies identified a C-terminal region of the *Ubx* protein that has evolved a specific amino acid sequence, a poly-alanine tract, that provides at least part of the limb repressing function of *Ubx*. This poly-alanine tract is absent in arthropods outside of the insects and these studies suggest that evolution of this specific region of the *Ubx* protein may have played an important role in the evolution of limb repression in the abdomen of insects.

Bias toward Studying Genes We Know About from Model Systems

In all of the studies discussed so far, the genes for comparative study were chosen on the basis of their known function in *D. melanogaster*. The success of this program has somewhat justified the approach; many patterning genes studied in *Drosophila* are in fact found to be expressed in distantly related taxa, often in similar expression patterns, and where this has been tested, they often have conserved functions. However, it is entirely possible that other taxa, even closely related taxa, in addition to using these conserved genes also rely upon novel sets of development genes. I provide one particularly telling example.

The *bicoid* protein acts very early in *Drosophila* embryogenesis to define the anterior–posterior (A-P) axis of the embryo. The mRNA for this transcription factor is placed in the anterior pole of the egg and the protein diffuses from the anterior pole to create a gradient of protein level from anterior to posterior. This *bicoid* gradient plays a key instructive role in defining the A-P axis. The absolutely central and early role of this gene in axis specification led many people to suspect that it would be strongly conserved in insects and perhaps throughout arthropods, just like the Hox genes and various segmentation genes. The surprising result, that took many years to be revealed, was that *bicoid* is a very rapidly evolving gene that is likely to have been derived from a duplicated Hox gene rather late in insect evolution. In fact, good *bicoid* orthologs have not been found outside of the higher Diptera. This example reveals the single largest limitation of relying upon model systems for clues to EDB; not all aspects of development are conserved and if novel genes guide development in divergent taxa then it is possible they may be discovered only by direct study of the novel taxa.

One point I will emphasize in the next section is that Comparative EDB has not, to date, provided an unbiased survey of the causes of developmental evolution. However, there is nothing inherent in the comparative method that prevents an unbiased survey for correlations. The power of Comparative EDB is derived from the fact that a random approach is explicitly not taken. Instead, genes are selected for study on the basis of their known role in development. In principle, a whole-genome approach to comparative questions can be taken, and comparative microarray studies fall in this category. In the future, we can imagine that the combination of whole-genome functional data combined with whole-genome comparisons of gene expression and gene function will provide hypotheses about the role of evolutionary changes across the genome in phenotypic evolution. The principle of such studies is much the same as I have outlined for Comparative EDB, which is that functional data dramatically improve the predictive power of what are otherwise rather simple comparisons.

GENETICAL EDB

Genetical EDB includes studies that utilize genetic recombination in some form to test the evolved function of a piece of DNA, such as transgenesis, genetic crosses, and association studies. Such studies can provide strong evidence of the evolved function of a particular piece of DNA. While transgenic assays can be used to transfer DNA between any two organisms, genetic crosses (including quantitative trait loci (QTL) analysis) are limited to individuals within a species or between closely related species. Association studies are limited to comparisons within species. These three methods can thus provide different levels of resolution of evolutionarily relevant molecular changes. I discuss and illustrate each of these approaches in turn.

Transgenesis

Conceptually, transgenesis is the most attractive method for testing the evolutionary relevance of a piece of DNA. In some model systems, such as mouse and increasingly *Drosophila*, the native gene of one species can be replaced precisely with the gene from a second species, or even with single nucleotide changes. Most transgenic assays are performed, however, by introducing a piece of DNA from one species into a random location in the genome of a second species. And finally, in some systems, pieces of DNA are tested using transient transfection assays, where the introduced piece of DNA does not integrate into the host genome (Box 15.2).

The first case study involves pigmentation amongst drosophilid species. Pigmentation is a highly variable feature amongst drosophilids, with some heavily melanic species and some that show little pigmentation anywhere on the body. Many of the genes involved in melanism have been well characterized in *D. melanogaster*, providing a set of candidate genes for examining pigmentation evolution in this genus. Wittkopp et al. (2002) showed that the distribution of the protein product of the pigmentation gene *yellow* is perfectly correlated with pigmentation patterns between *D. melanogaster*, which has a spatially restricted pattern of stripes in the posterior of abdominal segments, and *D. subobscura* and *D. virilis*, which show pigmentation throughout each abdominal segment. They then demonstrated that when the *yellow* genes from both *D. subobscura* and *D. virilis* were cloned and transformed into *D. melanogaster*, these genomic fragments were sufficient to produce *yellow* protein expression in the correct species-specific expression pattern. This study took the analysis one step further, by transforming the *D. melanogaster yellow* gene into *D. virilis*. This experiment demonstrated, surprisingly, that the *D. melanogaster* gene is not expressed in the correct spatial pattern; instead the pattern is similar to the expression of the *D. virilis, yellow* gene, but weaker.

BOX 15.2. Functional Assays in Nonmodel Organisms

Evolutionary developmental Biology (EDB) is moving quickly from comparisons of gene expression patterns to comparison of gene functions. There are two kinds of assays that are used. In the first, gene rescue is attempted by either transient or stable transformation of DNA. The second, and opposite, kinds of assay are attempts to reduce or eliminate gene function.

Stable integration is preferred for gene rescue experiments, since this provides a tool for further experiments. However, for many species that are difficult to rear in the laboratory, transient transfection is a more viable alternative. In transient transfection assays, pieces of DNA are injected or electroporated into embryos at an early stage of development. In stable transformation, the goal is to stably integrate a piece of DNA into the genome and assay function in later generations. The opportunity to perform stable transformation has been greatly enhanced by the use of transposable elements that can be used in a wide variety of organisms (Horn & Wimmer 2000). These transposable elements have been combined with dominant markers, typically variants of green fluorescent protein expressed in the eyes, to allow recovery of transformants in a wide variety of animal systems. Despite these improvements, the generation of stable transformants is still a formidable task, largely because it requires injection of individual embryos and manual screening for transformants. For these reasons, transformation is not widely used.

One means of reducing, altering, or eliminating gene function is to generate mutations in vivo. This approach has proved exceedingly powerful in several systems, notably bacteria, yeast, nematodes, flies, fish, and mice. Mutagenesis requires organisms that can be easily reared for multiple generations in the laboratory. Fortunately, there are two new approaches to reducing gene function that hold the promise of allowing tests of gene function in almost any organism. The first technology utilizes modified antisense oligonucleotides called morpholinos (Corey & Abrams 2001). These are short oligonucleotides in which the ribonuclease subunit has been converted to a morpholine ring structure, making these molecules highly stable in cells. Morpholinos bind to endogenous mRNA and block translation. One drawback of morpholinos is that they bind to the 5′UTR of genes and thus the entire 5′UTR of a target gene must first be cloned.

The second technology is RNA interference (RNAi), in which double-stranded RNA (dsRNA) is used to specifically reduce expression of a gene. This technique appears to hijack a strongly conserved mechanism found across eukaryotes that may have evolved to protect cells from RNA viruses. In any case, RNAi appears to have wide applicability in the study of gene function in embryos. In addition, in at least some cases, injection of dsRNA into adults can cause reduction of gene expression in eggs and embryos produced by these adults (Bucher et al. 2002). RNAi therefore holds great promise for allowing study of gene function in a wide variety of organisms.

Together, these experiments demonstrate that while there have been evolutionarily relevant changes in the *cis*-regulatory region of *yellow*, some *trans*-acting transcription factors must also have evolved to generate the spatially restricted pattern found in *D. melanogaster*. Most EDB studies do not utilize reciprocal transformation assays and such data may thus be open to misinterpretation.

Not all developmental genes show changed expression patterns between species, and this presumably reflects stabilizing selection on the function of the particular expression pattern. It is therefore interesting to know whether and how the regulation of such genes has evolved, because this provides some insight into how regulatory networks evolve under stabilizing selection. Ludwig and colleagues

(1998, 2000) have taken this approach by analyzing the sequence and functional evolution of a well-characterized *cis*-regulatory region, the *even-skipped* (*eve*) stripe-2 enhancer. *eve* is a so-called pair-rule gene, because at early stages of *Drosophila* segmentation it functions to define alternate segments along the A-P axis. It is expressed in seven stripes in the early embryo and discrete regulatory regions have been identified that control transcription in some individual stripes. After first determining that there had been considerable molecular evolution of the *eve* stripe 2 region in other drosophilid species, including the loss of known transcription factor binding sites, Ludwig et al. utilized transgenic assays to explore whether the function of the region had evolved. They cloned the stripe-2 enhancer from *D. pseudoobscura* and transformed this region into *D. melanogaster*. Careful analysis showed that the *D. pseudoobscura* enhancer functions correctly in the *D. melanogaster* genome. This was somewhat surprising, given the extensive sequence divergence between the regulatory regions from the two species. This result clearly indicates that the overall function of the enhancer has been under stabilizing selection and that the expression and function of all of the *trans*-acting factors required by the *D. pseudoobscura* enhancer are likely to be conserved. Ludwig et al. then asked whether the extensive molecular evolution in fact reflected the accumulation of compensatory mutations that altered the distribution of transcription factor binding sites throughout the enhancer. They tested this hypothesis by generating chimeric constructs containing the 5′ region from one species and the 3′ region from the second species, and vice versa, and transformed these constructs into *D. melanogaster*. In this case, the chimeric enhancer regions no longer operated correctly and both the position and the width of the stripe were altered. This observation suggests that enhancer regions that are apparently under stabilizing selection may in fact experience the turnover of binding sites, and the shuffling of individual transcription factor binding sites. This may result from weak selection on individual binding sites and the evolution of compensatory mutations to maintain the overall function of the enhancer.

Genetic Crosses

Genetic crosses provide an efficient tool for surveying the entire genome for the genes that have caused morphological divergence. The major limitation of such studies is that a large number of recombinants and markers are required to provide fine-scale resolution of the evolved genes. This limitation is often overcome either by selecting markers within candidate genes or by using other genetic tricks of model systems to identify the evolved genes. Below I discuss two examples that demonstrate the power of genetic crosses and the use of both candidate genes and other genetic tricks to identify evolved genes.

We return to pigmentation in drosophilids and recall that changes between some *Drosophila* species in the expression of the *yellow* gene were caused in part by changes in the *cis*-regulatory region of *yellow* itself. In a separate study, Wittkopp et al. (2003b) took a QTL approach with two closely related species that differ dramatically in pigmentation (*D. americana* and *D. novamexicana*). Since they knew of several good candidate genes, such as *yellow*, they used markers in some of these genes for the QTL mapping. The results of the QTL mapping identified at least four genomic regions that were responsible for the changed pigmentation pattern between these species, but none of the difference in pigmentation could be attributed to changes at the *yellow* locus. Therefore, similar changes in phenotype can be caused by evolutionary changes in different loci, even within a single genus of flies.

A second example illustrates how some tools of model systems have been used to identify evolutionarily relevant variation. In *Drosophila* species, the larval and adult cuticle is decorated with a complex pattern of nonsensory trichomes (the taxonomically correct term is microtrichia). We have explored the evolution of this pattern between *D. sechellia*, which has lost a broad swathe of trichomes on each segment of the dorsal cuticle of first-instar larvae, and other members of the *D. melanogaster* species subgroup, which all display the ancestral pattern of dense trichomes over the dorsum of most of each segment (Sucena & Stern 2000). We started by performing a full-genome screen for the evolved loci using genetic crosses between *D. sechellia* and *D. simulans*. Our initial studies indicated that the naked cuticle trait segregated as a single X-linked locus and that the *D. sechellia* allele was recessive to the alleles from *D. simulans* and *D. melanogaster*. We then surveyed most of the X chromosome using a set of *D. melanogaster* stocks that were deficient for different regions of the X chromosome. (This and other similarly powerful tools are the major benefits that are associated with performing EDB studies in species closely related to model systems.) This survey led us to a small region that included an excellent candidate gene, the transcription factor *shavenbaby/ovo* (*svb*). The *svb* gene acts like a switch; when it is expressed in a cell it causes the differentiation of one or more trichomes on the apical surface of this cell, presumably through the regulation of a battery of downstream genes that modify the distribution of actin to form an apical projection (Delon et al. 2003). When *svb* transcription is turned off, cells differentiate smooth cuticle. A diverse array of patterning genes are thought to regulate *svb* (Payre et al. 1999), thus *svb* provides a binary read-out of complex earlier patterning events. (Note that even at this fine scale of resolution of our mapping efforts, when we had narrowed the candidate region to about 10 genes, knowledge of gene function allowed us to focus our efforts to efficiently identify the causal gene.) Further analysis showed that evolution of the *cis*-regulatory region of *svb* had caused the morphological difference between species. Thus, evolutionarily relevant variation can be identified via whole-genome scans facilitated by genomics and other tools and we can expect an increasing number of such studies in the coming decade.

Association Studies

Within populations, there is a correlation between the causal genetic variation for traits and the phenotypic variation for such traits. This is the fundamental principle of association studies. I illustrate this approach with a set of studies that have identified the causal variants underlying phenotypic polymorphisms in natural populations.

In some species of vertebrates, adults of natural populations can be found in two morphs: a darker melanic morph and a lighter nonmelanic-morph. A remarkable series of recent papers has described the genetic causes of polymorphisms in melanism in pocket mice, bears, and three species of birds. In each of these species, some individuals have dark coloration and others have a lighter coloration. The lighter morphs are not cases of albinism but rather differences in the relative levels of eumelanin (black pigment) and pheomelanin (yellow or red pigment) that are produced by the melanocytes. In the three species of birds, the bear population and in one population of pocket mice, nucleotide differences causing changes in the amino acid sequence of the *melanocortin-1-receptor* gene (*Mc1r*) have been shown to be in perfect association with the melanism variation (Mundy et al. 2004; Nachman et al. 2003; Ritland et al. 2001; Theron et al. 2001). In the pocket mice, only one of four populations showed association of melanism variation and variation at *Mc1r* (Hoekstra & Nachman 2003). These cases of melanism variation are clearly examples of convergent phenotypic evolution, suggesting that the protein-coding sequence of *Mc1r* is a preferred mutational target for evolutionary changes in melanism in vertebrates. Why might this be? *Mc1r* is a G-protein-coupled transmembrane receptor that when activated by a peptide hormone, alpha-melanocyte-stimulating hormone, causes increased production of eumelanin in melanocytes. Another gene, *agouti*, acts as an antagonist of *Mc1r* signaling, causing reduced production of eumelanin and increased production of pheomelanin. Thus, *Mc1r* acts as a molecular switch integrating extracellular signals. In addition, *Mc1r* acts only in melanocytes, so mutational changes in the amino acid sequence of *Mc1r* are unlikely to have pleiotropic effects on other aspects of the phenotype. This is in sharp contrast to many of the approximately 80 other genes that also affect coat color in the laboratory mouse, mutations in which often cause various diseases (Price & Bontrager 2001).

In all of these cases of melanism evolution, current evidence strongly suggests that the relevant mutational changes causing melanism differences are changes in the amino acid sequence, which stands in stark contrast to the general consensus in the field of EDB that most developmental variation is likely to be caused by *cis*-regulatory variation in patterning genes. This example of melanism variation is, therefore, a cautionary tale that not all developmental variation results from evolutionary changes of transcriptional regulation. However, there is an even more important message inherent in these findings and this is that the precise structure of regulatory networks (be they regulating transcription or post-transcriptional events) may determine the distribution of adaptive mutations within the network. I return to this point below.

THE IMPORTANCE OF PARALLELISM

One important pattern has emerged from a wide diversity of recent studies in EDB: the prevalence of developmental parallelism underlying convergent morphological changes. The case of vertebrate melanism and *Mc1r* variation described above is, perhaps, the most compelling example, but other examples suggest that developmental evolution may be surprisingly predictable.

The first example is an extension of our studies of larval trichome patterning amongst drosophilids. We have recently shown that regulatory changes at the *svb* locus are likely to have caused similar changes in larval trichome patterning in at least three other lineages of drosophilids (Sucena et al. 2003). Other examples of similar developmental changes underlying convergent phenotype evolution have been described in other systems, including stickleback skeletal morphology (Colosimo et al. 2004; Cresko et al. 2004; Shapiro et al. 2004), cavefish eye degeneration (Yamamoto et al. 2004), and plant flowering architecture (Yoon & Baum 2004).

In each of these cases, evolutionary changes in the *cis*-regulatory region of the focal genes appear to have caused the morphological changes. That is, within the presumably complex genetic network that patterns particular organs there appear to be preferred mutational targets that cause evolutionary change. This can be illustrated with the example of *svb* evolution. The *svb* gene acts as a kind of molecular switch, integrating complex spatial information within the *cis*-regulatory region of *svb* and then regulating a diversity of genes downstream that are involved in the differentiation of a trichome. Therefore, changes in *svb* expression may cause fewer pleiotropic consequences than changes in any other genes in the pathway patterning trichomes. This is reminiscent of the argument I have offered for the prevalence of evolutionary changes at the *Mc1r* gene, above. Evolutionarily relevant mutations have tended to accumulate in positions in genetic networks that have minimal pleiotropic consequences.

Although I have stressed the general point that similar morphological changes can be caused by similar genetic changes in divergent taxa, there are also examples of divergent genetic changes underlying convergent morphologies (such as the evolution of convergent pigmentation phenotypes by divergent genetic mechanisms discussed above). Therefore, although there are strong indications that developmental parallelism is an important feature of morphological evolution, it is clearly not the only mode by which changes in development cause morphological evolution.

THE IMPORTANCE OF *CIS*-REGULATORY EVOLUTION

Transcriptional Gene Regulation

As reviewed above, many cases of morphological evolution appear to have resulted from evolutionary changes in *cis*-regulatory regions that have altered the pattern of transcription of developmental genes. There are several reasons why changes in *cis*-regulatory regions may be favored modes of morphological evolution (Stern 2000; Wray et al. 2003). To understand why this might be so, I first briefly review what is currently known about the structure of *cis*-regulatory regions and then return to this point. There are many excellent reviews of the general principles of transcriptional regulation (Davidson 2001; Ptashne & Gann 2002). Here I review only several features relevant to this chapter.

Cis-regulatory regions are stretches of DNA, usually several hundreds or thousands of base pairs long, that contain binding sites for transcription factors. Individual transcription factors bind to short specific stretches of DNA, on the order of 6–15 bp. These transcription factors may be activators or repressors of transcription, and a single region normally contains a combination of both. Binding of an individual transcription factor is often reinforced by the local binding of other transcription factors. Transcription factor binding sites that regulate the same gene are often clustered in the genome, presumably to allow interactions between transcription factors. This clustering sometimes results in experimentally identifiable "modules" of *cis*-regulatory action. These modules appear to be prominent (Davidson 2001), although not universal (Klingler et al. 1996), in the *cis*-regulatory regions of genes that are involved in pattern formation. The complex patterns of expression that are observed for genes involved in patterning embryos are often the result of output from multiple *cis*-regulatory modules.

How is all of the information encoded as the binding of transcription factors to regulatory regions integrated to generate a particular level of gene expression in a single cell? The best-studied example of the operation of a *cis*-regulatory region is for the *Endo-16* gene of the sea urchin, *Strongylocentrotus purpuratus* (Yuh et al. 1998). Several independent modules integrate inputs (the presence or absence of transcription factors) and output a quantitative response to a module located adjacent to the promoter. Transcriptional regulation is therefore the product of the binding of a combinatorial code of transcription factors present in a cell at a particular time to specific *cis*-regulatory modules. The required level of transcription is essentially computed by these modules and presented as an output to generate a particular level of transcription.

The Evolution of *Cis*-Regulatory Modularity

The modularity of *cis*-regulatory regions has several consequences for evolution. Perhaps most importantly, modular *cis*-regulatory structure allows a single gene to be recruited for multiple developmental functions without generating a conflict with its original role(s). For example, mutations that alter the transcription factor binding sites in one *cis*-regulatory module need not influence the transcription driven by other *cis*-regulatory modules. Such mutations will produce a highly specific change, an alteration only in the domain of expression determined by this module. In contrast, a mutation in the coding region of a transcription factor is likely to alter the transcription of all genes regulated by the protein, causing widespread changes to the phenotype.

This consideration leads to the identification of two advantages to *cis*-regulatory modularity. First, modularity limits the pleiotropic consequences of mutations that alter gene regulation. Mutations in *cis*-regulatory regions can have highly specific and nonpleiotropic effects. Second, altering the pattern of expression of a transcription factor often results in the immediate recruitment of an entire working set of downstream genes. That is, placing a similar organ in a new location does not require the invention of all the genetic circuitry to build the organ. For example, the entire working module that builds a trichome could be recruited to a new location simply through expression of *svb*. These two advantages, then, may explain why changes in the expression of transcription factors seem to play such a dominant mode in developmental evolution (Stern 2000; Wray et al. 2003).

An additional long-term evolutionary advantage of regulatory modularity is that this allows transcription factors to be easily coopted for patterning the same tissue at different times (Davidson 2001; Wilkins 2002). For example, the *Pax-6* gene is required for specification of eyes in divergent metazoans, even in taxa that are believed to have evolved complex eyes independently (Gehring & Ikeo 1999). However, *Pax-6* is also required to specify various elements of the complex eye, including the photoreceptors. One attractive model to explain the predominant role of *Pax-6* in specifying eyes across the metazoans involves the progressive cooption by *Pax-6* of additional regulatory roles associated with eye development (Davidson 2001; Wilkins 2002). The use of *Pax-6* for cooption would have been favored simply because this transcription factor was already expressed in the right place and time, originally simply to specify the production of photoreceptors in the head. In general, this model allows the evolution of increasing complexity without the invention of new transcription factors.

Finally, it is worth asking why metazoan genomes have evolved modular *cis*-regulatory regions, since such modules are presumably not the only way to regulate gene expression. One answer has been provided by Gerhart and Kirschner (1997), who proposed that organisms with modular regulatory circuitry may be more evolvable. This is an explanation due to selection above the level of the individual. The idea is that modular regulatory structure allows populations to respond more effectively to natural selection, perhaps for the two reasons discussed above: it limits pleiotropy and allows efficient cooption of sets of downstream genes.

FUTURE DIRECTIONS

What Can We Expect in the Future of EDB?

First, genomics is going to change everything. By genomics I mean both the acquisition of full-genome sequences for a large number of species and the assay of gene function on a genome-wide basis in these species. I do not mean to imply that genomic methodologies, especially as currently implemented, are necessarily going to supersede careful analysis of individual genes, although genomic methodologies are certainly making at least some tedious assays much more efficient. The real value of genomics will be that it forces us to consider the genome as an operational whole, which is probably the right level at which to consider developmental evolution.

Second, we will see an expansion of research on the evolution of genetic networks and how variation in these networks generates developmental evolution (Wilkins 2002). Again, I expect that genomic technologies will allow a more rapid accumulation of the relevant data. We have only begun to collect such data, one gene at a time, but the results (some of which were reviewed in this chapter) are promising.

Third, given a deeper understanding of the distribution of variation within genetic networks, we may be able to predict which genes within genetic pathways are more likely to vary to create phenotypic variation. The first observation in support of this hypothesis is the fact that in many cases changes in the pattern of transcription factor expression seem to be involved in alterations to development and morphology. If changes to morphology were distributed randomly amongst all genes involved in constructing the phenotype, it is unlikely that so much of the phenotypic variation could be explained by changes in transcription factor expression. While there appears to be considerable support for this hypothesis from Comparative EDB, there are currently only a small number of experimentally verified cases from Genetical EDB. Thus, the first prediction is that evolutionary changes to transcription factor expression are the most likely causes of changes in morphology. Beyond this, it also seems likely that the precise structure of regulatory networks is likely to influence where evolutionarily relevant variation is allowed to accumulate within networks. This, ultimately, may provide a predictive theory for the material basis of evolution.

SUGGESTIONS FOR FURTHER READING

Carroll et al. (2001) provides an accessible review of key concepts of animal development relevant to the study of EDB as well as a description of many examples. In particular, this book is a useful introduction to the so-called toolbox genes, patterning genes that are conserved throughout many taxa and are used in conserved and divergent ways in evolution. Wilkins (2002) is an excellent critical review of EDB that works hard to place EDB within a contemporary evolutionary framework. Wilkins uses the concept of developmental pathways to explore EDB. This book is not simply a review of the literature, but provides a number of fundamental insights and a useful conceptual framework for thinking about EDB. Wray et al. (2003) is a wonderfully comprehensive review of the data and concepts related to *cis*-regulatory evolution. The authors review a massive amount of data in a very clear format and provide a number of predictions about *cis*-regulatory evolution. This is an important contribution to the development of EDB as a predictive science. Coen (2000) is an introduction to developmental biology aimed at the lay reader, with its use of painting as an analogy for development, but there is much in here for scientists at any level. I strongly recommend this highly enjoyable book as a way for the reader to make their "developmental thinking" second nature. Since *cis*-regulatory evolution appears to be central to EDB, it is important to develop a deep understanding of transcriptional regulation. While there are several other good books on this subject, Ptashne and Gann (2002) provides a perspective that I have found particularly useful when considering evolutionary issues.

Carroll SB, Grenier JK & SD Weatherbee 2001 From DNA to Diversity: Molecular Genetics and the Evolution of Animal Design. Blackwell Science.

Coen E 2000 The Art of Genes: How Organisms Make Themselves. Oxford Univ. Press.

Ptashne M & A Gann 2002 Genes & Signals. Cold Spring Harbor Laboratory Press.

Wilkins AS 2002 The Evolution of Developmental Pathways. Sinauer Assoc.

Wray GA, Hahn MW, Abouheif E, Balhoff JP, Pizer M, Rockman MV & LA Romano 2003 The evolution of transcriptional regulation in eukaryotes. Mol. Biol. Evol. 20:1377–419.

Acknowledgments I thank Greg Davis for reading several versions of this chapter during its dramatic metamorphosis, all members of my laboratory for invigorating discussions, and two anonymous referees for being painfully and abundantly honest with critiques of a previous version of this chapter. E. B. Duffy graciously provided her empty office and her time, without either of which this chapter would not exist.

16

Canalization

MARK L. SIEGAL
AVIV BERGMAN

Canalization, a concept introduced by C.H. Waddington (1940, 1942) to explain the robustness of developmental processes in the face of environmental and genetic variation, remains one of biology's most confusing and contentious subjects. Perhaps it is a tribute to Waddington's prescience that his ideas continue to be provocative in the post-genome era. Still, a number of authors have lamented the conceptual muddle that has endured for so many years (Debat & David 2001; Gibson & Wagner 2000; Rutherford 2000). Some even pin the blame on Waddington himself, whose penchants for philosophical language and iconoclastic status (Wilkins 2002) clouded the debate from the outset. Nevertheless, Waddington's pioneering focus on developmental mechanism as central to understanding microevolutionary change cannot be overlooked, and indeed, we believe, is the reason why modern biologists still turn to him for inspiration and guidance as the much-heralded synthesis of evolutionary and developmental biology proceeds.

The goal of this chapter is to present a modern reading of the conceptual and experimental history of canalization, with particular attention paid to two recent case studies that illustrate the opportunities and challenges that lie ahead as we seek to understand the causes and consequences of developmental robustness. Thus, while we do not intend to provide an exhaustive review of canalization, we do recognize that current understanding of canalization requires tracing, at least in part, its history as a biological idea (for a comprehensive review, see Scharloo 1991; for perspectives on Waddington's life and legacy, see Slack 2002; Hall 1992; Gilbert 1991; Wilkins 2002; see the last reference in particular for an historical perspective that includes the views of contemporaries of Waddington, particularly Ivan Schmalhausen, who developed similar ideas). We will begin with a review of key concepts, starting with Waddington's early experiments and theoretical arguments. These will form the basis for a discussion of the subsequent experimental and theoretical developments, as well as the prospects for bringing modern perspectives and technologies to bear on the subject.

First, we must review the meaning, or meanings, of canalization. Waddington's original formulation of the concept was based on the observation that developmental processes seem to be structured such that they produce consistent end results in spite of minor perturbations, either environmental or genetic (Waddington 1942). This observation led to his now-famous visual metaphor of development as a ball rolling down an "epigenetic landscape" (Figure 16.1; Waddington 1957). Channels carved in the landscape represent alternative paths of differentiation; deeper channels signify developmental fates that, once chosen, are more resistant to perturbation. Over the years, canalization has taken on a number of meanings (Gibson & Wagner 2000; Debat & David 2001). Developmental biologists tend to equate canalization with developmental buffering (e.g., Wilkins 1997; Wilkins 2002). Evolutionary biologists tend to include its presumed causal basis, defining canalization as buffering that has evolved under natural selection (Gibson & Wagner 2000; Debat & David 2001). A distinction is also drawn between "environmental canalization"

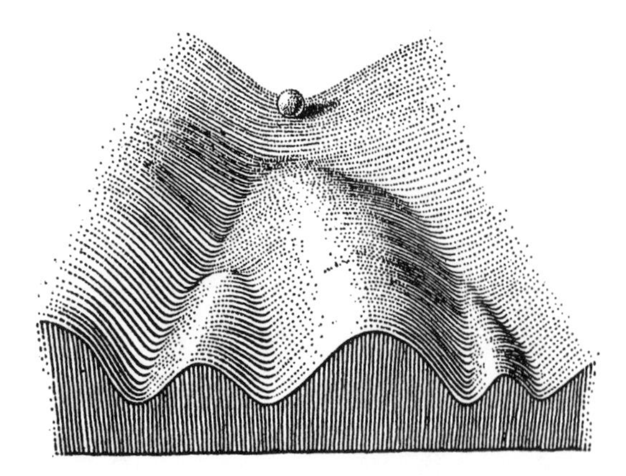

FIGURE 16.1. Part of an epigenetic landscape. The path followed by the ball as it rolls down the landscape corresponds to the developmental history of a particular part of the egg, which can adopt one of several discrete fates. Canalization is reflected in the depth and width of paths carved in the landscape. Minor perturbations in individual genes, or fluctuations due to environmental noise during development, would be buffered because they would be insufficient to force the ball out of its channel or to alter the gross topology of the landscape. Major perturbations, however, could either cause the ball to breach a threshold to an adjacent channel or alter the surface itself, leading to a new favored path. Reprinted from Waddington (1957) with permission.

(buffering of environmental variation) and "genetic canalization" (buffering of genetic variation) (Gibson & Wagner 2000; Debat & David 2001), although, as will be discussed, there is reason to believe that these are not completely separate phenomena (Gibson & Wagner 2000; Meiklejohn & Hartl 2002; Stearns et al. 1995; Bergman & Siegal 2003). Finally, as rightly pointed out by Wilkins (2002), canalization has sometimes been taken erroneously as synonymous with the related but distinct concept of "developmental constraint," or the restriction in possible paths of developmental change. Indeed, we will see in this chapter how canalization may actually play a role in opening certain avenues of evolutionary transformation.

We prefer the relatively unrestrictive definition of canalization as developmental buffering, for it leaves open the question of the ultimate reason that buffering exists and is consistent with Waddington's original formulation. Although Waddington himself introduced the idea of canalizing selection, it remains to be demonstrated that buffering is indeed a product of natural selection of the type he and others have envisaged, namely, "stabilizing" selection for optimum phenotypes. As we will see in the second case study, the assumption that buffering of key traits is the product of stabilizing selection requires critical re-examination (Siegal & Bergman 2002). That is, in the terminology of Gibson and Wagner (2000), canalization may be largely an "intrinsic," rather than "evolved," property of developmental systems.

Defining canalization as buffering raises the obvious question: What is buffering? For the purposes of this discussion, we consider buffering to be an umbrella term covering all developmental mechanisms by which the effect of genetic or environmental variation is dampened. Buffering thus includes relatively simple mechanisms, such as those that manifest as dominance (the capacity of a single wild-type allele at a diploid locus to cover for a nonfunctional allele at that locus) and redundancy (the compensation for loss of one gene's activity afforded by a paralogous gene) (Wilkins 1997). It also includes more complex mechanisms, such as dedicated proteins that ensure the proper functioning of other proteins, as will be discussed in the first case study, and complex nonlinear interactions among genes that attenuate the phenotypic effects of variation in the genes themselves or in environmental conditions (Nijhout 2002), as will be discussed in the second case study. In current parlance, the term "robustness" is often reserved for such complex mechanisms; we adopt this usage here. Whereas the simple buffering mechanisms are relatively easy to grasp, the buffering mechanisms of robustness and their effects on evolutionary trajectories are far less intuitive and will be the main focus of this chapter.

A fundamental question concerning robustness is whether it constrains or promotes evolutionary divergence. At first glance, it may seem natural to suppose that robustness only constrains divergence, because it prevents mutations from having phenotypic consequences. However, by providing for the buildup of stores of genetic variation, robustness may ultimately promote divergence. In grappling with this issue, Waddington developed the notion of genetic assimilation. Assimilation begins when an environmental perturbation induces a canalized system to reveal normally hidden genetic variation upon which selection then operates; under prolonged selection, the selected phenotype becomes independent of the perturbation and a new canalized system is established to consistently produce this phenotype never seen in the original environment

(Waddington 1961). Waddington argued that genetic assimilation provided a sound Darwinian explanation for apparent cases of Lamarckian evolution, in which environmentally induced ("acquired") characters appeared to be inherited (Waddington 1942). Moreover, if real, genetic assimilation would provide a mechanism for maintaining stasis under constant, favorable conditions, and generating phenotypic variation under challenging conditions. Mechanisms like this, which appear to invoke foresight in evolutionary processes, have historically generated much controversy (see, e.g., Foster 2000), and genetic assimilation is no exception; genetic assimilation was also assailed on the grounds that simpler models than the epigenetic landscape were sufficient to explain the pertinent experimental results (reviewed in Scharloo 1991). Without greater knowledge of the genetic architecture underlying complex traits, however, the debate over genetic assimilation lay relatively dormant for a few decades. Recently, though, the controversy was reignited when Rutherford and Lindquist (1998) declared the stress-responsive molecular chaperone Hsp90 to be an "evolutionary capacitor," buffering genetic variation under normal circumstances and revealing this variation when its function is overwhelmed by extreme stress.

The Rutherford and Lindquist (1998) study capped a period of a few years in which research on canalization and genetic assimilation was reinvigorated by modern molecular biology (Stearns & Kawecki 1994; Gibson & Hogness 1996; Wilkins 1997). These authors' work (1998), which dramatically raised important questions about canalization and assimilation, serves as the first case study in this chapter. These years also witnessed a resurgence of theoretical investigations into the evolutionary forces that mold canalized systems (reviewed in Siegal & Bergman 2002). Traditional population-genetic and quantitative-genetic models were complemented by computer simulations of complex gene networks, which have yielded important new insights. In the second case study, we discuss our own recent contributions that built upon this previous work. In particular we discuss the intriguing possibilities, raised by the results of our simulations, that complex gene networks may be canalized with or without stabilizing selection on the phenotypes they produce (Siegal & Bergman 2002), and that evolutionary capacitance may be a general feature of such networks (Bergman & Siegal 2003). Together the case studies illustrate that Waddington's legacy is strong, and that understanding the causes and consequences of developmental robustness will remain a major goal of modern biology.

CONCEPTS

Wild-Type Developmental Programs Appear Buffered Against Environmental and Genetic Variation

Mutant alleles can have dramatically different effects in different genetic backgrounds. This inconvenient fact has plagued geneticists for years. As noted decades ago by Waddington (1942), based on what was then known about the genetics of *Drosophila*, "the phenomenon is extremely obvious; there is scarcely a mutant which is comparable in constancy with the wild type, and there are very large numbers whose variability ... is so great that it presents a considerable technical difficulty." Waddington, however, looked beyond the nuisance of variability to see that it said something fundamental about developmental systems: "[t]he constancy of the wild type must be taken as evidence of the buffering of the genotype against minor variations not only in the environment in which the animals developed but also in its genetic make-up." This recognition that genetic as well as environmental variations must be buffered shaped Waddington's notion of the canalization of development, and led to a number of experiments by him and others in which major-effect mutations—or, more recently, specific alterations achieved by reverse genetic technology—were used to reveal genetic variation affecting a variety of traits (reviewed by Scharloo 1991; Gibson & Wagner 2000). *Drosophila* geneticists had also catalogued a number of cases in which a specific environmental treatment (e.g., heat shock or exposure to ether vapor during a critical developmental period) "phenocopied" certain mutational effects. Waddington championed the use of such phenocopies to study canalization as well, because they too manifested genetic background effects (also reviewed in Scharloo 1991).

Variation Revealed by Mutation or Environmental Challenge Is Heritable

The sensitivity of phenocopies to genetic background effects also formed the basis of Waddington's concept of genetic assimilation. Indeed, Waddington rarely separated canalization from genetic assimilation, a practice that frustrated his detractors (Scharloo 1991). Nevertheless his experiments and those of others were important in establishing that the variation induced by perturbing the wild-type system does indeed respond to selection (Scharloo 1991). An example here is useful. Waddington (1952) exposed pupae of a wild-type *Drosophila* strain to extreme temperature (40 °C) for 4 hours. As a result, approximately 40% of the treated flies had a specific defect in their wings: the absence or reduction of the posterior crossvein. Waddington then set up two selection lines, one in which flies exhibiting the crossveinless phenocopy upon heat treatment were chosen to found the next generation and one in which treated flies with apparently wild-type crossveins were so chosen. Selection in each line continued for 19 generations, at which point the crossveinless selection line produced over 90% crossveinless flies upon heat treatment, whereas the non-crossveinless selection line produced fewer than 20% crossveinless flies upon heat treatment. Thus, selection was successful in altering the genetic constitution of each line to either increase or decrease the response to heat treatment. Moreover, after generation 12, some flies from the crossveinless selection line developed the crossveinless phenotype even without heat treatment. In other words, the crossveinless phenocopy had become genetically assimilated, no longer requiring the inducing environmental treatment to be expressed.

The Mechanistic Basis of Canalization

While Waddington pursued experimental support for canalization and assimilation, he also sought to develop a theoretical framework for understanding them. This framework relied heavily on the notion of the epigenetic landscape (Figure 16.1), and was laid out most completely in a chapter on "The Cybernetics of Development" in his 1957 book, *The Strategy of the Genes*. The aspect of the book that is most striking to a modern reader is Waddington's all-out embracing of complexity. Waddington was unique in this respect, as other mid-century geneticists tended to focus—quite profitably it must be said—on simple biochemical pathways in simple organisms. Consider that the 1958 Nobel Prize was shared by George Beadle and Edward Tatum, for their pioneering work that established the "one gene—one enzyme" concept, and Joshua Lederberg, for his fundamental work in microbial genetics. Waddington's ideas were certainly shaped by the work of contemporary evolutionary biologists, particularly Sewall Wright, who introduced the notion of the "adaptive" landscape to conceptualize epistatic fitness interactions among genes (see Slack 2002). Nonetheless, Waddington's originality and farsightedness are evident in *The Strategy of the Genes*. In particular, his depictions of the complex gene interactions underlying the epigenetic landscape are remarkable to view even half a century later.

For the purposes of this chapter, we have already defined canalization as developmental buffering, making no reference to the metaphor of the epigenetic landscape. Thus, we do not wish to dwell on the metaphor, even though we personally concur with Waddington that "it has certain merits for those who, like myself, find it comforting to have some mental picture, however vague, for what they are trying to think about" (Waddington 1957, p. 30). Nonetheless, the spirit of the epigenetic landscape lives on in the consensus view that buffering must result from the complex and nonlinear processes of development (Nijhout 2002). Waddington relied on the metaphor because he understood that numerical methods had not advanced sufficiently to study complex systems in mathematical detail, and that the genetic mechanisms of development would stay mysterious for some time. Indeed, analysis of complex systems remains a challenge to this day, and methodologies for treating complex epistatic interactions among genes are still in their infancy. We are also still limited in our knowledge of the full complexity of developmental-genetic pathways underlying morphogenesis, despite enormous and ever-accelerating progress in their elucidation. Nevertheless, we have reached a point where complexity and actual genetic mechanisms are being considered in increasingly sophisticated ways. The two case studies exemplify this progress and, in so doing, vindicate Waddington's unconventional perspective.

The Evolutionary Cause of Canalization

Waddington endeavored to understand not just how canalized systems are built but why such systems should exist. That is, he sought the evolutionary forces favoring canalization. The explanation he proposed, independently forwarded by Schmalhausen (1949), was what is commonly termed "stabilizing selection," although Waddington (1957) preferred the term "canalizing selection." The basic premise is that selection for an optimum value of a particular trait will not only favor genotypes that produce the value on average but also those that reduce the variation around that value. Clearly, as canalization reduces variation by dampening the effects of environmental and genetic perturbations it should be favored by selection for optimum values of traits. A number of theoretical studies have supported this intuition (reviewed in Gibson & Wagner 2000; Siegal & Bergman 2002). However, an important complication was noted by G. Wagner et al. (1997), who showed using a population-genetic model that environmental canalization increases with increasing intensity of stabilizing selection but that genetic canalization increases only up to a point. The reason is that strong selection purges the population of the genetic variation that would drive the evolution of canalization. This result would seem to suggest that traits under strong stabilizing selection should not be genetically canalized, a conclusion that contradicts the experimental evidence of near-universal buffering of developmental systems. Gibson and Wagner (2000) offered two possible resolutions to this conundrum. The first, later formalized by Meiklejohn and Hartl (2002), is that stabilizing selection favors environmental canalization and that systems that are insensitive to environmental perturbation are also insensitive to mutations. The second is that canalization need not always be directly selected, but instead may be a by-product of other evolutionary processes or may be a relic of past canalizing selection. We (Siegal & Bergman 2002) provided the first demonstration that genetic canalization could evolve without stabilizing selection and that it likely represented an intrinsic property of complex developmental systems. We will return to this work in the second case study.

Models of Genetic Assimilation

To explain his experimental results demonstrating genetic assimilation Waddington largely relied on the metaphor of the epigenetic landscape. He argued (Waddington 1957) that the initial major environmental perturbation forced development out of its normal course and into an adjacent channel; through assimilation the environmental perturbation either was replaced by selection of a genetic variant that mimicked the perturbation, causing development to breach the same ridge, or was rendered unnecessary by selection of variants that changed the local topology of the landscape so as to make the adjacent channel the favored path. In either case, the result was to change a system that responded to the environment by adopting one of two fates into one that was insensitive to the environment. G. C. Williams, noting the pervasiveness in nature of cases of apparent phenotypic plasticity—that is, organisms with stereotyped and adaptive responses (reaction norms) to different environmental conditions—challenged that assimilation could therefore not be important in nature, because it would erase such flexibility (see Eshel & Matessi 1998). The relationship between canalization and plasticity is not so simple as to consider them exact opposites, however. While a full discussion of the connection between canalization and plasticity is beyond the scope of this chapter, suffice it to say that the stereotypy of response itself implies that plastic responses are canalized to some extent (for more on plasticity see Scheiner, Ch. 21 of this volume, and for its link to Hsp90-mediated buffering see Sangster et al. 2004).

The complex relationship between canalization and plasticity highlights the importance of models that consider all contributions to the mapping from genotype to phenotype in a less abstract form than the epigenetic landscape. Because of their explicit focus on genetic assimilation, we bring in two such efforts here: the early contribution of Rendel (1967) and the more recent development and expansion of his ideas by Eshel and Matessi (1998). Rendel (1967) introduced a model to explain canalization and genetic assimilation, based on the relationship between the phenotypic expression of a trait and a hypothetical underlying "entity," which he called "Make," that causes the trait to develop. Make is meant to represent in a single quantity the output of a complex interacting system that builds the trait. When the trait value is proportional to Make, the trait is not canalized, but when there is a plateau in the relationship such that a range of Make values yield the same trait value, the trait is canalized within the range of the plateau. Using this model, Rendel explained the experimental results of Waddington and others. For example, he explained the crossveinless experiment described above (Waddington 1952) by supposing that heat treatment lowered Make levels, sending those of some genotypes below the canalized plateau range, and thereby revealing phenotypic variation; continued crossveinless selection enriched for alleles in genes affecting the trait that led to lower amounts of Make, until eventually some genotypes produced insufficient levels of Make even without heat treatment.

Eshel and Matessi (1998) developed a mathematical formulation of Rendel's model and applied it to a population model in which in each generation individuals were assigned randomly to one of a finite number of environments, which differed in the optimum trait value they enforced. They found that selection favored a system that was canalized around the optimum corresponding to the most common environment, consistent with the predictions of Waddington (1957) and Schmalhausen (1949). They also built on the model by postulating that some extreme stresses might not merely affect levels of Make but might instead change the functional relationship between Make and the trait value such that canalization would entirely break down. They found that a system that responded to extreme stress in this way was favored. That is, they found evidence of adaptive inactivation of canalization. Eshel and Matessi (1998) carefully pointed out that cases of inactivation of canalization under extreme stress, if found in nature, are not necessarily adaptive and instead may be due to inevitable failure of the system, yet their conclusion that adaptive inactivation of canalization is plausible is relevant to the claim that Hsp90 is an evolutionary capacitor, as discussed directly below in the first case study.

CASE STUDIES

Hsp90 and the Conditional Release of Canalized Variation

A major obstacle to work on canalization and genetic assimilation was the absence of a sufficiently well defined mechanism for modulating the degree of buffering of a complex developmental system. The classical experiments clearly showed that developmental systems harbor genetic variation that is not normally expressed phenotypically, and that this buffering could be disrupted by large mutational or environmental perturbations. But how this occurred remained mysterious. Meanwhile, theoretical efforts demonstrated canalization and assimilation in models with reasonable assumptions about the evolutionary process, but these models characteristically considered the developmental mechanism in only abstract terms, if at all. Thus, the demonstration by Rutherford and Lindquist (1998) that Hsp90, a stress-induced molecular chaperone, buffers genetic variation under normal circumstances but reveals it when its function is overwhelmed, had immediate impact. For the first time, a molecular mechanism for the conditional release of buffered genetic variation could be studied in the laboratory. So what did Rutherford and Lindquist (1998) show?

Impairing Hsp90 Function Reveals Normally Buffered Polygenic Variation

Hsp90 maintains the proper conformation of a defined set of signaling proteins, keeping these developmentally important and inherently unstable proteins primed for activation. Its normally narrow target specificity distinguishes it from other stress-induced ("heat-shock") proteins, and places it uniquely at the interface between morphogenetic and stress-response pathways. Under conditions of cellular stress, Hsp90 becomes more abundant and its activity is partially redirected to misfolded proteins that are not normally its targets. When Rutherford and Lindquist (1998) looked at *Drosophila melanogaster* adults in laboratory stocks kept heterozygous for a mutation in the Hsp90 gene, they observed a surprisingly high proportion of flies with morphological defects (Figure 16.2; readers familiar with the idiosyncrasies of *Drosophila* gene nomenclature will not be surprised to learn that the gene encoding Hsp90 is called *Hsp83*—to avoid confusion, we refer to it here as the Hsp90 gene). To ensure that this was not an artifact of the particular stock they observed, they looked at multiple, independent mutant stocks, and, more importantly, outcrossed the mutants with a panel of wild-type stocks. The observation held: partial loss of Hsp90 function tended to induce morphological defects in a small but significant proportion of animals (on average 1–2%). Abnormalities were even seen when wild-type flies were fed geldanamycin, a specific pharmacological inhibitor of Hsp90.

An important trend noticed by Rutherford and Lindquist was that the morphological defects were not random with respect to genetic background. That is, siblings from the same cross tended to share the same defect, small wings for example (Figure 16.2A). This indicated that the effect of reducing Hsp90 function was not merely to make flies more sensitive to random microenvironmental fluctuations during development. Instead, compromising Hsp90 function uncovered pre-existing genetic variation in key developmental pathways.

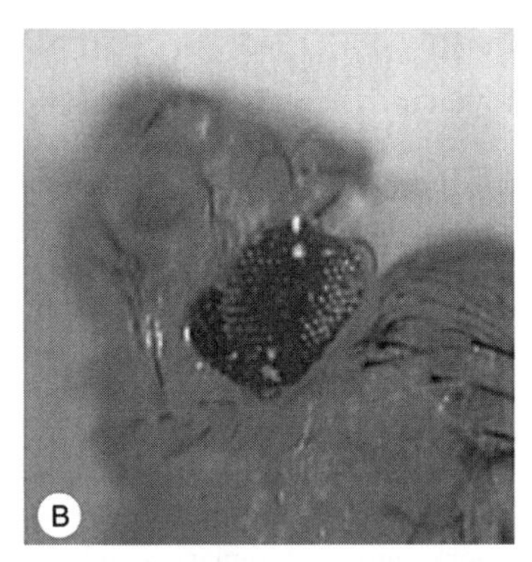

FIGURE 16.2. Developmental abnormalities associated with Hsp90 deficits. In addition to those shown here, Rutherford and Lindquist (1998) observed morphological defects in practically all external structures. Which defects emerged was dependent on the fly strain examined and the culture temperature. (A) Small wings; (B) deformed eye. From Rutherford and Lindquist (1998), reprinted with the permission of Nature Publishing Group.

The implication is that wild-type Hsp90 levels normally buffer this variation, thus making Hsp90 a contributor to developmental robustness (or, equivalently, a mechanism of canalization).

A more striking demonstration of the genetic basis of the phenotypic variation seen in Hsp90-mutant flies was provided by Rutherford and Lindquist in the form of artificial selection experiments on two separate Hsp90-mutant-induced phenotypes: deformed eyes and small wings. In both cases, fewer than 10 generations were required to establish lines in which a large proportion of flies manifested the selected phenotype (over 80% in some lines). The response to selection clearly indicated the heritability of the deformed-eye and small-wing traits, and the trajectory of the response suggested that the relevant variation was polygenic (i.e., it did not segregate in a simple Mendelian fashion).

Hsp90-Mutation-Induced Traits Can Be Assimilated

In the artificial selection experiments described above, selection was only for the induced traits, not for the continued presence of the Hsp90 mutation that originally induced them. Hsp90 mutants are lethal when homozygous, making the expected frequency of the mutation in adults of a selected line at most two thirds. As the frequency of the deformed-eye trait surpassed 80% in some selected lines, Rutherford and Lindquist surmised that the trait may have been assimilated. That is, some of the flies with deformed eyes may have been wild-type for Hsp90. To test this, they used polymerase chain reaction (PCR) to assay the presence or absence of the Hsp90 mutation. Remarkably, none of the flies they tested in generations 16–20 of the selection carried the mutation, confirming that the deformed-eye trait had been assimilated. The lack of (or very low frequency of) Hsp90-mutant flies in later generations may be due to pleiotropic fitness effects of the Hsp90 mutation, such that mutation-free flies with deformed eyes had higher fitness than their mutation-carrying siblings.

Does Hsp90 Provide a Mechanism for Evolvability?

Based on their experimental findings, Rutherford and Lindquist asserted that Hsp90 is an "evolutionary capacitor." Acting normally as a buffering mechanism, Hsp90 allows the accumulation of genetic variation in developmental pathways. Under conditions of environmental stress, Hsp90 activity is diverted away from its normal targets, thus resulting in developmental, and thereby morphological, variation. Natural selection may then act on the revealed variation, which in turn may be assimilated such that it no longer requires compromised Hsp90 function to be expressed. Under this model, Hsp90 modulates "evolvability," or the propensity to produce higher fitness variants (Gibson and Wagner 2000), in response to the environment.

Responses to Rutherford and Lindquist ranged, perhaps predictably, from the laudatory to the indignant (see references in Bergman & Siegal 2003).

From the latter end of the spectrum, some commentators zeroed in on the final, conjectural sentence of the paper, that "[t]he use of Hsp90 as a capacitor for the conditional release of stores of hidden morphogenic variation may have been adaptive for particular lineages ...." This statement appeared to suggest that Hsp90 was selected for its role in modulating evolvability. Some authors justifiably responded by cautioning that such arguments about adaptive "mutator" genes were difficult to defend. However, it remains an open possibility that Hsp90's role as capacitor is adaptive, and not merely a necessary product of its (adaptive) physiological properties. For example, similar wariness (Partridge & Barton 2000) met the suggestion by True and Lindquist (2000) that evolvability properties might have been responsible for the fixation in populations of *Saccharomyces cerevisiae* of a form of the Sup35 gene that makes its encoded protein capable of forming the prion [PSI^+]. Wild-type Sup35 functions in translation termination; the prion form leads to read-through of stop codons and therefore extra amino acids at the ends of a number of proteins. It has recently been shown by mathematical modeling that the ability to form the prion, and therefore to temporarily reveal variation in normally untranslated regions, is more likely due to a positive effect on evolvability than to chance alone (Masel & Bergman 2003). Thus, adaptive modulation of evolvability should be maintained as a viable hypothesis for both [PSI^+] and Hsp90, and should receive further experimental attention.

Doubts about the evolutionary "reason" for its existence notwithstanding, the demonstration of Hsp90's unique mechanism for translating environmental disturbance into morphological variation has reinvigorated research into Waddington's notions of canalization and, especially, genetic assimilation. One important question to be addressed is whether the types of variation produced by compromising Hsp90's function are plausibly adaptive. That is, does Hsp90 impairment always create "monsters" whose deformities render them virtually unfit in all but the most exceptional of cases? This question has been answered in part by Queitsch et al. (2002), who exposed *Arabidopsis thaliana* to the Hsp90 inhibitor geldanamycin. The defects seen in these plants tended to be more subtle than those seen in flies; the observed minor adjustments in leaf shape or pigment deposition, for example, seem less likely to be maladaptive than gross malformations of fly wings or eyes. Perhaps certain milder Hsp90 challenges in flies would produce fewer monsters and more potential adaptations. In any event, the *Arabidopsis* experiments established that Hsp90's ability to modulate phenotypic variation in response to the environment is conserved across kingdoms, and that the variants produced are at least in some cases conceivably adaptive.

Computational Modeling of the Evolution of Complex Gene Networks

The Hsp90 studies established that a dedicated buffering mechanism could exist and might have important evolutionary consequences. However, recall from the early experiments of Waddington and others that a number of mutations have been shown to increase phenotypic variation, suggesting that robustness may be an inherent feature of wild-type developmental programs. Traditional population-genetic and quantitative-genetic models are insufficient for addressing the causes and consequences of inherent robustness because they do not explicitly consider the developmental process. A. Wagner (1996) sought to overcome this problem by incorporating a computational model of a regulatory gene network into evolutionary simulations. The network model is conceptually simple, though its dynamics are complex (see Box 16.1). Each of N genes encodes a transcription factor that is in principle capable of regulating the expression of each of the other genes as well as itself. Whereas such a network of transcription factors is still quite far removed from a complete description of development, which takes place in both space and time and involves a diverse array of signaling and effector molecules, it certainly represents a type of regulatory structure that is prominent in developing biological systems (A. Wagner 1996; Davidson 2001; Wray et al. 2003).

It is useful at this point to consider the essential elements of a population-genetic model, and how these elements are represented in a model that explicitly includes a regulatory gene network. First, one must specify genotypes. In a standard multilocus population-genetic model, the genotype is notoriously arbitrary (e.g., $A_1A_2B_1B_2\ldots$). In a network model such as A. Wagner's (1996), the genotype is represented by all possible pair-wise interactions between the transcription factors and the genes encoding them, and includes the magnitudes of these interactions and their signs (negative interactions are

BOX **16.1.** Computational Modeling of the Evolution of Gene Regulatory Networks

As described in Stern (Ch. 15 of this volume), gene regulatory networks are central contributors to the mapping of genotype to phenotype. Thus, efforts have been made to incorporate gene-network representations into evolutionary models. Here we describe one such model, introduced by A. Wagner (1996), in some mathematical detail, for the benefit of those seeking a nuts-and-bolts understanding of the numerical simulations reviewed by us in this chapter.

Defining the Genotype

In the A. Wagner (1996) model, the regulatory network is envisioned as a set of N genes, each of which encodes a transcription factor. Each transcription factor is in principle capable of regulating the expression of each gene (including its own), via *cis*-regulatory sequences upstream of each gene's coding sequence (Figure 1). The effect of a transcription factor on a gene can be positive (activating), negative (repressing) or zero. The genotype is thus defined as an $N \times N$ matrix, W, whose elements, w_{ij}, denote the effect on gene i of the product of gene j. When a simulation is performed, certain w_{ij} elements are fixed at 0 and the rest are chosen randomly (e.g., from the standard Normal distribution). The proportion, c, of nonzero elements is a measure of the complexity of the network (its interconnectedness). Only regulatory, no protein-coding, mutations are allowed. That is, mutations are modeled as random replacements of individual nonzero w_{ij} elements.

Defining the Phenotype

In the model, factors regulating the same gene combine in a concentration-dependent, nonlinear way to influence the gene's expression level; when the net influence on a gene

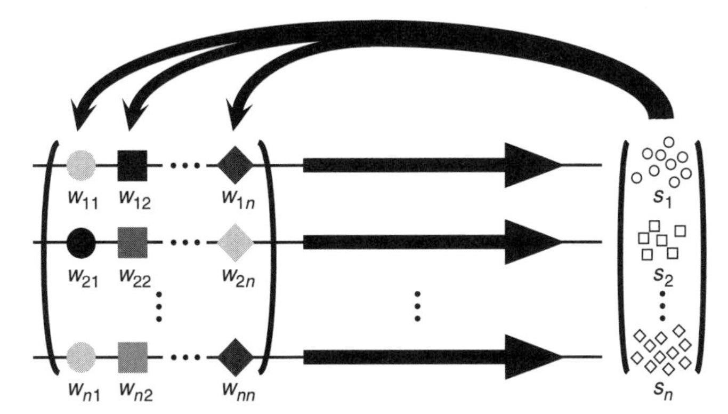

FIGURE **1.** Representation of a gene network. Each gene (horizontal arrow) produces a transcription factor (open circles, squares or diamonds). These factors influence the expression of each gene via upstream *cis*-regulatory elements (filled circles, squares or diamonds, shown with different gray levels to reflect inherent differences in affinity and in whether the interaction is activating or repressing). Genotype is represented as the matrix, W, of regulatory interactions. The regulatory interactions are iterated from an arbitrary starting point of gene-product levels, represented by the vector $S(0)$. Phenotype is the vector, $\hat{S}$, of gene-product levels at equilibrium.

(continued)

BOX 16.1. *(cont.)*

is positive, the gene is activated ("on") and when the net influence is negative, the gene is repressed ("off"). The system of interactions is iterated from an arbitrary starting point until it reaches a state of gene-expression equilibrium; an individual's equilibrium gene-expression profile is considered its phenotype. Specifically, each network is a dynamical system, with state vector $S(t) := (s_1(t),...,s_N(t))$ representing the expression levels of each gene at time t. The initial state, $S(0)$, is set by randomly choosing each $s_i = -1$ or $+1$, i.e., off or on, respectively. It is assumed that mRNA transcripts and their corresponding protein products are directly proportional in concentration (i.e., there is no post-transcriptional regulation), so $S(t)$ can be considered as either transcript or protein concentrations. The dynamics of $S(t)$ are modeled by the set of nonlinear coupled difference equations

$$s_i(t+1) = f\left(\sum_{j=1}^{N} w_{ij} s_j(t)\right), \tag{1}$$

where $f(x)$ is the sign function (−1 when x is negative, 0 when x is 0, +1 when x is positive). We (Siegal & Bergman 2002) generalized this step function to the sigmoidal function $f(x) = [2/(1 + e^{-ax}) - 1]$, in which a controls the rate of change from a state of repression to one of activation (the sign function is the special case when $a = \infty$). Equation. 1 is iterated a fixed number of times or until equilibrium. When $f(x)$ is the sign function, the equilibrium state, $\hat{S}$, is reached when $S(t) = S(t+1)$. When $f(x)$ is sigmoidal, the equilibrium state is defined to be reached when a measure analogous to a variance,

$$\psi(S(t)) = \frac{1}{\tau} \sum_{\theta=t-\tau}^{t} D(S(\theta), \overline{S}(t)),$$

is smaller than a threshold, ε, where $D(S^U, S^V) = (1/4N)\sum_{i=1}^{N}(S_i^U - S_i^V)^2$ and $\overline{S}(t)$ is the average of expression levels in the time interval $(t - \tau, \ldots, t)$. When this equilibrium criterion is satisfied, $\hat{S} = \overline{S}(t)$. If the system does not reach steady state within a predetermined large number of iterations, it is considered unstable.

Defining Fitness

The evolutionary model considers a finite population of constant size, M. All M individuals in the starting population are identical copies of a randomly generated founder individual. A. Wagner (1996) chose the founder individual by randomly selecting a $S(0)$ and $\hat{S}$ pair, then generating random W matrices until one was found that generated the chosen $\hat{S}$ from the chosen $S(0)$. We (Siegal & Bergman 2002) dispensed with this step, and instead merely chose a random $S(0)$ and a random W, then tested that the dynamical system so defined reached a steady state. In either case, the phenotype of the founder individual is arbitrarily defined to be the optimal one, S^{OPT}. The fitness of any other individual is evaluated as a function of the distance of its equilibrium state to the optimum:

$$F(\hat{S}) = e^{-\frac{D(\hat{S}, S^{OPT})}{\sigma}},$$

BOX 16.1. *(cont.)*

where σ determines the selection strength. That is, fitness drops off exponentially with increasing distance to the optimum (i.e., with increasing dissimilarity of the equilibrium gene-expression profiles). Note that when the sigmoidal parameter $a = \infty$, D becomes the so-called Hamming distance, the proportion of elements that are different between the vectors $\hat{S}$ and S^{OPT}. Individuals that do not have an equilibrium gene-expression state are assigned low fitness. In the A. Wagner (1996) model, such individuals have fitness corresponding to $D = 1$. In our modification of the model (Siegal & Bergman 2002), such individuals have fitness 0.

Evolutionary Simulations

The evolutionary model considers haploid individuals, pairs of which randomly mate to form a transient diploid that produces a single progeny consisting of randomly assorted rows of the W matrix (and perhaps one or more mutated w_{ij} elements). That is, there is free recombination among genes but complete linkage of cis-regulatory elements with their downstream coding sequence. A. Wagner (1996) also modeled the case of no recombination among genes, but found this had little effect on his results. An offspring is included as a member of the next generation if its fitness is greater than a random number generated from the Uniform distribution on the interval (0, 1). Note that in our simulations (Siegal & Bergman 2002), individuals with no equilibrium state never survived, whereas in those of A. Wagner (1996), such unstable individuals had fitness of approximately 5×10^{-5}, and therefore on rare occasion survived to the next generation. In each generation, new pairs of parents are chosen to produce an offspring until the number of surviving offspring equals M. Simulations are typically continued for hundreds of generations.

For more details see A. Wagner (1996) and Siegal and Bergman (2002).

repressing, positive interactions are activating). Mutation in the model occurs by random replacement of one of these interaction coefficients with a new value. Second, one must specify the relationship, or "mapping," between genotypes and phenotypes. Again, classical models have had arbitrary mappings, chosen as much to enable mathematical analysis as to reflect genetic reality. In the network model, phenotype emerges as the equilibrium gene-expression state of the network, arrived at by iterating the interactions between genes and transcription factors as specified by the genotype. Whereas this definition of phenotype clearly falls short of an actual morphological, physiological, or behavioral phenotype, it is reasonable to assume that differences in the expression states of key regulatory genes will somehow affect downstream differentiation programs so as to lead to actual phenotypic differences. Third, one must specify the relationship between an individual's phenotype and its expected reproductive success, or fitness. Note that in many traditional models, genotypes are mapped directly to fitnesses, with no intermediate specification of phenotype. This is natural, as the mappings are arbitrary anyway. In the network model, we specify an optimum equilibrium gene-expression profile (i.e., an optimum phenotype), and define fitness such that it decreases the more an individual's phenotype differs from this optimum. The strength of selection can be modulated by changing how quickly fitness drops off with increasing departure from the optimum phenotype—a steep curve amounts to strong selection, whereas shallow or flat curves amount to weak or absent selection, respectively. With genotypes, phenotypes, and fitnesses defined in the network model, it is nearly complete. All that remain to be specified are the population size, the mutation rate, and the mating regime, and then evolutionary simulations can be straightforwardly performed by computer.

A. Wagner (1996) used his network model to show that canalization (phenotypic insensitivity to mutation) evolved in randomly mating, finite populations subject to stabilizing selection, that is, selection for an optimum equilibrium gene-expression profile. This result was consistent with arguments going back to Waddington that such selection should favor canalization, and importantly showed that transcription-factor networks could in principle be altered to become less sensitive to mutation. However, there were still some lingering doubts that stabilizing selection was the ubiquitous cause of canalization. For instance, as mentioned previously G. Wagner et al. (1997), using a population-genetic model, had shown that if stabilizing selection is too strong, it cannot cause canalization to evolve; strong selection purges the variation that would itself provide the impetus for evolving buffering. So we (Siegal & Bergman 2002) reinvestigated A. Wagner's model and reached the surprising conclusion that stabilizing selection was not responsible for the evolved canalization, but that another type of selection was. In this case study, we discuss this work and subsequent work (Bergman & Siegal 2003), which links theoretical investigations to the evolutionary capacitance phenomenon introduced by Rutherford and Lindquist (1998).

Genetic Canalization Evolves in Simulated Gene Networks Without Stabilizing Selection

We (Siegal & Bergman 2002) built upon the model of A. Wagner (1996) to investigate the necessity of stabilizing selection in the evolution of genetic canalization. A. Wagner (1996) incorporated two forms of natural selection into his model, but did not examine each separately. That is, individuals were not only subject to phenotypic selection on their equilibrium gene-expression profiles, but also were subject to selection for reaching gene-expression equilibrium in the first place; the strength of each selection was set by the same parameter in the model, with individuals whose gene networks failed to reach equilibrium being assigned a fitness equal to that corresponding to the greatest possible distance from the optimum phenotype. We (Siegal & Bergman 2002) altered the model to decouple the intensity of stabilizing selection from the intensity of selection for gene-expression equilibrium. More specifically, we assigned fitness 0 (lethality) to "developmentally unstable" individuals (those whose gene networks failed to reach a stable equilibrium state) and then examined the consequences of varying the strength of phenotypic selection.

Under strong stabilizing selection, we (Siegal & Bergman 2002) observed results consistent with those of A. Wagner (1996). As shown in a typical simulation run (Figure 16.3A), phenotypic sensitivity to mutation decreased over the course of a few hundred generations. However, when the intensity of stabilizing selection was reduced by an order of magnitude, the effect on the trajectory of mutational sensitivity was imperceptible (Figure 16.3B). The same was true when stabilizing selection was eliminated, that is, when any phenotype was acceptable as long as it was developmentally stable (Figure 16.3C). We (Siegal & Bergman 2002) thus concluded that stabilizing selection was not necessary to evolve genetic canalization. Selection for the

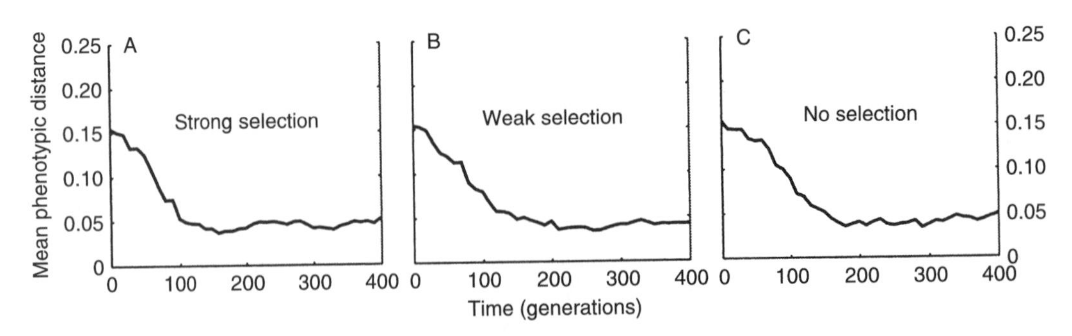

FIGURE 16.3. Typical time course of sensitivity to mutation (mean phenotypic distance of each individual in the population to phenotypes produced by single mutations of its genotype). The same founder individual evolves under strong stabilizing selection (A), intermediate (weak) stabilizing selection (B), and no stabilizing selection (C).

viability of individuals, not their particular phenotypic state, drove the evolution of canalization (Siegal & Bergman 2002). It is yet to be determined how genetic canalization emerges as a property of gene networks selected to have a stable equilibrium state, but further extensions and further analysis of the A. Wagner (1996) model are likely to yield clues, as are experimental attempts to understand the genetic basis and phenotypic consequences of developmental stability (Debat & David 2001).

The Revelation of Canalized Genetic Variation by Gene Knockout

Recall that one of the motivations for modeling abstract transcription-factor networks—as opposed to explicit, dedicated buffering mechanisms—was that a number of experiments through the years have demonstrated greater phenotypic variation in mutant than in wild-type strains. This suggests that buffering of minor perturbations is somehow inherent to wild-type organisms, and that major perturbations (i.e., mutations of large effect) in general disrupt this buffering. We have seen above that genetic canalization evolves in simulated gene networks. Do loss-of-function ("knockout") mutations in network genes reveal previously buffered variation?

We (Bergman & Siegal 2003) addressed this question using the same modeling framework described above. We founded simulated populations with 500 clones of an arbitrary, developmentally stable individual, then carried out 400 generations of mutation, reproduction, and selection. Selection was both for developmental stability (gene-expression equilibrium) and for the phenotype of the founder individual. Due to this strong stabilizing selection, the phenotypic distribution of a typical evolved population was very narrow (Figure 16.4). However, the population harbored a great deal of genetic variation, which is evidence that the gene network was indeed canalized. Importantly, this variation was revealed phenotypically when individual network genes were knocked out in turn, by zeroing interaction coefficients for a given gene and its encoded transcription factor (Figure 16.4). Out of 1000 comparisons of a wild-type evolved population with a derivative population in which each member had the same gene knocked out, only 4 showed higher phenotypic variation in the wild-type population. This is strong evidence that gene knockouts tend to reveal variation in interacting genes in these networks. It also suggests that there need not be anything "special" about either the knocked-out gene or the structure of the network: nearly all knockouts increased phenotypic variation. That said, it is important to note that the proportion of genetic backgrounds on which a gene knockout was lethal (i.e., caused the network not to reach a steady state) depended on which gene was knocked out, as did the exact distribution of nonlethal phenotypes. It is also noteworthy that the most frequent phenotypic class among the knockouts was often that corresponding to the optimum phenotype, suggesting that some of the evolved networks are even robust to major-effect mutations. These details of the behavior of networks upon major perturbation need to be connected to meaningful descriptions of network structure, and should be an area of active investigation.

Increased Environmental Sensitivity upon Gene Knockout

As noted earlier, the connection between genetic and environmental canalization has important evolutionary consequences. G. Wagner et al. (1997) showed using a population-genetic model that selective pressure for environmental buffering of a phenotype increases as the strength of stabilizing selection on that phenotype increases, whereas selective pressure for genetic buffering increases initially with the strength of stabilizing selection but drops off when stabilizing selection becomes so strong as to purge virtually all genetic variation from the population. This is not to say that traits under strong stabilizing selection cannot be genetically canalized, but instead that they must be so for other reasons. One such reason was laid out in detail by Meiklejohn and Hartl (2002), who argued that selection for environmental buffering would be strong and ubiquitous, and that developmental mechanisms that buffered against environmental perturbation would buffer against mutation as well. We (Siegal & Bergman 2002) suggested another possibility: that genetic canalization might be an inevitable property of complex gene networks.

Because of the importance of considering the link between environmental and genetic buffering, we (Bergman & Siegal 2003) explored the effect of gene knockouts on the sensitivity to environmental "noise." Noise was modeled by randomly perturbing an individual's initial gene-expression state, then iterating the network of gene interactions to

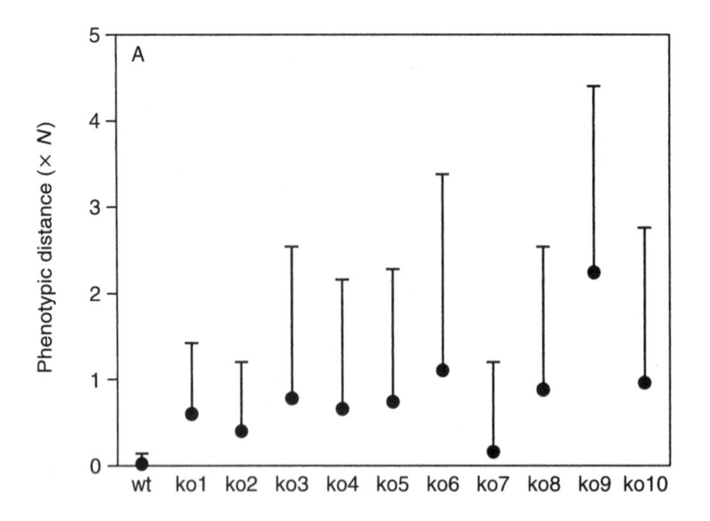

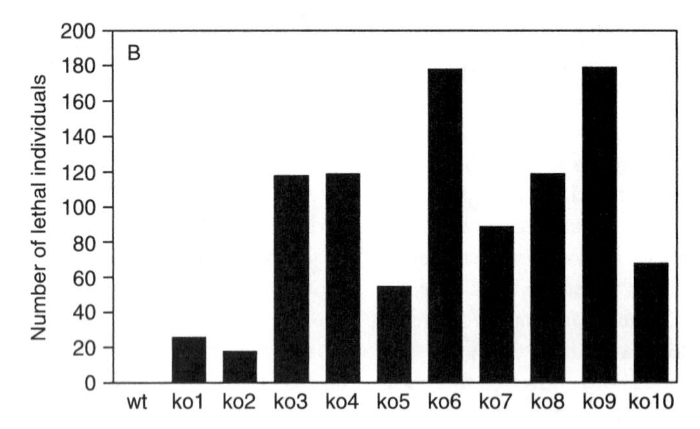

FIGURE 16.4. Greater phenotypic variation in single-gene knockouts than in the wild-type networks from which they derive. Shown are results of a typical simulation, in which a population of 500 individuals evolved for 400 generations under strong stabilizing selection. (A) Mean phenotypic distance to the selected optimum among viable individuals in the final wild-type population ("wt") and in derivative populations in which each individual has gene 1 knocked out ("ko1"), and so on. Phenotypic distances are multiplied by N, the number of genes in the network (10 for wild-type, 9 for knockouts). Error bars show the magnitude of the population standard deviation. (B) Number of lethal individuals in each population in (A). Data are from Bergman and Siegal (2003).

equilibrium as before. We found that phenotypic variance tended to be higher in networks with a single gene knocked out than in the wild-type networks from which they derived, and that this difference increased after a few hundred generations of the evolutionary simulations. Because the evolutionary simulations themselves did not incorporate noise, this is a clear demonstration that the evolution of one type of canalization (in this case, genetic) can in principle lead to the evolution of the other type (in this case, environmental). (Note that a similarly direct demonstration of the converse phenomenon, namely genetic canalization evolving as a result of the evolution of environmental canalization, would be difficult, in that any evolutionary simulation must include a mutation process, and mutations might provide selective pressure for genetic canalization independent of environmental variation.) This supports the argument for a single mode of canalization forwarded by Meiklejohn and Hartl (2002) and is consistent with the correlation between genetic and environmental canalization seen in experiments on life history traits in *Drosophila* by Stearns et al. (1995). It is also consistent with the possibility that complex, wildtype gene networks are inherently robust to both genetic and environmental perturbations, although clearly this robustness can be increased by selection against lethal disruptions in gene-expression equilibrium.

Evolutionary Capacitance and Evolvability in Gene Networks

Evolutionary capacitance requires, at a minimum, a mechanism for building up stores of genetic variation and a mechanism for revealing this variation phenotypically. As shown in *Drosophila* and *Arabidopsis*, Hsp90 clearly provides both these mechanisms (Rutherford & Lindquist 1998; Queitsch et al. 2002). However, to the extent that complex genetic networks are intrinsically canalized and respond to loss-of-function mutations by revealing normally buffered variation (Siegal & Bergman 2002; Bergman & Siegal 2003), they too provide both mechanisms. Hsp90 has two other attractive features as a candidate evolutionary capacitor: its pleiotropic effects on important developmental processes, and the sensitivity to environmental conditions of its buffering capacity. Thus, Hsp90 not only buffers variation likely to have morphogenetic consequences but it fails to buffer this variation in times of stress when alternative phenotypes may be advantageous (Rutherford & Lindquist 1998). Regulatory genes, as modeled by us (Siegal & Bergman 2002; Bergman & Siegal 2003), may each have more limited effects than Hsp90, but collectively are likely to affect a range of developmental traits as broad as, or broader than, that affected by Hsp90. Some may even individually approach the level of pleiotropy of Hsp90, as core genetic pathways tend to be reused at different developmental stages and in different developing tissues. This leaves the question of whether an evolutionary capacitor needs to respond to the environment to be effective, or more pointedly, whether unconditional loss-of-function mutations affect evolutionary trajectories by revealing otherwise hidden variation in interacting genes.

To begin to address the issue of whether evolutionary capacitance requires a mechanism for conditional release of variation only at opportune times, we (Bergman & Siegal 2003) performed evolutionary simulations in which knockout mutations occurred at low frequency relative to mutations in *cis*-regulatory elements. Reversion to wild-type occurred at even lower frequency (whereas knockouts that are actual deletions of genes are unlikely ever to revert to wild-type, other types of mutations, such as transposable element insertions, are expected to be reversible). We asked whether the existence of this knockout mutation process affected adaptation to a new optimum phenotype. Specifically, we asked whether knockout mutations led to faster arrival at the new optimum. Founder populations were allowed to evolve with and without knockouts. For 400 generations, the populations adapted to one optimum phenotype, and then the optimum was shifted such that the equilibrium expression states of three network genes flipped from on to off or vice versa. The populations evolving with knockouts arrived at the new optimum significantly faster than those evolving without knockouts (Bergman & Siegal 2003). Therefore, knockouts can affect the evolutionary trajectory without being conditional on the environment that enforces the new optimum. This simulation result, if borne out in real organisms, would dramatically widen the scope of evolutionary capacitance. As we pointed out (Bergman & Siegal 2003), the existence of a large set of capacitors with effects different from each other and from those of Hsp90, along with the lax requirement for occasional but not necessarily conditional release of variation, would mean that much "hidden" variation would actually be useful to evolving populations and could fuel adaptations even under conditions that do not involve chronic environmental stress.

FUTURE DIRECTIONS

Canalization, since its introduction by Waddington (1940, 1942), has attracted both theoretical and experimental attention, as well as a fair share of controversy. It is surprising, then, that it has also fallen below the radar of some biologists. For example, a major modern monograph on development and evolution (Gerhart & Kirschner 1997), which includes otherwise cogent, extensive chapters on "developmental flexibility and robustness" and "evolution and evolvability," does not mention canalization at all. We hope that our chapter contributes to a re-appreciation of Waddington's work and legacy, and to an excitement about future investigations of the causes and consequences of developmental robustness.

We find particularly exciting the prospect of bringing experimental and computational approaches together to a greater degree. Whereas the modeling approach introduced by A. Wagner (1996), and extended by us (Siegal & Bergman 2002; Bergman & Siegal 2003), incorporates a rather abstract representation of a gene-regulatory network, the growing body of knowledge of actual regulatory networks makes it possible both to construct more realistic models and to test the predictions of the models empirically. For example, it should be possible to model explicitly Hsp90 and its interaction with specific signaling pathways, so as to compare the relative effects on robustness and evolutionary capacitance of the buffering afforded by Hsp90 and the buffering intrinsic to complex networks. Such a model might build on that of Nijhout et al. (2003), who simulated one such signaling pathway, the mitogen-activated protein kinase (MAPK) cascade. On the experimental side, it is now feasible to use genetic engineering to alter individual components of a known complex network, and thereby to test predictions about how aspects of network structure contribute to phenotypic robustness. A guiding example in this respect is the work of Little et al. (1999), who used knowledge of the regulatory circuit governing the phage λ lysis/lysogeny switch to test in vivo the robustness of the circuit to targeted changes in its components. It is also now feasible to perform a Waddingtonian artificial selection experiment with a microorganism such as yeast, and to trace in the selected lineages not only the phenotypic changes that occur but the expression changes in every gene, some of which must underlie the phenotypic evolution. Such an experiment could be used to seek evidence of the phenomena seen in our evolutionary simulations. For example, do loss-of-function mutations increase in frequency (and increase phenotypic variance) as a population adapts to a new selection regime? With regard to this question, it is compelling to note the synergy of mathematical and experimental approaches with regard to a similar issue, that of adaptive "mutator" genes in bacteria. It has been shown by simulation, and confirmed by experiments in *E. coli*, that spontaneously arising alleles that increase the mutation rate can speed adaptation to a new environment (reviewed in Foster 2000). The simulations show that this may happen even when the mutator allele is maintained at very low frequency. It will be interesting to see whether loss-of-function mutations that reveal variation obey similar dynamics.

Modern genomic technologies and databases are also expected to yield insights into the genetic architecture of robust developmental systems and the capacity of these systems to change through evolution. In particular, the advanced state of functional characterization of the *S. cerevisiae* genome holds great potential. Already, genome-scale analyses have produced important results. For instance, by correlating the relative fitness effect of knocking out a yeast gene with the presence or absence in the genome of a closely related paralogous gene, A. Wagner (2000) concluded that robustness to mutation is largely due to epistatic interactions between unrelated genes. The strength of this conclusion has been challenged by Gu et al. (2003), who analyzed a larger data set and asserted that over one quarter of genes whose knockouts have no discernible phenotypic effect are compensated by redundant duplicate genes. Still, this leaves a large number of genes that are buffered by network interactions, and suggests that yeast may be an excellent system for studying buffering in particular network components. Our own analysis of genome-wide expression data for yeast single-gene knockout strains showed that mutants do indeed tend to have greater variance in gene expression, as predicted by our modeling (Bergman & Siegal 2003). Our analysis also supported a connection between gene-expression variation and phenotypic variation, in that knockout strains with lower growth rates relative to the other strains tend to have higher variation in the expression levels of the remaining genes (Bergman & Siegal 2003). However, the available data that we used were from experiments replicated at most two times, and therefore it was impossible to tease out the contributions of individual genes or subnetworks to the trends observed in the overall data. The existence of these trends, though, suggests that highly replicated experiments would be immensely valuable, allowing high-resolution analysis of the propagation of noise through wild-type and knockout strains' networks as well as investigation of connections between variation in particular genes' expression and phenotypic variation.

A final question that deserves greater attention concerns the relationship between canalization and regulatory complexity. The origins of regulatory complexity are of interest in their own right, and further thinking on the connection between regulatory complexity and canalization should be guided by our understanding both of the mutational processes by which genetic systems grow and change and of the possible selective forces that shape a system's ultimate structure (Duboule & Wilkins 1998; Zuckerkandl 2001). Further experimental work is also needed, specifically to test our hypothesis that canalization might be an inevitable property of complex regulatory systems. Testing this hypothesis directly is challenging because doing so would require proving a negative: that nothing is required for a genetic system to be canalized besides its being sufficiently complex and functional. However, certain evidence could certainly cast doubt on the hypothesis. For example, a correlation between the level of robustness of a system and the intensity of selection on the phenotype it produces would contradict our contention that stabilizing selection is not a major determinant of canalization. A well-chosen complex trait in a laboratory microbe could be used to seek such a correlation, in an experiment monitoring the trait and its variance under different artificial selection regimes.

There is clearly much work to be done before we achieve a truly satisfying understanding of canalization, its causes and effects. It is quite remarkable, however, that a concept introduced before the double-helical structure of DNA was known remains useful in an era of complete-genome sequences and routine analyses of entire transcriptomes and proteomes. Indeed, because canalization occupies a unique place at the interface of evolutionary and developmental concerns, and of theoretical and empirical approaches, it is one of the most exciting subjects in modern biology.

SUGGESTIONS FOR FURTHER READING

In *The Strategy of the Genes*, Waddington (1957) provides his most thorough explanation of canalization and the epigenetic landscape. The book provides a fascinating glimpse into the thought processes of one of the twentieth century's most intriguing and forward-thinking biologists. Waddington (1957) also touches on the notion of genetic assimilation, in a mostly theoretical way. Waddington (1961) provides a much more direct account of genetic assimilation. He reviews experimental evidence for its occurrence, and also proposes genetic mechanisms for the phenomenon and discusses the links between assimilation and canalization. Scharloo (1991) offers an excellent, comprehensive review of experimental work by Waddington and others on canalization and genetic assimilation. Much controversy surrounded this work, and Scharloo does an outstanding job of framing the debate and presenting competing arguments. The initial publication of Rutherford and Lindquist's study of Hsp90 impairment in *Drosophila* also generated much controversy, and the review by Sangster et al. (2004) is a measured, well-reasoned exposition on the cellular role of Hsp90, its effect of buffering against genetic and epigenetic variation, and its potential role in phenotypic evolution.

Sangster TA, Lindquist S & C Queitsch 2004 Under cover: causes, effects and implications of Hsp90-mediated genetic capacitance. BioEssays 26:348–362.

Scharloo W 1991 Canalization: genetic and developmental aspects. Annu. Rev. Ecol. Syst. 22:65–93.

Waddington CH 1957 The Strategy of the Genes. George Allen & Unwin.

Waddington CH 1961 Genetic assimilation. Adv. Genet. 10:257–293.

17

Evolutionary Epigenetics

EVA JABLONKA
MARION J. LAMB

Until recently, the Modern Synthesis of evolution forged in the 1930s and 1940s was the almost unquestioned basis of evolutionary thinking. At its core were two assumptions about variation. The first was that inherited differences between individuals are the result of genetic differences. The second was that the generation of new genetic variation is blind to function: variations do not arise as adaptive responses to environmental conditions. Ernst Mayr, one of the founders of the Synthesis, insisted repeatedly that inheritance is not "soft"—the hereditary material cannot be modified by environmental influences on the body, nor can it be changed by the use and disuse of body-parts (e.g., see Mayr 1982, p. 552). Lamarckism was totally rejected.

Molecular genetics at first hardened the assumption that the only variation that is relevant for evolutionary change is genetic variation. Evolution was seen in terms of changes in DNA, with new variants arising through chance alterations in DNA sequences. However, developments that began in the mid-1970s have led some evolutionary biologists to question the adequacy of this view of hereditary variation. They recognize that heredity involves more than genes and DNA. Studies of human and animal cultures have shown that variations can be transmitted behaviorally and culturally, and cell biologists have shown that genetically identical cells can transmit different structural and functional properties to their descendants. In other words, sometimes variations in phenotypes, as well as genotypes, are inherited. Moreover, since the generation of both cellular and cultural phenotypes is strongly influenced by the environment, some inheritance is soft. At first this realization had little impact on evolutionary thinking, and its implications still remain controversial. To some people, incorporating soft, nongenetic inheritance seems like an attempt to revive discredited Lamarckian ideas. Nevertheless, once it is accepted that heritable, inducible, nongenetic (epigenetic) variations exist, their potential influence on Darwinian evolution cannot be denied.

The term "epigenetic inheritance" is used both for nongenetic inheritance in cell lineages and for the transmission of cultural and behavioral variations, but in this chapter we are going to focus on cell heredity. We will begin by looking at the historical roots of the subject, and outline four types of epigenetic inheritance. Then, using three examples, we will show that epigenetic variants are transmitted not just in lineages of cells within organisms but also in lineages of organisms. Finally, we will consider the direct and indirect effects of epigenetic inheritance in adaptive evolution and speciation.

HISTORICAL OVERVIEW

The British embryologist Conrad Waddington introduced the term "epigenetics" in 1947 to describe "the branch of biology which studies the causal interactions between genes and their products which bring the phenotype into being" (see Waddington 1975, p. 218). He was interested in the processes through which the genetic information in a fertilized egg is used in the production of the visible phenotype, and he understood development in terms of intricate networks of regulatory interactions. Waddington depicted this view through his well-known visual metaphor, the "epigenetic landscape," which represents the developmental path from the fertilized egg to adult tissues or organs as bifurcating and deepening valleys running down from a plateau. The course, slopes, and cross-sections of the valleys are controlled by genes and their interactions, and selection can shape each of these features of the landscape. For example, it may deepen or broaden valleys, leading, respectively, to enhanced canalization (increased resistance to genetic and/or environmental perturbations) or enhanced plasticity (increased responsiveness to environmental factors). These ideas and the modern versions of them are discussed more fully in Siegal and Bergman (Ch. 16 of this volume).

"Epigenetics," the word that Waddington coined to capture his view of development, was derived from the much older term "epigenesis," which describes the influential Aristotelian theory that embryonic development involves the parts of the animal forming gradually and sequentially from an amorphous and unorganized egg. "Epigenetics" was clearly related to the old word, but since it incorporated the more modern term "genetics," it also pointed to the central role of genes in development. The "epi" part of the word, which means "upon" or "over," emphasized the need to study processes that go beyond the gene. Although Waddington's term did not catch on at first, its fortunes have revived in recent years, and it is now an essential part of the jargon of developmental biology. Its meaning has, however, drifted and diverged: "epigenetics" is still used in the original, Waddingtonian, general sense, but probably because the term incorporates the word "genetics," it is also now often used as a synonym for "epigenetic inheritance," which is a far narrower concept (Jablonka & Lamb 2002).

The study of epigenetic inheritance began in the last quarter of the twentieth century. About 20 years earlier it had become clear that the different cell types within an organism usually have identical genotypes, so the key to understanding differentiation was to discover which genes are expressed in a cell and what controls when and where they are expressed. One aspect of the problem that at first received little attention was how, once a cell becomes committed to a particular form or function, it transmits this determined state to daughter cells. A cell that is committed to becoming a liver cell, for example, "breeds true," and by division gives rise to a lineage of liver cells; similarly skin cells give rise to skin cells, kidney cells to kidney cells, and so on. One of the most illuminating examples of the way in which alternative states of gene expression are stably transmitted is found in female mammals (see Lyon 1999). Mary Lyon suggested in the early 1960s that although males have one X chromosome and females have two, they have the same effective dose of X-linked genes because in females one of the two X chromosomes in each cell becomes inactive in the early embryo. On the basis of genetic evidence, she argued that which of the Xs is inactivated is a random process—in some cells it is the chromosome inherited from the mother, in others it is that inherited from the father—but once inactivated, the state of inactivity is very stable. It is faithfully transmitted to all daughter cells. Such stable transmission of alternative phenotypes in cells with identical genotypes became known as "cellular heredity," "cell heredity," or "epigenetic inheritance," and investigations of its molecular basis began in the late 1970s. One of the pioneers in the field was Robin Holliday (see Holliday 1996), who not only recognized the importance of epigenetic inheritance but also suggested one of the mechanisms that might be involved.

At first it was assumed that epigenetic inheritance was significant only in development, but in the 1980s a surge of studies of genomic imprinting pushed the transmission of epigenetic variants from parents to offspring into the limelight. Genomic imprinting refers to situations in which the activity of a gene or chromosome depends on its parental origin. In some way, epigenetic modifications made to chromosomes in the parents affect gene expression in their offspring. Certain other curious non-Mendelian hereditary phenomena were also found to have an epigenetic basis (Urnov & Wolffe 2001). For example, the strange behavior of Barbara McClintock's jumping genes was found to be associated with epigenetic modifications. Similarly, paramutation, in which one allele in a heterozygote alters the heritable properties of the other allele, was shown to involve heritable epigenetic changes. Such studies led to the recognition that epigenetic information is transmitted not just between generations of cells but also between generations of individuals. As a result, biologists began to realize that epigenetic inheritance must be significant in evolution (Jablonka & Lamb 1989, 1995).

CONCEPTS

Epigenetic Inheritance in Cell Lineages

There is no single, well-defined mechanism of epigenetic inheritance. Instead, we have to think in terms of a whole spectrum of molecular processes that maintain different cell phenotypes and transmit them to daughter cells. Initially, three broad types of epigenetic inheritance system (EIS) were recognized: self-sustaining feedback loops, structural inheritance, and chromatin marking (Jablonka & Lamb 1995). Recently, a fourth type of EIS, that involving RNA-mediated gene silencing, has been added to the list (Jablonka & Lamb 2005).

The first type of EIS, self-sustaining feedback loops, is ubiquitous. At its simplest, it involves a gene positively regulating its own transcription. Once a gene has been turned on by some external inducing stimulus, the gene's own product induces its continued activity. Provided the concentration of the gene product within the cell is high enough and cell division is more or less equal, the gene's active state will be maintained after cell division in both daughter cells. The inheritance of the gene's state of activity is thus an automatic result of cell division. Consequently, if their inductive histories have been different, two genetically identical lineages of cells in the same environment can have different heritable states, one with the gene "on" and the other with it "off." Of course, most self-sustaining systems are more complicated than the one just described, with several genes and gene products involved in the feedback loop, but the regulatory principle is the same (Thieffry & Sánchez 2002).

The second type of EIS involves structural inheritance, in which pre-existing cellular structures act as templates for the production of new daughter structures (Grimes & Aufderheide 1991). Some types of cell membrane, for example, are formed only if similarly organized membranes are already present. Tom Cavalier-Smith (2004) refers to these as "genetic membranes," because they are part of a cell's hereditary endowment and are always transmitted to daughter cells. Structural inheritance is also seen with the reproduction of prions, the infectious proteins that are associated with degenerative diseases of the nervous system such as BSE (mad cow disease) and CJD (Creutzfeldt–Jakob disease) in humans. Once prions are present in a cell, they act as a kind of template that converts the normal form of the protein into their own abnormal conformation (Prusiner 1998). When cells divide, prions are automatically inherited by daughter cells.

The third type of EIS, chromatin marking, is based on the inheritance of DNA-associated molecules which can be proteins, RNAs, or small chemical groups attached to the bases. Collectively they are referred to as "chromatin marks." The chromatin marks associated with a gene's DNA sequence affect where and when it is transcribed. When DNA replicates, these chromatin marks are also reproduced. A relatively well researched example is the inheritance of patterns of cytosine methylation. The cytosines (C) of DNA sometimes carry a methyl group (C^m). This has no effect on its coding properties, but the presence of methyl groups does affect the likelihood of a DNA sequence being transcribed. Typically, highly methylated genes are inactive, whereas active genes have low methylation levels. During development, methylation levels are changed, so the same gene may have different patterns of methylation (marks) in different tissues. These alternative patterns can be transmitted to daughter cells. They are transmitted because methylation occurs symmetrically, usually in CG doublets (or in plants in CNG triplets, where N can be any base) in which C^mG on one strand is paired with GC^m on the other. Following semiconservative DNA replication, the old strand retains its methylated state (C^mG) but the complementary sequence on the new strand is unmethylated. These hemimethylated sites are then recognized by an enzyme that preferentially methylates the new strand, thus reconstructing the original pattern in the daughter molecules (see Holliday 1996).

There are still many gaps in our knowledge of how methylation marks are generated and reproduced (Bird 2002). How other types of marks, such as those involving patterns of histone modifications or different types of nonhistone proteins, are reproduced is even less well understood, although there are ideas about how it may happen (Henikoff et al. 2004). DNA methylation and histone modifications seem to be functionally related; for example, DNA methylation is associated with histone methylation and deacetylation (Weissmann & Lyko 2003). Nevertheless, there is still a lot to be found out about the way chromatin marks are established and reproduced. What is not in doubt, however, is that epialleles—the alternative marks that can occur on a DNA sequence—can be inherited for many cell generations.

The fourth type of EIS, which seems to be present in all eukaryotes, occurs through a recently discovered mechanism known as RNA-mediated gene silencing, or RNA interference (RNAi) (Mello & Conte 2004). The system is based on the silencing effect of small RNA molecules that originate from much larger RNA transcripts. These larger RNA molecules, which have certain topological peculiarities (usually, but not always, a double-stranded structure), are recognized by an enzyme that chops them into small RNAs. In some cases, these small RNAs are replicated, and copies of them are transmitted to daughter cells when the cell divides. Copies may also move horizontally from cell to cell (including to germ cells), so they behave like infectious agents or chemical signals. Crucially, the small RNAs associate with proteins to form complexes that silence or destroy copies of the large transcript from which they were derived and any other similar-enough RNAs. Sometimes they also prevent transcription of the gene that mothered their production by forming a stable chromatin mark (often a methylation mark), which can be transmitted to the future cell generations. The result of the activities of small RNAs is that genetically identical cells in the same environment may have different patterns of silent genes, depending on their history of RNA-mediated silencing.

All of the EISs that we have described show that history matters: what is inherited and passed on to the next generation depends on the conditions that a cell and its ancestors experienced. Although we have described the four types of EIS as if they were independent of each other, they are all interrelated and interact in various ways. For example, RNA interference may be closely associated with DNA methylation, and some chromatin marks may be generated through structural templating processes. The molecular details of all these processes have not yet been unraveled, but what is clear is that through the various interacting EISs, a package of acquired information is transmitted to daughter cells.

Transgenerational Epigenetic Inheritance

All organisms, including single-celled organisms, seem to have EISs. DNA methylation and other types of chromatin mark have been found in protozoans, and the ciliates provide some of the best examples of structural inheritance (Grimes & Aufderheide 1991). Even bacteria transmit epigenetic information through gene silencing and other mechanisms (Casadesús & D'Ari 2002).

Since EISs are found in all organisms, they were probably present well before the advent of multicellular organisms. One of the intriguing questions, therefore, is why early single-celled organisms should have evolved EISs at all. Of what advantage was cell memory to them? One possibility, suggested by Nanney (1960), is that epigenetic variants are beneficial in frequently and erratically changing conditions, because they allow more rapid adaptive adjustment than would be possible through genetic change. However, Lachmann and Jablonka (1996) believe that environments that fluctuate regularly (with the lunar cycle or with the seasons, for example) may have been even more important in promoting the evolution of reliable EISs in unicells. Their mathematical models show that in environments that cycle every 2–100 generations, natural selection should adjust the organisms' rate of switching between alternative adaptive states to the length of the different phases in the cycle. If organisms go through only a few generations in each phase, they suggest it is the epigenetic systems, with their high rates of variation production and reversion, that are most likely to be honed by natural selection into the efficient memory systems that would enable regular switching. Although switching every few generations by means of a DNA change is possible (and is known to occur in some pathogens), the adaptive flexibility of epigenetic regulatory systems makes them the prime targets for molding into the memory systems of unicellular organisms living in regularly cycling environments.

In complex multicellular organisms, EISs are essential for transmitting determined cell states during development, but, in addition, inherited epigenetic variants can in theory be the basis of adaptive evolutionary change. This is especially true for organisms that reproduce by asexual fragmentation or budding. If in response to changed environmental conditions a new heritable epigenetic variant arises in a tissue that will form part of a daughter individual, the acquired change may be inherited; that is, Lamarckian evolution may occur. Even if new variants arise accidentally, they can still be the basis of evolutionary adaptation, because Darwinian selection will occur between genetically identical but epigenetically different individuals in exactly the same way as between those differing genetically.

What about epigenetic variations in multicellular organisms that begin development from a single cell? At first glance it seems unlikely that they could be the basis of evolutionary change, because there are so many constraints on their transmission. For a new variant to be passed on, it has to be present in the germline and survive the massive changes that occur during meiosis and gamete formation. Moreover, the totipotency of the zygote depends on old epigenetic information being erased: development has to start from a clean slate. This must limit the range of epigenetic variations that can be transmitted by sexually reproducing organisms. Nevertheless, there is growing evidence that some epigenetic variations are transmitted between generations.

The first indications of this came from studies of genomic imprinting, a phenomenon that was recognized by Helen Crouse in 1960. She found that during the development of the dipteran fly *Sciara*, certain chromosomes that had been inherited from the male parent were regularly eliminated. For this to happen, she reasoned, the chromosomes must carry imprints that reflect their parental origin. Subsequently many other cases in which genes or chromosomes from the two parents behave differently were discovered in both animals and plants. They have been most intensely studied in mammals, where the expression of as many as 1% of all genes depends on whether they were inherited from the mother or the father.

Imprints are temporary, lasting only a single generation: in daughters the imprints inherited from the father will be replaced by female-specific marks before they are passed on, and in sons the mother's imprints will be replaced by male-specific marks. However, not all epigenetic marks are ephemeral. Some epialleles are transmitted to future generations in a way comparable with that of genetic alleles. There are now many examples of this, and in some cases the transmitted epialleles affect important and potentially selectable characters such as the sexual phenotype of plants (Janoušek et al. 1996), mouse coat color (Morgan et al. 1999), and various morphological traits. In what follows we will focus on three well-characterized examples of the latter type of transgenerational epigenetic inheritance: flower morphology in toadflax, tail morphology in the mouse, and eye morphology in the fruitfly *Drosophila*.

Case Study: Epigenetic Inheritance of Flower Morphology in Toadflax

Over 250 years ago, the great botanist Carl Linnaeus described a newly generated species of toadflax. This was remarkable, because fundamentally he believed that all species had been created by God at the beginning of the world, and had remained unchanged ever since. However, his system of naming and classifying plants was based on their reproductive parts, and the newly discovered variant had a floral structure that was very different from that of normal toadflax, *Linaria vulgaris*. The normal form is bilaterally symmetrical, with the upper and lower parts of the flower distinctly different, whereas the new variant was radially symmetrical, with five spurs instead of the single spur of the normal form (Figure 17.1). He named it "Peloria," which comes from the Greek for "monster."

Peloric variants of *Linaria* and other plants (e.g., *Antirrhinum*) fascinated many later biologists, including Goethe, Darwin, and de Vries (see Gustafsson 1979). Eventually, in the early days of genetics, it was recognized that the peloric form of *Linaria* is a Mendelian recessive character, not a new species, and in recent years botanists trying to understand the development of flower shape have investigated the molecular basis of peloric variants. Enrico Coen and his colleagues looked at *Lcyc*, the *Linaria* homolog of a gene known to control dorsoventral asymmetry in *Antirrhinum* (Cubas et al. 1999). Remarkably, they found that the DNA sequences of the normal and peloric forms were the same. What was different, however, was the pattern of methylation of the *Lcyc* gene. In the peloric variant the gene was heavily methylated and transcriptionally silent. Peloric strains are not totally stable, and occasionally branches with partially or even fully wild-type

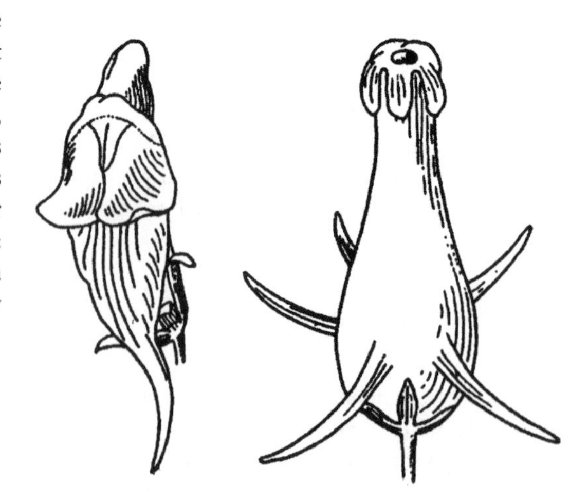

FIGURE 17.1. Goethe's illustrations of the normal form of *Linaria vulgaris* (left) and the peloric variant (right).

flowers develop on a peloric plant. In these the level of methylation of the *Lcyc* gene is intermediate and is correlated with the strength of the peloric phenotype.

Linnaeus's "Peloria" was therefore not a new species, nor was it the result of a mutation. It was caused by a heritable epigenetic modification—an epimutation. It is not known what initiated the methylation change. It could have been either an induced change or simply the result of developmental noise, but once the new pattern was established, it seems to have been transmitted for many generations. Over 200 years after Linnaeus's specimen was collected, the peloric variant of *Linaria* was still growing in the same locality, although there can be no certainty that it was a descendant of the ones growing there in Linnaeus's day.

Case Study: Epigenetic Inheritance in the Mouse: The Inheritance of Tail Morphology

The mouse mutant "Fused," which causes a kinked tail phenotype, was described in 1937. It has been maintained in the laboratory ever since. Fused is dominant, but its expression is very variable, with some individuals showing a very kinked tail, others having tails with only slight kinks, and some having a completely normal tail. In early studies it was noted that the degree of expression depends on the sex of the parent from which the Fused allele was inherited (i.e., it is imprinted), and also on how strongly it was expressed in the parent. Consequently, more than 20 years ago, a group of biologists in the former USSR concluded that the patterns of inheritance seen with Fused could be the manifestations of epigenetic, rather than purely genetic phenomena (Belyaev et al. 1981a). Molecular studies have now shown that they were right.

Using isogenic lines with very little genetic variability, so that differences between individuals could not be attributed to unidentified "modifier" genes, Rakyan et al. (2003) confirmed that the degree of expression of Fused is inherited through both male and female parents (Figure 17.2). When the tail phenotype is strongly expressed in the parent, the offspring are more likely to have kinked tails than when it is expressed weakly or not at all. The Fused allele (now known as $Axin^{Fu}$) has a transposon-derived sequence in one of the introns of the *Axin* gene, and Rakyan and colleagues found that the phenotypic expression of Fused is strongly correlated with the degree of methylation of this transposon. Heavy methylation leads to the development of a normal tail, whereas a demethylated transposon sequence leads to abnormal RNA transcripts and a kinked tail. The extent of tail kinkiness is negatively correlated with the extent of methylation, and since methylation marks tend to be inherited, the phenotype tends to be inherited too.

Transposon sequences are very common in mammalian genomes, and Fused is not the only

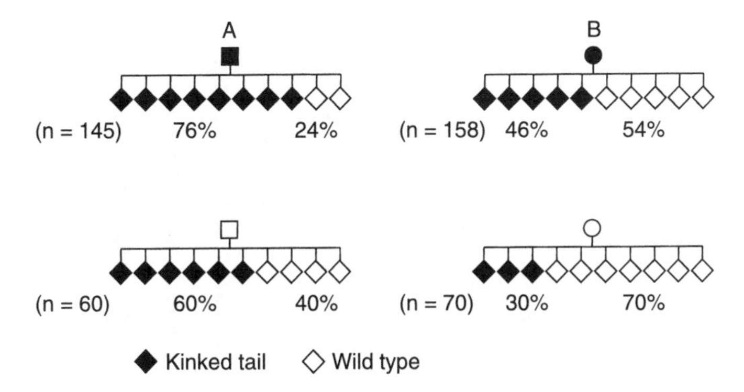

FIGURE 17.2. Epigenetic inheritance of tail morphology in the mouse. Mice that were all $Axin^{Fu}/+$ but either showed (filled symbols) or did not show (open symbols) the kinked tail phenotype were mated to +/+ animals. The proportions of kinked-tail and wild-type ($Axin^{Fu}/+$) offspring are shown. In (A), the male parents were $Axin^{Fu}/+$; in (B), the female parents were $Axin^{Fu}/+$; *n* is the total number of $Axin^{Fu}/+$ progeny from each type of cross. Redrawn from Rakyan et al. (2003).

example of a heritable variant that is associated with a transposon's level of methylation. Several workers in the field have suggested that such epiallelic variants may be quite common. One variant that is particularly interesting is the result of a transposon insertion in the agouti gene of the mouse. It causes the coat color to be yellow, but the expression of this allele (A^{vy}) is very variable: genetically identical mice can be yellow, yellow-agouti (mottled), or agouti (wild-type). The degree of expression tends to be inherited, and is correlated with the extent of methylation of the transposon sequence (Morgan et al. 1999). The reason this case is so interesting is that expression of the A^{vy} allele is influenced by the mother's diet. When pregnant females are fed with methyl-supplemented food, they have proportionally fewer yellow offspring and more agouti ones than mothers fed a normal diet (Waterland & Jirtle 2003). In other words, environmental conditions induce heritable changes in a gene's epigenetic marks that affect its expression in later generations.

Case Study: Epigenetic Inheritance of Eye Morphology in the Fruit Fly

Our final example of transgenerational epigenetic inheritance also illustrates how environmental conditions can have long-term effects on epigenetic marks and affect the variation that is available for selection. It comes from a study of selection in an isogenic strain of *Drosophila* that carried a mutant allele of the *Krüppel* gene (Sollars et al. 2003). Flies with this allele have abnormal, small and rough eyes, which in certain conditions are prone to form outgrowths. One way of increasing the production of outgrowths is to add geldanamycin, a drug that inhibits the activity of the heat-shock protein Hsp90, to the food on which they are raised. The researchers found that after adding the drug to the food for only one generation, six generations of selective breeding gradually increased the proportion of flies showing the anomaly from just over 1% to more than 60%, and it remained there until they ended the experiment at generation 13 (Figure 17.3). Since the strains used were isogenic, there were almost no genetic differences between the flies, so the response cannot have been the result of selecting genetic variations. The clue to the probable nature of the selected variations came from experiments in which the mothers of *Krüppel*-carrying flies had a defective copy of either the Hsp90 gene, or one of several genes that affect the maintenance and inheritance of chromatin structure. Some of the offspring of these mothers developed the eye outgrowth, even when they themselves did not inherit the defective

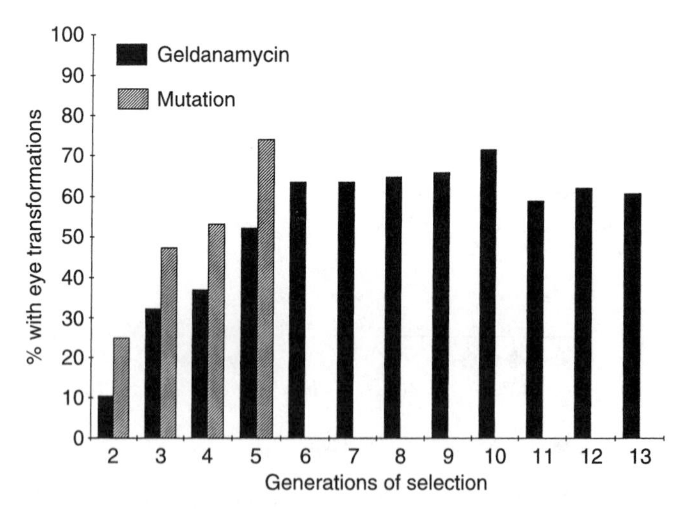

FIGURE 17.3. Selection for eye outgrowths in *Drosophila*. Grey bars, selection in a line in which the parental generation had a mutation (*vtd*^{3}) which affects chromatin structure; black bars, selection in an isogenic line in which the parental generation was raised on medium containing the drug geldanamycin. Note that the drug or *vtd*^{3} were present only in the parental generations, not during the subsequent generations of selection. Redrawn from Sollars et al. (2003).

Hsp90 or chromatin gene. As Figure 17.3 shows, selecting and breeding from these flies increased the proportion of offspring with the abnormality. From these and other results, Sollars and his colleagues concluded that the selectable variation in their isogenic lines must stem from new heritable differences in chromatin (epialleles), not from differences in gene sequences. The epialleles were probably heritable histone modifications, since inhibitors of histone deacetylation suppressed the development of outgrowths. When the newly induced chromatin modifications were transmitted to offspring, they affected when and where genes were expressed, and because the eye-development pathway had already been made unstable by *Krüppel*, they caused eye outgrowths.

How Epigenetic Inheritance Affects Evolution

The three examples that we have just described show that epigenetic variations can be transmitted between generations, and that they are selectable. But how abundant are they? And how important are they in adaptive evolution?

If we think about a single heritable cellular trait—a self-maintaining loop, one particular inherited three-dimensional structure, a chromatin mark, or an RNA-mediated state of silencing—the number of variant heritable states is usually small, sometimes only two. Nevertheless, when combinations of different active/inactive loops, cellular architectures, chromatin marks, and RNA-mediated silencing patterns are considered together, the amount of cell variation is vast. The evolutionary potential of EISs is therefore considerable.

EISs can affect adaptive evolution in very direct ways. First, as the *Drosophila* experiments that we have just described show, they provide an additional source of variation. This could be crucial if populations are small and lack genetic variability. Second, they can lead to very rapid adaptation, because most new epigenetic variants arise when conditions change, which is exactly the time when they can be most useful. Furthermore, when epigenetic variations are induced, many individuals in the population may acquire similar modifications at the same time, which will increase the rate of adaptation. Third, since epigenetic variations are usually more readily reversible than genetic variations, they have the advantage that little has been lost if they are selected when new conditions are short-lived, because the old variant can probably be re-established.

There is another way in which heritable epigenetic variations can affect evolution: they can bias genetic changes. They do so in two ways. One is by directly affecting the rate and type of genetic variation that occurs; the other is by influencing the selection of genetic variants.

The Effects of EISs on the Genetic System: Biasing Genetic Change

Genetic variation resulting from both mutation and recombination is affected by the structure of chromatin (Jablonka & Lamb 1995, Chapter 7). DNA sequences in condensed chromatin usually change less than those in more open, loose chromatin, which is associated with gene activity. The way chromatin marks influence DNA sequence variations can be seen most readily with transposition in plants. Demethylated transposable elements are usually very mobile, whereas when highly methylated the same elements rarely move (see Fedoroff & Botstein 1992). Since when transposable elements move to new locations they may introduce changes in coding or regulatory sequences, the marks they carry obviously affect their capacity to generate mutations. Methylation marks can also influence mutation rates more directly through their effects on point mutations: methylated cytosines are very much more prone to change to thymine than are nonmethylated cytosines (Yang et al. 1996).

The interplay between genetic and epigenetic variations is certainly complex. At present this is best seen in the evolution of cancer, where often the first sign of cellular abnormality is an epimutation (Jones & Baylin 2002). Commonly genetic changes follow the epigenetic ones, and sometimes they seem to be directly dependent upon them. For example, methylation of regulatory regions that are normally unmethylated may turn off one or more of the genes whose products are needed for DNA repair and cell division. As a result, damage and errors accumulate in DNA, and the control of cell division is disrupted. Since both the epimutations and the consequent genetic changes are inherited by daughter cells, the lineage accumulates more and more deleterious variations. According to this view of tumor evolution, genetic and epigenetic events interact, with epigenetic changes (such as increased methylation) leading to genetic changes, and genetic changes (such as mutations in genes coding for chromatin proteins) leading to further epigenetic changes. Simple models and computer simulations show that interactions between the epigenetic and genetic systems can also have profound effects on the rate of evolution when epigenetic marks are transmitted from parents to offspring (Jablonka & Lamb 1995, chapter 7).

Making Cryptic Genetic Variation Selectable

It has been known for many years that environmental stresses can alter epigenetic states and unmask variation that in normal conditions is cryptic. By altering patterns of gene activity, changed conditions upset developmental canalization, and previously hidden genetic differences between individuals are exposed and can be selected. Eventually, through selection, phenotypes that were initially produced only in the stressful conditions may become genetically fixed, no longer needing the environmental stimulus. Waddington (1961) called this process genetic assimilation. More recently, Rutherford and Lindquist (1998) have shown that reducing the amount of the molecular chaperone Hsp90 upsets canalization in *Drosophila*. Even a transient shortage of Hsp90 induces abnormal phenotypes such as eye and wing deformities. Moreover, Rutherford and Lindquist found that selecting and breeding from flies with these abnormalities led to partial genetic assimilation: the variant trait was produced even when Hsp90 levels were restored to normal, because the selected genes shifted development into a new pathway. Since the availability of Hsp90 and other molecular chaperones is influenced by environmental conditions, it has been suggested that these molecules may play a crucial role in revealing the hidden genetic variation that can be shaped into new adaptations when environments change (Sangster et al. 2004).

From both the old and new evidence it is clear that environmentally induced epigenetic changes can guide the selection of genetic variants. However, when induced epigenetic changes are themselves inherited, the effects on evolution can be even more interesting and significant. To understand why, we need to go back to the experiments of Sollar and his colleagues that we described earlier. Their results showed that inhibiting normal Hsp90 activity not only induces epigenetic changes that reveal genetic differences, but also that epigenetic changes are themselves heritable and can be selected. Therefore, even in the absence of genetic variation, adaptation can occur through the selection of heritable differences in chromatin structure. Although these epialleles may not be as stable as genetic alleles, adaptations based on such variation could do a "holding job" that would allow a population to survive until new mutations or combinations of mutations arise and make the adaptation more stable.

True and Lindquist's (2000) work with yeast strains carrying prions leads to a similar conclusion about epigenetic variants. They compared pairs of yeast strains differing only in whether or not they carried the prion form [PSI$^+$] of a protein that is involved in messenger RNA (mRNA) translation. By growing the pairs of strains in a variety of conditions, they found that the presence of the prion, an epigenetic structural variant, produced consistent, strain-specific effects on colony morphology and growth characteristics. The reason for this is that the normal form of the prion protein has a role in mRNA translation. It is needed for polypeptide chain termination. In its prion form, however, the molecules tend to aggregate, so it is not always available, and some stop codons in mRNA are ignored. Consequently, cells containing the [PSI$^+$] prion produce a variety of new protein products with potentially new functions (or malfunctions) because translation goes beyond the normal endpoint of functional genes, or because stop codons in the middle of nonfunctional genes are ignored. The result is that new phenotypes are produced without any change in DNA.

Since [PSI$^+$] is a prion, it is self-propagating. Daughter cells inherit prions, and with them the reduced fidelity of protein synthesis that leads to new phenotypes. In this way, the presence of the prion, an epigenetic variant, augments selectable variation by producing different heritable phenotypes from the same genome. This could enable a lineage to adapt and hold the adaptation until genetic changes take over. In other words, the heritable epigenetic variations could pave the way for genetic adaptation (True et al. 2004). Computer simulations have shown that the effects of such a prion system may lead to its evolutionary retention even if the response is adaptive only once in a million years (Masel & Bergman 2003). The variation generated through heritable epigenetic systems like this may be particularly important in asexual lineages, where the accumulation of mutational changes would be very slow.

The general conclusions that follow from the work described in this section are that when conditions change, epigenetic events can increase the rate of adaptive evolution (i) by revealing previously hidden genetic variation and making it available for selection, and (ii) by providing an additional type of selectable variation—heritable epigenetic variants. In addition, since epigenetic marks bias DNA sequence changes and usually increase the rate of change in active genes, heritable alterations in gene expression may also speed up evolution by increasing mutation and recombination in genes that are turned on by the new conditions. Together, the events initiated by an induced epigenetic change show that it is quite wrong to think of the environment as just a selector of heritable variation. The environment has a dual role in adaptive evolution—it does not just select among heritable phenotypic variations, it also induces them.

One of the most interesting long-term experiments in evolution—one that points to the dual role of the environment—was carried out by Dmitry Belyaev and his colleagues in the former USSR (Belyaev 1979; Belyaev et al. 1981b; Trut 1999). This is the same group that carried out some of the pioneering work on epigenetic inheritance using the mouse mutant Fused. For more than 40 years, they also selected for tameness in silver foxes, and managed to establish a population in which some animals were quite docile and dog-like in behavior. However, selection for tameness affected more than behavior. Within fewer than 20 generations the reproductive season of the females had become longer, the time of molting had changed, levels of stress hormones and sex hormones had altered, and several new morphological variations were present. The latter appeared quite early in the selection process, and although they affected only about 1% of the animals, they occurred repeatedly, often with more than one phenotypic change being found in the same animal. In addition, there were changes in the foxes' chromosomes, with some tame foxes having tiny additional heterochromatic microchromosomes.

The new phenotypes appeared too frequently for them to be the result of new mutations, and the mating scheme allowed very little inbreeding, so Belyaev rejected these genetic explanations of their presence. Instead, he interpreted them in what today we would call epigenetic terms. He believed that all animals have a large reservoir of "dormant" genes—genes that we would now describe as permanently inactivated. In stressful situations, he reasoned, some of these inactive genes become heritably active, and dramatically increase the amount of visible variability. What had happened with his silver foxes, he suggested, was that selection for domesticated behavior had altered their hormonal state. This in turn had affected chromatin structure, thereby activating normally silent genes and producing all the new phenotypes seen in the domesticated strain.

Belyaev's work with silver foxes and other mammals led him to conclude that in evolutionary theory the production of variation and its selection should not be treated as separate entities, because the environment that selects variation also contributes to its production by activating previously dormant genes. In other words, according to Belyaev, induced heritable epigenetic variations play an integral role in adaptive evolution.

The Role of EISs in the Evolution of Development

One of the most obvious and important roles of epigenetic inheritance is in the evolution of complex multicellular organisms. Reasonably reliable EISs were a precondition for the evolution of specialized cell lineages, because the cells in such lineages have to remember their determined state and transmit it to daughter cells, even when the conditions that initiated it are long past. However, effective EISs are not an advantage in all circumstances. In particular, acquiring and transmitting epigenetic information would usually be undesirable in potential germline cells, which need to retain or adopt an uncommitted state so that cells that give rise to the next generation are totipotent. Natural selection would be expected to favor mechanisms that counter the possibility of transmitting epigenetic variations through these cells.

There are at least three features of development that may be the outcome of selection to prevent cells with inappropriate epigenetic legacies from founding the next generation. First, it may be one of the evolutionary reasons why many epigenetic states are so difficult to reverse. As the problems that have been encountered when trying to clone animals make clear, somatic cells are very difficult to reprogram, so the chances that a rogue somatic cell will be able to become a germ cell and carry inappropriate epigenetic marks to the next generation are small. Second, the need for germ cells to be free from epigenetic legacies may be the evolutionary reason why in many animals primordial germ cells are separated off very early in embryogenesis and remain quiescent until sexual maturity. They have few epigenetic memories, and little chance of acquiring epimutations. Third, the massive changes that occur during meiosis and gamete production, when chromatin is restructured and the male gametes lose most of their cytoplasm, may, in part, have been an evolutionary result of selection against the transmission of epigenetic variations.

Although these features of development reduce the likelihood of transmitting irrelevant epigenetic information to the next generation, the evidence we outlined earlier shows that even after all the modifications that occur during gamete formation, some epigenetic information is nevertheless inherited. In fact, some differences—those we see as imprints—were probably originally incidental side-effects of the different ways that DNA is modified and packaged in male and female gametes. Usually such differences disappear during early development, but some persist and have phenotypic effects that can be exploited. Imprints seem to have evolved to have a variety of functions, depending on the taxon and trait studied (Hurst 1997). One function that has been found in several unrelated groups of insects is related to sex determination. Another is found in mammals, where parental imprints are sometimes significant in dosage compensation. In most mammalian cells, the X chromosome inherited from the father has the same chance of being inactivated as that inherited from the mother, but in marsupials the inactivated X is always that from the father; the same is true for cells in the extraembryonic tissues in other mammals (Lyon 1999).

A very influential idea about the way some organisms may have exploited parental imprints has been developed by Haig (see Haig 2002). He pointed out that when embryos are nourished by maternal tissues, as they are in mammals and flowering plants, there can be a conflict of interests between the two parents. Consider, for example, a pregnant female mammal that is carrying offspring from more than one father. She is equally interested in the welfare of all her offspring, and it is in her interests to invest equally in all of them. In contrast, it is in a father's best interests to help his own offspring to obtain as much nourishment as possible, even if they do so at the expense of their mother or their half-sibs. A father has no interest in the future welfare of the mother, because it is unlikely she will ever carry any more of his young, and his offspring's half-sibs have nothing whatsoever to do with him. Haig suggested that in such an asymmetrical situation, selection would favor any imprints on paternally derived chromosomes that make the embryos carrying them extract more than their fair share of nourishment from the mother. This will be countered, however, by selection for imprints on maternally derived chromosomes that overcome the paternal greedy-embryo strategy. What is therefore expected, and has been observed in mice and humans, is strong differential marks on paternal and maternal genes that are associated with embryonic growth. With some exceptions, paternal genes have growth-enhancing marks, whereas maternal genes have growth-suppressing marks.

Epigenetic Inheritance and Speciation

Darwin called the origin of species "the mystery of mysteries," and in spite of all the subsequent work, it remains a mystery. Recently, several biologists have concluded that some of the mystery could be dispelled by taking more notice of the epigenetic changes that might contribute to speciation.

West-Eberhard (2003) has outlined one simple and compelling way in which speciation could start from epigenetic rather than genetic processes. She argues that in polyphenic species, where individuals have alternative phenotypes that depend on the conditions in which they develop, speciation can be initiated when the expression of one phenotype is blocked. For example, several pierid butterflies have seasonal polyphenisms in which factors such as the rearing temperature determine which of the different morphs is produced. If the environment in part of their range changes, it might lead to only a single phenotype being produced in one subpopulation. Relaxed selection on the nonexpressed phenotype could then lead to genetic and epigenetic divergence of this subpopulation, and ultimately to its reproductive isolation. Notice that with this route to speciation, which depends on epigenetic switching between phenotypes, morphological divergence precedes reproductive isolation.

Most models of speciation assume that morphological divergence occurs after or at the same time as the development of reproductive isolation, and that speciation is initiated through a subpopulation becoming at least partially physically isolated. Although it is rarely considered, we believe that environmental and genomic stresses that affect heritable epigenetic variations may play a significant role in such speciation. Imagine, for example, that a few individuals arrive in a new region where they have to live in conditions to which they are not well adapted. This is likely to be both behaviorally and physiologically extremely stressful. As the experiments with *Drosophila* and silver foxes suggest, stress may induce heritable epigenetic changes that reveal selectable variation, and hence enhance the rate of adaptation to the new environment. Stress-induced epigenetic changes could also increase the activity of mobile elements and the number of mutations that they cause. The accumulation of such epigenetic and genetic changes could be the basis of new adaptations and the formation of a reproductive barrier.

In theory, epigenetic mechanisms alone could initiate reproductive isolation (Jablonka & Lamb 1995, Chapter 9). Nongenetic inherited behavioral differences could prevent mating taking place (Avital & Jablonka 2000), and differences in chromatin marks could result in hybrid offspring either failing to develop normally or being sterile, because normal development requires compatible marks on the two sets of parental chromosomes. We know this because when a mouse zygote is formed in the laboratory from two haploid nuclei from animals of the same sex, it does not complete development. For development to be normal, one set of chromosomes must carry female imprints and the other male ones. Incompatibility between the parental imprints is thought to be the reason for the abnormal development of hybrids between species in the rodent genus *Peromyscus* (Vrana et al. 1998). Disturbed epigenetic systems may also be responsible for hybrid sterility. In their studies of a wallaby hybrid, Waugh O'Neill and her colleagues (1998) found that its chromosomes were undermethylated and had structural modifications that they attributed to the movement of mobile elements. Such differences in chromatin and chromosome structure could cause failures in chromosome pairing and segregation during meiosis, and hence lead to the sterility that prevents species fusion.

Epigenetic inheritance may also play a significant role in speciation through polyploidization, which is a major feature of plant evolution (Pikaard 2001). In many naturally occurring and experimentally induced polyploids, DNA methylation patterns are dramatically altered, and genes in some of the duplicated chromosomes are heritably silenced. The genomic stress introduced by polyploidization also releases mobile elements, which increases the generation of genetic and epigenetic variations. Following polyploidization there is therefore a very rapid enhancement of selectable variation, with all the opportunities for adaptation that this provides. In general it seems that heritable epigenetic variations may play a large role in initiating the divergence between populations that leads to reproductive isolation, both by increasing selectable variation and by reorganizing genomes.

FUTURE DIRECTIONS

Evolutionary epigenetics is a new and exciting area of study. It impinges on every aspect of evolutionary theory, and suggests additions and alternatives to accepted ideas about evolutionary change. Although there is already some discussion of the evolutionary implications of epigenetic inheritance, especially in plants, and modeling of the processes involved has begun, most of the work belongs to the future. There are still very few epigenetic models of evolution, and even fewer empirical studies of epigenetic inheritance that are directly planned to answer evolutionary questions. The disciplines of ecological epigenetics (the study of heritable epigenetic variations in natural populations) and epigenetic epidemiology (the study of epigenetically inherited diseases in populations) have yet to be developed. The future of evolutionary epigenetics is very much in the hands of today's students of evolution.

SUGGESTIONS FOR FURTHER READING

The only detailed analysis of the nature of epigenetic inheritance and its role in evolution is the book by Jablonka and Lamb (1995), which was reprinted as a paperback edition with a new foreword and appendix in 1999. A short review of the evolutionary significance of nongenetic mechanisms of transmitting information, which includes behavioral and cultural transmission as well as epigenetic inheritance, is given in Jablonka et al. (1998). Criticisms of the idea that epigenetic inheritance has been important in evolution can be found in *Journal of Evolutionary Biology*, 1998, volume 11, pp. 159–260, which contains a target article by Jablonka and Lamb followed by commentaries from other authors. The August 10, 2001 issue of *Science* (volume 293) contains useful papers summarizing various facets of epigenetic inheritance, although it does not discuss evolutionary aspects of the subject. The volume edited by Van Speybroeck et al. (2002) provides some interesting papers on the historical, philosophical, and theoretical aspects of epigenetics, including epigenetic inheritance.

Jablonka E & MJ Lamb 1995 Epigenetic Inheritance and Evolution: The Lamarckian Dimension. Oxford Univ. Press,.

Jablonka E, Lamb MJ & E Avital 1998 "Lamarckian" mechanisms in darwinian evolution. Trends Ecol. Evol. 13:206–210.

Van Speybroeck L, Van de Vijver G & D De Waele (eds) 2002 From Epigenesis to Epigenetics: The Genome in Context. Ann. N. Y. Acad. Sci. 981.

IV

QUANTITATIVE GENETICS AND SELECTION

18

Evolutionary Quantitative Genetics

DEREK A. ROFF

Quantitative genetics developed from an attempt to explain the apparent "blending inheritance" observed in quantitative traits within the framework of Mendelian genetics. In this it has been extraordinarily successful. However, a change in perspective was needed to understand the inheritance of quantitative traits. Rather than being controlled by a single locus or a small number of loci, variation in quantitative traits needed to be understood as the aggregate action of many loci. The individual behavior of any given locus was subsumed under a statistical description of the whole, analogous to the prediction of the properties of a gas or the flow of traffic. A prime focus of interest to the evolutionary biologist is the change in trait means, but to understand and predict such changes we must take into account not only what the mean trait values are before and after selection but also how much of the variability is ascribable to genetic factors. If very little of the variability is due to genetic effects then even very strong selection, leading to large differences in trait means before and after selection, will not result in a rapid change in the population from one generation to the next. Thus the central conceptual perspective in quantitative genetics is that of variance, that is, the division of phenotypic variance into components due to genetic and nongenetic effects. Using this perspective a very rich field of theory was developed, which both accounted for the apparent blending inheritance and was capable of making predictions about the response to selection.

The initial application of quantitative genetics was primarily in the field of applied animal and plant breeding but theoretical developments were made and tested using model organisms such as *Drosophila* and *Tribolium*. A major transfer into the area of evolutionary biology began in the 1970s, largely spearheaded by the work of Lande. The last 30 years has seen an enormous growth in quantitative genetics research applied to natural populations. Much of this work has been setting the basic groundwork for further work. In this chapter I shall explain the theoretical basis for this work and present four case studies, which show that quantitative genetics can be applied to predict change in natural populations. However, there remain a number of significant unanswered questions and a need to both develop quantitative genetic theory further and to empirically test such developments by artificial selection experiments and studies of natural populations. I shall expand on this in the final section.

CONCEPTS

Partitioning the Variance

Mendelian traits (e.g., eye color) are typically discrete and analyzed by considering different states of the trait. In contrast, quantitative traits (e.g., body size) are, by definition, continuous and best characterized by parameters, generally the mean and variance, describing their statistical distribution.[1] Consider a single trait: by assuming that each locus contributes independently to the trait value (i.e., no interaction

[1]An interesting "exception" to these two categories is threshold traits, which are manifested as discrete traits but are determined by an underlying continuously distributed trait called the liability, which might, for example, be a hormone titer. Such a circumstance is discussed in the final case study of this chapter.

among loci; technically, no epistasis; see Wolf, Box 6.1 of this volume, for a discussion of epistasis), we can invoke the central limit theorem of statistics, which then tells us that the phenotypic values will be distributed as a normal distribution (Bulmer 1985). Simple extension of this idea leads to the conjecture that multiple traits will manifest a multivariate normal distribution. Whereas traits not under selection are expected to follow such a distribution, selection alters the symmetry of the distribution and introduces correlations between traits that make multivariate normality at best only a crude approximation (Turelli & Barton 1990).

The normal distribution is described by two parameters: the mean and the variance. The phenotypic variance, is itself composed of three major elements, each of which plays a different role in evolutionary change. These three elements are the additive genetic variance, σ^2_A, the nonadditive genetic variance, σ^2_{NA} and the environmental variance, σ^2_E. These three components are generally considered to be independent, in which case the phenotypic variance, σ^2_P is simply the sum of these three components:

$$\sigma^2_P = \sigma^2_A + \sigma^2_{NA} + \sigma^2_E. \tag{18.1}$$

The nonadditive genetic variance can itself be decomposed into the dominance variance, σ^2_D (which is a consequence of nonadditive interaction between alleles at a single locus) and the epistatic variance, σ^2_I (which is a consequence of nonadditive interactions among loci):

$$\sigma^2_P = \sigma^2_A + \sigma^2_D + \sigma^2_I + \sigma^2_E. \tag{18.2}$$

Under the assumptions given above it can be shown (Roff 1997) that the regression of mean offspring value, Y, on mid-parent value, X, is linear:

$$Y = \left(1 - \frac{\sigma^2_A}{\sigma^2_P}\right)\mu + \frac{\sigma^2_A}{\sigma^2_P}X = \left(1 - h^2\right)\mu + h^2 X, \tag{18.3}$$

where μ is the population mean, and h^2 is the narrow-sense heritability (the ratio of total genetic variance to phenotypic variance is known as the broad-sense heritability) or, more generally, simply heritability. Note that the important components in the offspring–parent relationship are the additive genetic variance and the phenotypic variance. This relationship can be used to derive the breeders' equation, as described below.

The Breeders' Equation

The fundamental equation of quantitative genetics is the equation predicting the response to selection on a single trait, commonly known as the breeders' equation. We can readily derive this equation by rearranging the offspring on parent regression in terms of the offspring (= population) means in successive generations:

$$\bar{Y}_{t+1} = \left(1 - h^2\right)\bar{Y}_t + h^2\bar{X}_t = \bar{Y}_t + h^2\left(\bar{X}_t - \bar{Y}_t\right). \tag{18.4}$$

Rearranging gives the breeders' equation:

$$\begin{gathered} \bar{Y}_{t+1} - \bar{Y}_t = h^2\left(\bar{X}_t - \bar{Y}_t\right), \\ R = h^2 S, \end{gathered} \tag{18.5}$$

where R is the response to selection (difference between generation means) and S is the selection differential (difference between the population mean and mean of the selected individuals). The equation is sometimes written as

$$R = \sigma^2_A i, \tag{18.6}$$

where i is called the selection intensity and is the selection differential scaled by the phenotypic variance. Because the selection intensity is a direct measure of the proportion selected, it is more convenient than the selection differential in comparing studies.

While the additive genetic variance clearly plays an important role in the breeders' equation, one should not forget that nonadditive genetic variance also plays a role via its effect on the phenotypic variance. Epistasis is actually ignored in the "simple" theory but does appear in the practical estimation of σ^2_P. The general defense for the omission of epistasis from the basic quantitative genetic model is that epistatic variance, as opposed to epistasis in the Mendelian sense of interaction among loci, is generally a small contributor to the phenotypic variance, though this is not always the case.

The breeders' equation has been tested on many different traits and organisms and has generally proved to be an adequate predictor of the response

BOX 18.1. Individual Fitness Surfaces and Multivariate Selection
Jason B. Wolf

Studies of multivariate selection tend to have two distinct but closely related goals. One goal is to understand the phenotype–fitness relationship. This may be part of the broader goal of understanding adaptation, or may be motivated by specific evolutionary hypotheses (e.g., morphological integration). The second goal is to characterize the form and strength of selection acting on traits in a population. In this case, selection is the process that changes the shape of the phenotype distribution within or between generations (e.g., changes the mean or variance of a trait) and is sometimes inferred indirectly from these changes.

The two goals are obviously closely related, and not surprisingly, there is a direct connection between the phenotype–fitness relationship and the strength/form of selection (see Phillips & Arnold 1989). Here I provide a brief introduction to the fitness surface and its relationship with the force of selection action traits in a population. I also briefly introduce the empirical approaches used to analyze selection, but do not discuss these in detail. For more detailed discussions of empirical approaches see the reviews by Brodie et al. (1995) and Philips and Arnold (1989).

The Individual Fitness Surface

The relationship between phenotypic values and expected individual fitness can be represented as a surface (Figure 1). We can regard the shape of this surface as a concise summary of the expected fitness associated with a particular set of phenotypic values. Here, we refer to "expected fitness" because, given the pattern of the phenotype–fitness relationship, this is the fitness value we would predict an individual to have given their particular set of phenotypic values. Of course, individuals may show stochastic deviations from this predicted value, but these deviations are assumed to be random with respect to phenotypic values. The shape of the surface is primarily defined by the interactions between organisms and their environment (which presumably determines expected individual fitness) and is, therefore, largely determined by ecology and sociobiology. Because of this, the surface shape may change as the ecology of the population changes. Therefore, it can be seen as a dynamic representation of the selective consequences of the current ecology, and one may study the shape of the surface to understand how various ecological factors lead to particular phenotype–fitness relationships.

The number of traits that affect fitness define the number of dimensions of space in which the surface exists. Since it is hard to conceptualize higher dimensional "surfaces" (i.e., hypersurfaces), it is often conceptually useful to focus on the three-dimensional surfaces formed by single pairs of traits. In theory, the fitness surface can be defined for all phenotypic values. However, a population exists only in a small local region of this surface. For example, in Figure 1 we can see two populations occupying different regions of a surface. Because the topography differs in these two regions, the populations have different patterns of the phenotype–fitness relationship. Thus, the pattern of the phenotype–fitness relationship is a function of phenotype distribution, which determines the location of the population on the surface. The way in which this phenotype–fitness relationship translates into a "force" of selection also depends on the current distribution of phenotypes in a population. Thus, the distribution of phenotypic values not only determines where a population lies on the surface, but also determines how the local geometry of the surface translates into selection. This also means that the pattern of selection can change as the population moves to a new region on the fitness surface.

(continued)

BOX 18.1. *(cont.)*

It is important to note the distinction between the adaptive landscape and the individual fitness surface since the two are sometimes confused, resulting in incorrect conclusions about evolutionary processes. The adaptive landscape relates population mean fitness to population attributes, such as mean phenotypic values and allele frequencies. The adaptive landscape is usually an abstraction that lies outside of the realm of studies of fitness and selection. It is used to visualize the evolution of populations as movement on the landscape surface, often in the context of adaptation and constraint. Phillips and Arnold (1989) provide a concise discussion of the relationship of the two.

Geometry of the Surface

To understand the importance of the fitness surface for evolutionary processes, we need to be able to describe the local geometry of the surface occupied by a population. There are two features of the surface that are of primary importance: the slope (or gradient) and curvature of the surface (although most selection studies focus only on the slope). The slope of the surface measures how "steep" a surface is in a particular

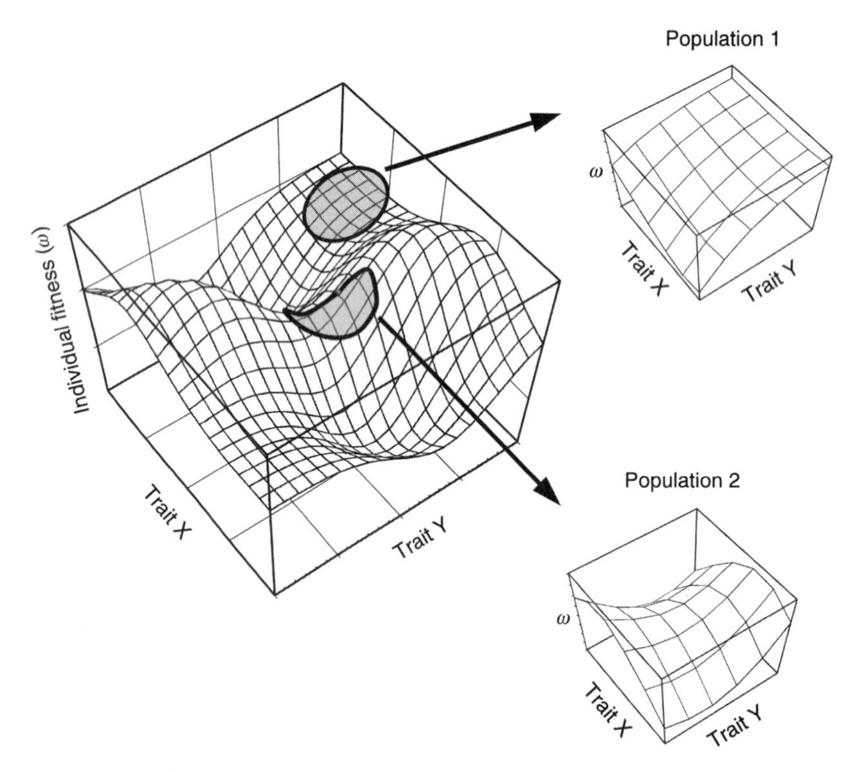

FIGURE 1. An individual fitness surface. The surface represents fitness as a function of the phenotypic values (z_i) of two traits (X and Y). The two shaded regions on the surface show the spread of phenotypic values of two different populations (1 and 2) with different mean phenotypic values for traits X and Y. The Local shapes of the surface in the region occupied by each of these populations are shown to the side of the surface.

BOX **18.1.** *(cont.)*

dimension and is related to directional or linear selection. The slopes of a surface in multiple dimensions are expressed as a gradient vector, which points in the steepest uphill direction on the surface. Curvature measures non-linearity of the surface (i.e., deviations from being a plane) and is associated with nonlinear components of selection, such as stabilizing or optimizing selection.

To better understand the conceptual connection between selection and the shape of the surface we can return to Figure 1. Population 1 lies in a region of the surface that is a very slightly curved plane. There is a positive relationship between the phenotypic value of trait Y and fitness, which indicates that selection "favors" individuals with larger phenotypic values of Y. However, there is almost no relationship between trait X and fitness, indicating that, in this region, trait X is not under selection. Population 2 lies in a very different region of this surface. There is, on average, no linear relationship between trait Y and fitness. Rather, there is a curved relationship, where fitness increases as we move away from the mean in either direction. This sort of curvature has been called disruptive selection since selection favors values that lie away from the mean. As with trait Y, there is no linear relationship between trait X and fitness since the average slope is zero (being positive one side of the mean and negative on the other). However, trait X shows a curved relationship between phenotypic values and fitness, with values closest to the mean being favored. This has been called stabilizing or optimizing selection.

Empirical Analysis of Multivariate Selection

Statistical approaches can be used to estimate the local shape (i.e., local average slopes and curvatures) of the fitness surface. There are a variety of ways in which this can be done (see Brodie et al. 1995), but by far, the most common approach is to approximate the individual fitness surface with a quadratic equation (Lande & Arnold 1983; Phillips & Arnold 1989). This quadratic equation is a fitness function that describes a surface with coefficients indicating slopes and curvatures. For a pair of traits with phenotypic values z_X and z_Y, the quadratic equation would have the form (assuming we standardize traits to a mean of zero):

$$w = \alpha + \beta_X z_X + \beta_Y z_Y + \tfrac{1}{2}\gamma_{XX} z_X^2 + \tfrac{1}{2}\gamma_{YY} z_Y^2 + \gamma_{XY} z_X z_Y + \varepsilon \qquad (1)$$

where α is a constant, β_X and β_Y are the linear or directional selection gradients, γ_{XX} and γ_{YY} are the quadratic terms describing curvature, which are associated with nonlinear selection, and γ_{XY} is an interaction term describing how fitness changes as a joint function or interaction between the two traits. The first four coefficients are associated with individual traits and therefore characterize slope and curvature along the individual trait axes while γ_{XY} characterizes fitness as a function of the joint values of the two traits. γ_{XY} has been called the "correlational selection gradient" because it acts to alter the covariance, or phenotypic association, of traits. Figure 2 shows a simple example of a surface where the only significant term is γ_{XY}, such that the fitness values associated with z_X depend on z_Y and vice versa. Here positive values z_X are favored only when in combination with positive values of z_Y and vice versa. This type of selection favors a positive correlation between the values of these two traits.

There are many ways in which one can empirically estimate the coefficients in Equation 1. The most common approach is to use multiple regression, where the

(continued)

BOX 18.1. *(cont.)*

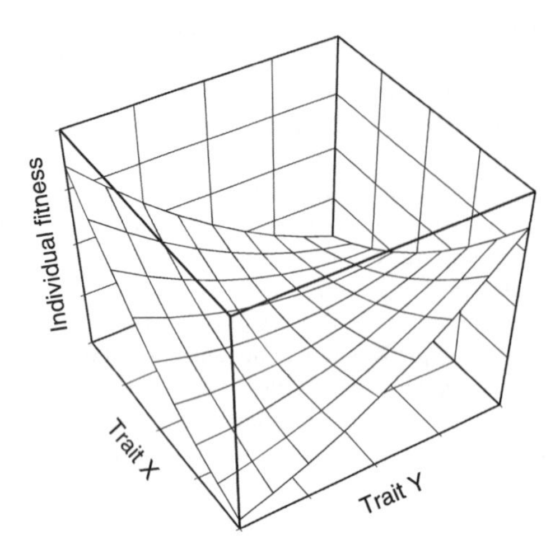

FIGURE 2. An individual fitness surface showing correlational selection. Fitness is shown as a function of the phenotypic values of traits X and Y. This surface represents the special case where the only nonzero coefficient from equation 1 is γ_{12} (i.e., linear terms, β_X, and β_Y and quadratic terms, γ_{XX} and γ_{YY} are all zero).

values of the βs and γs are partial regression coefficients corresponding to the *average* slope and curvature of the surface weighted by the phenotype distribution. The fitted surface described by these coefficients is a smoothed version of the actual surface (i.e., averaging removes the ruggedness) and represents the average shape in the local region. For example, if the distribution of traits in a population covered the entire surface in Figure 1, all of the coefficients would be near zero, but the surface is clearly not a flat plane.

The weighted average slopes are important because they can be combined with quantitative genetic models to predict the evolutionary response to selection. In those models, the βs are combined with estimates of additive genetic variances and covariances to predict the evolution of mean phenotypes. These models are presented and discussed in Roff (Ch. 18 of this volume) and Phillips and McGuigan (Ch. 20 of this volume). Less theory has been dedicated to understanding the evolutionary consequences of the non-linear terms from Equation 1 and fewer empirical studies have examined this form of selection (Kingsolver et al. 2001). Presumably, the quadratic terms (γ_{XX} and γ_{YY}) are associated with the evolution of trait variances while the interaction term (e.g., γ_{XY}) is associated with the evolution of the genetic covariance between traits. The effects of selection on genetic variances and covariances are discussed in detail by Phillips and McGuigan (Ch. 20 of this volume).

Acknowledgments I thank Allen Moore and Paula Kover for constructive comments on earlier drafts of this box.

to selection in the short term (approximately 15 generations) but tends to overestimate the response in the long term. This is entirely predictable as selection will change allele frequencies, thereby eroding additive genetic variance and reducing the heritability. Any attempt to extend the equation to the long term requires the addition of terms that account for the restoration of variance by mutation, genetic drift due to small population size, and a model for the changes in additive genetic variance due to the conversion of nonadditive genetic variance into additive genetic variance. It has been argued that in natural populations heritabilities will tend to remain constant, because selection is relatively weak and loss of variance due to selection is replaced by mutation. Estimates of selection intensities in free-ranging populations vary widely but can frequently be as high (20%) as used in artificial experiments (Kingsolver et al. 2001). There is undoubtedly a bias in the traits studied: given the statistical power of the approaches to measuring selection (Fairbairn & Reeve 2001), it is only pragmatic to select traits that one thinks, a priori, are under relatively strong selection.

The Multivariate Extension of the Breeders' Equation

Animal and plant breeders typically have their main focus on a single trait, for which the breeders' equation might be an adequate predictor, but evolutionary biologists are more likely to be interested in the joint evolution of many traits. For example, the evolution of body size may involve the evolution of growth rate, development time, and foraging behavior. Fitness is a single trait but is itself an amalgam of many traits and any analysis of fitness will necessarily involve the analysis of at least some of the component traits.

Genes do not typically have very precise localized effects, but rather influence several traits either because the genes are physically closely linked (i.e., close together on the same chromosome), are statistically associated (i.e., selection favors particular combinations of genes), or because the gene acts on several traits, a phenomenon known as pleiotropy. Genetic covariance via linkage may be important in some traits (e.g., mate choice) but pleiotropy is more important in general because its effects are not transitory as are covariances due to linkage (i.e., random mating cause linkage equilibrium). We can incorporate the effects of genetic and phenotypic covariance into a model of evolutionary change in two traits by a simple extension of the breeders' equation:

$$R_1 = \beta_1 h_1^2 + \beta_2 h_1 h_2 r_A, \\ R_2 = \beta_2 h_2^2 + \beta_1 h_1 h_2 r_A, \qquad (18.7)$$

where the subscripts identify traits, β is a measure of selection intensity (explained in more detail below and in Box 18.1) and r_A is the genetic correlation. Thus the response to selection is a consequence of two "forces": direct selection on the trait (the first term) and a response due to its genetic correlation with another trait (the second term). The above equations can readily be extended to more than two traits but a more convenient method of expression is to use matrix notation, which for two traits can be given as

$$\begin{pmatrix} R_1 \\ R_2 \end{pmatrix} = \begin{pmatrix} \sigma_{A11}^2 & \sigma_{A12} \\ \sigma_{A21} & \sigma_{A22}^2 \end{pmatrix} \begin{pmatrix} \beta_1 \\ \beta_2 \end{pmatrix}, \qquad (18.8)$$

where σ_{Aii}^2 is the additive genetic variance of trait i and σ_{Aij} is the additive genetic covariance between traits i and j. The vector on the extreme right is known as the selection gradient vector and can be decomposed into a phenotypic variance–covariance matrix and a vector of selection differentials:

$$\begin{pmatrix} \beta_1 \\ \beta_2 \end{pmatrix} = \begin{pmatrix} \sigma_{P11}^2 & \sigma_{P12} \\ \sigma_{P21} & \sigma_{P22}^2 \end{pmatrix}^{-1} \begin{pmatrix} S_1 \\ S_2 \end{pmatrix}. \qquad (18.9)$$

In shorthand the matrices can be written as

$$\mathbf{R} = \mathbf{G}\mathbf{P}^{-1}\mathbf{S}. \qquad (18.10)$$

This notation shows the relationship between the breeders' equation and its multivariate extension, the **G** matrix being the multivariate equivalent of σ_A^2, the inverse of the **P** matrix ($\mathbf{P}^{-1}$) being the equivalent of σ_P^{-2}, and the matrix **S** being the equivalent of the selection differential S. Note that the selection vector **S** measures the net effect of selection on a trait, whereas the gradient vector β measures the direct effect of selection on the trait (i.e., selection independent of selection on correlated traits; see Box 18.1). Exactly the same assumptions apply to the multivariate breeders' equation, with the additional assumption of multivariate normality. As noted above, selection on a single trait will lead to a change in heritability and hence only short-term predictions under strong selection can be made under the assumption of constant variances. Similarly, strong selection on multiple traits will alter covariances and cause deviation from prediction. Although some theory predicts that selection will change covariances more rapidly than variances, the empirical evidence suggests otherwise (see Roff 2002, p. 53).

Artificial selection experiments indicate that variances and covariances remain more or less constant for 15 to 20 generations. But this does not necessarily mean that the multivariate breeders' equation correctly predicts evolutionary trajectories. Given the importance of this equation for evolutionary biology one might suppose that it has been rigorously tested. Such is not the case. I have been able to locate only four experiments in which two traits were simultaneously selected and the response to selection compared with predictions (Roff 1997). In no case did results consistently match predictions (Figure 18.1). The overall finding is cogently summarized by Rutledge et al. (1973): "In contrast to single-trait-selection responses, the responses to index selection were not consistent with current

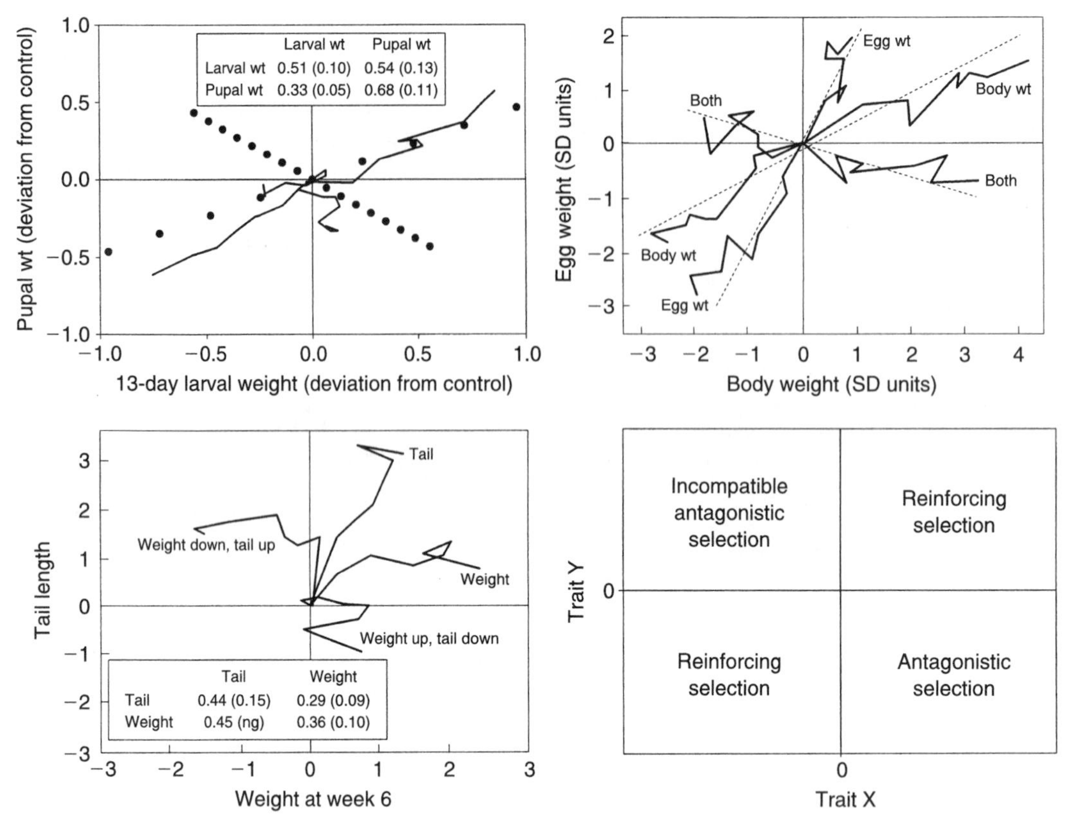

FIGURE 18.1. Three examples of joint selection on two traits. Upper left plot shows predicted (• = generation) response and observed (continuous lines) response in *Tribolium casteneum*. For clarity, the traces have been individually centered at zero. Heritabilities and genetic correlations (SE) are shown in the box. Data from Bell and Burris (1973). Upper right plot shows responses in the chicken. Data from Nordskog (1977). Lower left plot shows responses in mice, with heritabilities and genetic correlations (SE) shown in the box. Data from Rutledge et al. (1973). Lower right plot defines the type of selection in the four quadrants. Figure reprinted from Roff (2002) with the permission of Sinauer Associates, Inc.

theory...Our results indicate that in the dynamic situation of antagonistic selection, the genetic correlation may be more powerful in impeding component responses than predicted from presently available theory." One plausible reason for this breakdown between theory and experiment is that the multivariate model assumes multivariate normality, which is biologically inconsistent in many cases. In particular multivariate normality will break down when there are sets of functional constraints that limit the parameter space available to the traits. Suppose we have three traits, X, Y, Z, constrained to lie on a surface, such as a plane. The rate at which bivariate selection can move the combination of trait values, say X, Y, to some arbitrary combination will be contingent on the constraint imposed by the third trait (Figure 18.2). Under this hypothesis the failure of the quantitative genetic model lies in its failure to take into account functional constraints. Such constraints can be added to the model (Charlesworth 1990; Roff 2002) but their possible role in shaping evolutionary trajectories remains to be explored.

Nonadditive Components: Dominance

From the point of view of applied quantitative genetics nonadditive effects are a nuisance as they reduce the response to selection, at least in the short term. However, these effects are a "fact of life" and potentially important factors in the evolution of trait values within natural populations and hence cannot be ignored. There are three important subdivisions: dominance, epistasis, and maternal/ paternal effects.

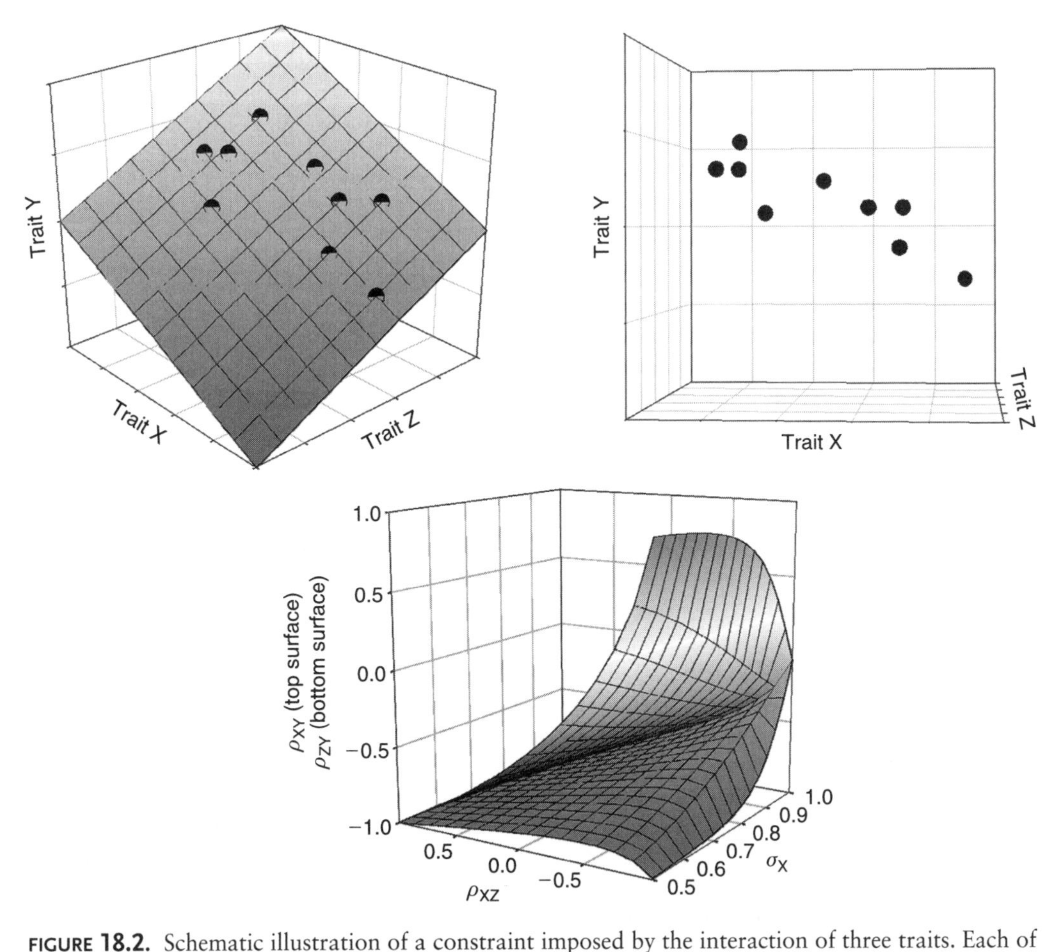

FIGURE 18.2. Schematic illustration of a constraint imposed by the interaction of three traits. Each of the three traits is a linear combination of the other two traits and hence trait combinations must fall on the indicated surface (top left panel). If only two traits are measured then there is apparent variation about the regression line due to the projection of the three-dimensional surface onto a two-dimensional surface (top right panel). This would suggest that there is no constraint on the evolutionary trajectory, when in fact the trait combinations can only move as shown in the left panel. The complexity in the set of possible correlations (ρ) between traits is illustrated in the bottom panel (for details of calculations see Roff, 2002, pp. 104–105.

Dominance variance arises as a consequence of interactions within a locus whereas epistatic variance arises from interactions among loci. In partitioning genetic variance, additive genetic variance is extracted first, followed by dominance and finally epistatic variance. It is this mode of variance partitioning that leads to a relatively small effect ascribable to epistasis even when epistasis may be ubiquitous among the genes controlling the trait. Dominance and epistatic variances are functions of allele frequencies: thus genetic drift, as might occur when a population passes through a bottleneck, can generate changes in the relative proportions of the three genetic variance components (Wolf et al. 2000). Most significantly, additive genetic variance can actually increase due to the "conversion" of nonadditive genetic variance to additive genetic variance. This can be seen using a very simple example in which the trait is determined by a single gene with two alleles, A_1 and A_2. Letting the three phenotypic values be $A_1A_1 = a$, $A_1A_2 = d$, $A_2A_2 = -a$ and the frequencies of A_1 and A_2 be p and q, respectively, the additive and dominance genetic variances are

$$\begin{aligned} \sigma_A^2 &= 2pq(a + d[q - p])^2, \\ \sigma_D^2 &= (2pqd)^2. \end{aligned} \qquad (18.11)$$

For complete dominance $d = a$ and for simplicity letting $a = 1$ we have $\sigma_A^2 = 8pq^3$. Now suppose in the original population $p = 0.75$, $q = 0.25$, giving $\sigma_A^2 = 8(0.75)(0.25)^3 = 0.09375$. If the population passes through a bottleneck it is quite possible for the frequencies to change by chance to $p = 0.7$, $q = 0.3$, giving $\sigma_A^2 = 8(0.7)(0.3)^3 = 0.1512$, which is more than a 50% increase in the additive genetic variance.

Inbreeding, the mating of close relatives, has long been recognized as having deleterious effects on fitness-related traits. Inbreeding depression is a consequence of dominance, which can be easily shown with the simple single-locus, two-allele model, used above. Let the genotypic values be $A_1A_1 = 1$, $A_1A_2 = d$, $A_2A_2 = -1$. For strictly additive action $d = 0$. From simple Mendelian inheritance, the mean trait value at generation t, μ_t, is given by

$$\mu_t = p_t - q_t + 2dp_tq_t, \tag{18.12}$$

where the allele frequencies have been subscripted with the generation. Due to drift the trait mean will decrease over time according to the relationship

$$\mu_t = \mu_0 - 2dp_0q_0\left[1 - \left(1 - \frac{1}{2N}\right)^t\right], \tag{18.13}$$

where N is population size. The above equation can be related more generally to inbreeding by use of the inbreeding coefficient (F = the probability that two genes are identical by descent):

$$\mu_t = \mu_t - 2dp_0q_0F_t. \tag{18.14}$$

Thus inbreeding depression will occur only if there is directional dominance ($d > 0$). Given directional dominance the rate of inbreeding depression increases with the degree of relatedness, with time and the inverse of population size. The effect of epistasis is to produce a curvilinear relationship between the mean and F (Crow & Kimura 1970).

Another important role played by dominance is in the preservation of additive genetic variance (as opposed to its generation by conversion after a bottleneck event) by antagonistic pleiotropy. Consider two traits genetically correlated with each other: if an increase in each trait increases fitness but the genetic correlation between them is negative we have a genetic trade-off, or antagonistic pleiotropy. As with inbreeding depression, theory shows that with only additivity allele frequencies at the two loci will become fixed. However, in the presence of dominance polymorphism is possible. Only the two-locus, two-allele case has been rigorously studied and in this circumstance the requirement for a polymorphic equilibrium is that on average the ratio of dominance to additive genetic variance should exceed 0.5 (Curtsinger et al. 1994). This ratio is rarely exceeded for morphological traits but quite frequently for life history traits (Roff 1997).

Fisher viewed adaptive evolution as proceeding largely through additive genetic effects whereas Wright developed a model in which nonadditive effects play a key role (Whitlock et al. 1995). Fisher assumed populations to be large and panmictic, with little variation in the genetic background. On the other hand Wright developed a model in which populations were small and connected by relatively infrequent migration. Further, he supposed that the genetic background was highly variable leading to considerable epistatic effects. The jury is still out on the relative importance of the two views but a key piece of evidence is the extent of nonadditive genetic variation among populations. The absence of nonadditive variance would favor the Fisherian view although its presence, while consistent with Wright's view, can also be explained as a consequence of founder events followed by selection. If nonadditive effects were common among populations we might reasonably expect that these effects would be at least as large between species.

Nonadditive Components: Epistasis

Epistasis may be an important component of evolutionary change in metapopulations and hybrid zones, because it can lead to the release of additive genetic variance, the creation of novel and more fit genetic combinations and disruptive selection favoring speciation by the production of incompatible interactions. Epistasis is without doubt ubiquitous in genetic systems but its appearance as a significant variance component might not be. Unfortunately detecting and measuring epistatic variance is technically very difficult. The most prevalent method of detecting epistatic effects is the analysis of hybrid breakdown, more commonly referred to as the analysis of line-cross means (Lynch & Walsh 1998, pp. 205–222). This method involves crossing several lines (which can be inbred lines for variation within a population, separate populations, or species) to produce an F_1 (this F refers to "filial" and is not to be confused with the inbreeding coefficient) from which is produced an F_2 and a backcrossed (= cross between parental and F_1) set of lines. Variation between two lines can then be partitioned as $mean = m + \alpha z + \beta h + \alpha^2 i + 2\alpha\beta j + \beta^2 l$, where α and β are coefficients dependent upon the type of cross (parental, F_1, F_2, backcross), z is the pooled additive effect, h is the pooled dominance effect, i is the pooled contribution due to additive × additive epistatic effects, j is the pooled contribution due to additive × dominance effects, and l is the pooled contribution due to dominance × dominance epistatic effects. Line-cross analysis has shown that both dominance and epistatic effects commonly occur (e.g., Table 9.5 in Lynch & Walsh 1998). However, these analyses refer to variation among trait means not the contribution of epistasis to genetic variance: they demonstrate that epistasis should be further investigated as a component of quantitative genetic variation. Unfortunately the nonlinearity introduced by epistasis destroys multivariate normality and makes analytical study of any quantitative genetic model very difficult. But this is not sufficient cause to ignore it.

Nonadditive Components: Maternal Effects, etc.

A parent can influence the fate of its offspring by providing care. I use "care" here in a very broad sense: it includes such things as resources directly provided by the parents, such as lactation in mammals, food in many bird species, egg provisioning in birds, reptiles, insects, etc., and the placement of eggs in locations that influence the growth and survival of offspring. Such care influences the phenotype of the offspring and hence alters the apparent heritability of the trait. The multivariate breeders' equation can be expanded to incorporate maternal effects, the general approach being to decompose the trait values into an additive genetic component, an environmental component (which includes nonadditive genetic effects), and a maternal effects component (Kirkpatrick & Lande 1989). The general approach and its consequences can be illustrated by the case of a single character that maternally affects itself, i.e., a trait such as body size that is determined in part by the individual's genes and in part by the phenotypic body size of the parent. The phenotypic value of the trait, X, in generation $t + 1$ is

$$X_{t+1} = A_{t+1} + E_{t+1} + c_m X_{m,t}, \quad (18.15)$$

where A is the additive genetic effect, E is the "environmental" effect, c_m is the maternal-effect coefficient, and X_m is the phenotypic value of the mother. The response to selection is given by

$$R_t = h^2 S_t + c_m(\beta_t + R_{t-1} - S_{t-1}). \quad (18.16)$$

The response to selection is equal to the usual response to selection (h^2S) augmented by the product of the maternal effect and the sum of the selection gradient ($\beta_t = S_t/\sigma_P^2$) and response in the previous generation and decremented by the product of the maternal effect and the selection differential in the previous generation. The important property of maternal effects is that they introduce a time lag into the evolutionary response. Two consequences of this are that there may initially be a reversed response to selection (when $h^2 S_t + c_m \beta_t < 0$), and that the response to selection changes each generation, only asymptotically reaching a constant value, which in the present case is approximately

$$R_\infty = \frac{2h^2 S}{(2 - c_m)(1 - c_m)}. \quad (18.17)$$

Maternal effects can significantly alter the response to selection. More complex patterns of maternal influence can occur but the same general effects hold (see Roff 1997, pp. 250–257).

CASE STUDIES

The Evolution of Genetic Architecture in the Oblique-Banded Leafroller

The general working hypothesis in theoretical studies of quantitative genetic studies is that the G matrix remains constant (obviously I exclude those studies specifically oriented toward examining change, such as studies on genetic drift or mutation effects). This assumption is not supported by the accumulating number of studies comparing G matrices among populations and species, although the relative importance of selection versus drift is still to be resolved (Roff 2000). In the first case study I present a theoretical prediction on the change in additive genetic variance resulting from a change in selection pressure, specifically the evolution of insecticide resistance in the lepidopteran, the oblique-banded leafroller (*Choristoneura rosaceana*).

Immune defense is metabolically costly (e.g., Demas et al. 1997; Veiga et al. 1998) and hence it is not unreasonable to hypothesize a negative trade-off between the two life history traits growth rate (X) and immune defense (Y). The genetic correlation between X and Y is the result of loci that affect both traits. Let the number of loci unique to X ("x" loci) be n_x the number unique to Y ("y" loci) be n_y and the number that are in common ("c" loci) be n_c. For simplicity I shall absorb the necessary "2" for the diploid condition into the n terms. Linkage disequilibrium is ignored and each locus treated as independent (i.e., no epistasis). The trait genotypic means are then

$$\begin{aligned} \mu_X &= n_x\mu_x + n_c\mu_c, \\ \mu_Y &= n_y\mu_y - n_c\mu_c, \end{aligned} \tag{18.18}$$

where μ_k is the mean at the kth type of locus (x, y, c and X, Y) and the trait genetic variances are

$$\begin{aligned} \sigma_X^2 &= n_x\sigma_x^2 + n_c\sigma_c^2, \\ \sigma_Y^2 &= n_y\sigma_y^2 + n_c\sigma_c^2, \end{aligned} \tag{18.19}$$

where σ_k^2 is the variance at the kth type of locus (x, y, c and X, Y). The only loci that contribute directly to the covariance are those that are in common. The expected value of the covariance, σ_{XY}, is thus

$$\begin{aligned} \sigma_{XY} &= E\{(x-\mu_X)(y-\mu_Y)\} \\ &= -E\{(c-n_c\mu_c)^2\} = -n_c\sigma_c^2, \end{aligned} \tag{18.20}$$

where $E\{\}$ is the expected value of the terms enclosed by {}. The **G** matrix is therefore

$$\mathrm{G} = \begin{pmatrix} n_x\sigma_x^2 + n_c\sigma_c^2 & -n_c\sigma_c^2 \\ -n_c\sigma_c^2 & n_y\sigma_y^2 + n_c\sigma_c^2 \end{pmatrix}. \tag{18.21}$$

Suppose that initially the environment is relatively benign and, as a consequence, immune defense is maintained only at a very low level under mutation–selection balance. Because of the negative trade-off, the immune defense (Y) loci, including those in common, would be maintained in the population at a relatively low frequency giving a low additive genetic variance in immune defense (trait Y). Now suppose there is a change in the environment such that selection favors a greatly increased allocation to immune defense, that is, the introduction of pesticide spraying. Alleles that favor defense will increase in frequency and hence the additive genetic variance of this trait will also increase. At the same time the increase in frequency of the shared loci will increase the additive genetic variance of the correlated trait, growth rate (X). This process can be readily visualized if we assume only a single common locus that is initially fixed giving a **G** matrix of

$$\begin{pmatrix} n_x p_x q_x & 0 \\ 0 & n_y p_y q_y \end{pmatrix}. \tag{18.22}$$

In the face of a challenge requiring increased immune defense a mutation arises that increases immune defense but negatively affects growth rate (trait X). The **G** matrix is now

$$\begin{pmatrix} n_x p_x q_x + \sigma_c^2 & -\sigma_c^2 \\ -\sigma_c^2 & n_y p_y q_y + \sigma_c^2 \end{pmatrix}, \tag{18.23}$$

where $\sigma_c^2 = 2p_c q_c(1 + d[q_c - p_c])^2$ (i.e., the additive genetic variance at a single locus). The second component in the parentheses is due to the dominance deviation, d. The influx of the mutation increases the additive genetic variances of both pesticide resistance and growth rate. Suppose that selection is sufficiently strong that it eventually drives the mutant allele to fixation. As the allele increases in frequency the variance contributed by this locus at first increases ($p_c q_c$ increases) but then decreases until at fixation it is zero. The actual turning point of the curve depends on the degree of dominance (Carrière & Roff 1995) but in all cases the components of the **G** matrix that are pleiotropically linked to the trait under selection will first show increases in magnitude followed by decreases as the alleles approach fixation. This process does not require the creation of new alleles by mutation, simply the increase in frequency of alleles previously at very low values.

The oblique-banded leafroller is a major pest of apple orchards in North America. The application of various types of pesticides led to an increase in pesticide resistance of the larvae in orchards in the area of Oka in Quebec, Canada (Figure 18.3). Populations from orchards not subject to spraying

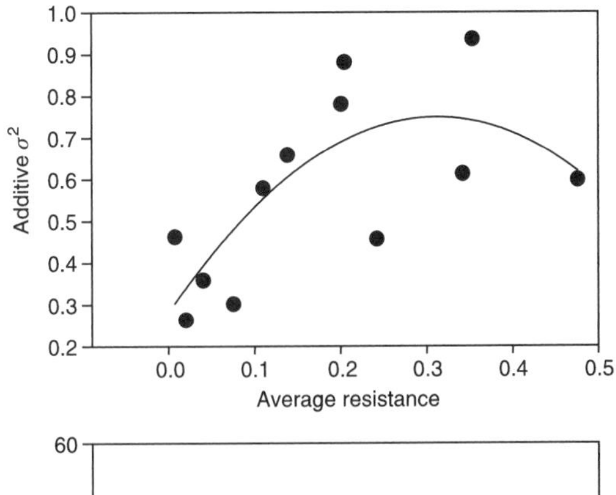

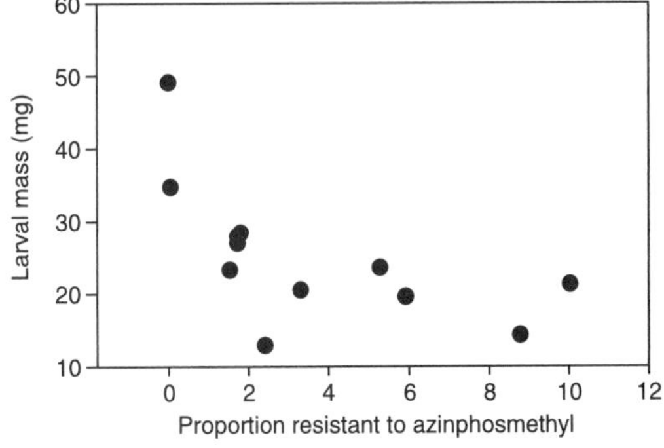

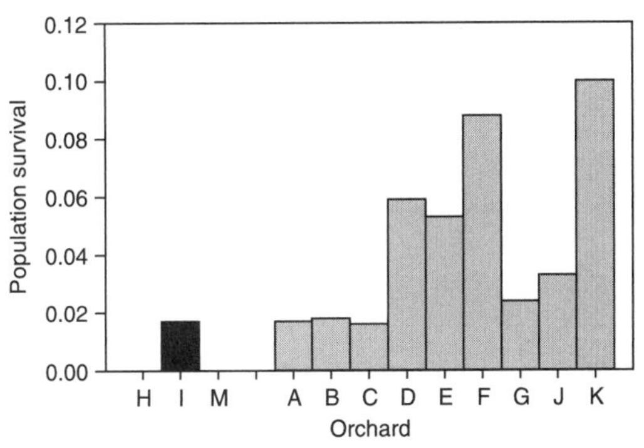

FIGURE 18.3. Lower panel: Survival of *Choristoneura rosaceana* larvae to the pesticide azinphosmethyl. Grey bars (A–K) show survival rates of larvae from populations within orchards that have a history of spraying, whereas the black bars (H, I, M) show survival for populations from unsprayed orchards (note that survival was 0 in H and M). Middle panel: Phenotypic relationship between larval mass after 16 days and resistance to azinphosmethyl. Each point is the mean for the populations shown in the lower panel. Upper panel: Additive genetic variance in larval weight as a function of the average resistance to three insecticides. Data sources: Carrière et al. (1994), Carrière and Roff (1995).

were highly susceptible to all three commonly used pesticides (Carrière et al. 1995). Test exposure of full-sib families to these three pesticides demonstrated that resistance was heritable ($0.4 < h^2 < 0.8$) and that resistance alleles were in low frequency in populations from untreated orchards (2.5% survival in untreated orchards vs. 20.2% in treated orchards; (Carrière & Roff 1995). There were highly significant negative correlations between 16-day larval weight, an index of growth rate, and pesticide resistance (Figure 18.3), between pupal mass, which is positively correlated with fecundity, and resistance; and a positive correlation between development time and resistance (Carrière et al. 1994). Growth rate, fecundity, and development time are all components of fitness and hence the foregoing correlations indicate a trade-off between resistance and other fitness components. Further evidence for trade-offs is the loss of resistance within five generations in a laboratory colony (Smirle et al. 1998). The mechanism underlying resistance and probably the trade-offs is the increased production of esterases, though the biochemical details remain to be resolved (Smirle et al. 1998).

To test the predictions outlined above Carrière and Roff (1995) estimated the additive genetic variances for the two life history traits larval weight and diapause incidence. Both traits are genetically correlated with pesticide resistance and hence were predicted to show changes in additive genetic variance as the resistance alleles increased in frequency in the populations. We were not able to measure these frequencies directly and so used average resistance to the three insecticides as a surrogate index. As predicted, the additive genetic variances of both traits showed a significant increase with average resistance (Carrière & Roff 1995; Figure 18.3). With larval weight there was also a significant quadratic term ($P = 0.034$, one-tailed test) while the quadratic was marginal for diapause ($P = 0.062$).

Predicting the Evolution of a Single Trait in the Wild: *Hyphantria cunea*

The lepidopteran *Hyphantria cunea* inhabits forests throughout North America. In Canada it is univoltine, the moths emerging in June and July, the larvae maturing in late summer and fall, and the pupae diapausing (the equivalent of hibernation) until the following spring. Diapause is broken after the pupae have been chilled and subjected to temperatures above 10.6 °C. Because in poikilotherms metabolic rates are strongly coupled to the external temperature, developmental periods are best described as a function of both real time (days) and the external temperature. This combination is termed degree-days and is equal to [Actual temperature − Threshold Temperature] × Number of days, where the threshold temperature is the temperature at which development ceases (Roff 2002, p. 384). The number of degree-days accumulated from the onset of chilling to adult emergence in *H. cunea* is defined as K_P (Morris & Fulton 1970). There is considerable variation in K_P, both across years and among different geographical sites, and mean offspring on mid-parent regression (Figure 18.4) indicates a heritability of 0.60 (Morris & Fulton 1970). In cold years the progeny from moths that emerge late in the spring (high K_P) fail to develop sufficiently to be able to pupate by the fall and hence die, while in warm years the progeny of moths that emerge early in the spring (low K_P) suffer high mortality because they pupate early and use up their fat reserves before the onset of chilling temperatures (Figure 18.4). The result is that each year the two tails of the distribution

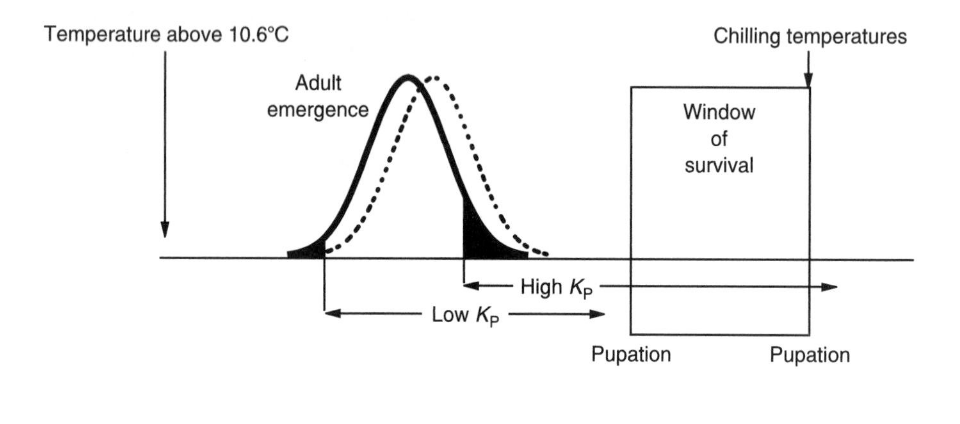

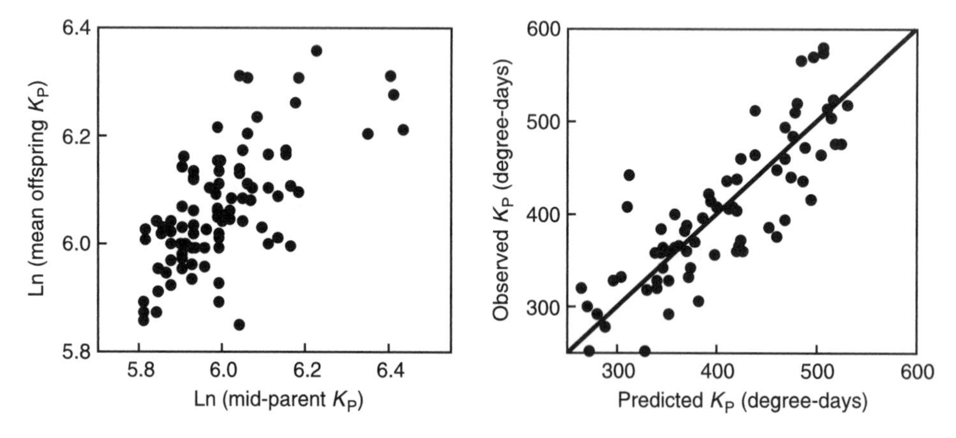

FIGURE 18.4. Upper panel: Schematic illustration of selection on *H. cunea*. Moths that eclose too early (low K_P) deplete their fat reserves before chilling temperatures induce diapause. Moths that emerge too late (high K_P) fail to reach the pupal stage before chilling temperatures. Both types are eliminated from the population. Changes in the average annual temperature favor shifts in the mean K_P, shown by the shift in the normal curve. Lower left panel: Regression of mean offspring K_P on the mid-parent K_P. Lower right panel: Regression of observed K_P on the value predicted from the simulation model incorporating the genetic relationship illustrated on the left. Model and analysis from Morris (1971); figure modified from Roff (1997).

will be eliminated and the mean K_P of the survivors, K_S, will differ from the K_P of the parents. Because temperature fluctuates from year to year, there will be overall directional selection each year but its sign will fluctuate (Figure 18.4).

From the above description we can infer that the developmental parameter K_P should vary from year to year as a result of fluctuating selection. To predict this variation we can use the breeders' equation, substituting in the relevant components:

$$K_{P,t+1} = K_{P,t} + R_t = h^2\left(K_{S,t} - K_{P,t}\right), \quad (18.24)$$

where $K_{S,t}$ is the K_P value of the survivors. Morris' approach differed somewhat from this and was actually incorrect, though, because of the high heritability of the trait, the error was not great (Roff 1997). Morris predicted changes in K_P for eight sites between the years 1958 to 1968, the prediction for each year being based on the predicted value from the previous year (omnibus prediction) not the observed value (stepwise prediction). This method of analysis is the most rigorous, as errors will tend to propagate through the predicted trajectory. After adjustment for a hypothesized effect due to temperature range, the match between prediction and observation was excellent (Figure 18.4). While the model can be faulted for its failure to incorporate an appropriate quantitative genetic model it is, so far as I am aware, the first attempt to take quantitative genetics into the real world and for that it deserves recognition.

The Evolution of Morphological Traits in Darwin's Medium Ground Finch, *Geospiza fortis*

The Galapagos islands are home to a wide range of finch species, the diversity of which, in part, inspired Charles Darwin in the development of his ideas on evolutionary change. The Galapagos finches have been the subject of a long-term research program seeking to measure the forces of selection acting on morphological traits (Grant & Grant 1989). Monitoring of climate and floristic conditions has shown that changes in rainfall are accompanied by changes in the type of seeds available, from large-hard seeds to small-soft seeds. These changes affect the survival of the Galapagos finches because the ability of an individual to handle each type of seed depends upon the characteristics of a bird's bill.

Birds are ideal subjects for the study of evolutionary change in natural populations, because they can be readily observed and their offspring measured and banded while in the nest. Such banding permits the estimation of the heritability of traits and the genetic correlations between traits by means of offspring on parent regression. Thus one can estimate the relevant genetic parameters within the same environment in which selection is acting. Several caveats must be considered: First, there may be bias introduced by the common environment within the nest and the effect of parental care. Both of these issues can be addressed by cross-fostering chicks. Such experiments with song sparrows, blue tits, willow tits, collared flycatchers, pied flycatchers, starlings and tree swallows found no evidence of common environment (nest and parental) effects on offspring morphological traits (reviewed in Weigensberg & Roff 1996). The second caveat is that there may be a substantial amount of extra-pair copulation, making paternity uncertain (e.g., barn swallows; Moller & Tegelstrom 1997): this issue can be addressed using molecular markers. In the absence of evidence that extra-pair paternity is insignificant the most conservative approach is to use a single offspring from each nest regressed on the trait values of the mother (Roff et al. 2004).

The genetic parameters for seven morphological traits have been estimated for *Geospiza fortis*, though these may be contaminated by common environment or uncertain paternity (Boag 1983). In a later study predicting the response to selection six of these traits were monitored (Table 18.1). The heritabilities ranged from 0.48 to 0.97 and genetic correlations from 0.67 to 0.94 (Grant & Grant 1995). Because genetic correlations between morphological traits are typically large and positive, predicting multivariate changes in morphological traits is not likely to be beset with the problems discussed earlier.

Over the two periods, 1976–1977 and 1984–1986, the Grants measured the survival of *G. fortis* on Daphne Major, a small island within the Galapagos archipelago. Both periods were characterized by a severe drought, though the direction of selection differed between episodes (Table 18.1). During the first period only 15% of birds survived and these were the larger birds. In contrast, during the second period survival was twice as large (32%) and selection favored smaller birds. A somewhat different picture emerges when the selection gradients are examined: in this case in both episodes direct selection favored larger size. Selection gradients

TABLE 18.1. Directional selection on morphological traits in Darwin's medium ground finch measured over two time periods Coefficients in bold face are significantly different ($P<0.05$) from zero.

	Standardized selection coefficients[a]			
	Selection differential, S		Selection gradient, β	
Trait	1976–1977	1984–1986	1976–1977	1984–1986
Weight	**+0.74**	−0.11	**+0.477**	−0.040
Wing length	**+0.72**	−0.08	**+0.436**	−0.015
Tarsus length	**+0.43**	−0.09	+0.001	−0.047
Bill length	**+0.54**	−0.03	−0.144	**+0.245**
Bill depth	**+0.63**	**−0.16**	**+0.528**	−0.135
Bill width	**+0.53**	**−0.17**	**−0.450**	−0.152
Sample size	634	556	632	549
Survival	15%	32%	15%	32%

Modified from Grant and Grant (1995).
[a]The selection differential measures the net effect of selection on a trait, whereas the selection gradient measures the direct effect of selection on the trait (i.e., effect independent of correlated effects).

show the direct effect of selection on a trait and hence indicate the focus of selection, whereas the selection differentials indicate the net effect of selection (Table 18.1). Difference between the two measures of selection thus indicate the effect due to correlations between traits.

The above data can be substituted in the multivariate breeders' equation and response to selection predicted. When both episodes are considered together, there is a highly significant correlation between predicted and observed evolutionary responses, though it is evident that the predictions for the second period do not fit as well as for the first period (Figure 18.5). The relatively poor fit for the second episode set is most likely a result of the relatively low selection intensity during this period compared to the first episode. The results are, nevertheless, very encouraging in demonstrating that, as in the previous two examples, it is possible to use the quantitative genetic model in a natural setting.

Predicting the Correlated Response in a Physiological Trait: Changes in JHE in the Sand Cricket, *Gryllus firmus*

Gryllus firmus, the sand cricket, occurs on sandy areas along the eastern coast of North America from Florida as far north as Connecticut. This species is wing dimorphic, meaning that some individuals possess well-developed wings and flight muscles and are capable of flight (macropterous) whereas others have reduced wings, highly reduced or nonexistent flight muscles and are flightless (micropterous). Although there are only two phenotypes, wing dimorphism is polygenic, and can be understood using the threshold model of quantitative genetics (Roff 1986, 1997). According to this model the dimorphic trait is determined by a continuously distributed underlying trait, called the liability, and a threshold of sensitivity: if the liability lies below the threshold one morph is produced whereas the alternate morph is produced when the liability is above the threshold (two examples are shown in the top panel of Figure 18.6). The liability is a quantitative trait and its heritability can be calculated by a modification of the statistical protocols to take into account that only two phenotypes are observed (see Roff 1997 for a description of the methodology).

Juvenile hormone esterase (JHE) is an enzyme that plays an important role in the determination of adult wing morphology. A significant genetic correlation between the liability and JHE has been demonstrated in *G. firmus* from both pedigree analysis (Roff et al. 1997) and directional selection on proportion macroptery (Fairbairn & Yadlowski 1997). The genetic correlation between JHE and the liability lead to the qualitative prediction that evolutionary changes in the proportion of macroptery will be accompanied by changes in JHE activity.

The purpose of the experiment described in this section was to use data collected from one population to predict how JHE would vary in another population as a result of selection for a change in the liability as indicated by an increase in the proportion of macropterous individuals in

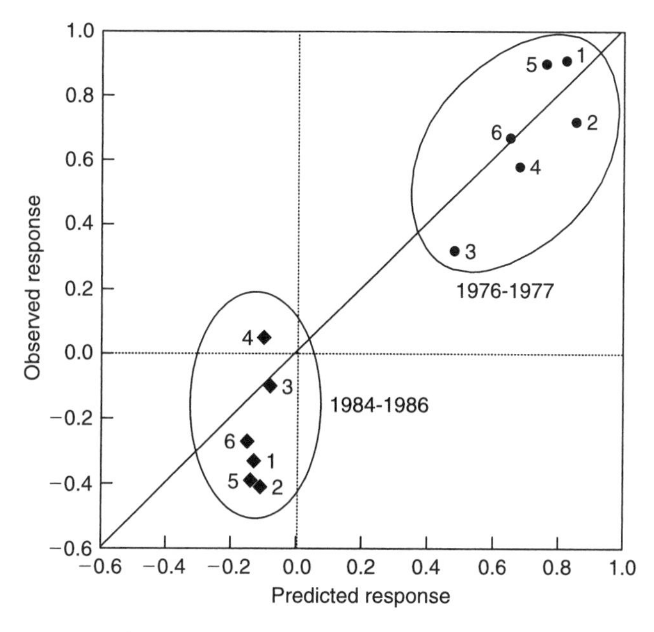

FIGURE 18.5. Comparison of observed and predicted response (in standard deviation units) to natural selection in Darwin's medium ground finch, *Geospita fortis*. Circles, 1976–1977, diamonds, 1984–1986. The actual values are those predicted for and observed in the generations born in 1978 and 1987, respectively. 1, weight; 2, wing length; 3, tarsus length; 4, bill length; 5, bill depth; 6, bill width. Data from Grant and Grant (1995), figure modified from Roff (1997).

the population. The data used to make the predictions were derived using a population collected in northern Florida. Using these data Roff and Fairbairn (1999) predicted JHE variation in a population collected from Bermuda. Using quantitative genetic theory Roff and Fairbairn (1999) predicted three differences between the Florida and Bermuda stocks. Here I shall describe two of these predictions. It is important to note that in the following analyses we are not assuming that the Bermuda population was derived from the Florida population: the equations predict the difference between the two populations using the Florida population as the reference population (this is mathematically equivalent to deriving the predictions for the two populations from some common ancestor and then simply comparing the two descendant populations). For the experiment, crickets from each stock were raised under two environmental conditions: (i) 28 °C and a photoperiod of 12L:12D, which corresponds approximately to early or late summer conditions; (ii) 30 °C and 14L:10D, which corresponds to midsummer conditions (we hereafter refer to these two environments simply by their temperature).

Prediction 1

An increase in proportion macroptery will be associated with an increase in the mean JHE activity (Figure 18.6). This prediction comes directly from the multivariate breeders' equation in which selection acts only on one trait, in this case the liability. Rearranging the equation to give the predicted (correlated) response of JHE, R_{JHE},

$$R_{\text{JHE}} = r_A R_L \frac{h_{\text{JHE}}\sigma_{\text{JHE}}}{h_L \sigma_L} \tag{18.25}$$

where r_A is the genetic correlation between JHE and the liability, R_L is the response to selection on the liability (measured indirectly by the change in proportion macroptery), h_{JHE} and h_L are the square root of the heritabilities, and σ_{JHE} and σ_L are the

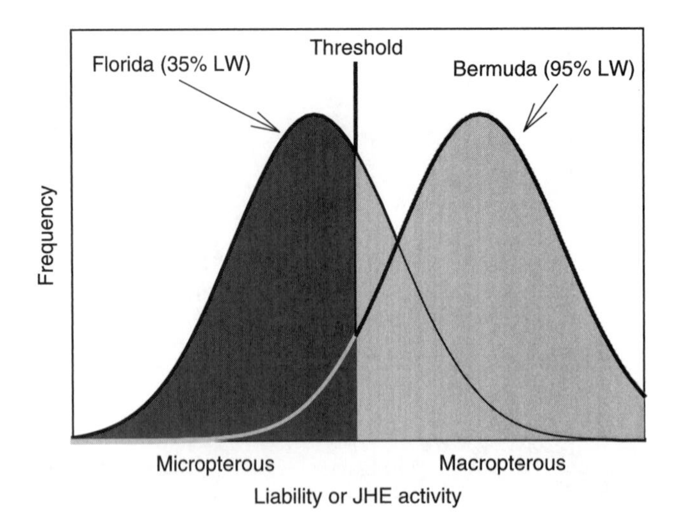

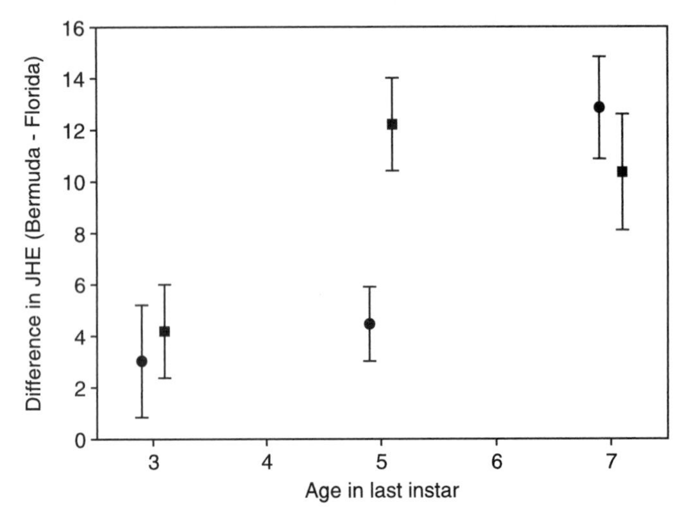

FIGURE 18.6. Prediction 1 of the quantitative genetic model for the change in juvenile hormone esterase (JHE) activity (nmol of JH converted to JH acid per minute per milliliter of hemolymph) accompanying a change in proportion macroptery in *Gryllus firmus*. Upper panel shows the distribution of the underlying trait, the liability, that determines the type of wing morph (LW, macropterous; SW; macropterous). This distribution is matched by the distribution of JHE, which is genetically correlated to the liability. Thus a shift in the mean liability, as observed in the Bermuda population, is predicted to be accompanied by an increase in the mean JHE activity. Lower panel shows the difference in mean JHE activity between the Florida and Bermuda populations at three ages and two environmental conditions (•, 28 °C, 12L:12D; ■, 30 °C, 14L:10D ±SE).

phenotypic standard deviations. The correlated change in JHE activity resulting from a change in proportion macropterous from 35% to 95% (Florida to Bermuda) is predicted to be 3.51 nmol/min (= nmol of JH converted to JH acid per minute per milliliter of hemolymph).

Because JHE activity varied with age Roff and Fairbain (1999) analyzed the data using residuals from the overall mean for each age and also using the standardized residuals, obtained by dividing the residuals by the standard deviation at each age. In accordance with the above prediction, the mean JHE activity at all ages was greater in the Bermuda than the Florida stock (Figure 18.6). A two-way analysis of variance on the residuals indicated no significant interaction between environment and population, nor an effect of environment alone. There was, however, a highly significant effect of population.

To estimate the 95% confidence region for the prediction Roff and Fairbairn used 1000 combinations of parameters within ranges estimated from a half-sib experiment: these gave an interval of 0.12 to 15.6 nmol/min, which includes the observed value. The relatively large interval is primarily a consequence of the large confidence region for the genetic correlation. Because of this large confidence interval the quantitative prediction of the model cannot be tested satisfactorily.

Prediction 2

Whereas an increase in the proportion macroptery will be associated with an increase in the mean JHE activity (prediction 1), the mean JHE activity within morphs will be decreased (Figure 18.7).

In addition to predicting changes in population mean values, the quantitative genetic model used here allows us to predict, and hence understand, complex patterns of coevolution such as the counterintuitive prediction given above that the decline in JHE activity within a morph occurs in spite of an overall increase in JHE activity (prediction 1). The expected JHE activity among macropterous individuals can be shown to be given (approximately, because it assumes full-sib families) by

$$\mu_{\mathrm{JHE}} + \sigma_{\mathrm{JHE}}\left(\frac{\int\limits_{-\infty}^{\infty}\int\limits_{f(x)}^{\infty}\left(r_A x\sqrt{0.5h_{\mathrm{JHE}}^2} + r_E y\sqrt{1-0.5h_{\mathrm{JHE}}^2}\right)\phi(x)\phi(y)\,dx\,dy}{\int\limits_{-\infty}^{\infty}\int\limits_{f(x)}^{\infty}\phi(x)\phi(y)\,dx\,dy}\right) \quad (18.26)$$

where

$$f(x) = \frac{T - \mu_{\mathrm{L}} - x\sqrt{0.5h_{\mathrm{L}}^2}}{\sqrt{1-0.5h_{\mathrm{L}}^2}},$$

$$\phi(x) = \frac{1}{\sqrt{2\pi}}e^{-0.5x^2},\ \phi(y) = \frac{1}{\sqrt{2\pi}}e^{-0.5y^2},$$

$$r_{\mathrm{E}} = \frac{r_{\mathrm{P}} - 0.5r_{\mathrm{A}}\sqrt{h_{\mathrm{L}}^2 h_{\mathrm{JHE}}^2}}{\sqrt{\left(1-0.5h_{\mathrm{L}}^2\right)\left(1-0.5h_{\mathrm{JHE}}^2\right)}}, \quad (18.27)$$

and μ_{JHE} is the mean JHE value in the reference (Florida) population, T is the threshold value (for details of its calculation see Roff & Fairbairn (1999), and r_{P} is the phenotypic correlation. Unlike the previous prediction in which the direction of change is independent of the parameter estimates, in the present case the directions of the two responses are contingent on parameter values. Roff and Fairbairn (1999) therefore examined the robustness of the prediction shown in Figure 18.7 by using the response to the 1000 combinations of parameters described above. In 84% of the combinations the mean JHE activity within the micropterous morph declined, whereas in the macropterous morph a decline occurred in 64% of cases. Thus a decline in mean JHE activity is more likely in the micropterous morph than in the macropterous morph.

Within each environment there are four observed and predicted JHE residuals. To compare observed and predicted values Roff and Fairbairn computed the regression of observed on predicted values (Figure 18.7). In both environments the model correctly predicted that macropterous individuals would have higher levels of JHE activity than micropterous individuals (Figure 18.7). For the 30 °C environment the correlation between observed and predicted was not significant ($r = 0.78$, $P = 0.22$) and, contrary to prediction, within each morph the Bermuda crickets had higher JHE levels than the Florida crickets, although in neither case was the difference statistically significant. The correlation between predicted and observed values was highly

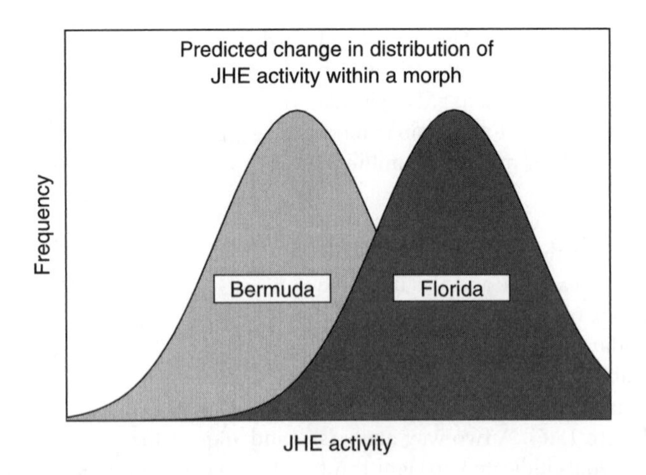

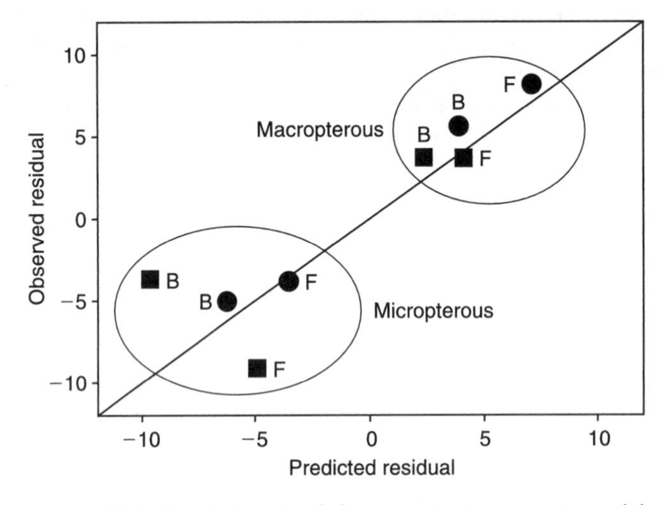

FIGURE 18.7. Prediction 2 of the quantitative genetic model for the change in JHE activity (nmol of JH converted to JH acid per minute per milliliter of hemolymph) accompanying a change in proportion macroptery in *Gryllus firmus*. Top panel shows the prediction that the mean JHE activity within morphs will decline as the frequency of macroptery is increased. Predicted changes were: micropterous, Florida (P = 0.35), $\overline{\text{JHE}}$ = 14.8; Bermuda (P = 0.95), $\overline{\text{JHE}}$ = 10.1; macropterous, Florida (P = 0.35), $\overline{\text{JHE}}$ = 23.9; Bermuda (P = 0.95), $\overline{\text{JHE}}$ = 22.1. Lower panel shows the observed and predicted JHE mean residuals within morphs of *G. firmus* (•, 28 °C, 12L:12D, ▪, 30 °C, 14L:10D ± SE).

significant in the 28 °C environment (r = 0.99, P = 0.007), and the values fell close to the 1:1 line (Figure 18.7). Further, the direction of difference in JHE activity within morphs between populations was as predicted.

FUTURE DIRECTIONS

Like the space shuttle the quantitative genetic model as exemplified by the multivariate breeders' equation was never meant to be the final product but rather an interim solution that served its original purpose but would be eventually replaced as need required. This process has been taking place over the last 20 years, as demonstrated by the inclusion of maternal effects in the model. However, there are still many weaknesses to be addressed. Theoretically, we still cannot effectively include epistasis in the models. As yet we are relatively ignorant of the effects of mutation in generating variance. Selection experiments show that additive genetic variances and covariances change under selection, which means that the **G** matrix will itself change. We have no reasonably comprehensive theory on what this change might be. There are two issues to be taken up: first, the development of a theoretical structure on the evolution of both the **G** and **P** matrices. Second, and equally important, the nature of the structure of variation in these matrices in natural populations: for example, latitudinal variation in the heritability of critical photoperiod has been observed in the pitcher-plant mosquito (Bradshaw & Holzapfel 2001) and in morphological traits of the *Allonemobius fasciatus/socius* complex (Roff & Mousseau 1999). We need much more information on how the variance–covariance matrices vary in natural populations to be able to formulate the questions to be addressed by theory.

The number of cases in which quantitative genetic predictions on multivariate evolution have been tested in either laboratory or natural populations is low. In part this lack of testing is due to the difficulties of estimating the parameter values with sufficient accuracy to make reliable predictions. However, the available data strongly suggest that the theory as we have it may be quite deficient in predicting evolutionary trajectories of many types of trait combinations. What we desperately need are experiments that combine multivariate selection with functional models that enable us to predict where the model should fail and where it should succeed.

19

Genetic Architecture of Quantitative Variation

JAMES M. CHEVERUD

The genetic architecture of quantitative variation plays an important role in evolutionary processes for complex traits. There are two aspects of the genetic architecture of quantitative variation that need to be considered. First, there is the genetic architecture of the quantitative traits themselves, including: (1) the number of loci that affect trait variation; (2) the typical size of genotypic effects; (3) whether alleles at a single locus display dominance interactions; (4) whether alleles at multiple loci engage in epistatic interactions; and (5) the range and pleiotropic patterns of gene effects. These features describe the genetic architecture of a trait and are specified in quantitative genetic theory by the genotypic values at a locus or set of loci (Falconer & Mackay 1996). Genotypic values are simply the mean phenotypic value of individuals carrying a specified genotype (see Box 19.1). These genotypic value definitions do not involve allele frequencies. The second aspect of the genetic architecture of quantitative variation is the genotype and allele frequencies at the loci in question. Genotypic values and allele frequencies are combined to define the population genic values that in turn specify the components of genetic variation and, hence, help to determine the response to evolutionary forces such as selection and genetic drift (Box 19.2; Falconer & Mackay 1996). It is through its manifold effects on heritable genetic variation that genetic architecture impacts evolutionary processes.

It is also important to remember that the genetic architecture of complex traits is not a fixed characteristic of the individual genes themselves. It can vary genetically within a population and also evolve under genetic drift and/or selection. Thus the number of genes affecting a trait, the magnitudes of their effects, levels of dominance and epistasis, and the range of pleiotropy are subject to evolution in addition to the evolution of trait means in populations. Although not treated in detail here, genetic architecture can also vary across different environmental contexts so that the evolution of genetic architecture can also be directly impacted by the environment.

Assumptions about the genetic architecture of complex traits play an important role in evolutionary models and concepts. However, for most of the last century there were few empirical data bearing on the five aspects of genetic architecture specified above. Attempts were made to investigate genetic architecture using line crosses (Lynch & Walsh 1998) between inbred strains. However, the results of these line-cross experiments are not diagnostic of genetic architecture at the level of individual loci. Some forms of line-cross analysis allow estimation of the number of gene effects separating lines but require the assumption that all segregating factors have the same effect. Furthermore, additivity in a cross does not guarantee additivity at the loci involved because dominance and epistatic effects may, collectively, cancel each other out. If a cross is not additive (F_1 hybrid not intermediate between parental strains), one can be certain that dominance and/or epistasis are present, but the placement of the F_1 mean midway between the parental strains does not preclude dominance and epistasis.

However, over the last 25 years advances in both quantitative and molecular genetic concepts and techniques have finally opened a window to the genetic

BOX 19.1. Genotypic Values: Additivity, Dominance, and Epistasis

Genotypic values (G_{ij}) are the average phenotypic value for individuals carrying a given genotype (Falconer & Mackay 1996). They are a basic concept in quantitative genetics (see Roff, Ch. 18 of this volume). In quantitative genetic models genotypic values are typically defined as follows at a single di-allelic locus (A) with alleles A and a:

- Additive genotypic value (a): half the difference between the genotypic values of the two homozygous genotypes:

$$a = (G_{AA} - G_{aa})/2.$$

- Dominance genotypic value (d): the deviation of the heterozygous genotypic value from the midpoint of the two homozygous genotypes:

$$d = G_{Aa} - (G_{AA} + G_{aa})/2.$$

- Note that genotype and allele frequencies at the locus in question play no role in defining genotypic values.

With a second locus, B, there are both single-locus and two-locus genotypic values (G_{ijkl}).

$$a_A = (a_{BB} + a_{Bb} + a_{bb})/3,$$

where each of the three additive genotypic values refer to genotypic values for locus A among individuals carrying each of the three B locus genotypes. A similar additive genotypic value can be defined for locus B.

$$d_A = (d_{BB} + d_{Bb} + d_{bb})/3,$$

where each of the three dominance genotypic values refer to genotypic values for locus A among individuals carrying each of the three B locus genotypes. Likewise, a similar dominance genotypic value can be defined for locus B.

Finally, with two loci there are also epistatic genotypic values (e_{ijkl}; Cheverud & Routman, 1995):

$$e_{ijkl} = G_{ijkl} - ne_{ijkl},$$

where ne_{ijkl} is the nonepistatic genotypic value:

$$ne_{ijkl} = G_{ij..} + G_{..kl} - G....$$

and

$$G_{ij..} = (G_{ijbb} + G_{ijBb} + G_{ijBB})/3,$$

$$G_{..kl} = (G_{aakl} + G_{Aakl} + G_{AAkl})/3$$

and

$$G.... = (G_{aabb} + G_{aaBb} + G_{aaBB} + G_{Aabb} + G_{AaBb} + G_{AaBB} + G_{AAbb} + G_{AABb} + G_{AABB})/9.$$

These genotypic values describe the genetic architecture of complex traits.

BOX 19.2. Genic Values and Genetic Variances

Genic values are a combination of genotypic values and their frequencies, or allele frequencies, when the population is in Hardy–Weinberg and linkage equilibrium (Falconer & Mackay 1996; Roff, Ch. 18 of this volume). For a single di-allelic locus there is the average, or additive, effect of a gene substitution (α):

$$\alpha_A = a_A + d_A(q - p)$$

where p is the frequency of the A allele and $q = (1 - p)$. The additive genic variance that determines response to selection is then

$$V_\alpha = 2pq\alpha^2. \quad (1)$$

There are three dominance deviations (δ), one for each genotype:

$$\delta_{AA} = -2q^2 d_A \; ; \; \delta_{Aa} = 2pqd \; ; \; \delta_{aa} = -2p^2 d_A$$

and the dominance variance is

$$V_\delta = (2pqd)^2.$$

When there are two di-allelic loci the average effect of a gene substitution (α_A) is

$$\alpha_A = a_A + d_A(q - p) + q(e_{aa..} - e_{Aa..}) + p(e_{Aa..} - e_{AA..})$$

where

$$e_{aa..} = r^2 e_{aabb} + 2rse_{aaBb} + s^2 e_{aaBB}$$

and s is the frequency of the B allele and $r = (1 - s)$ (Cheverud & Routman 1995). Similar equations define the other epistatic terms for $e_{Aa..}$ and $e_{AA..}$ and for the B locus. The additive genic variance is still measured by Equation 1:

$$V_\alpha = 2pq\alpha^2.$$

With epistasis the dominance variance at locus A becomes

$$V_{\delta A} = [2pqd_A - pq(e_{aa..} - 2e_{Aa..} + e_{AA..})]^2 .$$

Epistatic genic values (ε) are defined as

$$\varepsilon_{ijkl} = e_{ijkl} - e_{ij..} - e_{..kl} + e_{....}$$

and the epistatic variance is

$$V_\varepsilon = q^2r^2\varepsilon_{aabb} + 2pqr^2\varepsilon_{Aabb} + p^2r^2\varepsilon_{AAbb} + 2q^2rs\varepsilon_{aaBb} + 4pqrs\varepsilon_{AaBb} + 2p^2rs\varepsilon_{AABb} + q^2s^2\varepsilon_{aaBB} + 2pqs^2\varepsilon_{AaBB} + p^2s^2\varepsilon_{AABB}.$$

Note that the all three aspects of genetic architecture, additivity, dominance, and epistasis contribute to the additive genetic variance. The level of additive variance specifies response to selection and the extent of between-population differences produced by genetic drift.

architecture of complex traits at individual loci (Box 19.3). Studies measuring the effects of individual gene loci were made possible by the development of highly variable molecular genetic phenotypes, such as restriction fragment length polymorphisms (RFLPs) and short sequence length polymorphisms (SSLPs), in concert with quantitative mapping algorithms that allowed the unbiased estimate of gene effects at individual loci (Lander & Botstein 1989). These mapping studies are usually referred to as quantitative trait locus (QTL) studies.

CONCEPTS

Mapping Quantitative Trait Loci

In a QTL study, the segregation of quantitative phenotypes is correlated with a set of genotypes scored at molecular markers. There are two general strategies for measuring molecular markers in QTL studies: the candidate gene strategy and the whole genome strategy. In the candidate gene strategy molecular variants of a particular gene of interest

BOX 19.3. How To Perform a QTL Analysis

Lander and Botstein (1989) describe the basic strategy followed in quantitative trait locus (QTL) mapping experiments using the cross of inbred lines. First, two parental strains are crossed (SM/J × LG/J) producing a F_1 hybrid population (Figure 1). Given that the parental strains are fully inbred, all animals in the F_1 population will be genetically identical (except for the sexes). They will be heterozygous for all loci that are different between the parental strains and homozygous at all loci that have the same genotype in both parental strains. These F_1 hybrids are then interbred (intercrossed) with each other to produce a F_2 hybrid generation. Each animal in the F_2 generation is genetically unique, each being a different recombination of the parental haplotypes. Both the loci and their associated traits will segregate out in the F_2 generation. Phenotypes of interest are measured in the F_2 animals and the individuals are scored for molecular markers that varied between the parental strains.

Molecular markers used in the first QTL studies were restriction fragment length polymorphisms (RFLPs). However, in a few years large numbers of microsatellite loci with short sequence length polymorphisms (SSLPs) became widely available for model organisms and could be discovered for any species of interest. Most recently, with the advent of genome-wide sequencing, even larger numbers of single nucleotide polymorphisms (SNPs) are being detected. For an F_2 intercross experiment it is usual to space markers evenly throughout the genome at 10–20 cM intervals. Denser marker spacing is rarely of value because the number of recombinations available in a F_2 generation is limited.

The phenotypes are then subjected to interval mapping analysis (Lander & Botstein 1989). The interval mapping method provides unbiased estimates of QTL gene effect and position. In practice, genotypes between measured DNA markers are imputed using

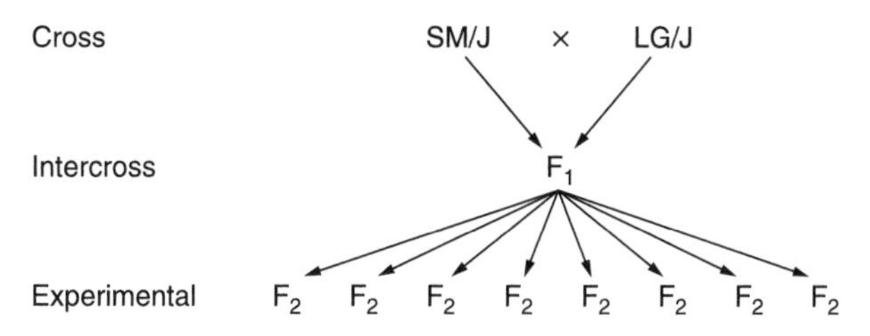

FIGURE 1. Mating design for a F_2 intercross QTL mapping experiment.

(continued)

BOX **19.3.** *(cont.)*

the flanking marker genotypes and the rate of recombination between these markers and the position of interest. Usually, positions of interest are identified every 2 cM through the interval bracketed by the flanking markers. The analysis proceeds by a separate regression at each point of interest (Haley & Knott 1992), with the phenotype of interest as the dependent variable and the imputed genotype as the independent variable. The additive and dominance genotypic values and an associated probability of no gene effect are obtained from this regression at each location of interest. It is usual to graph the base 10 logarithm of the inverse of the probability (LOD) against chromosomal position as in Figure 2. The peak of the LOD curve is the most likely position for the QTL. The confidence region for this position is usually given by the 1.0 LOD drop on either side of the peak. In an F_2 intercross the QTL confidence region is usually about 20 cM because of the strong linkage disequilibrium along a chromosome in an F_2 mapping experiment. This has been seen as a desirable feature of an F_2 intercross experiment because fewer markers need to be scored over the whole genome to assure that a marker will reside close enough to a gene with an effect on the phenotype to be detected.

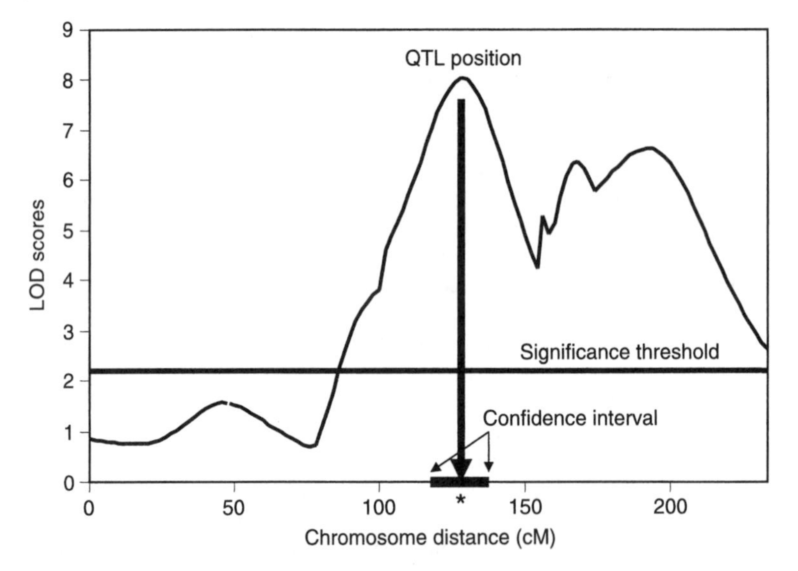

FIGURE **2.** Graph of LOD score versus chromosome position as developed in QTL mapping analysis. Higher LOD scores represent smaller probabilities of no QTL effect. The most likely position of the QTL is the peak of the curve. The confidence region for QTL position is given by the 1.0 LOD drop from the peak. Secondary peaks may, or may not, represent additional QTLs on the chromosome. A two-QTL model must be fit to evaluate this possibility.

are identified and are then tested for phenotypic differences. This approach can be quite fruitful but there is always the possibility that the gene at this locus in the study population is not variable, even though the gene product participates in trait development. The alternate strategy is to score molecular markers evenly spaced through the whole genome in families or specially constructed experimental populations so that correlations between markers and phenotype segregations can be evaluated. With these experimental populations it is also possible to impute genotypes at locations between flanking markers to identify more precisely the location of QTLs and obtain unbiased estimates of their effects (Lander & Botstein 1989; Haley & Knott 1992). This is referred to as interval mapping.

An especially powerful QTL experiment uses an intercross design involving the crossing of two genetically distinct populations. The analysis is simplest and most powerful when each of the populations crossed is inbred, although much can be gained by crossing distinct non-inbred populations (Haley et al. 1994). Using two discrete inbred populations, the parental strains are first crossed to form F_1 hybrids that are homozygous at all loci at which the parents carry the same allele and heterozygous at all loci different between the parental strains. This F_1 population is then intercrossed, F_1 hybrids mating with other F_1 hybrids, to produce an F_2 mapping population in which all variable loci and their associated phenotypes segregate. Association between segregating markers and phenotypes is used to map QTL locations and measure their effects. Using this breeding design all loci on different chromosomes are in linkage equilibrium in the F_2 generation while the distribution of linkage disequilibrium along individual chromosomes is enhanced relative to random mating. This allows one to use fewer markers to map QTLs on a chromosome but also limits resolution of QTL regions. Continued random mating of this F_2 and descendant populations results in an accumulation of recombination and, hence, finer scale mapping of QTLs in the genome. The randomly mated descendants of an F_2 population are referred to as an advanced intercross line (AIL; Darvasi & Soller 1995). The improvement in resolution at small genetic distances is approximately given by the number of generations since the parental strains divided by 2. Hence, an F_{10} AIL will have a 5-fold increase in mapping resolution.

QTL studies are carried out for several reasons but primarily to identify genes that affect quantitative phenotypes and to measure their phenotypic effects. Studies motivated by practical concerns of medicine or agriculture are often most interested in identifying the gene by name and determining its biochemical function. In evolutionary studies and some agricultural applications this remains a secondary goal while the primary goal is to measure the pattern and magnitude of gene effects. This second goal can be attained without specifying the direct biochemical function or role of the gene responsible for the QTL.

Most of the examples discussed below are drawn from collaborative research I have been involved in, evaluating the genetic basis for variation between the Large (LG/J) and Small (SM/J) inbred mouse strains (Chai 1956). I choose my examples primarily from this cross because of the extensive work done on these strains and my natural familiarity with them. This is not intended to be a comprehensive review of QTL studies. For an alternative system with very similar results see Mackay (2004). The LG/J and SM/J strains were generated in the first half of the last century by selecting for large (Goodale 1938) and small (MacArthur 1944) body size at 60 days of age in separate artificial selection experiments. After selection, the strains were inbred by brother–sister mating for over 100 generations. We first performed an intercross experiment generating an F_2 mapping panel of 535 animals using these parental strains. Each animal was genotyped at 76 microsatellite loci spread across the genome (Cheverud et al. 1996). This was followed a few years later by a second, replicate intercross composed of 510 individuals scored for 96 markers: 73 of the original 76 markers and 23 newly identified microsatellite polymorphisms (Vaughn et al. 1999). This second intercross population was used to establish an advanced intercross line by random mating and a series of recombinant inbred lines by brother–sister mating (Cheverud et al. 2004a). F_2 intercross animals have been measured for a wide variety of phenotypes, including weekly growth in body size (Cheverud et al. 1996; Vaughn et al. 1999), adiposity (Cheverud et al. 2001), and various internal organ weights. Individuals have also been macerated and aspects of skeletal morphology and asymmetry measured and mapped (Leamy et al. 1998, 1999, 2002; Ehrich et al. 2003. See Box 19.4). F_2 females were also scored for litter size and maternal effects on offspring growth and survival (Peripato et al. 2002, 2004; Wolf et al. 2002). No other QTL study has covered nearly the range of phenotypes investigated in the LG/J by SM/J intercross.

BOX 19.4. Evolutionary Morphometrics
Christian Peter Klingenberg

Morphometrics is the quantitative study of biological forms. It uses geometric and statistical methods to characterize morphology and quantify its variation. In the context of evolutionary biology, morphometrics is therefore intimately linked to quantitative genetics, development, and phylogeny.

Most morphometric studies in recent years have analyzed the shape of configurations of morphological landmarks, sets of points that can be located precisely on all specimens and that establish a clear one-to-one correspondence between them. The shape of a configuration of landmarks can be defined mathematically as all its geometric features except its size, location, and orientation—in essence, this is the geometric information by which you recognize objects in a photograph.

There is now a well-established methodology of shape analysis that combines aspects of geometry and multivariate statistics (e.g., Dryden & Mardia 1998). Therefore, the focus of morphometric research is on the application of these approaches to specific biological problems. In the context of evolutionary genetics, the primary purpose of morphometrics is to investigate the genetic and developmental basis of shape variation in populations and its implications for evolutionary change.

The most widely used method to extract information on shape from landmark data is *Procrustes superimposition* (Dryden & Mardia 1998). In this procedure, configurations of landmarks are scaled to unit size, shifted so that their centers of gravity are in the same location, and rotated so that a best overall fit to a common average configuration is achieved (Figure 1). The variation remaining in the landmark positions after

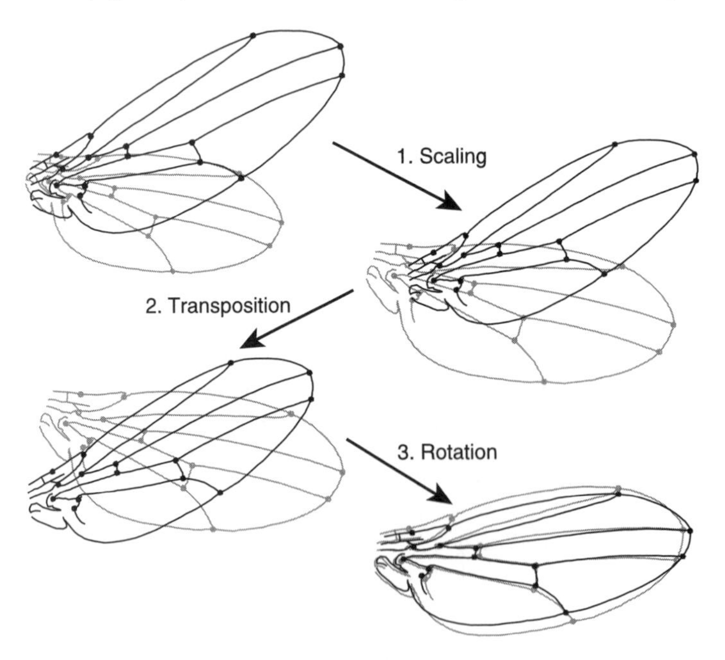

FIGURE 1. The Procrustes superimposition. Configurations of landmarks (dots on the outline drawings) are scaled to the same size, shifted to the same location, and rotated to an optimal fit. The remaining variation is information about shape, which can be used in further analyses.

BOX **19.4.** *(cont.)*

superimposition is the shape variation in the data. Because shape can vary in many different aspects, morphometric data need to be analyzed with the methods of multivariate statistics. The Procrustes methods have been extended for special cases such as landmark configurations that are internally symmetric, such as skulls (Klingenberg et al. 2002).

To examine the genetic basis of shape variation, geometric morphometrics has been combined with the multivariate theory of quantitative genetics (Lynch & Walsh 1998; Roff, Ch. 18 of this volume). Genetic and phenotypic covariance matrices of shape have been estimated in experiments using a standard breeding design (Klingenberg & Leamy 2001) and isochromosomal lines (Fernández Iriarte et al. 2003). The results of these analyses can be further analyzed with the multivariate breeders' equation to predict the potential responses to selection, for instance, to identify the shape features that respond most readily to selection (Figure 2; Klingenberg & Leamy 2001). This approach will be useful to determine the role of genetic constraints in the evolution of morphological traits.

A more detailed investigation of the genetic architecture of shape can be conducted by using experimental designs for mapping quantitative trait loci (QTLs; Lynch & Walsh 1998; Cheverud, Ch. 19 of this volume). Univariate analyses for single shape variables have located QTLs in both model and nonmodel organsims (Zimmerman et al. 2000; Albertson et al. 2003), but this type of analysis restricts itself to the particular aspect of shape that has been chosen a priori. A fully multivariate method for mapping QTLs for all aspects of shape is also available (Klingenberg et al. 2001). This approach estimates both the magnitude and direction of the QTL effects on shape, which can themselves be used in further investigations. For instance, analysis of the multivariate patterns of QTL effects can be used to test genetic modularity and genetic integration in a configuration of landmarks (Klingenberg et al. 2004; see also Mezey, Box 19.5 of this volume).

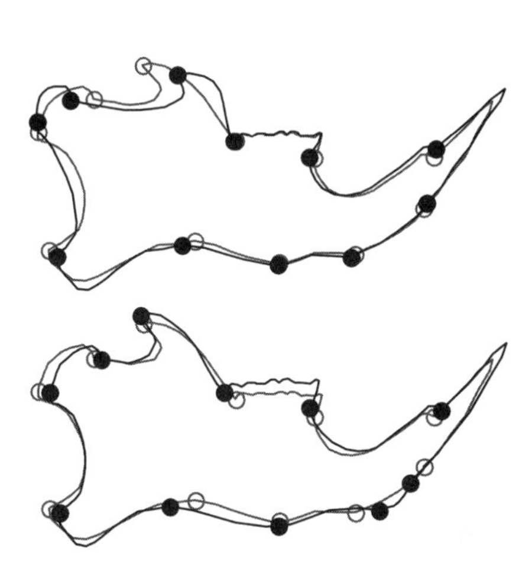

FIGURE 2. The shape feature with the maximum potential for response to selection in the mouse mandible. The two diagrams show shape changes from the mean shape (gray outline) in opposite directions along the dominant eigenvector of the matrix $\mathbf{GP}^{-1}$ (black outline). This is the shape variable with the greatest heritability. From Klingenberg and Leamy (2001, Figure 6); © Society for the Study of Evolution.

The developmental processes that build a morphological structure are among the primary factors that determine the patterns of integration among its component traits. Accordingly, these patterns of covariation provide an opportunity to investigate development by using morphometric methods. In particular, the specific nature of morphometric variation can be exploited to make specific inferences on developmental processes. For instance, fluctuating asymmetry can be used to infer direct developmental interactions between parts and therefore to delimit developmental modules (Klingenberg 2003). Because fluctuating asymmetry stems from random perturbations of development, there will be a systematic association between the asymmetries of two parts only if they are

(continued)

BOX 19.4. *(cont.)*

linked by direct developmental interactions that transmit the effects of perturbations between parts. Therefore, correlated asymmetries indicate developmental interactions between the developmental precursors of the parts under study (Klingenberg 2003). This information can be used to delimit developmental modules, because interactions should be localized primarily within modules rather than between them. In conjunction with QTL analysis, this approach can suggest a developmental basis for pleiotropic effects, that is, whether the joint effect of a QTL on two parts is due to changes that take place in the precursors of the two parts separately or whether it is the shared consequence of a single developmental effect (e.g., Klingenberg et al. 2004).

The methods of geometric morphometrics have become firmly established in the last decade, and the challenge now is to adapt them to the various experimental protocols used in different biological disciplines. For evolutionary genetics, this means that the various approaches of quantitative genetics (Lynch & Walsh 1998) can be applied to shape data. In combination with analyses that exploit the specific features of morphometric data, such as left–right asymmetry, these approaches provide a wide and mostly unexplored field for future research.

The mapping populations used in most QTL studies are certainly artificial and do not mimic a natural population. This is primarily due to the fact that allele frequencies are intermediate by design while natural populations are likely to have a wider range of allele frequencies across loci of interest. However, the alleles segregating in the intercross population can be considered as having been sampled from all the alleles available in the ultimate founders of the stocks. These alleles can be considered as randomly sampled in most instances. However, when the parental strains crossed were produced by selection, there will be some bias in the set of alleles affecting the selection criterion carried in the crossed population. We would expect the sample of alleles to be enriched for alleles with relatively large additive effects on the selection criterion. Alleles affecting other, uncorrelated traits would still represent a random sample fixed by inbreeding. Using selected populations as parental strains in a cross is advantageous in QTL studies because it increases the likelihood of detecting QTLs for the same reason as it produces bias in the genetic architecture of such traits. The genetic architecture of complex traits is concerned with genotypic effects of alleles, not their frequencies in populations. Thus, experimental populations are often unbiased with regard to their genetic architecture for most traits. Their complement of allele frequencies is that which allows the most statistical power for measuring genotypic values.

Various aspects of genetic architecture and their evolutionary consequences will be considered one at a time.

Number of Loci and Size of Effects

Most traits in natural populations are heritable. While morphological traits tend to have moderate to high heritability, physiological and behavioral traits have low to moderate heritability, and life history traits closely aligned to fitness have the lowest heritability (Mousseau & Roff 1987). This distinction in trait heritability has to do with high environmental variability for lowly heritable traits rather than low genetic variability, because most traits have comparable levels of genetic of variation when normalized by trait mean (Houle 1992). While genetic variation exists for most traits in most populations, we do not know whether that variation is due to a few genes of large effect or many genes of small effect.

Most of quantitative genetic selection theory is based on the assumption that genetic variation is due to many genes of small effect (Falconer & Mackay 1996). Deviation from this assumption, as in the house-of-cards models with few loci of large effect (Turelli 1988b), has important consequences for the level of genetic variation maintained in populations by the balance of new variation introduced by mutation and the removal of variation by stabilizing selection. The infinite alleles model requires a fairly high production of new genetic variation through mutation to account for the standing levels of genetic variance found in populations. The number of loci and size of effects also has important consequences for the evolution of additive genetic variance and covariance. Under the infinite alleles model this underlying variation structure is preserved during long periods of weak directional selection while under the house-of-cards models genetic variance and covariance evolve quickly in unpredictable ways when a population is exposed to directional selection.

QTL studies have produced mixed results when addressing the issue of whether variation is due to many small-effect genes or a few large-effect genes. In our own studies, we found substantial numbers of QTLs affecting morphological traits, including body size, growth, and skeletal morphology. The number of QTLs for these traits varies from about 10 to 30. Additive genotypic values tended to be small, on the order of 0.10 to 0.40 within genotype standard deviation units (Figure 19.1). The largest effect mapped for adult body weight resulted in a 3.6 g difference between homozygotes (Vaughn et al. 1999). An average mouse in this population weighs about 35 g. However, many other studies of natural populations or experimental populations have discovered a few QTLs of relatively large effect segregating in their populations (Lynch & Walsh 1998). The only traits that did not show many single-locus effects in our LG/J by SM/J intercross were those closely related to fitness, maternal effect on offspring survivorship (Peripato et al. 2002), and litter size (Peripato et al. 2004). Instead, the genetic architecture of these traits is dominated by pair-wise epistatic interactions.

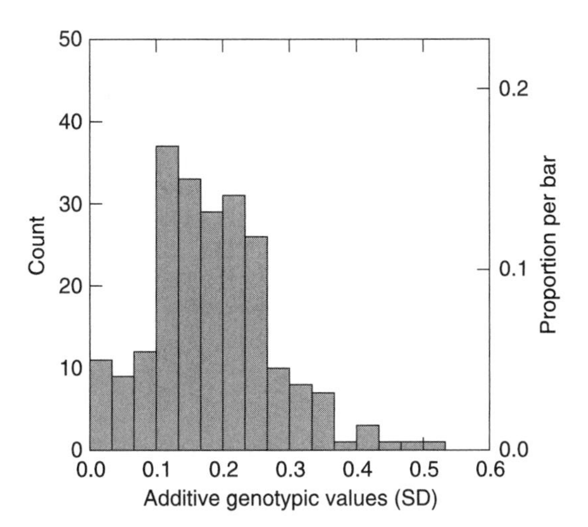

FIGURE **19.1.** Distribution of additive genotypic values for age-specific weights in standard deviation units based on Vaughn et al. (1999).

There are several reasons why the common finding of a few large-effect genes may be an artifact of the QTL experiments performed. When the purpose of the study is to identify named genes affecting a phenotype, greater statistical power is available for each single QTL when there are only a few of large effect. Therefore, researchers have often chosen specific mapping crosses that showed a priori evidence for this architecture in previous research. In contrast, the LG/J and SM/J strains were known from earlier crossing studies to differ at many loci of small effect (Chai 1956).

Second, there is a substantial bias toward finding only a subset of the segregating genes in a cross and attributing too large an effect to these genes when sample sizes and effect sizes are low (Beavis 1994). This bias only abates at sample sizes greater than 500 F_2 individuals and most QTL studies are performed with much smaller (n = 100 to 300) samples. A smaller sample size works fine if the study's purpose is to identify genomic regions affecting a phenotype for further study and finer mapping, because the subset of loci uncovered are likely to be true positive effects. However, a large proportion of QTLs will be missed in such studies because when there are actually many small-effect QTLs only a small random subset of them will be detected. For example, Beavis (1994) showed in a simulation with a sample size of 100 F_2 individuals segregating for 40 QTLs each accounting for only 1.6% of the variance that only one or two QTLs were detected in each population and their average effect was estimated at 16% of the total variance. At this sample size only 4% of the QTLs are found and their effects are overestimated by an order of magnitude. Thus small sample sizes can lead to substantial bias toward finding a genetic architecture of few, large-effect genes even in populations with many genes of small effect segregating for the phenotype in question.

Finally, as mentioned above, F_2 intercross gene mapping is often fairly crude, indicating the presence of QTL effects on a chromosome and limiting the QTL region to a broad 20 cM map interval containing hundreds of different genes. If multiple QTLs with similar effects reside within 20 cM of each other, their effects will merge and appear to represent only a single effect mapping at a location between the true QTLs. The 20 cM confidence region in most F_2 intercross experiments occurs due to the typical sample of recombinations and density of markers available. Alternatively, if the linked QTLs have opposite effects they will cancel each other out so that neither is discovered. In either case, too few QTLs are discovered and when their effects reinforce each other, the apparent effect size will be exaggerated. Determining whether an initial QTL finding represents one or more linked loci requires finer scale mapping with accumulated recombination. In our own QTL experiments mapping genes affecting body weight and obesity, our largest-effect F_2 QTLs have calved into three or more linked QTLs when mapped in the F_{10} generation of the advanced intercross line. This has been a common finding when F_2 mapping studies are followed up with studies involving more recombinations (Mackay 2004). Thus there are many reasons for QTL studies to underestimate the number of segregating genes affecting a trait and to overestimate their effects.

Dominance

Dominance is a property of a pair of alleles at a single locus (*A*,*a*) in relation to a specific phenotype. It is defined in terms of the relationship between the average phenotype of the heterozygote (*Aa*) and the midpoint of the homozygote genotypic values (Figure 19.2 and Box 19.1). If the heterozygote lies at the midpoint of the homozygote means, there is no dominance. Deviations from the midpoint can be classified as *A* dominant when the heterozygote is equivalent to the *AA* homozygote, *a* dominant when the heterozygote is equivalent to the *aa* homozygote, overdominant when the heterozygote exceeds both homozygotes, and underdominant when the heterozygote is below both homozygotes.

Dominance can be difficult to measure precisely in QTL studies because the dominance genotypic value, being defined by all three genotypes, has a relatively high standard error compared with the additive genotypic value. Even so, some broad distinctions can be made. In our QTL studies we have observed all kinds of dominance relations between LG/J and SM/J alleles. For age-specific body sizes there were 120 QTLs (Vaughn et al. 1999). The most common dominance relation was codominance ($-0.5 < d/a < 0.5$), present at 57% of the QTLs. Twenty-seven percent of the loci showed dominance ($0.5 < d/a < 1.5$ LG/J dominant to SM/J; $-1.5 < d/a < -0.5$ SM/J dominant to LG/J), LG/J being dominant to SM/J in 75% of these instances. Finally, 16% of the loci had heterozygotes far outside the range of the homozygotes. Sixty percent of these loci showed overdominance ($d/a > 1.5$) while the remainder were underdominant ($d/a < -1.5$). Age-specific weights between 1 and 3 weeks were responsible for most of the overdominant loci while weights at later ages were responsible for the few underdominant loci.

Most other morphological traits that have been studied in the LG/J by SM/J intercross, including skeletal traits and organ weights, show a pattern of dominance similar to that observed for post-weaning age-specific weights, most loci are codominant, and when dominance exists the LG/J allele is usually dominant to the SM/J allele (Leamy et al. 1999; Ehrich et al. 2003). There are some instances where the SM/J allele produces a larger phenotype and is dominant to LG/J. However, there are also some other traits that display dominance relationships like those affecting early age-specific weights, with the heterozygote value being beyond those of homozygotes,

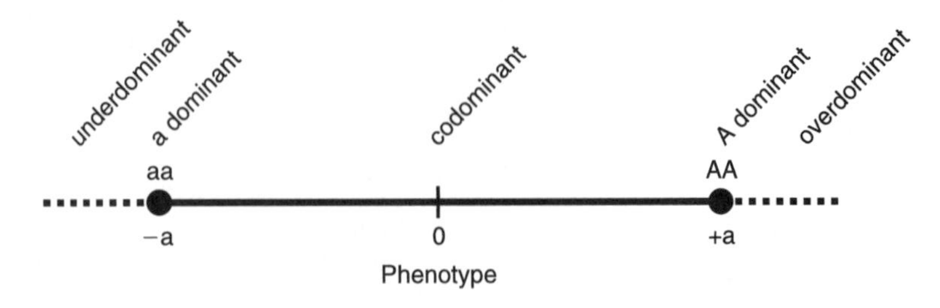

FIGURE 19.2. Classification of dominance relations on a genotypic value number line centered on 0.0. Appropriate labels are provided for different positions of the dominance genotypic value (*d*) in relation to the additive genotypic value (*a*) (Falconer & Mackay 1996).

including the development of skeletal nonmetric traits (Leamy et al. 1998), fluctuating asymmetry (Leamy et al. 2002), maternal care (Peripato et al. 2002), and litter size (Peripato et al. 2004). Thus it seems that different kinds of traits may display different dominance patterns.

Most QTL effects on individual traits do not display overdominance but rather tend to be codominant or have the allele resulting in a larger size dominant to the allele producing the smaller size. Therefore, it would appear that overdominance may be relatively rare or restricted to certain classes of traits (see above). However, when the manifold effects of a QTL on several traits are considered together, there are very likely to be linear combinations of those traits which are overdominant (Falconer & Mackay 1996; Figure 19.3). Whenever a gene affects many phenotypic traits and there is variation among the traits in the dominance relations between alleles at a locus, there will be a set of linear combinations of the traits that display overdominance even when no single trait displays overdominance itself (Ehrich et al. 2003). This so-called differential dominance appears to be a common finding in QTL studies. With differential dominance, when selection occurs orthogonal to the direction of additive gene effects, fitness will display overdominance at the locus and it will be subject to balancing selection. Directional phenotypic selection can result in balancing selection at the individual loci affecting a trait if the traits vary in their dominance relations.

The best-worked example of differential dominance and the associated multivariate overdominance involves QTLs affecting mouse mandibular morphology (Ehrich et al. 2003). In this study there was substantial variability in dominance among the many traits affected by individual QTLs. The typical direction of additive effects for multiple traits was (– –) versus (++) for the SM/J and LG/J alleles and represents size variation in local mandibular regions. Therefore, local shape variation (–+ vs. +–) displayed overdominance. Directional selection on shape would result in balancing selection at most of the loci affecting mandibular morphology in this cross. This balancing selection would act to maintain

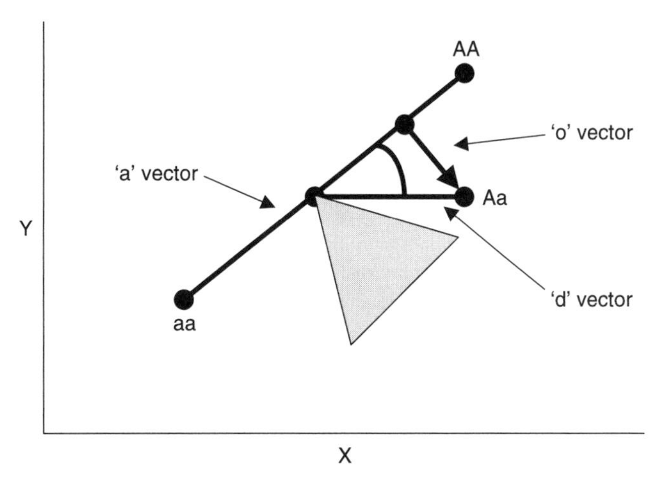

FIGURE **19.3.** Differential dominance at a locus produces overdominant linear combinations of traits. The three genotypic values G(*aa*), G(*Aa*), and G(*AA*) are plotted for two traits X and Y. The vector connecting the two homozygotes is the additive genotypic value vector (a) while the dominance genotypic vector (d) connects the midpoint between the homozygotes with the heterozygote genotypic value. The overdominance (o) vector specifies the linear combination of X and Y with the greatest extent of overdominance. The range of overdominant dimensions is indicated by the shaded region. Any linear combination of X and Y that falls within the shaded area will show overdominance at this locus (Cheverud et al. 2004b).

Additive by additive epistasis

		aa	Aa	AA	Genotypic values (B) a	d
	bb	−1	0	1	1	0
	Bb	0	0	0	0	0
	BB	1	0	−1	−1	0
Genotypic values (A)	a	1	0	−1		
	d	0	0	0		

Dominance by additive epistasis

		aa	Aa	AA	Genotypic values (B) a	d
	bb	−1	0	1	1	0
	Bb	2	0	−2	−2	0
	BB	−1	0	1	1	0
Genotypic values (A)	a	0	0	0		
	d	3	0	−3		

Additive by dominance epistasis

		aa	Aa	AA	Genotypic values (B) a	d
	bb	−1	2	−1	0	3
	Bb	0	0	0	0	0
	BB	1	−2	1	0	−3
Genotypic values (A)	a	1	−2	1		
	d	0	0	0		

Dominance by dominance epistasis

		aa	Aa	AA	Genotypic values (B) a	d
	bb	1	−2	1	0	−3
	Bb	−2	4	−2	0	6
	BB	1	−2	1	0	−3
Genotypic values (A)	a	0	0	0		
	d	−3	6	−3		

FIGURE 19.4. Four canonical forms of epistasis: additive by additive epistasis where the additive genotypic value at locus *A* varies depending on the genotype present at locus *B* and vice versa, additive by dominance epistasis where the additive genotypic value at locus *B* depends on the genotypes present at locus *A* and the dominance genotypic value at locus *A* depends on the genotypes present at locus *B*, dominance by additive epistasis where the dominance genotypic value at locus *B* depends on the genotypes present at locus *A* and the additive genotypic values at locus *A* depend on the genotypes present at locus *B*, and dominance by dominance epistasis where the dominance genotypic values at locus *A* depend on the genotype present at locus *B* and vice versa.

genetic variation at the individual QTLs displaying differential dominance. Perhaps this is one way in which genetic variation for morphological traits is maintained in populations.

Epistasis

Epistasis occurs when the genotypic values at one locus vary across genotypes at a second locus (Figure 19.4; see Wolf, Box 6.1 of this volume). This phenomenon is of critical importance because with epistasis it is possible for genetic architecture to vary in a heritable fashion and itself evolve. Figure 19.4 illustrates four canonical types of two-locus epistasis:[1] additive by additive, additive by dominance, dominance by additive, and dominance by dominance epistasis. As one scans across the rows and columns of each graph one can see that the genotypic values (*a* and *d*, see Box 19.1) at locus *A* vary across the genotypes at locus *B* and vice versa. Epistatic relations are always reciprocal. If locus *A* affects the genotypic values at locus *B*, locus *B* will always affect the genotypic values at locus *A*. In this sense, epistasis does not involve directional effects between modifiers and major genes but is mutual. The distinction between major genes and modifiers depends on allele frequencies at the loci involved, not genotypic effects. When epistatic interactions occur, the effects of an allele are not a property of that allele but vary depending on the genotypes present at another locus. Hence, with epistasis the effects of a gene are context-dependent rather than absolute.

A special form of epistasis involves the interaction of the Y chromosome, or other sex-determining chromosome, with loci on the autosomes.

[1]Higher order interactions are certainly possible and likely but the study of such interactions is difficult due to the very large sample sizes needed to evaluate three-locus and higher order interactions.

These interactions are the basis for genetic variation in and evolution of sexual dimorphism in populations. The genetic correlation between the sexes is a critical parameter in the evolution of dimorphism. A very high correlation between trait manifestations in males and females constrains the evolution of sexual dimorphism (Lande 1980a). Sexual dimorphism in gene effects due to genotype by sex interactions results in intersex genetic correlations of less than 1 and allows for evolution of dimorphism. Most instances of sex chromosome by autosome interaction uncovered in QTL studies have indicated that some QTLs are sex-limited, with effects restricted to a single sex (Mackay 2004). Rarely does a single QTL have opposite effects in the two sexes. For age-specific weights 25% of the 21 QTLs had sex-specific effects (Vaughn et al. 1999), while 33% of QTLs affecting tail length had sex-specific effects (Cheverud et al. 2001). A much larger proportion (63%) of QTLs affecting adiposity had sex-specific effects (Cheverud et al. 2001). Some of these sex-specific effects may be related to the females having been bred and weaning their litter prior to adiposity measurement. Relatively strong epistasis between the sex chromosomes and autosomes for adiposity provides heritable variation in sexual dimorphism, so that adiposity sexual dimorphism could evolve more readily than dimorphism for body weight or skeletal characteristics (Cheverud et al. 2001).

The frequency and strength of epistatic interactions have been controversial in evolutionary studies. Wright (1980) considered that epistatic interactions of relatively small effect were very common and had important effects in evolution. In contrast, Fisher (1930) felt that epistatic interactions were rare and played little role in evolutionary change. A consideration of physiological genetics leads one to believe that the potential for epistasis is present in the direct and indirect pathways involved in various physiological and developmental processes. However, traditionally it has been difficult to detect epistasis because of the very low statistical power for measuring epistatic variance in most quantitative genetic designs. QTL studies allow for a direct consideration of epistasis at the loci involved so that epistatic contributions to all the variance components (see Box 19.2) can be included in tests for epistasis (Cheverud & Routman 1995; Routman & Cheverud 1997).

We have discovered large amounts of epistasis in our QTL studies of the LG/J by SM/J intercross. Routman and Cheverud (1997) found that 25% of pair-wise body weight QTL combinations involved epistasis. Furthermore, all 19 QTLs considered were involved in at least one interaction with another QTL. Forty-three percent of the paired adiposity QTLs and 39% of the paired tail length QTLs exhibited epistasis and, as for body weight, all of the QTLs were involved in at least one interaction (Cheverud et al. 2001). The magnitudes of these epistatic effects were of the same order as the magnitudes of the additive and dominance effects at these loci. Therefore even when QTLs show significant single-locus effects they are involved in multiple small-effect epistatic interactions.

In contrast to the morphological traits discussed above, variation in fluctuating asymmetry (Leamy et al. 1998, 2002), and fitness-related traits, maternal care and litter size (Peripato et al. 2002, 2004), are dominated by epistatic effects. Leamy et al. (1998) found scant evidence for the effects of single loci on fluctuating asymmetry in mandibular morphology but found substantial evidence for epistasis (Leamy et al. 2002). Epistasis also dominated in QTL studies of maternal performance for offspring survival (Peripato et al. 2002) and litter size (Peripato et al. 2004). Unlike the situation for body weight, adiposity, and tail length, these loci are not fully interconnected by a network of epistatic interactions but rather form a sparse network where each QTL typically interacts with only one or two other QTL.

Just as dominance is different for different phenotypes at a locus, epistatic patterns can differ from trait to trait for a pair of loci (Cheverud et al. 2004b). This differential epistasis results in variation in the range and strength of pleiotropic effects exhibited at a locus. In the example in Figure 19.5 we see that traits X and Y can be affected by loci *A* and *B*. However, when the *AA* genotype is present at the *A* locus, the effects of locus *B* are restricted to trait X, there being no effect on trait Y. Therefore, variation at the *B* locus controls the range of pleiotropy expressed by the *A* locus. This phenomenon can also be observed when the phenotypic relationship between two or more traits is affected by the genotype at a locus. Such loci can be referred to as "relationship QTLs" because they affect the relationships between traits, not necessarily affecting the means of the traits themselves. Examples of relationship QTLs between portions of the mandible and mandibular length are provided in Figure 19.6. In this study we mapped 23 mandibular relationship QTLs. One third of these QTLs occurred in the same location as trait QTLs but the remainder occurred at locations without direct effects on the mandible. In these cases, the effect of the allele on the size of a mandibular region depends on the size

No epistasis — Trait X

	aa	Aa	AA
bb	−2	−1	0
Bb	−1	0	1
BB	0	1	2

Epistasis — Trait Y

	aa	Aa	AA
bb	−2	−1	0
Bb	−1	0	0
BB	0	1	0

FIGURE **19.5.** Differential epistasis for traits X and Y at loci *A* and *B*. Trait X shows a simple additive architecture with no epistasis while for trait Y both additive and dominance genotypic values at locus *A* depend on the genotypes present at locus *B* while the additive genotypic value at locus *B* depends on the genotype present at locus *A*. Therefore, epistasis for trait Y at loci *A* and *B* is a combination of additive by additive and dominance by additive epistasis. Note that when the *aa* genotype is fixed in the population, locus *B* affects both traits X and Y while when the *AA* genotype is fixed locus *B* only affects trait X. Thus, differential epistasis produces genetic variation in pleiotropy at locus *B*.

of the mandible. For example, the SM/J allele leads to relatively large mandibular corpus height in a short mandible but leads to a smaller mandibular corpus height size in a long mandible (Figure 19.6). The effect of the gene on one trait depends on the value of another.

Epistasis can have important consequences for evolution by either genetic drift or natural selection. Researchers have shown (Meffert, Ch. 27 of this volume; Goodnight 2000a; Routman & Cheverud 1997) that epistasis can result in an increase in genetic variance as populations undergo genetic drift due to population bottlenecks. Cheverud et al. (1999) showed that epistasis dramatically limited the loss of additive genetic variance in body size as populations of mice passed through a series of bottlenecks, despite loss of heterozygosity at the loci involved. Also, with epistasis, founder effects and genetic drift will result in evolution of the genetic background or in Mayr's genetic revolution (1963). Mayr discussed how the effects of alleles at a locus would evolve when allele frequencies at epistatically interacting loci were modified by random processes. These changes in allelic effects can then interact with selection so that even when divergent populations are under the same selection regime, an allele will be favored in one population but not in another depending on the random frequencies of genotypes at epistatically interacting loci. In an experiment testing the effects of genetic drift on the evolution of genetic background, de Brito et al. (2005) found that the effects of migrant haplotypes differed dramatically between populations that had diverged by genetic drift. The phenotypes considered in this experiment were those that had many small single-locus and epistatic effects, such as body size and adiposity, so this result is robust in involving traits with substantial direct genetic effects.

Pleiotropy

Pleiotropy occurs when a single gene affects multiple phenotypes. Wright (1980) suggested that all loci have strong pleiotropic effects on a multitude of characters. Pleiotropy is a major source of genetic correlation between traits and thus helps define the coevolution of traits under both natural selection and genetic drift. Earlier studies had found that there are variable levels of genetic correlation among phenotypes such that functionally and developmentally related traits tend to be more highly correlated than unrelated traits. This pattern of morphological integration has been seen in a variety of species and for various morphological systems. However, the genetic architecture behind this pattern remained obscure until the availability of molecular markers and QTL analyses.

Genetic correlations between traits (r_G) due to pleiotropy are the weighted average of the number of segregating loci demonstrating positive pleiotropy (k), negative pleiotropy (n), and independent non-pleiotropic effects (m):

$$r_G = \left\{(+1)\sum_{i=1}^{k}\sqrt{(V_{AXi}V_{AYi})} + (-1)\sum_{i=1}^{n}\sqrt{(V_{AXi}V_{AYi})} + (0)\sum_{i=1}^{m}\sqrt{(V_{AXi}V_{AYi})}\right\} / \sqrt{V'_{AX}V'_{AY}},$$

where V_{AXi} is the additive genetic variance at the locus in question and V'_{AX} is the total additive

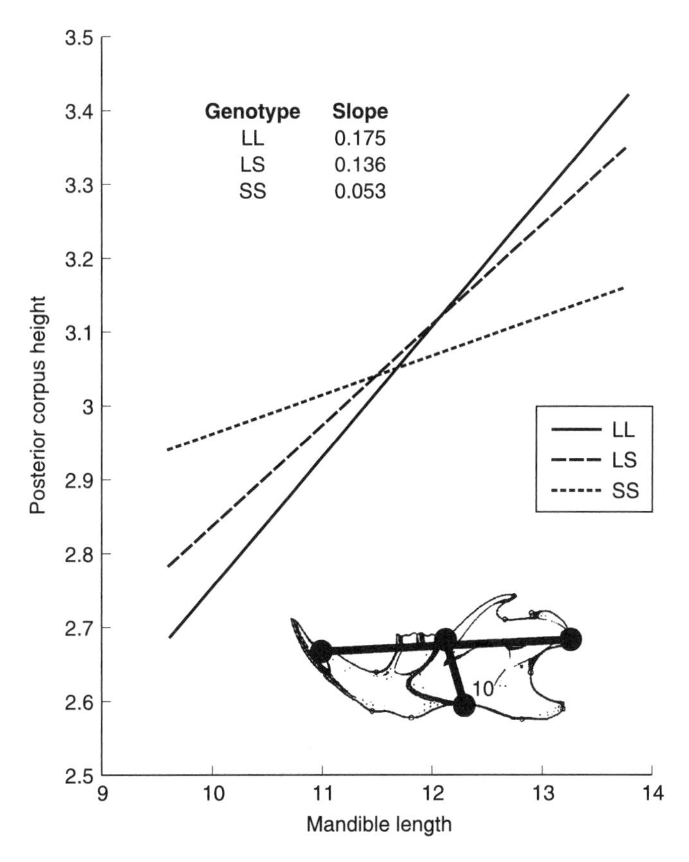

FIGURE **19.6.** Relationship QTLs near the proximal end of murine chromosome 17. The relationship between posterior corpus height and mandibular length varies depending on the genotype present at the QTL. This is likely due to differential epistasis for these two traits in interactions involving proximal chromosome 17.

genetic variance for the trait (Cheverud 1984). The weights are the geometric mean of additive genetic variation for the two traits at each locus. A lack of genetic correlation between a pair of traits can arise by two different mechanisms. If every locus affecting the two traits is pleiotropic, then the level of correlation observed in a population will be the evolved balance between positive and negative pleiotropy. If the number and relative effects of positively and negatively pleiotropic loci balance, the correlation will be zero despite a large number of pleiotropic effects. Alternatively, a low genetic correlation could be achieved by a balance of positive pleiotropic and nonpleiotropic loci. In this case, pleiotropic effects leading to low correlations would be modular, restricted to a subset of functionally and developmentally related traits (Box 19.5). As discussed above, loci may evolve their pleiotropic effects when there is differential epistasis for a pair of traits, changing pleiotropic categories (positive, negative, null) depending on genotype frequencies at other loci.

Our QTL studies on morphological traits have shown that pleiotropy tends to be modular, restricted to subsets of developmentally and functionally related traits (Figure 19.7). We found modular pleiotropic effects on body size growth in mice (Cheverud et al. 1996; Vaughn et al. 1999), separate sets of genes affecting early growth (1–3 weeks) and later growth (6–10 weeks) reflecting the different physiological pathways known to have time-specific effects on growth (Riska et al. 1984). We also found modular pleiotropy for QTLs affecting body composition (Cheverud et al. 2001) with largely independent QTLs affecting adiposity and skeletal size. In several

BOX 19.5. Modularity
Jason G. Mezey

Modularity refers to the autonomy displayed by parts of the phenotype when considering a process such as evolution or development (Schlosser & Wagner 2004). The definition of a module varies depending on the process under consideration, but in all cases a module is a part of the phenotype that can act as a relatively independent unit. The existence of modules is a function of the connectivity of the phenotype.

In evolutionary quantitative genetics, we are largely concerned with genetic modules. A genetic module refers to traits of the phenotype where genetic variation is statistically independent of variation at other traits (Wagner & Altenberg 1996; Mezey & Houle 2004). These are also called variational or evolutionary modules. To be a genetic module, the distribution of gene effects must produce statistical independence that is a nontransient property of the module. For example, the module will be statistically independent even as allele frequencies in a population change (Mezey & Houle 2004). Genetic modules are just now beginning to be identified (Mezey et al. 2000; Klingenberg et al. 2004). These modules are of particular interest in evolutionary genetics because evolution of a module is not genetically tied to other aspects of the phenotype. For example, under genetic drift, modules would be expected to evolve independently. If traits outside of the module are under directional selection, this will not cause a correlated response of traits within the module, and vice versa. Modules therefore reflect aspects of the phenotype where the genetic and consequently evolutionary properties need not depend on other aspects of the phenotype.

It might seem intuitive that genetic modules would also be modules of development. A developmental module is an aspect of the phenotype that can develop outside of its normal context (Raff 1996; Wagner & Mezey 2004). Classic examples of developmental modules are the imaginal discs of *Drosophila*. These are discrete tissues in the larvae

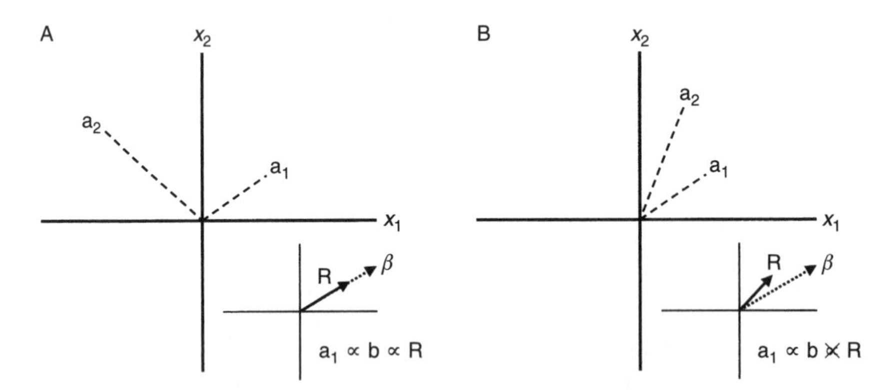

FIGURE 1. Measured traits (x_i) are represented by the axes and the additive effects of segregating alleles at two loci are represented by the dashed vectors ($\mathbf{a}_i$). (A) is a case of two modules and (B) is a case of no modules. The lower graphs show the change in mean of x_1 and x_2 in a population (the response vector **R**) when a selection gradient β has the same direction as $\mathbf{a}_1$ (see Roff, Ch.18 of this volume). In (A), $\mathbf{a}_1$ defines a module so the response is entirely in the direction of $\mathbf{a}_1$, regardless of the allele frequencies at the two loci. In (B), the response is at an angle to $\mathbf{a}_1$ and this will always occur for this nonmodular case when there is a selection gradient in the direction of $\mathbf{a}_1$.

BOX **19.5.** *(cont.)*

which develop into the major adult structures (wings, legs, eyes, etc.). When presumptive disc cells are transplanted from a donor embryo to the different location in a host embryo, adult structures develop in the host in these new locations (Simcox & Sang 1983), that is, legs can develop in the wrong location in the host. However, independence of developmental context does not necessarily mean that genetic variation of the developmental module in its *normal* context will display nontransient statistical independence. Developmental modules are therefore not necessarily genetic modules. The degree to which genetic and developmental modules correspond is an area of continuing research.

The existence of a genetic module depends on the distribution of pleiotropic effects of genes. Modules are easiest to define in the case of two traits where genetic variation is determined by an additive two-locus, two-allele system. In this case, if the gene effects at each locus affect either one trait or the other, the two traits are modules. For example, Figure 1 contrasts a modular and a non-modular case. In Figure 1A, the axes may be rotated to line up with the allelic effect vectors a_1 and a_2 producing two new traits. After this rotation, the effects of alleles at each locus are entirely limited to one of the two traits. These two traits represent "perfect" modules in the sense that genetic variation in traits corresponding to a_1 and a_2 are completely independent of one another. What is more, even as allele frequencies at these two loci change, variation in the a_1 and a_2 traits will continue to be independent (Mezey & Houle 2004). These traits may therefore change in an evolutionarily independent manner. For example, if the trait corresponding to a_1 were placed under directional selection, the trait would respond without a correlated response in the trait corresponding to a_2. Even if both traits are under directional selection, while both traits will evolve, genetic variation in these two traits will continue to be independent, that is, the potential to evolve as independent units is retained. The case diagrammed in Figure 1A contrasts with the case in Figure 1B. In Figure 1B there is no rotation that will produce two independent traits. For this case, whatever traits are defined, most allele frequencies at the two loci cause genetic variation in the traits to be correlated (Mezey & Houle 2004), that is, independence of any traits in this case will be transient.

When considering many traits, a "perfect" genetic module occurs if it is possible to identify a set of traits where the effects of genes responsible for variation in the set *only* affect traits in the set. For example Figure 2 diagrams a case of three traits where variation is determined by four loci. We can define two modules for these three traits, one corresponding to the vector A and the second for two traits that span the plane B. These modules will be statistically independent even as allele frequencies change. A modular organization is therefore defined in terms of a nonrandom distribution of the pleiotropic effects of genes when considering a specific set of traits.

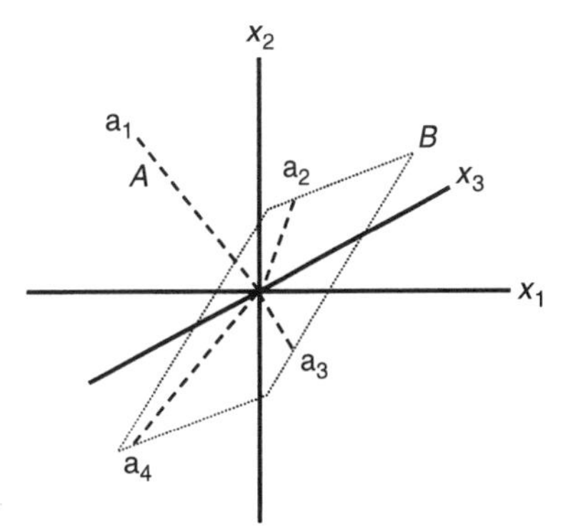

FIGURE 2. A case of two modules (A and B) for three traits.

There are currently two methods that can be used to identify genetic modules: analysis of quantitative trait loci (QTLs) and comparison of additive genetic covariance matrices (**G**) matrices (see Cheverud, Ch. 19 of this volume; and Roff, Ch. 18 of this volume). The goal of a QTL analysis is to quantify the effects associated

(continued)

BOX 19.5. *(cont.)*

with individual loci, although in practice, QTLs tend to summarize the composite effects of alleles at multiple, linked loci (Lynch & Walsh 1998). However, when there are modules, we still expect to be able to partition the effects of identified QTLs into sets that affect different groups of traits (Mezey et al. 2000). The second approach for identifying modules is based on the observation that a (one-dimensional) modular structure causes the principal components (PCs) of the **G** matrix of a population to be constant, even as the structure of the **G** matrix evolves (Mezey & Houle 2004). If two populations have the same PCs, this indicates the existence of a module. Comparison of the PCs of **G** matrices has been an important quantitative genetic approach in recent years (Steppan et al. 2002).

We would almost never expect to find cases of "perfect" modules in nature. However, cases which approximate a perfect scenario occur. For example, analyses of QTLs affecting the mouse mandible (Figure 3) have found a statistically significant tendency for QTLs to affect either the ascending ramus or the alveolar region (Mezey et al. 2000) and that the covariation of QTL effects between the regions is weaker than arbitrary sets of traits (Klingenberg et al. 2004). While the effects of each QTL are not entirely isolated to one region or the other as in the perfect case, the effects of individual QTLs tend to have larger effects on one region or the other. These two regions therefore reflect genetic modules of the mandible.

In the future, studies that estimate the effects of individual loci and individual polymorphisms (association studies) will become increasingly feasible. These studies will provide additional resolution for quantifying pleiotropic effects and for identifying genetic modules. This approach will tell us a great deal about the distribution of pleiotropic effects on the phenotype, the dynamic structure of genetic variation, the evolutionary potential of traits, and how genes act through developmental processes to produce variation at the level of the phenotype.

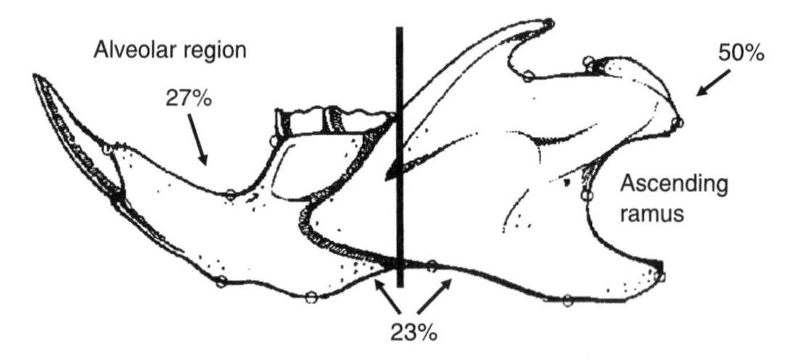

FIGURE 3. Genetic modules of the mouse mandible. The percentages show the number of QTLs that were found to affect only traits in either the ascending ramus or the alveolar region and QTLs found to affect traits in both regions (Mezey et al. 2000; figure from Wagner et al. 2004, Figure 15.2 © University of Chicago Press).

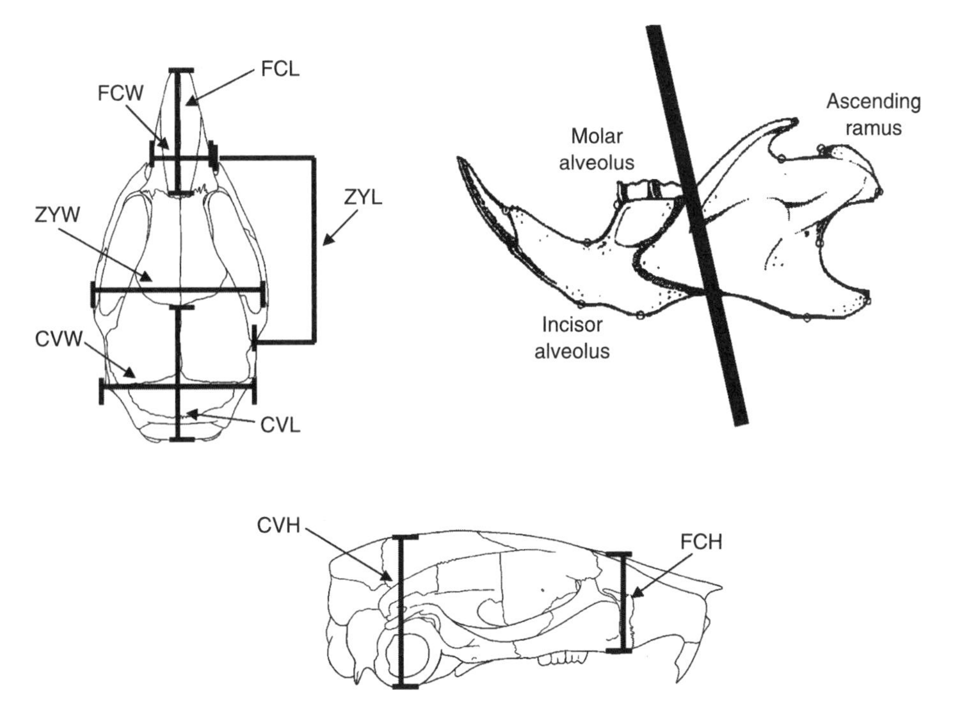

FIGURE 19.7. Modular pleiotropy in the skull (Leamy et al. 1999) and mandible (Ehrich et al. 2003). QTL effects are typically restricted to separate functional or developmental regions such as the face versus the braincase of the skull or alveolus versus the ascending ramus of the mandible.

detailed studies of mandibular morphology we found that most genes with pleiotropic effects had those effects restricted to specific developmental and functional regions, such as the alveolus or ascending ramus (Ehrich et al. 2003). Furthermore, QTL studies of skull morphology found that a majority of the genes affecting the skull had effects restricted to either the face or the braincase (Leamy et al. 1999). These two regions of the skull grow at different times and apparently do so under largely separate developmental pathways.

Reidl (1978) predicted that the structure of the genetic system would mimic that of the phenotype so that if the phenotype was modular, the genotype would evolve such a structure. Wagner and Altenberg (1996) also predicted modular gene effects because they facilitate adaptive evolution, allowing for the mosaic evolution of different developmental and functional systems. From our QTL studies it appears that morphological integration is due to modular gene effects. We propose that the range of pleiotropic effects associated with a locus becomes selectively restricted so that developmental and functional modules are represented in the genome by modular pleiotropic effects. Variation in pleiotropic effects due to differential epistasis forms the genetic basis for the evolution of pleiotropy. Modular pleiotropy also has different effects on correlated response to selection relative to situations where most loci are pleiotropic. Gromko (1995) showed that variation in correlated response was quite high in modular pleiotropic systems compared to populations where the genetic correlation was due to ubiquitous pleiotropy. Rice (2004) has also recently shown that hidden pleiotropic effects can have dramatic effects on the evolution of traits, whether they are correlated or not.

SUMMARY AND FUTURE DIRECTIONS

Over the past 15 years quantitative trait locus analysis has provided an unprecedented, if imperfect, view of the genetic architecture underlying complex traits. While much has been learned over the years, several issues remain to be addressed. First the number of segregating genes affecting trait variation in natural populations and the magnitudes of their effects remain unclear. QTL studies are not especially well suited for answering this question because the populations crossed are often preselected to have relatively few segregating factors of large effect or to have many segregating factors of small effect. The number and effect sizes of segregating genes are part of the experimental design rather than an outcome of the study. Furthermore, many studies have insufficient samples for accurate measures of gene effects. Smaller samples produce biased estimates of gene effect and number so that even if the populations considered differ by many genes of small effect there will appear to be only a few segregating genes of large effect.

It is perhaps possible to estimate how many loci can affect a complex trait by combining the results of many QTL studies of the same trait in different populations. For example, Snyder et al. (2004) tabulated the 133 obesity QTLs detected in 31 mouse intercrosses. Some regions appear repeatedly in this compilation and the distribution of the number of times regions have been detected is given in Figure 19.8. Fifty-six percent of the 133 QTLs are unique. However, these QTLs are mapped to rather broad, approximately 20 cM regions. These unique regions account for nearly all of the 1600 cM length of the mouse genome. It is also possible that many of these 20 cM wide QTLs represent the effects of several linked genes, so that the 75 unique obesity

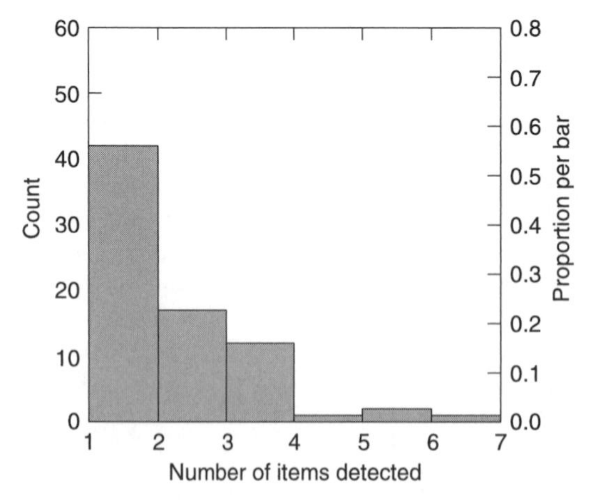

FIGURE 19.8. Frequency distribution of the number of times various obesity QTLs have been detected in a large variety of mouse crosses.

QTL regions should be considered as a minimum for the number of loci potentially affecting obesity.

In order to determine the number and size of gene effects on complex traits in natural populations it will be necessary to pursue QTL studies of closely related natural populations or linkage disequilibrium analysis in families within a single population. Sample sizes will need to be quite large if an accurate picture of the number and distribution of gene effects is to be obtained. Furthermore, highly polymorphic mapped markers are needed in order to detect QTL locations. To date, large samples of mapped markers are available only for humans and model systems. While in some instances markers can be transferred across closely related species this still leaves only a few species available for study. Perhaps as genomic approaches become more efficient in the future it will be possible to quickly develop polymorphic molecular markers in a wider variety of species.

Unlike the distribution of the number and size of gene effects, other aspects of genetic architecture can be approached more directly. Even when only a few large-effect genes are detected due to biases resulting from linkage and inadequate sample size, their characteristics, such as dominance, epistasis, and pleiotropy, may be examined. Most studies of complex traits find that gene effects are nearly codominant. Where dominance exists for morphological and body size traits, the heterozygote is most likely to be similar to the larger homozygote. Overdominance at single traits is relatively rare, occurring in specific situations such as early postnatal growth and for developmental plasticity. However, when a QTL has pleiotropic effects on multiple characters and the dominance values vary from trait to trait, combinations of those characters will display overdominance. Thus overdominance for combinations of traits may be common. Overdominant dimensions subjected to directional selection result in balancing selection and maintenance of genetic variation at the pleiotropic locus involved. This differential dominance could be a mechanism for maintaining genetic variability in populations.

Genetic interactions of small effect appear ubiquitous for body size and morphology so that, in addition to many small single-locus effects, small epistatic interactions are also common for these traits. Some fitness-related traits, such as litter size, maternal care, and developmental stability, show primarily epistatic interactions. Collectively, these small epistatic interactions are sufficient to limit the loss of additive genetic variance as populations pass through a bottleneck and to be responsible for the evolution of genotypic effects in populations due to genetic drift.

There are still problems in measuring the extent and influence of epistatic interactions in populations. Cheverud and Routman (1995) showed that epistasis contributes to all three population genetic variance components—additive, dominance, and epistatic—and that epistatic genotypic values can be defined independently of the allele frequencies at the loci considered. However, these methods were fully described only for experimental populations. Applications of these concepts to populations with natural allele frequencies are needed before the importance of epistasis for complex traits can be generally addressed in natural populations. While there has been substantial work on the interaction between epistasis and genetic drift, less is known about the interaction of selection and epistasis. Goodnight (2000a) was concerned with situations in which an allele or its alternative will be favored by selection, depending on genetic background. It is possible that differential epistasis (different epistatic patterns for different traits at the same loci) may result in overdominance for some combinations of traits resulting in balancing selection at the loci involved, as in differential dominance.

The most important aspect of epistasis as a component of genetic architecture is that it permits the evolution of aspects of genetic architecture itself, such as the presence and magnitude of single-locus additive and dominance effects and the range of pleiotropy. The concept of epistasis opens the door to a variety of theoretical and empirical studies of the role of epistasis in evolution.

Several studies of complex morphologies have indicated that pleiotropy is modular and hierarchical in nature. The effects of genes are often limited to nested sets of developmentally and functionally related traits. The existence of relationship QTLs and differential epistasis among traits affected by a pair of loci indicates that the range of pleiotropic effects displayed by a locus is subject to evolution by selection and drift. Much more work is required on the theoretical aspects of the interaction of selection with differential epistasis before we understand the full significance of this phenomenon.

Overall, our studies of the genetic architecture of complex traits suggest that variation in populations is due to many genes of relatively small effect with variable dominance, ubiquitous epistasis, and pleiotropic effects largely restricted to sets of functionally and developmentally related traits. The presence of epistasis indicates that the genetic architecture, by which we refer to the genotypic effects underlying a trait, is subject to evolution through genetic drift and natural selection. Models for the evolution of genotypic effects are relatively poorly developed, especially concerning the effects of selection on genetic architecture. Detailed studies of genetic architecture are also relatively rare and currently restricted to experimental rather than natural populations. There is substantial scope for future work in this field.

SUGGESTIONS FOR FURTHER READING

The original paper on interval mapping is by Lander and Botstein (1989) but Haley and Knott's 1992 paper discusses interval mapping in a multiple regression framework and are very clear on how to impute genotypes between markers. Mackay (2004) provides an overview of the tremendous scope of work by Mackay's group on QTLs in *Drosophila*. This paper provides entry to a large set of relevant literature.

Haley CS & SA Knott 1992 A simple regression method for mapping quantitative trait loci in line crosses using flanking markers. Heredity 69:315–324.

Lander ES & D Botstein 1989 Mapping Mendelian factors underlying quantitative traits using RFLP linkage maps. Genetics 121:185–199.

Mackay TFC 2004 The genetic architecture of quantitative traits: lessons from *Drospohila*. Curr. Opin. Genet. Dev. 14:253–257.

20

Evolution of Genetic Variance–Covariance Structure

PATRICK C. PHILLIPS
KATRINA L. McGUIGAN

One of the features of organisms that makes the study of biology so compelling is their apparent complexity. Molecular, cellular, developmental, physiological, neurological, and behavioral systems are each fascinating in and of themselves, but it is their interaction that generates what we see as the organism as a whole. By necessity, biologists have a tendency to break down organisms into their component parts to see how they work. Yet, no trait is an island (Dobzhansky 1956). In particular, we need to consider the organism in its entirety when looking at evolutionary change. This perspective is necessitated simply by the fact that it is the whole individual, not specific traits or particular alleles, that lives and dies. The whole individual is therefore the central unit of selection (Lewontin 1970). How can we deal with the complexity inherent in trying to integrate biological function and evolutionary change across the whole organism? Quantitative genetics provides an attractive potential framework because it attempts to summarize important features of the genetic system while focusing firmly at the level of the phenotype. The multiple trait (multivariate) formulations of quantitative genetics that we will discuss here can be seen both as a conceptual framework for integrating multiple systems at many levels and as a tool for making quantitative predictions about evolutionary change. Whether it succeeds on either count is the subject of active debate (Barton & Turelli 1989). Ways of successfully building upon the traditional statistical framework of quantitative genetics is one of the exciting areas of future growth within the field of evolutionary genetics.

A preoccupation of many evolutionary biologists is the prediction of future response to selection and/or the retrospective estimation of the selection pressure responsible for contemporary phenotypes. In trying to reconcile experimental simplicity with the reality that selection acts on the whole organism, we are generally concerned with multiple traits, each of which has a polygenic basis and varies continuously (not discretely). For these suites of traits, past and future evolution can be explored using the multivariate extension of the breeders' equation (Lande 1979; Lande & Arnold 1983; Roff, Ch. 18 of this volume):

$$\Delta \mathbf{z} = \mathbf{G}\boldsymbol{\beta}. \qquad (20.1)$$

Response to selection (the vector $\Delta\mathbf{z}$) depends jointly on the strength and direction of selection (the vector $\boldsymbol{\beta}$), and the matrix of additive genetic variances and covariances, G (Roff, Ch. 18 of this volume) (Those unfamiliar with the concept of covariance should consult the brief exposition in Box 20.1.) The contribution of G to evolution is nontrivial, influencing not just the response of traits directly under selection but also the rate and direction of evolution in traits that genetically covary with selected traits (Figure 20.1).

As the name suggests, the breeders' equation has been used extensively by plant/animal breeders to select for stock with particular attributes. Typically, breeders are interested in response to selection in the short term. In contrast, evolutionary biologists are frequently interested in long-term responses.

BOX 20.1. What Is a Covariance?

Before getting carried away discussing the importance of genetic covariances for the evolutionary process, it is necessary to understand what a covariance is. Statistically, a covariance describes the association between two variables. It is closely related to the more familiar concepts of regression and correlation, which are covariances normalized on different scales. The covariance between X and Y is given by:

$$\sigma_{xy} = \mathrm{E}\left[\left(X - \overline{X}\right)\left(Y - \overline{Y}\right)\right]$$

and can be estimated as

$$\mathrm{Cov}(X, Y) = \sum_{i=1}^{n}\left(X_i - \overline{X}\right)\left(Y_i - \overline{Y}\right)/(n-1).$$

The correlation between X and Y is the covariance normalized by the standard deviation of each trait $\rho_{xy} = \sigma_{xy}/\sigma_x\sigma_y$, while the regression of Y on X is the covariance normalized by the variance of X ($\beta_{XY} = \sigma_{XY}/\sigma_X^2$). The covariance of a variable with itself is simply its variance ($\sigma_{XX} = \sigma_X^2$). Covariances, rather than correlations, are used to describe the evolution of traits because they are not normalized, existing on the same scale as the traits themselves.

At the genetic level, a covariance between two traits is generated when alleles affecting both traits tend to be found within the same individual (Box 20.2). Genetic covariances can be estimated using standard quantitative genetic approaches (Roff, Ch. 18 of this volume), for instance, from the regression of offspring values for trait X on the parental values for trait Y (and vice versa). As second-order statistics, covariances require large sample sizes to be accurately estimated. As genetic covariances are subject to sampling variation at more than one level (genetic and environmental), the accuracy of their estimation is even more troubling. Only a few studies in the evolutionary literature are of sufficient sample size to provide more than the crudest estimates of the pattern of genetic covariance among traits.

With only two traits, the genetic covariance is easy to interpret, but as we consider more traits it becomes more difficult to interpret the pattern of genetic covariances. Principal components analysis is one statistical tool used to interpret patterns of genetic covariation among traits (Figure 20.3; see Phillips & Arnold 1999). Principal components analysis generates new variables (principal components), which are linear combinations of the "traits" (genetic variances and covariances) in **G**. As vectors, principal components have direction and length. The first principal component describes the trait combination (direction) for which there is the most genetic variance (length). Subsequent principal components will describe less variation, but unique directions. G is then interpreted by its orientation (the directions of its principal components) and size (the length of principal components). For example, two matrices that are proportional to one another (an important prediction for genetic drift, for instance) would have identical principal component orientations, but the dimensions along each component would be proportionally expanded or contracted (see Figure 20.3).

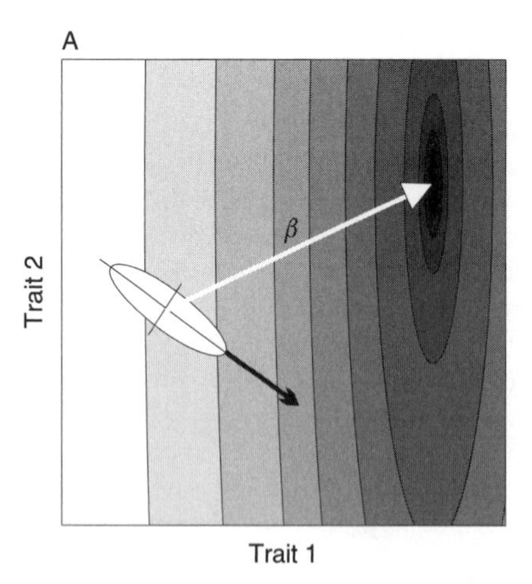

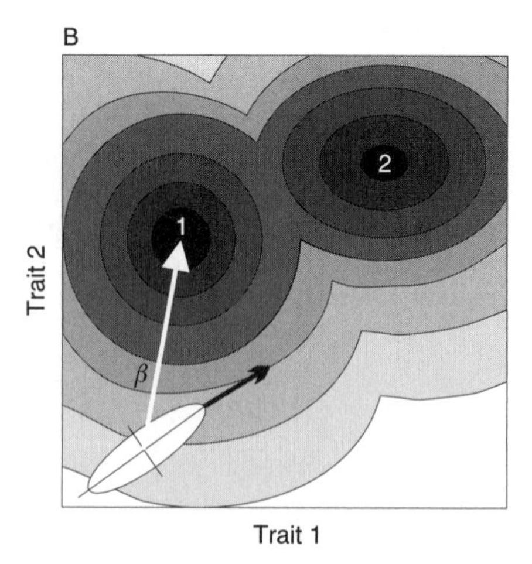

FIGURE **20.1.** The response of two genetically correlated traits to correlational selection. The orientation and size of **G** is described by the white ellipse, with the narrow oval shape indicating most of the genetic variance is common to both traits, rather than independent (i.e., covariance). The cross-hairs of the ellipse indicate the current population mean for the two traits; the position of the population on the fitness surface. The darker the region of the fitness landscape, the higher the fitness, with black areas indicating local adaptive peaks; regions shaded the same color have equal fitness. **β** (white line) describes the most direct approach of the population to the nearest fitness peak. (A) Traits are negatively genetically correlated, but there is selection for an increase in both traits; the initial direction of evolution will be away from the fitness peak. The population will eventually ascend the fitness peak, but this will take longer than if **G** were oriented in the same direction as the fitness surface (i.e., along **β**). (B) Traits are positively correlated and the population is in the neighborhood of two fitness peaks, both of which generate selection for an increase in both traits. However, because of the genetic covariance among traits, the population will evolve toward peak 2 rather than peak 1.

The utility of the breeders' equation in evolutionary biology depends on **β** and **G** remaining relatively stable over the time frame of interest. Gain and loss of traits through evolutionary time is sufficient to demonstrate that **G** changes. However, we know very little about *how* **G** evolves, which makes it difficult to predict *when* **G** will evolve, and thus when we can or cannot assume constancy and employ Equation 20.1 in the analysis of evolution.

The G matrix evolves through processes that alter allele frequencies: mutation, migration, selection, and drift (Arnold 1992; Barton & Turelli 1987; Lande 1980b). Conditions under which **G** might remain relatively constant have been predicted both from theory (Turelli 1988a) and computer simulations (Jones et al. 2003), but we do not know how often these conditions are met in natural populations. Empirically, the data suggest **G** might remain relatively stable over the short term, but does not always do so (reviewed by Jones et al. 2003). Overall there is little support for researchers assuming a stable **G** in their system. Thus, understanding the evolution of **G** remains an outstanding issue in quantitative genetics.

Quantitative geneticists will continue addressing these issues by conducting computer simulations of evolution and by estimating **G** from populations under a range of conditions. An emerging mechanistic approach to tackling these same issues is built upon the fact that **G** summarizes large amounts of information about the physical genetic and developmental bases of traits. **G** is estimated from observations on the phenotypic similarity among relatives, and we have historically been unable to directly observe what underlies **G**. Recent advances in developmental and molecular genetics are bringing us to a position where we can break down a **G** matrix so as to consider the contributions of each locus to each trait. This interface between quantitative and molecular genetics may provide insight into the evolutionary processes shaping G, thereby expanding the framework within which we can understand the evolution of organisms as an integrated whole.

CONCEPTS

What Makes G?

In many ways, the G matrix can be thought of as a nexus connecting genetic information with evolutionary processes. On one hand, the structure of G is determined by the functional and developmental interactions that generate pleiotropy and by the genomic features that characterize linkage relationships among loci. On the other, G determines the rate and direction of the response to selection, as well as the pattern of divergence among populations. G is a statistical abstraction stuck between these levels, seeking to sufficiently summarize the molecular details at one end so that it can serve as an adequate predictor of evolutionary response at the other. Lofty goals. The elements underlying these aims bear closer scrutiny.

The essential genetic underpinnings G can be appreciated by a careful study of Lande (1980b)—which is somewhat akin to saying that the essential underpinnings of sub-atomic physics can be appreciated by a careful study of a piece of fruitcake.[1] Things are not always as self-evident as they could be. Notational difficulties aside, the fundamental concepts are actually relatively straightforward. The most important step toward understanding what G summarizes at the genetic level is to take a very general view of allelic effects at a given locus (Box 20.2). First, imagine that a large set of alleles at a single locus might generate a continuous distribution of effects on a given trait. Next, note that each allele can also affect more than one trait (pleiotropy), such that different alleles at a locus will have different levels

[1]With apologies to Douglas Adams—and Russ Lande.

BOX 20.2. Pleiotropic Effects

We might have a tendency to think about pleiotropic effects in a very diffuse way, perhaps calling forth the traditional definition of a single locus that affects more than one trait. To understand the genetic basis of covariances we need a much more precise view of pleiotropy—one that focuses at the level of the allele rather than the locus as a whole. In particular, some alleles at a given locus might have pleiotropic effects while others do not. More subtly, the direction and magnitude of the pleiotropic effects on a suite of traits might vary from allele to allele. This requires that every locus be represented by the range of pleiotropic effects generated by each allele, as in Figure 1.

The array of all possible allelic effects at a given locus will generate a distribution of effects, as shown in Figure 2. Each locus might have its own distribution, and, in

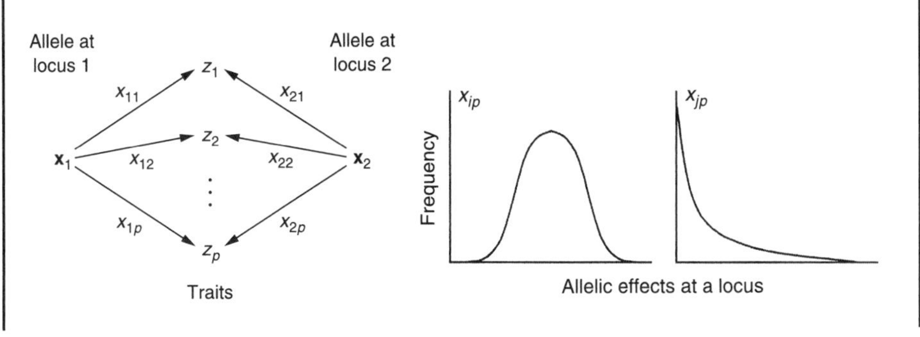

(continued)

BOX 20.2. *(cont.)*

general, we do not really know what these distributions tend to look like. Popular possibilities are fairly "normal" symmetrical distributions (Figure 2, left) and "L"-shaped, highly skewed distributions (Figure 2, right). To assess the pleiotropic contributions of each allele, these distributions must be combined in a bivariate (and ultimately multivariate) view. Alleles at a given locus can potentially have any pattern of pleiotropic effects, including no pleiotropy at all. The effect of correlational selection (Figure 20.1) might have on the pattern of pleiotropic effects of the underlying alleles can be visualized by laying a fitness contour on top of the distribution of allelic effects. Correlational selection occurs when fitness depends on a combination of trait values rather than on the individual traits themselves (e.g., on how the upper and lower mandibles fit together to make a single fruit-cake-cracking jaw). Given unconstrained effects, the distribution of allelic effects should evolve to match the pattern of selection. The actual translation between the alleles is mediated through the mapping of the allelic effects onto the traits under selection (Lande 1980b). Mutational input can disturb the pattern of pleiotropy in each generation, so if there is bias in the pattern of mutational input, the distribution of effects will evolve to match that bias to some degree Figure 3. The equilibrium distribution of effects will therefore depend on a balance between selection and mutation, as well as other factors (Box 20.3).

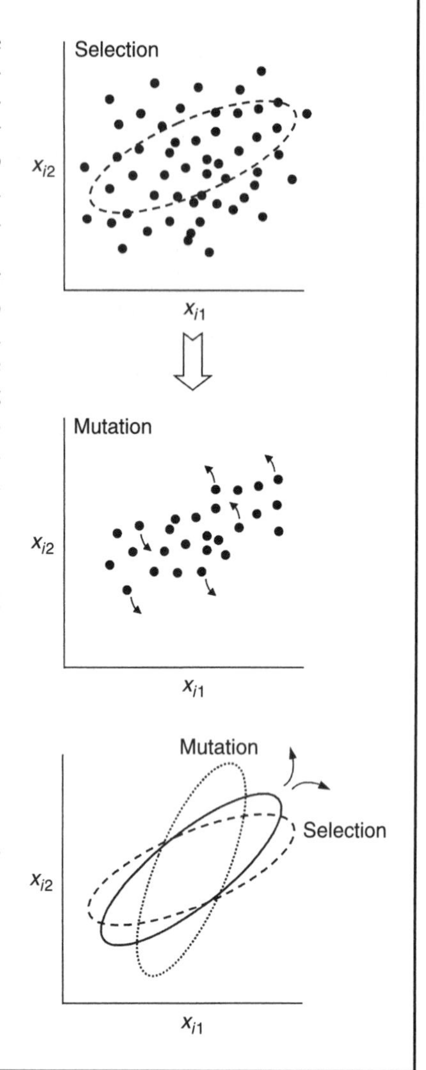

of direct and pleiotropic effects on the suite of traits under study. We can represent the effect an allele at locus i has on trait p as x_{ip}. The whole array of pleiotropic effects at the locus is then represented by the vector $\mathbf{x}_i$ (Box 20.2). Lande (1980b) in fact uses only this vector notation for compactness. If we really wanted to proliferate subscripts, we could have included another subscript (say k) to indicate that x_{ikp} is the effect of the kth allele from locus i on trait p. Instead, x_{ip} serves more like a random variable that indicates the particular value that we have drawn from the infinite set of possible allelic effects at locus i. In practice, we often want to average over all of these effects. We represent this averaging using the expected-value function, E[], as in $E[x_{ip}] = \bar{x}_{ip}$. This expectation is taken over all of the individuals in the population and so is dependent on the underlying frequencies of the alleles.

The pattern of association of alleles both within and between loci will generate the pattern of genetic variance and covariance for the traits. For now, we will use a system with only two traits for simplicity.[2]

[2]It will be rough going for a bit here. Try to stay with the big picture point of view at first. You can always come back for the details if interested.

Following Lande (1980b), there are three possible sets of allelic association that need to be considered. First, there are associations among the effects of alleles segregating at a *single locus*. Each allele at locus i can have a particular influence on each of the two traits (Box 20.2), so we will collect their effects together in a single matrix, $\mathbf{C}_{ii}$:

$$\mathbf{C}_{ii} = E\left[\begin{pmatrix} (x_{i1} - \bar{x}_{i1})^2 & (x_{i1} - \bar{x}_{i1})(x_{i2} - \bar{x}_{i2}) \\ (x_{i1} - \bar{x}_{i1})(x_{i2} - \bar{x}_{i2}) & (x_{i2} - \bar{x}_{i2})^2 \end{pmatrix}\right] = \begin{pmatrix} \sigma_{i1}^2 & \sigma_{i1i2} \\ \sigma_{i1i2} & \sigma_{i2}^2 \end{pmatrix}, \quad (20.2)$$

where σ_{i1}^2 is the usual per-locus additive allelic or genic variance affecting trait 1 and σ_{i1i2} is the additive covariance between traits 1 and 2 contributed by pleiotropic effects of alleles at locus i (Box 20.2). These latter terms will be generated when alleles at a given locus tend to affect more than one trait, either directly (e.g., a transcription factor turning on a similar set of genes within two different tissues) or indirectly (e.g., a gene that affects metabolic efficiency influencing both growth rate and egg production). *Big picture*: $\mathbf{C}_{ii}$ give the single-locus-affecting-multiple-traits perspective that is the hallmark of pleiotropy and that is frequently the focus of most functional interpretations of genetic covariance structure.

Second, there can be an association between alleles at *different loci* within the same gamete, summarized by $\mathbf{C}_{ij}$:

$$\mathbf{C}_{ij} = E\left[\begin{pmatrix} (x_{i1} - \bar{x}_{i1})(x_{j1} - \bar{x}_{j1}) & (x_{i1} - \bar{x}_{i1})(x_{j2} - \bar{x}_{j2}) \\ (x_{i2} - \bar{x}_{i2})(x_{j1} - \bar{x}_{j1}) & (x_{i2} - \bar{x}_{i2})(x_{j2} - \bar{x}_{j2}) \end{pmatrix}\right] = \begin{pmatrix} \sigma_{i1j1} & \sigma_{i1j2} \\ \sigma_{i2j1} & \sigma_{i2j2} \end{pmatrix}. \quad (20.3)$$

Each of the terms in this matrix is a covariance of effects across loci within a gamete (i.e., *cis*-acting effects) and are the associations that we would expect to be primarily generated by physical linkage along a chromosome. *Big picture*: linkage between loci (or between genetic elements within genes) contributes to genetic correlations because these linked effects tend to be inherited together. As one breaks down the elements of a gene (say to the nucleotide level), the distinction between linkage and pleiotropy becomes somewhat arbitrary.

Third, the effects from alleles of loci located on *different gametes* (i.e., *trans*-acting effects) can be correlated with one another, as in $\mathbf{C}'_{ij}$:

$$\mathbf{C}'_{ij} = E\left[\begin{pmatrix} (x_{i1} - \bar{x}_{i1})(x_{j'1} - \bar{x}_{j'1}) & (x_{i1} - \bar{x}_{i1})(x_{j'2} - \bar{x}_{j'2}) \\ (x_{i2} - \bar{x}_{i2})(x_{j'1} - \bar{x}_{j'1}) & (x_{i2} - \bar{x}_{i2})(x_{j'2} - \bar{x}_{j'2}) \end{pmatrix}\right] = \begin{pmatrix} \sigma_{i1j'1} & \sigma_{i1j'2} \\ \sigma_{i2j'1} & \sigma_{i2j'2} \end{pmatrix} \quad (20.4)$$

The prime indicates the allele is from the other gamete. How can this be? How can loci from different gametes with effects on different traits possibly matter? If the loci have a tendency to be inherited together, say by assortative mating, then they can indeed influence the correlation across seemingly unrelated traits. Terms in this matrix are therefore especially important in models of sexual selection that depend on a build-up of genetic covariance between traits such as male display and female preference. *Big picture*: correlations across gametes (contributed separately by a mother and father) can contribute to a genetic correlation if the effects of assortative mating can overcome the natural tendency for independent segregation among gametes. In most quantitative genetic models, except those explicitly concerned with sexual selection, these effects are usually assumed to be negligible.

The G-matrix itself is the sum of these three potential sources of covariance of allelic effects,

$$\mathbf{G} = 2\sum_{i=1}^{n}\sum_{j=1}^{n}(\mathbf{C}_{ij} + \mathbf{C}'_{ij}), \quad (20.5)$$

with the sum extending over allelic covariances for all n loci (Lande 1980b). This is a little easier to interpret by writing the elements of $\mathbf{G}$, say for traits p and q, after separating the terms from Equation 20.5 based on their definitions from Equations 20.2–20.4:

$$G_{pp} = 2\sum_{i=1}^{n}\sigma_{ip}^2 + 2\sum_{i=1}^{n}\sum_{j=1}^{n}\sigma_{ipjp} + 2\sum_{i=1}^{n}\sum_{j'=1}^{n}\sigma_{ipj'p}, \quad (20.6a)$$

$$G_{pq} = 2\sum_{i=1}^{n}\sigma_{ipiq} + 2\sum_{i=1}^{n}\sum_{j=1}^{n}\sigma_{ipjq} + 2\sum_{i=1}^{n}\sum_{j'=1}^{n}\sigma_{ipj'q}. \quad (20.6b)$$

Here G_{pp} is the additive genetic variance for trait p, and G_{pq} is the additive genetic covariance between traits p and q. This breaks the composition of the **G** matrix into three separate (and fairly comprehensible) pieces. In each case, the first term refers to the pleiotropic effects of each locus, the second term refers to the variance or covariance generated by linkage disequilibrium, and the third term refers to the covariance among alleles that can be generated by nonrandom (or assortative) mating.

G matrices thus represent covariances of covariances. Pleiotropic effects of alleles at a single locus, linkage of alleles with similar effects at multiple loci, and associations of alleles across gametes generated by nonrandom mating can all generate covariance in genic effects, which in turn generate the genetic covariances observed in **G**. The actual values of the covariances in **G** will depend on the frequencies of the underlying alleles. Like all quantitative genetic models, this approach sums over the individual per locus effects so that we deal solely with the summed variances of genic effects. The genetic covariances therefore average over many individual genic effects, which can even potentially cancel each other out (e.g., Gromko 1995). Ultimately, it is the genic effects themselves that harbor the interesting information regarding the functional basis for the genetic coupling between traits (Box 20.2), but it is the overall genetic effects that we can observe in the resemblance between relatives.

Evolution of G

Although the pattern of genetic covariation among traits can play an important role in the evolutionary response of those traits, as exemplified by Equation 20.1 (Figure 20.1), it is equally interesting to ask how **G** itself will evolve. Extrapolating from Lande (1980b), the deterministic changes in **G** can be modeled using the equation shown in Box 20.3. From a theoretical perspective, we can consider both how deterministic processes, such as selection and mutation, and stochastic processes, such as drift, change **G**.

BOX 20.3. Evolution of the **G** Matrix

$$\Delta G = \underbrace{G(\gamma - \boldsymbol{\beta}\boldsymbol{\beta}^{T})G}_{\text{Selection}} + \underbrace{2M}_{\text{Mutation}} - \underbrace{2\sum_{i=1}^{n}\sum_{j=1}^{n} r_{ij}(C_{ij} - C'_{ij})}_{\text{Recombination}}$$

This is a composite equation that shows how different evolutionary forces influence the evolution of the genetic variance–covariance (**G**) matrix (mostly based on Lande 1980b). The first term describes the influence of natural selection, where γ is the average curvature and orientation of the individual selection surface and β is the average slope of the individual selection surface (Lande & Arnold 1983). Taken together, these two elements describe the local curvature and orientation of the adaptive landscape (Phillips & Arnold 1989). The response to selection is generated through a balance between the tendency of directional selection to erode genetic variation and the ability of stabilizing and correlational selection to reorient the pattern of among-trait correlation. In this term, the influence of selection is reflected back on change in **G** through the lens of the existing pattern of genetic covariance. The second term describes the role that new mutations play in structuring **G**. This is summarized by the mutational covariance matrix **M**, where $\mathbf{M} = \Sigma_i \mathbf{M}_i$ is the sum of the mutational covariances generated by the pattern of pleiotropic mutation at each locus. In the absence of other evolutionary forces, **G** will tend to match any biases induced through mutational effects guided by, say, developmental processes. Finally, the third term describes the rate of degradation of linkage-induced genetic covariance generated by recombination, where r_{ij} is the recombination rate between loci i and j ($r_{ii} = 0$). Additional terms are needed when the distribution of allelic effects is not normal (e.g., Barton & Turelli 1987).

Selection

Directional selection ($\boldsymbol{\beta}$) alone is expected to cause an erosion of genetic variation, but the pattern of correlational selection (γ) can potentially mold the pattern of genetic covariation. The term ($\gamma - \boldsymbol{\beta\beta}^{T}$) in Box 20.3 is actually the curvature of the adaptive landscape (Phillips & Arnold 1989), so one interesting prediction from this model is that **G** should evolve to match the orientation of the adaptive landscape (Box 20.2; Cheverud 1984; Jones et al. 2003). However, the actual pattern of **G** will depend on the balance of selection with other evolutionary forces.

Mutation and Recombination

The potential importance of mutation and recombination in the evolution of **G** remains relatively unexplored. The influence of recombination is usually ignored because selection ordinarily needs to be fairly strong in order to overcome the ability of even small amounts of recombination to eliminate linkage disequilibrium. Intermittent admixture of isolated populations can generate substantial amounts of linkage disequilibrium, however, as will strong assortative mating among similar genotypes. The relative importance of linkage disequilibrium will ultimately depend on the genomic organization of genes influencing suites of correlated traits. Traditional population genetic models have tended to treat loci as equivalent to distinct genes, which are usually assumed to be loosely linked. Emerging insights into molecular genetics may require changes in these assumptions. For instance, if one treats the regulatory regions of genes separately from their translational products, then these parts of a "gene" need to be treated as two distinct, tightly linked loci. In more explicit models of developmental regulation, considering the influence of tight linkage between separate factors will become more important.

There is an even more pressing need to consider the influence of mutational covariance on the evolution of **G**. Over the last few decades, evolutionary geneticists have speculated extensively on the potential role of genetic correlations as an evolutionary constraint (Barton & Turelli 1989). The true nature of these constraints is most likely to be revealed in the pattern of mutational covariance, since it is against this background of variance that selection can act and the structure of the **G**-matrix is determined (Box 20.2; Jones et al. 2003). It remains an open question whether or not patterns of developmental interactions among genes will influence new mutations to adhere to constrained pleiotropic pathways or whether most loci are instead capable of a wide array of possible phenotypic effects (Box 20.2). Any such mutational biases are bound to have an important influence on the long-term evolution of **G**. As difficult as accurately estimating the elements of **G** can be, estimating the components of **M** is much harder and unfortunately still leaves us far removed from the distribution of the mutational effects themselves.

Genetic Drift

Stochastic variation in gametic frequencies generated by genetic drift will influence both the evolution of the **G**-matrix itself (Phillips et al. 2001) and the direction of phenotypic evolution, which is mediated by **G** (Lande 1979). Drift can change **G** both through sampling alleles with differing pleiotropic effects and through a build-up (or change in the pattern) of linkage disequilibrium (Equations 20.6 and 20.7). The former effects might be expected to be more persistent than the latter, although linkage can potentially decay fairly slowly (Whitlock et al. 2002). Unfortunately, we do not have a firm theoretical handle on the importance of drift through either of these processes (Phillips et al. 2001). Within a population, drift is expected to influence all genetic variances and covariances similarly; they should decline at a rate of $(1 - 1/2N_e)$ per generation, where N_e is the effective population size. If mutation restores some of this variation every generation, then we would expect the long-term mutation–drift equilibrium for **G** to be $2N_e\mathbf{M}$ (after Lynch & Hill 1986). Thus, we expect both the short-term and long-term effects of drift to lead to proportional changes in **G** (Box 20.1; Phillips et al. 2001). As we discuss below, however, there can be tremendous sampling variation around this expectation, such that for any given population the orientaion of the **G**-matrix can change substantially due to drift (Phillips et al. 2001), especially over extended periods of time (Whitlock et al. 2002).

When populations diverge through drift, it is predicted that the orientation of population means will match the orientation of **G** (Lande 1979; Phillips et al. 2001). This is the same expectation as for the correlated response to selection on a set of traits (Lande 1979), which unfortunately means the observation that divergence is aligned with the major direction of genetic covariation (Schluter 1996) cannot be used to distinguish whether drift and selection has been responsible for the divergence among populations (although see McGuigan et al. 2004).

Assumptions and Complications

The equations and predictions given above are only first approximations of the true complexity underlying the evolution of genetic associations among traits. First, the equations describing the evolution of the G-matrix are derived under the assumption that the effects of alleles (both within and between loci) are additive. This is the only way to easily write the genic covariances in terms of the deviation of the effects of a particular allele from effects of other alleles at the locus. If that deviation depends on genetic context, either because of dominance of other alleles at the same locus or by interactions between that allele and alleles at other loci (epistasis), then things can get much more complicated. Epistasis, particularly when combined with varying levels of linkage, could potentially have a large impact on the apparent pattern of pleiotropy at a given locus. Imagine, for example, an enhancer region that modulates expression of a given gene in different tissues. The pattern of pleiotropy will depend on changes in the enhancer, on changes in the structural gene or, most likely, on some coordinated set of changes shared between them. To some extent, complications caused by nonlinear interactions among genetic factors can be statistically accounted for, especially for predictions of changes in trait means (Roff, Ch. 18 of this volume), but their effects on variances are complex (Whitlock et al. 1995) and their influence on genetic covariances is virtually unexplored (Lopez-Fanjul et al. 2004).

An equally complex but more subtle complication is caused by differences in the distribution of allelic effects (Box 20.2). These distributions do not affect the definition and interpretation of **G** in a static sense (Equations. 20.2–20.6), but even if we assume all effects at a locus are additive, the distribution of allelic effects can have an important influence on the evolution of the elements of **G**. When allelic effects are normally distributed, as assumed by Lande (1980b), Equation 20.7 can be used to describe changes in **G**. The essential underlying result here is that even as the elements of the allelic covariance matrices change, under normality the linear relationship between the allelic effects and the phenotypic traits does not (Lande 1980b). When the distribution of effects is non-normal, however, there is a complex interplay between the higher moments of the allelic effects and changes in the moments of the traits themselves (Barton & Turelli 1987). For example, if there is skew in the distribution of allelic effects, changes in the trait mean will lead to changes in the genetic variance of that trait, irrespective of changes in variation in the underlying alleles (Barton & Turelli 1987). More complex approaches that explicitly include allelic dynamics provide a strong direction for further work (e.g., Kirkpatrick et al. 2002), although a general set of results, especially with regard to genetic covariances, remains elusive. In the end, all of these complications suggest that we need to know much more about the specific nature of the genetic effects of interest before we will be able to determine what level of theoretical abstraction is appropriate.

CASE STUDIES

There are two broad empirical approaches to determining how **G** evolves: manipulative laboratory experiments and comparative field experiments. Each approach has its advantages and disadvantages (Table 20.1). Generally, laboratory studies can be used to generate predictions about which evolutionary process generate what patterns in **G**, while comparative studies of natural populations can identify what patterns (and therefore processes) occur in the wild. Table 20.1 suggests the choice of natural populations is more difficult than the choice of taxa for laboratory systems. Meeting the stringent criteria for natural populations makes it possible to extrapolate results of the study to other natural systems. Here, we use case studies to discuss the attributes and methods of comparative field investigations of **G** versus manipulative laboratory studies of the evolution of **G**.

Comparative Study: Butterfly Wing Patterns

Butterflies have the short generation time and limited space requirements that, when coupled with extensive natural variation in size, shape, color, and pattern of their wings (Figure 20.2), make them attractive for quantitative genetic studies. Despite this variation, basic pattern elements are relatively easy to identify on most wings, facilitating comparison

TABLE 20.1. Relative attributes of comparative and manipulative experiments to investigate the evolution of **G**

	Natural comparative	Manipulative laboratory
Utility in determining which evolutionary processes act on **G** and what patterns they generate	• Can infer which processes operate in natural populations, and how processes interact • But, difficult to estimate parameters of selection, drift, migration, or mutation	• Cannot infer which processes operate in natural populations, nor how they interact • But, can estimate and/or control the strength of selection, mutation, migration, and drift, and therefore determine the cause of observed patterns
Limitations on choice of taxa	• Quantitative genetic parameters frequently estimated in the laboratory due to difficulties in estimating relationships in wild populations (although the latter is greatly preferable). Thus, usually limited to organisms with easy husbandry and ability to conduct controlled matings/crosses • Generation time less critical because evolution has already occurred • Relationships between the populations should be known from well-supported phylogenies or historical data • Differences (or lack thereof) in selection pressure between populations should be known or inferred • Population parameters that affect drift or response to selection (e.g., population size and generation time) should be estimated	• Need to house in laboratory restricts choice to taxa with easy husbandry and limited space requirements • Generally conducted on taxa with short generation times to keep experimental duration short when populations need to be taken through multiple generations of evolution • Generally begin with a single stock, and generate several populations (lines) which are subject to known evolutionary process • Population parameters known and controlled
Replication/sample sizes	• Replication limited by time/space of researcher, but also by the availability of wild populations/individuals • Require high replication of populations due to the multitude of evolutionary processes acting (e.g., many selective forces) • Require high replication within populations due to variability in natural habitats	• Replication/sample size usually limited by space and manpower rather than availability • Control the evolutionary processes acting on the population, so do not need sample multiple populations to have the power to estimate processes • Frequently, lower environmental variance, so more accurate estimates of **G** for a given sample size
Genetic tools	• Need to develop genomic resources for interesting natural systems	• Often, molecular genetic and developmental tools are also available for taxa that are tractable for laboratory quantitative genetic analyses

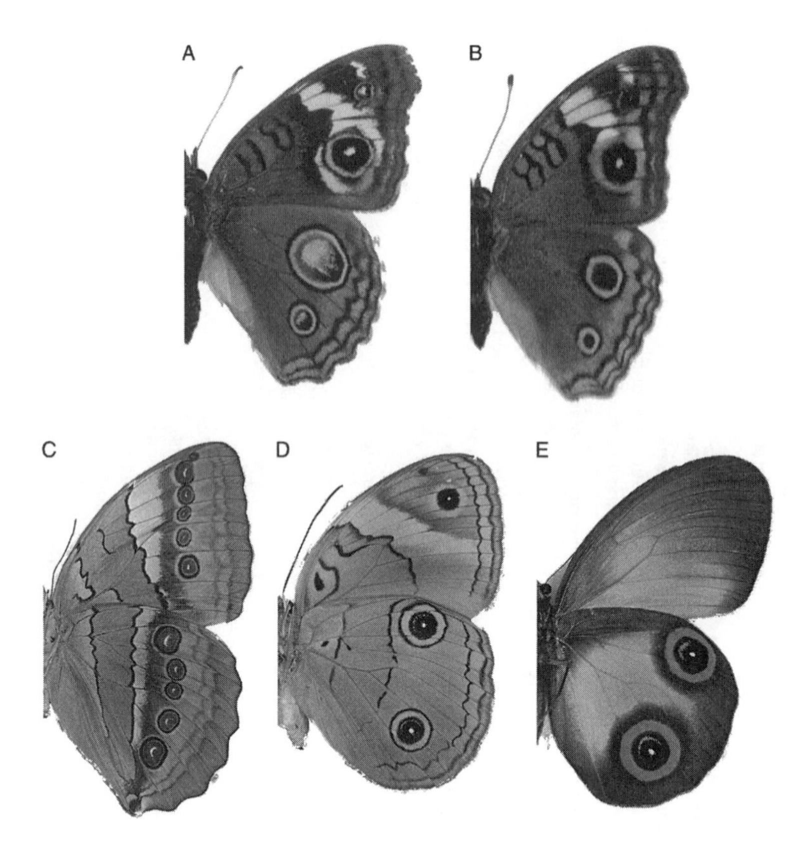

FIGURE 20.2. Variation in dorsal eyespots of nymphalid butterflies. (A) *Junonia coenia* (Nymphalinae); (B) *J. evarete* (Nymphalinae); (C) *Stichophthalma camadeva* (Morphinae); (D) *Faunis menado* (Morphinae) and; (E) *Taenaris macropus* (Morphinae). Note the difference in size between the two hindwing eyespots in *J. coenia,* but not *J. evarete.* Pictures kindly provided by F. Nihout based on those originally presented in Nijhout (1991) (cf. Nijhout's Figure 2.2 and Figure 5.18A and C).

among taxa (reviewed by Nijhout 1991). Wing pattern divergence (Δ**z**) has been investigated by trying to determine the causative selective force (**β**) acting upon them, as well as by studying their genetic basis (**G**). It has often proven difficult to determine the function of some wing pattern elements (e.g., Lyytinen et al. 2003) and therefore the nature of their selection, but the genetic basis of the traits is amenable to investigation.

Artificial selection to increase or decrease the size of one eyespot in *Bicyclus anynana* resulted in evolution (in the selected direction) of other eyespots (Monteiro et al. 1994). By inference, eyespot size genetically covaries positively among eyespots. This inference was supported by the calculation of genetic correlations among eyespots in two species of butterfly: *Junonia (Precis) coenia* and *J. evarete* (Paulsen 1996). All bivariate genetic correlations among eyespot sizes were positive. Paulsen (1996) also included other wing pattern traits in the **G**, with an overall conclusion that the size of similar pattern elements (e.g., eyespots) tended to positively covary, whereas size among different pattern elements (e.g., wing bands) showed no covariation of size. Similarly, there was little correlation between size and position of the pattern elements. The pattern of genetic correlations appeared similar between Paulsen's two species, and a similar pattern of correlation among replicates of the same element, but not among different elements, has been observed in other butterflies (e.g., Kingsolver & Wiernasz 1991; Monteiro et al. 1994). Is the genetic basis of wing pattern in butterflies similar across taxa and can we use the relationships in the breeders' equation to determine the nature of selection responsible for current patterns of diversity?

Using both matrix-wide and an element-by-element approaches, Paulsen (1996) tested the hypothesis that **G** did not differ between the closely related *J. coenia* and *J. evarete*. No differences between matrices were detected; we can conclude that **G** has not significantly diverged between these *Junonia* species. Can we then assume the genetic relationships among traits are the same for all other butterfly species? To ascertain how broadly applicable the result is we need to consider why **G** remained stable. The first hypothesis to consider is that *J. coenia* and *J. evarete* began diverging too recently for differences in allele frequency (due to mutation, selection or drift) to accumulate. Putative divergence time is positively correlated with divergence in phenotypic covariance matrices (**P**) (e.g., Baker & Wilkinson 2003; Steppan 1997), suggesting this as a plausible hypothesis. Since we do not know how long *J. coenia* and *J. evarete* have been diverging, we cannot infer what a reasonable time frame for assuming stability of **G** might be. The best way to determine whether the observed stability is due to recent divergence is to compare these **G**-matrices with those of other butterflies within the framework of a known phylogeny. Using a phylogenetic comparative approach, if we observe more recently speciated butterflies to have divergent **G**, we can ask whether stabilizing selection has maintained the ancestral pattern of genetic covariation among traits in our two species.

What generates/maintains the genetic interrelationships among these traits? The orientation of **G** will be determined by an interaction between genetic variation, selection, and drift (Box 20.3). If functional relationships among traits generate patterns of correlational selection, then the **G** should reflect these relationships. Similarly, developmental relationships among traits will influence the pattern of mutational input into **G**, and selection should also shape development itself. **G** may therefore be an interesting place to look for a signature of the interaction between function and development. Kingsolver and Wiernasz (1991) estimated **G** for melanin patches on the wings of the butterfly *Pieris occidentalis*. Using data from previous studies, they generated a matrix that described the hypothetical thermoregulatory functional relationships among melanin patches and a matrix that described the hypothetical developmental relationships among patches. Comparing each of these with **G**, they concluded the observed patterns of genetic covariaton were due to both functional and developmental relationships among traits (Kingsolver & Wiernasz 1987).

Hypotheses about the function of eyespots suggest different roles for dorsal eyespots, which are not exposed at rest, and ventral eyespots, which are exposed at rest (e.g., Breuker & Brakefield 2002; Lyytinen et al. 2003). The existence of a positive genetic covariation of eyespot size across both wing surfaces therefore cannot be generated by a functional relationship between these traits. Although developmental hypotheses have not been formally tested, similar developmental mechanisms (namely response to a morphogen gradient) appear to operate in the formation of all eyespots (reviewed by Brakefield 2001), suggesting **G** is influenced by the developmental relationships. If this assessment of the relative influences of development and function is correct, we can predict change in the function of eyespots will have little impact on **G**, whereas changes in the underlying developmental relationships among traits will result in evolution of **G**.

Thus far, we have interpreted similarity of **G** between species as evidence of stability. However, similarity of **G** in *J. coenia* and *J. evarete* might be due to convergent or parallel evolution: similar selection regimes generating the same **G** independently in each species. This hypothesis cannot be addressed in a two-species system, but instead requires a broader phylogenetic context to infer the direction of evolution. An alternative hypothesis is that phenotypic divergence between the species was accompanied by divergence in **G**, but that this divergence was not maintained (see Agrawal et al. 2001; Reeve 2000). Agrawal et al. (2001) demonstrated genes of major effect could dramatically, but transiently, change **G** during directional evolution. Mutagenesis experiments with *B. anynana* has indeed revealed some loci with alleles that dramatically affected eyespot development (Monteiro et al. 2003). Questions about transient instability of **G** could be addressed with independent natural populations that have colonized the same habitat at different times, or through temporally fine-scale sampling of **G** during experimental evolution.

Experimental Studies

Effect of Drift on Genetic Covariation of Drosophila Wing Shape

Phillips et al. (2001) designed a laboratory experiment to test the theoretical prediction that random genetic drift does not change the genetic covariation among traits (i.e., the orientation). Rather, because it erodes genetic variance, drift will scale **G** by 1 minus the inbreeding coefficient (see above). Changes in the orientation of **G** will affect the direction in which evolution proceeds (Figure 20.1), whereas changes in the variance will affect the rate. Exploiting the experimental benefits of *Drosophila melanogaster*, Phillips et al. (2001) generated a large data set consisting of six wing-shape traits measured for eight daughters from about 90 families in each of 52 inbred lines (4680 families total) and from 1945 families across six control (outbred) lines. The large sample generated here facilitated accurate estimation of **G**. This laboratory data set is equivalent to estimating **G** for 52 neutrally diverging populations (founded by one female and her brother), as well as estimating a known ancestral **G**.

At first glance, the results of Phillips et al. (2001) support the theoretical prediction that drift reduces genetic variance but does not change the genetic covariation of traits. Common principal components analysis supported proportionality of the average inbred **G** (across the 52 lines) to the outbred (ancestral) **G** (Figure 20.3). However, using proportionality of **G** as a signature of drift depends on how drift operates within single populations rather than on the average behavior of drifting populations. In this case, the 52 inbred lines varied considerably in both the orientation of **G** and the level of genetic variance (Figure 20.3). Therefore, although theory predicted the average outcome across all populations, it could not predict what would occur within any individual population. Implicit in this result is the conclusion that **G** can diverge through drift over very short periods of time, although whether this occurs frequently in nature has yet to be determined. Following these populations for an additional 20 generations after the initial bottleneck demonstrated that the drift-induced changes in **G** were not transient (Whitlock et al. 2002).

Phillips et al. (2001) noted that the impact of drift on genetic covariance varied from trait to trait. Changes induced by drift evidently depended on the underlying genetic details. Differences in cell lineage and gene expression patterns have led researchers

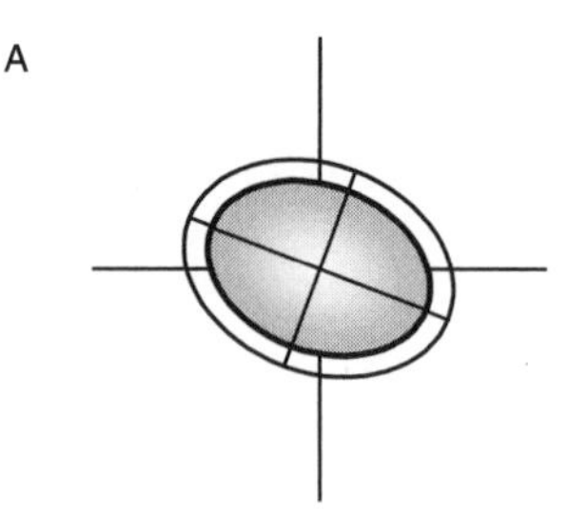

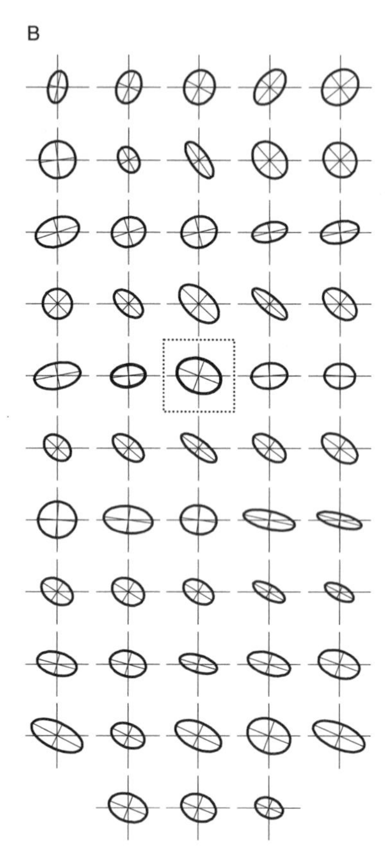

FIGURE 20.3. Effects of genetic drift on the genetic covariance between two wing characteristics in *Drosophila melanogaster.* (A) A graphical representation of the genetic covariance between two wing-vein angles. The outer ellipse shows the orientation of covariance in the outbred population whereas the grey ellipse shows the average covariance of 52 inbred lines created by brother–sister mating. Note that the ellipse for the inbred lines has shrunk proportionally relative to the outbred population as predicted by theory. (B) Estimates of genetic covariance for these two traits for each of the 52 inbred lines. Note that there is a great deal of variation around the average shown in (A) (center box). After Phillips et al. (2001).

to propose that anterior and posterior portions of *D. melanogaster* wings are separate developmental units, although this hypothesis has been questioned by Klingenberg and Zaklan (2000) based on their observation that only a small proportion of variation among wings is due to region-specific effects. Unfortunately, it is difficult to determine how traits measured by Phillips et al. (2001) fall across these morphogenic regions, making it difficult to assess whether **G** changed more within or between developmental units. In separate analyses of anterior and posterior regions of *D. melanogaster* wings, Zimmerman et al. (2000) detected multiple quantitative trait loci (QTLs) associated with wing shape, a first step toward more finely dissecting the underlying genetic basis of these traits. There appears to be significant epistatic genetic variance for wing shape (Gilchrist & Partridge 2001), suggesting that interactions among loci contributing to shape might be complex. Knowing the number of loci contributing to traits in **G** and the interactions among loci is a step toward understanding whether variation among traits in these genetic parameters causes the genetic relationships among some traits (as summarized in **G**) to change rapidly while others remain stable. This also highlights the fact that the choice of a particular set of traits and the way in which those traits are measured might strongly influence both the interpretation of results and our ability to correlate causal factors across levels of genetic organization.

Distinguishing Pleiotropy from Linkage

If not all aspects of **G** respond in the same way to a specific process evolutionary processes, the proximal cause must be differences in the underpinnings of **G**. Observed genetic correlations are usually expected to be caused by pleiotropy because correlations due to linkage should break down relatively quickly (Lande 1980b). That is, theory predicts that genetic covariance through pleiotropy will be more stable than covariances generated by linkage, which is also consistent with computer simulations that indicate a strong stabilizing effect of pleiotropic mutations on the structure of **G** (Jones et al. 2003). Similarly, Mezey and Houle (2003) have shown that shared similarity in **G** matrix structure is likely to be preserved only in cases in which developmental modules have a shared underlying pleiotropic basis.

Pleiotropy is considered ubiquitous, but is necessarily difficult to distinguish from tight linkage. For example, floral traits in plants are tightly coupled both functionally and developmentally and also show substantial genetic correlations. Are these correlations caused by pleiotropy or linkage? Conner (2002) enforced nine generations of random mating in the wild radish (*Raphanus raphanistrum*) and observed no change in covariance of six highly correlated floral traits. Forced random mating should have moved the population closer toward linkage equilibrium, especially if the linkage was generated by either correlational selection or assortative mating (Box 20.3). This study therefore strongly suggests a pleiotropic basis to the genetic covariances. Of course, very tight linkage may take substantially longer to decay (Conner (2002) showed that the genes influencing these traits would need to have average recombination rates of less than 0.01–0.05 to retain substantial linkage disequilibrium over this time period.) When one considers extremely tight linkage, say between different control regions within a gene, then the difference between linkage and variability in pleiotropic effects becomes somewhat semantic (Box 20.2).

FUTURE DIRECTIONS

Comparative Approaches

The recent, rapid acquisition of data on the evolution of **G** (Steppan et al. 2002) places us in a position to refine experimental design and identify approaches that will yield the most information. We now know that although **G** can diverge greatly over few generations (e.g., Phillips et al. 2001), it does not always do so (e.g., Paulsen 1996). Therefore, comparisons of natural populations must contribute more than "yes, **G** has evolved," or "no, **G** has not evolved." Comparative analyses are still the only way we can estimate the direction and rate of evolution of **G** in natural populations. However, comparisons must be made within a robust phylogenetic framework to achieve this aim.

Although still in its infancy, computer simulations have the potential to generate expected distributions of **G** matrices under particular evolutionary scenarios (e.g., Jones et al. 2003). Empirical data from laboratory experiments (e.g., Phillips et al. 2001) are also expanding our understanding of the patterns generated by particular evolutionary processes. This information on relationships between pattern and processes will assist in the interpretation of patterns observed in phylogenetic comparative studies. At the very least, we need to develop null models (e.g., drift) against which other evolutionary hypotheses (e.g., selection) can be tested. The phylogenetic framework serendipitously facilitates identification of taxa whose G matrices have evolved more or less than expected based on information from the other taxa. We can then address the questions of why and how these taxa differ in population parameters, selection pressure, etc.

Comparisons within a rigorous phylogenetic framework will also help us assess whether functional or developmental relationships among traits affect stability of G. Integration theory, which predicts developmentally or functionally related traits will be pleiotropically controlled (Cheverud 1984), and computer simulations, which show that pleiotropic mutations can stabilize **G** (Jones et al. 2003), lead to the prediction that the portion of the G matrix that describes functionally and/or developmentally related traits will be more stable than the portions describing unrelated traits. For example, in the case of butterfly wings, Paulsen (1996) detected a marginally insignificant difference ($P = 0.056$) between **G** matrices from *J. coenia* and *J. evarete* for the full set of 29 traits, including eyespot size, position and wing-vein measurements, whereas a comparison of the submatrix of **G** including just the developmentally related eyespot diameters provided no support for divergence between the species ($P = 0.273$). Further comparisons within a robust phylogenetic framework can determine: (a) whether functionally or developmentally connected traits are characterized by strong genetic covariation, and (b) whether **G** matrices of these traits are more stable than **G** of unconnected traits.

There are still significant methodological issues to be overcome in order to conduct the proper analysis in a comparative framework (Phillips & Arnold 1999; Steppan et al. 2002). Further, there is little value in conducting studies with insufficient power (Box 20.1), so the total level of effort required to conduct a complex comparative analysis among a number of different taxa will be substantial—perhaps even daunting.

Manipulative and Experimental Approaches

Although there is no substitute for comparative natural experiments to determine what processes shape **G** under natural evolution, the valuable role of laboratory experiments is clearly established (e.g., Phillips et al. 2001; Shaw et al. 1995). As we accumulate more data on the behavior of **G** under particular evolutionary scenarios, we can increase the sophistication of experimental studies to include simultaneous action of more than one process and thus toward the consideration of how different processes interact. We can then generate more realistic predictions of the behavior of **G** under natural conditions.

The Genetic Basis of Genetic Covariances

There are effectively two levels at which we can attempt to dissect the molecular genetic basis of **G**: detailed analysis of the function and variation of individual genes versus broad analysis identifying all loci contributing to a particular suite of traits. By considering the action of individual genes, researchers investigating butterfly eyespots are moving toward understanding which genes contribute to eyespots, as well as the nature of the interactions among those genes (e.g., Beldade et al. 2002). Taking this approach generally requires a priori information about the system (for example, information from related taxa about genes affecting particular traits or processes) but enables very precise consideration of variation at those loci.

Quantitative trait locus (QTL) analyses represent a method taking the other, top-down approach to identifying loci associated with phenotypic variation. There are certain advantages to using QTL analyses to dissect **G**: they require no a priori knowledge of the genetic basis of traits; they can be used to investigate multiple traits simultaneously, facilitating identification of pleiotropic/linked loci; they can potentially be used to estimate the total number of loci contributing to any one trait; and can be used to identify (candidate) loci for further study. Currently, there are also disadvantages to using QTLs: a QTL covers considerably more of the chromosome than a gene does, making it impossible to ascertain whether a QTL reflects pleiotropic effects of a single locus or linked loci; and, with small genetic differences and small sample sizes, QTL analyses are not very powerful. Both of these problems might result in underestimation of loci and overestimation of pleiotropy. Lack of power could also result in detection of different QTLs in different populations, leading us to the erroneous conclusion that the underlying basis of **G** has evolved when in fact it has not (Gibson 2002). Ultimately, we will need the combination of functional specificity coupled with global analysis that is promised by the genomics revolution. In reality, this will still end up demanding a great deal of fairly traditional genetic analysis, but the potential for rapid advance is definitely on the horizon.

The need for this molecular dissection is inevitable because we ultimately need to know much more about the distribution of allelic effects underlying quantitative traits before we can move on to a more sophisticated understanding of the evolution of those traits. Not all mutations (i.e., not all alleles) have the same effect on all traits to which that locus contributes (Box 20.2; Stern 2000). This highlights a fundamental aspect of **G**, one that is accessible only through developmental genetics. **G** is determined by allele frequencies and, if not all alleles are equally pleiotropic, changes in allele frequency will change the orientation of **G**. The similarity of pleiotropic affects among alleles at a locus might depend on whether the locus codes for structural or regulatory proteins (Stern 2000). Variation at structural loci is likely to impact equally all traits to which that the protein contributes, resulting in similar pleiotropy of all alleles. In contrast, variation at regulatory loci might affect when or where proteins are expressed, thus having different effects on different traits, and thereby generating different levels of pleiotropy among alleles. The different contributions of regulatory versus structural genes to evolution is a subject of ongoing debate in molecular evolution and has implications for our understanding of the evolution of **G**.

G stands at the center of quantitative models that help to describe the pattern of phenotypic change generated by a wide variety of evolutionary processes. **G** also serves as a metaphor for the role integrated genetic systems play in understanding the evolution of the whole organism. It is perhaps fitting that matrices usually have very box-like representations, because for the last several decades the G matrix has been used as a black box into which tremendous amounts of underlying genetic complexity could be placed while research focused squarely at the level of the phenotype. To move from metaphor to reality, we need to open that box and see what lies inside.

SUGGESTIONS FOR FURTHER READING

Steppan et al. (2002) provide a more extensive review of a number of studies that have compared G matrices. Lande (1979) is the most important paper understanding the multi-trait representation of the evolution of quantitative characters. Barton and Turelli (1987), though a difficult paper, creates a general approach for examining the role that allelic distributions play in structuring the pattern of genetic variation and other features of phenotypic evolution. This approach has been expanded in a number of other papers by this group. Cheverud (1984) is an interesting paper that strongly asserts the value of a union between developmental and evolutionary genetic thinking for understanding the pattern genetic covariation among traits. Stern (2000) presents a strong statement of the relationship between developmental gene regulation and the generation of phenotypic variation; a theme that is put in a more explicitly quantitative genetic framework by Johnson and Porter (2001).

Barton NH & M Turelli 1987 Adaptive landscapes, genetic distance, and the evolution of quantitative characters. Genet. Res. 49:157–174.

Cheverud JM 1984 Quantitative genetics and developmental constraints on evolution by selection. J. Theor. Biol. 110:155–171.

Johnson NA & AH Porter 2001 Toward a new synthesis: population genetics and evolutionary developmental biology. Genetica 112:45–58.

Lande R 1979 Quantitative genetic analysis of multivariate evolution, applied to brain:body allometry. Evolution 33:402–416.

Steppan SJ, Phillips PC & D Houle 2002 Comparative quantitative genetics: evolution of the G matrix. Trends Ecol. Evol. 17:320–327.

Stern DL 2000 Evolutionary developmental biology and the problem of variation. Evolution 54:1079–1091.

Acknowledgments We thank all of the members of the UO-OSU G-matrix reading group for influential discussions and helpful comments. Critical comments from the editors and a reviewer were most helpful. This work was supported by an NSF IGERT grant in Evolution, Development, and Genomics and grants from the NSF and NIH to P.C.P.

21

Genotype–Environment Interactions and Evolution

SAMUEL M. SCHEINER

The link between genotype and phenotype is complex and that complexity can have profound effects on the course and nature of evolution. Chapters 15 and 16 described how the developmental system can control evolutionary outcomes. But developmental systems do not operate in isolation. All individuals are embedded in an environment. That environment plays two roles in determining the course of evolution: it determines which phenotypes are most fit (natural selection) and it interacts with the genome to determine the phenotype.

One of the central goals of evolutionary theory is to explain why traits of organisms have the properties they do. These properties can be studied from four perspectives—adaptational, functional, historical, and developmental—which define four classes of constraints on trait evolution. For example, a small beetle, *Stator limbatus*, lays it eggs on the seeds of a variety of trees in the pea family (Fox et al. 1994, 1997, 1999). On the blue palo verde (*Parkinsonia florida*) females lay just a few, large eggs, while on the catclaw acacia (*Acacia greggii*) they lay many, small eggs. Considered from an adaptational perspective, these differences in laying behavior match differences in the optimal phenotype in each environment. On blue palo verde, larval survivorship increases with egg size. Thus, females maximize their fitness by laying larger eggs. In contrast, on catclaw acacia survivorship does not depend on egg size, so higher fitness comes from laying as many eggs as possible. These differences can also be explained from a functional perspective. The seed coat of blue palo verde is harder and has a greater chemical resistance than that of catclaw acacia; a larger egg leads to larger larvae that have greater survivorship. The history of the species also plays a role, in this case the amount of time that a population has been exposed to a host species. Texas ebony (*Ebenopsis ebano*), as the name implies, is native to the gulf coast regions of Texas and Mexico. Recently it was planted as an ornamental in Phoenix, Arizona, where *S. limbatus* is found. While *S. limbatus* can survive on this species, it does not do particularly well because the two species have not been in contact long enough for adaptation to occur. Finally, development can play an important role. The size of egg that a female lays can be manipulated by the female. However, this alteration of egg development does not occur instantaneously. The developmental system requires the female to be in contact with the seeds for 1 to 2 days before egg size changes.

This chapter explores the interaction among these different perspective. Debate over the importance of different processes in evolution is often portrayed as a dichotomy of the external (historical or adaptational) versus the internal (functional or developmental). Such a dichotomy is a simplification because the developmental program is at least partially shaped by selection over very long time periods (Siegal & Bergman, Ch. 16 of this volume). In this chapter I examine the nature of this interaction, how the interaction of genes and environment shapes evolution, and how that interaction itself evolves.

CONCEPTS

Linking Genotype to Phenotype

All traits exist within some sort of developmental system. Although we typically think of development as a property of complex organisms such as eukaryotes with multiple cell and tissue types, we can broaden that concept to encompass the entire process of translating genotype into phenotype. Even a trait such as the ability of a bacterium to metabolize lactose, which may be determined by a single gene, still depends on the expression of that gene: going from DNA to messenger RNA to protein. Thus, the manifestation of any trait, from a single enzyme to a complex morphology, depends on the interaction of many genes and the environment plays a role in this complex dance.

The environment can mediate trait expression in two ways (Table 21.1). First, the environment can act as a signaler, initiating a cascade of events. When an *S. limbatus* female arrives on a seed, she senses the species that she is crawling over. Depending on the species, she may alter the size of the egg that she will lay. When the environment acts as a signaler, the type, nature, and strength of the signal can be uncorrelated with the magnitude of the induced phenotype change. The environment can also act directly on the expression of a trait. For example, plants raised in low light conditions have a lower biomass than those grown in high light because they receive less energy with which to fix carbon dioxide into biomass. In this instance, differences among environments are proportional to phenotypic changes. This distinction, the environment as signaler (active plasticity) versus direct actor (passive plasticity), is related to the likelihood that the plasticity is adaptive.

As a general rule, if the environment acts as a signaler, it is highly likely that the resulting pattern of plasticity will be adaptive. Because such a signal–response system requires a complex set of environmental detectors and changes in gene expression, it is unlikely to be maintained without positive selection for the system. However, adaptation should be taken only as a prima facie hypothesis that needs to be tested. The system might have been selected for in the past but no longer be selected for currently. For example, females from *S. limbatus* populations in Texas change egg size in response to seeds from blue palo verde, a species that does not grow in Texas (Fox et al. 1999). The snail *Physa* develops a shell that is rounder in shape in the presence of fish, an adaptive response because a rounder shell is harder for fish to crush (DeWitt 1998). However, this change occurs in the presence of many species of fish, not just those that eat snails. In this case, the signal that the snail perceives does not discriminate among fish species. Thus, not all active plasticity is necessarily adaptive.

In contrast, if the environment is a direct actor, it is less likely that the plasticity will be adaptive. In many of these cases the environment acts directly on the phenotype without involving changes in gene expression. Such effects can include changes in the rate of enzyme reactions as a function of temperature or changes in the size of individuals as a function of the amount of food they ingest. For example, the herbaceous plant *Brassica rapa* is shorter when grown in the presence of other plants than when grown alone (Poulton & Winn 2002). The change in height is a direct effect of root competition for limiting resources. But selection actually favors

TABLE 21.1. The relationship between the type of plasticity (active or passive) and the type of phenotypic response (discontinuous or continuous).

	Type of plasticity	
Type of response	Active	Passive
Continuous	Usually adaptive (e.g., spines on water flea in response to phantom midge)	May be adaptive (e.g., ratio of wing length to thorax length of fruit fly in response to temperature)
Discontinuous	Usually adaptive (e.g., form of geometrid moth in response to phenology of oak tree)	Rare to nonexistent

Active plasticity exists when the environment acts as a signal, initiating a cascade of developmental changes, and is usually adaptive. Passive plasticity exists when the environment acts directly on a trait, and may or may not be adaptive. While active plasticity can result in both continuous and discontinuous responses, passive plasticity results only in continuous responses. I do not know of any examples of passive plasticity resulting in a discontinuous response.

taller plants in the presence of neighbors, so this plastic response is maladaptive. Sometimes the adaptive nature of passive plasticity can be seen only in the context of trait complexes, such as the ratio of wing area and body mass in *Drosophila melanogaster* (Figure 21.1).

Active and passive plasticity are sometimes confused with the form of the response: discontinuous or continuous. Passive plasticity will almost always result in a continuous response, such as shown in Figure 21.1. In contrast, active plasticity can result in a continuous or discontinuous response. An example of a continuous response can be found in the water flea *Daphnia pulex*. In response to the presence of predaceous larvae of the phantom midge, *Chaoborus americanus*, juveniles increase the size of their body size and tail spine, which results in decreased predation rates. This response is triggered by chemicals given off by *C. americanus* and the magnitude of the response is proportional to the concentration of the chemical in the water (Parejko & Dodson 1991).

A discontinuous response is exemplified by the geometrid moth *Nemoria arizonaria*, which grows on oak trees (Greene 1989). In the spring, newly hatched larvae feed upon the inflorescences (catkins) and grow to look like catkins. Presumably such mimicry results in lower predation. Larvae that hatch later in the year, after flowering is over and the catkins are gone, feed upon newly emerging leaves. These leaves have much higher levels of tannin which causes a switch in the developmental form; the larvae now look like twigs. These two forms look so different that they were once classified as members of different genera.

The form of the plastic response is called a norm of reaction. In the first graphical presentation of the reaction norm by Woltereck (1909), both the abscissa (nutrient level) and the ordinate (relative head height) were continuous variables (Figure 21.2A). More often, though, studies are limited to only two environmental conditions. In such a case, although the data are graphed using continuous, linear functions (Figure 21.2B), there is no guarantee that the response is really continuous.

The form of the overt phenotypic response is not necessarily indicative of the underlying cause or what is occurring at the cellular level. Discontinuous traits may have a continuous underlying cause but a reaction norm with a sharp threshold. For example, adult

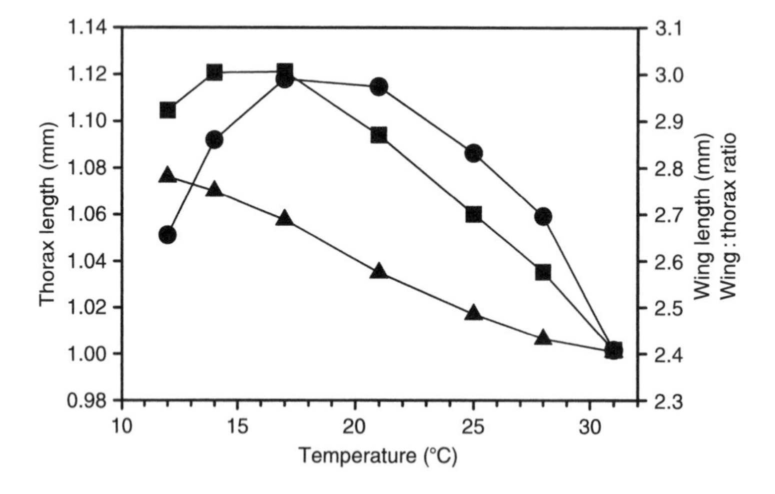

FIGURE 21.1. In *Drosophila melanogaster,* both body mass (measured here as thorax length; circles) and wing area (measured here as wing length; squares) are plastic in response to temperature, and the temperature of maximal size differs. An individual's flight ability depends on the ratio of wing area to body mass (wing loading) and the optimal ratio is lower at higher temperatures (Pétavy et al. 1997). The combination of increasing then decreasing temperature responses in body size and wing size with different maxima results in a smoothly decreasing ratio (triangles) that matches the change in the optimal ratio. Data from David et al. (1994).

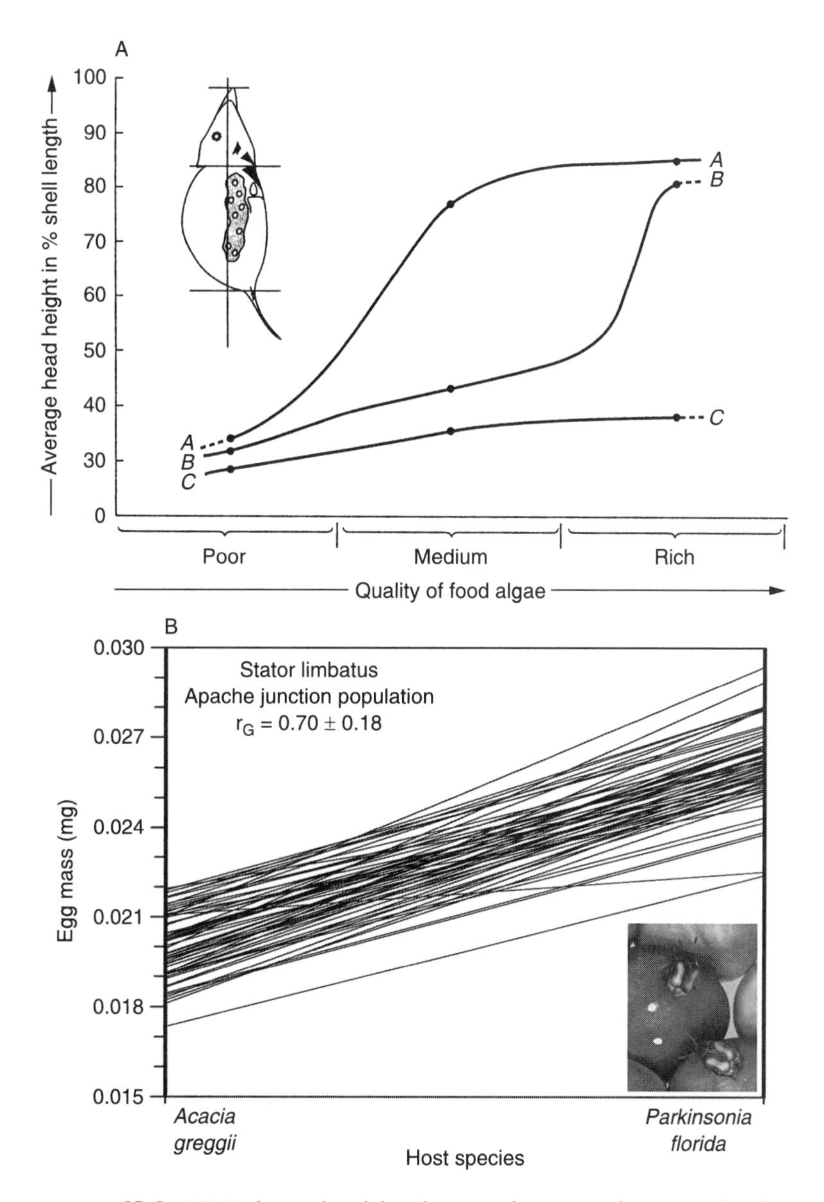

FIGURE 21.2. (A) Relative head height as a function of nutrient level in *Hyalodaphnia cucullata*. Each line is a single clonal lineage. Reaction norms as continuous functions can be described mathematically using various polynomials (Gavrilets & Scheiner 1993; Gibert et al. 1998) or continuous-value functions (Kirkpatrick & Heckman 1989; Meyer 1998). From Woltereck (1909) (B) The size of eggs laid by *Stator limbatus* females on acacia and palo verde seeds. Each line represents one full-sib family, half raised on one seed type and half raised on the other. Although the reaction norms are shown as a continuous function (straight lines), these are actually discrete environments. An alternative method for conveying these data is a cross-environment correlation (r) where $r = |1|$ indicates no $G \times E$ variation and $r = 0$ indicates maximal $G \times E$ variation. Data from Fox et al. (1999).

male beetles of the species *Onthophagus taurus* vary in whether they have horns; larger individuals are more likely to have them (Emlen & Nijhout 1999). Adult size is determined by the amount and quality of dung with which a larvae was provisioned. This continuous variation in the environment and overall body size leads to differences in the amount of juvenile hormone in large and small larvae during a critical window during larval development. Larvae below a critical mass experience a spike in juvenile hormone levels and lack horns as an adult.

What Is the Environment?

In the previous discussion I spoke of "the environment" and gave a few specific examples. But what is meant by "the environment?" The definition typically encompasses that which is external to the organism. Often this is the abiotic, physical environment, but can also include the biotic environment such as interactions among individuals in a population (Frank, Ch. 23 of this volume), parental interactions, and competitors or predators. Plasticity responses can extend over multiple generations so that the environment experienced by the mother can determine the phenotype of the offspring (e.g., Groeters & Dingle 1987; Schmitt et al. 1992; Case et al. 1996).

Environmental responses (plasticity) are not the only nongenetic source of phenotypic variation. Phenotypic variation can also come from random variation during development: developmental noise. Developmental noise is often studied by looking at variation in matching left- and right-side structures such as limbs or wings or repetitive structures such as leaves. While one can make a theoretical distinction between plasticity and developmental noise, in the context of an actual experiment this is not easy to do. In practice, we typically lump together developmental noise, microenvironmental variation, and errors of measurement.

Individual Variation and G × E

Egg size in *S. limbatus* varies among individuals due to differences in genotype, environment, and errors of development (Fox et al. 1997, 1999). The percentage of phenotypic variation due to genetic variation is heritability. Mathematically, we can express this idea as follows. First, consider the total phenotypic variation (σ^2_P). In the simplest case, we could partition the phenotypic variation into that part due to genetic variation (σ^2_G) and that part due to all other causes (σ^2_e): $\sigma^2_P = \sigma^2_G + \sigma^2_e$, and the heritability is $H^2 = \sigma^2_G/\sigma^2_P$. (For simplicity of discussion in this chapter, I am not distinguishing between total genetic variation and additive genetic variation. For these and other issues involving heritability see Roff, Ch. 18 of this volume).

How does the environment influence heritability measures? Heritability estimates are often treated as though they are the same regardless of the environment within which they are measured. Such treatment is the equivalent of assuming that if an *S. limbatus* individual lays eggs that are 10% larger than another when reared on acacia seeds, the eggs will be 10% larger when reared on palo verde seeds. But what happens when this assumption does not hold? For example, when *S. limbatus* is reared on palo verde, all individuals lay eggs that are about the same size, but when reared on acacia some of those individuals lay much smaller eggs than the others due to genetic differences (Fox & Mousseau 1996). In other words, the genetic differences are apparent in some environments but not in others. The heritability of egg size measured only on palo verde would be very low because the amount of phenotypic variation would be low. On the other hand, heritability measured only for individuals reared on acacia would be larger. Thus, heritability is always context-dependent. Differences in genetic expression as a function of the environment are referred to as genotype–environment (G × E) interactions.

In the previous example, we considered the environment in a coarse-grained fashion: individuals experienced only one seed type. But consider a situation in which an acacia and a palo verde are growing side-by-side and all of the individuals in the population lay eggs on seeds from both plants. This population experiences the environment in a fine-grained fashion. A full partition of the phenotypic variation in eggs size of this population is: $\sigma^2_P = \sigma^2_E + \sigma^2_{G\times E} + \sigma^2_G + \sigma^2_e$, where σ^2_E is variation due to the environment (the presence of more than one type of seed), $\sigma^2_{G\times E}$ is variation due to genotype–environment interactions (differences in how individuals react to those seeds), σ^2_G is averaged over all of the environments, and σ^2_e now refers only to variation due to errors of development. Since σ^2_P is the denominator in calculating heritability, changes in any of its components will affect the heritability of the trait. Because fine-grain environmental heterogeneity can effect both the numerator and denominator of heritability, changes are potentially complex and heritability could either increase or decrease. Thus, genotype–environment interactions can affect how evolution proceeds, accelerating or retarding the rate of evolution and affecting its direction.

How G × E Affects Evolution

What is the effect of G × E variation on evolution? This debate stretches back to the early days of evolutionary theory. Wright (1931) argued that the ability of a single genotype to express a variety of phenotypes would allow it to adapt to a range of environments. Within a population, this plasticity could result in more rapid evolution toward the optimal phenotype, but would slow genotypic differentiation and speciation. Baldwin (1896, 1902), in contrast, argued that G × E variation would speed up differentiation through a three-stage process now referred to as the Baldwin effect or genetic assimilation (see Simpson (1953) for a summary of the early history of this idea). In the first stage, plasticity allows a species to express a wider range of phenotypic variation and occupy a wider range of environments. G × E variation results in previously unexpressed genetic variation to manifest in the new environment. Even if the mean phenotype of the population does not shift in the direction of the new optimal phenotype, some genotypes might have a higher fitness in the new environment. In the second stage, there is selection for those genotypes, together with additional selection on other modifier loci, moving the population toward the optimum in the new environment. In the third stage, there is selection against plasticity such that the new, optimal phenotype becomes fixed in the population.

Waddington (1953), who coined the term "genetic assimilation," demonstrated this phenomenon by selecting on a shortening of the crossvein in the wings of *Drosophila melanogaster*. This phenotype appears when the pupae are exposed to a brief period of high temperature. Initially about one third of the population used in his study exhibited this phenotype, but after about 20 generations of selection the frequency increased to nearly 100%. In addition, the crossveinless phenotype had gone from being a plastic trait to being nonplastic such that even flies not exposed to the high temperature also lacked the crossvein.

The Baldwin effect was little studied for many years. Recently, though, a number of authors have emphasized the potential importance of the Baldwin effect in trait evolution and the promotion of speciation (e.g., Pigliucci & Murren 2003; West-Eberhard 2003; Schlichting 2004). Alternatively, Wright may have been correct: plasticity itself may be favored thereby slowing down differentiation.

The Evolution of Plasticity

When faced with environmental heterogeneity, a species may adapt in one of three ways: (1) multiple genotypes that optimize on different aspects of the environment, (2) a single jack-of-all-trades genotype with a fixed phenotype that is not optimal in any single environment but has a high average fitness, or (3) a single plastic genotype that matches the optimal phenotype in each environment (Scheiner 1993a). Which of these outcomes, or combination of outcomes, prevails is a complex function of the organism's genetic architecture and developmental capacities and the environment's temporal and spatial structure. The pathway of evolution can be complex. The evolution of specialist genotypes, for example, could proceed through a process like the Baldwin effect, or through the more direct route of selection on nonplastic genetic variants.

Natural selection will favor plasticity over fixed strategies (specialist or jack-of-all-trades) when the average fitness of individuals with the plastic strategy exceeds that of individuals with the fixed strategy. Very generally this requires: environmental heterogeneity, reliable cues, benefits that outweigh the costs of plasticity, and a genetic basis to plasticity (Scheiner 1993a).

Environmental Heterogeneity and Cue Reliability

The environment can be heterogeneous in time or space. Consider first temporal variation that happens during the lifetime of an individual and a trait that can continue to change. If the physiological or developmental system of the organism can respond to such environmental variation quickly enough, a plastic genotype will be favored. For example, an *S. limbatus* individual can alter the size of her eggs within a day or two. The evolution of this egg-size plasticity is constrained by the developmental system (e.g., range of egg sizes that can be expressed by a single individual and the speed of that response), the functional system (e.g., the ability to correctly sense the environment), and how fast the developmental system can respond to a change in the environment.

Evolution operates slightly differently if the trait is fixed at a single time during development and selection occurs at a later time. In this instance, cue reliability is critical for a plastic genotype to be favored. Cue reliability decreases with an increase in the time between when a trait becomes fixed during development and when selection occurs. An unreliable cue will always result in a fixed strategy; as reliability increases a plastic strategy becomes more favored.

Spatial heterogeneity can be similar in effect to temporal heterogeneity. Migration acts as a form of environmental uncertainty; by moving, the environment that an individual develops in may not be the same as the environment in which selection occurs. In the case of *S. limbatus*, adults move among seed pods and trees, but offspring remain on a single seed from the egg stage through pupation. Thus, different life history stages experience different amounts of environmental heterogeneity. In this instance adults migrate just before they reproduce and the offspring experiences only one environment while developing. Such spatial heterogeneity with adult migration is similar in effect to a system where individuals never migrate but the environment changes from one generation to the next. Environmental heterogeneity, migration, and cue reliability interact so that greater migration rates or temporal heterogeneity favor plasticity only when cues are reliable.

The frequency distribution of environments also plays a role in whether a plastic genotype is favored (e.g., de Jong 1999; Sultan & Spencer 2002). If one set of environmental conditions predominates, then plasticity will not be favored because individuals will experience mostly one environment even as they move from place to place. With *S. limbatus*, plasticity is expected to be favored where there is a mixture of acacia and palo verde trees. The actual result is more complex, possibly depending on rare, long-distant dispersal events (see Case Study).

Costs and Limitation of Plasticity

Environmental heterogeneity occurs on a variety of spatial and temporal scales, yet not all organisms change their phenotypes to match this heterogeneity. One possible reason for incomplete matching is costs of plasticity (DeWitt et al. 1998). A cost is a decrease in fitness even if an individual has the optimal phenotype in a given environment. Costs include: the costs for maintaining the sensory and regulatory mechanisms of plasticity, the costs of acquiring information about the environment, increased developmental instability in other traits due to being plastic, and other forms of pleiotropic or epistatic interactions. Costs of plasticity are often confused with the cost of producing a given phenotype. For example, in response to herbivory a plant may produce secondary chemicals to deter future herbivory. These chemicals can be costly to produce: for example alkaloids are rich in nitrogen, which is often in short supply. But another individual that produces those alkaloids all of the time, even in the absence of herbivores, would incur the same nitrogen costs. Rather, a cost of plasticity is one incurred above and beyond those incurred by individuals with fixed phenotypes, such as the costs to the organism to retain the ability to change even if no change occurs. Currently, there is mixed evidence for costs, with only a few studies finding evidence for them (e.g., DeWitt 1998; Scheiner & Berrigan 1998; Donohue et al. 2000; Poulton & Winn 2002; Relyea 2002).

Limitations of plasticity are distinct from costs. A limitation is any factor that results in a failure to match the optimal phenotype. For example, cue reliability is a limitation, not a cost. Other limitations include: the lag-time between an environmental change and a change in the phenotype, limits to the developmental range of a plastic genotype in contrast to a fixed genotype, and epiphenotypic problems so that structures created through a plastic response are less well developed than similar structures produced early in development.

Finally, for any trait to evolve there must exist genetic variation for that trait. Genotype–environment interactions are the genetic component of phenotypic plasticity. They represent the extent to which individuals in a population differ in their plasticities. One potential limitation on the evolution of plasticity is a lack of genetic variation. Although genetic variation for plasticity has been found in an extensive array of species and traits, very often the amount of variation is small, usually much lower than genetic variation for the trait itself (Scheiner 1993a).

The Genetics of Plasticity

We can model evolution at two levels: individual loci and whole genotypes. Both approaches are valuable, although a failure to distinguish between them has led to some vigorous debates concerning the evolution of plasticity. Nearly all empirical work on the genetics and evolution of plasticity has been done at the genotypic level using the standard tools of statistical genetics, such as measures of genetic variances, covariances, and heritabilities. Theoretical work, in contrast, has been devoted to a wide range of conceptual and mathematical models, including both quantitative genetic and single-locus models. In the 1980s a disagreement arose concerning the range of gene types one needs to consider to sufficiently model the evolution of plastic traits.

This debate was encapsulated in a trio of articles published in 1993 Via (1993) and the subsequent replies by Scheiner (1993b) and Schlichting & Pigliucci (1993); see also Via et al. (1995). Via's contention was that the only type of genes that need be considered are those whose expressions are environmentally sensitive, often termed plasticity genes. All parties agree on this general definition of plasticity genes. Schlichting and Pigliucci, though, further constrained plasticity genes to regulatory loci. In contrast, I contended that to understand trait evolution, one must also include genes that are not environmentally sensitive (Figure 21.3; Scheiner & Lyman 1989). This disagreement about genetic architecture was related to additional arguments over the form of selection on traits and plasticity.

Everyone agrees that selection always acts on individuals and traits within environments. What is at issue is whether selection can also act directly on plasticity. Unfortunately, the words and descriptions used for such multilevel selection have an unfortunate history, including a failure to recognize the mathematical equivalence of forms of selection given different labels (e.g., Frank, Ch. 23 of this volume; Goodnight, Ch. 6 of this volume). Population structuring in space, time, or by behavior (e.g., deme or kin groups) can provide an opportunity for selection on plasticity (Scheiner 1998; Goodnight, Ch. 6 of this volume). It is a matter of semantics whether such selection is on plasticity, per se, or on the total context that an allele experiences over many individuals, environments, and generations. Although the argument over the form of selection became tangled up in the argument over the genetic basis of plasticity, the form of selection is a function of the ecology of plasticity, and independent of its genetics.

Bradshaw (1965) was the first to suggest that the plasticity of a trait could evolve independently of its mean value. Confusion has arisen over the years by a failure to clearly articulate whether the mean refers to a within-environment measurement,

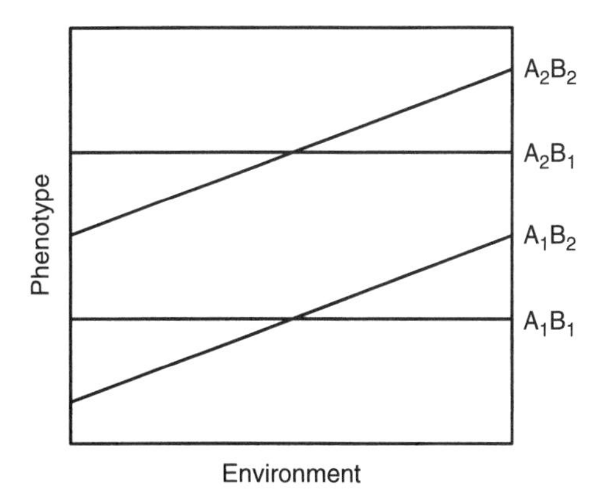

FIGURE **21.3.** A two-locus haploid model of plasticity. The expression of locus *A* differs among alleles, but not in a way which is environmentally sensitive. It affects the elevation of the reaction norm (the mean expression of the trait across environments) such that the marginal difference between individuals with genotype A_1 and A_2 is constant across all environments. Locus *B* is a plasticity gene and its environmentally sensitivity differs among alleles. It affects the slope of the norm such that individuals with genotype B_1 do not change phenotypes across environments, while those with genotype B_2 do change. The differences among *B* alleles create genotype–environment interaction (G × E) variation. The reaction norm phenotype is determined by a three-way interaction among the two loci plus the environment. In a plasticity-gene-only model, only locus *B* type genes would exist. After Scheiner (1993a).

or an across-environment measurement. Subsequent work clarified that it is the across-environment mean (Via et al. 1995).

Whether the mean of a trait and the plasticity of a trait will evolve independently depends, in part, on how one parameterizes the reaction norm. The coefficients of a reaction norm are a mathematical description of the shape of the reaction norm, how the phenotype changes across environments. For example, a quadratic reaction norm can be written as: $p = P_m + g_2(e - E_m)_2$, where p is the phenotype, and e is the environmental index. This reaction norm is defined by three values: the environment of the minimum/maximum (E_m), the phenotypic value in that environment (P_m), and the curvature (g_2). If parameters are standardized to an environmental midpoint defined by the minimum/maximum of the curve, then the three parameters are mathematically independent (Gibert et al. 1998).

Such independence refers only to the coefficients of the reaction norm. It does not imply independence between a reaction norm coefficient and a trait value within an environment. Any change in plasticity, except under very narrow circumstances, will always result in a change in trait values within a given environment. Confusion over which elements (reaction norm coefficients vs. trait values within environments) had the potential to evolve independently fostered ambiguities in the disagreements among myself, Via, Schlichting and Pigliucci.

While the form of selection is a function of ecology, the evolution of plasticity depends on its genetic basis. Evolution will proceed in a different fashion if a genome contains only environmentally sensitive alleles than if it also contains alleles whose expression does not depend on the environment. In a simulation of plasticity evolution along a gradient, with the plasticity-gene-only model the metapopulation matches the local optimum everywhere by achieving the optimal plasticity (Scheiner 1998). In contrast, when both types of loci are included in the model, demes are locally adapted but their plasticity is less than optimal for adaptation elsewhere on the gradient. It is the intersection of genetic architecture with selection ecology that determines the evolutionary outcome.

CASE STUDY

G × E and Beetle Evolution

I have laid out the potential effects of G × E variation on the evolutionary process; now I examine how these effects have played out in the evolution of the beetle *Stator limbatus*, work carried out by Fox and his colleagues (e.g., Fox et al. 1994, 1997, 1999; Fox & Mousseau 1996; Fox & Savalli 2000; Czesak & Fox 2003). This insect lays its eggs on the seeds of species in the pea family. Over its geographic range—northern South America to the southwestern United States—it has been found on more than 70 species in nine genera. In the state of Arizona, the focus of these studies, it is abundant on many species of acacia, particularly catclaw acacia (*Acacia greggii*), and the two common species of palo verde, the blue palo verde (*Parkinsonia florida*) and the foothill palo verde (*P. microphylla*). However, in any one location only one or a few hosts may be present. (Taxonomic note: The genus *Parkinsonia* has recently been revised and the published papers refer to these species as members of the genus *Cercidium*.)

Stator limbatus is a seed parasite. Females lay eggs directly onto host seeds, and thus are restricted to seed pods that have already opened or been damaged. Upon hatching, the larvae burrow into the seed, where they complete development and pupate. They emerge from seeds as adults, the only dispersing stage of this species. Thus, the larval environment is determined by the choice of the mother. An individual requires only the resources inside a single seed to complete development, and multiple larvae can develop in larger seeds.

To understand whether and how G × E variation has affected the evolution of *S. limbatus*, we need to answer a series of questions. First, does this species live in a heterogeneous environment? Second, what trait is being differentially selected in each environment? Third, is there genetic variation for that trait and for plasticity in that trait?

Is the environment of *S. limbatus* heterogeneous? The answer is a clear yes. The host species differ substantially in quality, as measured by egg-to-adult survivorship. On the blue palo verde, fewer than half the individuals survive, with most of this mortality occurring when the first instar larvae attempt to penetrate the seed coat. In contrast, on catclaw acacia survivorship is greater than 90%. These environmental differences are most likely factors of seed-coat hardness and defensive chemicals.

What trait is being differentially selected in each environment? The key response of the beetle to these differences in survivorship is variation in egg size. On the blue palo verde, larvae that hatch from larger eggs have a greater chance to survive (Figure 21.4). In contrast, egg size has no effect on survivorship on catclaw acacia. The larger eggs come at a cost, though, of lower total fecundity. Females actively respond to these environmental differences by adjusting the size of eggs that they lay. If you allow a beetle to lay eggs on acacia seeds, and then move it to palo verde seeds, the beetle recognizes the change in its environment and after 1 to 2 days egg size increases by about 30%. And the reverse happens moving from palo verde to acacia seeds. When females are not given any contact with seeds, they lay small eggs.

Third, genetic variation exists for both the trait and trait plasticity (Figure 21.2B; Fox & Mousseau 1996; Fox et al. 1999). This system therefore contains all of the necessary components for evolution by natural selection: phenotypic variation for

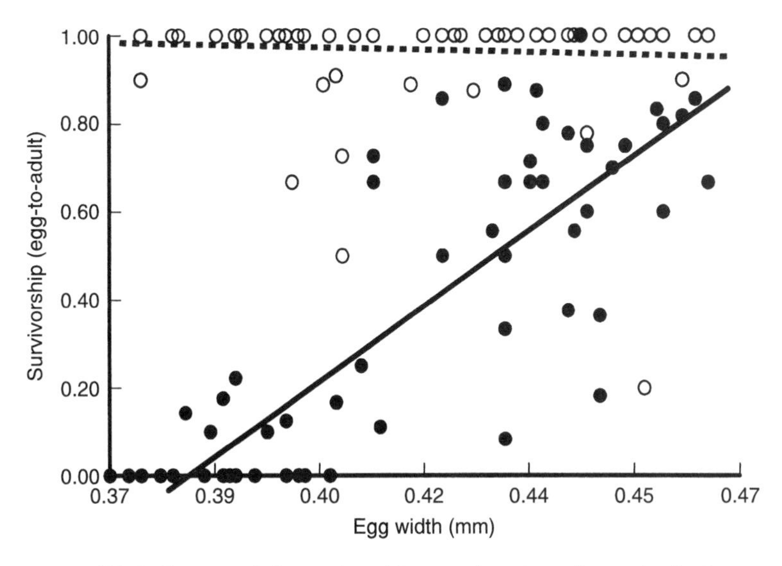

FIGURE 21.4. Egg-to-adult survivorship as a function of egg size in *Stator limbatus* for individuals reared on acacia seeds (open circles, dotted line) or on palo verde seeds (closed circles, solid line). (Data from Fox & Mousseau (1996).

egg-size plasticity, a relationship between egg-size plasticity and fitness, and genetic variation for egg-size plasticity.

G × E and Population Divergence

Given that this system has the necessary ingredients for evolution by natural selection, has it led to differences among populations? One study (Fox et al. 1999) compared two populations: one from Texas in an area with a dense stand of *Acacia berlandieri*, and one from Arizona where only palo verdes were growing. Palo verdes do not grow in Texas; in contrast, acacias are found in Arizona, although not in the exact location where these Arizona beetles were collected. Females from both populations laid larger eggs on palo verde seeds. The populations did not differ in this plastic response, somewhat surprising given the lack of palo verdes in Texas. Both populations showed genetic variation for egg size on both host species, and genetic variation for egg size plasticity (G × E variation). In this study all of the genetic variation was within, rather than between, populations.

Is this pattern of plasticity in response to palo verde seeds found across all populations of *S. limbatus*? Throughout the United States, including populations in California and Arizona where palo verde are found, and in Texas, where palo verde does not occur, all populations show similar amounts of egg-size plasticity. On a larger scale, there are four distinct clades of *S. limbatus* throughout the New World. A comparison of two of the clades (one from the United States and one from Colombia) found that beetles from Colombia respond to a native Colombian species by laying larger eggs on *Pseudosamanea guachapele* (relative to the size of eggs laid on acacia). This suggests that egg-size plasticity in *S. limbatus* is ancestral (because it is present in two extremes in the clade). The Texas populations may simply retain this evolved plasticity, or there might be sufficient gene flow between Arizona and Texas to maintain the plasticity.

The presence of plasticity in the Texas populations also suggests that plasticity has little or no costs. If costs existed, then there would be selection against plastic genotypes in areas were the plasticity did not increase fitness and was not being selected for.

G × E and Continued Trait Evolution

What about the potential for further evolution of egg-size plasticity? Texas ebony (*Ebenopsis ebano*) grows within the range of *S. limbatus*, but *S. limbatus* has never been found on that species in Texas ebony's native range. This is particularly surprising because Texas ebony is used as a host plant by a sister species of *S. limbatus*, *S. beali*. About 30 years ago, however, Texas ebony was introduced as an ornamental tree in the Phoenix metropolitan area of Arizona and *S. limbatus* has since colonized this plant in that area. Four populations of *S. limbatus* were compared for their survivorship on ebony, one population growing on ebony from the heart of the Phoenix metropolitan area, one from the edge of the metropolitan area growing on palo verde, one from about 90 km away also growing on palo verde, and one from Texas growing on acacia. Females from all populations were reared in the laboratory from a common environment, catclaw acacia seeds, and allowed to lay eggs on ebony seeds. In general, survivorship on ebony was very low (ca. 3–4%) and differed little among the populations. However, this survivorship was substantially lower than that observed for the Phoenix population in the field (ca. 10%). Perhaps some aspect of the environment was enhancing survivorship.

The shift to the use of Texas ebony in the Phoenix area may have been facilitated by the existence of palo verdes. In another experiment, beetles that were reared on acacia seed but exposed to palo verde seeds as adults, were then moved to ebony seeds. These palo-verde-induced beetles continued to lay larger eggs on ebony for a day or so and survivorship of their larvae was higher and positively correlated with egg size (mean survivorship of nearly 50%). Interestingly, ebony does not induce *S. limbatus* females to lay larger eggs, even though it would be adaptive to do so. Females reared on ebony seeds that did not experience palo verde seeds while maturing eggs had larvae with survivorship rates of only 4–8%. Thus, there is selection for *S. limbatus* to lay larger eggs on ebony seeds. The existence of egg size plasticity has allowed *S. limbatus* to expand its host range from acacia to palo verde, and this plasticity may explain the generally wide host range of this species. It will be very interesting to see whether eventually *S. limbatus* manages to adapt directly to Texas ebony by evolving a cue to lay larger eggs on that species.

Artificial Selection on Plasticity

While G × E variation can be important in the evolution of a species, as demonstrated by host-range evolution in *S. limbatus*, we can also directly examine the process of evolution of plasticity. One way to explore this issue is to perform a selection experiment, which I did with *Drosophila melanogaster* (Scheiner & Lyman 1991). The targeted trait was thorax length and the environmental variable was temperature. Over the range of temperatures used (19 °C and 25 °C) the reaction norm is approximately linear and negative (Figure 21.1). The experiment involved selection to both increase and decrease plasticity using a family selection protocol.

We found a significant response to selection on plasticity with a realized heritability of 0.088 ± 0.027 SE (Figure 21.5), establishing that plasticity of thorax length was a selectable trait. This heritability was much lower than that for the trait itself (0.246 ± 0.027 SE), as is the usual pattern for the heritability of plasticity.

Demonstrating the selectability of plasticity was not the most important finding, though, as others had previously demonstrated the selectability of plasticity. More important were our conclusions about the genetic architecture of thorax-length plasticity. We demonstrated that plasticity was not the result of overdominance as had been theorized for many decades (Lerner 1954). Plasticity had been posited to be a function of homozygosity, with increasing heterozygosity resulting in decreasing plasticity. The overdominance model predicts that the response to selection will plateau as maximal heterozygosity or homozygosity is reached. We found no such plateau when selecting for increased plasticity. The overdominance model predicts an increase in plasticity when selecting on the mean of the trait, regardless of the direction of selection, because directional selection causes fixation of loci and, thus, homozygosity. Instead, selection on the mean of the trait resulted in plasticity increasing in some lines and decreasing in others. Finally we failed to find a predicted correlation between the amount of genetic variation for thorax length and the plasticity of thorax length.

We demonstrated that plasticity and the response to selection was due to a complex interaction among a number of different types of loci. Various pieces of evidence led to this conclusion. One critical piece was a lower bound to selection on plasticity. During the first 11 generations we succeeded in decreasing plasticity so that flies raised at 19 °C were the same size, on average, as flies raised at 25 °C. That is, plasticity was reduced to zero. In theory, plasticity should have been reversible: larger flies at 25 °C than at 19 °C. However, during the next 11 generations we failed to push the lines

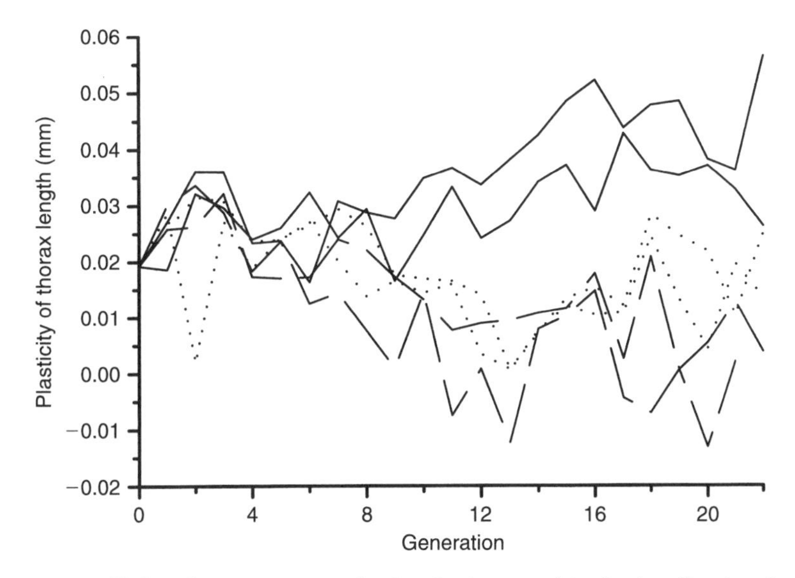

FIGURE 21.5. The response to selection for increased (unbroken lines) and decreased (dashed lines) phenotypic plasticity of thorax length in response to temperature in *D. melanogaster.* Controls shown as dotted lines. For downward selection, the response halted after approximately 11 generations at a zero slope of the reaction norm. Data from Scheiner and Lyman (1991).

below zero plasticity. We postulated that thorax length plasticity has two components: a response/no response component and a magnitude of response component. This, admittedly ad hoc, model explains a number of puzzling aspects of the experiment. First, estimates of the heritability of plasticity based on sib-analyses (Scheiner & Lyman 1989) were much higher than the realized heritability. In particular, the overestimates were much greater than those for the mean of the trait. Such a consistent overestimate could result if selection operated primarily on only one of the two components of plasticity—response or magnitude—while measures of genetic variation included both components.

FUTURE DIRECTIONS

In the past 20 years we have gone from simply recognizing that phenotypic plasticity might be an important factor in evolution, to measuring the magnitude of G × E variation in numerous of populations and species, to now determining when that plasticity is adaptive. However, advances in theoretical models continue to outpace data. In just the past few years the importance of epistasis in the evolutionary process has received renewed recognition. The first steps have now been taken to create a general evolutionary framework that unites epistasis and G × E variation (Wolf et al. 2004).

Plasticity studies are a gateway to understanding the importance of developmental processes in evolution. Until now the evolution of development has focused primarily on large-scale patterns (e.g., comparisons of mice, hydra, roundworms, and fruit flies) with each species considered to have a single, fixed developmental trajectory. Now more attention is being given to within-species variation and the importance of environmental influences on development.

Genomics, the study of whole genomes from a molecular perspective, is leading to a convergence of quantitative genetics and molecular genetics (e.g., Stearns & Magwene 2003). Quantitative genetics deals with phenotypes and serves as the bridge between the ecology of natural selection and evolutionary responses. Molecular genetics tells us about the constraints of genetic architecture. The importance of such constraints is an open question. Is a quantitative genetic framework a reasonable first-order approximation of the evolutionary process? Theory shows that genetic architecture *can* matter for the evolution of plasticity (e.g., Scheiner 1998), but we do not know whether it *does* matter. This is very much an open, empirical question.

All of these topics—G × E variation, the evolution of plasticity, epistasis, the evolution of structured populations, the importance of development, and genomics—are rapidly uniting. They form a natural nexus for the studies in evolutionary genetics and its wider importance in understanding the evolutionary process.

SUGGESTIONS FOR FURTHER READING

Both Pigliucci (2001) and DeWitt and Scheiner (2004) provide comprehensive reviews and syntheses of the concepts and literature of phenotypic plasticity. West-Eberhard (2003) presents her theory of genetic accommodation, a theory of the evolution of the appearance of new phenotypic variation by environmental change and the plastic capacity of genomes, wrapped within a massive literature review. More focused reviews can be found in the papers of Agrawal (2001; the importance of plasticity in species interactions), Kassen (2002; experiments on the evolution of specialists), and Scheiner (2002; selection experiments and plasticity). Schmitt et al. (1999) present what is arguably one of the best case studies of adaptation and plasticity in their summary of work on shade-response in touch-me-nots.

Agrawal AA 2001 Phenotypic plasticity in the interactions and evolution of species. Science 294:321–326.

DeWitt TJ & SM Scheiner (eds) 2004 Phenotypic Plasticity: Functional and Conceptual Approaches. Oxford Univ. Press.

Kassen R 2002 The experimental evolution of specialists, generalists, and the maintenance of diversity. J. Evol. Biol. 15:173–190.

Pigliucci M 2001 Phenotypic Plasticity: Beyond Nature and Nurture. Johns Hopkins Univ. Press.

Scheiner SM 2002 Selection experiments and the study of phenotypic plasticity. J. Evol. Biol. 15:889–898.

Schmitt J, Dudley SA & M Pigliucci 1999 Manipulation approaches to testing adaptive plasticity: phytochrome-mediated shade-avoidance responses in plants. Am. Nat. 154:S43–S54.

West-Eberhard MJ 2003 Developmental Plasticity and Evolution. Oxford Univ. Press.

Acknowledgments I thank Chuck Fox for making sure that I got the *Stator limbatus* story correct and generously supplying figures. Jason Wolf and two anonymous reviews provided many useful comments. This chapter is based on work done while serving at the National Science Foundation. The views expressed in this chapter do not necessarily reflect those of the National Science Foundation or the United States Government.

22

Genetics of Sexual Selection

ALLEN J. MOORE
PATRICIA J. MOORE

Sexual selection was defined by Darwin (1871) as selection that "depends on the advantage that certain individuals have over other individuals of the same sex and species, in exclusive relation to reproduction." Sexual selection was developed separately from natural selection because Darwin wished to explain the evolution of elaborate characters, particularly secondary sexual characters (traits "not directly connected with the act of reproduction"; i.e., traits that influence but are not required for copulation). Such traits apparently have no benefit outside of influencing the outcome of male–male competition or female mate choice. Secondary sexual traits might even be counter to survival and natural selection by drawing the attention of predators or limiting physiological performance. Why should traits be so elaborate when their function is apparently limited to influencing the competition for mates? Darwin's solution was sexual selection with two main mechanisms: female mate choice and male–male competition. Darwin suggested that male combat or female preferences resulted in particularly strong selection. He saw these as fundamentally different from the mechanisms leading to the process of natural selection because they involved social interactions and influenced mating success rather than survival or fecundity and so might have no limits to the evolved outcome or expression. There need not be an optimum expression, as the "best" was that which was most extreme.

Sexual selection was not nearly as popular as natural selection. Wallace (1889) dismissed the idea outright while later influential evolutionists either accepted sexual selection but rejected aesthetic preferences of females as unlikely (e.g., Huxley 1938) or produced verbal arguments in support of mate choice that went unheeded (Fisher 1930). Not until almost 100 years after it was proposed was there an explosion of research showing that females prefer males that provide superior resources, increase the fitness of their mates or offspring, provide protection or help with the offspring, or were simply more attractive (Andersson 1994; Bakker 1999; Zeh & Zeh 2003). Thus, sexual selection is well established as an important concept and research area in evolutionary biology. There are, however, a number of important issues associated with sexual selection that remain unresolved.

Darwin's solution to the evolution of extravagant characters is still only a partial one. Why should mate choice or male competition result in a different form of selection, especially a particularly powerful one? Sexual selection should be (all else being equal) only half as strong as natural selection because sexual selection acts on only one sex at a time, and so "misses" half of the population (see Shuster & Wade 2003). Is there a genetic constraint to elaboration? Genetic variation is expected to erode under such strong selection, ultimately limiting the evolutionary outcome. Are the different mechanisms of sexual selection equivalent or is only mate choice capable of evolving elaborate traits that are not weapons or agonistic displays? We still have few studies comparing male–male competition with female mate choice. Does sexual selection typically result in sexual conflict, the evolution of traits that increase the mating success of one sex while reducing the fitness of the other sex (Parker 1979)? The evolution of traits that are harmful to one sex while benefiting the other has attracted considerable attention, but documenting true sexual conflict requires explicit measures of the fitness consequences of a trait, which is often difficult in this context (Chapman et al. 2003; Pizarri & Snook 2003). Finally, do sexually selected traits reflect aspects of fitness outside of attracting a mate or influencing mating (Kokko et al. 2003; Mead & Arnold 2004)? Addressing the genetics of sexually selected traits can help address all of these issues. We believe that what is needed more than ever is to develop a basic understanding of how mechanisms of sexual selection result in genetic changes in a population—that is, evolution. It is this point that we concentrate on in this chapter.

Throughout this chapter we will suggest areas that need more data or models. We promote the empirical agenda for students of sexual selection, and emphasize the unknown over the known. We therefore highlight a few topics that we are especially interested in seeing pursued. Our goal is to illustrate why a genetic approach, and especially diverse genetic approaches, will improve our understanding of sexual selection and sexual conflict. Our theme is simple: we need to quantify variation in fitness and genetic architecture.

There are many areas of research in sexual selection that can benefit from a genetic approach (Mead & Arnold 2004), but we have chosen to emphasize two. First, we will suggest that research in the field has been hampered by a lack of techniques for quantifying fitness. Mate choice or male–male competition, or the outcome of these interactions, are typically scored as all-or-none phenomena. Manipulative experiments allow us to show that mate choice does or does not occur, or that male traits do or do not influence the outcome of intrasexual competition, but it is variation in these traits that is required for testing models of sexual selection (Mead & Arnold 2004) and for examining the process of evolution resulting from sexual selection (Blows et al. 2003, 2004). Very few researchers have documented the nature of the distribution of the within-population variation in mating preference—that is, the preference function (Ritchie 1996, 2001; Ritchie et al. 2001; Brooks 2002; Blows et al. 2003, 2004). Quantifying variation in male success in male–male competition is equally difficult although not impossible (Hoffman 1994; Horne & Ylonen 1998; Moore et al. 2002). Demonstrating sexual conflict requires documenting the sex-specific fitness consequences of the traits influencing mating success (Chapman et al. 2003; Pizzari & Snook 2003). The second area we will emphasize is the need to quantify genetic architecture. This is more difficult than it sounds—a lack of a quantifiable phenotype complicates quantitative genetic studies. Furthermore, the very property that makes sexual selection so fascinating to students of animal behavior, the social interactions, hampers our ability to study genetics. Mate choice and male–male competition not only affect sexual selection, they are themselves subject to evolution. Thus, they are evolving (social) environments. This gives rise to indirect as well as direct genetic effects (Wolf et al. 1997, 1999; Moore et al. 1998) and creates difficulties for separating selection and genetics (Moore et al. 2002). The joint evolution of the trait and the mechanism influencing the evolution of the trait, however, may well lead to the linkage between mate choice and the preferred character and the extreme elaboration that results (Mead & Arnold 2004). It may take some creativity to disentangle phenotypic and genetic influences and correlations, but creativity is a hallmark of research in evolutionary studies of behavior.

CONCEPTS

Mechanisms of Sexual Selection

Mate Choice

In the 1970s researchers began to accept that females play an active role in mating outcomes, and compelling verbal and mathematical models supported the notion that female preferences for some males over others can have dramatic evolutionary consequences (Andersson 1994; Mead & Arnold 2004). Coevolution of the preference and the trait was one of the driving features of the models, highlighting the unique nature of preferences—they both cause evolution and evolve themselves as a result of being genetically influenced. The net result is a "runaway" process, where male traits and female preferences evolve ever more elaborate expressions, checked only when opposing natural selection overcomes the process. Soon empiricists caught up and found that mate choice was widespread and that outlandish ornaments, elaborate weapons, melodic songs, and odoriferous emissions influence female discrimination of potential mates (Andersson 1994). However, genetic studies were rare, and many of the genetic aspects of sexual selection remain controversial, such as the genetic correlation between male traits and female preferences and the reason for the existence of considerable genetic variation underlying sexually selected traits (Mead & Arnold 2004).

One of the attributes of the models that stimulated research into sexual selection was that they could be classified into two competing theories. The debate became whether females chose males based on traits that indicate fitness benefits outside of sexual selection to the offspring ("good genes") or characters important only in sexual selection ("arbitrary characters"), and stimulated both empirical and mathematical investigations (Kokko et al. 2002, 2003; Mead & Arnold 2004). Currently, this dichotomy appears to be convenient but false (Kokko et al. 2002). However, whether evolution as a result of sexual selection is "different" or not remains unresolved. Recent models of sexual selection suggest that condition dependence, where the male trait reflects differences in male condition rather than genetic differences among individuals, may help explain why elaboration is possible (Rowe & Houle 1996). Of course, males may vary genetically in their ability to produce a "condition," but the only models of this involve maternal effects on male traits (Wolf et al. 1997, 1999). It is not always clear how one measures condition.

Male–Male Competition

Some of the most spectacular displays in the animal kingdom involve interactions between males that are competing for access to females (Andersson 1994; Shuster & Wade 2003). Agonistic interactions, fights, territorial displays, and the associated weapons and structures are often fantastically elaborate. Not surprisingly, early research in animal behavior focused more on male–male interactions than mate choice and the diversity of male mating tactics is reflected in the classification of mating systems among animals (Shuster & Wade 2003). Yet models of sexual selection are almost always concerned with mate choice rather than male–male competition. It would be surprising if all the forms of male–male competition have equivalent evolutionary consequences, but the importance of this variation for the outcome of sexual selection is poorly documented. The similarity or difference in the evolutionary process resulting from male–male competition and female mate choice are equally unknown although different sorts of traits may evolve under these two mechanisms. There are a few models that suggest there is no difference in evolution resulting from inter- or intrasexual selection (e.g., Moore et al. 1997) but we know of no model that addresses this explicitly.

In practice it is difficult to separate the different mechanisms of sexual selection without manipulative experiments. Male–male competition is often obvious and female mate choice is often cryptic or contingent upon male–male competition. The relative importance of male–male competition versus female mate choice is therefore unknown: we cannot even agree if male–male competition and female mate choice are complementary or conflicting processes (Moore & Moore 1999). What happens when male–male competition and female mate choice oppose each other—is there balancing selection as suggested by Moore and Moore (1999)?

Sexual Conflict

Studies of sexual selection have grown to include sexual conflict, which occurs when the sexual selection advantage to one sex confers a selective disadvantage to the other sex. Although this aspect of sexual selection was first defined and modeled several decades ago by Parker (1979), research in this area did not really take off until this century (Chapman et al. 2003; Pizzari & Snook 2003). Investigations of sexual conflict have now become one of the most popular areas of sexual selection research. We now recognize that sexual conflict can involve both male–male competition (e.g., sperm competition or manipulative substances in ejaculates) and female mate choice (e.g., mate choice to avoid manipulative males). In large part this newer theory arises as a result of the recognition that females often mate with more than one male, raising the question of why females should mate multiply if they can choose their mates. One possibility is that "cryptic" mate choice occurs after mating by biasing the use of sperm (Eberhard 1996) or by inciting male–male competition through sperm competition (Zeh & Zeh 2003). There is a fascinating interplay between selection reflecting "pure" sexual selection (male–male competition and female mate choice for traits that provide a mating advantage to one sex without harming the other sex) and the finer details of sexual conflict (male traits that manipulate female reproduction to limit opportunities for other males to fertilize eggs but at a fitness cost to the female, mate choice to avoid manipulation rather than receive benefits from males; Chapman et al. 2003). Nevertheless, why and how females should choose continues to dominate much of animal behavior research. Females can discriminate among males, and do, but which factors inspire their discrimination? Are females avoiding manipulation or choosing a superior mate? The answer is often best found by adopting the perspective of evolutionary genetics and quantifying fitness variation and genetic architecture.

Unlike other areas of sexual selection, genetic approaches have long been a component of empirical research into sexual conflict research. Surprisingly, there are few genetic models of the process of evolution arising from sexual conflict (Moore & Pizzari 2005). Most models of sexual conflict have followed Parker and examined evolutionary outcomes using a phenotypic (optimality) approach. However, like other areas of sexual selection, the phenotypic models can be contentious and there is not universal agreement on how to test the models or even what sort of data to collect. These phenotypic models will benefit by developing quantitative genetic models. In addition, genetic models in this area will provide powerful lines of research for empiricists as long as we pay attention to the definitions of fitness and make the assumptions as explicit as possible.

Documenting Variation in Fitness and Quantifying Selection

The choice of method for demonstrating natural (or sexual) selection depends entirely on the question of interest. As we have suggested, simply documenting sexual selection or any of the mechanisms giving rise to sexual selection provides evidence for the occurrence but not the form of selection and is unlikely to further the field although it may well provide valuable insights into the evolution of specific organisms. Data that can be used to compare species, or add to our general understanding of the nature of sexual selection, will typically need to be quantitative. The most popular method is to measure selection differentials, gradients, or opportunities for selection by looking at covariances between phenotypes and fitness (Brodie et al. 1995; Shuster & Wade 2003). However, all quantitative approaches require a reliable measure of fitness, and that can be problematic in studies of sexual selection and sexual conflict (Wolf & Wade 2001; Pizzari & Snook 2003).

There appears to be a conflict between verbal or phenotypic models of sexual selection and genetic models of sexual selection (e.g., McNamara et al. 2003). It is dangerous to assume without evidence that phenotypes and genotypes are equivalent (Cheverud & Moore 1994; Boake et al. 2002). Perhaps the main area of disagreement is the way in which fitness is described (Wolf & Wade 2001; McNamara et al. 2003; Pizzari & Snook 2003). Part of the problem is in finding a suitable measure of fitness. As all empiricists know, we often require surrogate measures, such as survival, mating, or number of offspring produced (Fairbairn, Box 4.1 of this volume). Phenotypic models often use fitness surrogates that extend beyond a single generation (e.g., "number of surviving and reproducing offspring" or future descendants, e.g., Weatherhead & Robertson 1979; or "offspring value," e.g., McNamara et al. 2003). Quantitative genetic models, in contrast, assign fitness to just one individual (i.e., survival is the offspring's fitness and never contributes to parental fitness; Cheverud & Moore 1994). Wolf and Wade (2001) provide a discussion of the problem and limitations of each approach.

How and why fitness is measured is especially important when testing theoretical models. As McNamara et al. (2003) argue, phenotypic models (whatever their flaws) have great intuitive appeal and, especially in sexual selection and sexual conflict, have generated considerable enthusiasm among behavioral ecologists. Such models have also provided considerable insight into what may happen. Still, empiricist should treat phenotypic models as potentially flawed hypotheses, not as something to be supported with data.

One advantage of quantifying fitness is that fitness functions relating phenotypes to fitness, and therefore defining selection (individual fitness functions) or possible evolutionary trajectories (population fitness functions), make explicit the form and function of selection (Brodie et al. 1995; see Ritchie 1996; Brooks 2002 for applications to mating preferences). Perhaps the main limitation in identifying fitness functions for sexually selected characters, let alone the behavioral mechanisms of sexual selection and sexual conflict, is that quantifying variation in behavior can be difficult.

Another area that needs further study is to document how selection associated with mate choice, male–male competition, or sexual conflict may vary over space and time, and with social conditions (Moore & Moore 1999). As we know from our own research on the cockroach *Nauphoeta cinerea*, dominant males may not be dominant for even a significant proportion of their life (Moore & Moore, unpublished data), females may change their choice of mates in response to age or other factors

(Moore & Moore 2001), and the expression of sexual conflict changes as a function of life history (Montrose et al. 2004). In addition, as others have shown in species such as guppies, mate choice can vary both within and among populations (e.g., Brooks 2002). The importance of variable expression in mate choice is unknown, but temporal variation in selection can lead to the maintenance of genetic variation depending on gene flow and the temporal nature of the changes (Turelli & Barton 2004). Of course, gene flow among populations and variation in the expression of mate choice over time can also slow evolution. Once again, empirical information is needed, and a combination of population genetics with quantitative studies of sexual selection would be extremely valuable.

Quantitative Genetics of Sexual Selection

Testing models of sexual selection always require genetic information—heritabilities, genetic correlations, and quantitative measures of selection. Is there a genetic correlation between mating preferences and preferred characters? Is genetic variation underlying sexually selected traits less than that underlying traits experiencing only natural selection? Is there a relationship between fitness and sexually selected characters or mate choice? Is there an increased opportunity for selection resulting from sexual selection? Are selection gradients operating on sexually selected characters stronger than selection gradients operating on naturally selected characters? Initially it was thought that answering these questions would discriminate between good genes and the runaway process associated with arbitrary characters, but it is increasingly obvious that there is not a simple test to distinguish these models. In hindsight this makes sense; both must involve heritable characters, a genetic relationship between preferences and the preferred character, and elaborate evolution.

By far the greatest effort in the study of the genetics of sexual selection has been expended in looking at quantitative genetics of sexually selected (secondary sexual) male traits using a quantitative genetics approach (Bakker 1999 and references therein). Fewer studies have documented the inheritance of mating preferences, perhaps again limited primarily by an inability to score individual female preferences on a continuous scale allowing the calculation of variances (Bakker 1999; Brooks 2002). Nonetheless, perhaps stimulated by the good genes versus arbitrary trait arguments of sexual selection, we do know that females vary and their preferences can be heritable (Bakker 1999; Jang & Greenfield 2000; Ritchie 2001; Brooks 2002; Blows et al. 2004).

One of the problems of measuring quantitative genetic parameters of sexual selection is that traits such as mate choice and social dominance are interacting phenotypes, that is, their expression depends on the social context (Moore et al. 1997, 2002). This is also true of traits influencing sexual conflict (Moore & Pizarri 2005). Traits such as mate choice, male manipulation, and male social dominance cannot exist outside of an interaction with another individual. How can a female express a choice without a social interaction with a potential mate? How can a trait manipulate unless there is an individual that is manipulated? The concept of dominance is meaningless outside of the specific interaction between individuals. Interacting phenotypes complicate our understanding of inheritance, but do not make it impossible to study (Moore et al. 2002).

We can accommodate the unusual character definition of interacting phenotypes by recognizing that there will be indirect genetic influences on such traits (Moore et al. 1997, 2002; Moore & Pizarri 2005). The genetic influences on interacting phenotypes reflect both direct additive genetic influences (effect of genes inherited from their parents) and indirect genetic influences (heritable genetic variation influencing the trait in the other individual that is necessary for the expression of the trait of interest in the focal individual). A simple analogy is that it is the equivalent of a genotype–environment interaction but where the environment is a social environment that can be genetically influenced and therefore evolve. Because the social environment may reflect genetic differences among the interacting partners, the genetics of that partner have to be measured (or controlled for) as well as the genetics of the focal individual. Therefore more attention should be paid to the interacting nature of sexual selection phenotypes; after all, this is the initial fascination with and unique aspect of mate choice. The evolution of a preference changes the nature of selection on the preferred trait because preferences are both selection and an evolving phenotype. This evolution of selection can lead to the genetic covariance between preferences and traits that is thought to be integral to sexual selection.

Quantitative genetic studies have dominated our studies of the genetics of sexually selected traits because these are complex characters and are likely to have a multifactorial inheritance. The complex and continuous nature limits our ability to examine the molecular genetics of traits like preferences (e.g., Kyriacou 2002) although advances in the mapping and detection of quantitative trait loci (QTL studies) suggest that, at least for morphology, progress can be made (e.g., Gleason et al. 2002; Shaw & Parsons 2002). However, we will need to eventually study the genes that influence traits in order to fully understand how sexually selected traits evolve (see below). Understanding the molecular nature and genetic architecture underlying traits will also facilitate unifying phenotypic and genetic models.

Genetic Architecture

As emphasized by the multivariate quantitative genetic models of evolution, evolution rarely acts on traits in isolation (see Brodie et al. 1995; Brooks 2002). This makes it especially important to study genetic architecture—not only interactions within and between loci, and the number of loci, but interactions and joint genetic influences on multiple traits. All of these affect evolution. Boake et al. (2002) provide a detailed argument for the rationale behind studying genetic architecture of behavioral traits, and these arguments hold equally well for genetic studies of sexual selection. Ultimately a comparison of the genetic architecture of traits influenced by natural selection and sexual selection will help us uncover any fundamental differences between these evolutionary processes.

One basic aspect of genetic architecture that needs to be addressed is the number of genes involved and genetic relationship (e.g., pleiotropy) among traits—that is, the **G** matrix for sexually selected characters. Multivariate selection may well be more complicated given the interacting nature of sexually selected phenotypes (Moore et al. 1997; Wolf et al. 1999), and as Blows et al. (2003) show there may be more than one way to be attractive. Blows et al. (2004) provide a particularly powerful and elegant analysis of how selection and the variance–covariance matrix may interact, and conclude that the apparent genetic variation underlying sexually selected characters is misleading. In fact, by comparing patterns of selection (i.e., the fitness surface) and the genetic variance–covariance matrix (i.e., **G**), these researchers show that potential evolutionary responses are greatly limited despite significant standing genetic variation. Unfortunately, for most organisms, data on genetic correlations between sexually selected characters and between the characters and the behaviors causing selection remain elusive. It remains likely that this is where sexual selection and natural selection differ.

Molecular tools are increasingly accessible, and details of genetic architecture for sexually selected traits will soon be available (Boake et al. 2002). This means that we can now begin to identify the specific regions of a chromosome (QTLs) that influence sexually selected characters. Although QTL studies are not the same as identifying the genes involved, it is a step closer (Boake et al. 2002). The number, location, and action and interaction of QTLs should be compared with QTLs underlying naturally selected traits in the same organism. This will require the molecular toolbox to be opened to species other than *Drosophila*, mouse, or other domesticated animals where such comparisons are likely to be uninformative either because we know little about selection on these organisms in natural populations, or because sexual selection either does not occur or is complicated by the laboratory environment. Jumping from model organisms to interesting organisms is increasingly common, and we expect this will expand to sexual selection studies.

One of the possibilities Boake et al. (2002) suggest be addressed at the molecular level is the distribution of allelic effects for naturally and sexually selected traits of equal complexity (although the latter is a caveat that may be difficult to meet in practice). Are major genes involved more often in sexually selected characters? Is there a difference in coding and control region evolution between the two processes? It is here where we may begin to get a clue as to how sexually selected characters are elaborated.

CASE STUDIES: HOW TO UNDERSTAND THE COMPLEXITIES OF SEXUAL SELECTION

Genetics of Traits That Make Males Successful in Attracting Females and the Genetics of Female Mate Choice

Considerable experimental effort has been directed at determining whether or not traits elaborated by sexual selection are heritable. In short, the answer is yes. However, Blows et al. (2004) provide a cautionary tale concerning simply measuring genetic variation in sexually selected traits. Simply noting genetic variation may not be enough; we also need to know the nature of selection and compare both selection and the genetic variance–covariance matrix. The genetics of female mating preferences poses additional difficulties. While measuring variation in (typically) male (typically) morphological characters that are sexually dimorphic and elaborate poses few technical difficulties; measuring variation in female preferences that are contingent on the males presented to the female is more problematic when males cannot be reused. Nonetheless, we need to know more about the genetics of female preferences as both traits and preferences are expected to evolve under sexual selection.

Greenfield and his colleagues working with the lesser wax moth, *Achroia grisella*, provide one of the most complete stories with regard to female mating preferences. Females of this species use sequential comparisons of males (Greenfield et al. 2002) to discriminate among potential mates based on ultrasonic calls. This means that females can be presented with artificial calls rather than the males themselves. Females prefer males that produce pulses of sound more quickly (i.e., have high values of pulse rate), thus exerting directional selection (Jia et al. 2000). Moreover, there is genetic variation in both the male character and the female preference, and for both of these there is a considerable genotype by environmental interaction—success of calls or expression of specific mate choice depends on the environments in which they are made—that may help maintain genetic variation in this system (Jia et al. 2000; Rodriguez & Greenfield 2003). This phenotypic plasticity in female preferences also leads to variation in the selection that males experience, further influencing genetic variation in the male characters through variable selection. Genotype by environment interactions may work in nature to help maintain genetic variation in this system because wax moths are honeybee symbionts and live inside honeybee hives, and there may well be considerable environmental heterogeneity among honeybee hives.

Studies of mate choice based on male cuticular hydrocarbons in *Drosophila serrata* by Blows and coworkers provide another model for how a genetic approach provides extraordinary insight into evolution by sexual selection. Cuticular hydrocarbons in this species are under both natural and sexual selection (Blows 2002). Females gain genetic benefits from their choice as there is a positive genetic correlation between offspring fitness and female preferences (Blows 2002; Hine et al. 2002). Despite very strong direction selection and the benefit to mate choice, significant genetic variation remains in both male characters and female preferences (Blows et al. 2004). This may reflect an orientation between **G** and the fitness surface for male cuticular hydrocarbons that has arisen because of selection, and there is little scope for further evolutionary change despite selection and genetic variation (Blows et al. 2004).

We have chosen to highlight these two studies because they are remarkably complete, illustrate that measuring genetic variation in female preferences as well as the male character is possible, and show how genetic information provides unique insights into evolution by sexual selection. There are other systems that are well studied, such as the guppy (see Brooks 2002 for a review). Nevertheless, more studies of the genetic architecture underlying female preferences (in particular) are needed. We are especially lacking in molecular studies of the genetics of mating preferences (Kyriacou 2002), although QTL studies of sexually selected signals are now appearing (Gleason et al. 2002; Shaw & Parsons 2002).

Genetics of Traits That Influence Male Success in Male–Male Competition

One complication in the study of the genetics of male–male competition is that the phenotype, male success, only exists in the context of the social interaction with another male (Moore et al. 2002). One way of approaching this problem is to study the genetics of traits that can be used to reliably predict the outcome of male–male competition, as many traits important in precopulatory male–male competition, such as body size, have an underlying genetic contribution. For example, in a study of the heritability of dominance in male bank voles, Horne and Ylonen (1998) used urine-marking behavior and preputial gland weight as indicator traits for dominance status. These traits reflecting social dominance had relatively high heritabilities compared with traits related to male body size.

Given that underlying genetic variation in social behaviors is notoriously difficult to measure (Moore et al. 2002), why is it necessary to undertake such experiments? As discussed above, a purely phenotypic approach to studies of sexual selection in general, and male contests in particular, can lead to misleading predictions regarding response to selection. In an elegant series of studies on territoriality in *Drosophila melanogaster*, Hoffmann (1994) showed that larger males have an advantage in male contests, with heavier males more likely to win escalated encounters and gain the mating advantage associated with territoriality. Thus, taken alone, phenotypic studies predict a change in body size in response to selection for increased territoriality. In genetic analyses of territoriality, however, Hoffmann and coworkers found that factors unrelated to male body size contributed to variation in territorial success at a genetic level. While territorial behavior and success evolved rapidly under artificial selection, there was no correlated increase in body size. Thus, other traits must influence the fighting abilities of males and their success in male–male competition.

In reality, qualities important to success in male–male competition are likely to be composed of a number of different behavioral and morphological attributes, influenced by a multitude of genes that may or may not overlap. Thus, the overall trait of "success in male–male competition" is likely to require the integration of multiple characters. Genetic architecture will provide a signature of this integration. Our studies on the evolution of the male pheromone in the cockroach *Nauphoeta cinerea* provide an example of how patterns of genetic variation in a multiple component trait, a pheromone blend, can provide insight into the evolution of functionally related characters that might ultimately lead to a complex phenotype such as social dominance (Moore & Moore 1999; Moore et al. 2002).

Males may also be required to compete for fertilizations following mating if a female mates with more than one male (Parker 1979). Genetic variation exists between males in traits important in their success during sperm competition. Variation in both the offensive and defensive components of sperm displacement have been observed (reviewed in Clark 2002). Sperm competitiveness depends on a male's diploid genotype, not the haploid genotype of a sperm. Ejaculates are complex characters, and are likely to involve the integration of several traits (Moore et al. 2004). Several studies examining the quantitative genetics of traits thought to be important in sperm competition (see Simmons & Kotiaho 2002; Moore et al. 2004), including sperm morphology, sperm size or length, sperm viability, quantity of associated accessory compounds, and testis size have revealed a genetic basis for these traits. In addition, selection experiments demonstrate changes in sperm morphology and size, indicating underlying genetic variation. For example, in the nematode *Caenorhabditis elegans* larger sperm are at a competitive advantage as they have greater motility and exclude smaller sperm from the spermatheca (LaMunyon & Ward 2002). The social environment is again an important consideration. When sperm competition is experimentally increased, sperm volume produced by individual males increases.

One motivation for studying the genetics of sperm competitive traits is as a mechanism for delineating the selective pressures arising through postcopulatory sexual selection. In order for sperm competition to occur, females must mate with multiple males. In species in which females gain direct benefits from males and mating, it is easy to explain polyandry that results in sperm competition. The evolution of polyandry is less easy to explain in species in which females do not benefit from males or mating. One hypothesis to explain the evolution of polyandry in the absence of direct benefits to females is the "good-sperm" hypothesis. Under this hypothesis, if fertilization success by males is correlated with viability, females will incite sperm competition in order to gain genetic benefits for their offspring. The "good-sperm" hypothesis predicts a positive genetic correlation between sperm competitive ability and offspring viability, as well as female mating frequency. Simmons and Kotiaho (2002) used a quantitative genetic approach to address this question in the dung beetle, *Onthophagus taurus*. In addition to demonstrating significant genetic contributions to traits thought to be important in sperm competitiveness, such as testis weight, ejaculate volume, and sperm length, they demonstrated that a significant genetic correlation exists between male condition and testis weight, as well as male condition and sperm length. These genetic correlations provide the link between competitively superior ejaculates and increased offspring quality required by the "good-sperm" hypothesis of sexual selection.

One of the most contentious issues facing researchers interested in the mechanism of sexual selection is whether differences in male fertilization success following multiple mating by females results from sperm competition or from cryptic female mate choice (Eberhard 1996). Quantitative genetics provides a powerful approach for unraveling these two mechanisms of sexual selection because the predicted patterns of genetic covariation that would arise through evolution by postcopulatory male–male competition versus cryptic female mate choice differ and can be tested (Simmons 2003). Polyandry in species where females derive no direct benefit from mating multiply is hypothesized to arise through indirect benefits to females. As discussed above, the "good-sperm" hypothesis may be one mechanism by which females benefit indirectly from mating with multiple males. An alternative mechanism by which females may benefit indirectly from mating with more than one male through sperm competition has been proposed. In this "sexy-sperm" hypothesis, females mate with more than one male in order to ensure that competitively superior sperm will fertilize their eggs and therefore produce sons with competitive sperm (Simmons 2003). The difficulty in testing this hypothesis through phenotypic studies is that it is impossible to determine whether a bias in fertilization success occurs because a particular male has highly competitive sperm or because a female is biasing fertilization toward that male on the basis of features of his ejaculate. However, the "sexy-sperm" hypothesis generates specific predictions relative to genetic architecture. Thus, if postcopulatory sexual selection occurs via sperm competition rather than cryptic female mate choice we would predict a positive genetic correlation between female mating frequency and male fertilization success.

Simmons (2003) examined phenotypic and genotypic variation and covariation in sperm competition traits, investment in testes and accessory glands, as well as variation in female investment in ovaries and female mating frequency in the field cricket. He found no genetic correlation between polyandry and fertilization success. Thus, the prediction for the "sexy-sperm" hypothesis was not met. Instead, a genetic correlation between female investment in oogenesis and male investment in accessory glands was found. This fits a model of indirect selection through cryptic female mate choice, where alleles for the trait, male accessory gland proteins, should become positively associated with alleles for the preference, increased oogenesis. Thus, in the field cricket, polyandry may have evolved as a mechanism for females to sample multiple males rather than to incite sperm competition. Patterns of genetic covariation can thus differentiate between alternative mechanisms of postcopulatory sexual selection that would be difficult to distinguish through phenotypic studies.

Even in systems where we accept that the sperm of multiple males compete for fertilization, it is becoming increasingly clear that the role of the female reproductive tract as the arena for sperm competition, and thus a potential source of variation in male success, cannot be ignored. As discussed above, it has been found that there are genetic factors that influence the competitive ability of sperm. So why do alleles that lead to highly competitive sperm fail to become fixed in the population? Genetic variation in sperm competitiveness may be maintained because the relative competitive ability of a male of a given genotype depends on the genetic context of the female with which he mates (reviewed in Clark 2002). Thus, in *Drosophila melanogaster* there is genetic variation between populations in the tendency of females to store sperm from the first or second male with which they mate. In addition, the success of a particular male genotype in competition with a second genotype depends on the genotype of the female within which the sperm are competing. Sperm competitive ability thus depends on the genetic context of the female and this needs to be accounted for both in empirical work and in theoretical considerations of how sperm competition will shape evolution through sexual selection.

Does Sexual Selection Affect the Location or Expression of Genes?

Defining the location of genes that influence sexually selected traits may help explain the apparent rapidity of evolution under sexual selection. Sexually selected traits are often sex-limited in expression, and one way to generate sex-specific expression is sex linkage. A great deal of attention, both empirical and theoretical, has focused recently on the X chromosome and sex linkage of sexually selected and sexually antagonistic traits (Reinhold 1998; Lande & Wilkinson 1999; Ritchie 2001; Gibson et al. 2001; Lindholm & Breden 2002; Pizzari & Birkhead 2002; Reeve & Pfennig 2002; but see Rowe & Houle 1996; Fitzpatrick 2004). It would seem reasonable to suppose that genes influencing sexually selected traits might be sex-linked. For example, because males are the heterogametic sex in mammals, and because genes on the Y are in the nonrecombining region, the Y chromosome in mammals is passed unaltered from father to son. The Y chromosome could therefore carry genes for competitive traits even if they would be disadvantageous to females because they never exist in the female. This could allow for the evolution of traits beneficial to males without a cost to females. There are few genes on the mammalian Y chromosome, but these include genes for spermatogenesis, which could influence sperm competitive ability, genes for adult size, and perhaps surprisingly, genes for tooth development (Roldan & Gomendio 1999). Many mammalian fights involve teeth, and sexual dimorphism in tooth size is common, suggesting an area where sexually selected and naturally selected traits might be compared.

Empirical research provides ambiguous results. Reinhold (1998) finds support for a biased distribution of sexually selected traits, with more genes influencing sexually selected traits being sex-linked. Sex-linked sexually selected traits are common in poecilid fish (Lindholm & Breden 2002). Perhaps the most compelling evidence comes from the work on stalk-eyed flies (Lande & Wilkinson 1999; Wolfenbarger & Wilkinson 2001). Evidence contrary to the hypothesis also exists. Sexually selected male characters in Hawaiian crickets tend to be located on autosomes (Shaw & Parsons 2002). Most recently, Fitzpatrick (2004) argues that we cannot just examine the genes that influence sexually selected traits without regard to pleiotropic effects of those genes. He finds that sexually selected genes are not preferentially sex-linked in *Drosophila* and argues this may be because of pleiotropy. Furthermore, female mating preferences and preferred male traits appear not to be linked when the male trait is disproportionately influenced by the X chromosome (Ting et al. 2001; Ritchie 2001). Clearly the data available for testing this hypothesis are limited—we have only limited information on genes that influence sexually selected male traits or mating preferences, and it is likely that X- or Y-linked genes will be easier to detect than autosomes simply because detecting sex linkage is easier than locating or detecting other single-gene effects. Additional genomic data on organisms with sexually selected traits are therefore needed.

Reeve and Pfennig (2002) argue a related idea: that genetic systems predispose organisms to showy traits evolving as a result of sexual selection. Their argument is that some genetic systems (e.g., male homogamety) are better protected against random loss of rare alleles influencing male ornaments or mate choice, which they support with a model. While their models are intriguing, in addition to the contrary evidence provided above, they support their ideas with a comparison of sexual dimorphism across taxa. Equating sexual dimorphism and sexual selection is incorrect as sexual dimorphism may result from a number of evolutionary processes of which sexual selection is only one of many possibilities (Shine 1989). This again highlights the problem for testing these ideas; it is hard to obtain data that have no ascertainment bias or are a true random sample. Reeve and Pannig's model may be correct, but it needs additional testing. This is also an area that could profitably be pursued with more genomic data and fewer indirect measures of genetics of showiness.

One of the problems with the sex-linkage hypothesis is that not all sexually selected species have sex-specific chromosomes such as the Y or W chromosome. In many insects, for example, the males lack one of the two sex chromosomes (i.e., are X/0). A virtually unexplored alternative is epigenetic mechanisms that result in sex-specific gene expression, such as imprinting. There is too little theory addressing epigenetics and sexual selection, and even less empirical work, to make many generalizations. The nature of imprinting and the distribution of sex-specific gene expression in relation to sexual selection are only now beginning to be explored (Lande & Wilkinson 1999). Given the prevalence of sexual conflict when sexual selection influences traits, and that sexual conflict is expected to result in sex-specific gene expression, it is not difficult for us to predict that examining epigenetics associated with sexually selected characters should become an important area of research for both fields.

In both the sex-linked and genetic systems hypotheses relating sexual selection and genomics, the problem lies in a lack of quality data. Where we have excellent genomic data (e.g., *Drosophila* or mice) we have poor data on which traits are sexually selected. Where we have high-quality data on genetics of sexually selected traits (e.g., Jia et al. 2000; Brooks 2002; Blows et al. 2004) we rarely have genomic data. Collecting such genomic information should be a priority for students of sexual selection.

FUTURE DIRECTIONS

The vast majority of research in sexual selection has focused on documenting that this form of selection occurs. It is now time to focus on variation—variation in mate choice, in fitness associated with mate choice, in selection itself. We also need to address why there is so much variation underlying such an important process. Is this because environments where mating occurs, especially social environments, are more variable than we have suspected?

We need to have more G × E studies, especially where E is a social environment. Variation in selection over space and time has rarely been addressed in studies of sexual selection, and we therefore cannot say how common it is for this type of variation to occur. Temporal, spatial, and social context variation is more likely to be the norm than the exception. The main technique we need developed here is to find ways to provide quantitative measures of the social behaviors integral to sexual selection. Shuster and Wade (2003) is an excellent starting point for students wishing to pursue this course of study.

Genetic architecture has been investigated more often, particularly with a quantitative genetic approach and a description of the G matrix underlying elaborate male morphology, but these studies are still too rare. The rise of genomics provides further data on genetic architecture that students of sexual selection need to assimilate and examine. The old arguments over good genes and arbitrary characters are giving way to where and how many genes influence sexually selected traits. All of these questions are valuable, but all still need new data. The important thing, then, is for students of sexual selection, whether it is red-winged blackbirds or guppies or moths or even a live-bearing African cockroach that are being studied, to ensure that wherever possible genetic data and especially genomic data are collected. Model genetic systems focusing on domesticated or cosmopolitan species such as *Drosophila* are very unlikely to be the best systems for studying sexual selection, and therefore it is important that model behavioral systems are examined genetically. Borrowing information and approaches developed in the genetic model species and applying these to well-studied behavioral systems is where the breakthroughs in sexual selection research are likely to occur.

SUGGESTIONS FOR FURTHER READING

In this chapter we focused on the models and evidence for genetic aspects of sexually selected traits and behaviors. Our review did not provide a particularly detailed discussion of the basics of sexual selection. To begin to get a foothold in the field, there are a number of excellent reviews of sexual selection available, and the starting point for anyone should be Andersson (1994), which is both comprehensive and accessible. Shuster and Wade (2003) provide a genetically oriented theoretical perspective on sexual selection and mating system evolution. Brooks (2002) and Kokko et al. (2003) provide some of the finer details of mate choice research. Mead and Arnold (2004) review the quantitative genetic models of sexual selection. Research on sexual conflict is newer, but Eberhard (1996), Kokko et al. (2002, 2003), Chapman et al. (2003), Pizzari & Snook (2003) and Zeh & Zeh (2003) provide an introduction to the focus and controversy in this area. All of these should provide insights into the areas where we need measures of how sexually selected phenotypes correspond to fitness and underlying genetic attributes.

Andersson M 1994 Sexual Selection. Princeton Univ. Press.

Brooks R 2002 Variation in female mate choice within guppy populations: population divergence, multiple ornaments and the maintenance of polymorphisms. Genetica 116:343–358.

Chapman T, Arnqvist G, Bangham J & L Rowe 2003 Sexual conflict. Trends Ecol. Evol. 18:41–47.

Eberhard W 1996 Female Control: Sexual Selection by Cryptic Female Mate Choice. Princeton Univ. Press.

Kokko H, Brooks R, McNamara JM & AI Houston 2002 The sexual selection continuum. Proc. R. Soc. Lond. B 269:1331–1340.

Kokko H, Brooks R, Jennions MD & J Morley 2003 The evolution of mate choice and mating biases. Proc. R. Soc. Lond. B 270:653–664.

Mead LS & SJ Arnold 2004 Quantitative genetic models of sexual selection. Trends Ecol. Evol. 19:264–271.

Pizzari T & RR Snook 2003 Sexual conflict and sexual selection: chasing away the paradigm shifts. Evolution 57:1223–1236.

Shuster SM & MJ Wade 2003 The Evolution of Mating Systems and Strategies. Princeton Univ. Press.

Zeh JA & DW Zeh 2003 Toward a new sexual selection paradigm: polyandry, conflict and incompatibility. Ethology 109:929–950.

23

Social Selection

STEVEN A. FRANK

Individuals sometimes give up their own resources to benefit their neighbors. Such altruistic traits posed a difficulty for the original Darwinian formulation of natural selection, which emphasized the spread of individually advantageous characters. So how have altruistic traits become common in some populations? This is an important question because a purely individualistic world would look very different from the one that we see. With no altruism, there would be no multicellularity with specialized nonreproductive tissues, no social insects with specialized worker castes, and nothing at all like complex human societies.

In this chapter, I discuss three processes that can promote altruism and the evolution of social cooperation. I start with kin selection, in which an individual may give up some of its own reproduction to aid relatives, or an individual may coordinate its behavior with phenotypically similar neighbors to promote the good of the group. Altruism toward kin helps to explain patterns of parasite virulence, sex ratios, and complex sociality with division of labor between different individuals.

Some groups build up a high level of social cohesion in spite of little relatedness and low opportunity for kin selection. In the second section, I discuss how repression of competition can be a powerful force integrating the interests of individuals. With no opportunity to compete against neighbors, an individual can only increase its own success by increasing the success of the whole group. Meiosis provides the classic example, in which the strict control of chromosomal segregation into gametes prevents competition between different chromosomes. It is only through such repression of competition between chromosomes that the genome developed into a highly integrated and cohesive unit.

In the third section, I turn to another key theme in the history of life—the evolutionary innovations of cooperative symbioses between different species or different kinds of genomes. The first genomes near the origin of life probably evolved by biochemical synergism between different replicating molecules; eukaryotic cells arose by symbioses between different species; and lichens, mycorrhizal–plant systems, and many other symbioses have contributed greatly to the complexity of modern life. The evolution of symbioses concerns the same social tensions between conflict and cooperation as the more familiar problems from kin groups and animal societies.

Cooperative symbioses may evolve by positive feedback between partners. In such synergistic relations, one party gives up some of its resources to enhance the success of its partner, and the partner does the same. The vast majority of cooperative symbioses arose as biochemical synergisms between organisms without complex behavioral flexibility. By contrast, the exchange of benefits between partners with the capacity for memory and the potential for strategy leads to issues of cheating, detection of cheaters, and strategic assessment of partner behavior and quality. Such problems of reciprocal altruism (Trivers 1971) pervade many aspects of vertebrate sociality, in which individuals remember particular partners and their past behaviors, and individuals can assess the complex strategies of others and form their own strategies in response. I do not cover behavioral reciprocity in this chapter, in order to focus on the more genetically relevant aspects of social selection that fit the themes of this volume.

My three topics of kin selection, repression of competition, and synergistic symbiosis all play fundamental roles in the evolution of genetic systems.

KIN SELECTION AND CORRELATED BEHAVIORS

Two different kinds of problems often arise when studying altruistic behavior in social interactions (Frank 1997b, 1998). First, an individual may give up some of its own reproduction to help kin increase their reproduction. For example, a young bird may remain with its parents and help raise its siblings rather than leave to set up its own nest and reproduce directly. This is the classic problem of altruism and kin selection.

Second, an individual may live in a group and face a tension between selfish and cooperative behaviors. For example, an individual may gain by selfishly grabbing a larger share of local resources but at a cost to the efficient use of those resources by the group. This trade-off between selfish gains for the individual and prudent benefits for all arises in problems such as the evolution of sex ratios and parasite virulence. Cooperation increases with the correlation in behaviors between group members. Correlation in behavior can arise for various reasons, of which kinship is often the most important.

The history of these subjects has turned on how to calculate when an altruistic behavior toward kin will increase, or how to calculate the optimum mixture of selfish and cooperative behavior in groups that share a common resource. The following sections give a sense of those calculations. I simplify the mathematics to emphasize the essential concepts. After developing the concepts, I illustrate the main ideas with a few examples.

Hamilton's Rule for Kin Selection

How can we determine whether a trait for self-sacrifice spreads in a population? Hamilton (1964a, b, 1970) took a population genetics approach by calculating when the allele frequency for an altruistic character increases. Suppose, for example, that an individual gives up its own opportunity to reproduce and instead uses its resources to help its sister. Do genes that reduce individual reproduction in this way increase or decrease over time?

Hamilton's calculation proceeds roughly as follows. The altruistic individual gave up the chance to make C babies and instead helped its sister to make an extra B babies. We need some measure of exchange to figure out how to weight the loss of the individual's own offspring and the gain in its sister's offspring. Our ultimate scale is change in allele frequency.

What effect does a loss of C babies by an individual have on the allele frequency of altruistic traits? This depends on the difference between the frequency of those alleles carried by the individual, q, and the allele frequency in the population, $\bar{q}$. Let us write this difference as $\delta = q - \bar{q}$. If the individual has the same allele frequency as occurs in the population, $\delta = 0$, then no matter how many babies this individual makes, there will be no consequence for the population allele frequency. In general, loss of C babies by our individual has allele frequency consequences in proportion to $-\delta C$, where the minus sign arises because this term represents a loss in reproduction.

What effect does a gain of an extra B babies by the individual's sister have on allele frequency change? If our focal individual has a allele frequency deviation δ, then on average the sister has a deviation $r\delta$, where r is the coefficient of relatedness of the focal individual to its sister (Box 23.1). Thus, a gain of an extra B babies by the individual's sister has allele frequency consequences in proportion to $r\delta B$.

We get the total effect on allele frequency change by combining the two terms, $r\delta B - \delta C = \delta(rB - C)$. If the total effect is greater than zero, then the frequency of alleles causing the altruistic behavior increases, and altruism spreads in the population. This gives us the famous result

$$rB - C > 0 \tag{23.1}$$

known as Hamilton's rule for the spread of an altruistic behavior, where B is the benefit of the behavior directed toward kin related by r, and C is the cost of the behavior.

For example, an individual is typically related to its mother by $r = 1/2$. Thus an individual would be favored to forgo reproduction and stay with its mother if, for every lost baby of its own, $C = 1$, the altruistic individual added more than two offspring to its mother's reproduction, $B > 2$.

In this problem, individuals exchange direct transmission of genes for transmission by indirect routes via the extra reproduction of kin. The coefficient of relatedness, r, is the exchange rate that scales direct and indirect reproduction to obtain the ultimate

BOX 23.1. Coefficients of Relatedness

There is a vast literature on coefficients of relatedness in social interactions (Frank 1998). The two different kinds of social problems—self-sacrifice and correlated behaviors—are rarely distinguished in a clear way, causing the literature on this topic to be confusing and difficult to read. I will not attempt to review the literature or provide technical details here. Rather, I will outline the main concepts of relatedness in an intuitive way, using simple equations. Mastery of the subject requires deeper study, but the points here highlight the essential differences between the two types of social problems for which relatedness has been used.

The study of relatedness can be developed most naturally through the basic equation of linear regression. Simple regression predicts the value of a variable given a measurement on another variable. For example, we may predict the amount of rain given a measurement of cloudiness, or we may predict weight given a measurement of height.

The basic linear regression equation is

$$y = a + bx + \varepsilon,$$

where x is a value we measure, and y is the value we wish to predict given the measurement on x. In regression analysis, we predict y given a measurement of x as $\hat{y} \mid x = a + bx$. From regression theory, $a = \bar{y} - b\bar{x}$ and $b = \text{cov}(y,x) / \text{var}(x)$, where b is called the regression coefficient, and overbars denote average values.

We can use linear regression to predict expected values of y whenever we have paired observations (x, y). The values of a and b come from minimizing the average distance between the actual values of y and the predicted values, $\hat{y}$. No assumptions are required about the distributions of x and y; for example, they do not have to be normally distributed. Requirement of normality arises only in tests of statistical significance, not in developing predicted values of y that minimize the distance between prediction and observation.

The regression coefficient, b, provides a measure of conditional information about the variable we wish to predict. We can see this by rewriting the basic regression equation. First, define the deviation of x from its average value as $\delta = x - \bar{x}$. Then, using the above details about the standard form of regression, we can rearrange the terms as

$$\text{E}(y - \bar{y} \mid \delta) = b\delta,$$

which can be read as: the expected deviation of y from its average value, $\bar{y}$, given the deviation, δ, of the predictor variable x from its average value, equals the regression coefficient multiplied by the deviation of the predictor from its average, $b\delta$. In other words, the regression coefficient tells us how much y is expected to deviate from its average given a certain deviation of x from its average.

Now let us take the first kin selection problem of self-sacrifice, in which we need to measure allele frequency deviations from the population average. Suppose the average frequency of an altruistic allele is $\bar{q}$, and the frequency of the allele in the actor who may behave in a self-sacrificing way is q. What information about allele frequency in recipients of the altruistic act is contained in the fact that the actor has allele frequency deviation $\delta = q - \bar{q}$? Suppose the allele frequency in recipients is q'. Then the expected allele frequency deviation in recipients is

$$\text{E}(q' - \bar{q} \mid \delta) = r\delta,$$

BOX 23.1. *(cont.)*

where r is the relatedness coefficient, which is the regression coefficient from standard linear regression

$$r = \frac{\text{cov}(q',q)}{\text{var}(q)}.$$

There are more general ways of analyzing such problems that avoid assuming a single locus controls self-sacrifice. For example, one can formulate the theory in terms of breeding value from quantitative genetics (Frank 1997b, 1998). But in the more general theory, the basic use of regression remains the same. For self-sacrifice, relatedness measures the conditional prediction about genetic deviation of social partners from the population average given the deviation of the actor from that average. This provides the scaling needed to measure gains and losses in allele frequency in different classes of individuals.

The second type of problem concerns correlated behaviors in groups where all members express a behavior. This is more of a game-like situation in which behaviors are strategies that determine payoffs to individuals. The question is how an individual should adjust its behavior, such as its sex ratio, in order to maximize its own direct payoff.

Consider a group game, such as the tragedy of the commons or the sex ratio. Here, our focal individual has a behavior y, the actor's group has average behavior z which includes the contribution of the actor, and the population average is $\bar{z}$. Let the focal individual's deviation from the population average be $\delta = y - \bar{z}$. Then

$$E(z - \bar{z} \mid \delta) = r\delta,$$

that is, the expected deviation of social partners from the population average, given the actor's deviation, is $r\delta$. Here, r measures conditional information about the behavior of social partners given the actor's own behavior. In this case, the regression coefficient is

$$r = \frac{\text{cov}(z,y)}{\text{var}(y)},$$

the regression (slope) of average group phenotype on actor phenotype. This gives r entirely in terms of phenotypes, which is what we need if we are interested in the immediate payoff to an actor when playing a game in which partners have correlated strategies. In evolutionary analysis, we are more interested in what the actor transmits to progeny, so we may choose to focus on the genetically transmitted value (breeding value) for the behavior, g, where g is roughly the expected contribution of the actor to the value of y in progeny (Frank 1998). We can do the analysis using g in place of y, giving

$$r = \frac{\text{cov}(z,g)}{\text{var}(g)},$$

which is the slope of partners' phenotype on actor's genotype. Partners will often have correlated behaviors because they are genealogical relatives. But genetic relatedness is not required, only an association between partner phenotype and the breeding value of the actor.

In the first model of self-sacrifice, the proper measure of relatedness is the regression of recipient genotype on actor genotype. This provides a measure of information

(continued)

BOX 23.1. *(cont.)*

about the genetic value transmitted by the recipient given the actor's genetic value (or allele frequency), allowing one to measure the total gains and losses in the transmission of genetic value. In the second model of correlated behavior, the proper measure of relatedness is the regression of the partners' phenotype on the focal individual's genetic value. This provides a measure of how partners' behaviors affect the transmission of the focal individual's genetic value.

In more complex social situations, the particular trait of an individual can affect reproduction by different kinds of recipient individuals. For those complex situations, it is often best to consider how the trait of an actor affects the fitness of the recipients in different classes. For example, a helper that remains with its parent may affect the fitness of its parent and of its siblings of different ages in the extended family.

If we would like to study the evolution of helper behavior, we must follow the effect of the helper's phenotype on the transmission of genetic value for the helping trait in the different classes of recipients. One can usually use the method for self-sacrifice discussed above, but it is often more natural conceptually and mathematically to formulate the problem in terms of the direct fitness method (Taylor & Frank 1996). With that method, the flow of effects goes from the actor's phenotype to the transmitted genetic value of the different classes of recipients. Consequently, the direct fitness coefficient of relatedness measures the regression of actor phenotype on the recipients' transmitted genetic value (Frank 1998).

The general conclusion is that relatedness measures for studying social evolution take on different forms of regression coefficients according to the flow of effects in particular analyses.

consequences for allele frequency change. The coefficient r can be thought of as a measure for the fidelity of transmission of genetic information via different pathways of direct and indirect reproduction—an extended form of the standard heritability coefficient of quantitative genetics (Frank 1997b, 1998).

Partitions and Scaling

In the previous section, I considered how an altruistic behavior may increase the transmission of genes in nondescendant lineages. Hamilton's rule for that problem partitions the total effect of a behavior on allele frequency change into two components: the effect, C, on the actor's reproduction and the effect, B, on the recipient's reproduction. The coefficient r provides the proper scaling so that rB and C give allele frequency effects on the same scale.

Whenever we wish to partition the total effect of some behavior into direct and indirect components, we will end up with three factors. First, we must measure the direct effect on the scale of interest. Second, we need the indirect effect, usually measured on some different scale. Third, we must render the indirect effect on the same scale as the direct effect so that we can obtain the total effect. For example, we used C for the direct effect, B for the indirect effect, and r for scale translation.

Given the general structure of partitioning into direct and indirect effects, we should not be surprised to find that different partitions all end up looking like Hamilton's rule. The next section provides an example of another partition that looks exactly like our first Hamilton's rule, yet has a very different meaning.

Correlated Behaviors and Cooperation in Social Groups

Consider an individual that interacts with a partner or with a small number of others in an isolated group. Each individual faces the essential tension of sociality. On the one hand, it can act selfishly to grab a larger share of the limited resources in its group, but selfish behavior causes inefficient use of local resources and lowers the total output of the group.

On the other hand, an individual can act altruistically, taking a smaller share of local resources and raising the total success of the group.

In this case, we are concerned only with how changes in an individual's behavior affect its own success. Let the individual's level of cooperation be y, such that larger values mean better cooperation with neighbors and smaller values mean greater selfishness. Let the average value of y in our focal individual's group be z, and let the average level in the population be $\bar{z}$. Thus, our focal individual's difference from the population average is $\delta = y - \bar{z}$. Higher δ means our focal individual is more cooperative than average and gives up a greater share of personal gain. The direct loss in success caused by cooperative behavior is $-\delta C$. Here, δ is the deviation from average phenotype and C scales between phenotype and a measure of success such as the number of offspring.

The group's deviation from the population average can be measured as $r\delta = z - \bar{z}$, where r is the slope of group behavior, z, on individual behavior, y. Here r measures phenotypic similarity—from an individual's point of view, it is a measure of information about the behavior of social partners (Box 23.1). When individuals in the group all act in the same way, then $r = 1$. When there is no association between individuals in a group of size N, then $r = 1/N$, because our focal individual is identical to itself and contributes a part $1/N$ to the group average.

The higher the group's level of altruism, z, the greater the benefit to our focal individual for living in a cooperative and efficient group. The expected group deviation from the population average is $r\delta$. Let the benefit per unit deviation be B, so the total benefit from group-level altruism is $r\delta B$. Thus, the total gain to an individual for increasing y, its level of altruism, is $r\delta B - \delta C$, which is positive when $rB - C > 0$.

This is the same expression as Hamilton's rule, but this expression must be interpreted differently from the prior result about allele frequency. Here, C is the direct cost of the individual's own altruism on its success. The term B is the indirect effect of the group's altruism and efficiency on our focal individual's success. The scale in this case is the individual's own success.

For each unit change in the individual's own phenotype, the group phenotype changes by r. Thus, I like to think of r as a measure of information about social partners. If r is high, then individuals acting altruistically have a high chance of associating with similar, altruistic partners. If r is low, then altruistic individuals often associate at random and will often have selfish partners that take advantage of them.

Loosely speaking, if one knows that partners will behave similarly to oneself, then acting altruistically and promoting group efficiency provide a direct benefit to oneself. By contrast, if one has little information about partners, then altruistic behavior will often be taken advantage of by selfish partners.

It is easy to be misled by the identical $rB - C > 0$ form of the expressions for the increase in altruism under self-sacrifice and under correlated behaviors (Frank 1997b, 1998). Whenever we partition total effects into two components measured on different scales, we end up with results that have the same $rB - C$ structure. In the two cases, the terms have different meanings. The first applies to allele frequency change when an actor behaves toward a neighbor, with the term r translating allele frequency deviations between actor and recipient. The second applies to an individual's success when actors and neighbors both act mutually in a social game of cooperation and selfishness, with the term r translating between individual- and group-level deviations from the average level of altruism.

We can use the second model for any sort of mutual interaction between individuals, including interactions between different species (Frank 1994b, 1998). Most often, however, the phenotypic similarities measured by r will arise from genetic similarity. By contrast, the first model makes reasonable sense only when applied to the behavior of individuals toward genetically similar recipients of the same species.

Applications to Self-Sacrifice

Suppose an offspring gives up its own opportunity to reproduce and instead helps its parents to raise more of its siblings. The most extreme case arises in sterile workers of social insects, but there are many examples in birds and mammals in which offspring spend at least part of their adult life aiding their parents. This is a clear case of the first model: self-sacrifice for a genetic relative.

Our theory tells us that we need to evaluate $rB - C > 0$ to determine when selection favors sacrifice of direct reproduction to help parents. At first glance, this seems to make a clear, simple, and testable prediction about whether or not offspring will stay to help their parents. We do indeed have the right pieces, because our partition identifies the three key factors: direct reproduction, C, indirect reproduction, B, and scale translation, r. But quantitative evaluation can be difficult.

Consider a young bird that may either try to reproduce on its own or remain in its natal territory to help the current breeders (Emlen 1984; Brown 1987). How do we estimate loss in direct reproduction, C? This cost depends on demographic opportunities to obtain a territory, a mate, sufficient food to raise progeny, and the bird's vigilance to defend against predators. In addition, individual vigor and competitiveness vary, so we must account for each individual's particular attributes. And we must compare the value of an offspring raised alone with the value of an offspring raised on the natal territory, taking account of such issues as the potential for offspring to inherit their parents' territory.

In studying such problems, observers sometimes try to measure all possible factors needed to tally $rB - C$ and determine whether the balance for a young individual tips toward helping or going it alone. In my opinion, while the theory is valuable in calling attention to the relevant issues, there is little hope of measuring all factors that contribute to costs and benefits. Thus, consistency checks that seek to match the sign of $rB - C$ with individual behavior in a particular setting can be difficult to interpret.

I prefer simple comparative hypotheses and tests that respect the general, abstract nature of the theory. Comparative use of the theory can explain much of observed behavior, without trying to explain more than we are really able to do. For example, the greater the value of r, the more likely an individual will remain on its natal territory as a helper rather than try to reproduce on its own. The more severe the competition for establishing new territories or taking over existing ones, the more likely an individual will help rather than try to reproduce.

Applications to Correlated Behavior in Groups

The "tragedy of the commons" problem captures the tension between individual selfishness and group efficiency (Hardin 1993; Frank 1995b). Suppose each group has a common, renewable resource. The more each individual takes from the common pool, the greater its success. However, greater exploitation of local resources lowers total yield, so group productivity rises if individuals restrain their selfish tendencies. For example, a group of parasites may share a common host. The faster a parasite consumes resources and reproduces, the greater its share of the host. Rapid consumption may, however, overexploit the host, reducing host vigor or survival and lowering the total yield of the parasite group. In this tension between individual success and sustainable yield, parasite virulence may be shaped by a tragedy of the commons (Frank 1996).

A simple model for this problem can be written as

$$w(y, z) = \frac{y}{z}(1 - z) \tag{23.2}$$

where an individual's fitness, w, depends on its selfish tendency to grab local resources, y, and the average selfish tendency in the local group, z (Frank 1994c, 1995b). The term y/z is the relative success of an individual within its group; for example, $y/z = 2$ means that the individual gets twice the average share. The term $1 - z$ is group productivity—the greater the average selfishness, z, the lower the group productivity.

Here the trait y is selfishness, so we may say that altruism increases as y declines. We know by the general theory of partitioning effects that altruism increases (y declines) when $rB - C > 0$, that altruism decreases when $rB - C < 0$, and that the system comes to equilibrium when $rB - C = 0$. So all we need to do is solve $rB - C = 0$ to determine what level of altruism tends to evolve in this situation. However, it is not obvious from looking at Equation 23.2 how to determine costs and benefits. Remember, in this situation we focus on an individual and measure the direct cost to its fitness of becoming more altruistic, the benefit to the focal individual from living in a group that shares its altruistic tendency, and the measure r that gives the translation between our individual's level of altruism and the tendency of its neighbors also to be altruistic.

There is a simple technique to extract the cost and benefit terms (Frank 1995b, 1998; Taylor and Frank 1996). The cost is the change in fitness, w, with respect to change in individual behavior, y, holding group behavior z constant. In mathematical terms this is $\partial w/\partial y = -C$, where the minus sign arises because costs enter into the total in a negative way. The benefit term is the change in w with respect z, holding y constant, $\partial w/\partial z = B$.

What we have done is step through the mathematical expression for the total change in fitness,

w, with respect to individual behavior, y, which can be expanded as

$$\begin{aligned}\frac{dw}{dy} &= \frac{\partial w}{\partial y} + \frac{\partial w}{\partial z}\frac{dz}{dy} \\ &= -C + Br\end{aligned} \tag{23.3}$$

where $r = dz/dy$ shows that the relation between group behavior and individual behavior is simply given by the slope from the derivative.

From all this we want the equilibrium behavior to which the population settles. At equilibrium, small changes in behavior do not increase fitness, otherwise the population would continue to evolve; thus we want $\delta w/\delta y = 0$, which is equivalent to $rB - C = 0$. At this equilibrium point all individuals have converged to the same, optimal behavior, so we evaluate the condition at $y = z = z^*$. Checking our calculus book to get B and C by applying Equation 23.3 to Equation 23.2, and solving $dw/dy = 0$, gives the equilibrium $z^* = 1 - r$ (Frank 1994c, 1995b). As the similarity in behavior within groups, r, rises, individuals lower their selfish tendencies, z^*.

I went through the steps in some detail, but the overall approach to understanding social selection is very simple. From the fitness expression in Equation 23.2, we take the slope of fitness on individual behavior in Equation 23.3 and solve at the equilibrium where the slope is zero.

Now let us analyze a sex ratio problem, focusing on the biology rather than the method. Sex ratio has been the most important problem for developing and testing the theory of social selection in group-structured populations (Hamilton 1967; Charnov 1982; Godfray & Werren 1996).

Suppose that several mated female insects (foundresses) land in a patch of resources and lay their eggs. The progeny emerge and mate among themselves. The males die, and the mated females fly off to find a new patch in which to lay their eggs. The problem concerns the ratio of sons and daughters produced by foundresses. To express the sex ratio in a consistent way, I use the number of males divided by the total number of progeny—the frequency of males. Making sons is a selfish act because males do not contribute to group productivity. Instead they compete in the mating arena with the sons of other foundresses for the local resource, which is unmated females. Making daughters is an altruistic act because it increases the pool of the local resource available for mating by males.

Let a focal foundress's sex ratio be y and the average sex ratio in a group be z. Our focal foundress's relative share of matings through sons is y/z, and the pool of available females is in proportion to $1 - z$, so total success through sons is proportional to $(y/z)(1 - z)$. Success through daughters is proportional to $1 - y$, the fraction of progeny that are female, assuming females do not compete in the local patch for food or space but instead compete after mating and dispersal.

Putting the terms together, fitness is

$$w(y,z) = \frac{y}{z}(1 - z) + 1 - y. \tag{23.4}$$

Setting the derivative in Equation 23.2 to zero and evaluating at the equilibrium condition $y = z$ gives the optimal sex ratio as $z^* = (1/2)(1 - r)$ (Frank 1986). Note the similarity to the model for the tragedy of the commons, both in the expression for the male component of fitness and in the result.

If there are n foundresses in each patch, and a female's sex ratio is uncorrelated with her neighbors' sex ratios, then $r = 1/n$ and $z^* = (n - 1)/2n$, which is Hamilton's (1967) famous result for sex ratio under local mate competition. The result here based on r is more general because it shows how the sex ratio evolves when neighboring foundresses have correlated sex ratio behavior, perhaps because the foundresses are genetically related and sex ratio behavior is influenced by genotype.

Summary

In this section, I discussed two distinct processes. In the first case, individuals sacrifice their own reproduction to aid nondescendant genetic relatives. Hamilton's rule, $rB - C > 0$, partitions the consequences of self-sacrifice into a cost in direct reproduction, a benefit in the reproduction of relatives, and a scaling factor r that measures the genetic relatedness between altruist and recipient. The factor r can also be thought of as a scaling for the heritability of traits via nondescendant lineages compared with the heritability in direct reproduction—a measure of information about the transmission fidelity of characters via different pathways.

In the second case, individuals face a tension between their share of group resources and the efficiency of the group. Members of the group interact symmetrically, each having its own behavior that affects its success and the success of the group.

The consequences of each individual's behavior on its own reproduction can be partitioned into three terms. The first is a cost that measures the direct effects of the behavior on the individual's success, holding constant the group-level efficiency. The second is an indirect benefit that measures the effects of the average behavior in the group on group efficiency, holding constant individual behavior. The third is a scaling r that measures the similarity between an individual's behavior and the behavior of its neighbors—a measure of information about the behavior of social partners. Thus, r measures the extent to which a more cooperative individual tends to have more cooperative partners. In this case, the partners are not required to be genetic relatives, because payoff is measured entirely in terms of effects on our focal individual's total success.

The three terms of the second case can be combined into an expression for the increase in cooperative behavior, $rB - C > 0$. The similarity to Hamilton's rule for self-sacrifice is obvious, and indeed those of us who have developed this expression have called it a form of Hamilton's rule (Queller 1992; Taylor & Frank 1996; Frank 1997b, 1998). However, the two distinct processes of social selection—self-sacrifice for genetic relatives and cooperation in groups—give rise to similar expressions because the best method of analysis partitions success into costs, benefits, and a scaling factor, r. The similarity is both instructive and misleading; it is important to recognize the distinct biological interactions in the two cases and the different interpretations of r (Frank 1997b, 1998).

OTHER SOCIAL PROCESSES

I discussed the concepts of kin selection and correlated behaviors at length because they are the major forces of social selection. In the remainder of this chapter, I briefly introduce two additional topics. These topics build on the foundation of natural selection and kin selection to show how additional processes have shaped social interactions. I continue to use the word "social" in the broadest way, to cover all aspects of evolutionary change that deal with the tension between conflict and cooperation. Thus, social selection is important for understanding how different genetical elements came to be integrated into genomes, and how different cells became integrated into complex multicellular organisms.

REPRESSION OF COMPETITION

It would not make sense to speak of the genetic relatedness between genes on the Y chromosome and genes on chromosome 7 of the human genome. Those different genes are different kinds of things, almost like different species.

The genes on different chromosomes are integrated into a cohesive, cooperative group, yet they may also conflict over transmission to the next generation. For example, genes on the Y chromosome pass to the next generation only through sons, whereas those on chromosome 7 pass equally to sons and daughters. A Y chromosome that biased the sex ratio toward sons would increase in frequency. The genes on chromosome 7 in the same genome with the biasing Y would end up in sons. If a male bias developed in the population, sons would have lower fitness than daughters because the excess of males would be competing for relatively few females, and each male would on average be the father of less than one brood. So chromosome 7 would lose in transmission as the Y gained from causing a male bias.

In general, a chromosome can gain by increasing its transmission to offspring above the standard Mendelian fraction of one half (Lyttle 1991). Such segregation bias often imposes a fitness cost on the entire group of genes in the genome, either by sex ratio distortion or because the "driving" chromosome typically carries deleterious effects. Thus, the integration of the genome into a cohesive unit requires repression of the selfish transmission gains by subunits of the genome. The standard Mendelian segregation ratio of one half probably depends on mutual suppression of drive between chromosomes, repressing the potential for internal competition within the genome. With repression of the opportunities for gain against neighboring genetic elements, the only way that parts of the genome can increase their own success is by increasing the success of the entire group.

The Mendelian segregation ratio of one half for chromosomes is sometimes called fair meiosis, to emphasize that each chromosome has an equal or fair chance of being transmitted. Leigh (1971, 1977) was perhaps the first to emphasize that fair meiosis may have arisen to repress competition in the genome, thus integrating the reproductive interests of genomic subsets into a cohesive unit. Leigh (1977, p. 4543) expressed this idea rather colorfully when he said that the many genes of the genome repress biases of individual chromosomes "as if we had to do with a parliament of the genes, which so regulated itself as to prevent 'cabals of a few' conspiring for their own 'selfish profit' at the expense of the 'commonwealth.'"

Leigh (1977) noted that alignment of individual and group interests shifts selection to the group level. However, meiosis was the only compelling case known at that time. Without further examples, there was no reason to emphasize repression of internal competition as an important force in social evolution. From the conceptual point of view, it may have been clear that repression of internal competition could be important, but not clear how natural selection would favor such internal repression.

Alexander and Borgia (1978) joined Leigh in emphasizing the potentially great potency of internal repression in shaping interests and conflicts in the hierarchy of life. From this, Alexander (1979, 1987) developed his theory of human social structure. In that theory, intense group-against-group competition dominated the success of humans and thus shaped societies according to their group efficiencies in conflicts. Efficiency, best achieved by aligning the interests of the individual with the group, favored in the most successful groups those laws that partially restricted the opportunities for reproductive dominance. For example, Alexander (1987) argues that socially imposed monogamy levels reproductive opportunities, particularly among young men at the age of maximal sexual competition. Those young men are the most competitive and divisive individuals within societies, and also the pool of warriors on which the group depends for its protection and expansion.

In the late 1970s, the concept of internal repression remained limited to meiosis and perhaps some aspects of human social structure. The concept could not gain attention as a potentially important process in the history of life without further examples. In the 1980s, three independent lineages of thought developed on social insects, cellular competition in metazoans, and domestication of symbionts. These different subjects would eventually contribute to a fuller understanding of the conceptual issues and biological significance of internal repression of competition (Frank 2003). Here, I briefly summarize only one of those subjects: cellular competition in metazoans.

Many multicellular animals are differentiated into tissues that predominantly contribute to gametes and tissues that are primarily nonreproductive. This germ–soma distinction creates the potential for reproductive conflict when cells are not genetically identical. Genetically distinct cellular lineages can raise their fitness by gaining preferential access to the germline. This biasing can increase in frequency even if it partly reduces the overall success of the group.

One way to control renegade cell lineages is with policing traits that enforce a germ–soma split early in development (Buss 1987). This split prevents reproductive bias between lineages during subsequent development. Once the potential for bias has been restricted, a cell lineage can improve its own fitness only by increasing the fitness of the individual. This is another example of how reproductive fairness acts as an integrating force in the formation of units.

Maynard Smith (1988) agreed with Buss's logic about the potential for cell lineage competition, but he argued that metazoans solved their problems of cell lineage competition by passing through a single-celled stage in each generation. When an individual develops from a single cell, all variation among subsequent cell lineages must arise by de novo mutation. In Maynard Smith's view, such mutations must be sufficiently rare that the genetic relatedness among cells is essentially perfect. Thus, the soma sacrifice reproduction as an altruistic act in favor of their genetically identical germline neighbors. Buss recognized the need for de novo mutations within an individual and argued that these would be sufficiently common to favor significant cell lineage competition and policing. Arguments on this topic continue (Michod and Roze 2001).

In summary, Leigh (1971, 1977) may have been the first to emphasize how repression of internal competition aligns individual and group interests. However, meiosis provided the only good example at that time, so the idea did not lead immediately to new insight. Alexander (1979, 1987) used the idea and the example of meiosis as the foundation for his novel theories about human social evolution. Buss's (1987) argument followed on the role of cellular competition and repression in the evolution of metazoans.

Buss stimulated Maynard Smith (1988) to consider how social groups became integrated over evolutionary history. Maynard Smith disagreed with Buss's particular argument about the importance of the germ–soma separation in metazoans. But in considering the general issues, Maynard Smith had in hand several possible examples, including meiosis and genomic integration, limited cellular competition in metazoans, and the social insects. From these examples, Maynard Smith (1988, pp. 229–230) restated the essential concept in a concise and very general way:

> One can recognize in the evolution of life several revolutions in the way in which genetic information is organized. In each of these revolutions, there has been a conflict between selection at several levels. The achievement of individuality at the higher level has required that the disruptive effects of selection at the lower level be suppressed.

Disruptive effects may be repressed by high relatedness and kin selection, which favors self-restraint, or by repression of competition among unrelated or distantly related members of a group. Together, kin selection and repression of competition define the key evolutionary processes that have driven the major revolutions in the organization of genetic information (Maynard Smith & Szathmáry 1995).

SYNERGISTIC SYMBIOSIS

Gene products act within complex biochemical networks. From a social perspective of conflict and cooperation, biochemical networks usually pose no difficulties when the genes reside within a single genome. Each gene gains or loses in transmission along with its genomic neighbors—the group is bound by its common timing of replication and its common pathways of transmission. Adaptation has to do with engineering of biochemical networks for increased performance.

The smooth integration of genomes into biochemical networks obscures a great evolutionary puzzle. The earliest replicating molecules in the history of life probably did not live in integrated genomes with synchronous replication and common pathways of transmission (Maynard Smith & Szathmáry 1995). How did those individual replicators evolve to make complex, well-integrated biochemical networks? Put another way, how did those different species of early replicators come to act synergistically in symbiotic biochemical networks? In modern organisms, how do genes that reside in different species evolve synergistic symbioses?

The general problem of synergistic symbiosis can be studied by focusing on the joint evolution of two genetic loci (Frank 1995c, 1997c). The two loci may be in the same genome or in different species. I divide aspects of social selection into two parts. The first occurs when symbiont and partner have mutually beneficial effects on each other—a positive synergism between loci. The use of "locus" to describe partner and symbiont may seem a bit strange; it would seem more natural to say "a positive synergism between species." I use "locus" to emphasize that the symbiont and partner could be two different replicating molecules (genes or chromosomes) in a primitive genome or an insect and its bacterial symbiont.

The second part of social selection concerns various processes that bind together the reproductive interests of the two loci. The most obvious form of binding is physical, in which two separate replicators are joined together chemically to form a longer chromosome. The joined pair of loci may always be transmitted together, in which case their reproductive interests are completely aligned and they form a single evolutionary unit, as if they were a single locus. Or the loci may be shuffled occasionally by recombination, in which case they "codisperse" with a probability of one minus the recombination fraction. I have used standard genetical language, but physical binding might just as well cause a host locus and a symbiotic bacterial locus to codisperse, with shuffling defined by a parameter analogous to recombination.

Physical binding is easy to understand, but other types of association between pairs of loci have similar evolutionary consequences. Reproductive synchrony prevents competition and binds reproductive interests via the common timing of replication. Reproductive entrainment among chromosomes is certainly one of the outstanding features of mitosis and meiosis. These orderly cellular processes are derived conditions from the primitive state of scramble competition among a pool of unconstrained replicators.

Loci that have a positive synergism on reproductive success can develop statistical correlations between genetic variation at the loci (Frank 1994b). These correlations can arise even when there is limited codispersal. Such conclusions are well known in standard Mendelian population genetics. A pair of loci on separate chromosomes, recombining freely, will develop a statistical association when there is a positive or negative interaction between loci (epistasis). This statistical association is called linkage disequilibrium. Thus, synergism creates associations between loci, and statistical association may have consequences similar to physical linkage.

This discussion emphasizes that symbiotic genetics shares many properties with standard, Mendelian genetics. But a generalization is required, removing the standard assumptions of meiotic reproductive synchrony and rigid patterns of codispersal.

Many models of cooperative symbiosis start with the assumption that each individual donates a fraction of its energy to aid partners. For example, hypercycle models assume mutual enhancement of replication by separate species of replicators and then study the conditions under which complex genomes can evolve (Eigen & Schuster 1979; Maynard Smith & Szathmáry 1995). Models for the origin of chromosomes start with the assumption of positive synergism between separate replicators and then ask when selection favors those separate replicators to become biochemically linked on chromosomes (Maynard Smith & Szathmáry 1993).

I studied the prior step in the evolution of cooperative symbiosis (Frank 1994b, 1995c): How do different loci first evolve to aid partner loci? This step must be passed before one can invoke synergism to study hypercycles, genomic integration, and the evolution of chromosomes. I emphasized the early evolution of genetic systems, but the models apply to any kind of cooperative mutualism with behaviorally inflexible traits (e.g., biochemical mutualism).

Two processes influence the origin of synergistic traits. First, both partners must have a minimal level of expression for their mutualistic trait. Second, pairs that develop positive synergism must be associated in space so that benefits conferred to a partner are returned to the initial donor. These spatial associations have two components: selection creates spatial association (linkage disequilibrium) in trait values between symbiotic partners (Frank, 1994b), and the benefits of cooperation, returned from partners, must be provided to relatives of the original donor (Hamilton 1972; Wilson 1980).

The initial level of trait expression and the spatial associations determine threshold trait values that are required for the origin and evolution of synergistic symbiosis (Figure 23.1A). Locus 1 has a trait, T_1, that enhances the reproductive rate of species 2 but reduces its own fitness. Likewise, locus 2 has a trait, T_2, that enhances the reproduction of locus 1 at a cost to itself. Larger values of T provide more benefit to the partner at a higher cost to the donor. When both loci have low trait values, as would be expected when the partners first meet, selection pressure continually pushes the traits to lower values. If, however, the pair of traits is above a threshold upon first meeting, then cooperation can increase because of synergistic feedback (Frank 1995c). Statistical association between loci increases the probability that a particular group will have a pair of symbionts above the threshold.

An example of how the benefit:cost ratio affects cooperative evolution is shown in Figure 23.1B. The benefit:cost ratio defines a scaling for the positive effect an individual has on its partner relative to its own cost. In this example, both partners start with the same trait value, T. If the benefit:cost ratio

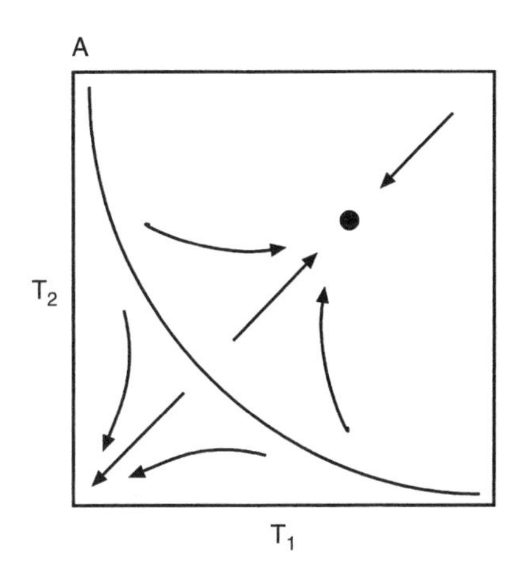

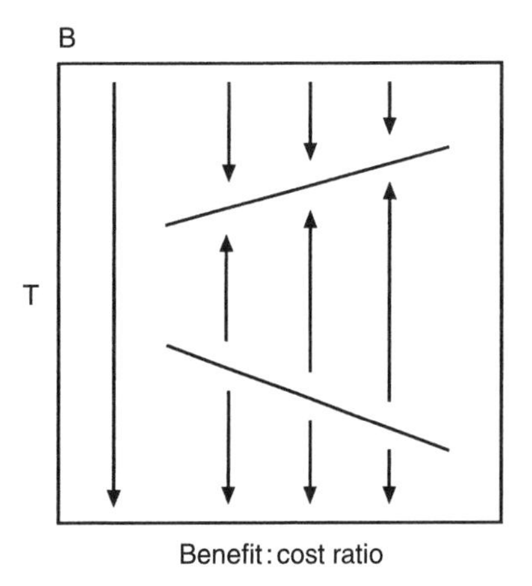

FIGURE 23.1. The threshold model for the evolution of synergistic symbiosis. For details see text. Reprinted from Frank (1997c).

is low, then selection reduces trait values from any starting point. As benefits increase relative to costs, the potential for positive feedback increases: lower trait values are needed to get over the initial threshold, and the traits evolve to higher equilibrium values.

This threshold is a key step in the origin of synergistic traits and cooperative symbiosis. Once the threshold is passed, symbionts may evolve through an irreversible stage, leading to an obligate relationship in which neither partner can live alone (Frank 1995c; Maynard Smith & Szathmáry 1995). By this process, genes with different phylogenetic histories can become integrated into complex biochemical networks.

DISCUSSION AND FUTURE DIRECTIONS

Genes act in biochemical networks. These social aggregations of genes define the environment in which reproductive interests play out in terms of replication and transmission—and in terms of conflict and cooperation. In this way, social selection dominates the evolutionary history of genetic systems.

At first glance, I may seem to be overstating the case. In the normal Mendelian system of classical population genetics, the rigid rules of replication, segregation, and transmission bind the interests of most genomic components into a unit that shares a common interest. The biochemical network can be studied from an engineering perspective of function without regard to social phenomena. However, the Mendelian condition derives from a social history of conflict and cooperation, in which replicating molecules developed systems of cotransmission, replicative synchrony, and reproductive fairness. The social processes of kin selection, repression of competition, and synergistic symbiosis melded the Mendelian genome into a cohesive unit.

Even modern Mendelian genomes fail to achieve complete unity (Werren et al. 1988; Hurst et al. 1996). Transposons violate the strict controls on replication and transmission. Uniparentally inherited mitochondria and other symbiont genomes often distort sex ratios. Chromosomes may gain transmission advantage by violating the normally fair segregation ratios. Each case shows the conflicting and common interests of different genomic components—the normally repressed social processes seething just below the surface of Mendelian rigidity.

Bacteria and viruses do not follows Mendel's rules. Their genetic systems remain more openly social, probably more like the primitive condition of genetic systems. For example, plasmids sometimes cotransmit with their bacterial hosts' genomes, joining plasmid and chromosome into a cooperative group. At other times, plasmids transmit horizontally as parasites (Levin & Lenski 1983). Multiple plasmids in a bacterial cell can be related as genetic kin; biochemical synergisms between different plasmids or between host and plasmid may favor combinations of genes to form more stable associations. Reproductive synchrony of plasmid and host genomes represses competition and tends to align the interests of genomic components.

Viral genetics also varies in non-Mendelian ways. For example, when multiple RNA viruses infect host cells, then there are multiple copies of each viral gene. Some viral genomes lose part of their genes through deletions. Those defective genomes can be copied when they coinfect with full genomes that provide the needed viral gene products (Holland 1990). The shortened genomes often replicate faster than the full genomes, probably because there is less RNA to copy. Thus, the infected cell produces proportionately more of the defective viruses than the wild-type viruses. This scenario matches the tragedy of the commons. When viruses coinfect with correlated genomes, then full genomes tend to go with full genomes, and defective genomes tend to end up with defective partners. The correlation between partners determines whether defective genomes match with full genomes sufficiently often to increase—in other words, the coefficient of relatedness plays a key role in the social evolution of viral genomes.

Other viruses have multipartite genomes, in which different components of the genome are packaged into separate particles (Matthews 1991). Viral success requires coinfection by all the different genomic pieces. This is a synergistic symbiosis in which the genes have become obligately entwined into cohesive biochemical networks, yet genomic components retain separate identities and interact as mutualistic symbionts.

Social selection also continues to be a dominant force in the more familiar types of sociality. But the powerful role of social selection in the history of life and in the evolution of genetic systems is sometimes overlooked.

SUGGESTIONS FOR FURTHER READING

Frank (1998) provides detailed summaries and references on kin selection and inclusive fitness along with methods to solve problems of sex ratios, dispersal, and altruism. Maynard Smith and Szathmáry (1995) emphasize the role played by repression of competition in shaping the cohesiveness of social groupings and the creation of new evolutionary units in the history of life. The ongoing conflicts within groupings at the genomic level may have important consequences for several aspects of genomic organization (Hurst et al. 1996). Conflict within groups also influences parasite virulence, which Frank (1996) framed within the general theory of kin selection and life history evolution. Many aspects of evolutionary units, genome evolution, and transitions in the history of life depend on the tension between conflict and cooperation among symbionts (Frank 1997).

Frank SA 1996 Models of parasite virulence. Q. Rev. Biol. 71:37–78.
Frank SA 1997 Models of symbiosis. Am. Nat. 150:S80–S99.
Frank SA 1998 Foundations of Social Evolution. Princeton Univ. Press.
Hurst LD, Atlan A & BO Bengtsson 1996 Genetic conflicts. Q. Rev. Biol. 71:317–364.
Maynard Smith J & E Szathmáry 1995 The major transitions in evolution. Freeman.

Acknowledgments National Science Foundation grant DEB–0089741 and National Institutes of Health grant AI24424 support my research. I used material from Frank (1997c) and Frank (2003) with permission from the original sources.

V

GENETICS OF SPECIATION

Species Concepts

James Mallet

> *One should never quarrel about words, and never get involved in questions of terminology. One should always keep away from discussing concepts.*
>
> —Karl Popper, *Objective Knowledge: An Evolutionary Approach*

Darwin (1859) believed he had disproved the need for a species "concept" by demonstrating that evolution could account for the diversity of life. He showed that species were part of a continuum from local varieties, geographic races and subspecies, through species to genera and higher taxa. All we need are practical criteria to distinguish varieties from species: "Varieties have the same general characters as species, for they cannot be distinguished from species,—except, first, by the discovery of intermediate linking forms ...; and except, second, by a certain amount of difference, for two forms, if differing very little, are generally ranked as varieties, notwithstanding that intermediate linking forms have not been discovered."

Species can be delimited broadly and inclusively, or narrowly, and there has been a long-running conflict between groups of taxonomists known as "lumpers" or "splitters." Somewhat apart from this argument about how actual species should be split is the argument about the true nature or "reality" of species, in other words, about species concepts. By the mid-twentieth century, a post-Darwinian reconceptualization of species was under way, as evolutionary biologists increasingly adopted the view that species were real and fundamental units of nature, qualitatively distinct from lower and higher taxonomic ranks.

Unfortunately, opinions today differ rather strongly on the correct underlying reality of species, leading to a variety of species concepts (partially listed in Table 1). "No one definition [of species] has as yet satisfied all naturalists; yet every naturalist knows vaguely what he means when he speaks of a species." This statement is perhaps more true today than it was when Darwin wrote it. Below, I cover a few of the major alternative concepts and definitions, while classifying their results in terms of species delimitation, the most important practical effect of species concepts in taxonomy. For more detailed discussions and critiques of various species concepts, see Claridge et al. (1997), Howard and Berlocher (1998), Wheeler and Meier (1999), Hey (2001), Mallet (2001), and Coyne and Orr (2004).

TABLE 1. A partial listing of species concepts and other ideas about species

Name of species concept (alphabetically arranged)	Brief definition	Reference
"Biological" or reproductive isolation concept	Taxa possessing reproductive isolation with respect to other species. Characterized by reproductive isolating mechanisms	Poulton 1904, Mayr 1970
Cladistic Concept	Species are unbranched segments or lineages in an organismal phylogeny	Hennig 1968, Ridley 2004
Cohesion Concept	A taxon characterized by cohesion mechanisms, including reproductive isolation, recognition mechanisms, ecological niche, as well as by genealogical distinctness	Templeton 1998

(continued)

Species Concepts *(cont.)*

TABLE 1. *(cont.)*

Name of species concept (alphabetically arranged)	Brief definition	Reference
Darwin's morphological concept	"Varieties" between which there are no or few morphological intermediates	Darwin 1859
Diagnostic ("phylogenetic") Concept	A species "is an irreducible (basal) cluster of organisms, diagnosably distinct from other such clusters, and within which there is a parental pattern of ancestry and descent"	Cracraft 1989
Ecological Concept	"A lineage which occupies an adaptive zone minimally different from that of any other lineage..."	Van Valen 1976
Evolutionary concept	A lineage evolving separately and "with its own unitary evolutionary role and tendencies"	Simpson 1951
Genealogical concept	Species are mutually monophyletic in the genealogies at all (or at a consensus of) gene genealogies in the genome	Baum and Shaw 1995
General lineage concept	Species are independent lineages. According to De Queiroz: all other species concepts agree on this fundamental principle; conflict about species concepts refers mainly to criteria applying to different stages of lineage divergence	de Queiroz 1998
Genotypic (genomic) cluster criterion	Sympatric species are clusters of genotypes circumscribed by gaps in the range of possible multilocus genotypes between them	Mallet 1995, 2001
Phenetic concept	Clusters of individuals circumscribed using multivariate statistical analysis	Sokal and Crovello 1970
Polytypic Species	Taxa having many "types," i.e., geographic subspecies. Geographic populations are part of the same species if they intergrade in areas of overlap	1890 onward, reviewed by: Mayr 1970, Mallet 1995, 2004
Population concept	Populations are the real units of evolution, not species, because gene flow is generally weak. Morphological and genetic uniformity of species is explained by stabilizing selection acting separately in each population	Ehrlich and Raven 1969
Recognition concept	Species possess a shared fertilization system, known as "specific-mate recognition systems"	Paterson 1985
Taxonomy without species	Species are no more real than any other hierarchical level in the tree of life. Species and other taxonomic ranks should be replaced either by "rank-free taxonomy" (which can name each node in a bifurcating phylogeny—Mishler), or by genotypic clusters described according to their genetic divergence from other clusters (Hendry et al.)	Mishler 1999, Hendry et al. 2000

Species Concepts *(cont.)*

Polytypic Species Criteria: Species Concepts for Lumpers

Large collections of specimens had been amassed by the late nineteenth century, and it began to be realized that morphologically divergent forms in different areas could often be united via intergradation in intervening regions. Geographically differentiated forms began to be recognized formally as subspecies within more inclusive *polytypic species*, rather than as separate species, as earlier. "True" species, it was argued, were more inclusive; they consisted of taxa that could remain distinct in regions of overlap. In this inclusive formulation, species are the lowest-ranking taxa capable of contributing to local biodiversity.

In 1904, E.B. Poulton suggested that inclusive species delimitation in sexual taxa could be justified by appealing to reproductive continuity ("syngamy") within species, and reproductive isolation ("asyngamy") between species (for history, see Mallet 2004). Poulton's view later became formalized into the *biological species concept*, in which "Species are groups of interbreeding natural populations that are reproductively isolated from other such groups" (Mayr 1970).

The biological species concept achieved a rather broad consensus, at least from the 1950s onward, but the act of specifying the "reality" of species, rather than merely letting species remain groups of individuals to be delimited by taxonomists, eventually led to dissent. Critics felt that other species traits were more important than reproductive isolation. For example, species that hybridized frequently, but remained distinct due to ecological factors, could be classified under the *ecological species concept* (Van Valen 1976). The *recognition concept* argued that species should be defined by "specific-mate recognition systems" (Paterson 1985). Paterson's concept is in a sense a mirror image of the idea that reproductive isolation defines species, but includes only those processes leading up to fertilization. The *cohesion concept* argues that species are defined by post-mating and pre-mating "cohesion" processes, including mate recognition systems, reproductive compatibility and incompatibility, and ecological selection, as well as via gene-genealogical monophyly (Templeton 1998).

Several authors have questioned the need to invoke processes maintaining separateness when delimiting species. Sokal and Crovello (1970) argued that statistical clustering algorithms should delimit *phenetic species*, which would avoid worrying about the ontological status of such entities; however, Sokal and Crovello did not explicitly deal with geographically divergent populations. I have suggested that genetic data can be used to minimize Hardy–Weinberg and linkage disequilibria within species, in order to distinguish species as *genotypic clusters* in zones of overlap (Mallet 1995). This character-based methodology could allow for polytypic species, without the theoretical overhead of requiring that any particular process is most important.

Recently, genotypic "partitioning" has been used in exactly this way to detect species-level subdivisions within local populations (Pritchard et al. 2000; Anderson & Thompson 2002), and to identify hybrids between such species (e.g., Cianchi et al. 2003). Polytypic species with multiple geographic subspecies are justified in this framework by investigating regions of overlap: if there is free intergradation in the hybrid zone or region of overlap, divergent forms should be considered members of the same species because all morphs and genotypes form part of a single cluster. Geographically isolated populations are hard to classify, but this is true in all species concepts. A pragmatic "null hypothesis" approach might name such taxa as separate subspecies within the most suitable species, until other evidence (e.g., laboratory breeding or phylogenetic studies) indicates whether overlap without fusion would be likely.

(continued)

Species Concepts *(cont.)*

Phylogenetic Criteria: Species Concepts for Splitters

An early attempt to take account of history in species definitions resulted in the *evolutionary species concept*, in which a species is considered to be "a lineage (an ancestral-descendant sequence of populations) evolving separately from others and with its own unitary evolutionary role and tendencies" (Simpson 1951). Beginning in the 1950s with Hennig, phylogenetic principles began to be applied in systematics, particularly an increasing emphasis on using shared derived characters to establish monophyly in classification; in the previous evolutionary systematics tradition, all characters, including ancestral traits, had been used in classification, and paraphyletic groups were recognized as valid taxa. It seemed natural that these phylogenetic principles should apply at the species rank, as well as at higher levels of classification. Hennig (1968) distinguished between "tokogenetic" relationships (between individuals within species) and "phylogenetic" relationships (between species or separate lineages, Figure 1). Under this view, "species reside at the boundary between reticulate and divergent genealogy" (e.g., Baum & Shaw 1995). In Hennig's *cladistic concept* (see also Ridley 2004), a pair of new species (B and C in Figure 1) is formed whenever a species lineage splits; the original species (A) becomes technically extinct to avoid the problem of such a species becoming paraphyletic.

While evolutionary and cladistic species concepts seem satisfying philosophically, they are hard to use in practice. For example, lineages B and C, diverging from lineage A in Figure 1 would all be separate species under this phylogenetic criterion, even if lacking morphological or genetic character differences; in fact even if the phylogenetic divergence itself were undetectable by means of taxonomic characters. Such a phylogenetic concept can allow a great deal more splitting than under the family of inclusive concepts of the 1900s onward.

A practical phylogenetic species concept, perhaps best termed the *diagnostic concept*, is now used widely in species delimitation: "A phylogenetic species is an irreducible (basal) cluster of organisms, diagnosably distinct from other such clusters, and within which there is a parental pattern of ancestry and descent" (Cracraft 1989). This concept can allow delimitation of species by one or more fixed differences, such as base pairs in a mitochondrial DNA sequence. However, species diagnosed in this way may include paraphyletic species, because species can be diagnosed via ancestral as well as derived character states. Curiously, although Cracraft justifies the need for a phylogenetic species concept instead of a reproductive isolation concept on the grounds that reproductive compatibility is an ancestral trait (Cracraft 1989, pp. 34, 46), he accepts that his own species concept, depending on traits other than reproduction, might also lead to paraphyletic entities (Cracraft 1989, p. 35).

Many recent dicussions about species include diagrams similar to Figure 1 and derive from Hennig's conceptualization of speciation. However, we must remember Figure 1 is actually a cartoon or caricature of a much more complex, underlying process of gene genealogical divergence. Each sexual individual consists of a phenotype determined by multiple genes. Genes at one locus will typically be inherited from different ancestors (and therefore have a different genealogy) from genes at other loci, because of recombination between genes. Baum and Shaw (1995) therefore argue that species should be defined on the basis of underlying genealogies and coalescence, and argue for a *genealogical species concept* requiring reciprocal monophyly in all (or perhaps a consensus of) gene genealogies, rather than merely monophyly at the population level as in Figure 1. This concept would not be able to delimit recently

Species Concepts *(cont.)*

reproductively isolated species, because under the neutral theory such species should retain polyphyletic and paraphyletic genealogies at some genes long after genetic isolation (Hudson & Coyne 2002). Indeed many cases are now known of ancestral polymorphisms shared between species.

The cartoon of Figure 1 is overly simplistic in other ways, as well. For example, it depicts individuals within a single species as if they were all in contact at any given time. In fact, spatial separation between individuals within continuous, but viscous populations and between isolated and semi-isolated populations will ensure that many temporary and some permanent lineages will form in each generation, even in the absence of any evolutionary divergence. Systematists cannot regard all separate lineages as separate species, or they would overburden nomenclature with trivial local populations or variants. Conversely, if a single lineage speciates in the inclusive sense by developing an ability to overlap with its ancestor, it will typically leave a large and diverse paraphyletic remnant of multiple such sublineages. After speciation, there may be hybridization, leading to exchange of some genes or introgression between lineages. Nonetheless, such taxa largely "evolve separately from others" and have "separate evolutionary roles and tendencies" (Simpson 1951) in other respects, so that almost everyone prefers to call them separate species. Thus, the question of how to use phylogeny in species concepts is primarily a practical, not theoretical issue. While the idealized theory of Figure 1 is appealing, it cannot easily solve the problem of where to delimit real species.

Phylogenetic species concepts, particularly the diagnostic concept, have been widely used recently in cataloguing the diversity of life. Diagnostic species do not have to be reproductively isolated and may intergrade at range boundaries. Taxa previously classified as subspecies can become recognized as separate species. This has led to taxonomic inflation compared with earlier taxonomies, and to a wave of taxonomic splitting, particularly in charismatic vertebrates such as birds and primates (Isaac et al. 2004).

Because the evolution of a new trait, leading to a new "diagnostic species," is hardly distinguishable from any other evolution within species, evolutionists studying speciation tend to employ a more inclusive, polytypic species concept (see above), which demands that a pair of lineages should be classified as separate species only if they can remain distinct when overlapping. Phylogenetic concepts are, however, more suitable in biogeography and phylogeography, where there is a need to understand the evolution of all lineages rather than just of the species rank.

Attempts at Consensus

Little consensus on species concepts has yet been reached. Some even argue that named Linnean ranks, including species, are no longer useful in taxonomy at all (Mishler 1999; Hendry et al. 2000). However, attempts at consensus have been made. Poulton (1904), Simpson (1951) as well as Templeton (1998) have argued that a combination of morphological, ecological, phylogenetic, and reproductive criteria should be used. Sokal and Crovello (1970) and Mallet (1995) attempted the reverse argument: that one could arbitrate between conflicting "concept" arguments by using the results of clustering processes on phenotype or genotype, rather than by specifying the processes themselves. de Queiroz (1998) has argued that conflict between species concepts is illusory, because different concepts represent criteria applicable to different stages in the lineage-splitting process; they are horizontal slices at different levels near the nexus between a divergent pair of lineages like those in Figure 1. According to de Queiroz,

(continued)

Species Concepts *(cont.)*

all these concepts agree implicitly on a single, underlying concept, the *general lineage concept*, in which species are independent lineages as in Figure 1. However, this attempt at consensus does not help with the practical question of whether to use inclusive or diagnostic criteria in taxonomy; that is, whether to be a lumper or a splitter.

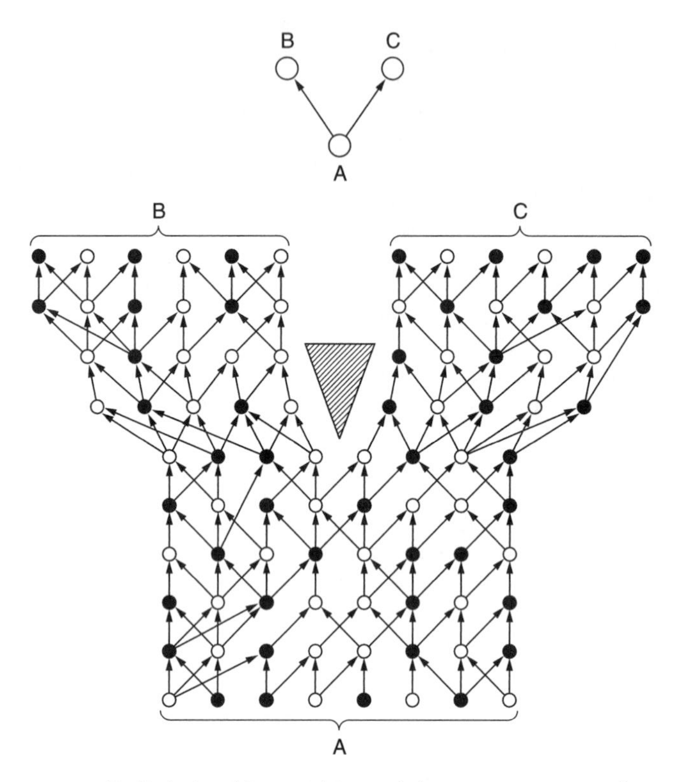

FIGURE 1. Relationships within and between species (after Hennig 1968). Above: phylogenetic relationships between species. Below: microscopic examination of individuals in successive generations. "Tokogenetic" relationships, representing the genealogies of individuals, are shown as arrows. A shaded wedge, consisting of factors that lead to reproductive isolation, is shown splitting species A into two new species B and C. Speciation is considered to occur when there is a break in tokogeny between two groups of individuals because of genetic or intrinsic isolation, in other words at the point at which tokogeny gives way to phylogeny between species. In both figures, time travels upward and arrows represent relationships. Individuals are shown as hollow (female) and filled (male) circles.

Species Concepts *(cont.)*

Conclusion. Which Species Concept?

It seems likely, therefore, that species concepts and criteria will continue to be debated for some time. Until a practical solution is widely agreed, we need to develop nomenclatural databases for comparative biology and biodiversity, as well as conservation, that can continue to provide useful information while fashions in the taxonomic rank considered species fluctuate (Isaac et al. 2004).

In evolutionary studies such as those discussed in this book, we are generally concerned about evolution of reproductive isolation of various kinds. Nonetheless, we must be aware of the uncertainty of the species rank (Hey 2001); taxonomists currently may mean by the term "species" different types of evolutionary entity in different taxonomic groups. Species counts on different continents or in different organisms will give only a roughly comparable idea of biodiversity. We should make sure that evolutionary hypotheses to be tested are either independent of the precise taxonomic rank (reproductive isolation can be measured via its effect on gene flow without specifying taxonomic level, for instance), or take into account the uncertainty about that rank, particularly in comparative studies of speciation (Barraclough & Vogler 2000) or adaptive radiation (Schluter 2000).

A polytypic, "lumper's" delimitation of species is implicit in most evolutionary literature on the topic of speciation (e.g., Howard & Berlocher 1998; Coyne & Orr 2004; Chapters 24–26 in this volume). Under this view, the study of speciation then reduces to the problem of understanding how reproductive isolation, ecological, or other factors diverge to the level where populations are stable enough in the face of potential gene flow so as to remain separate when in contact. Under this view of species, the most important thing about speciation is that it can contribute to local, sympatric biodiversity. In contrast, many parapatric or allopatric "phylogenetic species" found in taxonomic lists contribute to global biodiversity but are not reproductively isolated, and are therefore not considered to have speciated under this evolutionary biology point of view.

24

The Evolution of Reproductive Isolating Barriers

NORMAN A. JOHNSON

Despite the title of his famous book, Darwin was notably unsuccessful in solving the real problem of organic diversity: why plants and animals in a habitat fall into discrete, nonoverlapping packages. It is widely accepted that his failure came from his inability to conceptualize species as noninterbreeding groups.

Coyne (1994, p. 19)

Much of evolution occurs on time scales that are too long for humans to observe. Thus, as in many other branches in science, evolutionary biologists often must infer process from observed patterns. The study of speciation, how one species becomes more than one species, is no exception as the process usually takes place over thousands of generations. In this chapter, I will highlight some of the recent progress in uncovering patterns about speciation and discuss what inferences can be drawn from those patterns.

In this chapter, I will define species based on the Biological Species Concept (BSC) (Dobzhansky 1937; Mayr 1963). Following the BSC,[1] two populations are separate species if they lack the potential to interbreed and are the same species if they can interbreed. Thus, by this definition, reproductive isolation (non-interbreeding) is the essence of species. Based on the BSC and its variants, the process of speciation consists of diverging populations accumulating genetic changes that result in reproductive isolation. Note that such reproductive isolation is intrinsic to the organisms; two geographically isolated (allopatric) populations that do not exchange genes solely owing to their geographic isolation would be of the same species. As speciation is a process, there is a continuum between one interbreeding species and two completely reproductively isolated species. This continuum has made it possible for geneticists to do crosses between taxa at various stages along the way to them becoming completely reproductively isolated.

This chapter is not meant to be a comprehensive review about speciation. The multi-authored volume edited by Howard and Berlocher (1998) and the recent book by Coyne and Orr (2004) provide such reviews. In addition, Michalak and Noor (Ch. 25 of this volume) discuss the genetic basis of traits related to reproductive isolation. They address such questions as how many genes are involved in speciation and phenotypic species differences and whether

[1]Species concepts/definitions and discussions about them proliferated during the 1980s and 1990s as shown by Coyne (1994) and several chapters in Howard and Berlocher (1998). Despite this proliferation, the biological species concept remains the one most used by biologists interested in the process of speciation. Further, most of those biologists studying speciation who do not subscribe to the BSC still consider reproductive isolation to be important in speciation (Avise & Wollenbug 1997). Coyne and Orr's (2004) recent book includes a long appendix devoted to discussions of alternatives to the BSC. See also Mallet, Species Concepts, p. 367.

certain types of genes contribute disproportionately to these traits. They also address questions pertaining to the function of these genes, the size of their phenotypic effects, and the evolutionary forces that have caused these genes to diverge. Such topics will not be discussed here, except as in how they relate to general patterns and processes of speciation.

In this chapter, I will highlight some important recent findings about observed patterns of speciation and what inferences about process can be drawn from these patterns. After a brief review about how evolutionary biologists typically divide reproductive isolating barriers (RIBs), I will discuss those barriers that occur between mating and zygote formation, a class of RIBs that has received increasing attention during the 1990s. The major portion of this chapter will deal with questions regarding general patterns of speciation, derived from studies that examine many species from a given taxon. Here I will examine patterns of speciation such as whether there is a speciation clock and the relationship between sex chromosomes and speciation. I then discuss the role of sexual selection in speciation, including but not limited to the process of reinforcement. Throughout the chapter, I discuss how process can be inferred by the observed patterns. I close with a discussion of challenges and suggestions for future directions in the study of the patterns of speciation.

Division of Reproductive Isolating Barriers

Ever since the evolutionary synthesis of the 1930s and 1940s, biologists have typically divided reproductive isolating barriers (RIBs) into two main classes. There are RIBs that prevent gene exchange before mating (pre-mating isolating barriers) and there are those that prevent genetic exchange after mating has taken place (post-mating isolating barriers) (see Table 24.1 for a brief catalogue of RIBs). An obvious example of a pre-mating RIB is behavioral incompatibility. One hypothetical example of behavioral incompatibility would occur if males from one damselfly population display a type of courtship dance to which females of that population respond readily but females from another population do not respond as well. Differences in timing of the life cycle (phenology) are another possible form of a pre-mating RIB. Apple maggot flies that emerge from hawthorn apples emerge 3 weeks later

TABLE 24.1. Catalogue of some reproductive isolating barriers (RIBs)

A. RIBs that act before copulation (pre-mating)
Mating occurs at different times of year
Mating occurs at different times of day
Mating occurs on different host plants
Differences in olfactory cues diminish mating
Differences in acoustic signals diminish mating
Behavioral differences diminish mating
Incompatibilities between genitalia diminish copulation success
B. RIBs that act after copulation but before fertilization (post-mating, pre-zygotic)
Foreign sperm/pollen do not transfer well into female reproductive tract
Foreign sperm/pollen have reduced viability in female reproductive tract
Foreign sperm cause "insemination reaction" sterilizing females
Foreign ejaculate fails to properly stimulate oviposition
Conspecific sperm/pollen outcompete foreign sperm/pollen (conspecific sperm/pollen precedence)
C. RIBs that act after fertilization (post-zygotic)
F_1 have reduced viability
F_1 have reduced fertility
F_1 are behavioral intermediates (accepted by neither parental species)
F_1 have defective behavior
F_2 or backcross offspring have reduced viability
F_2 or backcross offspring have reduced fertility

than do apple maggot flies that emerge from apples. This difference in phenology contributes to the reproductive isolation between these host races (detailed in Feder 1998).

Post-mating reproductive isolating barriers include hybrid inviability and hybrid sterility. The sterility or inviability need not be complete in the hybrids for it to be a barrier. In addition, hybrid could be viable and fertile but the hybrids could be less successful at mating. Reduced hybrid mating success could arise because the hybrids are defective or because the hybrid behavior is intermediate between that of the two parental species and thus not recognized as well by potential pure-species mates. Hybrid behavioral dysfunction has been found in hybrids between species of parrots, species of neotropical butterflies, and host strains of fall army worms (references in Noor et al. 2001). Noor's group has recently genetically mapped the dysfunction seen in male hybrids between *Drosophila pseudoobscura* and *D. persimilis* (Noor et al. 2001).

Among other things, the founders of the evolutionary synthesis have given us a framework to understanding how post-zygotic RIBs can accumulate in allopatry (Dobzhansky 1937; Muller 1942; Mayr 1963). Dobzhansky and Muller[2] each realized that post-zygotic reproductive isolation based on single genetic changes would be very unlikely because the first individuals to harbor these hybrid incompatibility genes would have low fitness (Michalak & Noor, Box 25.1 of this volume). The answer they independently reached was that hybrid incompatibility involved changes at least two genetic loci. Genetic studies, mostly from *Drosophila*, have supported this Dobzhansky–Muller model: hybrid fitness reduction results from interactions of regions of the genome (reviewed in Johnson 2000 and Coyne & Orr 2004). Moreover, as anticipated by Muller (1942), the genetic interactions that underlie hybrid sterility are usually complex (Johnson 2002).

An alternative way to classify RIBs is to divide them into two general categories, not based on whether the barriers occur before or after mating, but whether they occur before (pre-zygotic) or after (post-zygotic) hybrid zygotes are formed. Often, this difference in division is not made clear as pre/post-zygotic and pre/post-mating are used synonymously. This distinction becomes important, however, when considering the middle category of RIBs: those that come after mating but before zygote formation.

Pre-mating, Post-zygotic RIBs

Consider a female animal with internal fertilization that has been mated by both a conspecific male and a heterospecific male. In between mating and zygote formation, the female reproductive tract may be more conducive to sperm of conspecific males than those of the heterospecific males. Alternatively, the conspecific sperm may incapacitate the heterospecific sperm. Such phenomena would be examples of this middle category of pre-mating but post-zygotic RIBs. Because the conspecific sperm have an advantage, this phenomenon has sometimes been called "conspecific sperm precedence" (Howard 1999). Although this class of RIBs was virtually ignored until the 1990s, the emerging pattern is that these pre-mating, post-zygotic barriers may be much more prevalent and important than previously thought (see Howard 1999). Below, I discuss some case studies.

Flour beetles of the genus *Tribolium* have been used in laboratory studies of population ecology, quantitative genetics, and evolutionary genetics. The two most used species in the genus are *T. castaneum* and *T. confusum*, which are both human commensal species with worldwide distributions. These species not only do not interbreed but also have different karyotypes and are classifed in different subgenera. A species of flour beetle, *T. freemani*, was rediscovered in the late 1970s. It had been previously described in the 1940s but had not been seen until its rediscovery. *T. freemani* is morphologically similar than *T. castaneum* but two to three times larger. There is no detectable pre-mating reproductive isolation between these species as interspecific matings occur freely. Matings between the species will produce an abundant number of sterile or quasi-sterile hybrids; the number of hybrids will vary depending upon the population or strain of *T. castaneum* used (see Wade et al. 1994 and references within for details). There is strong conspecific sperm precedence between these species. When a female of either *T. castaneum* or *T. freemani* is paired with both a heterospecific male and a conspecific male simultaneously, the conspecific male will sire the vast majority of the offspring

[2]Bateson actually came up with a similar explanation for the evolution of reproductive isolation three decades before Dobzhansky and Muller. Bateson's contribution was not well known until Orr (1996) rediscovered it.

(Wade et al. 1994; Robinson et al. 1994). Moreover, as soon as the female mates with a conspecific male, nearly all her offspring are sired by that male regardless of how long she is paired with a heterospecific male. In contrast, females that are paired first with a heterospecific male and then with a conspecific male, will first produce hybrids and then will rapidly shift to producing all pure-species offspring (Robinson et al. 1994).

Dan Howard and his colleagues found a similar phenomenon between two species of ground crickets, *Allonemobius fasciatus* and *A. socius*. Even when females are paired with one conspecific male and several heterospecific males, the conspecific male sires the majority of offspring (Howard et al. 1998). This conspecific sperm precedence appears to be the major RIB maintaining the mosaic hybrid zone between these cricket species, as other RIBs such as pre-mating isolation or hybrid dysfunction appear to be absent or weak (Howard et al. 1998).

In both the *Tribolium* and *Allonemobius* cases of conspecific sperm precedence, the isolation is competitive. The barrier to fertilization of the heterospecific sperm occurs only in the presence of conspecific males. There are also examples of noncompetitive post-mating, pre-zygotic RIBs such as when the heterospecific sperm have reduced fertilization efficiency. One example is the much-reduced transfer of sperm in matings between *Drosophila simulans* females and *D. sechellia* males resulting in few hybrids produced even in the absence of *D. simulans* males (Price et al. 2001). Mating with a *D. simulans* male further reduces the number of hybrids produced by a *D. simulans* female mated to a *D. sechellia* male, showing that noncompetitive and competitive forms of post-mating, pre-zygotic isolation can work together. Coyne and Orr (2004, pp. 233–235) present other examples of noncompetitive post-mating, pre-zygotic RIBs.

In nearly all these cases, little or nothing is known about the mechanism of conspecific sperm precedence. Most likely, the mechanism will likely vary across taxa. The best worked out mechanism of conspecific sperm precedence is in the *Drosophila simulans* system. Here, Price et al. (2000) found two different barriers to fertilization by the heterospecific males in crosses between *D. simulans* and *D. mauritiana*. When a male *D. mauritiana* mates with a *D. simulans* female that has already mated with a conspecific male, his sperm are incapacitated by the seminal fluid of the *D. simulans* male. This incapacitation does not occur when the conspecific matings are aborted after only 5 minutes, suggesting that the incapacitation requires a substance that is transferred late in copulation. When a *D. simulans* female first mates with a *D. mauritiana* male, seminal fluid from a subsequent mating with a conspecific male will displace the heterospecific sperm. Such displacement can occur even after only relatively short copulations with the conspecific male. Price et al. (2000) argue that conspecific sperm precedence in these species is likely a by-product of within-species sperm competition due to the similarity of mechanisms in conspecific sperm precedence and within-species sperm competition.

Although conspecific sperm precedence often is among the first class of reproductive barriers to form, it is not always. Some species pairs may form other RIBs before showing detectable conspecific sperm precedence. For instance, Dixon et al. (2003) found no detectable conspecific sperm precedence between the Zimbabwe and United States races of *D. melanogaster*, in contrast to the strong interracial pre-mating isolation. Dixon et al. also found only weak conspecific sperm precedence between *D. pseudoobscura pseudoobscura* and *D. pseudoobscura bogotana*. In contrast to the weak conspecific sperm precedence, these subspecies display complete hybrid male sterility in one of the reciprocal crosses.

GENERAL PATTERNS OF SPECIATION

The broad patterns characterizing intrinsic postzygotic isolation—its gradual evolution, the ubiquity of Haldane's rule, the rapid evolution of sterility versus inviability—suggest that common themes pervade animal speciation. Nevertheless, some details differ among groups.

Presgraves (2002, p. 1176)

In Search of a Speciation Clock

What do we know about the time course of speciation? How long does it take RIBs to evolve? Do certain classes of RIBs evolve faster than others? Is there such a thing as a speciation clock, where reproductive isolation evolves in a more or less clockwise fashion, like the molecular clock? Coyne and Orr took a major step toward answering these questions about the patterns of speciation during the late 1980s by taking advantage of two different large and growing data sets. The first was on the hybridization patterns between pairs of species within *Drosophila*. The other was on allozyme differences between species pairs, which could be translated into genetic distance information. Fortunately, there was a sizeable intersection of the two data sets—for many species pairs, both hybridization and genetic distance data had been collected. Eight years after their first paper, Coyne and Orr published a second paper updating and expanding their data sets.

Coyne and Orr's papers have sparked the growth of a cottage industry of studies looking at the patterns of reproductive isolation in different taxa. Large-scale studies of the patterns of speciation have been conducted in such diverse animal groups as frogs (Sasa et al. 1998), Lepidoptera (butterflies and moths; Presgraves 2002), birds (Price & Bouvier 2002; Lijtmaer et al. 2003), and a genus of freshwater fish (Mendelson 2003). Although the *Drosophila* data set is the most extensive and probably will remain so, the other data sets are sufficiently large to enable a reasonable degree of statistical power with which to make comparisons. The existence of such studies from diverse taxa allows us to address other questions, such as whether there is a universal speciation clock or whether RIBs accumulate at different rates in different taxa.

How is reproductive isolation assayed? Coyne and Orr's (1989, 1997) post-zygotic index is the number of sexes from reciprocal cross that are inviable or sterile divided by four (the total number of possible sexes). Thus, if males from one reciprocal cross but not the other are sterile and the females are both fertile, the post-zygotic index is 1/4. Note that this is a rather insensitive assay as it fails to account for partial reduction in hybrid fertility or viability. It also fails to consider hybrid fitness reduction beyond the F_1 (in the F_2 and backcrosses). Most studies of the pattern of speciation have used Coyne and Orr's index. Sasa et al. (1998) used several other indices in addition to Coyne and Orr's index. The pre-zygotic mating index used by Coyne and Orr is based on the relative numbers of heterospecific and conspecific matings. It is 0 when there is random mating and 1 when there are no heterospecific matings.

The most striking general pattern is that reproductive isolation increases with time (Figure 24.1). In all of the cases, there is a strong and significant positive correlation of the post-zygotic isolation index with time as measured by genetic distance (or in fewer cases, measured by DNA sequence divergence). Moreover, reproductive isolation increases in an apparently linear fashion until the point at which many species pairs are completely reproductively isolated. As genetic distance is often used as a proxy for the amount of time separating taxa, finding a strong linear correlation between the extent of post-zygotic isolation and time would imply that at least rough "speciation clocks" exist for these large taxa. Note that these clocks may be very sloppy and may differ substantially among the taxa.

Drosophilid flies, frogs, and Lepidoptera (butterflies and moths) appear to evolve post-zygotic RIBs at similar rates when scaled to the rate of divergence in allozyme frequencies (Coyne & Orr 1997; Sasa et al. 1998; Presgraves 2002). Based on standard calibration of allozyme divergence in *Drosophila*, Coyne and Orr (1997) placed the completion of speciation in allopatric taxa to be about 2.7 million years. Because less is known about the rates of allozyme divergence in other taxa, it is more difficult to establish absolute dates for them. Hybrid incompatibilities appear to evolve much slower in birds than they do in the other taxa (Price & Bouvier 2002). It is difficult to make direct comparisons between the bird study and the others because the genetic distance data come from differences in melting temperatures in DNA–DNA hybridization in the bird sample (Price & Bouvier 2002) and allozyme frequency differences in the others. Nonetheless, the evolution of post-zygotic RIBs in birds is apparently several-fold lower than it is in the other taxa. Moreover, Lijtmaer et al. (2003), who used sequence divergence as their measure of genetic distance, also found that pigeons and doves evolved post-zygotic RIBs more slowly than non-avian taxa.

Hybrid sterility usually evolves well before hybrid inviability. This pattern is seen in *Drosophila* (Coyne & Orr 1997), in which male are the heterogametic sex (sex with heteromorphic sex chromosomes), and frogs (Sasa et al. 1998), which are mostly male heterogametic. The same pattern occurs in female heterogametic taxa: Lepidoptera (Presgraves 2002) and birds (Price & Bouvier 2002). This faster evolution of hybrid sterility than hybrid inviability is interesting and cannot be explained by sterility being easier to obtain than inviability. In fact, within *Drosophila* mutations that cause inviability greatly outnumber those that cause sterility (Wu & Davis 1993, and references within).

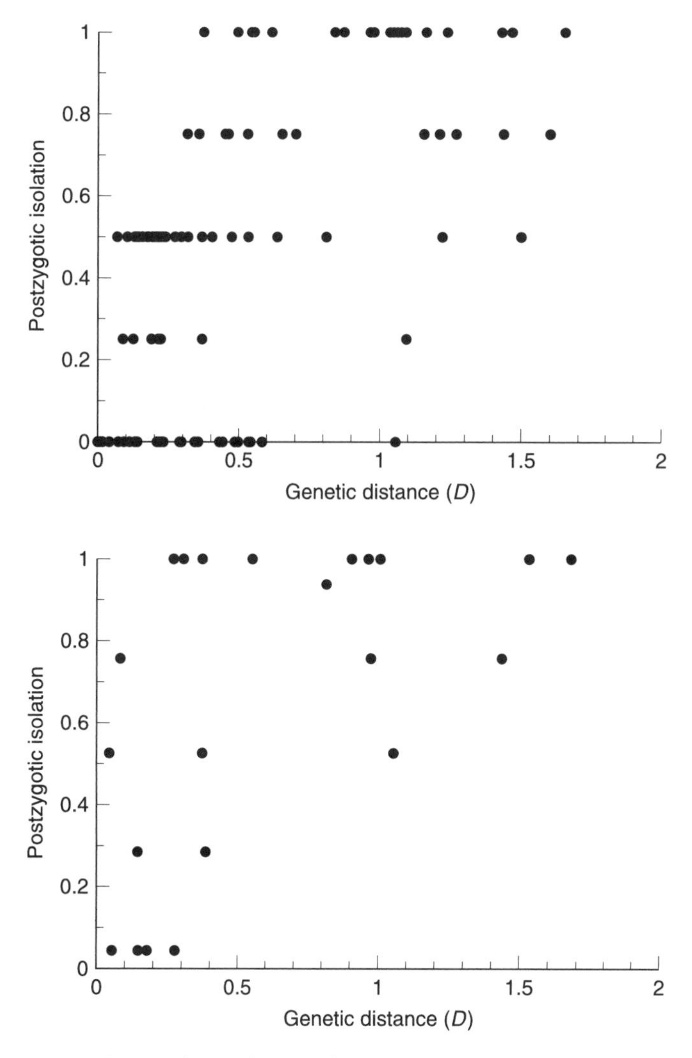

FIGURE 24.1. The evolution of post-zygotic reproductive isolation with respect to genetic distance. The upper panel shows results from *Drosophila* (Coyne & Orr 1997, Figure 2B). The lower panel results from frogs (Sasa et al. 1998, Figure 2D). Note how similar the results are.

What about Plants?

Plants and animals differ in so many fundamental ways that one would suspect they would differ greatly with respect to patterns of speciation. One of the biggest differences is that polyploidy (doubling of chromosome number), known to result in near-instantaneous speciation, is much more common in plants than it is in animals (Coyne & Orr 2004, Chapter 9). Other potentially important differences between plants and animals include a greater tolerance of inbreeding and the lack of a sequestered germline in plants. Until 2004, there had been no large-scale study investigating patterns of speciation in plants.

Moyle et al. (2004) examined three genera of flowering plants (*Glycine, Silene,* and *Streptanthus*) for patterns of the evolution of reproductive isolation. They investigated both "post-pollination," pre-zygotic reproductive isolation and the proportion of sterile pollen in interspecific F_1 hybrids.

As Coyne and Orr (2004, p. 79) note, Moyle et al.'s index of pre-zygotic isolation could include cases of early-acting post-zygotic isolation (the abortion or early inviability of the zygotes before seed was set). Moyle et al. (2004) found a strong, positive correlation between reproductive isolation and genetic distance in the genera *Glycine* and *Silene*, consistent with the patterns observed in a number of animal taxa (see above). On the other hand, they found no support for such correlation in *Streptanthus*. A possible reason for the lack of correlation in this genus is that species pairs tested within it are younger than those tested in the other genera and the relative dearth of older species pairs may have decreased the power. In the cases of *Glycine* and *Silene*, near-complete reproductive isolation was achieved at genetic distances much lower than those seen in animals. Because we lack good calibration for molecular clocks in angiosperms, it is not certain whether the differences in rates of reproductive isolation per genetic distance reflect true differences in the rates of the evolution of reproductive isolation.

Moyle et al. (2004) found some evidence that changes in ploidy level did affect the patterns of reproductive isolation within *Silene* (closely related species with different ploidy levels almost always had high levels of reproductive isolation). This effect, however, was not enough to obscure the general pattern of a steady increase in reproductive isolation with increasing genetic distance.

Pre-mating Versus Post-mating RIBs: Which Evolve Faster?

There is no solid evidence yet to address whether, as a general rule, pre-mating isolation evolves faster, slower, or at the same rate as post-mating isolation. Only a few studies have directly compared rates of evolution of both pre-mating and post-mating. Second, the relative rates of the evolution of these RIB classes could vary among the different taxa.

Another confounding problem is that indices used to examine the rates of evolution of these different RIBs are not directly comparable. It is often claimed that Coyne and Orr found that pre-mating and post-mating reproductive isolation evolve at approximately equal rates between allopatric species of *Drosophila*. However, justification for that statement is lacking because Coyne and Orr's indices for pre-mating and post-mating reproductive isolation are not comparable. Specifically, the pre-mating isolation index can be much more sensitive than the post-mating isolation index. Why is that? The pre-mating index is based on relative mating success while the post-mating index is based on the number of hybrid sexes in reciprocal crosses that are viable and fertile.

In Mendelson's (2003) study of the *Etheostoma* genus of fish, the indices for pre-mating and post-mating isolating barriers are comparable. The index for behavioral isolation (pre-mating) is the number of conspecific spawning events minus the number of heterospecific spawning events divided by the total number of spawning events. Likewise, the index for hybrid inviability (post-mating isolation) is the conspecific hatching success rate (measured in percent) minus the heterospecific hatching success rate divided by the total hatching success rate. Mendelson found that behavioral isolation in this group evolves faster than hybrid inviability because the slope of the regression for behavioral isolation is significantly greater than that for hybrid inviability. It should be noted that other barriers, such as hybrid sterility and conspecific sperm precedence, exist between species in this genus but have yet to be quantified.

The Missing Snowball?

According to Orr's theoretical treatment of the Dobzhansky–Muller model (Orr 1995), hybrid incompatibilities should not accumulate in a linear fashion but should instead accelerate with time. Specifically, if hybrid incompatibility involved the interaction of one allelic change in each species, hybrid incompatibility should accumulate with the square of time. If hybrid incompatibility generally involves more complex genetic interactions, its accumulation would accelerate even faster. Orr (1995) called this faster-than-linear accumulation "snowballing" (Figure 24.2).

However, there is no apparent sign of snowballing from the empirical data, which show a basically linear accumulation of post-mating RIB with time. There are several nonexclusive explanations for this apparent discrepancy. First, as was suggested by Sasa et al. (1998), the discrepancy could reflect the asymptotic relationship between the number of genetic incompatibilities between the species and the post-zygotic index. Second, the analyses thus far have been simple. Perhaps, deviations from linearity might be detected with more sophisticated analyses. (T. Mendelson, unpublished

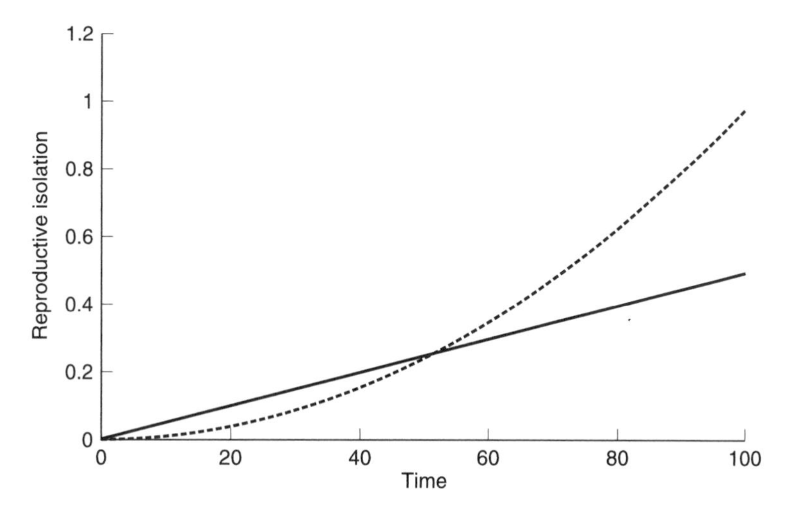

FIGURE 24.2. The accumulation of post-mating reproductive isolation with time. The dashed line shows the expected faster than linear "snowball" pattern under the Dobzhansky–Muller model. The solid line shows the general observed linear pattern.

manuscript.) Third, the bulk of the genetic distance data thus far are derived from differences in the frequencies of allozyme variants. Perhaps more DNA-based data will provide better resolution.

Haldane's Rule

According to Haldane's (1922) rule, the heterogametic sex (one with heteromorphic sex chromosomes) should be more adversely affected by hybridization than the homogametic sex. This pattern has been well documented in *Drosophila* and mammals, where the heterogametic sex is male, and in birds and Lepidoptera, where the heterogametic sex is female (reviewed in Wu et al. 1996; Orr 1997). Moreover, it is not the case that there are two distinct pathways to complete reproductive isolation, one in which Haldane's rule is the first step and one in which both sexes become inviable or infertile at the same time (Coyne & Orr 1989). There are very few cases in which both hybrid sexes are sterile or inviable early in the speciation process. Instead, the patterns demonstrate that Haldane's rule is a nearly unavoidable first step in speciation.

During the 1990s evolutionary geneticists studying Haldane's rule reached a consensus that Haldane's rule is a composite phenomenon—several forces act to produce the general pattern (Wu et al. 1996; Orr 1997; Coyne & Orr 2004). The relative strengths of these forces vary in different taxa and different researchers have emphasized different forces. There is consensus that two forces—the dominance theory and the faster male theory—are major contributors to Haldane's rule and these will be discussed below. A third force, the faster X theory, is more controversial. It will be discussed in the following subsection Sex Chromosomes and Speciation.

The dominance theory begins with the postulate that the deleterious affects of alleles that contribute to hybrid dysfunction are at least somewhat recessive in hybrids (Turelli & Orr 1995). These effects will thus be masked to some degree by alleles from the other parent in the homogametic hybrids. In contrast, there is no masking in the homogametic hybrids. The deleterious effects are manifest in full. Although there is fairly strong evidence that the deleterious effects in hybrids are partially recessive (Orr 1997), exactly why this is the case remains a mystery.

The faster male theory states that some force has accelerated evolution of hybrid incompatibilities that are manifested in hybrid males. The faster rate of hybrid male sterility over hybrid female sterility in both mammals and birds is strong proof that the faster male theory explains at least part of the pattern of Haldane's rule (Wu & Davis 1993;

Wu et al. 1996; Orr 1997). Why is there this accelerated evolution in males? Wu and Davis (1993) proposed that hybrid male sterility may be accelerated due to features of spermatogenesis. This proposition, while intriguing, has not received much in the way of rigorous testing. A more commonly argued reason for the faster male evolution is sexual conflict and sexual selection (Wu & Davis 1993; Wu et al. 1996; Holland & Rice 1998; Arnqvist et al. 2000). The role of these forces will be discussed in Sexual selection, Reinforcement, and Specification below.

These two major theories could explain differences in rates of accumulation in the different taxa. A plausible reason why post-zygotic RIBs might evolver slower in birds is that "fast-male" and dominance are working in different directions in birds. Hence, those deleterious interactions are not coming out as quickly in one sex. It must be noted that these two forces will act at cross-purposes in Lepidoptera also, and yet this taxon evolves post-zygotic RIBs at least as quickly as *Drosophila* does.[3]

Sex Chromosomes and Speciation

Because the heterogametic sex has only one copy of the X chromosome, rare recessive beneficial X-linked alleles will have an immediate advantage whereas rare recessive beneficial autosomal alleles are effectively neutral. This difference would permit faster a rate of X-linked evolution if beneficial alleles tend to be recessive (Charlesworth et al. 1987). Note that there is a difference in type of dominance involved in this model compared with that in the "dominance model" of Haldane's rule. Here dominance refers to the beneficial effect of alleles in pure-species background while that of the dominance model of Haldane's rule refers to their deleterious effects in hybrids.

Does the X chromosome accumulate more hybrid sterility genetic factors than the autosomes? Although early studies gave ambiguous results, a more recent and more comprehensive study by Tao et al. (2003) argues that the X chromosome has accumulated disproportionately more of these factors. They estimated that the density of hybrid male sterility factors was 2.5 times higher than that of the autosomes in hybrids of *D. simulans* and *D. mauritiana*. Tao et al. (2003) discuss several of the caveats associated with this estimate but conclude that the accelerated rate of evolution of X-linked versus autosomal hybrid sterility factors is a real and biologically significant phenomenon. These studies have serious limitations in that they are very time/effort-consuming and can only be done with pairs of species that are extremely well genetically characterized.

Turelli and Begun (1997) derived and tested two predictions about the dominance theory of Haldane's rule and sex chromosomes. The first prediction from the dominance theory is that pairs of species that have relatively large X chromosomes should, on average, evolve Haldane's rule sooner than those pairs that have smaller X chromosomes. The second prediction of the dominance theory is that provided the X chromosome evolves at a similar rate to the autosomes, the size of the X should not influence the rate at which homogametic hybrids become inviable or sterile.

Using Coyne and Orr's data and other data from *Drosophila*, Turelli and Begun (1997) found that pairs of taxa with large X chromosomes (about 40% of the nuclear genome) exhibit Haldane's rule for sterility at significantly smaller genetic distances than those pairs with smaller X chromosomes (about 20% of the genome). This is in accordance with the first prediction of the dominance theory. In accordance with the second prediction, the genetic distances between taxa that exhibited female hybrid inviability or sterility did not significantly differ between the "large X" and the "small X" pairs (Turelli & Begun 1997).

The X chromosome[4] does seem to have a disproportionately large effect in Lepidoptera (reviewed in Prowell 1998). Even though the X chromosome in most Leipdoteran species comprises only 3–5% of the genome, many important intraspecific variations

[3]If anything, Lepidoptera evolve RIBs at lower genetic distances than *Drosophila* (but there is no calibration of the molecular clock in Lepidoptera). Recently, Tao and Hartl (2003) have proposed that genetic factors that cause incompatibilities in hybrids of the heterogametic sex may accumulate faster than any other factors. They argue that there would be genomic conflict between different parts of the genome over the sex ratio (e.g., the Y chromosome in heterogametic male taxa would favor a male-biased sex ratio). Such conflicts would be more prevalent in the heterogametic sex. See Tao and Hartl (2003) and Coyne and Orr (2004, p. 297) for further detail.

[4]Recall that females are the heterogametic sex in Lepidoptera. Some researchers have used the terms "Z" and "W" chromosomes in place of X and Y when referring to the sex determination in Lepidoptera (males ZZ, females ZW). See Prowell (1998) and references within about the large X effect in Lepidoptera.

and interspecific differences have been linked to the X. These differences are not just associated with hybrid incompatibility but include traits associated with morphology, mating behavior, ovoposition behavior, and physiology.

Presgraves (2002) has argued that the large X effect for hybrid dysfunction in lepidopterans may be owing to butterflies and moths evolving Y-linked incompatibilities sooner than the other taxa. He further suggested that the reason lepidopteran species evolve more Y-linked incompatibilities is that they have a less degenerate Y chromosome than do species in *Drosophila*.

SEXUAL SELECTION, REINFORCEMENT, AND SPECIATION.

Acceptance of reinforcement has resembled stock market fluctuations.

Noor (1999, p. 503)

Reinforcement

One of the most fascinating results from Coyne and Orr's (1989, 1997) studies is the acceleration of pre-mating isolation in sympatry as compared with allopatry in *Drosophila*. These authors observed that species pairs currently in sympatry had substantially higher pre-mating isolation index scores on average than those of comparable genetic distance currently in allopatry. In particular, they found several cases of sympatric species pairs that had moderate to high pre-mating isolation but little genetic distance separating them. In contrast, there were no such allopatric species pairs (Figure 24.3).

Coyne and Orr (1989, 1997) proposed that reinforcement was the mechanism driving that pattern of faster evolution of pre-mating isolation in sympatry. Reinforcement, first championed by Dobzhansky (1940), is the direct selection for pre-mating RIBs as a means to prevent matings that would produce hybrid offspring that are less fit than pure-species offspring. Females from species that frequently were in contact with males of closely related species or subspecies would be under selection to reduce mating with those males, but females from species that did not contact other close relatives would not experience such selective pressures. Thus the pattern observed by Coyne and Orr (1989, 1997) is consistent with the reinforcement model being an influential contributor to patterns of speciation.

Although Coyne and Orr's (1989, 1997) pre-zygotic results are consistent with reinforcement, other processes could conceivably lead to the same pattern. For instance, suppose incompletely formed species re-established contact. If insufficient reproductive isolation had evolved, those proto-species would fuse back into one. This differential fusion hypothesis would also produce the pattern observed by Coyne and Orr: few species currently in allopatry with large amounts of pre-zygotic isolation and little genetic distance. Mitigating against the likelihood of differential fusion being responsible for the pattern is the fact that this hypothesis, unlike the reinforcement hypothesis, predicts that post-zygotic isolation should also follow the same pattern—and it does not. Species pairs that are currently allopatric have that are statistically indistinguishable from those of patterns of the evolution of post-zygotic isolation species pairs that are currently sympatric.

At the time of Coyne and Orr's (1989) first paper, acceptance of reinforcement was at a recent nadir. Like the United States stock market, its acceptance rose markedly during the 1990s (Noor 1999; Coyne & Orr 2004). This rise started with Coyne and Orr's results, which would spark renewed interest in both theoretical and empirical studies of reinforcement (reviewed in Noor 1999). The new theoretical studies would demonstrate that the conditions for obtaining reinforcement were not as restrictive as previously thought. Empirical studies would provide case examples of the probable action of reinforcement.

Reinforcement is an evolutionary force that is expected to often generate the pattern known as "reproductive character displacement" (RCD). Consider a species A whose range overlaps with another closely related species, B. RCD occurs when females of species A that are from the region of overlap discriminate more against species B than do females of species A from outside the region of overlap. Howard (1993) showed that RCD was a rather common pattern in nature, thus providing support for the notion that reinforcement often takes place. Howard (1993) agreed with the critics of reinforcement that RCD does not prove the action of reinforcement and provided a list of requirements for attributing the pattern of RCD to the actions of reinforcement. These requirements have been largely satisfied in several case studies in

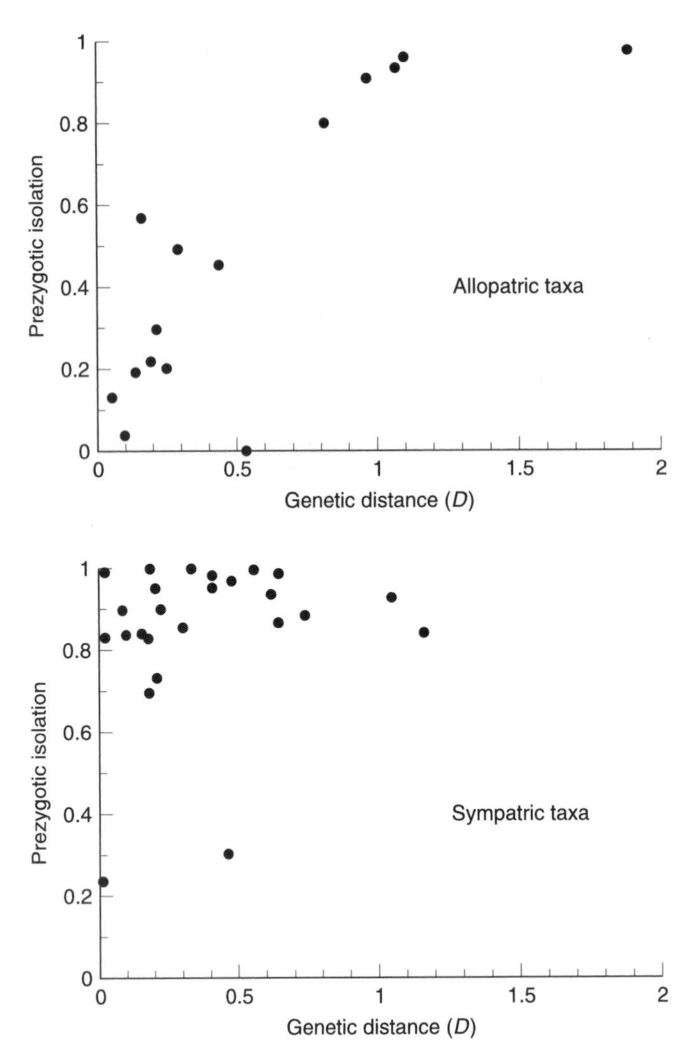

FIGURE 24.3. Pre-zygotic reproductive isolation with respect to genetic distance in *Drosophila*. The upper panel shows results for pairs that are currently allopatric. Note the absence of pairs that have evolved substantial pre-mating isolation at low genetic distances. From Coyne and Orr (1997, Figure 3B). The lower panel shows results for pairs that are currently sympatric. Here there are considerably more pairs that have substantial pre-mating isolation at low genetic distances. From Coyne and Orr (1997, Figure 3A).

taxa as diverse as *Drosophila*, flycatchers, and sticklebacks (reviewed in Noor 1999).

Sexual Selection and Speciation

As discussed above, sexual selection can accelerate speciation in the form of reinforcement. Pre-mating isolation evolves from the action of sexual selection acting to reduce matings with heterospecific males provided that the hybrids are already less fit (pre-existing post-mating reproductive isolation).

Mendelson (2003) has suggested that sexual selection for exaggerated male recognition characters has driven the rapid evolution of pre-mating isolation in the fish genus *Etheostoma*. Consistent with this hypothesis is the strong degree of sexual dimorphism seen in this genus. An alternative hypothesis is that because many species of this genus lack sex chromosomes, the evolution of post-mating RIBs has been retarded. Many of the incompatibilities that cause F_1 sterility in flies involve heterozygous autosome changes interacting with a hemizygous X. Such incompatibilities will not arise in taxa without sex chromosomes, and sterility/inviability will necessarily evolve much later.

In recent years, several authors have argued that sexual selection and/or sexual conflict have been accelerating the evolution of some forms of post-zygotic reproductive isolation in some taxa (Wu & Davis 1993; Wu et al. 1996; Holland & Rice 1998). There are several variants to these hypotheses but the basic idea is that within species, sexual selection drives the evolution of new genetic variants that act primarily within the male reproductive tract. The new variants of one species are incompatible with those of the other species and the incompatibilities also primarily affect the male reproductive trait.

Consistent with these sexual selection models are the observations that genes expressed in the male reproductive system evolve faster than those expressed in the female reproductive system or in other tissues. For instance, Torgerson et al. (2002) examined 43 mammalian genes that are expressed only in the spermatozoa and 473 genes that are expressed in all other tissues. Considering only those nucleotide changes that change the amino acid (nonsynonymous substitutions), the sperm protein genes have evolved more than twice as fast as the nonsperm genes. Yet, there is no significant difference between these classes of genes in the rate of synonymous changes. Further analysis shows that the accelerated rate of evolution of the sperm protein genes reflects positive selection rather than reduced selective constraints.

Further evidence that sexual selection and sexual conflict drive the evolution of RIBs comes from studies investigating the correlation between mating system and speciation. Mitra et al. (1996) found an association between mating system and species richness within birds. Those taxa that had promiscuous mating systems had on average more species than those that were monogamous. Arnqvist et al. (2000) found that insect taxa in which females mate with multiple males have 4-fold higher speciation rates than related groups in which females mate but once. Although correlation does not imply causation, this finding is consistent with sexual selection driving speciation, as one would expect sexual selection to be more pronounced in the taxa with promiscuous species.

CHALLENGES FOR THE FUTURE

Until recently, the genetics of speciation remained a form of classical genetics: "speciation genes" were black boxes of unknown function that resided in poorly defined chromosome regions and that interacted with other poorly defined chromosome regions.

Coyne and Orr (2004, p. 313.)

Although much has been learned about speciation from the recent large-scale studies on patterns of speciation in different taxa, more can be learned still from further such studies. No large-scale studies of the patterns of speciation have yet been done with mammals despite extensive work on both hybridization and genetic distance in this class. As mammals are homeothermic vertebrates with male heterogameity, it would be interesting to compare the mammalian results with those of both birds (homeothermic vertebrates with female heterogamety) and *Drosophila* (invertebrates with male heterogamety). Hymenopetera would be another interesting group in which to study the patterns of speciation, given the complications of haplodiploidy and thus the entire genome being effectively a sex chromosome. Unfortunately, the requisite data on both genetic distance and hybridization are probably not available and thus would have to be collected.

Recently researchers have begun studying the genetics of speciation in the context of gene regulation and genetic pathways. Several of the genes that have been identified as being involved in hybrid incompatibility appear to have regulatory functions and genetic pathways are good source of generating the epistastic gene interactions necessary for hybrid incompatibility (reviewed in Johnson & Porter 2000; Coyne & Orr 2004; Michalak & Noor, Ch. 25 of this volume). Using microarrays and other tools of genomics, Michalak and Noor (2003) have shown that a considerable number of genes are misregulated in interspecific hybrids between the closely related species *Drosophila simulans* and *D. mauritiana*. Johnson and Porter (2000) have modeled hybrid incompatibility as a consequence of natural selection driving divergence of genetic pathways in allopatric populations. Moreover, Porter and Johnson (2002) have shown that selection operating on these regulatory genetic pathways can result in speciation even in the face of considerable gene flow. Of course, examining gene regulation is just the first step; both proteomic and extensive functional studies will be necessary in the longer term for determining the real impact of changes in gene expression (or gene expression regulation) between species.

Can differences in the patterns of speciation in different taxa be explained in terms of differences in gene regulation in those taxa? For instance, could the slow rate of evolution of post-zygotic RIBs found in birds be a consequence of features of avian gene regulation? Although only speculative at present, answers to such questions will be plausible in the near future.

SUGGESTIONS FOR FURTHER READING

Coyne and Orr (2004) have recently published an excellent, comprehensive book on speciation that should be required reading for all graduate students interested in evolutionary biology. I also highly recommend Howard and Berlocher's (1998) symposium volume for all who are interested in current work being done on speciation. Noor (1999) provides a short, understandable review of the exciting work investigating the process of reinforcement. The most comprehensive review of that venerable pattern of speciation, Haldane's rule, is by Orr (1997).

Coyne JA & HA Orr 2004 *Speciation.* Sinauer Assoc., Inc.

Howard DJ & SH Berlocher 1998 *Endless Forms: Species and Speciation.* Oxford University Press.

Noor MAF 1999 Reinforcement and other consequences of sympatry. Heredity 83: 503–508.

Orr HA 1997 Haldane's Rule. Ann. Rev. Ecol. Syst. 28:195–218.

Acknowledgments I first thank Charles Fox and Jason Wolf for inviting me to write a chapter for this volume. I thank Tamra Mendelson and Mohamed Noor for comments and discussion of an earlier version of this chapter. Sean Werle provided assistance with figures. This work has been supported in part by NSF award DEB 0075451.

25

Genetics of Reproductive Isolation and Species Differences in Model Organisms

PAWEL MICHALAK
MOHAMED A. F. NOOR

The two attributes most commonly used to identify and define species are distinct phenotypic differences and reproductive isolation between groups of organisms, respectively. Because of this usage, understanding the genetic basis of these attributes is of fundamental importance in evolutionary genetics. The two phenomena are not independent of each other: phenotypic differentiation may cause reproductive isolation and reproductive isolation may allow further phenotypic differentiation (Orr 2001). Despite our realization of the importance of these attributes and that they interact, we have only begun to scratch the surface of their underlying genetic complexity. We still attempt to answer the old questions: What portions of the genomes contribute to species differences, and how many genes are involved? What is the function of these genes, their molecular nature (regulatory or coding), and where are they located? How large are their phenotypic effects? Are the same loci responsible for intraspecific variation and interspecific differences? What is the role of epistatic and dominance effects? Is their divergence driven by natural selection or genetic drift? Finally, are the genetic patterns underlying species differences common across taxonomic groups?

In this chapter, we focus on the genetics of traits related to reproductive isolation or phenotypic differentiation between species. Johnson (Ch. 24 of this volume) discusses the different types of reproductive isolation as well as the evolutionary forces that may have spawned them. He especially focuses on major patterns that can be derived from studies of multiple taxa, and the processes inferred to bring those about.

Our progress in identifying genes related to species differences and reproductive isolation (note that we do not imply that most species differences play a role in preventing gene flow between species) has been discouragingly slow despite decades of intensive study in such model systems as *Drosophila*. Just in the past few years, studies have identified a handful of major developmental gene effects in some phenotypic differences between species (e.g., Gompel & Carroll 2003; Sucena et al. 2003), primarily in *Drosophila*. Similarly, no more than three or four hybrid dysfunction genes are identified and partly characterized so far in *Drosophila* (Barbash et al. 2003; Presgraves et al. 2003; Ting et al. 1998) and just one outside *Drosophila* (Schartl 1995). One might therefore conclude that little more impressive progress has been made to identify genes underlying species differences since J. B. S. Haldane's "The nature of interspecies differences" that was published in 1938. The main obstacle to isolating individual gene effects is reproductive isolation itself: crosses that fail to produce viable and fertile progeny are of little use in classical genetic approaches. This problem is apparent in the model *Drosophila* species, *D. melanogaster*, which does not produce progeny beyond the F_1 generation when crossed to any of its sibling species. The second difficulty, far from unusual in experimental biology, is the complexity of the genetic system. The pattern that emerges from existing data shows that the number of genes involved in species differences and speciation is substantial, even in sibling species or races within species (Orr 2001). Also, extensive gene interactions may often underlie species and speciation traits (Palopoli & Wu 1994; Cabot et al. 1994; Schartl 1995), and such interactions complicate simple mapping studies. Traits are often further subject to disparate evolution rates, which contribute to the complexity; a remarkable example is provided by genetic studies of hybrid sterility and inviability (Johnson, Ch. 24 of this volume).

Recent biotechnological advances yield cause for optimism. Quantitative Trait Loci (QTL) mapping and associated statistical development is accompanied by fast growth of DNA sequence databases, thus expanding the body of molecular markers available for denser mapping strategies. Thus far, QTL studies have their power limitations and have not resulted in pinning down and cloning a species/speciation gene (Orr 2001). The power effect is exemplified by the analysis of posterior lobe differences in *Drosophila simulans–mauritiana*. Liu et al.'s (1996) first estimate indicated eight QTLs. Later, Zeng et al. (2000) used a larger sample size and found 14 QTLs via composite interval mapping and 19 QTLs via multiple interval mapping. However, various genetic engineering innovations, such as targeted mutations and saturated mutations and construction of deficiency libraries in *D. melanogaster*, are all tools whose value is difficult to overestimate in the genetics of speciation. Genome-wide profiling of gene expression via DNA microarrays has already been employed to investigate species differences and interspecies hybrid dysfunction (Michalak & Noor 2003; Rifkin et al. 2003; Ranz et al. 2004). Such technologies as serial analysis of gene expression (SAGE) and RNAi will almost certainly be incorporated into this research field.

The literature on speciation genetics is heavily *Drosophila*-biased, and inevitably so is the perspective adapted in this chapter. Although there are certain regularities of postzygotic reproductive isolation shared among animal taxa, such as Haldane's rule and reinforcement (Wu & Palopoli 1994; Orr 1997; Coyne & Orr 1998; Servedio & Noor 2003; Johnson, Ch. 24 of this volume), there is little information about applicability of results from *Drosophila* to other systems, and thus caution in generalization may be required. Similarly, we in no way mean to imply that the findings in *Drosophila* are more significant than the excellent research on the genetics of such differences that has begun in nonmodel systems, but none would argue that *Drosophila* has not held an advantage because of the decades of research in the genus and availability of genetic tools. Even within *Drosophila*, there are examples of sharply contrasting results (see e.g., Ting et al. 2001; Doi et al. 2001). There is a growing number of exciting reports on the genetics of reproductive isolation between *Saccharomyces* yeast species (e.g., Greig et al. 2002a; Delneri et al. 2003), and we also present an incomplete review of those in this chapter. There are a relatively large number of excellent reviews about the genetics of species differences and there are even whole issues of *Trends in Ecology and Evolution* (16, no. 7, 2001) and the *Journal of Evolutionary Biology* (14, 2001) dedicated to this subject. It was not our intention to create a comprehensive review of reviews; rather we focus on a functional genomics approach to speciation and those few study cases that have led to molecular depiction of genes.

GENETICS OF SPECIES DIFFERENCES

When we contemplate interspecies differences, morphological and behavioral characters typically occur to us first. In recent years, two distinct approaches have been pursued to determine how changes at the genetic level translate into phenotypic diversity. First, developmental biologists have concentrated on correlating macroevolutionary changes in morphology, such as those related to body plans, with patterns of gene regulation (Warren et al. 1994; Averof & Akam 1995). Although fruitful for understanding gross patterns of development, these broad-scale taxonomic analyses have thus far given little insight into differences between closely related species and rapidly evolving morphological traits such as secondary sexual characters. Second, biometric and genetic analysis has been employed to estimate the number and characterize genes involved in morphological differentiation among species. Unfortunately, QTL mapping and QTL-related approaches, most commonly used in these studies, have thus far provided little consistency in the genetic architecture patterns. The estimated numbers of genes range from one to many, even for apparently similar systems. For example, in *Drosophila* multiple loci underlie differences in the shape of the male genital arch, bristle numbers, male and female pheromones, and acoustic mating signal (see Orr 2001 for a review). In contrast, changes in hydrocarbon composition between species often involve a few "major genes." For example, the difference in cuticular hydrocarbon profile between African and Caribbean cosmopolitan *D. melanogaster* is due to a single locus (Takahashi et al. 2001). Another example of a single-gene polymorphism underlying a difference is that in larval morphology between *D. simulans* and *D. sechellia* (Sucena et al. 2003). Comprehensive reviews of numerous studies can be found elsewhere (see Suggestions for Further Reading), and many biases exist in estimates of gene numbers. Below we focus on the genetic basis of pigmentation, illustrating the hunt for species-specific genes contributing to morphological differentiation.

Adult *Drosophila* Pigmentation

Examples of convergent evolution that involve changes in the same developmental pathways, giving rise to similar phenotypes in different species and taxa, provide evidence that there may be a limited number of developmental changes available to evolution. It remains unclear to what extent this convergence is caused by orthologous genes. In this context, the evolution of pigmentation has attracted much interest, and such phenomena as melanic forms, observed in most animal taxa, were subject to scrutiny for years. Melanism in birds and mammals can be caused by amino acid changes in a melanocortin receptor protein (Theron et al. 2001). Pigmentation in vertebrates and insects is known to develop through different cellular mechanisms (e.g., Urabe et al. 1993). In *Drosophila*, the pigmentation of adult abdominal segments is one of the most variable characters and is sometimes a sexual-dimorphic trait. For example, in the subgenus *Sophophora*, sexual dimorphism is present in species of the *melanogaster* subgroup and the closely related subgroups, whereas the *ananassa*e and *montium* subgroups include both sexually dimorphic and sexually monomorphic species (Kopp et al. 2000).

The plethora of genetic tools in *D. melanogaster* has helped enormously in describing genetic and developmental pathways underlying pigment patterning and their within- and between-species differentiation (Kopp et al. 2000). *D. melanogaster* has non-sex-specific pigmentation in the form of posterior stripes on abdominal segments 1–6 in females and 1–4 in males and male-specific dark coloration of segments 5 and 6. The nonspecific pigmentation, which is typical of most *Drosophila* species, is under control of a pleiotropic transcription factor, *optomotor-blind* (*omb*). Omb null mutants (*omb*⁻), both males and females, lack the segment striping, but males retain pigmented segments 5 and 6. Conversely, loss of a homeotic gene expression, *Abdominal-B* (*Abd-B*), eliminates male-specific pigmentation of segments 5 and 6 but has no influence on the non-sex-specific stripe patterning. In addition to *Abd-B*, *abdominal-A* (*abd-A*) also contributes to the pattern of male segments 5–6.

Interestingly, male-specific pigmentation appears to be expressed by default, and the development of female pigment pattern requires the presence of the suppressors *doublesex* (*dsx*) and *bric-a-brac* (*bab*). The *dsx* transcript is alternatively spliced, producing a distinct male-specific product (*dsxM*) and female-specific product (*dsxF*). Loss of *dsx* expression in females results in the development of male-like pigmentation. The *bab* locus contains two adjacent genes (*bab1* and *bab2*) that encode a product with transcription factor activity involved in sex differentiation. At the time of writing, FlyBase records (http://flybase.bio.indiana.edu) referred to 19 known *bab1* and 13 *bab2* mutants, including 3 in vitro constructs, and more than 65 references, reflecting a considerable interest in this gene. Loss of one *bab* copy results in the development of male-like pigmentation in females, but has no effect on the male abdomen pigmentation. Both *bab1* and *bab2* are strong repressors of male-specific pigmentation in females and their transcription is downregulated by the homeotic genes *Abd-B* and *abd-A*, and upregulated by *dsxF* in females (Figure 25.1). Thus, *bab* integrates the homeotic and sex determination regulatory inputs (Kopp et al. 2000).

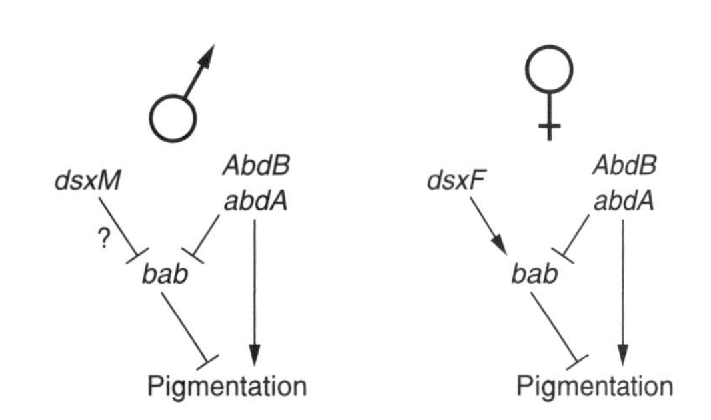

FIGURE 25.1. Regulation of sexually dimorphic abdominal pigmentation in *Drosophila* through *bab*, *abd-A*, *Abd-B*, and *dsx*. Blunt pointer ends indicate suppression, arrows indicate upregulation. Adapted from Figure 3h in Kopp et al. (2000).

The *bab* locus is consistently downregulated in *Drosophila* species that are sexually dimorphic with respect to pigmentation of segments 5 and 6 (Figure 25.2). In species with no sexual dimorphism, such as *D. subobscura*, *D. affinis*, *D. willistoni* or *D. virilis*, *bab* exhibits equal expression in both sexes. But exceptions to this phylogenetic pattern are found among *montium* subgroup species, such as *D. kikkawai*, in which *bab* expression is strongly downregulated in males despite the lack of sex-specific pigmentation (Kopp et al. 2000). Even within the *Sophophora* group, *D. santomea* and *D. serrata* have *D. melanogaster*-like pattern of Bab2 expression with no correspondence to pigmentation, as *D. santomea* both sexes are devoid of pigmentation and *D. serrata* females are darker than males (Gompel & Carroll 2003). It seems that in the ancestral condition *bab* expression was largely independent of *Abd-B* and *dsx*, resulting in sexually monomorphic pigmentation, while sexual dimorphism evolved through acquisition of two novel gene regulations. The *bab* locus is also known to regulate segment shape, and patterns of bristles and trichomes (small nonsensory hairs covering cuticle)—again in an antagonistic way relative to *Abd-B*.

Llopart et al. (2002) observed a significant association between the *bab* genotype and the degree of abdominal pigmentation in backcrosses of closely related *D. santomea* and *D. yakuba*. *D. santomea*, a newly discovered species from the island São Tomé, is the only species in the *D. melanogaster* subgenus lacking distinct abdominal pigmentation in both sexes. Five other genetic markers were nonrandomly associated with pigmentation: *Abd-B*, *yellow*, *vermilion*, *Annexin X* and *twinstar*. Interestingly, Bab2 expression perfectly correlates with the abdominal pattern of trichomes in *D. santomea* (Gompel & Carroll 2003). The *yellow* locus, which contains five independent *cis*-regulatory elements, has been long known to influence cuticle pigmentation in *D. melano*gaster. For *D. melanogaster*, FlyBase lists *yellow*'s interactions with 13 other genes, along with its 826 alleles and 562 references. Abdominal pigmentation pattern also varies dramatically among Caribbean species of the *Drosophila dunni* subgroup: lighter *D. arawakana* are sexually dimorphic, with males darker than females, while almost completely pigmented *D. nigrodunni* are sexually monomorphic. No specific genes responsible for the species differences have been identified, but a genetic analysis showed that both the X chromosome and the autosomes contributed (Hollocher et al. 2000) in addition to a small maternal effect.

Another pair of sister species exhibiting a striking difference in pigmentation is *D. americana–D. novamexicana* from the *virilis* group: *D. americana* is heavily melanized and *D. novamexicana* is predominantly yellow with little melanization (Wittkopp et al. 2003b). At least four pigmentation QTLs contribute to this difference, but neither the relative contribution of these nor the proportion of the phenotypic differences explained by them has been identified (Wittkopp et al. 2003b). One of the QTLs identified contained the *ebony* locus, encoding the enzyme *N*-β-alanyldopamine (NBAD) synthetase that converts dopamine to NBAD, which is subsequently oxidized to produce a yellowish pigment and reduce the amount of black melanin. The *ebony* expression was correlated with pigmentation differences: There was a higher concentration of Ebony in the abdominal tergite of *D. novamexicana* than *D. americana*, with an intermediate level in hybrids (Wittkopp et al. 2003b). No association between the *yellow*, *ddc*, *omb* and *bab* loci and QTLs were observed in *D. americana–D. novamexicana* by Wittkopp et al. (2003b), which is in contrast with the findings in other species. Thus, similar pigment patterns may have evolved through regulatory changes in different loci in different lineages.

GENETICS OF REPRODUCTIVE ISOLATION

Traditionally, prezygotic reproductive isolation is distinguished from postzygotic reproductive isolation. The former is most often studied in the context of behavioral mate discrimination, though increasing emphasis in the last decade has been laid upon both ecological differentiation and diminished interspecies fertilization. Postzygotic reproductive isolation is caused by hybrid dysfunctions, frequently observed in the form of sterility or inviability (Coyne & Orr 1998). Dobzhansky (1937) and Muller (1940) independently formulated a theory to explain how such hybrid incompatibilities may have evolved, which is now known as the Dobzhansky–Muller model (Box 25.1). This model predicts that different alleles evolve in different populations during and after speciation. When alternative alleles from different populations are brought together in hybrids then deleterious interactions

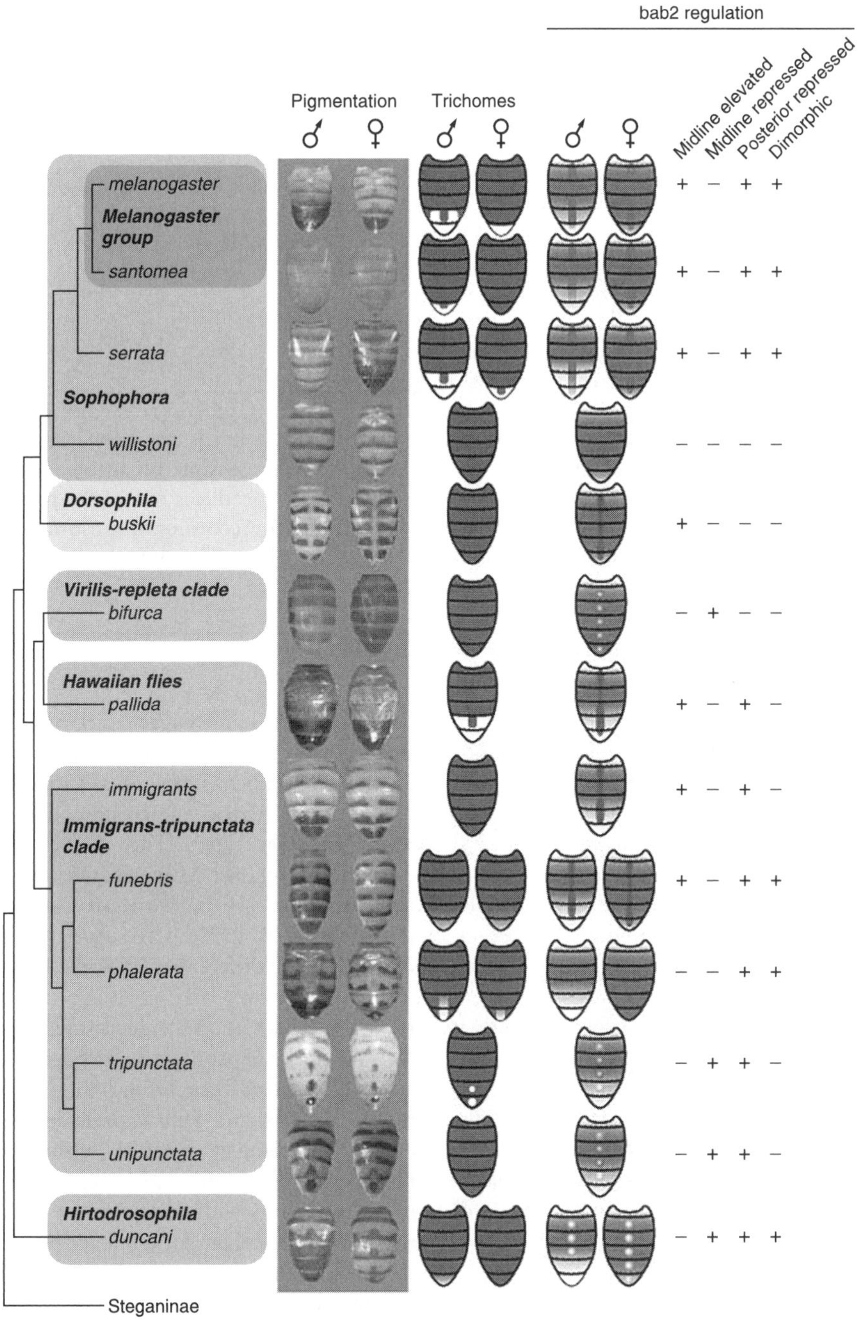

FIGURE 25.2. Relationship between phylogeny of the *Drosophilinae*, Bab2 expression (+, presence; –, absence), abdominal pigmentation, and trichome patterning (gray shading). Adapted from Figure 2 in Gompel and Carroll (2003) with the permission of Nature Publishing Group.

BOX 25.1. The Dobzhansky–Muller Model

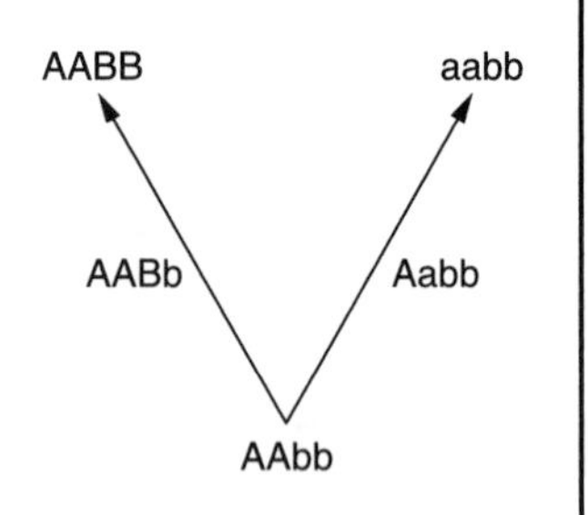

Let us consider the Dobzhansky–Muller scheme in its original formulation involving a two-locus, two-allele diploid population, initially monomorphic for a genotype *AAbb*. Assume further that this population is fragmented into two geographically isolated parts. In one part, a substitution of *B* for *b* occurs and a subpopulation *AABB* is formed. In the other part, there is a substitution of *a* for *A*, thus giving rise to a subpopulation *aabb* (Figure 1). Assume that there is no reproductive isolation among genotypes *AAbb, Aabb*, and *aabb* and among genotypes *AAbb, AABb*, and *AABB*, that is, individuals within these two groups freely interbreed and produce fully viable and fertile offspring. On the contrary, let the cross of *AABB* and *aabb* be difficult or impossible because alleles *a* and *B* are incompatible in the sense that their interaction "produces one of the physiological isolating mechanisms" (Dobzhansky 1937, p. 282). Accordingly, substitutions of *B* for *b* in one subpopulation and *a* for *A* in the other subpopulation will lead to the emergence of two reproductively isolated populations.

between them can be exposed because they were each selected in different genetic backgrounds. Here we focus only on molecular and genetic studies resulting in fine-scale characterization of genes responsible for reproductive isolation.

We overemphasize studies of postzygotic isolation in this review because of the more detailed genetic studies performed on these traits, not because they are necessarily more directly involved in the speciation process. Indeed, many studies have found that prezygotic isolation both evolves earlier in evolutionary divergence and plays a more significant role in preventing gene flow between coexisting taxa. Nonetheless, postzygotic isolation also potentially limits introgression, and many cases are known wherein hybridizing taxa form sterile or inviable hybrids in nature (see Coyne & Orr 2004). We highlight this work in the context of the "genetics of potential barriers to gene flow," not as genetic studies of those traits that actually caused speciation or are currently effective at preventing gene flow between species.

Genetics of Prezygotic Isolation in *Drosophila*

Despite numerous studies of the genetic basis of prezygotic isolation, again with an overwhelming majority conducted on *Drosophila*, almost no associated genes are characterized at the molecular level (Powell 1997). Within *D. melanogaster*, there are two distinct behavioral races: Z (from Zimbabwe and Southern Africa) and cosmopolitan M (worldwide common type). The Z females strongly discriminate against M males, whereas M females mate with males from both races indiscriminately (Hollocher et al. 1997). This suggests, together with limited genetic differentiation at the DNA level and incomplete postzygotic isolation, that these two races are in fact at an incipient stage of speciation. The third chromosome accounted for more than 50% of the total genetic variation in both male mating success and female preference (Hollocher et al. 1997). Ting et al. (2001), using chromosome balancers and a multimarker line, generated a series of recombinants with respect to the third chromosome and tested them for a linkage between genetic markers and male mating success and plus female preference. All remaining chromosomes were made M-type.

First, Ting et al. (2001) found that the distal part of the right arm of chromosome III significantly affected male mating success: males with two copies of the Z race chromosome at the distal chromosomal segment mated better with Z females than males with only one or no Z copy. They subsequently decomposed this segment into three smaller regions in order to test whether this genetic effect was due to one major effect locus or several minor effects loci. No major effect locus was found and instead a nonadditive interaction between at least two loci producing "Z maleness" was detected. Overall, they found at least four loci contributing to the mating success of Z males and at least two loci responsible for Z female mating preference, all located on chromosome III. Fang et al. (2002) suggested that the *desat2* locus located on IIIR, and involved in cuticle hydrocarbon biosynthesis, is correlated with "Z femaleness." However, Caribbean *D. melanogaster* bear the putative Z allele at *desat2* and yet do not exhibit a Z female mating preference. Hence, the exact role of this locus in mate discrimination remains uncertain. Greenberg et al. (2003) replaced *desat2* of M race with *desat2* of Z race using site-specific gene disruption. Although the cuticle hydrocarbon profile was changed, the manipulation failed to produce a Z-like profile.

D. ananassae is a cosmopolitan species found in sympatry with *D. pallidosa* in the Tongan and Fijian islands of Melanesia (Doi et al. 2001). Similarly to *D. melanogaster* M and Z, these two species do not show postmating isolation such as hybrid inviability or sterility, but instead mating isolation mostly due to female preferences occurs. Using surgical treatments, Doi et al. (2001) showed that female preference is oriented at acoustic courtship elements produced by males. The absence of the song of *D. pallidosa* males largely increased interspecies mating with *D. ananassae* females but reduced intraspecies mating with *D. pallidosa* females. Moreover, chromosomal introgressions by repeated backcrosses to *D. pallidosa* males possibly identified a small region on the middle of IIL (near the *Delta* marker) disproportionately contributing to females' preference for *D. ananassae* males: *D. pallidosa* females with the introgressed segment from *D. ananassae* readily mated with *D. ananassae* males. This predominant effect exerted by a single segment of a few centimorgans contrasts with the more complex genetics of female preference divergence in *D. melanogaster* Z–M.

Hybrid Lethality in *D. melanogaster*

D. melanogaster females produce lethal F_1 males and viable but sterile F_1 females (inviable in higher temperatures) when crossed with *D. simulans*, *D. mauritiana*, or *D. sechellia* males. These three species are estimated to have diverged from *D. melanogaster* approximately 2–3 million years ago (Powell 1997). *D. melanogaster* hybrids have a long history as a model to investigate hybrid dysfunctions. For example, Sturtevant (1920) showed that this sex-specific inviability of the hybrids could be attributed to the X chromosome. Muller and Pontecorvo (1940) created F_2 equivalents from crosses between triploid *D. melanogaster* and heavily irradiated diploid *D. simulans* males and found that autosomes may also be involved in hybrid lethality. However, the early studies could not go to a finer scale than whole chromosomes because of the complete sterility of F_1 hybrids, which prevented genetic mapping via backcrosses. Instead, researchers focused on mutant alleles in a locus called *Hybrid male rescue* (*Hmr*) that suppress F_1 male hybrid inviability and the high-temperature female inviability, as well as partially suppress female infertility (Barbash et al. 2003 and references therein). Increased dosage of the wild-type Hmr^+ has been observed to reduce hybrid viability more severely.

Barbash et al. (2003), using transgenic constructs, characterized a single gene located on the X chromosome that was responsible for the *Hmr* effect. First, they obtained complete clone coverage across the *Hmr* region by chromosome walking and defined its boundaries by using breakpoints of deficiencies previously known as Hmr^+ and Hmr^-. The *Hmr* region turned out to contain five predicted genes. Next, transgenic constructs covering the four candidate genes were made and assayed in hybrid males for suppression of rescue function by *Hmr*. The principle behind the usage of transgenic constructs was as follows. If a hybrid male is Hmr^- and a cloned copy of Hmr^+ on a *P* element is added then eventually the Hmr^+ phenotype should be obtained. Only one construct out of four, characterized by a frameshift mutation in the *CG1619* coding region, had no Hmr^+ activity. All remaining constructs retained the Hmr^+ phenotype.

Additionally, Barbash et al. (2003) confirmed that the original *Hmr* rescue allele, Hmr^1, could be mimicked by a P element insertion near the 5′ end of *CG1619*. The *CG1619* region encodes the HMR protein containing homology to the MADF domain that is the first sequence-specific transcription factor (ADF1) found in *Drosophila*. The MADF domain of ADF1 is similar to the DNA-binding motifs of the MYB oncoproteins and SANT domain, the latter commonly found in DNA-binding proteins in eukaryotes. This *Hmr* region also characterizes a high level of sequence divergence between *D. melanogaster* and its sibling species: the average replacement rate (Dn) is 0.077. Also, the *D. melanogaster Hmr* contains between 8 and 11 in-frame indels compared with the sibling species (Barbash et al. 2003).

Another gene responsible for hybrid inviability between *D. melanogaster* and *D. simulans* has been identified on chromosome III by Presgraves et al. (2003) (Figure 25.3). They used hybrids between *D. melanogaster* females heterozygous for an autosomal deficiency and a balancer chromosome and *D. simulans* males carrying *Lethal hybrid rescue* (*Lhr*). Alternating deletions, they scanned approximately 70% of the *D. simulans* autosomes for recessive alleles producing lethal interactions with the *D. melanogaster* X chromosome gene(s). Two complementation groups containing 12 loci were narrowed down and the loss-of-function mutations were individually tested in each of those. Only mutants in one locus, *CG10198*, failed to complement their counterpart in the *D. simulans* genome, thus producing hybrid lethality. This locus is the *Drosophila* homolog of a dicistronic gene encoding two distinct nucleoproteins, Nup98 and Nup96, and the hybrid inviability is specifically associated with the N-terminus of Nup96. Sequence polymorphism analysis of *Nup96* showed a significant excess of replacement changes fixed between *D. melanogaster*, *D. simulans*, *D. mauritiana*, and *D. yakuba*, suggesting the occurrence of adaptive evolution in this locus.

Hybrid Sterility in *Drosophila*

In an extensive review of reproductive isolation genetics, Wu and Palopoli (1994) concluded that sterility and inviability effects are due to multiple genetic elements of weak direct effects and pervasive epistatic interactions rather than to single loci of major effects. However, a single locus conferring almost complete sterility, *Odysseus* (*Ods*), has been identified via introgressions of small segments of the X chromosome from *D. mauritiana* into *D. simulans* (Perez et al. 1993). Repeated backcrosses over more than 20 generations were conducted, and the extent of introgressions that produce male sterility was contrasted with those that did not cause it. No gradual decline in fertility with the insert size was observed. The "sterile" introgressions produced 100% sterility and the "fertile" introgressions were on average more than 90% fertile, thus indicating the existence of one major gene. Perez and Wu (1995) further refined the mapping of this region and narrowed down *Ods* to an interval of 20 kb. The more recent analysis also showed that *Ods* is not sufficient to produce sterility and instead it interacts with some other *D. mauritiana* genes to cause hybrid sterility in *D. simulans* (Perez & Wu 1995).

What is the *Ods* gene? Chung-I Wu's group explored this question further; Ting et al. (1998) created 190 new recombinants with progressively shorter introgressions. The location of *Ods* was defined by the longest fertile and shortest sterile introgressions (delineated using genetic markers) and narrowed down to a genomic region of 8.4 kb. The exons putatively identified in this region belonged to a homeobox that is well conserved across distant taxa. Surprisingly, there were as many as 15 amino acid differences found between *D. simulans* and *D. mauritiana* in the homeodomain. For a comparison, it is divergence similar to that between *Drosophila Ods* and its mouse ortholog, and larger than that between *Caenorhabditis. elegans* and mouse. The rate of nucleotide substitution in the homeodomain between *D. simulans* and *D. mauritiana* is 8 times higher than that in intron, suggesting positive selection. The ratio of amino acid replacements to silent substitutions in the homeodomain is 10:1 between *D. simulans* and *D. mauritiana* but 10:9 between their branch and that leading to *D. melanogaster*. Also, it seems that the selective pressure in the homeodomain relative to nonhomeodomain is stronger in *D. mauritiana* than in the other two species (Ting et al. 1998). Curiously, *Ods* seems a "dispensable" gene because its experimental deletion barely affects male fertility (Sun et al. 2004).

Speciation Genetics of Yeast

Most of the experimental work on *Drosophila* described above, with the exception of *Hmr* mutations, relies on manipulations revealing rather than suppressing reproductive isolation. The latter experimental interventionism, however, can provide more compelling evidence of particular genes' contribution to speciation because it involves a preexistent phenomenon rather than giving a chance of creating a new convergent phenomenon. Genetic engineering to "undo" reproductive isolation between species has been successively applied to yeasts of the *Saccharomyces sensu stricto* group (Delneri et al. 2003).

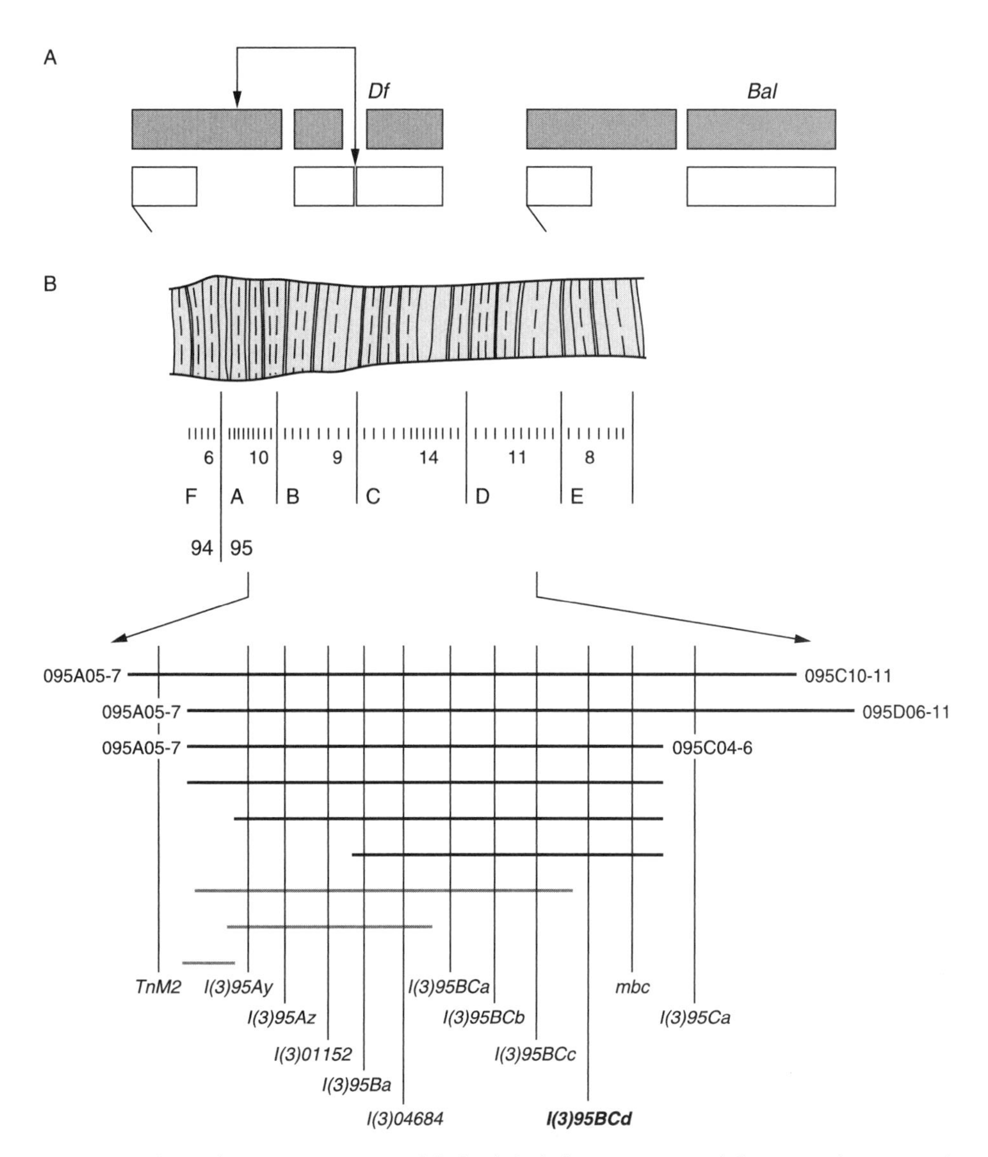

FIGURE 25.3. Deficiency mapping of hybrid lethality in *Drosophila.* (A) Schematic of hybrid male genotypes, with the sex chromosomes on the left (Y chromosome shorter with hook) and one autosome on the right of each genotype. The right genotype represents a male heterozygous for *D. simulans* genotype (white) and a balancer (Bal). *D. melanogaster* chromosomes shown in gray, *D. simulans* chromosomes shown in white. The deficiency depicted as a slot exposes a hemizygous autosomal region (black) that interacts (arrows) with chromosome X. (B) Interspecific complementation tests using deficiencies (horizontal lines) in cytological region 95 of chromosome 3R. Gray, complements hybrid lethality; black, fails to complement hybrid lethality. Adapted with permission from Figure 1 in Presgraves et al. (2003) © 2003 AAAS.

The yeast species group consists of six species: *S. cerevisiae*, *S. paradoxus*, *S. bayanus*, *S. cariocanus*, *S. mikatae*, and *S. kudriavzevii*. The species readily hybridize, but hybrids are mostly sterile (<5% viable ascospores). Three of these species are characterized by the presence of specific reciprocal chromosomal translocations believed to play a role in reproductive isolation: four in *S. bayanus* and *S. cariocanus*, and one and two, respectively, in the two strains of *S. mikatae*—IFO1815 and IFO1816. Delneri et al. (2003) reconfigured *S. cerevisiae* chromosomes to produce the two reciprocal translocations typical of *S. mikatae*. These reconfigured *S. cerevisiae* strains were used in intraspecies and interspecies crosses to assess the effect of the translocations in reproductive isolation. Intraspecific crosses exhibited a significant decrease in hybrid fertility. Remarkably, restoring chromosomal collinearity between *S. cerevisiae* and *S. mikatae* increased hybrid fertility from near zero (typical of noncollinear crosses) to approximately 30%. The analysis of viable spore products of interspecific hybrids indicated widespread aneuploidy involving almost all chromosomes, but it was unclear whether chromosome loss occurred during spore formation or the hybrid zygote formation itself. The hybrids are not tetraploids—the pattern of spore karyotypes is rather compatible with duplication of one of the parental species having occurred in hybrids. Producing spores that include the complete genome of one parent is plausibly a mechanism required to overcome the meiosis problems. Interestingly, enforced allotetraploidy from crosses among the *S. cerevisiae sensu stricto* species results in fully viable spores (Greig et al. 2002a).

In another study, Greig et al. (2002b) managed to obtain fertile F_1 and F_2 hybrids between *S. cerevisiae* and *S. paradoxus*. The F_2 hybrids had high fertility when crossed with themselves, but they had low (~7.5%) fertility when backcrossed with either parental species. Although 10 independent F_2 strains were fully fertile (100%), their fertility decreased dramatically when intercrossed to generate F_3 hybrids. Extensive tetrasomy was observed in the hybrids, but this pattern of sexual isolation was not easily explained by either aneuploidy or chromosomal incompatibility. Tetrasomes are not expected to have unbalanced meiotic segregation, while chromosomal incompatibilities can be excluded because F_2 hybrids are fully homozygous, having originated from single autofertilized gametes. Instead, Greig et al. (2002a) estimated that genic incompatibilities between interacting *S. cerevisiae* and *S. paradoxus* genes account for 50% of the fertility variation.

These studies prove that yeast species may provide excellent systems to investigate speciation. Genomes of four yeasts (*S. cerevisiae*, *S. mikatae*, *S. paradoxus*, and *S. bayanus*) have already been sequenced completely (Kellis et al. 2003), and the yeast system offers even a broader battery of genetic tools than the *Drosophila* system.

FUTURE DIRECTIONS: EPISTASIS AND GENOME-WIDE ANALYSIS

It is premature to generalize about speciation genes based on the scarce data available. Nevertheless, there are certain patterns becoming increasingly persuasive. First, there is a very broad versatility of genes responsible for species differences and reproductive isolation with respect to their function, number, and genome location. Even for sibling species, apparently identical features such as pigmentation patterning in *Drosophila* may be due to regulatory changes in different loci. There are, of course, examples of conserved genetic effects across species as well as independent origin of the same genic effect, such as *svb* expression affecting *Drosophila* larval morphology (Sucena et al. 2003). There is no consistent pattern with respect to the number of genes involved in reproductive isolation and the magnitude of their effect. The only fairly safe point we can make is that neither of the long-contrasted views on the genetic background of reproductive isolation, namely numerous genes of additive effects versus infrequent major effect genes, consistently explains all observations. A common feature of the recently discovered speciation genes (*Hmr*, *Nup96*, *Ods*) is that they exhibit signatures of strong positive selection, but more data are needed to confirm this generalization.

Nearly all study cases described in this chapter invoke some sort of epistatic interactions. Dobzhansky (1933) was probably the first to see the importance of dominant gene incompatibilities in F_1 hybrid sterility and inviability. He observed that sterility in a *Drosophila* hybrid was caused by "a profound disturbance of spermatogenesis, including failure of chromosome pairing at meiosis." He reasoned this observation was either due to "chromosomes of one species finding no complete homologues among the chromosomes of another species or race" or "the action of complementary genetic factors contributed by parent species." The observation that islands of tetraploid spermatocytes still failed to pair made him argue that the sterility was caused by genetic incompatibilities rather than the lack of chromosomal complementation. The latter should have been furnished by the doubling of chromosomes in tetraploids.

In F_2 and backcross hybrids, loci may be homozygous for incompatible alleles, allowing for hybrid dysfunctions due to expression of recessive genic incompatibilities. An example is provided by crosses of swordtails (*Xiphophorus helleri*) with platyfish (*X. maculates*), in which lethal tumor develops in hybrids (Schartl 1995). F_1 hybrids are viable despite the fact that they develop benign melanomas. Fifty percent of backcrosses $F_1 \times X.$ *helleri* are melanoma-free, 25% develop benign melanomas, and 25% develop malignant melanomas. A two-locus system with a sex-linked tumor locus and an autosomal suppressor was demonstrated to underlie this pattern (Schartl 1995). Recessive incompatibilities have also been proposed to explain Haldane's rule (Orr & Presgraves 2000; Johnson, Ch. 24 of this volume).

There are also theoretical reasons to believe that epistasis in general is a strong force driving reproductive isolation. Johnson and Porter (2000; also Porter & Johnson 2002) modeled gene regulation as a matching function between the product of one locus and the promoter of another locus in a genetic pathway, with binding strength determining the amount of product. They found that if the regulatory system is subject to parallel selection in two independent populations then hybrid fitness may readily decrease leading to reproductive isolation. In this context, it has been very instructive to find an empirical counter-example of the importance of genic epistasis in creating reproductive isolation. Greig et al. (2002a) observed restoration of yeast hybrid fertility in tetraploids, demonstrating that at least dominant gene incompatibilities play no role in reproductive isolation of yeasts. Also, the significance of recessive incompatibilities in yeast hybrids turns out to be enigmatic. As yeast gametes express their genes, recessive incompatibilities should be manifested in hybrid gametes. Therefore, other mechanisms, such as the mismatch repair system, have been suggested to underlie speciation modes in yeasts (Greig et al. 2002a).

To assess the role of gene expression deregulation in hybrids at the genome-wide scale, transcriptome profiling with the use of DNA microarrays has been recently applied (Michalak & Noor 2003). These expeditions into hybrid transcriptomes have been preceded by microarray analysis of interpopulation variation in teleost fish of the genus *Fundulus* (Oleksiak et al. 2002) and interspecies variation in *D. simulans*, *D. yakuba,* and *D. melanogaster* (Rifkin et al. 2003). The latter study addressed changes in gene expression patterns in relation to a dramatic phenotypic transformation during puparium formation. Of 12,866 genes scanned, approximately 50% significantly changed expression level between the two time points (puparium formation and 18 h before puparium formation). About 50% of these genes (27% across the genome) significantly differed between species, and these differences followed phylogenetic relationships (Rifkin et al. 2003). Michalak and Noor (2003) used oligonucleotide DNA chips to compare expression profiles of sterile F_1 males from *D. simulans* × *D. mauritiana* crosses with both their parental lineages (Figure 25.4). A dozen transcripts were significantly downregulated in hybrids compared with parental lineages, and misexpression of three of these was independently confirmed with a quantitative real-time PCR (polymerase chain reaction) analysis. We also found that statistically significant misexpression tends to occur much more often among male-specific genes than others (Michalak & Noor 2003). Ranz et al. (2004) found even more pronounced misregulation of genes in hybrid females of *D. melanogaster* and *D. simulans*: nearly 70% of all transcripts covered by their microarray study were eiter up- or downregulated relative to the parental species.

Ultimately, we believe that genome-wide functional analyses will and already do provide an extraordinary opportunity to assess the importance of gene regulation, and more generally epistasis, in speciation and species evolution. A logical next step would be to use proteomics tools and assay entire proteome changes associated with speciation processes—a step for which we will probably not have to wait too long.

In this chapter we have largely ignored the ecological aspects of speciation, but it by no means implies that genetics is more important in speciation than ecology. Conversely, we recognize that the genetics of reproductive isolation makes little sense when detached from its ecological context. Future genetic studies must no longer neglect this aspect of speciation.

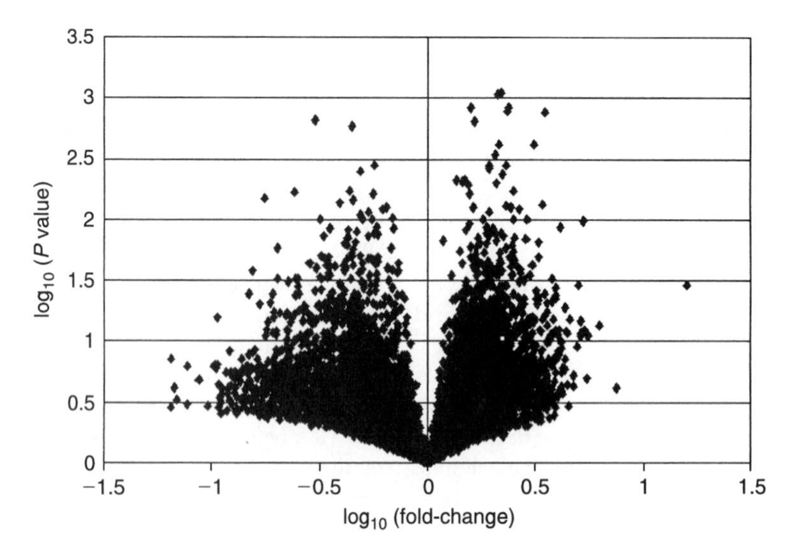

FIGURE 25.4. Volcano plot of gene expression difference between pure and F_1 *D. mauritiana* × *D. simulans* hybrids and pure *D. mauritiana* lineage. Each point represents a transcript from the microarray assay. The log 10 fold-change (ratio of mean expression values of pure species and F_1 hybrids) is shown on the x-axis and the –log 10 P values from ANOVA significance tests are shown on the y-axis. Positive values of x correspond to underexpression and negative x values represent overexpression in hybrids relative to nonhybrids. Points located over the line $y = 2$ are statistically significant at a p value lower than 0.01. Adapted from Figure 1B in Michalak and Noor (2003).

SUGGESTIONS FOR FURTHER READING

The most comprehensive reading on speciation is provided by Coyne and Orr's *Speciation* (2004). Orr and Coyne (1992) and Orr (2001) provide extensive reviews of information on the genetic background of species differences. The latter review is accompanied by others on speciation in a special issue of *Trends in Ecological Evolution.* One may find a more comprehensive review on the genetics of pigmentation polymorphism in *Drosophila* in Wittkopp et al. (2003a). Wu and Palopoli (1994) is a classical (and therefore slightly outdated) review about the genetics of reproductive isolation. Wu (2001) is a target review followed by commentaries, some of them largely critical, in a special issue of *Journal of Evolutionary Biology.*

Coyne JA & HA Orr 2004 Speciation. Sinauer Assoc.

Orr HA 2001 The genetics of species differences. Trends Ecol. Evol. 16:343–350.

Orr HA & JA Coyne 1992 The genetics of adaptation revisited. Am. Nat. 140:725–742.

Wittkopp P, Carroll SB & A Kopp 2003a Evolution in black and white: genetic control of pigment patterns in *Drosophila.* Trends Genet. 19:495–504.

Wu C-I 2001 The genic view of the process of speciation. J. Evol. Biol. 14:851–865.

Wu C-I & MF Palopoli 1994 Genetics of postmating reproductive isolation in animals. Annu. Rev. Genet. 28:283–308.

26

Natural Hybridization

MICHAEL L. ARNOLD
JOHN M. BURKE

The focus of this chapter is natural hybridization, that is the natural interbreeding of individuals from wild populations that can be distinguished on the basis of one or more heritable characters (Harrison 1990; Arnold 1997). Over the years, the evolutionary importance of natural hybridization has been a topic of considerable debate. Indeed, the potential outcomes ascribed to natural hybridization range from the extinction of one or both hybridizing taxa to the formation of new species (reviewed in Arnold 1997; see also Box 26.1).

Because crosses between genetically divergent lineages often result in offspring with reduced levels of fertility and/or viability, natural hybridization has often been viewed as an evolutionarily unimportant phenomenon (e.g., Mayr 1963). In those cases where natural hybridization has been implicated directly in the evolutionary process, it has often been for the role it may play in the finalization of speciation (reviewed in Howard 1993). Thus, hybrid zones generally have been viewed as transient, with selection on mating preferences leading to the production of reproductively isolated taxa. Alternatively, in cases where hybrid zones have been shown to be stable over time, they have often been considered impediments to continuing evolutionary divergence, rather than evolutionarily important in their own right (Arnold 1997).

A less common perspective, but one that has gained increasing support in recent years, is that of hybridization as a relatively widespread and potentially creative evolutionary process (e.g., Lotsy 1931; Anderson 1949). Support for this view comes from studies documenting the occurrence of introgression (reviewed in Arnold 1997), as well as from those suggesting that hybridization may serve as a source of adaptive genetic variation (e.g., Lewontin & Birch 1966; Grant & Grant 2002; Rieseberg et al. 2003). In addition, there are a number of well-documented cases of speciation resulting from hybridization in a number of animal and plant taxa (e.g., Dowling & DeMarais 1993; Rieseberg 1997). Taken together, these findings suggest that, rather than serving as a roadblock along the pathway of divergent evolution, natural hybridization may play a direct and important role in evolutionary diversification.

We begin this chapter by providing a brief history of investigations of natural hybridization, beginning in the eighteenth century, continuing through the Modern Synthesis, and concluding with the present state of research. We then review several models of hybrid zone evolution that provide a conceptual framework for modern-day analyses of the role of natural hybridization in evolution. This is followed by a discussion of the possible outcomes of natural hybridization, with particular reference to its creative potential. Finally, we close with a brief discussion of the types of research on natural hybridization that demonstrate great promise for expanding our understanding of this important evolutionary process.

HISTORY OF INVESTIGATIONS

Pre-Modern Synthesis

Though mythological "hybrid" organisms abound in Greek, Egyptian, and Hindu writings and art,

BOX 26.1. Potential Outcomes of Natural Hybridization

Extinction (Merging) of Lineages

One of the conservation biology issues arising from natural hybridization is that of extinction through the genetic assimilation of one species, subspecies, etc., by another. Rhymer and Simberloff (1996) described this process, in the context of exotic species, as follows: "Nonindigenous species can bring about a form of extinction of native flora and fauna by hybridization and introgression either through purposeful introduction by humans or through habitat modification, bringing previously isolated species into contact. These phenomena can be especially problematic for rare species coming into contact with more abundant ones."

Reinforcement

This process was first defined clearly by Dobzhansky (1940) and first termed "reinforcement" by Blair (1955). Reinforcement may occur when individuals from populations, which have diverged in allopatry, come into secondary contact and mate. If the hybrids are less fit (i.e., lower viability or fertility) than offspring produced from matings between conspecific individuals, selection will favor those individuals that choose conspecific mates. The frequency of individuals that mate "assortatively" with conspecifics will increase leading to the "reinforcement" of reproductive isolation, and possibly ending in complete reproductive isolation.

Adaptive Introgression

The formation of natural and experimental hybrids with fitnesses that equal or exceed that of their parents, or conspecific offspring, is now well documented (e.g., Arnold 1997; Campbell & Waser 2001). A predicted outcome from hybrids that demonstrate high fitness is the formation of new, hybrid evolutionary lineages. An additional consequence, if the suite of adaptations present in the hybrids differs from that in the parents, may be their occupation of novel habitats relative to their parents. Some of the best potential examples of this process come from introgression and hybrid speciation in the sunflower genus *Helianthus* (e.g., Kim & Rieseberg 1999; Rieseberg et al. 2003) and in Darwin's finches (e.g., Grant & Grant 2002). However, this process has also been implicated in additional plant and animal species groups (e.g., charr: Doiron et al. 2002; irises: Cruzan & Arnold 1993).

Hybrid Sink

Whitham (1989) introduced the concept that natural hybrids might act as pest sinks, preventing adaptive evolution of the pests to the parental populations and thus protecting the parental populations from infestations. He defined the hybrid sink concept using data from the interactions between aphids and parental and hybrid cottonwood trees. He found that "the more susceptible trees acted as pest sinks supporting most of the aphid population ... The concentration of aphids on ... [the hybrid trees] ... suggested that susceptible plants [i.e., hybrids] not only acted as sinks in ecological time, but may also have prevented aphids from adapting to the more numerous hosts [i.e., parental, nonhybrid trees] in evolutionary time" (Whitham 1989).

BOX 26.1. *(cont.)*

Hybrid Speciation

The identification of evolutionary lineages resulting from hybridization is known from every kingdom (e.g., see Arnold, 1997, 2004 for reviews). The formation of lineages recognized as new "species" is also recognized across all taxonomic groups. Such species can be grouped into two subheadings based upon the new lineage's chromosome number. If the new lineage has the same, or very close to the same, number of chromosomes as the parental lineages, they are said to be a diploid or homoploid hybrid species (Grant 1981, p. 256). If, on the other hand, the new hybrid species has a chromosome number that is higher than that found in the parentals by one or more sets of haploid chromosomes, it is termed an "allopolyploid" hybrid species. An allopolyploid is thus formed by crosses between two individuals that come from divergent evolutionary lineages (Soltis et al. 2004). In contrast, an "autopolyploid" arises from crosses between individuals from the same evolutionary lineage (Soltis et al. 2004). In animal taxa, the formation of polyploid hybrids normally results in unisexual forms that reproduce asexually (e.g., via parthenogenesis; Moritz et al. 1989). In contrast, naturally occurring plant polyploids are normally capable of sexual reproduction (see Soltis et al. 2004 for a review).

the history of scientific investigations into the process and products of natural hybridization can be traced to the mid-eighteenth-century work of Linnaeus. Linnaeus held to a creationist model for the origin of species, but also believed that species could arise through hybridization between previously created forms. Thus he wrote in his *Disquisitio de Sexu Plantarum* (1760; as cited by Grant 1981, p. 245), "It is impossible to doubt that there are new species produced by hybrid generation." However, shortly thereafter, Kölreuter (as referenced by Darwin 1859, pp. 246–247) showed that hybrids from heterospecific crosses are often sterile, a result which cast doubt on the possibility of hybrid species formation. Like Kölreuter, Darwin (1859) was impressed by the fact that heterospecific crosses were difficult to form and that the offspring from such crosses were generally highly infertile. These observations led him to conclude that the most common outcome of natural hybridization would be the production of offspring with low fitness (Darwin 1859, pp. 276–277).

The viewpoint of Kölreuter and Darwin contrasted sharply with subsequent work by Lotsy (1931), who saw natural hybridization as the primary mechanism for evolutionary change. Specifically, Lotsy (1931) hypothesized that the origin of new taxa was due to the interbreeding of individuals from different syngameons ("an habitually interbreeding community"; p. 3), rather than from the gradual accumulation of heritable differences in the absence of reproductive contact between populations.

The Modern Synthesis

Scientific investigations into the process and outcomes of natural hybridization were profoundly influenced by the period of intense conceptualization and data gathering known as the Modern Synthesis. Indeed, the evolutionary literature from the late 1930s onward reveals two divergent viewpoints concerning the role of natural hybridization in evolution. The first, voiced mainly by evolutionary zoologists, considered natural hybridization to be a mechanism by which species barriers are finalized, or perhaps as a tool for understanding the process of speciation (e.g., Dobzhansky 1937). Otherwise, natural hybridization was assumed to be of little long-term evolutionary importance, an assumption that was incorporated into the ever-hardening conceptual foundation of the Modern Synthesis.

In contrast to the above was the viewpoint that natural hybridization could be a pervasive and important evolutionary process. Botanists (e.g., Anderson 1949) were the primary advocates of this paradigm. Results from numerous studies provided evidence of (i) the widespread occurrence of natural hybridization and (ii) significant evolutionary effects of this process through the production of new species and novel adaptations (e.g., Stebbins 1959).

Postmodern Synthesis

Between the end of the Modern Synthesis and the early 1990s, the most crucial difference between studies of natural hybridization involving either plants or animals involved the differences in worldview described above. In studies of animal species, natural hybridization was likely to be considered evolutionarily unimportant (but see e.g., Lewontin & Birch 1966), while for plants this process was considered a potential source for evolutionary novelty. The writings of Mayr (1963) and Stebbins (1959), in particular, highlight these divergent viewpoints. This difference in paradigm is perhaps the best explanation for the different types of studies undertaken by zoologists and botanists subsequent to the Modern Synthesis. Zoological studies during this time period emphasized microevolutionary processes that were ongoing in parental and hybrid populations (i.e., those processes associated with "biological speciation"; see Barton & Hewitt 1985; Harrison 1990 for reviews). This emphasis resulted in numerous, wonderfully detailed cases of animal hybridization (e.g., *Lepomis*: Hubbs 1955; *Allonemobius*: Howard 1986). In contrast, relatively few botanists examined population-level phenomena associated with natural hybridization during this same time period (but see Anderson 1949; Levin 1963; Stebbins 1959; Grant 1981). This is somewhat surprising given the relative ease (compared with most animal groups) of studying the population biology of plants, combined with the general acceptance of hybridization as an evolutionarily creative process by the botanical community. It has been suggested (Arnold 1997) that the numerous process-oriented analyses of animal hybrid zones and the few such analyses in plant groups reflected the zoologists' focus on the mechanics of speciation (i.e., the evolution of reproductive barriers) and the botanists' focus on systematic relationships. It has also been argued that, since allopolyploidy (see Box 26.1) was viewed as the most common and important outcome of hybridization between plant taxa (e.g., Stebbins 1959), a disproportionate number of studies during this time tested for polyploidy in plants, rather than focusing on population-level phenomena (Arnold 1997).

In spite of the different conceptual frameworks previously held by zoologists and botanists, the past 15 years have seen somewhat of a convergence in both the underlying paradigm and the approach of natural hybridization studies. Thus, zoologists have begun to accept natural hybridization as a potentially creative evolutionary process (see Dowling & DeMarais 1993 for a review) and botanists have completed numerous process-oriented studies (e.g., *Helianthus*: Rieseberg et al. 2003; *Ipomopsis*: Campbell & Waser 2001; *Iris*: Johnston et al. 2001).

Predicting the Outcome of Natural Hybridization

Numerous models have been proposed that predict the outcome of natural hybridization (e.g., Dobzhansky 1937; Endler 1977; Barton & Hewitt 1985; Howard 1986; Harrison 1990; Moore & Price 1993; Arnold 1997). Each of these models incorporates assumptions concerning the effects of natural selection and/or dispersal. For example, assumptions regarding the relative fitness of hybrids are always central. Process-oriented studies prior to the 1990s most often adopted a model in which selection was assumed to act against hybrids, regardless of the environment in which they occurred (e.g., Darwin 1859; Dobzhansky 1937; Barton & Hewitt 1985). This decrease in hybrid fitness was attributed to the disruption of the parental gene combinations (Barton & Hewitt 1985) or to differences in the number and/or structure of chromosomes in the parental taxa (Grant 1981).

Of course, not all studies of natural hybridization from this era assumed that environment-independent selection acted against hybrids. In fact, several models assumed that interactions between the genotype of hybrid individuals and their environment determined the genetic structure of hybrid populations (Figure 26.1; Endler 1977; Howard 1986; Harrison 1990; Moore & Price 1993; Arnold 1997). In addition, some of these environment-dependent models included dispersal as a major factor in determining the genetic structure of hybrid populations (Endler 1977; Moore & Price 1993), while others assumed that dispersal played a minor role (Howard 1986; Harrison 1990). Another important distinction between the various environment-dependent models related to the fitness of hybrid genotypes compared with their parents. For example, Moore and Price (1993) argued that hybrids could be more fit than their parents in certain "hybrid" environments, while the parental individuals would be most fit in their own microhabitats.

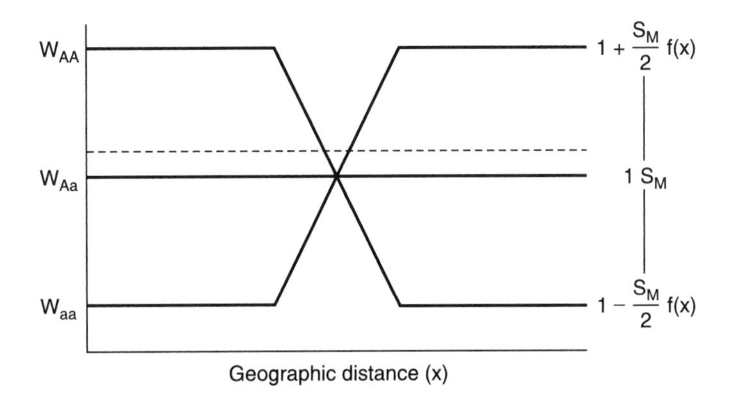

FIGURE 26.1. Model of selection in a hybrid zone. The unbroken horizontal line indicates heterozygote fitness. W_{AA}, fitness of the AA homozygote, W_{Aa}, fitness of the heterozygote; W_{aa}, fitness of the aa homozygote. S_M is the selection coefficient as defined by May et al. (1975). The dashed line indicates heterozygote fitness in the area of transition in the selection gradient. From Moore and Price (1993), reprinted with the permission of Oxford University Press.

Arnold (1997) reflected a similar viewpoint, but also suggested that some hybrid genotypes could exceed the fitness of the parental types, even in the parental habitats. In the following sections, we discuss the major environment-independent and environment-dependent models of hybrid zone evolution, with particular reference to their underlying assumptions.

Dynamic Equilibrium or Tension Zone Model

The Tension Zone or Dynamic Equilibrium model has been the most widely applied, environment-independent framework for hybrid zone analyses. Defined by Barton and Hewitt (e.g., Barton & Hewitt 1985), its use led to numerous excellent mathematical and empirical studies of evolution in animal hybrid zones. Prior to the 1990s, it was the work of Barton and Hewitt and their colleagues that served most often as a conceptual framework for process-oriented analyses of hybrid zone evolution (but see Endler 1977; Moore & Price 1993).

Barton and Hewitt (1985) defined a Tension Zone (TZ) in the following way: "We will argue that most of the phenomena referred to as hybrid zones are in fact clines maintained by a balance between dispersal and selection against hybrids." The critical assumption was that stable hybrid zones were the result of intrinsic selection against hybrids balanced by the continual dispersal of the parental types into the region of overlap (and their subsequent regeneration of new hybrid individuals). Barton and Hewitt (1985) went on to argue that the steepness of transitions (i.e., clines) in allele frequencies across hybrid zones reflected the intensity of selection against hybrid genotypes, and that transitions for different markers should be coincident and concordant (i.e., the clines should be centered in the same geographic region, and should have the same shape). Finally, TZs will be characterized by significant linkage disequilibrium. In other words, selection against the disruption of parental gene combinations, coupled with the continued migration of parental individuals into the zone of contact, will result in statistically significant associations between markers from each parental type (i.e., parental gene combinations will be more common than expected by chance).

The TZ concept remains a useful null model upon which to base tests of the factors that affect the genetic structure of hybrid zones. However, most (if not all) hybrid zones now appear to be best explained by models that incorporate environment-dependent selection (Arnold 1997).

Mosaic Model

In the mid- to late 1980s, Howard and Harrison suggested that the usual depiction of hybrid zones as areas of smooth transitions between alternate forms was insufficient. Instead, Howard (1986) and Harrison (1990) recognized that genotypes were distributed across the landscape in a "mosaic" pattern, and that the distribution of these genotypes mirrored an underlying mosaic of microhabitats. The adaptation of the parental taxa to different habitats was thought to govern the genetic structure of these Mosaic Hybrid Zones (MHZs). Howard (1986) described the MHZ concept in the following manner: "The mosaic pattern…may form between two populations which have become adapted to different environments in allopatry and which meet in a region where the two environments, as well as intermediate environments, are patchily distributed." There are numerous zoological examples of MHZs. Indeed, we would argue that the vast majority of instances of hybridization result in a mosaic distribution of genotypes. For example, this is reflected by numerous cases where fish species come into patchily distributed zones of secondary contact through the connection of previously separated drainages (see Hubbs 1955 for a review).

Natural hybridization among plant species also often reflects the formation of MHZs. One such example has been identified for two Louisiana iris species, *Iris fulva* and *I. brevicaulis* (Cruzan & Arnold 1993; Arnold 1997). This example also illustrates the common observation that an environment-dependent hybrid zone can be characterized by (TZ-like) clinal variation (Arnold 1997). Figure 26.2 illustrates the pattern of genetic variation and the genetic–environmental associations in this *I. fulva* × *I. brevicaulis* hybrid zone. The pattern of genetic variation in this zone (Figure 26.2, upper panel) demonstrates coincident and concordant clinality for the species-specific chloroplast DNA (cpDNA) and all but one of the nuclear markers. The clinal pattern demonstrated by the genetic markers could have resulted from environment-independent selection against all hybrid types, balanced by dispersal of the parental forms into the hybrid zone (Barton & Hewitt 1985). However, in contrast to the predictions of the TZ model, the clinality did not reflect a changeover from one parental genotype to the other (i.e., *I. fulva* to *I. brevicaulis*), but rather a transition from one class of hybrid to another (Figure 26.2, lower panel). Indeed, the four genotypic classes represented in Figure 26.2 occurred throughout the transect. Furthermore, three of the classes were significantly associated with specific environments scattered throughout the hybrid zone (Cruzan & Arnold 1993). A recent analysis found similar genetic structure and genotypic class–environment associations in a second *I. fulva* × *I. brevicaulis* hybrid zone (Johnston et al. 2001). Clearly, these hybrid zones are best defined within the environment-dependent subset of models.

Evolutionary Novelty Model

It has been suggested that none of the earlier models explained adequately the patterns of genetic structure and habitat associations in most hybrid zones (Arnold 1997). In particular, none of these models explicitly recognized the role of both intrinsic and extrinsic selection, as well as variable, environment-dependent fitness values for hybrid genotypes. This shortcoming prompted Arnold (1997) to propose the Evolutionary Novelty model (Figure 26.3), which incorporates aspects of both environment-dependent and environment-independent selection, and allows parental and hybrid genotypes to take on a wide array of fitness values. The diversity in possible evolutionary outcomes predicted by this model results from the assumptions of recurrent hybridization in regions of range overlap, as well as the production of hybrid genotypes that exhibit a range of fitness values in either novel or parental habitats.

NATURAL HYBRIDIZATION: TWO OUTCOMES

There are numerous possible outcomes when individuals from two species meet, mate, and reproduce in the wild (e.g., Figure 26.3; Box 26.1). Perhaps the most common and evolutionarily significant of these are introgressive hybridization and hybrid speciation. In the following sections, we detail the occurrence and evolutionary effects of these two outcomes in several species complexes.

Introgressive Hybridization

For both plants and animals, one of the most common outcomes of natural hybridization involves the transfer of genomic segments between the hybridizing taxa (i.e., "introgressive hybridization" or "introgression"; Anderson & Hubricht, 1938). In fact, this process is known to occur in every major taxonomic group of both animals (e.g., fish: Doiron et al. 2002; mammals: Vilà et al. 1997; amphibians: Malmos et al. 2001; birds: Grant & Grant 2002) and plants (e.g., ferns: Kentner & Mesler 2000;

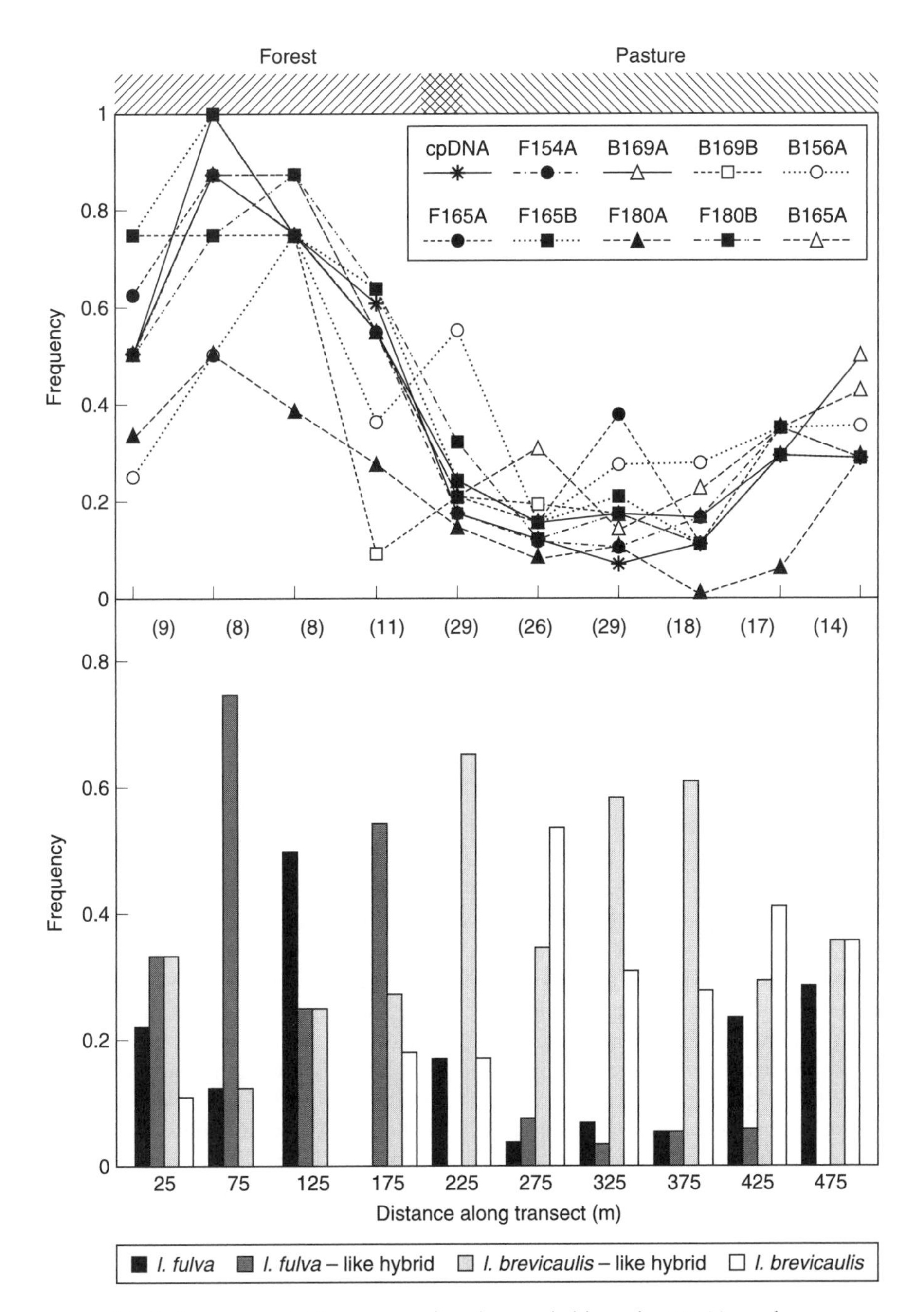

FIGURE **26.2.** Upper panel: Frequency of nuclear and chloroplast DNA markers across a spatial transect in an *Iris brevicaulis* × *I. fulva* hybrid zone. Filled shapes and open shapes represent *I. fulva* and *I. brevicaulis* markers, respectively. Lower panel: Frequency of four genotypic classes (i.e., *I. fulva*, *I. fulva*-like hybrid, *I. brevicaulis*-like hybrid, *I. brevicaulis*) along the same transect. Sample sizes are given in parentheses. From Cruzan and Arnold (1993), reprinted with the permission of *Evolution* and the Society for the Study of Evolution.

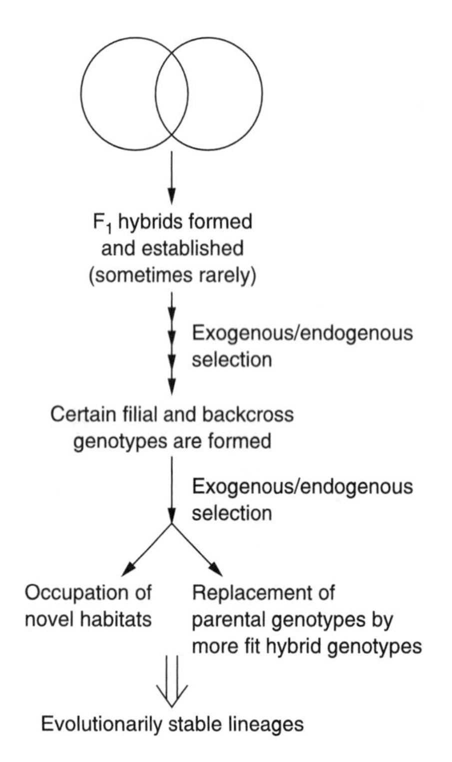

FIGURE 26.3. Evolutionary Novelty model of hybrid zone evolution. From Arnold (1997), reprinted with the permission of Oxford University Press.

gymnosperms: Wheeler & Guries 1987; monocots: Johnston et al. 2001; dicots: Sweigart & Willis 2003).

In 1949, Edgar Anderson defined the process of introgression using the Louisiana iris species complex as his "typical example." In general, Anderson concluded that introgression was evolutionarily important because it "greatly enriches variation in the participating species ... and ... would far outweigh the immediate effects of gene mutation" (Anderson, 1949, pp. 61–62). Nearly two decades after the publication of Anderson's classic work, L.F. Randolph published his own findings in a paper entitled "Negative evidence of introgression affecting the stability of Louisiana iris species" (Randolph et al. 1967). Randolph's 56-page monograph reviewed morphological and pollen fertility data collected for *I. fulva*, *I. brevicaulis*, *I. hexagona*, and natural hybrid populations. Although the data presented by Randolph and his colleagues were similar to those used by Anderson and others, Randolph et al. (1967) concluded that there was only localized hybridization between the Louisiana iris species, and that this localized gene exchange had not resulted in significant introgression among the various species.

A series of studies dating from the mid-1960s, involving the Australian fruit fly species *Dacus tryoni* and *D. neohumeralis* (now *Bactrocera tryoni* and *B. neohumeralis*), show a similar ontogeny to the Louisiana iris work. First, Lewontin and Birch (1966) suggested that introgression from *B. neohumeralis* into *B. tryoni* had facilitated the spread of the *B. tryoni*-like introgressants into habitats different from those of either parent, arguing that "introgression of genes from *D. neohumeralis* into *D. tryoni* has provided the needed variability for rapid evolution. . . . " However, like the Louisiana iris example, subsequent studies cast doubt on the occurrence of introgression. For *Bactrocera*, what had been seen as evidence for introgression was subsequently claimed to be evidence of a retained ancestral polymorphism (see Arnold 1997 for a discussion).

Over the past 15 years, reanalyses of both the Louisiana iris and the Australian tephritid fly data have pointed to a lack of resolution afforded by the genetic/morphological markers available to the earlier workers as the cause for their contradictory conclusions. Indeed, Anderson and Hubricht (1938)—decades before any of the later studies—recognized the limitations of quantitative (e.g., morphological) characters for detecting introgression between species. They concluded that the effect of repeated backcrossing into one or both hybridizing taxa was to transfer a small amount of genetic material that might "be of tremendous biological import" but that "might well be imperceptible" (Anderson 1949, p. 102). In other words, the lack of resolution resulting from their relatively "crude methods" (Anderson 1949, p. 102) made it impossible for the authors of these studies to test definitively for the occurrence of introgression in either the Louisiana irises or *Bactrocera*. This deficiency, however, disappeared with the advent of techniques for assaying DNA variation.

Case Study: Introgressing Irises

In the case of the Louisiana iris species, both nuclear (ribosomal RNA, allozyme, RAPD) and cytoplasmic (chloroplast DNA; i.e., cpDNA) loci have been employed to test for introgression between *I. fulva*, *I. hexagona* and *I. brevicaulis* (Arnold 1997). Furthermore, the comparison of population genetic variation in maternally inherited cpDNA markers and biparentally inherited nuclear markers allowed inferences concerning the initial establishment of hybrid zones via seed or pollen transfer. Contrary to the conclusions of Randolph and his colleagues, molecular analyses revealed a pattern of genetic variation consistent with the effects of both localized and dispersed introgression. More specifically, patterns of clinal variation for species-diagnostic nuclear and cpDNA markers across contemporary hybrid zones provided evidence of localized introgression, whereas the presence of the diagnostic markers in allopatric populations of the three species indicated the effect of dispersed introgression (Arnold 1997).

Molecular methods also allowed a determination of whether introgression between the Louisiana iris species was initiated by the transfer of pollen or seeds. The biological characteristics of the Louisiana iris species make plausible both of these avenues for gene exchange. Animals capable of long-distance movements (i.e., bumblebees and hummingbirds; e.g., Wesselingh & Arnold 2000) pollinate these species, and each iris species produces seeds that can float long distances in the waterways associated with their wetland habitats. Because the cpDNA of Louisiana iris species is inherited maternally (Arnold 1997), the cpDNA haplotype of seeds will identify the female parent. Thus, if hybridization were the result of the immigration of seeds from one species into allopatric populations of the alternate species, the hybrid populations should possess cpDNA haplotypes from both species. Alternatively, if only pollen was transferred between the allopatric populations, no cpDNA mixtures would be expected. Patterns of cpDNA variation in naturally occurring Louisiana iris populations have revealed no evidence for seed-mediated introgression between allopatric populations (Arnold 1997). Furthermore, seed-mediated gene flow within hybrid zones, and thus over very small spatial scales, was extremely limited (Arnold 1997). These results are consistent with the hypothesis that introgression between the various Louisiana iris species was initiated by pollen, rather than seed, transfer.

Case Study: Fraternizing Flies

A test of earlier hypotheses concerning introgression in the Australian tephritid fruit fly species was also made possible by the application of molecular techniques. Morrow et al. (2000) examined patterns of genetic variation within and between *Bactrocera tryoni*, *B. neohumeralis* and a third species, *Bactrocera aquilonis*. This study, based on two nuclear and two mitochondrial loci, revealed numerous shared polymorphisms, resulting in extremely low levels of genetic differentiation between *B. tryoni* and *B. neohumeralis* (Figure 26.4; Morrow et al. 2000). In contrast, *B. aquilonis* (a sibling species to *B. tryoni*) showed clear mitochondrial differentiation from both of the other species (Morrow et al. 2000). Taking into consideration findings from morphological, molecular, and behavioral studies, these authors concluded that introgression had, in fact, influenced the genetic differentiation (or lack thereof) between *B. tryoni* and *B. neohumeralis*, lending further support to the earlier conclusions of Lewontin and Birch (1966).

Hybrid Speciation

Instances of hybrid speciation can be placed into two categories—allopolyploidy and homoploidy—as defined by the number of chromosome complements in the hybrid taxon. The first category reflects those cases in which the hybrid derivative possesses an increase over the parental chromosome number by multiples of the haploid chromosomal set. In contrast, homoploid hybrid speciation reflects those instances in which the hybrid derivative possesses the identical, or nearly identical, chromosome number found in the parental taxa.

Allopolyploidy

Allopolyploidy has several important evolutionary consequences. First, it can result in instantaneous reproductive isolation between a hybrid lineage and its parents—the triploid offspring of a tetraploid × diploid cross are at least partially sterile due to the presence of unpaired chromosomes at meiosis. Second, it has been argued that genome duplication can generate biochemical, physiological, and developmental changes that give polyploids quite different ecological tolerances compared with their parents (e.g., see Soltis et al. 2004 for a review). Altered ecological preferences can increase the likelihood of successful establishment of an allopolyploid because it need not compete directly with its diploid parents. Third, genome duplication provides a means for stabilizing the hybrid vigor that is often associated with first-generation hybrids. Finally, genome duplication can promote a series of genetic and chromosomal changes that increase the differences between the polyploid species and its

```
                             00000000000111111111111111111222222222222222223333
                             03344688999001112234667778899000111122245678990001
                             80659247049230373624081498904124036706964435176892

                             CTCCGTAATATACTAATTCCAATTCTACTCCACTACCTCATAACCACTTG
B. tryoni
Cairns          TC1          G.........................................G.......
                TC2          ...................G......G...............G.......
                TC3          .......G...........G..................A.C.G.......
                TC4          ..................................G........T.....A
                TC5          ..................................................
                TC6          ..................................................
Tamborine       TT1          ..................................G..............A
                TT3          ...A......................................G.......
Nambour         TN1          .....G.................C..........G...............
                TN4          ..........................G...............G.......
                TN5          ..................................................
Brisbane        TB2          ...T....CC..T......G.......T.....................A
                TB4          ..................................................
                TB5          .......G...........G.............................A
                TB6          ..................................................
                TB7          ..........................................G.......
B. neohumeralis
Cairns          NC1          ..................................G...G.C..T......
                NC2          ....A.....................................GT......
                NC3          ........................G...............C.GT......
                NC4          ..................................G.....C.GT...CA.
                NC5          ..................................G.......GT......
                NC6          .................................CG........T.....A
Nambour         NN1          ...T....C...T..............T.....................A
                NN3          ..........................................G.......
                NN6          ...T....C...T..............T..............G......A
                NN7          ..................................G...............
                NN8          ........C............G......................G.....
B. aquilonis
W.A             A325.1       .......................................G..........
                A363.1       .......................................G..........
                BagD5.1      .......................................G..........
B. jarvisi
Nambour         JN1          .CTT..GC....TCCTC.TTC....G..CTTCT..TTCT..G.TTTT.CA
                JN2          .CTT..GT..CGT..TCCTT..C...G..TTCT..TTCT..G.TTTT.CA
```

FIGURE 26.4. Polymorphic nucleotide positions found for the mitochondrial gene cytochrome oxidase II isolated from the Australian tephritid fruit fly species *Bactrocera tryoni*, *B. neohumeralis*, *B. aquilonis*, and *B. jarvisi* (outgroup taxon). Names of collecting localities are given to the left of the sequence data and below each of the specific names. The numbers above the sequence data indicate the polymorphic sites, which appear below. From Morrow et al. (2000), reprinted with the permission of *Evolution* and the Society for the Study of Evolution.

diploid progenitors. These include loss of DNA, silencing or divergence of duplicated genes, and an increase in frequency of alleles that perform best in a polyploid genetic background (Soltis et al. 2004).

Allopolyploidy has long been recognized as a significant factor in plant evolution (see Grant 1981 for a review). In fact, it is estimated that a majority of flowering plants have an allopolyploid event somewhere in their phylogenetic history (Grant 1981). In contrast to plants, animal allopolyploids are typically unisexual, clonal species that are generally assumed to be evolutionarily short-lived. Maynard Smith (1992) reflected this viewpoint when he stated that the estimated 100,000 years of existence of such a taxon was "but an evening gone." Other evolutionary biologists studying animal allopolyploids have, however, emphasized the longevity of certain hybrid, unisexual lineages (e.g., Hedges et al. 1992). Regardless of how long an asexual lineage may persist, it is clear that the recombination associated with ongoing sexual reproduction in plant allopolyploids will serve to elevate the initial level of variation afforded by hybridization; this will not be the case for unisexual animals.

The establishment of allopolyploid taxa in both plants and animals reflect similar modes of origin and results in a similar population genetic structure. For example, the origin of allopolyploid species often occurs repeatedly (e.g., Moritz et al. 1989; Soltis et al. 2004), and the genetic variability of allopolyploid taxa reflects variation segregating within the parental populations.

An excellent example of an allopolyploid taxon that has arisen multiple times is the parthenogenetic gecko, *Heteronotia binoei* (Moritz et al. 1989). In this species, bisexual populations occur across most of the Australian continent, while all-female, triploid populations occur from the "center" to the western coast. Moritz et al. (1989) found a minimum of 52 genotypes among 143 triploid individuals (see Table 26.1 for a partial list of genotypes). Using chromosome banding and allozyme data, they were able to demonstrate that most of the triploid genotypes were made up of combinations of alleles found to be segregating in populations of two bisexual cytotypes (CA6 and SM6; Moritz et al. 1989). Thus, the triploid parthenogens apparently arose through multiple, independent hybridization events between the two diploid cytotypes (Moritz et al. 1989). Furthermore, the allopolyploid *Heteronotia* possess a range of genotypes due to segregating variation in the parental populations.

The plant genus *Draba* represents a botanical example showing very similar patterns to those found in the Australian geckos. Repeated origination of polyploid *Draba* taxa apparently involved crosses between different combinations of parental genotypes. As in *Heteronotia*, this resulted in high levels of genetic diversity among the allopolyploid derivatives (see Soltis et al. 2004 for a discussion). Furthermore, in *Draba* (and in *Heteronotia* and other allopolyploids), variation among the hybrid derivatives has apparently increased further by subsequent mutations, recombination, and/or gene flow (Moritz et al. 1989; Soltis et al. 2004).

Homoploidy

Reports of homoploid hybrid speciation in plants are relatively widespread, albeit often controversial (Rieseberg 1997). In animals, on the other hand, there are few proposed homoploid hybrid species. One such case involves the cyprinid fish genus *Gila*; a genus typified by numerous examples of introgressive hybridization (Dowling & DeMarais 1993). In addition, one species, *Gila seminuda*, has been

TABLE 26.1. Occurrence of cytotypes and multilocus allozyme genotypes found in 15 (of 43 sampled) populations of parthenogenetic geckos.

Locality	*N*	Cytotype/genotype
1	6	A-e, A-g, A-h, A-t, A-v, A-y
2	4	A-w, A-y
3	4	A-b, A-k
4	2	A-f, A-x
5	5	A-u, A-y, C-j
6	3	A-o, C-l, C-j
7	2	A-a, A-c
8	1	A-k
9	1	A-y
10	3	B-l, B-k
11	2	B-k, B-l
12	3	B-n, C-k
13	3	B-m, B-n, C-l
14	2	B-m
15	7	B-m, B-n

Uppercase letters indicate chromosomal cytotype and lowercase letters indicate multilocus genotype. From Moritz et al. (1989).

identified as a diploid, bisexual hybrid derived from crosses between *G. robusta* and *G. elegans* (Dowling & DeMarais 1993). *Gila seminuda* thus possesses an intermediate morphology, segregates for diagnostic isozyme alleles from *G. robusta* and *G. elegans*, and is characterized by a mitochondrial haplotype that is nearly identical to that found in *G. elegans* (Dowling & DeMarais 1993).

If we accept, as has been argued elsewhere (e.g., Arnold 1997), that the formation of hybrid taxa is evolutionarily significant regardless of the taxonomic level, then we can also use hybrid subspeciation to illustrate the process of homoploidy. For example, Franck et al. (2000) detected patterns of genetic variability in the honeybee subspecies *Apis mellifera ligustica* and *A. m. sicula* that were consistent with an hypothesis of hybrid origin. The nuclear genomes of these Italian taxa (*A. m. ligustica* from continental Italy and *A. m. sicula* from Sicily) were characteristic of other "southeastern European" subspecies. In contrast, their mitochondrial genomes were more similar to those found either in distant European populations or in Africa. This pattern of discordance between cytoplasmic and nuclear markers is stereotypical for cases of introgressive hybridization, where it is often found that the cytoplasmic markers introgress while the nuclear markers do not (Harrison 1990). Because additional subspecies of *A. mellifera* also demonstrated genetic patterns indicative of homoploid origin, Franck et al. (2000) concluded that natural hybridization "played an important role in the evolutionary history" of this complex.

Returning now to homoploid hybrid speciation in plants, the best-characterized examples belong to the sunflower genus *Helianthus*. This genus contains three well-defined homoploid hybrid species: *H. anomalus*, *H. deserticola* and *H. paradoxus*. The origin of each of these hybrid species has apparently involved recombinational speciation (i.e., the fixation of a set of chromosome rearrangements resulting in the reproductive isolation of the hybrid derivative; Grant 1981; Rieseberg 1997). As with each of the other two homoploid species, *H. anomalus* originated from hybridization between *H. annuus* and *H. petiolaris* (Rieseberg 1997). Genetic map-based analyses have demonstrated that *H. anomalus* chromosomes are mosaics of those of the two parental taxa (Figure 26.5). This suggests that the origin of *H. anomalus* involved extensive recombination among the parental linkage groups. Furthermore, as predicted by the recombinational speciation model, genomic mapping of *H. anomalus* discovered a unique combination of the chromosomal structural differences (i.e., translocations and inversions) that distinguish *H. annuus* and *H. petiolaris* (Rieseberg 1997).

As noted above, at least two other homoploid speciation events (along with occurrences of homoploid raciation) have been identified in the annual *Helianthus* group. The numerous occurrences of hybrid taxa led Rieseberg (1997) to conclude that evolution in *Helianthus* reflects a reticulate evolutionary history. In addition to exemplifying the predictions of the model of recombinational speciation, these taxa also reflect the outcome of ecological speciation. Thus, each of these taxa possesses novel ecological adaptations that likely arose as a consequence of their hybrid origin (e.g., Rieseberg et al. 2003).

THE EVOLUTIONARY NOVELTY MODEL AND EVOLUTIONARY OUTCOMES

As noted earlier in this chapter, the Evolutionary Novelty model assumes that hybrid genotypes will demonstrate a range of fitness values; while some hybrid genotypes will be less fit than their parents, others will possess levels of fitness equivalent to their progenitors. Some hybrid genotypes will even surpass their parents in fitness (Figure 26.3). The relative fitness of the various hybrid genotypes may be endogenously determined (as in the TZ model), or governed by environmental variables. We will now examine how well these expectations fit patterns observed in cottonwoods, Louisiana irises, Darwin's finches, and sunflowers.

Whitham and his colleagues have developed the Fremont (*Populus fremontii*) and Narrowleaf (*Populus angustifolia*) cottonwood species complex into a model system for investigating natural hybridization (e.g., Schweitzer et al. 2002). One of their findings, consistent with an assumption of the Evolutionary Novelty model, is that hybrid individuals demonstrate various levels of fitness relative to their Fremont and Narrowleaf parents. Significantly, some hybrids possess a combination of the life history traits of both parental species (Schweitzer et al. 2002). This amalgamation of life history traits results in the hybrids possessing a higher fitness than either parent across a variety of habitats. A similar finding has also been made in the Louisiana iris species complex. Burke et al. (1998) discovered the

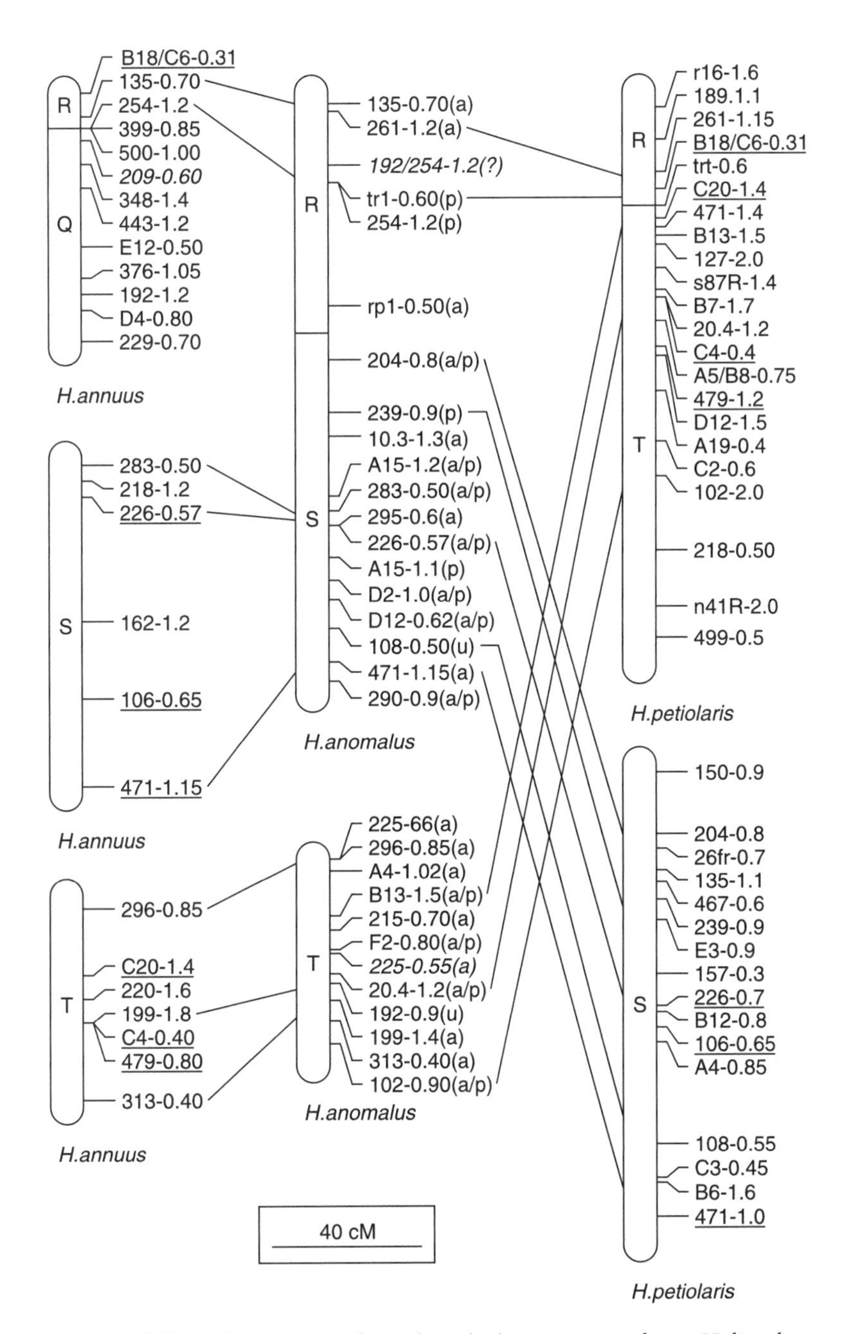

FIGURE 26.5. Linkage maps for selected chromosomes from *Helianthus annuus*, *H. petiolaris* and their hybrid derivative, *H. anomalus*. Letters within each linkage group indicate chromosomal blocks and their relationship with homologous blocks in other species. Numbers and letters to the right of the *H. annuus* and *H. petiolaris* chromosomes indicate locus designations. The designations "a" and "p" to the right of the *H. anomalus* chromosomes indicate a locus found in either *H. annuus* or *H. petiolaris*, respectively. From Rieseberg et al. (1995).

highest sexual and asexual reproductive output for F_1 hybrids rather than for their parents, *I. hexagona* and *I. fulva*. As with the *Populus* study this was a reflection of the F_1s possessing a unique combination of traits associated with the alternate life history strategies of these two species (i.e., the high sexual and asexual reproductive output of *I. fulva* and *I. hexagona*, respectively). The following quote from Schweitzer et al. (2002) summarizes well the findings from both the *Populus* and *Iris* studies: "It appears that hybrids benefit from both the introgression of ... traits for larger seed size and ... traits, for increased asexual reproduction."

Another example that illustrates patterns predicted by the Evolutionary Novelty model comes from studies of Darwin's finches. For 30 years Peter and Rosemary Grant have detailed the ecological and evolutionary trajectories of these species in their Galapagos Island habitats (Grant & Grant 2002). Over the past three decades, and associated with major vegetation changes, these authors have documented a fitness transition for hybrid offspring of *Geospiza fuliginosa*, *G. fortis*, and *G. scandens* (e.g., Grant & Grant 2002). The transitions in habitat and relative fitness of parental and hybrid individuals were caused by an El Niño event. Prior to the El Niño, hybrid individuals were rarely hatched and did not reproduce. Subsequent to the El Niño perturbation, F_1 and backcross individuals demonstrated equivalent or higher fitness relative to their parental species for survivorship, recruitment, and breeding. The specific factor causing this switch in fitness for parental and hybrid individuals appeared to be a changeover in the array of available seeds. Hybrid beak morphologies facilitated more efficient use of the new array of seeds, resulting in higher fitness. Hybridization thus provided novel genetic variation (Grant & Grant 2002), giving rise to hybrid phenotypes that were more fit in a newly created habitat. The *Geospiza* studies exemplify well one prediction of the Evolutionary Novelty model: introgression can lead to hybrid genotypes that have higher fitness, relative to parental genotypes, in novel habitats (e.g., Lewontin & Birch 1966).

As with the cottonwood and Darwin's finch species, natural hybridization between annual sunflower species has given rise to some hybrid genotypes that demonstrate increased fitness. In addition, as detailed above, natural hybridization between *Helianthus annuus* and *H. petiolaris* has led to the production of three hybrid species (Rieseberg 1997). Yet, hybridization between *H. annuus* and *H. petiolaris* also demonstrates an apparent paradox of hybrid species formation; unfit early generation hybrids are often the foundation for new evolutionary lineages. Though seemingly counterintuitive, this makes sense; stronger reproductive isolation between parental lineages typically translates into stronger isolation between a hybrid lineage and its parents. Thus, although more difficult to form, hybrids between strongly isolated lineages stand a much greater chance of surviving in the face of ongoing reproductive contact with their parents. In this regard, experimental crosses between *H. annuus* and *H. petiolaris* result in F_1 individuals with pollen fertilities of 0–30% and F_2 and first backcross generation individuals that produce a maximum of 2% viable seed (see Rieseberg 1997 for a review). In spite of the extremely low fitness of the initial hybrid generations, crosses between these species are ongoing, resulting in numerous, widespread, contemporary hybrid zones (Rieseberg 1997). Most significantly, however, Loren Rieseberg and colleagues have confirmed the formation of stable hybrid lineages that possess novel ecological adaptations (Rieseberg et al. 2003). The evidence for these conclusions comes from (i) genomic mapping of experimental hybrid populations and the naturally occurring hybrid species and (ii) QTL mapping of traits associated with the novel habitats occupied by the hybrid species (Rieseberg et al. 2003). Thus, hybrid sunflower genotypes have formed that are adapted to novel habitats, habitats in which the parental forms are less fit (Rieseberg et al. 2003)—a factor which further reduces the likelihood that new hybrid lineages will be driven to extinction by parental gene flow.

FUTURE DIRECTIONS

Although great progress has been made in the investigation of the evolutionary importance of natural hybridization, our understanding of the role of hybridization in the evolutionary process is far from complete. For example, estimates of hybrid fitness under natural conditions are still lacking for all but a few systems, yet models that predict the outcome of hybridization rely heavily on assumptions concerning hybrid fitness. Along these same lines, there is presently a dearth of mathematical models that incorporate both genetic and ecological components of reproductive isolation. In view of the importance of both of these types of reproductive barrier, it seems likely that such models will prove to be an absolute necessity if we ever hope to make realistic predictions regarding the ultimate fate of hybridizing taxa. Finally, we are beginning to see a transition from the more or less descriptive studies that have dominated the hybrid zone literature to an increasingly mechanistic approach, wherein researchers seek to elucidate the molecular basis of various aspects of reproductive isolation. Indeed, a number of recent studies have begun to shed light on the genetic architecture of traits that may limit or promote gene flow (e.g., Ting et al. 1998; Fishman et al. 2002), and the increasing availability of the molecular tools necessary for such investigations should result in a dramatic increase in both the frequency and level of detail of these sorts of studies in the coming years.

SUGGESTIONS FOR FURTHER READING

Barton and Hewitt's (1985) review of hybrid zone theory has served as the jumping-off point for many of the more recent investigations into the evolution of hybrid zones. In particular, the authors argued that "the great majority [of hybrid zones] are maintained in a stable balance between dispersal and selection," thereby making their Tension Zone model the de facto null hypothesis in a great many studies. Arnold (1997) provides a comprehensive review of investigations into the evolutionary importance of natural hybridization, and argues forcefully that it can play a significant (and creative) role in evolutionary diversification. Rieseberg (1997) evaluates the evidence for species formation via hybridization in plants, and reviews the cases for which there exist convincing evidence of this mode of origin. Finally, Burke and Arnold (2001) review what is known about the genetics underlying variation in hybrid fitness in plants and animals.

Arnold ML 1997 Natural Hybridization and Evolution. Oxford Univ. Press.
Barton NH & GM Hewitt 1985 Analysis of hybrid zones. Annu. Rev. Ecol. Syst. 16:113–148.
Burke JM & ML Arnold 2001 Genetics and the fitness of hybrids. Annu. Rev. Genet. 35:31–52.
Rieseberg LH 1997 Hybrid origins of plant species. Annu. Rev. Ecol. Syst. 28:359–389.

Acknowledgments We wish to thank A. Bouck and S. Cornman for reviewing this chapter. M.L.A. was supported by NSF grants DEB-0074159 and DEB-0345123 and J.M.B. was supported by awards from the USDA (#03-35300-13104) and NSF (#DBI-0332411).

27

Population Bottlenecks and Founder Effects

LISA MARIE MEFFERT

Patterns of geographic variation arise from the combined effects of migration, selection, mutation, and genetic drift. One measure of the evolutionary potential for such geographic differentiation is the level of additive genetic variance (i.e., the linear resemblance between relatives, hereafter denoted V_A; see Roff, Ch. 18 of this volume; Cheverud, Ch. 19 of this volume). Migration has relatively straightforward effects that increase genetic variation within populations. In contrast, directional selection should exhaust V_A for fitness traits, with the residual genetic variance hidden within the nonadditive components of genetic variance (see below) and maintained by some balance between selection and mutation (see Roff, Ch. 18 of this volume). Genetic drift, however, can have less predictable effects on levels of V_A, particularly in the cases of bottleneck or founder events (i.e., involving severe reductions in population size). In a purely additive genetic model (i.e., no nonadditive genetic effects), a population bottleneck should reduce V_A, on average, by $1/2N_e$, where N_e is the number of founders. Models that address nonadditive genetic effects, however, have shown that bottlenecks can increase V_A, in contrast to purely additive theory (Goodnight 1988; Cheverud & Routman 1996; and see below).

A major component in the study of geographic differentiation involves the bottleneck consequences of colonization. For example, serial founder-flush events are thought to be critical in the cycles of glacial spread and retreat that have generated geographic variation in plants (e.g., Cwynar & MacDonald 1987) and animals (e.g., Hard et al. 1993). There are several basic models for the kinds of altered selection pressures that could occur through these colonization events (see Box 27.1). First, small population size effectively affords neutrality to alleles at loci that would have strong selective effects in large outbred populations. This selective neutrality results in an increased potential to fix alleles, and, as noted above, reduces V_A, on average. When considering those loci that retain genetic variation, particularly those with nonadditive effects, bottlenecks can increase V_A with both adaptive and maladaptive consequences (detailed below). This kind of genetic reorganization of intra- and inter-locus interactions alters selective pressures in the internal genomic environment (Box 27.1), potentially opening evolutionary pathways that were inaccessible to the ancestor. With regard to external selective pressures, the act of colonizing new habitat is likely to introduce a population to a different suite of pressures from the environment. Moreover, shifts in external selection pressures can occur even in habitat that was previously occupied, particularly when the reduction in population size reduces intra- and/or interspecific competition. In such situations, reduced selection, particularly during the flush phase, could promote increased recombination and sorting among alleles, again with the potential to open novel evolutionary trajectories.

Bottleneck models of speciation extend these mechanisms of geographic differentiation to the stimulation of speciation, particularly through shifts in reproductive isolating mechanisms (see Johnson, Ch. 24 of this volume). The classical proposed scenario of founder-induced speciation involves the radiation of the Hawaiian *Drosophila*. Over 25% of all *Drosophila* species worldwide are endemic to the Hawaiian islands, although the islands are known to

BOX 27.1. Models of the Shifts in Selection Experienced by Bottlenecked Populations

The most fundamental genetic consequence of a population bottleneck is that there will be an increased effect of drift, and thus an increased probability of alleles being driven to fixation (Falconer & Mackay 1996). In the simplest scenario, the maximum number of alleles that can be retained for a single-pair bottleneck is four (in a diploid outcrossing organism). This process reduces average heterozygosity and levels of additive genetic variance by a function of $1/2N_e$, where N_e is the number of founders (Falconer & Mackay 1996). Importantly, alleles that would have deleterious effects in a large, outcrossing population can be fixed (e.g., see Hedrick 2005). Thus, a crash in population size effectively causes alleles to evolve as though they had neutral effects on fitness. Importantly, relatively moderate bottlenecks may also promote the fixation of beneficial alleles (Day et al. 2003). In particular, deleterious recessive alleles are difficult to purge in large populations because they can remain hidden in the heterozygote carriers (Falconer & Mackay 1996). Thus, the selective neutrality that bottlenecks confer can have some combination of adaptive and maladaptive effects, depending upon the severity of the bottleneck and the amount of genetic load in the ancestral population.

Quantitative genetic models generally focus on the loci that retain allelic diversity. In such approaches, the most relevant parameter is the level of additive genetic variation (V_A), since it is a measure of the evolutionary history and evolutionary potential of a trait. For traits structured by dominance, bottlenecks can cause increased frequencies of detrimental recessive alleles (Willis & Orr 1993)—those alleles that are harbored at rare frequencies in the outbred ancestor (see text). The increased frequencies of these alleles thus increases V_A, and thus the evolutionary potential for the trait. Such shifts are not so much adaptive in promoting geographic differentiation or speciation, but can threaten a population with extinction (Barton & Charlesworth 1984). For traits structured by epistasis, bottlenecks can increase the frequencies of allelic combination across loci that were once rare. Depending upon the form of epistasis, such shifts can facilitate evolution along novel, viable trajectories or increase inbreeding depression effects (Goodnight 1988; Charlesworth 1998).

These various mechanisms that can fix alleles or increase V_A can cause significant alterations in the architecture of internal selective pressures within the genome. In particular, networks of pleiotropic and epistatic interactions dictate genetic intercorrelation structure, and thus define the patterns of evolutionary potential and constraints. More specifically, genetic correlations are responsible for correlated shifts in selection responses across traits (e.g., Bohren et al. 1966), reducing the theoretical universe of the multivariate phenotypes that can evolve. Bottleneck reorganizations of the genome can alter genetic intercorrelations (e.g., Bryant & Meffert 1988), and thus promote differentiation along evolutionary pathways that were inaccessible to the ancestor, which, in turn, alters the interactions among alleles (see Templeton 1996). In this way, alleles are responding to the shifts in the selective pressures of the genomic environment.

Bottleneck and founder-flush models also address the effects of shifts in the external environment. As suggested for the Hawaiian *Drosophila*, the process of colonizing new habitat is likely to expose the founding population to a new suite of selection pressures from the external environment, such as those that influence development or feeding (Kaneshiro 1988). For example, many species Hawaiian *Drosophila* have adapted specialized feeding phenotypes in response to the novel selective pressures of the tropical rainforest (White 1978). The genetic perturbations of founder events are thought to have promoted the rapid evolution of these specializations. External selection pressures can shift even in habitat that is similar to the ancestral environment. In particular, a dramatic decrease in population size could alter intraspecific and interspecific

(continued)

BOX 27.1. *(cont.)*

competition, such as for food or space. Kaneshiro (1988), for example, proposed that the competitive pressures of the "sexual environment" can shift during founder-flush, altering patterns of mate discrimination.

The mechanisms of shifting internal and external selection pressures are not mutually exclusive. For example, Carson (1982) suggested that reduced external selection pressure during the flush phase would promote increased recombination and sorting of alleles across loci. This, in turn, would alter the internal genomic selection pressures, generating new epistatic associations, with the potential to promote adaptive responses along novel evolutionary trajectories.

Detractors of bottleneck models note that theoretical work shows that a single bottleneck can only rarely stimulate a viable evolutionary shift, and that large populations are more responsive to selection pressures than those that have undergone bottlenecks (Barton & Charlesworth 1984; Turelli et al. 2001). Moreover, empirical work has found that natural selection alone can account for patterns of geographic differentiation and speciation, without invoking bottleneck effects (e.g., Clegg et al. 2002). I suggest that the more salient issue concerning the adaptive effects of population bottlenecks may not be how common they are in nature, but whether or not founder events involve special mechanisms for the rapid evolution of evolutionary novelties.

be relatively young (i.e., the youngest is less than 500,000 years old; White 1978). In founder-flush speciation, a very small number of individuals successfully establish a population by either colonizing a new environment or repopulating the original habitat after some catastrophe (i.e., undergoing a flush phase of population growth after a founder event; Giddings et al. 1989). For the Hawaiian *Drosophila*, it has been proposed that a few founders from the mainland colonized the islands, such as by riding on storm winds or floating on debris, followed by a number of founder events among the islands (see Kaneshiro 1988). Figure 27.1 shows one network of founder-flush events proposed to have occurred after colonization from the mainland an estimated 5 million years ago (White 1978). Using karyotype analyses, Carson et al. (1970) postulated that a minimum of 22 such founder-flush events could account for the radiation of the Hawaiian Drosophilidae. The serial founder-flush events are thought to have opened up evolutionary avenues for divergence in the behavioral and ecological traits that characterize these distinct species (Box 27.1; Kaneshiro 1988).

Founder-flush speciation models vary in their specifics, but all describe some form of genetic reorganization that opens up avenues for the evolution of complex traits, such as those involved in reproductive isolation (Mayr 1954; Templeton 1980; Carson 1982). The simplest model uses the purely additive expectation that, on average, V_A, will be reduced by the function of $1/2N_e$, where N_e is the number of founders (see Goodnight, Ch. 6 of this volume). In the original founder-flush model, Mayr (1954) expanded this principle by noting that a

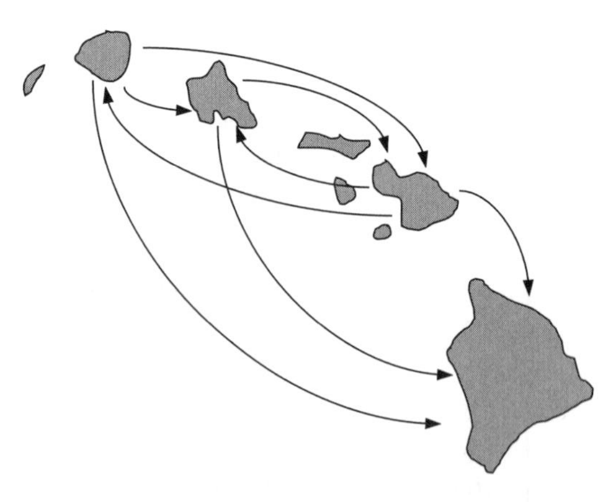

FIGURE 27.1. A proposed scenario for the founder-flush colonization events of the Hawaiian *Drosophila* (adapted from Figure 10 in White 1978). Carson et al. (1970) concluded, based upon karyotype analyses, that a minimum of 22 colonizations occurred.

small random sampling of individuals can found a population composed only of alleles and allelic combinations across loci that were once rare, promoting geographic differentiation. The opportunity for speciation has been enhanced if this derived genetic composition alters reproductive or ecological traits that can define new species. Carson (1982) developed this model further to explain how the evolutionary lineages of Hawaiian *Drosophila* species can be deciphered from their karyotypes. In particular, he showed how the formation of novel karyotypic conformations in colonizing populations could influence the evolution of reproductive isolation (e.g., through hybrid inviability). In the same vein, Kaneshiro (1980) suggested that the evolutionary histories of the Hawaiian *Drosophila* could be surmised through the patterns of lost behavioral elements in courtship repertoire (but see Meffert & Bryant 1991).

Later models of founder-flush speciation (Templeton 1980; Carson 1990) focused more on bottleneck effects on the genetic variance components of quantitative genetic traits (i.e., those affected by more than one gene; see Goodnight, Ch. 6 of this volume). While additive quantitative genetic models also predict a reduction in V_A in proportion to the size of the founder population, it is unlikely that all of the loci that influence a trait will lose all of their genetic variation. These models thus focus more on the shifts in the frequencies of alleles at the loci that retain genetic variation, rather than those loci that simply become fixed for one allele. In these models, a founder event (or bottleneck; see details on the distinctions below) can result in increased rather than decreased V_A, especially when the nonadditive genetic processes (i.e., dominance or epistasis; see below) are involved (Goodnight 1988; Willis & Orr 1993; Cheverud & Routman 1996). More specifically, theory predicts that V_A for fitness traits should be minimized by selection and camouflaged under the nonadditive genetic components of genetic variance. A bottleneck that increases the frequency of the rare alleles or allelic combinations then actually increases V_A, and, thus, evolutionary potential, by making the frequencies of the alleles or allelic combinations more equal (e.g., Willis & Orr 1993).

Founder-flush speciation is perhaps the most experimentally tractable mode of speciation because the major genetic disruptions are expected to occur in only a few generations (see Chippindale, Ch. 31 of this volume). Speciation via ploidy (or parthenogenesis; see Templeton 1979) can be even faster but does not apply generally to obligately sexual organisms. The studies that have tested founder-flush theory used model experimental systems, such as flies, because they are easily reared in the laboratory and have generation lengths of only a few weeks (Table 27.1). Additionally, small space requirements permit large-scale experiments on multiple populations. In these studies, a few founders, such as one or two male–female pairs, are randomly drawn from a stock population. These founders then "colonize" a new cage, and the populations are allowed to flush to a large population size over the course of a few generations. In some studies, the populations are subjected to additional founder-flush events (e.g., Meffert & Bryant 1991). Replicate founder-flush populations are then tested for evidence of emergent speciation through: (i) shifts in the genetic variances (and covariances, see below) of complex traits, (ii) assortative mating preferences (i.e., females preferentially mating with their own males), or (iii) divergent courtship repertoires (Table 27.1). There is substantial theoretical and empirical support for increased evolutionary potential and reproductive isolation in bottlenecked populations (e.g., see Wolf, Box 6.1 of this volume, and below), but appreciable controversies exist as to whether such evolutionary shifts constitute the catalysis of speciation proposed by founder-flush models (Box 27.1).

CONCEPTS

Quantitative Genetic Models of Bottleneck Effects

Figure 27.2 depicts the theoretical shifts in V_A under basic quantitative genetic models of pure additivity (i.e., the alleles at a locus sum linearly in their effects), pure dominance (as defined by Falconer & Mackay 1996; i.e., one allele at a locus completely masks another), and additive-by-additive epistasis (i.e., two or more additive loci interact in their net effects; see Wolf, Box 6.1 of this volume). In each case here, two representative loci are modeled, with each locus having two alleles. The surfaces represent the levels of V_A for the range of all possible allele frequencies with these two-locus, two-allele systems (see Wayne & Miyamoto, Ch. 2 of this volume). Under pure additivity, the level of V_A at each locus is given simply by $2pq$, where p and q are the alternate allele frequencies (Falconer & Mackay 1996).

TABLE 27.1. Conclusions found in experimental tests of founder-flush speciation

Species	Increased V_A for complex traits?	Reproductive isolation?	Divergent mating behavior?	Reference
Drosophila melanogaster	–	Yes	–	Hay 1976
Drosophila melanogaster	Yes	–	–	Lints and Bourgois 1982
Drosophila melanogaster	Yes	–	–	López-Fanjul and Villaverde 1989
Drosophila mercatorum	–	Yes	–	Templeton 1979
Drosophila pseudoobscura	–	Yes	–	Ehrman 1969
Drosophila pseudoobscura	–	Yes	–	Powell 1978
Drosophila pseudoobscura	–	Yes	–	Galiana et al. 1993
Drosophila simulans	–	Yes	Yes	Ringo et al. 1986
Drosophila spp.	Yes	–	–	Carson 1990
Musca domestica	Yes	Yes	Yes	Meffert 2000 for a review
Tribolium spp.	Yes	–	–	Wade 1991
Drosophila melanogaster	–	No	–	Averhoff and Richardson 1974
Drosophila melanogaster	–	No	–	Rundle et al. 1998
Drosophila pseudoobscura	–	No	–	Powell and Morton 1979
Drosophila pseudoobscura	–	No	–	Galiana et al. 1993

"—" indicates that the study had neither supporting or refuting evidence of the particular founder-flush phenomenon. The authors' conclusions on the presence or absence of the founder-flush predictions are identified by "Yes" and "No," respectively.

The total effects across the two loci are then summed to yield the net V_A (Figure 27.2A). For loci with pure dominance, V_A at each locus is $8pq^3$ (where p and q are the frequencies of the dominant and recessive alleles, respectively; Falconer & Mackay 1996), again with the effects across the two loci being summated (Figure 27.2B). Finally, V_A for a simple form of epistasis (i.e., additive-by-additive epistasis) is calculated as $(V_{A\,1}\,(|q_1 - q_2|)^2 + V_{A\,2}\,(|p_1 - p_2|)^2)/4$ (see Cheverud & Routman 1996; Figure 27.2C). In this formula, $V_{A\,1}$ and $V_{A\,2}$ are the additive genetic variances for the first and second loci, respectively, as defined independently by the purely additive model for a single locus (i.e., $2pq$, see above). The alternate allele frequencies for the first locus are p_1 and q_1, with the alternate allele frequencies for the second locus being p_2 and q_2. Models of epistasis that incorporate dominance (i.e., dominance-by-additive and dominance-by-dominance epistasis) have been expanded in a similar manner by Cheverud and Routman (1996).

In a purely additive system, the maximum V_A occurs when the two alleles at each locus are in equal frequency ($p = q = 0.5$; Figure 27.2A). A bottleneck

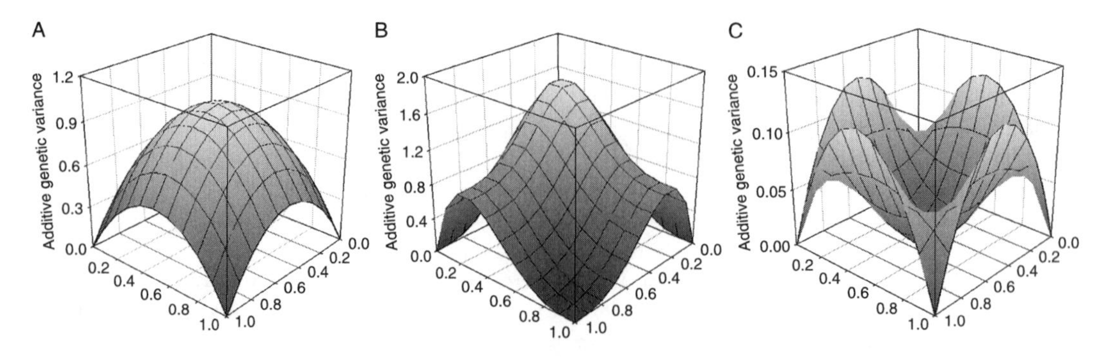

FIGURE 27.2. Levels of additive genetic variance under the models of (A) pure additivity, (B) pure dominance (where $p = 1.0$ for the fixation of the dominant allele), and (C) additive-by-additive epistasis. (A) and (B) are adapted from Falconer and Mackay (1996), and (C) is adapted from Cheverud and Routman (1996). The models depict the additive genetic variances for the range of allele frequencies for a two-locus system, with each locus having two alleles.

that shifts the allelic frequencies away from parity, toward fixation of one allele or the other, will decrease V_A. For a neutral trait, there is no a priori expectation for the allele frequencies in the ancestral population. Thus, all possible initiation points for the allele frequencies are equally likely in a pre-bottlenecked population. However, the concave shape of the V_A space (Figure 27.2A) shows that the characteristic steep declines at any one locus, decreasing V_A, will overcompensate for relatively shallow inclines at another locus, increasing V_A. Therefore, the average expectation is that a bottleneck will decrease V_A for additive quantitative traits.

With pure dominance, the maximum V_A at each locus occurs for a ratio of 1:3 for the allele frequency of the dominant allele relative to that of the recessive allele ($p = 0.25$, $q = 0.75$; Figure 27.2B). If the recessive alleles have negative fitness consequences (i.e., deleterious states for homozygous recessives) then the a priori expectation is that a non-bottlenecked population will have low frequencies of these recessive alleles. More specifically, selection on a large outbred population will purge the deleterious recessive alleles through time up to the point of a mutation–selection equilibrium (see the front corner of Figure 27.2B). Purging the genetic load in large populations, however, is hindered by the way that the recessive alleles are masked and passed on by the heterozygote individuals (Falconer & Mackay 1996). Consequently, recessive alleles can remain in a large population but will be rare. A bottleneck that increases the frequencies of the rare recessive alleles thus increases V_A (Figure 27.2B). Relatively slight increases in the frequencies of rare recessive alleles, increasing V_A, overcompensate for the reduced V_A that results from the decreased frequencies (or elimination) of the recessive alleles at other loci (Falconer & Mackay 1996). Thus, the average expectation is that V_A will increase for quantitative traits that are structured by directional dominance (when the recessive alleles are deleterious). This process of extracting V_A from a nonadditive genetic variance component has been termed "conversion" (see Wolf, Box. 6.1 of this volume). Conversion of purely dominant traits, however, causes an overall reduction in the average fitness of a population. Therefore, this kind of conversion does not so much catalyze speciation, as it more likely threatens a population with extinction (Barton & Charlesworth 1984).

The conversion of epistatic genetic variation is considered to be more effective in stimulating founder-flush speciation. With epistasis, the combination of alleles across loci determines the phenotype, so different genetic combinations can produce comparable fitness levels (Wright 1977; Goodnight 1988; Cheverud & Routman 1996). Selection toward a fitness optimum will fix certain allelic combinations, causing the frequencies of equally viable genotypes to become rare. In the simplified model presented here (Figure 27.2C), a population under such selection pressures could equilibrate at one of the "corners" of the V_A space. A bottleneck can increase V_A in an evolutionarily adaptive way by increasing the frequencies of the rare, but viable, genotypes (to promote a jump to a new corner, Figure 27.2C; see also Wolf, Box 6.1 of this volume). The conversion of all three components of di-genic epistatic variance (e.g., additive-by-additive, additive-by-dominance, and dominance-by-dominance epistasis) can contribute to increased V_A in bottlenecked populations (Cheverud & Routman 1996). Epistasis can be particularly important to the process of speciation when the merging of favorable epistatic combinations results in outbreeding depression (i.e., reduced fitness of the hybrids; see Michalak & Noor, Ch. 25 of this volume). Note that increased epistatic inbreeding depression is also a potential consequence of a bottleneck (Charlesworth 1998). Thus, partitioning the influences of the various genetic variance components is critical for interpreting the evolutionary significance of inflated V_A in bottlenecked populations.

Complex traits, such as those important to speciation, are intercorrelated due to linkage and pleiotropy (see Phillips & McGuigan, Ch. 20 of this volume). Besides increased V_A for individual quantitative traits, bottlenecked populations can thus exhibit increased evolutionary potential in a multifactorial sense. In particular, a non-bottlenecked population that is at fitness optima for a suite of genetically correlated traits will have reduced evolutionary accessibility to the multifactorial phenotypic space. Bottlenecks that increase the V_A of one trait can cause shifts in correlated traits and thus open up avenues for the evolution of novel multifactorial phenotypes while also closing down others (see Phillips & McGuigan, Ch. 20 of this volume; Regan et al. 2003; Box 27.1). It is important to consider the gamut of potential bottleneck effects for evaluating the efficacy of founder events to promote geographic differentiation or speciation.

CASE STUDIES

Nonadditive Genetic Effects in the Geographic Differentiation of the Pitcher Plant Mosquito, *Wyeomyia smithii*

An impressive amount of work has been conducted by Bradshaw, Holzapfel, and colleagues on the post-glaciation recolonization of the pitcher plant mosquito, *Wyeomyia smithii* (e.g., Hard et al. 1993). These insects are tightly affiliated with their host, the purple pitcher plant, which is found in isolated bogs ranging from the Gulf of Mexico to Canada. Abundant geographical, morphological, behavioral, and physiological evidence supports the thesis that *W. smithii* recolonized North America on a latitudinal gradient from south to north (Hard et al. 1993). Adaptive differentiation has been identified along this gradient for life history traits such as photoperiodism (i.e., the shift in day length that triggers pupation; Hard et al. 1993). The genetic variance for neutral molecular genetic markers decreases along this gradient, in support of the model of serial founder-flush events from south to north, as predicted by a purely neutral model of bottleneck effects (Box 27.1; Armbruster et al. 1998). The V_A for the fitness traits, however, increases along this cline, in support of the model of conversion of V_A from the nonadditive components of genetic variance (Armbruster et al. 1998). Using line-cross analyses, Hard et al. (1993) identified the influence of epistasis directly. Thus, both the neutral and nonadditive models of bottleneck effects are implicated in the patterns of geographic differentiation in this system, depending upon the trait in question.

Experimental Support for Founder-Flush Theory

A number of studies have used the housefly, *Musca domestica* L., as an experimental model to test bottleneck theory, especially with regard to the process of speciation. In each experiment, replicate populations were derived from a few founders and allowed to flush to the size of parallel non-bottlenecked populations (i.e., the controls). In several experiments, founder-flush populations exhibited increased V_A for morphometrics and mating behavior relative to the controls (see Table 1 in Meffert 2000 for a review). Additionally, serial founder-flush cycles caused significant shifts in the genetic variance–covariance structure among morphological traits, which were related to patterns of phenotypic divergence along evolutionary trajectories that were ostensibly inaccessible to the ancestor (see Table 1 in Meffert 2000). Selection experiments on morphological shape (see Meffert 2000) and assortative mating (Regan et al. 2003) also showed that founder-flush populations could have comparable or greater selection responses relative to their non-bottlenecked controls. As in the morphometric studies, the mating behavior of founder-flush populations diverged along the minor multifactorial axes of courtship "shape" (see Meffert 2000; Regan et al. 2003), suggesting that covert V_A had been released from the nonadditive structure of the ancestor (see Meffert 2000). These experiments also identified significant assortative mating biases in founder-flush populations (see Meffert 2000). Moreover, dominance and epistasic contributions have been confirmed for the morphometric and mating behavior characters targeted by these studies (see Meffert 2000; Meffert et al. 2002).

Figure 27.3 shows how bottleneck increases in V_A can translate to increased evolutionary potential for housefly mating behavior. In this experiment, six replicate founder-flush populations were derived from the offspring of exactly two randomly selected male–female pairs (a different set of founders was used to create each founder-flush population). As noted above, the populations were allowed to flush to several thousand individuals, the size of the two non-bottlenecked control populations. In each of the bottlenecked and non-bottlenecked populations, V_A's were estimated for five courtship traits in approximately 50 families per population (by parent–daughter covariances; see Meffert 1995 and Meffert et al. 2002 for details). Four traits were assayed in terms of the intensity of the male's display (BUZZ, HOLD, LIFT, and LUNGE; see Meffert 1995), along with one display performed by the female (WING OUT; see Meffert 1995). Meffert and Regan (2002) showed that assays of the intensities of the male's displays include a component of the female's cryptic preferences for his performance. While negative V_A estimates are often set to zero under the assumption of sampling error, the negative values found here (Figure 27.3) are most likely biologically real. In particular, genotype-by-environment interactions, where the environment is the mating partner, can generate negative V_A's (see Meffert 1995 and Meffert et al. 2002). Bootstrap simulations confirmed that the power was sufficient to identify significant V_A coefficients in these

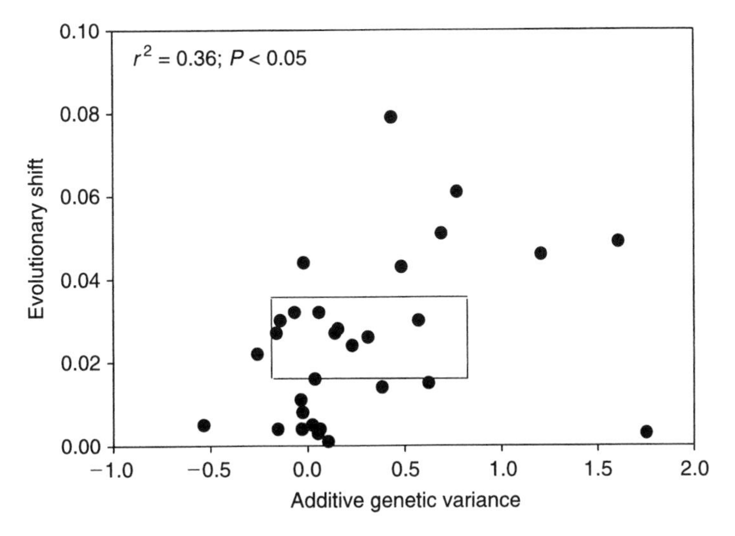

FIGURE 27.3. Evolutionary shifts in courtship components relative to the levels of additive genetic variance in founder-flush populations of the housefly. The data depict the pooled analyses across five courtship traits in six bottlenecked lines (i.e., after a two-pair founder-flush treatment). Additive genetic variances were determined by parent–daughter covariances following the founder-flush event, and evolutionary shifts were assessed by the regression of the courtship phenotypes onto the 26 generations over which populations were monitored in laboratory (see Meffert 1995; Meffert et al. 1999). The box identifies the 95% confidence boundaries for the expected shifts from the pooled control (non-bottlenecked) treatment based upon purely additive theory (i.e., a reduction of $1/2N_e$, where N_e is 4, the number of founders). The correlation coefficient is given, along with its significance value.

populations (Meffert 1995), such that the founder-flush populations had up to 550% higher V_A than the controls (Figure 27.3).

All of the populations showed significant evolutionary potential in terms of the phenotypic shifts in courtship repertoire over the course of 26 generations in the laboratory (estimated by the absolute value of the coefficient for the regression of courtship phenotype over time; see Meffert et al. 1999). Note that evolutionary potential was measured conservatively in terms of the phenotypic shifts due to genetic drift and the "natural" selection pressures of the laboratory. An artificial selection protocol would have assessed the maximum evolutionary potential due to shifts in V_A (see Box 27.1 and below). The founder-flush populations had up to 300% greater evolutionary potential relative to the controls, yielding a significant correlation with the estimates of V_A ($P < 0.05$, Figure 27.3).

Several studies on other systems have supported founder-flush theory (Table 27.1). Increased genetic variances for morphological (Lints & Bourgois 1982), viability (López-Fanjul & Villaverde 1989), and karyotypic (Carson 1990) traits have been reported for founder-flush or inbred populations of *Drosophila* (Table 27.1). Similarly, inbred *Tribolium* populations have shown increased genetic variances for life history parameters (Wade 1991; Table 27.1). Additionally, significant patterns of reproductive isolation have been identified for experimentally inbred or founder-flush *Drosophila* populations (Ehrman 1969; Hay 1976; Powell 1978; Templeton 1979; Ringo et al. 1986; Galiana et al. 1993). Moreover, Ringo et al. (1986) were able to relate differentiation in courtship behavior to the patterns of mating success among founder-flush lines of *D. simulans*. Thus, empirical support for founder-flush theory has been replicated in other systems.

Counter-Evidence of Founder-Induced Speciation

One concern about bottleneck theory involves the relatively low probability that a founder event will increase levels of evolutionary potential or reproductive isolation (e.g., see Florin & Ödeen 2002). In particular, theory has shown that the probabilities of viable genetic and phenotypic shifts after a single bottleneck are negligible (Barton & Charlesworth 1984; Turelli et al. 2001). With regard to the experimental work, Rice and Hostert (1993) suggested that assays for assortative mating tend to have inadequate corrections for multiple testing, such that the conclusions about generating reproductive isolation (see Table 27.1) would be invalid. In the same vein, Galiana et al. (1993) concluded that their cases of significant isolation were too rare to be meaningful to speciation theory (thus this study is reported as positive and negative evidence of founder-flush speciation in Table 27.1). Importantly, other experiments have failed to find significant assortative mating among founder-flush or inbred lines of *Drosophila* (Averhoff & Richardson 1974; Powell & Morton 1979; Rundle et al. 1998, see Table 27.1). In general, arguments about the probability of creating a new species via bottlenecks may remain moot, since all speciation models, founder-flush or otherwise, hold that such radical shifts in genetic architecture should be rare (Slatkin 1996; Templeton 1996). Founder-flush models also invoke additional bottlenecks and/or shifts in selection pressures (Box 27.1) to boost such low theoretical probabilities to more biologically meaningful levels.

Another concern about founder-flush experiments involves whether or not the observed genetic and phenotypic shifts constitute viable shifts in evolutionary potential or reproductive compatibility. As noted above, increased V_A may simply reflect the increased frequencies of deleterious alleles or epistatic combinations. In such scenarios, bottlenecked populations would have generally reduced fitness. The observed deviations in mating behavior and mate preferences, for example, would be attributable to increased inbreeding depression effects (e.g., see Barton & Charlesworth 1984; Meffert & Bryant 1991). In many experiments, some inbred or founder-flush populations did indeed exhibit inbreeding depression in general mating activity (Ehrman 1969; Ringo et al. 1986; Meffert & Bryant 1991), as predicted by a model of pure dominance (see above). In such cases, the populations would be expected to simply restore the ancestral equilibrium as natural selection again took its course against the pool of deleterious recessive alleles (and epistatic combinations; see Figures 27.2B, C). Nevertheless, inbreeding depression effects were not ubiquitous, and some bottlenecked populations actually had higher fitness than the controls (e.g., Meffert & Bryant 1991; see Meffert 2000). As noted above, dominance and the three forms of di-genic epistasis have been identified for the morphological and behavioral traits targeted by the housefly experiments (see Meffert 2000; Meffert et al. 2002), but teasing apart the separate contributions of the nonadditive components has not been feasible. For most complex traits, especially those important to speciation, the combined influences of both dominance and epistasis are likely.

Even for the experiments that found increased evolutionary potential and assortative mating biases, evaluating the formation of stable incipient species via founder events has been difficult. In an experiment to select on housefly morphological shape, bottlenecked populations showed significant rebounds toward the original phenotype after selection was suspended (Bryant & Meffert 1995). In a separate housefly study, patterns of significant reproductive isolation among founder-flush lines decayed as the populations apparently evolved in response to selection pressures of the laboratory environment (Figure 27.3; see Meffert et al. 1999 for details). Importantly, three *Drosophila* experiments also showed that patterns of reproductive isolation can dissipate over time (Ehrman 1969; Dodd & Powell 1985; Moya et al. 1995). Thus, inflated V_A in founder-flush populations, while contradicting purely additive genetic predictions, can permit too much evolutionary volatility for the long-term establishment of differentiated populations (Meffert et al. 1999). The question that still remains is whether or not the evolutionary trends in these populations simply represented purging of genetic load in order to restore the ancestral mutation–selection balance. In order to force a population to establish new stable optima, additional founder-flush events (Meffert & Bryant 1991) and/or diversifying selection pressure may be critical (e.g., Regan et al. 2003; Box 27.1).

Merging the Predictions for Bottleneck Effects

As noted above, any and all aspects of bottleneck theory may apply to the various components of a complex phenotype, and they can be difficult to decipher. That is, some loci may lose V_A at the same time that others gain V_A. Therefore, some trajectories of the evolutionary universe may be closed off while others are opened up. In Regan et al. (2003), non-bottlenecked populations exhibited concordant evolutionary trends in courtship behavior over the course of 18 generations of selection for assortative mating, but the bottlenecked lines exhibited a more variable set of evolutionary responses. It is difficult to determine whether the initial founder-flush events increased V_A (e.g., Templeton 1980) or locked up evolutionary potential in accordance with more neutral bottleneck expectations (Falconer & Mackay 1996). Thus, the greater variability in evolutionary trajectories for the bottlenecked lines could have been due to either increased access to the multifactorial phenotypic universe or, more simply, reduced potential to follow the convergent trajectories of the non-bottlenecked lines. In the same vein, Whitlock et al. (2002) found that some bottlenecked populations of *D. melanogaster* had stable alterations in the genetic intercorrelation structure of morphological traits. In this experiment, the internal selection pressures appeared to be shifted due to the founder events (see Box 27.1), but some of the shifts involved reduced, rather than increased, genetic variance. Most likely, all of the potential bottleneck processes occur for such complex traits. In this way, a combination of the earlier and later versions of founder-flush theory (i.e, focusing on the losses and increases in V_A, respectively; see Carson 1990 and above) can still explain the increased number of evolutionary opportunities for bottlenecked populations.

QUESTIONS FOR FUTURE DIRECTIONS

What Ancestral Genetic Structure Is Optimal for Maximizing Evolutionary Potential?

Theory has described the kinds of genetic compositions that are optimal for founder-induced increases in evolutionary potential, but direct experimental examinations have been elusive. In particular, an epistatic structure that conceals appreciable levels of V_A and is less prone to inbreeding depression (and more prone to outbreeding depression) is the most favorable (see Michalak and Noor, Ch. 25 of this volume). For example, Templeton (1996) suggested that the lack of success in generating assortative mating in some *Drosophila* experiments (e.g., Rundle et al. 1998) was because some species lack the requisite genetic architecture for founder-induced speciation. In his model of genetic transilience, the ancestral architecture most likely to promote founder- flush speciation would have a relatively small number loci, with many pleiotropic effects on the target traits, which, themselves are affected by many epistatic modifying loci (Templeton 1996). Thus, he concluded that experiments on *D. mercatorum* were more relevant to founder-flush theory than those on *D. melanogaster* because they had the requisite genetic architecture (see Table 27.1). In practice, the relative roles of nonadditive genetic effects and the magnitudes of influence by modifying loci have been difficult to partition out. An ideal experimental approach would be to examine the predictive power of a particular ancestral genetic architecture in testing founder-flush theory.

What Factors for the Severity, Number, and Frequencies of Founder Events Promote Differentiation?

Up to this point, the terms "bottleneck" and "founder event" have been used almost interchangeably, but the distinctions are not always subtle. In general, bottlenecks are those restrictions in population size that dramatically increase the effects of genetic drift. A founder event is then an extreme bottleneck. In principle, there should be an optimal founder size that would constitute a significant genetic disruption while still not threatening a population with imminent extinction (Bryant et al. 1986; Galiana et al. 1993). More importantly, the term "founder effect" in speciation theory usually invokes the subsequent flush phase of reduced inbreeding and relaxed or altered selection pressures (see Box 27.1). A period of enhanced recombination potential should be valuable for the kind of catalysis of speciation envisioned by the models. In contrast, prolonged bottlenecks, or serial inbreeding events, are more likely to simply purge genetic load or increase the risk of extinction (Day et al. 2003; see Frankham, Ch. 32 of this volume). A more complete understanding of founder-flush theory requires tests on the rates of inbreeding and recombination that promote geographic differentiation or speciation.

How Can Founder-flush Populations Maximally Exploit Increased V_A?

Artificial selection experiments are valuable for assessing the maximum evolutionary potential in founder-flush populations (see Box 27.1), but few systems have been amenable to such approaches (Bryant & Meffert 1995; Regan et al. 2003). In particular, tests on variable founder sizes and rates of inbreeding have proven to be quite feasible (e.g., Day et al. 2003), but selection experiments pose significantly more logistical constraints. The major logistical conflict of any selection experiment is attempting to maximize the sizes of the experimental populations (i.e., to maximize N_e) while also maximizing the number of replicate lines (i.e., for greater statistical power). Very few selection protocols can thus avoid the inherent confound of inbreeding, so that subsequent inbreeding could simply negate any evolutionary "advantage" of an initial founder- flush event. Experiments on how founder-flush populations can exploit increased V_A are necessary for understanding how founder-induced geographic variation or speciation might occur in nature.

What Are the Implications for Conservation Biology?

Conservation projects generally assume that maximizing genetic variation is the best strategy for rescuing an endangered population (see Frankham, Ch. 32 of this volume), but founder-flush studies have suggested that such strategies may sometimes be counterproductive (e.g., see Meffert 1999). In particular, some of the increases in V_A appear to involve the increased frequencies of detrimental recessive alleles (e.g., Willis & Orr 1993; Figure 27.2B). Under such scenarios, the conventional captive breeding strategy to balance founder contributions in the descendant generations could increase extinction risks by inflating the frequencies of deleterious alleles (see Meffert 1999). Nevertheless, naturally bottlenecking populations (e.g., Hard et al. 1993), pest invasions, and highly inbred laboratory strains all attest to the ability of founder events to produce robust populations (e.g., see Meffert 1999). Founder-flush experiments that can elucidate the influences of components of genetic architecture are thus necessary for understanding how natural or captive populations escape the threat of extinction.

SUMMARY

Founder-flush theory holds that bottlenecks can open evolutionary avenues for geographic differentiation and speciation. Specifically, bottlenecks can alter evolutionary constraints by increasing the additive genetic variances of traits that are structured by nonadditive genetic effects with consequent perturbations of the genetic intercorrelation structure. Geographic surveys and experiments on model systems have shown that bottlenecks can indeed increase the additive genetic variances of fitness traits, alter courtship repertoires, and generate reproductive isolation. However, theoretical work has suggested that viable genetic and phenotypic shifts in bottlenecked populations should be rare. In particular, inbreeding depression is a common consequence of a bottleneck, and several experiments have shown that a single bottleneck was insufficient to generate reproductive isolation. Additional founder-flush events and/or diversifying selection pressure may be crucial for establishing differentiated populations in the long term. Additional research on the types of genetic architectures, sizes and durations of the bottlenecks, as well as influences of subsequent selection pressures is critical for a more complete understanding of how founder events could increase evolutionary potential and, more generally, how populations can recover from bottlenecks.

SUGGESTIONS FOR FURTHER READING

Meffert (2000) provides a review of housefly bottleneck experiments, with supporting evidence of the role of epistasis in founder-induced increases in the additive genetic variances of behavioral traits. Meffert (1999) summarizes the results of several bottleneck studies with regard to the potential for the inflated frequencies of deleterious recessive alleles, suggesting that some efforts to preserve genetic diversity in conservation programs may actually be counterproductive. Barton and Charlesworth (1984) provide theoretical evidence against proposed models of founder-induced speciation, and Florin and Ödeen (2002) extend Rice and Hosterts' (1993) critique of founder-flush experiments.

Barton NH & B Charlesworth 1984 Genetic revolutions, founder events, and speciation. Annu. Rev. Ecol. Syst. 15:133–164.

Florin AB & A Ödeen 2002 Laboratory environments are not conducive for allopatric speciation. J. Evol. Biol. 15:10–23.

Meffert LM 1999 How speciation experiments relate to conservation biology. BioScience 49:701–715.

Meffert LM 2000 The evolutionary potential of morphology and mating behavior: the role of epistasis in bottlenecked populations. pp. 177–193 in JB Wolf, ED Brodie & MJ Wade, eds. Epistasis and the Evolutionary Process. Oxford Univ. Press.

Acknowledgments This research was supported by grants from the National Science Foundation (BSR-910651 and DEB-0196101). Many thanks go to Charles Fox, Autumn Hardin, Sara Hicks, Toartou Kargou, Sanjeet Patel, Jennifer Regan, Kara Stabler, and Amaris Swann for their comments on earlier drafts of the chapter.

28

Theory of Phylogenetic Estimation

ASHLEY N. EGAN
KEITH A. CRANDALL

The genetic makeup of an organism (genotype) provides a blueprint for its physical existence (phenotype)—from morphological characteristics to biochemical constituents and processes—characterizing the organism as a unique individual, unlike any other in the world. Yet there are obvious biological similarities among different organisms. Scientists have noted these similarities and differences and endeavored to use them to reconstruct historical relationships between extant species. The observations and endeavors of past scientists to understand the evolution of life have led to the scientific discipline of phylogenetics.

Evolutionary theory implies that all species are historically related, each species occupying a branch on the tree of life united through shared characteristics due to common ancestry, yet independent through unique characteristics found only in their lineage. Evolutionary, hierarchical relationships are portrayed by a phylogeny, a branching diagram that represents hypothetical lines of descent or paths of inheritance among organisms or their traits. Phylogenies are built upon the shared and contrasted traits of a group of organisms or their parts. Morphological, chemical, physiological, demographic, and genetic characteristics have all been used to reconstruct phylogenies. The advent of molecular biology has provided a powerful tool for unraveling evolutionary relatedness. Modern biotechnology has allowed the estimation of phylogenies based on the detection of genetic variation among different lineages through DNA sequence analysis. By tracing backward along the phylogeny, a common ancestor for a group of alleles, traits, or organisms can be found.

In its infancy, molecular phylogenetics was mostly applied to evolutionary systematics and species relationships. As technology and methodology have expanded, phylogenetics has been applied to diverse issues and scientific disciplines. Evolution, ecology, conservation biology, systematics, comparative biology, behavior, genetics, molecular biology, development, forensics, epidemiology, and infectious disease are just a few of the fields now incorporating phylogenies and phylogenetic methods into their research. Phylogenies provide essential knowledge concerning evolutionary relationships, establishing a foundation for hypothetical reasoning. The evolutionary information portrayed by a phylogeny allows scientists to formulate and test hypotheses or draw inferences concerning historical events. Phylogenetic analyses allow more enlightened choices in matters such as conservation planning and resource management, drug development and disease treatment, or even matters of criminal guilt or innocence.

A phylogeny represents paths of descent that depict the transmission of characters from ancestral lineage to descendent lineage. It consists of terminals (the species or genes sampled), internal nodes (ancestral state reconstructions of the associated characters), and branches (which define the relationships of the units and can represent the relative divergence among the terminals and nodes). Branch lengths may depict the time between speciation events according to the mutation rate or the number of mutations along a lineage, depending on the operational taxonomic units (OTUs). The tree topology may also have a root (the common ancestor to all involved taxa). A tree can be rooted using the outgroup method in which one or more taxa known to share a distant common ancestor are specified as the root. Alternatively, the tree can be rooted based on the assumption of a molecular clock in which character evolution is said to assume a certain rate along the branches of the tree. The topology is rooted at a point that effectively splits the amount of character evolution in half. OTUs may be hierarchical groups or parts thereof, such as species, individuals, or even genes (Figure 28.1). A distinction must be made between phylogenies that portray organismal relationships and those that depict the evolutionary history of a part of an organism (e.g., a gene).

Gene Trees Versus Species Trees

A gene tree depicts the evolutionary history of a specific gene shared by a group of organisms, such as the 18S ribosomal subunit. Analysis of this gene provides a phylogeny that proposes a hypothesis for the paths of inheritance of the 18S ribosomal subunit among the involved taxa. Different genes may exhibit different paths of inheritance, thereby providing a different hypothesis for species relations. A species tree can be inferred by tracing the evolutionary history of many genes from a group of species, delimiting the evolutionary events that mark routes of inheritance between the involved taxa.

Species are usually delineated by reproductive boundaries, which help mold the course of evolution. A bifurcation on a species tree represents a speciation event in which an ancestral species split into two descendant species. The position of a bifurcation on a phylogeny represents the time of speciation. The topology of gene trees versus species trees may or may not differ depending on where the mutation took place along the tree relative to the speciation event. Genetic divergence among the genes may have occurred before the speciation event. In the case of Figure 28.2A, the gene split came relatively close to the species split, resulting in no major problems with phylogeny estimation. The gene split in Figure 28.2B greatly preceded the species split. While this does not change the topology, it can cause an overestimation of branch lengths—the time between evolutionary events. The branching pattern of the tree may differ between gene and species trees when genetic polymorphism exists in the ancestral species, causing paths of inheritance to be misconstrued. This effect, known as incomplete lineage sorting, arises from the persistence of gene lineages present before speciation, some of which may not be inherited by all species. In the case

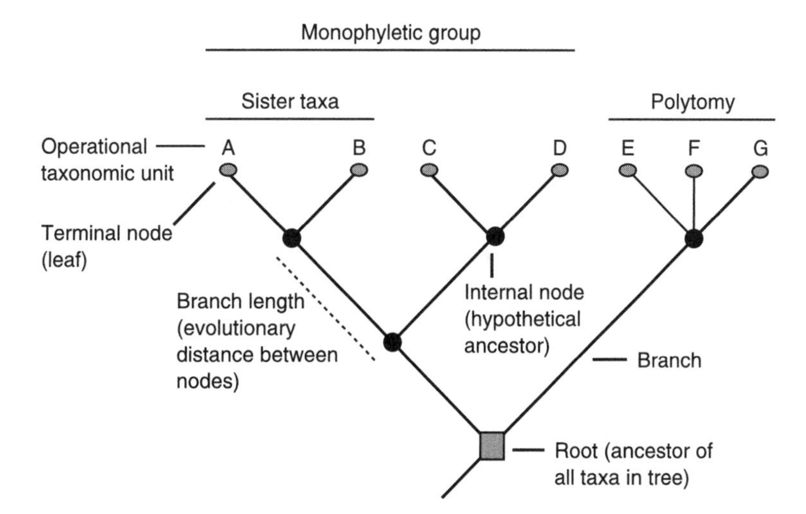

FIGURE 28.1. Example of a phylogeny with included terminology. Sister taxa are two taxa derived from a common ancestral node, they are each other's closest relatives; a monophyletic group includes an ancestor with all its descendants; a polytomy is an internal node with more than two immediate descendants, representing either simultaneous divergence or an unresolved node.

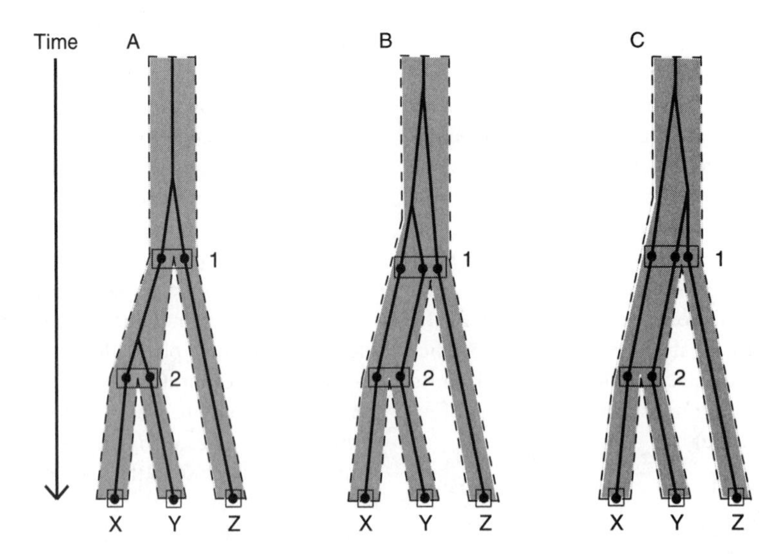

FIGURE 28.2. Possible relationships between gene trees (unbroken lines) and species trees (shaded area). The gene and species tree are the same in A and B with divergence times between genes and species similar in A. Genes X and Y diverged before the respective species in B. The topologies of the gene tree and species tree differ in C due to incomplete lineage sorting. Modified from Nei (1987).

of Figure 28.2C, the gene split prior to speciation event 1 was sorted between the two resultant species, creating a polymorphism within one species just after the event. This group then continued to speciate through event 2, with each gene lineage comprising a new species. This misconstrued the true tree by leading to an inference that genes X and Y are related via speciation event 2 when in actuality genes Y and Z are related via the gene split prior to speciation event 1. Other evolutionary processes such as recombination, horizontal transfer, hybridization, selection, and gene duplication may also create incongruence between gene and species trees.

SEQUENCE ALIGNMENT

Alignment between nucleotide positions in a group of sequences provides a statement of homology, or shared ancestry, upon which historical inferences of evolution can be made. The manner in which nucleotide or amino acid sequences are aligned has a profound effect on phylogenetic analysis. Sequence alignment endeavors to maximize similarity or minimize inferred changes between nucleotide or amino acid character states within an alignment. Alignments can be straightforward for closely related individuals or conserved coding genes, but can be very problematic for distantly related taxa and/or noncoding sequences. Alignment is complicated by length differences between sequences due to insertion or deletion events (indels). Alignments involving indels can be determined through the use of cost functions and gap penalties. Determination of the globally best alignment is based on minimizing the cost function associated with each alignment, ascertained by assigning penalty scores for nucleotide or amino acid changes and indels or gaps. Different theories behind cost functions can result in varying optimal alignments. For example, transitions and transversions may be assigned different costs. Gaps can be assigned one overall cost, regardless of position. Alternatively, affine gap costs—which endeavor to incorporate biological processes into gap cost functions—institute a gap initiation cost and gap continuance cost of a lesser value. Some scientists oppose this method because the complexity of the gap cost function cannot be incorporated into the phylogenetic analysis (Phillips et al. 2000). In addition, the use of cost functions for sequence alignment is criticized by some because the cost functions are often user-defined, making the procedure less objective. Needleman and Wunsch (1970) first suggested a dynamic programming algorithm for optimal pair-wise sequence alignment. This algorithm elucidates the optimal pair-wise alignment by traversing a matrix plot of sequences and determining the least costly path, resulting in the optimal pair-wise alignment. Computationally, this process is limited to aligning only a few sequences. Phylogenetics deals with multiple sequence analysis.

Multiple Sequence Alignment

Theoretically, the Needleman and Wunsch algorithm can be used to align multiple sequences through an *n*-dimensional matrix. However, this becomes nearly impossible above three sequences. The MSA program (Lipman et al. 1989) expanded this number to five to seven sequences, depending on length. The algorithm of MSA also traverses an alignment matrix, but uses a branch and bound technique to pare down the area traversed, essentially removing cells that will not contribute to the optimal alignment by calculating and imposing an upper bound on the alignment cost. As a cost measure, the algorithm minimizes the sum of pairs by adding the alignment costs between each pair of sequences in the multiple alignments. However, the sum of pairs measure does not take into account common ancestry, requiring correction through use of a weighting scheme based on relatedness. In spite of the MSA program algorithm allowing up to seven sequences, the fact that the weighting scheme requires phylogenetic information introduces a measure of circularity if the alignment is used to infer evolutionary relationships within phylogenetics. In addition, optimality is not guaranteed because the upper bound calculated by the algorithm is usually high. MSA uses a neighbor-joining algorithm to create a guide tree for the alignment. The neighbor-joining method creates a tree by determining the pair of taxa with the shortest branch length, in this case, with the least sequence divergence. Those sequences are used to create a composite sequence and combined with the least divergent remaining sequence and so on (Saitou & Nei 1987).

The alignment of many sequences can become an NP (nondeterministic polynomial) complete problem in that no known algorithm can guarantee a correct solution in polynomial time. Therefore, like phylogenetic analysis, heuristic methods for multiple sequence alignment must be used. Various software exists to tackle the task of aligning multiple sequences, most of which uses a progressive alignment algorithm. Clustal (Thompson et al. 1997) is a popular program that begins by computing a distance matrix of all pair-wise comparisons between nucleotide or protein sequences. It then uses a cluster method, such as the neighbor-joining method, to create a guide tree from the matrix that specifies the order in which sequences are to be aligned. The least costly pair of sequences are aligned and subsequently locked. The next sequence is then aligned to the pair, and so on. In addition to gap and mismatch penalties, Clustal also allows other penalty considerations, such as penalties for breaking secondary structure. The standard approach for multiple sequence alignment is to use Clustal for an overall initial alignment and then adjust by eye. The advantage of this progressive algorithm is speed. The process, however, does not guarantee the minimum cost alignment. In addition, different parameters will result in differing alignments.

The program MALIGN (Wheeler & Gladstein 1994) adds an additional alignment optimization technique beyond the Clustal framework. After creating an initial guide tree via methods similar to Clustal, MALIGN then endeavors to find an optimal guide tree through random addition and branch swapping on the guide tree. The guide tree that gives the least cost alignment is then used to progressively align the sequences. MALIGN does not guarantee the optimal alignment, but often finds less costly alignments than Clustal. It requires greater computation time, however. MALIGN can find multiple guide trees of the same least cost, an attribute that can be both advantageous and disadvantageous depending on research objectives.

Although multiple sequence alignment algorithms produce a good alignment, anomalies can still be created. Therefore, the researcher should always check the alignment for possible incongruencies. If the sequence is coding, the alignment can be translated into amino acid sequence to check for erroneous nucleotide homology statements based on the reading frame. Two examples of sequence alignment editing programs that can aid in this process are Se-Al (Rambaut 2002) and MacClade (Maddison & Maddison 2000). The alignment can then be adjusted to ensure correct amino acid translation and reading frame. Coding regions are often highly conserved and can often be completely aligned by eye. Noncoding genes can be more problematic.

A messy region in the alignment may adversely affect estimation of the phylogeny. For this reason, areas of questionable alignment are often deleted prior to phylogenetic analysis. Still more quandaries exist. Given the sequence data, several alignments of equal score can be inferred. In addition, each equal-cost alignment may produce different trees. Therefore, determining which alignment will give the optimal tree topology is problematic. This has led to the creation of methods that integrate alignment and phylogeny estimation together, thus eliminating several stumbling blocks often encountered in phylogenetic reconstruction.

Simultaneous Alignment and Phylogeny Reconstruction

Because phylogeny reconstruction is based on aligned sequences, many researchers optimize the alignment separate from the tree-searching portion of phylogenetic analysis. However, the ideal method would be one in which sequence alignment and phylogeny estimation take place simultaneously, using the same optimality criterion for both. TreeAlign is a program that creates alignments and phylogenies based on a combination of distance matrices and approximate parsimony. A pair-wise distance matrix is used to create a guide tree for use in constructing ancestral sequences. The algorithm seeks to minimize the sum of user-defined weights of indels and substitutions in an effort to optimize the alignment and phylogeny (Hein 1994). Like the multiple alignment algorithms, the use of a guide tree by TreeAlign introduces circularity into the analysis, an aspect seen as unfavorable by many. In addition, the program is limited in use due to the amount of memory needed for the analysis. TreeAlign is best used on a Unix workstation or mainframe.

Direct optimization (DO), also known as optimization alignment, combines multiple sequence alignment and tree reconstruction into a single step (Wheeler 1996). It avoids the problems involved with instituting gaps and determining the gap cost function by incorporating indels as transformation events that link ancestral and descendant nucleotide sequences through heuristic determination of hypothetical ancestral sequences. Unique nucleotide homologies are dynamically determined for a given topology and are optimized at every node using a Fitch optimization down-pass and an alignment algorithm that minimizes the weighted union/intersection cost. Putative ancestral sequences are formulated using an up-pass. This method can use a variety of cost minimization functions including parsimony and maximum likelihood models and is implemented in the program POY (Wheeler & Gladstein 2000). Like other heuristic alignment and phylogeny algorithms, DO does not guarantee the optimal alignment or phylogeny. One drawback to DO is that it does not produce a sequence alignment for review. However, a synapomorphy-based implied alignment can be obtained from the results (Wheeler 2003).

DO provides promising potential for simultaneous sequence alignment and phylogeny reconstruction, but can only work within certain limits. Some researchers disagree with this approach because it seems to compound the difficulty of alignment and phylogeny reconstruction into one larger NP complete problem. Therefore, multiple sequence alignment is the prevailing alignment method, followed by phylogeny estimation.

MODELS OF EVOLUTION

Molecular evolution is based on change within a sequence over time. The observed difference between two sequences may underestimate the amount of evolutionary change because of the possibility of unseen mutational events or multiple hits. For example, during the course of evolution a nucleotide at a specific site may mutate from the ancestral state of "A" through an unobserved intermediary state of "G" and finally to the extant state of "T." The intermediate mutation was not observed. The prospect of unobserved reversals can also occur: the ancestral nucleotide "A" can mutate to a "C" and experience a reversal back to "A." Thus, a mathematical model that describes and corrects for the difference between observed and expected change is needed. All phylogeny reconstruction methods, including alignment and topology estimation, make use of evolutionary models that describe the underlying process of nucleotide substitution, whether implicit or explicit. The model of sequence evolution must consider both the pattern and rate of change. In addition, models can account for biases stemming from differing nucleotide frequencies, substitution rate heterogeneity across a sequence, invariable sites, codon position, and secondary and tertiary structure.

Numerous models of evolution have been created to account for various evolutionary scenarios, moving from simple to complex. The Jukes–Cantor model (JC69) is the simplest (Figure 28.3), assuming that equilibrium nucleotide frequencies are equal and that any nucleotide can change to any other with equal probability (Felsenstein 2004). The Kimura 2-parameter model (K2P) allows for differences in transition and transversion rates while keeping equal base frequencies. The Felsenstein 1981 model (F81) allows some bases to be more common than others while keeping substitution probabilities equal. F81 assumes that base frequencies are similar across sequences. Combining F81 and K2P gives the Hasegawa, Kishino, and Yano model (HKY85) which allows variation in base frequencies and transition/transversion bias. The general time reversible model (GTR) builds on HKY85 by permitting each of the six substitution pairs to have different rates. GTR contains the other models nested within it (Figure 28.4). More models can be created by allowing irreversible substitution rates to enter the hierarchy (Felsenstein 2004).

The above models make several assumptions: nucleotide sites change independently; the substitution rate is constant between lineages and over time; nucleotide frequencies are in equilibrium; and nucleotide substitutions are equal across sites and time. These assumptions may not hold for every situation. More evolutionary parameters have been proposed to deal with various situations. The LogDet transformation deals with varying base composition by approximating an additive distance between sequences. Rate heterogeneity within the sequence can also be observed. The gamma distribution (Γ) shape parameter (α) describes the range of rate variation among nucleotide sites. A small α gives an L-shaped curve in which most sites have low substitution rates while a few have high substitution rates. A larger α conveys less rate heterogeneity among sites (Yang 1994). The proportion of invariable sites (I), those with zero probability of change, also incorporates rate heterogeneity (Steel et al. 2000). These parameters combined with models already discussed provide an array of evolutionary scenarios to model the underlying substitution patterns in a given data set. By adding parameters to a model, it more closely approximates the observed pattern underlying the data. However, overparameterization can become a problem through increased sampling error and computation time and inconsistency. Therefore, determining the simplest model that adequately fits the data is paramount.

Model Optimization

Given the range of evolutionary models, there must be a way to choose the simplest model that best fits the data. Maximum likelihood provides a statistical framework in which to determine the best model for the data (Huelsenbeck & Crandall 1997). The likelihood ratio test (LRT) uses a chi-square distribution to determine whether two nested models are significantly different across the same topology,

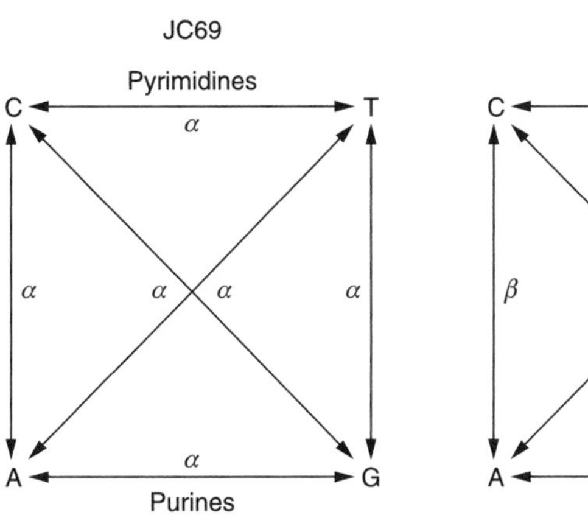

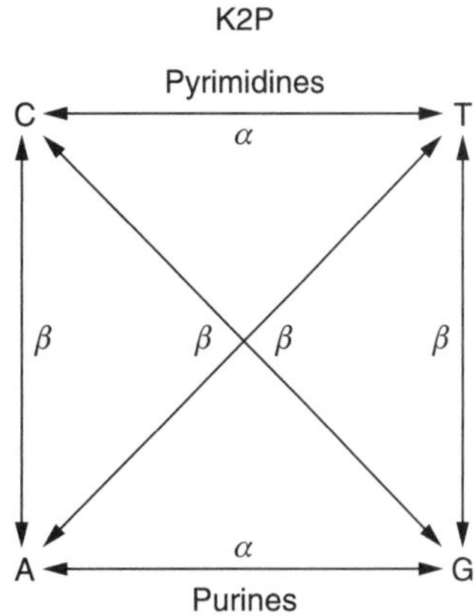

FIGURE 28.3. The Jukes–Cantor model (JC69) and Kimura 2-parameter model (K2P). α, transition rate; β, transversion rate.

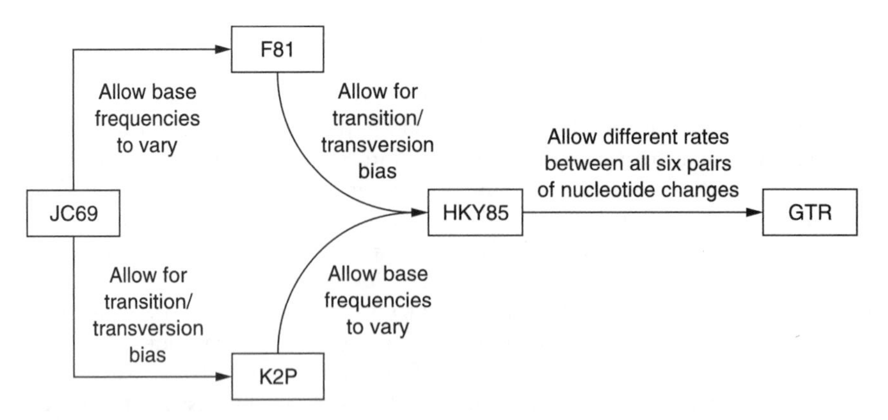

FIGURE 28.4. The nested hierarchy of five models of nucleotide substitution. Model abbreviations: JC69, Jukes–Cantor; F81, Felsenstein 81; K2P, Kimura 2-parameter; HKY85, Hasegawa, Kishino, and Yano 1985; GTR, general time reversible. See Felsenstein (2004) for a thorough discussion of models of evolution as well as citation to the primary literature on models.

one providing a better fit to the data than the other. The alternative hypothesis is the more complex model. A chi-square test of the simple versus complex model can be tested using the difference in the number of free parameters as the degrees of freedom (Felsenstein 1988). ModelTest, a program for determining the best model of evolution for a data set using maximum likelihood, uses the LRT to find the simplest scenario that models the underlying process of evolution (Posada & Crandall 1998). A starting tree topology is inferred from the data using parsimony or neighbor-joining. The program then compares the likelihood estimates of 56 nested evolutionary models inferred on the topology using a hierarchical LRT. This determines the model with the least parameters that adequately fits the data, thus eliminating unnecessary parameters. The chosen model and estimated parameters can then be used for phylogeny inference. Drawbacks include the need for a starting tree which introduces a measure of circularity. In addition, the outcome may be sensitive to the starting topology, but simulation results suggest that this is not a great concern (Posada & Crandall 2001).

DISTANCE METHODS

Once an alignment is determined, historical inferences concerning the course of evolution can be made. This is done by reconstructing a phylogeny which outlines the branching pattern and branch lengths linking taxa from ancestor to descendant. Several methods exist to reconstruct the tree topology. Distance methods transform the data into a matrix of evolutionary distances between all pairs of sequences, usually defined by the number of nucleotides that differ between pair-wise comparisons. Phylogenetic relationships are then reconstructed by algorithmically combining taxa in a manner that reflects the distance values. For distance methods to be accurate, the distance between two sequences needs to correspond to the branch lengths between the species, that is, evolutionary distances need to be additive. For long distances this is not usually the case due to multiple hits. Unless corrected for, the possibility of multiple hits can distort the phylogeny estimated from evolutionary distances. Differing rates of substitution can also impact the outcome. Models of nucleotide substitution can be applied to correct for these differences.

Unweighted pair-group method of arithmetic means (UPGMA) is a clustering algorithm that assumes a molecular clock, that is, that nucleotide substitutions accumulate at a specified rate across time. The phylogeny is built in a stepwise manner, joining the least divergent sequences first. A new distance matrix is computed treating the combined sequences as a composite individual. This is repeated until two composite individuals exist, becoming sister to each other in the phylogeny. This method is fast, but often inaccurate (Hillis et al. 1994).

The neighbor-joining method (NJ) uses arithmetic means for distances between nodes with multiple taxa. Starting with a star phylogeny, each node is resolved by adding the one pair of neighboring taxa which minimize the branch length of the topology overall. Like UPGMA, this method also uses composite taxa to create subsequence distance matrices. The addition of "neighbors" continues until all taxa are included. The algorithm of the NJ method allows it to make a good approximation of the minimum evolution tree. One advantage of the NJ method over UPGMA is that it does not assume a molecular clock, an often unrealistic assumption. The NJ method is computationally fast and correct if the tree is additive (Saitou & Nei 1987). It is the most popular distance method for phylogeny reconstruction and is used in heuristic searching strategies as well.

Distance methods as a whole have both advantages and disadvantages. One advantage over parsimony is the ability to institute different models of nucleotide substitution. In addition, distance methods are faster than other methods of phylogeny reconstruction. The major drawback to distance methods is the loss of information. The transformation of nucleotide sequences to evolutionary distances makes it impossible to derive ancestral sequences using these methods. In addition, branch lengths can often be difficult to interpret due to negative or fractional distances. Another drawback of the UPGMA and NJ methods is that they do not compare alternative topologies in an effort to find optimal trees. The final trees are simply a product of the algorithmic functions used to create the tree.

OPTIMALITY CRITERIA

Algorithmic methods provide one tree when in actuality the number of possible tree topologies for a given number of taxa can be exponentially large. Optimality criteria, such as maximum parsimony, maximum likelihood, or minimum evolution, provide a means of determining which tree topology best illustrates the evolutionary history between involved taxa by maximizing or minimizing an optimality function on all or a heuristic subset of possible tree topologies. Each optimality criterion makes explicit assumptions about the processes of evolution. Differing opinions tend to cause heated debates concerning which optimality criterion is best (see Box 28.1).

Minimum Evolution

Minimum evolution is an optimality criterion applied to distance data using exhaustive or heuristic search strategies. This criterion uses the minimum sum of branch lengths and is based on the assumption that the topology with the least total branch length score best represents the true evolutionary relationships of the taxa. The data are fit to a possible tree topology and branch lengths are computed across the topology using the unweighted least squares method. Branch lengths can be interpreted as the evolutionary distance between ancestor and descendant, assuming a specific model of evolution. Minimum evolution can produce many trees of equal score, an advantage or disadvantage depending on the study. However, it is not guaranteed to provide the true tree. Some believe that minimum evolution is not a good optimality criterion because of its foundation of minimizing branch lengths, asserting that there is no biological basis that says the true phylogeny must have minimum branch lengths. Of course, there is no *biological* reason why the true phylogeny should be the most parsimonious or maximum likelihood either.

Maximum Parsimony

Maximum parsimony (MP) is an optimality criterion based on Occam's Razor—that the simplest solution is the best. There are several ways to explain the criterion for MP: the best tree is the one with the fewest evolutionary events or nucleotide substitutions, the tree which minimizes homoplasy, or the tree that minimizes the number of ad hoc hypotheses. Homoplasy is defined as the existence of similar character states derived independently rather than from common ancestry. Homoplasy requires ad hoc hypotheses to explain ill-fitting data on the phylogeny. These ad hoc hypotheses include homoplasy attributed to parallelism, convergent evolution, or reversal events.

MP determines the best tree by counting the minimum number of nucleotide changes required to fit the data to a topology. Based on the search method chosen, MP will assess the number of changes on all possible trees or a heuristic subset thereof. The tree with the lowest tree score or tree length, as defined by the number of changes summed along all the branches, becomes the most parsimonious tree and is taken as the tree that best represents the evolutionary history patterned

BOX 28.1. Philosophical and Methodological Differences in Phylogenetics

Phenetics Versus Cladistics

From its infancy, phylogenetics has been divided between different philosophical and methodological factions, each school of thought opposing the other in almost warlike proportions. It began in the context of taxonomy with numerical phenetics. In the early 1960s, scientists began to apply numerical clustering methods to morphological data in an effort to establish classification schemes based on similarity. Pheneticists used an algorithmic, distance-based approach and purported that their methods were only used for taxonomic reasons. Indeed, they claimed that evolutionary history could not be inferred. Systematists interested in phylogenetic relationships began to oppose numerical phenetics, contending that evolutionary history could be inferred. Thus began the cladistic school of thought in the late 1960s. Willi Hennig was influential by defining the aim of systematics to be monophyletic classification in which all taxa were related by common ancestry into nested sets of sister taxa, or clades. In time, cladistics became the prevailing school of thought.

Cladistic Parsimony Versus Statistical Phylogenetics

As cladistics became more popular, there began to be a schism in thought concerning methodology. In the beginning, parsimony analysis was the method of choice, in which the best phylogenetic hypothesis is that tree with the least number of evolutionary events needed to group taxa. A few phylogenetic systematists advocated the use of statistical methods. But cladists of the day fervently denied the application of statistics to phylogenetics, claiming that statistics has no bearing on history and phylogeny therefore was not an inference problem. The advent of molecular data in the 1970s would add fuel to the fire, however. Parsimony was still strong in the molecular arena with Fitch's optimization of characters on phylogenies, but molecular data introduced more controversies in the form of conflicting data couched as unseen mutational events or multiple hits. In the early 1980s, Felsenstein proposed his phylogenetic method of maximum likelihood, which uses a probabilistic framework to determine the most likely phylogeny given the data and a model of nucleotide substitution which accounts for multiple hits. Statistical phylogenetics had gained enough popularity by the 1990s to see the introduction of several statistical phylogenetics software programs, greatly boosting the use of statistics for phylogeny inference. The middle 1990s saw the application of Bayesian statistics to phylogenetics as well. While the current decade still suffers from the schism between cladistic parsimony and statistical phylogenetics, the fire has died down slightly (Felsenstein 2001).

within the data. MP does not offer branch lengths for each node, instead providing an overall length of each topology in terms of the total number of changes required. Often, MP finds two or more trees that are equally parsimonious—that have the same number of minimum changes. This can be good in some cases. However, when a single evolutionary hypothesis is desired, this becomes problematic because there is no definite way to determine which of the set of most parsimonious trees best represents the actual underlying history of the organisms. Often a strict (or majority-rule) consensus tree of the MP trees is used to summarize the common associations found in all (or 50%) of the MP trees.

MP works well for sequences with little or moderate divergence but may be inaccurate when sequences of large divergence are analyzed. This stems from the assumption that the tree with the fewest evolutionary changes is the best, that is, the tree that minimizes homoplasy. When sequence divergence is small, the number of homoplasies is likely to be small and MP will accurately reflect that. However, when sequence divergence is large, MP may draw erroneous phylogenetic inferences. This occurs when some branches evolve faster than others, and thus are more likely to accumulate homoplasies. MP tends to groups these sequences together. Thus, MP may converge on the wrong tree as more data are added, causing the method to suffer from inconsistency. This phenomenon is known as long-branch attraction or the Felsenstein zone (Felsenstein 2004). Inconsistency may also stem from the assumption that nucleotide substitutions are equally likely. If this assumption is violated, MP may produce faulty phylogenies. However, there are ways to institute weighting schemes (e.g., transversions are weighted more than transitions)—a method called weighted parsimony. Using different substitution rates may better reflect the underlying biology (Hillis et al. 1994).

Because of the ramifications of MP's underlying assumption, many phylogeneticists are dissatisfied with this optimality criterion. Others see the fact that MP does not rely on an explicit model of evolution to be an advantage of the method (although it does rely on an implicit model). MP uses only those nucleotide sites or character states that are shared among taxa and derived from the ancestral state. These sites are deemed parsimony informative. Some researchers criticize MP for not using all the available information in a data set. While MP requires more computation time than distance methods, it is comparatively faster than more computationally intense methods such as maximum likelihood.

Maximum Likelihood

Maximum Likelihood (ML) optimizes the likelihood of observing the data given a tree topology and a model of nucleotide evolution. In other words, ML finds the tree topology that explains the observed data with the greatest probability under a specified model of evolution. In statistical terms, the likelihood, L, is stated as $L = \Pr(\text{data}|\text{tree})$, which is the probability of observing the data given the tree.

Various models of nucleotide substitution can be applied to describe the underlying processes of nucleotide change. The model parameters and branch lengths can be specified by the researcher or estimated from the data. Those parameters that give the highest likelihood are considered the maximum likelihood estimates of the model parameters. The likelihood of observing the data on a given tree with specified branch lengths can be computed by multiplying the individual probabilities of each nucleotide site, requiring the assumption that each site is evolving independently. The likelihood of a site on a given topology with specified branch lengths is calculated by summing the probabilities of all the possible ways the nucleotide could have evolved. The overall likelihood of the tree is the product of each site's likelihood over all the sites, a number which is often exceptionally small, for example 9×10^{-29}. For mathematical reasons, the likelihood is often reported as the natural log of the likelihood score: $\ln L = -66.88$ for the above number. When comparing alternative tree topologies, the tree with the highest log likelihood is the maximum likelihood tree. Remember, this is not the probability that the ML tree is the true tree, but the probability that the tree has produced the observed data. Also keep in mind that the ML tree under one substitution model may not be the ML tree under another model.

Perhaps one of the biggest advantages of ML is the ability to make statistical comparisons between topologies and data sets, allowing the testing of different hypotheses. ML can return several equally likely trees—a pro or con depending on the study, but because likelihood is measured in real numbers instead of integers, fewer equally likely trees are found compared with the number of equally parsimonious trees. ML makes explicit assumptions about the process of evolution through the use of a model of nucleotide substitution. If the model of evolution does not accurately represent the underlying processes of the data set, the method may become inconsistent. However, ML is postulated to be fairly robust to violations of model assumptions (Felsenstein 2004).

Perhaps the biggest drawback to ML is the extreme computational intensity required. In addition to computing the likelihood of each nucleotide on each topology, branch lengths must also be estimated from the data. Depending on the model of evolution used, other parameters may also require estimation. The main arguments against ML stem from the use of statistics to infer history. Some opponents purport that history should only be inferred from induction and that statistics has no bearing on history.

Bayesian Inference

Bayesian Inference (BI) is a statistical method that builds on maximum likelihood (Huelsenbeck et al. 2001). Whereas likelihood is stated as the probability of observing the data given the hypothesis, BI is based on the probability that a certain hypothesis is true given the data. This notion is summarized in the posterior probability (PP) of a given phylogenetic hypothesis. BI combines a researcher's prior beliefs, in the form of a prior distribution, with the likelihood of the data under a specified evolutionary model in order to obtain the posterior probability of each possible topology. In the case of phylogenetic analysis, a prior distribution may give equal weight to all tree topologies, or may incorporate a more complex set of prior beliefs, such as monophyly of a particular group of taxa. The flat or equal prior distribution, however, is often far from the truth.

In mathematical terms the PP is defined as

$$\mathrm{PP}(T, \tau, \theta \mid \mathrm{D}) = \frac{\Pr(\mathrm{D} \mid T, \tau, \theta) \times \Pr(T, \tau, \theta)}{\Pr(\mathrm{D})},$$

which states the posterior probability of the tree T, with branch lengths τ, and model parameters θ, given the data D, is equal to the likelihood (the probability of the data given the tree, branch lengths, and model parameters) multiplied by the prior probability of the tree with branch lengths and model parameters divided by the unconditional probability of the data. BI has an advantage in the straightforward interpretation of the PP as the probability that the tree is correct.

The PP of the data can be difficult to estimate. BI approximates the PP of phylogenies by sampling trees from the PP distribution using an iterative search strategy known as the Markov chain Monte Carlo (MCMC). This method starts with an initial tree, either random or specified, with accompanying branch lengths and model parameters. A new tree is proposed and accepted with a probability proportional to its likelihood. This is repeated thousands, even millions of times. The PP of a tree is approximated by the proportion of times that tree is sampled during the MCMC search. Typically, there is a burn-in phase of the MCMC process in which the search wades through suboptimal trees. After this phase, MCMC is said to converge on a set of optimal trees. The Bayesian estimate of the phylogenetic tree is often the topology that maximizes the PP. A variant of MCMC called Metropolis-coupled Markov chain Monte Carlo (MCMCMC) allows a greater searching of the PP distribution by escaping local optima. One problem with these search strategies is knowing when the process has converged onto a set of probable trees. It is not known whether a longer analysis time would have converged on a new PP distribution. In addition, MCMC requires the assumption of independence among search samples, a highly violated foundation. The nuances of searching strategies will be discussed in detail later.

MrBayes 3 (Ronquist & Huelsenbeck 2003) provides an incredible tool for analyzing heterogeneous data sets under BI. Most researchers incorporate data sets from various partitions into a single analysis—mitochondrial, nuclear, chloroplast, and morphological data, for example. Data from differing partitions may be evolving under completely different evolutionary models. The MrBayes 3 program allows the data partitions to be simultaneously analyzed under varied evolutionary models, providing a powerful tool for phylogenetic inference. Although somewhat faster than maximum likelihood, BI is still computationally intensive.

SEARCHING THE TREE SPACE

The next step in phylogenetic analysis is to search the tree space, an area defined by all possible tree topologies. The tree space is analogous to a three-dimensional landscape with peaks representing local optima. Tree plateaus or islands exist that contain numerous trees of similar class. These islands may represent different sets of locally most parsimonious trees (Maddison 1991). The globally optimal tree, or best tree, is located at the top of the highest peak in the tree landscape. The ultimate goal is to find the best tree defined by an optimized tree statistic, the tree with the highest likelihood for example. Several search methods exist, depending on the size and complexity of the tree space.

Exact Searches

Ideally, every tree in the tree space should be examined. An exhaustive search does just that. Unfortunately, this method becomes computationally untenable when dealing with more than about 12 taxa due to the exponential growth of possible trees upon addition of taxa. The branch and bound technique reduces computation time somewhat, but still guarantees the optimal tree (Hendy & Penny 1982). This method creates an upper bound optimality score based on a reasonable phylogenetic hypothesis (e.g., UPGMA). As trees are created, scores are estimated for each taxon addition. If a subtree, one without all the taxa added, has an optimality score worse than the upper bound, it is immediately discarded. For trees with scores better than the upper bound, an exhaustive search is instituted. However, this method is still unfeasible for data sets above 20 taxa. For problems involving larger data sets, heuristic methods must be used.

Heuristic Searches

Heuristic "hill-climbing" techniques search the tree space by trial and error and are thus faster than exhaustive methods, but do not guarantee that the optimal tree has been found. Because multiple islands of locally optimal trees exist, the best heuristic method repeatedly searches the tree landscape with a randomly selected tree topology in an effort to escape local optima and find the globally optimal tree. Heuristic searches may endeavor to find the best tree by swapping branches on an initial tree topology. Three types of rearrangement strategies are employed: nearest neighbor interchange (NNI), in which interior tree branches are exchanged; subtree pruning and regrafting (SPR), where a sister group is pruned off and grafted elsewhere on the tree; and tree bisection and reconnection (TBR), in which an internal branch is split and the subtrees grafted in a different place. TBR branch swapping is the most thorough and is typically used in standard heuristic searches (Page 1993b).

The most computationally expensive process of phylogenetic analysis is searching for the optimal tree. Various algorithms and computational methods have been formulated in an effort to decrease the computational time while increasing the accuracy of tree searching methods. Several different algorithms used within different optimality criteria now allow quicker, more thorough searching of the tree space. The neighbor-joining method (NJ) is often used as the starting point for heuristic algorithmic searches. It is a fast method for determining a starting tree to be improved upon by other criteria. The beam-bootstrap algorithm (BB) builds on the NJ algorithm by widening the search space. An initial NJ tree is used for a bootstrap analysis. In the search process, the putative pair to be added is analyzed in terms of its bootstrap score: the percentage of times the pair shows up within a certain number of resampled replicates. If the score is 95% or higher, the pair is accepted and the search strategy proceeds to the next node. If not, several topologies corresponding to the higher bootstrap values are saved and evaluated. The BB algorithm is computationally fast and allows the search path to examine wrong paths which may allow movement between local optima or tree islands, thus widening the search path (Rodin & Li 2000).

The parsimony ratchet is an iterative search algorithm that cuts down search time using a few different strategies (Nixon 1999). The algorithm starts by finding a local optimum using random addition with TBR swapping. A subset of the characters are then perturbed through reweighting and used for a second TBR branch swapping event on the first local optimum until a new optimal tree is found. Finally, the characters are reset to their original weight and TBR branch-swapped on the second optimal tree until a third and final optimum is found. This round of optimization events is performed between 50 and 200 times, expanding the tree space searched. In addition, Vos (2003) has modified the parsimony ratchet algorithm for use in maximum likelihood, significantly decreasing the time required to find optimal topologies.

Heuristic algorithms built on the idea of grafting also aid in tree searching. Both tree fusing (TF) and tree drifting (DFT) search for the optimal topology by exchanging subgroups on a tree. In TF, the subgroups must be composed of the same taxa. All possible exchanges between two trees are quickly evaluated, optimizing each subgroup individually, thereby determining the optimal topology. This process is repeated several times, changing the initial tree before each iteration. The DFT algorithm allows suboptimal solutions to be accepted conditional upon the raw length and relative fit differences between the optimal and less optimal trees. DFT differs from TF by swapping subgroups of conflict between optimal and less optimal trees, instead of areas of consensus. Both TF and DFT dramatically reduce computation time while widening the area of the tree space searched (Goloboff 1999).

Within maximum likelihood, several divide-and-conquer methods of heuristic searching have been developed recently. Quartet puzzling allows a comparatively fast heuristic tree searching method for large data sets. Instead of tackling *n* taxa as a whole, quartet puzzling constructs all possible four-taxon maximum likelihood trees. In the puzzling step, the four-taxon subtrees are randomly combined to create an overall *n*-taxon topology, one sequence at a time. The puzzling process involves a voting procedure in which a new sequence is placed in an edge position according to scores in the four-taxon subtrees. The puzzling procedure is repeated several times, the resulting trees being summarized in a majority rule consensus tree that includes a measure of nodal support (Strimmer & von Haeseler 1996). Some researchers claim that the use of quartets does not completely represent the original data. This method may also have difficulties when subtrees are incongruent with each other, presenting conflicts within the data. The disc-covering method does not require all combined subtrees to be compatible, allowing it to deal with noise in the data. It also allows the taxa to be partitioned into subtrees larger than quartets. This method divides the taxa into subsets according to a user-defined threshold value representing the allowable distance between taxa. Like quartet puzzling, the subtrees are combined into a consensus supertree (Huson et al. 1999). One flaw exists: the disc-covering method is often unable to determine any discs that will cover portions of the tree when under clocklike conditions.

Stochastic Searches

Stochastic searches introduce a random element into the search strategy, allowing a greater search of the tree landscape. In terms of searching within maximum likelihood, the simulated annealing algorithm locally rearranges topologies, accepting those that increase the likelihood score. The algorithm also accepts suboptimal topologies with a probability proportional to the decrease in likelihood. The introduction of a less likely topology allows the search to jump between local optima in a more efficient manner than heuristic reiterations used in hill-climbing techniques. This random reintroduction is accompanied by a decrease in computation time (Salter & Pearl 2001).

A new and promising method of searching the tree space lies in genetic algorithms—algorithms that use biological processes metaphorically to guide tree searching. These algorithms simulate evolution to guide the heuristic tree search. The genetic algorithm of Lewis (1998) allows "individuals" within a "population" to evolve via "natural selection" acting on variation contributed by "mutation" and "recombination." Individuals are defined by tree topologies, branch lengths, and parameter values specified within the model of evolution. Those individuals with greater fitness values, defined by its log likelihood score, contribute more "offspring," or topologies, to the next generation of searching. "Mutations" are simulated by randomly changing one component that defines an individual. "Recombination" occurs through branch swapping between offspring and parent topologies. Each generation consists of (1) evaluation of fitness; (2) the ranking of individuals by fitness; (3) the most fit individuals being allowed to leave *k* offspring in the next generation with every one but the first going through a round of mutation; and (4) recombination performed between offspring and parental generations. This tree searching method is exceptionally fast and efficient. The metapopulation genetic algorithm (metaGA) of Lemmon and Milinkovitch (2002) expands on the genetic algorithm by creating parallel searches, several "populations," that interact to find the optimal tree, speeding up the process even more. Using interpopulation consensus information, populations cooperate to find the optimal solution. The authors have incorporated metaGA into their computer program METAPIGA. The random component introduced by mutation allows the genetic algorithms to escape local optima, moving the search into the global landscape. However, the landscape is not guaranteed to be exhaustively searched and the algorithm may waste time by considering the same topology more than once.

CONFIDENCE ASSESSMENT

Phylogenetic analysis endeavors to reconstruct evolutionary relationships via an optimality criterion applied to data. Each node within a phylogeny is a hypothesis unto itself, carrying potential information about evolutionary history. However, application to evolutionary history comes only with a measure of support or confidence expressed for each node of the phylogeny, providing strength for each hypothesis. This can be done with the bootstrap or the jackknife using any optimality criterion, Bremer support within parsimony, or posterior probabilities under the Bayesian inference criteria. It is essential to note that all of these methods perform a confidence assessment of sorts in terms of the confidence in the estimated tree relative to the data at hand. They say nothing about the confidence in the tree relative to the true underlying evolutionary history.

The Bootstrap and the Jackknife

The bootstrap procedure was first applied to phylogenies by Felsenstein (1985) and has become the most widely used measure of support in phylogenetics. Bootstrapping produces a measure of support by randomly resampling columns of the original data matrix with replacement, building a pseudoreplicate data set in which some characters or nucleotide sites are represented multiple times or not at all. This pseudoreplicate is then analyzed using the same optimality criterion as the original data set. The process is repeated between 100 and 1000 times, creating a distribution of trees. The percentage of replicates in which a particular node or grouping exists becomes an estimate of how strongly the data support that group, called bootstrap proportions (BPs).

BPs can estimate relative support of groups based on any method of analysis or optimality criterion. However, this very advantage can also be disadvantageous in the light of computation time. If using neighbor-joining methods, bootstrapping can be a fast, effective method of support analysis. When using a criterion such as maximum likelihood, however, 1000 iterations can become overwhelming computationally. Some studies suggest that BPs are overly conservative, purporting a value of 70% to be roughly equal to a support measure of 95% (Hillis & Bull 1993). High BPs can give a good indication of strong phylogenetic signal, provided there is no evidence for misleading signals within the data. Low BPs do not necessarily mean false hypothetical relationships, just poorly supported ones.

The bootstrap method is not without opposition. Objections to the use of the bootstrap within phylogenetics involve sampling issues and statistical inference. The bootstrap method assumes that characters within a data set must be identically and independently distributed, an assumption that some say is unrealistic. Others have combated this objection with the theory that random sampling will create those very conditions. Statistical objections include the application of probabilistic models to phylogenetics, claiming that a phylogeny is a single historical event and that probability cannot be applied to a single trial. Answers to this objection include the view that phylogenetics is a sampling problem including parameter values and character choice, two situations in which probability can be applied (Sanderson 1995).

Jackknifing differs from bootstrapping only in the character resampling scheme; a proportion of the columns in the original data set are selected and deleted, instead of resampled with replacement. For example, deleting 50% of the data columns is known as the delete-half jackknife. Each pseudoreplicate is analyzed and the results are interpreted in the same manner as the bootstrap. Jackknifing and bootstrapping tend to produce comparable results.

Bremer Support

In parsimony, the length difference between the shortest consensus tree containing a clade and the shortest consensus tree without the clade is known as the decay index or Bremer support (Bremer 1988). This provides a measure of support for that clade. Total support for the tree is calculated by summing all clade decay indices. Unlike BPs, Bremer support is not scaled and is therefore much more difficult to interpret and offers no statistical power. Because this value is based on tree length, it is restricted to the parsimony criterion. Finally, because tree length changes with the number of characters and taxa, Bremer support values are not comparable across different data sets.

Posterior Probability

Posterior probabilities (PP) assess confidence within Bayesian inference. The PP builds on prior probabilities, which reflect the researcher's knowledge about the hypothesis before the data are seen. The PP is proportional to the prior probability multiplied by the likelihood of the data. Most researchers will use a flat prior, allowing the likelihood of the data to control the PP. The direct interpretation of the PP as the probability of monophyly and the faster computation of support relative to BPs make this method appealing.

Method Comparisons

There is controversy concerning the applicability and relationship between PPs and BPs. Several empirical and simulation studies have found that PPs and BPs are not equivalent. Alfaro et al. (2003) suggest that PPs are less likely to reject a true monophyletic grouping, have greater sensitivity to phylogenetic signal, and estimate accuracy better than BPs. Other studies purport that PPs are excessively high and may strongly support false phylogenetic hypotheses, suggesting the conservative nature of BPs to be preferable to the overestimation of PPs (Douady et al. 2003).

HYPOTHESIS TESTING IN A PHYLOGENETIC FRAMEWORK

A phylogenetic hypothesis suggested by the analysis of one data set will often conflict with the topology determined by a different data set. Statistical methods have been formulated to test whether the topologies of two trees are significantly different and which better fits the data. In testing tree topologies, two scenarios exist: a priori and a posteriori hypothesis testing. A priori testing involves pitting two alternative topological hypotheses defined before phylogeny estimation against one another. For example, say we have two hypothetical sister taxa to birds: bats and reptiles. To test the a priori hypotheses ((bird, reptile) bat) versus ((bird, bat) reptile) we obtain molecular data and fit the data to the phylogenetic hypotheses, enabling the determination of the sister group that best fits the data. A posteriori hypothesis testing involves testing the best estimate of phylogeny against alternatives (be they defined a priori or a posteriori); one example is testing the trees with the highest and next highest likelihoods. Several tests for each situation will be discussed.

Tests Within the Parsimony Framework

Hypothesis tests have been formulated within the parsimony framework that test a priori hypotheses: the Templeton test (Templeton 1983), the winning sites test (Prager & Wilson 1988), and the Kishino–Hasegawa test modified for use with parsimony (KH; Kishino & Hasegawa 1989). All of these approaches test the null hypothesis of no significant difference in the number of character differences between tree A and tree B. The winning sites test assumes that each character is equally likely to support either tree. This simple signed test sums the number of differences supporting one tree over the other and uses a binomial distribution to determine the probability of observing the difference in the number of winning sites. The Templeton test is similar to the winning sites test but differs by taking into account the magnitude of differences in the supporting characters for each tree; it is a nonparametric Wilcoxon Signed Ranks Test of the differences of character fit to the trees. The KH test is a parametric test that uses a *t*-test to compare the difference in steps between the two trees with a normal distribution. If the difference in steps is significantly different from zero, the null hypothesis (that the trees are the same) is rejected and the suboptimal tree can be declared worse than the optimal tree.

Tests Within Maximum Likelihood

The KH test was originally formulated to test the difference between log-likelihoods of two trees using a *t*-test. It compares both topology and branch length and is one of the most widely used tests of phylogenetic hypotheses. As mentioned above, this test is for a priori hypotheses. Many studies have used the KH test erroneously by applying it to a posteriori hypotheses, which dramatically violates the underlying assumptions of the test (Goldman et al. 2000). The Shimodaira and Hasegawa (1999) test (SH) is similar to the KH, but is applicable to both a priori and a posteriori situations and allows the comparing of several topologies. Shimodaira recognized a tendency of the SH test to be somewhat conservative and to be less accurate when comparing large numbers of trees. The weighted Shimodaira test (WSH) is more suited to comparing larger groups of trees. To overcome the biased nature of the KH test and the problems with the SH test, Shimodaira developed the approximately unbiased test (AU), which uses a multiscale bootstrap technique and can accommodate large numbers of candidate trees (Shimodaira 2002).

The Swofford–Olsen–Waddell–Hillis (SOWH) test is a parametric maximum likelihood test applicable to a posteriori hypothesis comparisons. Parametric tests differ from nonparametric tests by simulating replicate data sets according to parameter values estimated from the original data set; analysis of the replicate data sets is according to the specified model (Goldman et al. 2000). Because parametric tests allow the use of complex and specific models, they may provide more power and be less prone to type 1 error than nonparametric tests. Many real data sets, however, do not conform to the underlying assumptions of complex models, making the SOWH less applicable. Likelihood ratio tests also provide a simple form of testing both a posteriori and a priori hypotheses (Huelsenbeck & Bull 1996).

COMBINING DATA

In reconstructing phylogenies, most researchers have access to more than one kind of data set—a morphological and molecular data set, or two or more genes from different partitions. The use of multiple data sets with different underlying evolutionary models requires special attention in phylogenetic analysis, as the hypothesis from each data set may differ from each other as well as from the combined analysis.

The issue of combining data sets is a hot topic in phylogenetics; various perspectives exist. The "total evidence" approach argues that data should always be combined into a single analysis to maximize explanatory power, and that the combined analysis results make individual results irrelevant. Opponents of the "total evidence" approach claim that combining the data masks information held within individual data sets, including the need for different evolutionary models. Thus, the method of analysis for one data set may not be the best for another. Some proponents of individual analyses prefer to summarize results from each data set using consensus trees, finding confidence in those nodes found in each of the different hypotheses. Others prefer to use conditional combination in which data sets are analyzed individually and compared for incongruence between resulting hypotheses using various statistical tests. If incongruence is low, the data sets may be combined for a total evidence hypothesis (Huelsenbeck et al. 1996). Some statistical tests for incongruence within parsimony include the Templeton test discussed above, and the incongruence length difference (ILD) test, also known as the partition homogeneity test (Farris et al. 1994). The ILD test looks at the difference between the number of steps required for the individual versus combined analyses. This test, however, has been criticized as weak. Simulation tests have shown that the ILD test lacks power to detect incongruence when caused by differing topologies, small number of informative sites, and large heterogeneity of among-site substitution rates (Darlu & Lecointre 2002). Within maximum likelihood, the likelihood heterogeneity test compares the likelihood scores obtained if the same phylogeny resulted under all data sets versus if different phylogenies were allowed (Huelsenbeck & Bull 1996).

CASE STUDY

The Origin of the Freshwater Crayfish

This case study is based on Crandall et al. (2000a).

Objective

(1) To investigate the origin of freshwater crayfish; (2) to test hypotheses of crayfish monophyly versus inclusion of the Nephropoidae between the Northern and Southern Hemisphere crayfish.

Sampling

Infraorder Astacidea includes three superfamilies: Astacoidae (Northern Hemisphere), with two families Cambaridae and Astacidae; Nephropoidea (clawed lobsters); Parastacoidae (Southern Hemisphere). DNA was extracted and amplified for 18S, 28S and 16S mitochondrial DNA genes via polymerase chain reaction and subsequently sequenced.

Phylogeny Reconstruction

Sequences were aligned using ClustalX and edited by eye; phylogenetic analysis was carried out using PAUP*; the model of evolution was determined using Modeltest; a starting tree was obtained via neighbor joining; the tree was rooted using the outgroup method; maximum likelihood and maximum parsimony optimality criteria were used; heuristic tree searches were done using random sequence addition; confidence assessment was carried out via 1000 bootstrap replicates; and a partition homogeneity test was conducted on three separate data sets to assess combinability.

Results

The partition homogeneity test found no significant heterogeneity among the three gene partitions, justifying a combined analysis ($P = 0.618$). Modeltest showed the transversion model of evolution with a gamma-distributed rate heterogeneity model and an estimated proportion of invariable sites to be the best model for the combined data. A 10-replicate heuristic search using random sequence addition under the above model found one maximum likelihood tree with a score of –ln 11,820.57, shown in Figure 28.5.

Discussion

The combined analysis strongly supports monophyly of freshwater crayfish with Nephropoidae as outgroup. Astacidae and Cambaridae grouped with 100% bootstrap support. Parastacidae is monophyletic. However, the position of the genus *Cambaroides* as sister to *Pacifastacus*, an Astacidae member, inhibits monophyly of both the Astacidae and Cambaridae families. The monophyly of freshwater crayfish and the geographical distribution suggests that crayfish originated in Pangaea by the Triassic period. Pangaea's split into Laurasia and Gondwana fits with the hypothesis. In the application of phylogenetics to geographical history, the phylogeography of the Southern Hemisphere crayfish found in South America, Madagascar, Australia and fossils in Antarctica follow the pattern of Abelisauridae, a predatory dinosaur group, providing further evidence for the hypothesis of contact between these land masses via Antarctica.

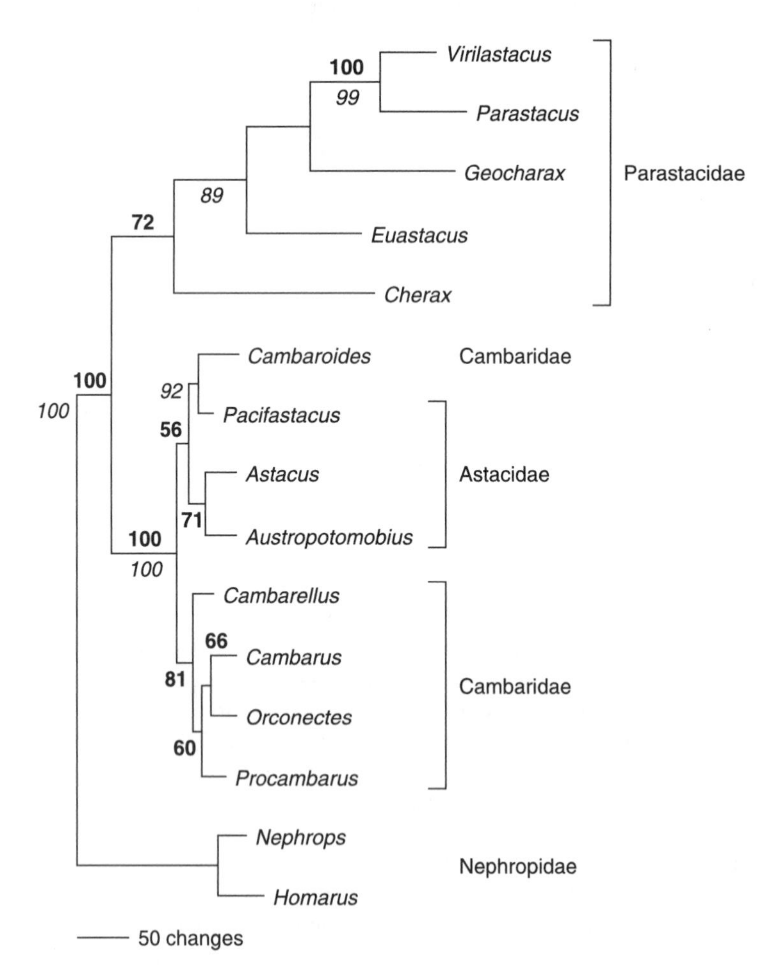

FIGURE 28.5. Maximum likelihood tree for the phylogeny of freshwater crayfish. Bootstrap scores are shown for maximum likelihood (italics) and maximum parsimony (bold) based on 1000 bootstrap replicates.

PHYLOGENETIC SOFTWARE

All the above tests except WSH, AU and SOWH can be implemented in PAUP* 4.0 (Swofford 2002). The AU test and WSH can be applied using the computer program CONSEL downloadable from http://www.is.titech.ac.jp/~shimo/prog/consel/index.html (Shimodaira & Hasegawa 2001). SOWH can be obtained from the authors at http://www.ebi.ac.uk/goldman/tests/index.html. Felsenstein keeps a website that lists most of the currently available software packages for phylogenetic inference (freeware, shareware, and commercial packages) at: http://evolution.genetics.washington.edu/phylip/software.html

SUGGESTIONS FOR FURTHER READING

The best text on phylogeny reconstruction is the recent book by Felsenstein called *Inferring Phylogenies* (Felsenstein 2004). It is without parallel in terms of depth and completeness of coverage. An easier introduction to phylogenetics including programs and techniques is Barry Hall's book *Phylogenetic Trees Made Easy: A How To Manual* (Hall 2004). For the more adventuresome and mathematically inclined, *Phylogenetics* offers a mathematically rigorous treatment of phylogeny reconstruction (Semple & Steel 2003).

Felsenstein J 2004 Inferring Phylogenies. Sinauer Assoc.

Hall BG 2004 Phylogenetic Trees Made Easy: A How To Manual. Sinauer Assoc.

Semple C & M Steel 2003 Phylogenetics. Oxford Univ. Press.

VI

EVOLUTIONARY GENETICS IN ACTION

29

Evolutionary Genetics of Host–Parasite Interactions

PAULA X. KOVER

Parasites[1] include a broad group of organisms: from viruses to parasitic fish, including a few plant species. While some parasites have little or no effect on their hosts, most have been shown to significantly affect host traits such as morphology, life history, and mating system (Kuris 1974; Clay & Putten 1999). Consequently, parasites have the potential to significantly affect host fitness and exert strong selective pressure on host traits. Likewise, parasite fitness is dependent on the defense strategies deployed by hosts. Thus, evolutionary changes in the host can also have an effect on pathogen fitness. This interdependence between host and parasite fitness can lead to a coevolutionary process, where reciprocal, adaptive changes occur in both host and parasite.

The degree to which parasites reduce host fitness is called *virulence*, and the degree to which hosts reduce pathogen growth, or their exposure to pathogens, is called *resistance*. For coevolution to occur, both resistance and virulence traits must be genetically based and genetically variable. While empirical studies have confirmed that resistance and virulence are genetically based and variable in many natural systems, there are still very few well-documented cases of coevolutionary interaction between host and parasites. In part, the paucity of well-documented cases is due to the fact that demonstrating coevolution requires studies across generations and geographic areas, and at the molecular level (further discussed below). It is also due to the fact that the genetic basis of traits underlying resistance and virulence is much more complex than initially thought.

Parasites have caused severe human epidemics, such as the 1918 influenza pandemic. They have also been responsible for drastic food shortages, such as the potato famine of 1846 in Ireland. In addition, new emerging diseases are constantly threatening human health, agriculture, and conservation efforts. Consequently, the genetics and evolution of host–parasite interactions have been under investigation for more than a hundred years. Furthermore, because the interaction between hosts and parasites has been shown to affect species distribution and community-level diversity (Kareiva 2000), understanding their interaction has also become of interest to conservation biologists. While much of the research on host–parasite interactions has focused on identifying genes that determine resistance and virulence, extensive work has also been carried out on their coevolutionary dynamics. Evolutionary modeling of host–parasite interaction has been particularly focused on the following questions: Why is there genetic variation for resistance and virulence? and How is this variation distributed and maintained?

The reason why the observed widespread polymorphism in resistance and virulence is puzzling is that both traits are expected to be favored by selection (but see Box 29.1), and should therefore quickly lose genetic variation as alleles that confer resistance and virulence goes to fixation. Two main hypotheses have been proposed to

[1]The term "pathogen" is also used to describe microparasites such as viruses, bacteria, and fungi. In this chapter, the more general term "parasite" will be used equivalently to "pathogen."

BOX 29.1. The Coevolutionary Consequences of Tolerance Versus Resistance

Host response to parasite selection is usually equated with the evolution of resistance traits such as biochemical or physical barriers that prevent or contain parasite attack. However, an alternative response to parasite selection is to evolve tolerance traits. Tolerance traits are those that reduce the effect of infection on host fitness, without directly affecting pathogen growth. While tolerance has been primarily investigated in interactions between plants and herbivores (reviewed in Strauss & Agrawal 1999), genetic variance in tolerance to a bacterial parasite has been demonstrated in plants (Kover & Schaal 2002) and animals (Hansen & Koella 2003). Tolerance traits in plants often involve augmentation of resources to compensate for losses due to parasites and herbivores, such as an increase in chlorophyll concentration in leaves, nutrient uptake, and vegetative branching.

While resistance and tolerance can equally improve host fitness upon parasite attack, the evolutionary outcome of host–parasite interactions can be fundamentally different depending on whether hosts evolve resistance or tolerance (Roy & Kirchner 2000). Disease resistance in the host population places strong selection on pathogens to evolve new genotypes that can avoid plant defenses, generating the kind of frequency-dependent selection that favors complex coevolutionary dynamics and maintains genetic polymorphism in both host and pathogen populations. In contrast, tolerance traits decouple the association between traits that confer resistance and host fitness, reducing selection for resistance. Furthermore, alleles that confer tolerance are expected to increase in frequency until they go to fixation due to a positive feedback. An increase in frequency of tolerance within the host population can lead to an increase in disease incidence (since infection is not being reduced by resistance traits), causing an increased selection for tolerance. Thus, when hosts respond to parasites by evolving tolerance, frequency-dependent cycles should not be observed and genetic diversity is not maintained. On the other hand, because tolerance traits arrest the coevolutionary "arms race" between host and parasites, they might be the ideal types of trait to provide long-term protection against crop loss due to parasites and herbivores. Currently, new virulent races of parasites and herbivores can overcome the resistant genes introduced into crops within a few years. Thus, a constant search for new resistant genes to be introduced is needed.

Very few studies have specifically investigated the occurrence of tolerance to parasites in natural populations. This is clearly an area that needs more research since the occurrence of tolerance can affect many aspects of the interaction between host and parasites. Besides the factors mentioned above, tolerance can also affect the empirical estimates of resistance in natural populations. Because it is commonly assumed that pathogen effects on host fitness are a direct result of pathogen growth in host tissues, many empirical studies estimate plant resistance by measuring a single trait such as symptoms or seed production. However, if tolerance traits are present these measurements are not equivalent. Studies of tolerance may also provide insights into the pathway for the evolution of mutualism.

explain polymorphisms in resistance and virulence. One possibility is that polymorphisms are maintained through frequency-dependent selection. Alternatively, it has been suggested that polymorphisms are maintained by a balance between costs of resistance and virulence and their selective advantage. In this chapter I will first introduce the two main coevolutionary models that underlie the two hypotheses outlined above. The assumptions and predictions of each model are also discussed. This will be followed by a discussion of the empirical evidence that supports or complicates these models.

COEVOLUTIONARY MODELS

Many theoretical models of the evolutionary interaction between hosts and parasites have been developed since the 1940s (Haldane 1949; Jayakar 1970; Hamilton 1980; Frank 1994a; Parker 1994). These models vary in the assumptions they make about the genetic basis of disease resistance and virulence. Factors considered in these models include: the ploidy of hosts and parasites, the number of alleles and loci involved in their interaction, and whether epistasis and pleiotropy are present. Other factors that have been considered are: random versus assortative mating, discrete versus continuous generations, the relative generation times of hosts and parasites, and the amount of gene flow between neighboring populations. These models have shown that the interaction between hosts and parasites can lead to cycling in allele frequencies that can explain the maintenance of genetic diversity and sexual reproduction. However, many other outcomes are possible depending on the assumptions made. One of the most significant factors affecting the outcome of these models is the assumed mechanism of resistance and virulence. While the exact genetic mechanisms remain elusive for most organisms, two main approximations have been developed to capture the empirical results available: the "matching-allele" and "gene-for-gene" mechanisms. Below I discuss the assumptions of these two mechanisms and their consequences for the coevolutionary dynamics of hosts and parasites.

Matching-Allele Models

The matching allele mechanism (MAM) is based on the "self/nonself" recognition system of animals (Grosberg & Hart 2000). With this mechanism, hosts are hypothesized to have self-recognition genes that can prevent parasite infection by detecting parasites as "nonself." However, if a parasite matches exactly the host self-recognition genotype, it will go undetected, causing infection in the host (see Figure 29.1). When interactions between hosts and parasites are modeled assuming MAM, it is easy to see that parasites capable of "matching" the most common host genotype will have a higher success of infecting hosts, and will consequently enjoy higher fitness. Conversely, rare host genotypes will have a selective advantage because they will have a lower probability of becoming infected. However, as rare host genotypes increase in

"Gene-for-gene"

Pathogen genotype: / Host phenotype:	VV avirulent	Vv avirulent	vv virulent
rr	+	+	+
Rr	–	–	+
RR	–	–	+

"Matching–allele"

Pathogen genotype: / Host phenotype:	AA	AB	BB
AA	+	–	–
AB	–	+	–
BB	–	–	+

FIGURE 29.1. Comparison of the matching-allele and gene-for-gene models of interaction between host and parasite. A (+) indicates infection was successful, and a (–) indicates infection was unsuccessful and the host is resistant.

frequency, matching parasites will also increase in frequency, leading to a frequency-dependent cycling of allele frequencies in both host and parasite (Figure 29.2A).

Frequency-dependent models have been of particular interest because they support the "Red Queen theory" championed by William Hamilton and colleagues (Hamilton 1980; Hamilton et al. 1990). According to this theory, the negative frequency-dependent selection exerted by parasites on host populations will favor genetic recombination. Thus, selection exerted by parasites may explain the widespread occurrence of sex despite its 2-fold cost relative to asexual reproduction. Models that explicitly test this hypothesis have shown that recombination is favored in this scenario because it breaks unfavorable genetic combinations created by previous selection and recreates rare genotypes that have been selected against in the past (Hamilton et al. 1990; Peters & Lively 2000). While other antagonistic species can cause frequency-dependent selection, parasites are viewed as a particularly powerful candidate to explain the maintenance of sex because parasites are ubiquitous and can exert strong selection on specific host genotypes. More importantly, parasites tend to have a much shorter generation time and larger population sizes than their hosts. Thus, parasites should be able to adapt quickly to the most common host, leaving genetic variation over time as the best strategy for the hosts.

While empirical tests of the Red Queen hypothesis are hard to perform, it is possible to test some of its main assumption and predictions. A central assumption of the theory is that parasites exert frequency-dependent selection on hosts. Support for this assumption has been found in the interaction between snails and trematodes (Lively & Dybdahl 2000), where rare host genotypes are less targeted by sympatric parasites than common hosts. Frequency-dependent selection exerted by parasites was also observed in an experimental population of wheat, where seed production of different lines was correlated with their frequency in the presence of the rust parasite, but not in its absence (Brunet & Mundt 2000). While these results are encouraging, it would be preferable to have data from more systems before this result can be generalized. Nevertheless, other indirect tests (discussed below) have been performed and provide evidence for the occurrence of frequency-dependent selection in a wider group of organisms.

Gene-for-Gene Models

The gene-for-gene (GFG) mechanism was proposed by Flor (1947) to describe the outcome of his empirical experiments in the interaction between races of flax rust fungus and flax plants. He showed that their interaction was highly specific, and proposed that the outcome depended on the interaction between the products of a resistance locus in the plant and a virulence[2] locus in the fungus. In the GFG model, hosts are resistant when they have a dominant resistant allele R at the resistance locus, unless the parasite has a recessive genotype vv at the virulence locus (see Figure 29.1). Flor's results have been used sometimes to support the Red Queen hypothesis, because they provide evidence that host–parasite interactions are genotypic-specific and could, therefore, lead to frequency dependent selection. However, Parker (1994) has shown that a strict GFG interaction does not lead to frequency-dependent selection, and therefore does not support a role for parasites in the maintenance of sexual recombination. Instead, with GFG, the coevolutionary dynamics are better described as a series of "selective sweeps" (see Figure 29.2B). In the GFG mechanism, the resistant locus in the host is usually envisioned as a molecular receptor capable of recognizing the product of a specific virulence locus. When specific recognition occurs, the defense responses are triggered. Thus, a new R allele capable of recognizing an elicitor previously undetectable should be selectively favored and sweep through the population until it becomes fixed. As the new R allele becomes common, any mutation in the virulent locus that renders the elicitor unrecognizable will produce a new virulent allele for the parasite. The new virulent allele should also sweep through the parasite population and become fixed. Therefore, with GFG, polymorphism at resistant and virulent loci is transient, and new resistant genotypes can evolve through the appearance of new alleles or new loci.

Disease resistance in plants is typically described as following GFG (but see Frank 1993 and Innes 1995), and many resistance and avirulent genes have

[2]Although this gene present in the parasite ultimately determines infectivity or "host range"(i.e., the ability to infect or not a given host), it was originally named by Flor (1947) as a "virulent gene." Since all of the plant pathology literature still calls it a "virulent gene," I will keep with the original nomenclature. But it is important to understand that the role of this gene is different from the commonly accepted definition of virulence in the evolutionary literature (i.e., the effect of the parasite on host fitness).

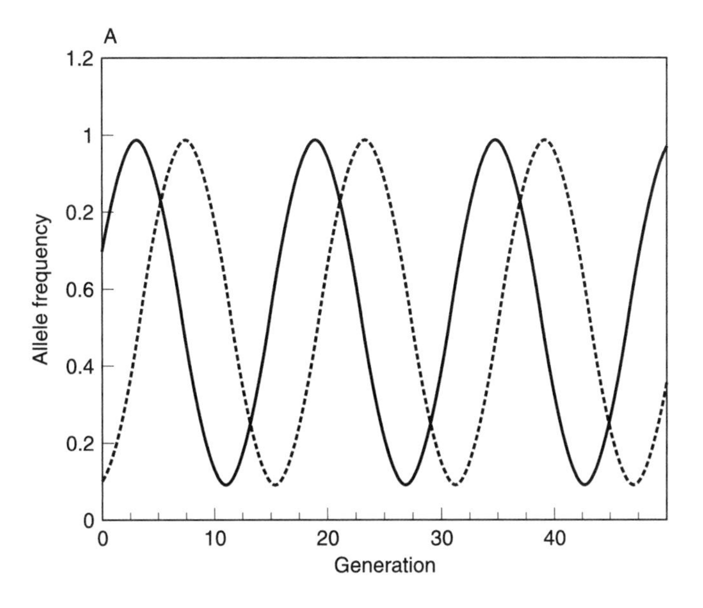

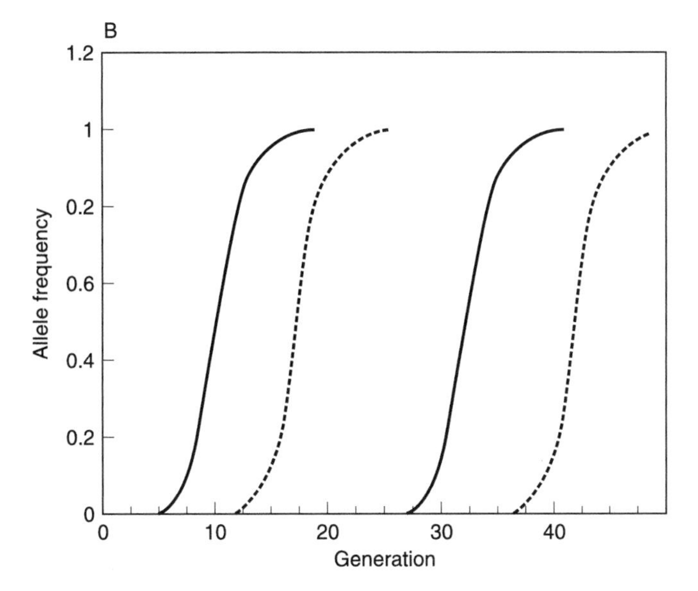

FIGURE 29.2. Changes in allele frequency as a result of coevolution between host and parasite. (A) The cyclic changes in allele frequency in host and pathogen expected under frequency-dependent selection (Red queen dynamics). (B) The "arms race" expected under the gene-for-gene mechanism of resistance, with successive selective sweeps. Each line represents a new resistant (black) or virulence (gray) allele.

been identified and cloned (reviewed in Staskawicz 2001). Thus, we would expect that most polymorphisms in resistance and virulence in natural populations are transient. However, stable polymorphisms for resistance are often observed. How can this be explained? One possibility is that genes that confer resistance and virulence bear some fitness costs that counterbalance the selective advantage of the resistance loci in the presence of parasites.

Studies on the existence of a cost of resistance have been numerous (reviewed by Bergelson & Purrington 1996). A cost of resistance is indicated by a resistant genotype having lower fitness than a susceptible genotype in the absence of infection. While many studies have shown a cost for resistance, an almost equal number failed to find evidence of a cost. Thus, the issue remains controversial. One of the difficulties in testing for costs of resistance is the separation of the effect of the resistance gene from its genetic background. In other words, a true cost of resistance comes from pleiotropic effects of genes that confer resistance. Costs that are due to genes being in linkage disequilibrium to resistance traits can be separated from resistance genes through recombination and should not be considered true costs. Backcrossing of resistant lines into susceptible lines has been used to reduce the effect of the genetic background, but this procedure does not completely eliminate the effect of tightly linked genes. Model organisms that can be genetically transformed with a single resistance gene offer a good solution to this problem. In a recent study, Tian et al. (2003) used this approach to estimate the cost of a single *R* gene in *Arabidopsis thaliana* (See Box 29.2). Multiple transformations were used to control for the possible effect of the position where the gene was inserted. They found that plants carrying the *Rpm1* gene produced on average 9% fewer seeds than susceptible plants. This is a considerable cost for a single resistance gene, considering that each individual is expected to have many *R* genes. While the cost of resistance has typically been estimated through changes in host survival or reproduction, other aspects of host fitness that are not usually measured could also be affected. For example, it has been shown that resistance can affect the competitive ability of hosts (Kraaijveld & Godfray 1997). Thus, before the issue of cost of resistance can be resolved, we need more experiments in which the effect of resistance genes can be properly isolated and a large number of traits in the host are measured under different environmental conditions.

Similar issues plague research on the cost of virulence (i.e., the cost of carrying virulence genes). In addition, this literature is further muddled by the fact that the cost of virulence varies depending on whether the gene in question affects the host fitness, the pathogen infectiveness, the pathogen range, or a combination of these traits. Nevertheless, a recent review supports the existence of virulence costs (Zhan et al. 2002 and references therein) and suggests that trade-offs with transmission might be a possible cause of the costs. For example, Thrall & Burdon (2003) found that, in a metapopulation of flax plants, flax rust strains containing virulent genes were more common in flax populations with high resistance than in populations that were mostly susceptible. In addition they demonstrated the existence of a trade-offs between spore production (needed for transmission) and the presence of virulence genes. Similar trade-offs between virulence and transmission have been observed in other systems (e.g., the protozoan parasite that causes malaria; Ferguson et al. 2003), suggesting this might be a general mechanism for the cost. The presence of a cost and the properties of the coevolutionary dynamic can also affect the way in which parasites evolve to become more or less virulent (see Box 29.3).

Quantitative Models

In both GFG and MA models, resistance and virulence are qualitative. What this means is that hosts are either resistant or susceptible, and parasites are either virulent or avirulent. However, in natural populations hosts often do not show categorical variation in whether they are susceptible or resistant, but rather in the degree to which they are susceptible (Burdon 1987). While in the past the study of the genetic resistance has been limited to genes of large effect, the availability of polymerase chain reaction (PCR)-based molecular markers and powerful statistical techniques has allowed the development of methods to investigate the genetic architecture of continuous traits (see Cheverud, Ch. 19 of this volume). A recent survey of the genetic architecture of disease resistance, which included 115 quantitative trait loci (QTL) studies, found that disease resistance in plants often involves more than one locus (Kover & Caicedo 2001). In contrast, there are relatively few coevolutionary models that explicitly consider resistance and virulence being determined by more than three

BOX 29.2. *Arabidopsis* as a Model Organism in Evolutionary Genetics
Kentaro K. Shimizu and Michael D. Purugganan

Arabidopsis thaliana (L.) Heynh. (family Brassicaceae) is an annual weedy plant (Figure 1), with a native range that covers Eurasia and Northern Africa, and is naturalized widely in the world, including in North America and Japan. It occupies disturbed habitats including the margins of agricultural fields. Natural populations usually germinate in the fall and overwinter as vegetative rosettes (a fall annual strategy), and flower in spring. A spring annual strategy, with germination and flowering in spring, is also observed. This plant is predominantly selfing, with a reported outcrossing rate of approximately 1%. The plant is small in stature and, for the spring ephemerals, has a generation time from 30 to 90 days, depending on the specific genotype. The size, ease of cultivation, and short generation time have made *A. thaliana* a model species for genetic studies in plant biology, including investigations in plant development, pathogen resistance, and environmental response. In 2000 the complete genome sequence, including the annotation of over 25,000 genes, was reported, making it the first plant genome to be completely sequenced (Arabidopsis Genome Initiative 2000).

FIGURE 1. *Arabidopsis thaliana* in Johston County, North Carolina, USA.

Over 400 *A. thaliana* accessions (commonly referred to as ecotypes) have been collected by research workers across its growing range, and seed material from the majority of these accessions are readily available from stock centers in the United States, Europe and Japan. The accession collections show various aspects of phenotypic variation, including natural quantitative differences in flowering time, pathogen resistance, and morphology. These accession collections have been extremely useful in ecological and evolutionary studies, including quantitative and molecular population genetic studies. The level of silent site nucleotide diversity for these accessions, π, is on average 0.7%, and within-species linkage disequilibrium extends from 50–250 kb (Nordborg et al. 2002). As *A. thaliana* is an inbreeding species, most of the variation is distributed between rather than within populations, although polymorphic lines are found within populations, possibly as a result of population admixtures. Analysis of neutral loci reveals evidence of recent population expansion. The disruption of the self-incompatibility *SCR* gene is thought to have facilitated the evolution of selfing in this annual species, and thus the rapid expansion from glacial refugia (Shimizu et al. 2004; see below).

Recent taxonomic studies recognize ten species in the genus *Arabidopsis*, many of which are also utilized in ecological and evolutionary studies. *A. lyrata* is often used as an outgroup in evolutionary studies of *A. thaliana*. An allopolyploid,

(continued)

BOX **29.2.** *(cont.)*

A. suecica, is providing a unique opportunity to study the mechanism of speciation by polyploidization. Research tools and conclusions from *A. thaliana* can readily be transferred to these other wild Brassicaceae species (Mitchell-Olds 2001).

QTL Mapping in *A. thaliana*

Many phenotypic traits of evolutionary interest are quantitative in nature, and mapping quantitative trait loci (QTLs) has been a major objective in evolutionary and ecological genetic studies (Cheverud, Ch. 19 of this volume). QTL mapping in *A. thaliana* is facilitated by the availability of several recombinant inbred lines (RILs), a number of which have already been genotyped with 100–600 markers (see Koornneef et al. 2004 for details). QTL analyses of several traits have been reported, including plant size, life history traits such as the time to flower induction, levels of secondary compounds to avoid insect herbivory, and fitness components such as seed number. Moreover, the sessile nature of plants allows QTL mapping in natural field conditions, providing the basis for studying the genetic architecture of phenotypic traits under ecologically relevant environments (Weinig et al. 2003).

Linkage disequilibrium (LD) or association mapping is another approach to identify genes underlying natural variation in ecological and evolutionary traits. This approach has been used to map a number of different flowering time genes, and future LD mapping studies will be facilitated by the availability of high-density sequence data for approximately 3000 gene fragments in a worldwide collection of 96 accessions (Nordborg et al. 2002).

Isolation and Analysis of Evolutionary and Ecological Gene Functions

Molecular geneticists traditionally used laboratory-induced mutants to elucidate the function of new genes. They are currently utilizing naturally occurring variation to isolate new genes and new alleles in developmental and physiological pathways. In cloning a QTL, small genomic regions spanning the particular QTL are introgressed into a parental background, creating near-isogenic lines (NILs). In NILs, the QTL behaves as a quasi-Mendelian trait, with one defined locus segregating for the quantitative phenotype, and the gene can be isolated using traditional map-based cloning techniques.

The *Arabidopsis* research community has developed a large number of genetic and genomic tools to facilitate map-based cloning, including the whole genome sequence, genetic maps, BAC libraries, more than 30,000 SNP markers, and comprehensive gene-disruption lines. Alleles of three life history genes—*FRIGIDA*, *FLC*, and *CRY2*—which influence flowering time (see below) have been identified and/or isolated from natural alleles segregating in *A. thaliana* populations using map-based cloning procedures. In some cases, the gene can also be easily isolated without constructing a NIL, because only one major QTL often accounts for most of the variation; an example is the disease resistance gene *RPM1* (see Stahl et al. 1999 and below).

The large number of resources available to the *Arabidopsis* genetics community facilitates studies of the genetic basis of evolutionary and ecological phenomena. A large body of information is available for many genes, and gene disruption lines of most genes based on T-DNA insertions are available. Transcriptome analysis is readily undertaken, with DNA chips, cDNA and long-oligo microarrays available.

BOX 29.2 *(cont.)*

Gene expression profiles for a large number of genes are already available from a large number of studies, including data from massively parallel signature sequencing (MPSS).

Adaptive Molecular Variation: Genes to Ecology

Understanding the molecular basis of adaptation remains a critical area of research in evolutionary biology. Molecular population genetic studies can detect the action of historical selection, but such an approach alone cannot identify the relevant phenotype that is the target of selection. On the other hand, molecular genetics alone cannot demonstrate the evolutionary and ecological importance of any particular locus. Recent ecological and evolutionary studies in *Arabidopsis* have taken explicitly interdisciplinary approaches to dissect the evolutionary forces acting on genes responsible for phenotypic evolution. Here, we discuss a few examples from the recent literature.

Host–Pathogen Interaction

The *RPM1* gene confers resistance to the pathogenic bacteria *Pseudomonas syringae avrRpm1*. Many accessions have a non-functional allele of *RPM1* gene, and both functional and nonfunctional alleles have been maintained for much longer time than neutral expectation, which is incompatible with a traditional "arms race" hypothesis that predicts a rapid turnover of alleles (Stahl et al. 1999). It has been suggested that a trade-off between resistance and fitness costs results in balancing selection on functional and nonfunctional alleles of this gene (Tian et al. 2003).

Herbivory Genes

Secondary metabolites play major roles in plant resistance to insect herbivores. QTL mapping studies have detected a major locus that underlies the diversity of glucosinolate compounds and resistance to generalist insect herbivores. This locus harbors a small gene family encoding MAM enzymes, and analysis of 25 accessions reveals complex patterns of molecular variations at this locus and the possible action of balancing selection (Kroymann et al. 2003).

Flowering Time

The timing of flowering is a major life history transition in plants. Recent molecular analyses have identified alleles at three genes—*FRIGIDA*, *FLC*, and *CRY2* (Caicedo et al. 2004; Koornneef et al. 2004)—that underlie natural variation in reproductive timing and its response to ecological cues. An epistatic interaction between *FRI* and *FLC* has also been shown to be associated with an observed latitudinal cline in flowering time (Caicedo et al. 2004).

The Evolution of Selfing

In 1876, Darwin proposed that selfing in plants can be advantageous if mates or pollinators are scarce, and recent analysis of *A. thaliana* provides the first molecular evidence for the hypothesis. Selfing in *A. thaliana* has evolved by the inactivation of the self-incompatibility recognition genes, and a molecular evolutionary analysis indicates that directional selection has driven the evolutionary fixation of nonfunctional alleles of the self-incompatibility gene *SCR* (Shimizu et al. 2004).

BOX **29.3.** Evolution of Virulence

Until recently, the medical community believed that parasites should evolve toward reduced virulence because high virulence could decimate the host population, and therefore could also drive parasites to extinction. However, evolutionary theory predicts that natural selection should favor genotypes that have the highest fitness at any point in time without regard to the long-term consequences. Thus, evolutionary models of virulence propose that optimal virulence level is the result of a trade-off between parasite reproduction and transmission (Anderson & May 1992; Ewald 1983). Maximization of pathogen reproduction should cause an increase in virulence, if the reduction in fitness experienced by an infected host is due to the growth and/or multiplication of the parasite. However, under some circumstances transmission requirements may favor reduced virulence. For example, sexually transmitted parasites will only accrue fitness if their hosts stay alive through their reproductive age. In contrast, parasites that are transmitted independent of whether their hosts are alive or dead (e.g., fungi that sporulate on plant leaves) should maximize reproduction independent of their effect on host fitness. Thus, according to the trade-off hypothesis there is an optimal solution between these two extremes, where parasite transmission is maximized over the lifetime of the infection (box figure). The predictive value of this hypothesis depends strongly on the relationship between virulence, transmission, and parasite fitness. While a recent review provides support for a generally positive correlation between these traits (Lipsitch & Moxon 1997), there are many exceptions. For example, the bacteria that cause meningitis mainly grow in the nasal passages of humans where is transmitted through droplets, without causing any symptoms. The actual disease occurs when bacteria enter the cerebrospinal fluid symptoms from where it cannot be transmitted. Symptoms can also be the result of host immune response and not the pathogen growth per se. The decoupling of transmission and virulence can significantly affect the outcome of the evolution of virulence as predicted by the trade-off hypothesis (Figure 1).

The trade-off hypothesis has been lauded as a revolutionary idea for medical practice, because it suggests that human management of disease could directly influence parasite virulence (Ewald 1994). This notion, however, has strong critics (e.g., Ebert & Bull 2003). At issue is the fact that most of the empirical support for the trade-off hypothesis has been from experiments performed in very controlled situations, or comparing extremes (such as exclusively vertical versus exclusively horizontal transmission). More importantly, experiments that tried to directly affect virulence by selecting on transmission had very limited success, suggesting that evolutionary pathways in parasites are more complex that anticipated. Further theoretical developments have also called for caution in applying the trade-off model to manage disease, by showing that many departures from the basic assumptions can lead to different outcomes. Among the most noteworthy is the presence of multiple strains of parasites within species. In this case, the within-host competition is expected to favor higher parasite reproduction and consequently an increase in virulence.

Further caution on the application of the trade-off hypothesis comes from some technical difficulties in estimating virulence, which affects our ability to test the available models. By definition, the virulence of a parasite is the reduction it causes in host fitness. Thus, while virulence is defined as a parasite trait, its expression depends on host traits. Furthermore, the effect of the parasite on host fitness is most commonly estimated as differences in host mortality. This may constitute a problem, since it ignores the fact that parasites can affect many other aspects of host life history, such as developmental and reproductive schedule. For example, many parasites castrate

BOX 29.3. *(cont.)*

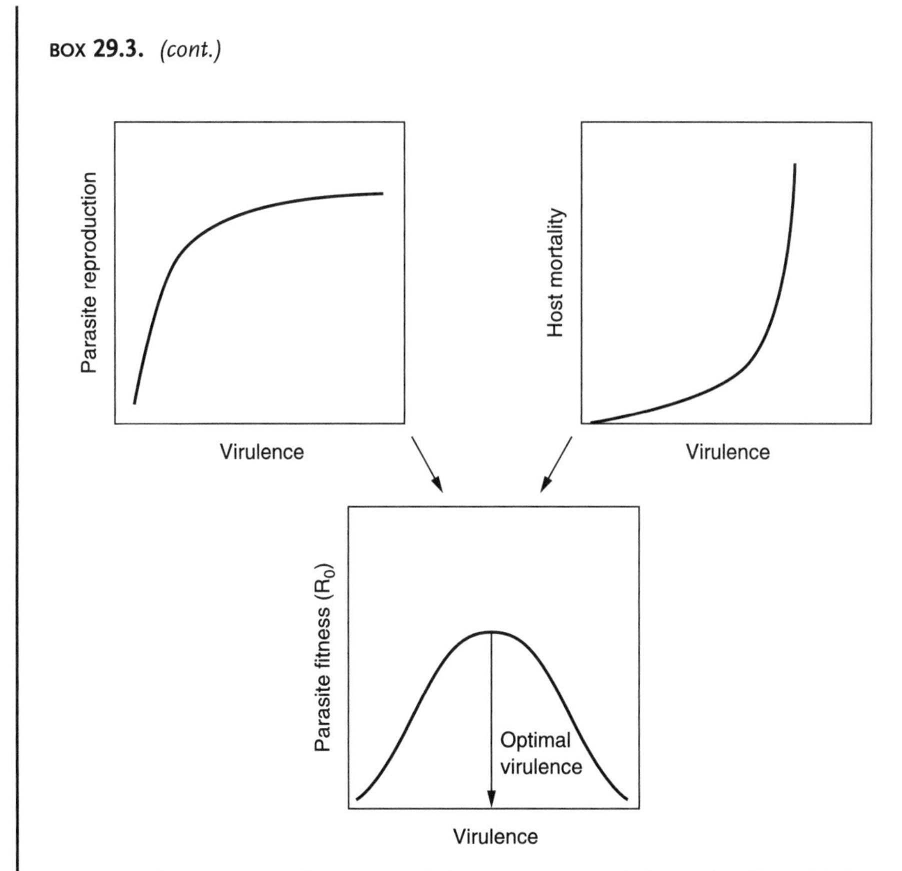

FIGURE 1. Graphical illustration of the assumptions of the trade-off model for the evolution of virulence (upper panels) that leads to the prediction of an optimal intermediate virulence (lower panel).

their hosts, which often increase their lifespan but reduces host reproductive output. In contrast a parasite can kill its host after reproduction and have no effect on that individual's fitness.

loci and varying quantitatively. Notable exceptions are Frank (1994a) and Gavrilets (1997). Frank's model was built within an ecological framework, and includes density-dependent and frequency-dependent forces, as well as costs for being more resistant or virulent. He found that ecological factors such as the stability of population size are very important in the coevolutionary dynamics of hosts and parasites. While the maintenance of genetic variability depended on a combination of factors, the relationship between cost and benefit of resistance and virulence traits played a particularly important role. Overall, he found that maintenance of genetic variation and allele cycling is more likely with quantitative than qualitative traits.

Gavrilets used an explicit quantitative genetic model to investigate the coevolution between two antagonistic species. While the model is not specific for host and parasites, it is directly applicable. With his model, Gavrilets observes coevolutionary allelic cycling under broad conditions, particularly when selection is strong and hosts ("victims") have either a stronger incentive or a larger genetic variance than the parasites ("exploiters"). While no specific assumption about the mechanism of resistance is made, loci are assumed to be additive and the result of the species, interaction depends on the difference between the value of the trait in the two species (which is akin to MAM). It is also assumed that the fitness of a given genotype is frequency-dependent. These results show that polymorphism can be maintained by dynamics caused by the genetic properties of resistance and virulence only. In other words, the maintenance of polymorphism does not depend exclusively on population dynamics.

Other Models

GFG and MA models have been important heuristic cornerstones, which generated many important concepts and insightful studies. However, it is likely that the actual genetic mechanism of resistance and virulence will not strictly match either the MAM or GFG model. Frank (1993), for example, suggest that GFG and MA are part of the same continuum, and that most systems would fall somewhere in between. More detailed analysis of GFG systems has indeed shown that many do not actually follow the strict definition of GFG (Innes 1995). Furthermore, the cloned *R* genes are only the beginning of a long signaling pathway that leads to the defense system, which contains numerous other genes (Staskawicz 2001). Since the genetic basis of resistance and virulence is likely to vary among organisms, a better approach might be to understand the effect of deviations from these extremes instead of analyzing each particular case separately. To determine how much of the parameter space leads to cycling in allele frequency, Agrawal and Lively (2002) modeled a continuous space between GFG and MA. They found that very small departures from a strict GFG can lead to cycling of allele frequencies, even in the absence of costs. Moreover, they found that while the coevolutionary dynamics strongly depend on the genetic assumptions, allele cycling is observed over a large proportion of the parameter space. Thus, their result suggests that coevolution between host and parasites should be a likely outcome in a wide range of host–parasite interactions.

EMPIRICAL EVIDENCE OF COEVOLUTION BETWEEN HOSTS AND PARASITES

The models above make clear that there are number of requirements for coevolution to occur. Among the most important is that resistance and virulence are genetically determined traits and variable within populations. Also, resistance and virulence need to be genotype-specific and reciprocally affect host and parasite fitness. Extensive research has provided strong support for these requirements (this subject has been extensively reviewed elsewhere: e.g., Fritz & Simms 1992). The more challenging task is to demonstrate that coevolution has occurred or is occurring in natural populations. For obvious reasons, empirical studies are generally limited to interactions where host and parasite have short generation times. Even when infection leads to host death, it can still take 100 generations of the host for resistant alleles to go to fixation (Levin et al. 1996). Consequently, most studies rely on detecting patterns that would be expected if coevolution has occurred. The two main types of evidence collected are either tests of whether the outcome or assumptions of the models are met, or tests for historical patterns from which coevolution can be inferred. In all cases, it is more likely to actually detect coevolution when the parasite is specialized, prevalent, and has a large effect on host fitness.

Spatial Evidence

One of the predictions from the Red Queen hypothesis is that parasites will be locally adapted to their sympatric hosts. In other words, it is expected that parasites will have a higher success at infecting hosts from their own population (with whom they coevolve) than from other populations. This outcome is expected because frequency-dependent selection should cause parasites to track the most common host genotype, reducing its performance in other host populations which are unlikely to be at the some point of the coevolutionary cycle. Local adaptation of the parasites can also result from the fact that parasites usually have smaller generation times and larger population sizes than hosts, and therefore evolve at a faster pace. If the opposite were true and hosts evolved faster, hosts would be expected to be locally adapted to their parasites. In other words, hosts would be better at avoiding infection by sympatric parasites than allopatric ones.

Local adaptation has been studied in a large number of systems, including animals and plants (reviewed in Kaltz & Shykoff, 1998). Local adaptation of parasites is commonly observed, and compatibility seems to decrease with distance as expected. However, many cases of maladaptation (i.e., parasites being more successful at infecting allopatric hosts) are also observed. This result is not completely unexpected, since it is possible that, by chance, a more suitable host may occasionally be found in an allopatric population. Maladaptation is actually predicted under some conditions (Gandon 2002). For example, the relative rate of host versus parasite migration can strongly affect the pattern of the local adaptation. When hosts have higher migration rates than pathogens, hosts are predicted to be locally adapted to their parasites, which is equivalent to maladaptation for the pathogens.

Whether pathogens are locally adapted is important not only as a test of the Red Queen hypothesis, but also to medicine and conservation. In medicine, parasite local adaptation is important because it could cause the efficacy of some treatments to be area-specific. Furthermore, it can affect the emergence of new diseases through anthropogenic movement of parasites. Geographic variation in host–parasite interactions can determine the success of invasive species and biological control. This type of evidence, however, does not provide a very strong test of coevolution. The fact that parasite adaptation is often observed can be interpreted as support for the Red queen hypothesis. However, as discussed above, other genetic mechanisms can cause coevolutionary cycling and consequently local adaptation of parasites. Thus, local adaptation is better seen as broad evidence for coevolution. Still, care must be taken with this broad interpretation since alternative hypotheses exist to explain local adaptation (e.g., Thompson 1999). The other empirical tests discussed below provide stronger evidence for coevolution.

Temporal Evidence

Demonstration of coevolution ultimately requires showing reciprocal changes in allele frequencies in both host and pathogen. Thus, one of the strongest pieces of evidence of coevolution is the tracking of alleles between host and parasites over time. However, these sorts of data are very hard to collect and, as a result, only a few long-term field studies have been performed. Obtaining this type of data is particularly difficult because in most natural systems the loci underlying the host–parasite interaction are not known. A particularly good example of this type of evidence is provided by Dybdahl and Lively (1998), which is also further discussed in the case study below. Their results support the idea that parasites can adapt to the most common host genotype in nature, and can cause frequency-dependent selection that favors rare genotypes. If similar results are observed in more systems there will be strong support for models that show coevolutionary cycling of host and parasite alleles.

Case Study: The Interaction Between a Snail and its Trematode Parasite

One of the best examples of temporal data is from Lively and colleagues, who studied the interaction between the snail *Potamopyrgus antipodarum* and its trematode parasite *Microphallus* sp. Dybdahl and Lively (1998) found a clever solution to track unknown resistance alleles by following changes in frequency of asexual snail clones. Because they were monitoring allele frequencies in asexual clones they were able to assume that isozyme markers were linked to the relevant loci that determine the outcome of the interaction between the snails and their common trematode parasite. They tracked isozyme genotypes for 5 years (approximately 10 to 15 snail generations) and found that clone frequencies changed over time, as expected if allele cycling due to frequency-dependent selection was occurring. Furthermore, they found that common clones were overrepresented in their sample of infected snails, and that these common clones decreased in frequency after being targeted by the trematodes.

Molecular Evidence

An alternative way to study temporal patterns of coevolution is to investigate whether patterns of molecular variation are compatible with the occurrence, in the past, of coevolution. If coevolution has occurred, a likely candidate to show the molecular signature of coevolution would be the genes that underlie host–parasite interactions. Because genes have to be well characterized and cloned before molecular variation can be studied, investigations have focused particularly on *R* genes from *Arabidopsis* and close relatives. The Red Queen hypothesis would predict that patterns of molecular variation should be consistent with frequency-dependent models, while strict GFG models would predict strong directional selection and selective sweeps. To determine whether *R* genes are under selection and, if so, what kind, two types of analysis have been performed: phylogenetic analysis, and analysis of rates of substitution.

Phylogenetic analysis is used to build the evolutionary history of the alleles of resistance genes, also called "gene genealogies." The fewer differences between the sequences of two alleles the more closely related they are, and the more likely it is that they have diverged recently (see Rodriguez-Trelles et al., Ch. 8 of this volume; Thornton, Ch. 11 of this volume;). Gene genealogies with mostly closely related alleles are expected when most new mutations sweep through the population, "erasing" old alleles from the phylogeny. In contrast, observed gene genealogies of resistance genes show long branches, indicates that the existing alleles are old, and most likely maintained by frequency-dependent selection (e.g., Caicedo et al. 1999; Tian et al. 2002). Phylogenetic data have been complemented with analysis of the rate of base substitution, which tests whether resistant and virulence genes have been under selection. If hosts and parasites are in a constant coevolutionary chase, with cycling in allele frequencies due to reciprocal selection, then the genes that underlie the interaction should show the signature of selection. The signature of selection is a higher ratio of nonsynonymous (amino-acid-changing) to synonymous (silent) substitutions (see Nachman, Ch. 7 of this volume, for more details of this type of analysis). Evidence for selection at the molecular level has been collected for several genes involved in resistance and virulence (Endo et al. 1996; Rose et al. 2004, and references therein). While evidence for selection supports the idea that coevolution is shaping the molecular variation in host and parasite populations, further analysis is needed to understand what type of selection underlies the coevolutionary dynamics. For example, Tian et al. (2002) demonstrated that the RPS4 gene of *Arabidopsis thaliana,* a gene that confers specific resistance to a strain of the bacterium *Pseudomonas,* has been under balancing selection as predicted by the Red queen model. This is a particularly interesting result because *RPS4* is a resistance gene that behaves as a classical *R* gene from a GFG model. However, a strict GFG model will predict strong positive selection that would lead to allele sweeping.

While the current results do suggest that molecular variation support the occurrence of frequency-dependent selection, studies have been performed only in a handful of model organisms. Data on genes that underlie the interaction between vertebrates and their parasites have been lacking. This is because much of the antigen diversity observed in vertebrates is due to recombination outside the germ cells, and therefore does not reflect true coevolution. Thus, caution is recommended before this result can be generalized. Furthermore, it would greatly improve our understanding of coevolutionary dynamics if studies of molecular variation considered simultaneously the interacting genes in both hosts and parasites.

Phylogenetic Evidence

It has been proposed that coevolutionary dynamics between hosts and parasites should also leave a historical signature at the macroevolutionary level. The "Farhenholz rule" states that coevolution can lead to cospeciation. Therefore, the phylogenetic trees of host and parasite groups that are associated have been predicted to be congruent (Figure 29.3). While phylogenetic congruence seems a straightforward testable hypothesis, there are many difficulties with the methodology (summarized in Brooks 1988). While congruence has been observed in some cases (e.g., pocket gophers and their obligate ectoparasitic lice; Hafner & Page 1995), most often hosts and parasites do not show phylogenetic agreement (Desdevises et al. 2002 and references therein). This incongruence is often interpreted as an indication that host switching is common and that ecological factors are more important than historical events in determining host–parasite association. This is further supported by the fact that the groups in which congruence has been observed include very

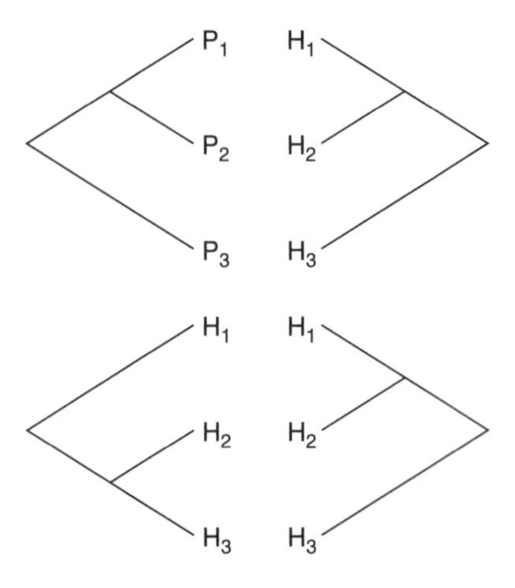

FIGURE 29.3. Examples of a congruent (top) and an incongruent (bottom) phylogeny of associated groups of host species (H) and parasite species (P).

specialized host–parasite relationships, with little opportunity for host switching. However, Rannala and Michalakis (2002) showed that significant discrepancies in host and parasite phylogenies can be observed even when they have a common tree of speciation. What this means is that phylogenetic congruence can provide evidence for cospeciation, but that phylogenetic incongruence does not necessarily mean that cospeciation did not happen. Another problem with interpreting macroevolutionary data is that phylogenetic congruence can be the result of different processes, only one of which is coevolution in the strict sense (i.e., through reciprocal selection; Rannala & Michalakis 2002). In summary, given the current limitations of the analysis of phylogenetic congruence this type of data cannot be used as unequivocal evidence about coevolution.

FUTURE DIRECTIONS

Clearly, progress in understanding the evolutionary dynamics of hosts and parasites has been much more advanced theoretically than empirically. In part, the lack of adequate empirical data is due to the fact that the experiments needed require systems with very particular qualities (e.g., short generation time, strong selection, knowledge of the genetic basis). Thus, it is not surprising that many of the current empirical studies have been limited to model organisms. Further effort in developing natural systems where knowledge about the genetics and ecology can be accumulated is fundamental. Conceptual areas that also deserve special attention are outlined below.

Multiple Parasites

While most of the theoretical and empirical work on host–parasite interactions has focused on single pairs of interacting species, hosts in natural populations are often challenged by multiple strains of one parasite and multiple species of parasites. Although some work has been done in modeling coevolutionary dynamics when multiple strains coinfect a host (e.g., May & Nowak 1995), further work is needed. In particular, it is important to determine whether there is a correlation between resistance and virulence to different species. Negative correlations can prevent response to selection and, consequently, a lack of coevolution.

Genetic Architecture

Resistance to pathogens is often more complex than a single pair of interacting loci. QTL studies investigating the genetic architecture of disease resistance in plants have found that the genetic architecture of resistance varies considerably among different species (Kover & Caicedo 2001). Unfortunately, all the coevolutionary models discussed here clearly demonstrate that the coevolutionary dynamics between host and parasites are sensitive to the specifics of the genetic basis of resistance and virulence. Thus, models that consider a specific genetic architecture will often have limited application. Studies that investigate wide parameter space, such as the model developed by Agrawal and Lively (2002), might offer a better alternative. It would be particularly useful if we could determine what types of genetic architecture could lead to allele cycling versus sweeping. The advantage of such information is that while specific knowledge of the genetic basis of resistance and virulence will always be limited to a few systems, studies of natural populations can provide estimates of the shape and strength of natural selection. Thus, if this information could be used to determine the potential for coevolution, we might be able to obtain information from a wider group of organisms and perhaps reach a more general conclusion about the evolutionary dynamics of hosts and parasites.

Coevolution in Metapopulations

There is mounting evidence that coevolution between hosts and parasites does not occur in isolated populations but in metapopulations. Metapopulations are groups of distinct demes linked through gene flow, where each individual deme can go extinct or be recolonized. Recently, models have been developed to investigate the effect of extinction and gene flow on the maintenance of genetic variation in host–parasite interactions within a metapopulation framework. These models have shown that metapopulation dynamics can substantially affect the prediction of the single population models. For example, Thrall and Burdon (2002) showed that spatial variation coupled with migration can maintain polymorphism for resistance and virulence without costs even when a strict GFG mechanism is considered. Such models represent an important step toward building models that are more relevant to natural populations. While some empirical data on host–parasite dynamics have begun to appear in plant systems, more data on animal systems are needed.

Case Study: The Interaction between Linum marginale *and* Melampsora lini *in a Metapopulation*

Most coevolutionary models assume host and parasites occur in large populations with random mating. Theoretical models have shown that spatial subdivision and level of gene flow can have a significant effect on the coevolutionary dynamics. However, only a few systems have been studied empirically. One exception has been the long-term studies spearheaded by Burdon and colleagues on the interaction between a perennial herbaceous plant native to Australia (*L. marginale*) and its host-specific fungal parasite (*M. lini*). The genetic basis of resistance and virulence is well understood in this system (Burdon 1994), and follows a typical GFG model. In addition, Burdon et al. have been able to follow the frequencies and diversities of resistance and virulence genotypes over time and space. Thus, this is an ideal system in which to combine genetic, temporal, and spatial data to better understand the coevolutionary dynamics of host–parasite interaction.

To determine the structure of resistance and virulent genotypes across 16 demes of *L. marginale* spread over a 1 ha area in the Kiandra plains (Australia), Thrall et al. (2001) collected seeds and fungal spores, and monitored disease prevalence (percentage of host population infected), in each of these demes. The seeds collected were germinated and used to test for their resistance to the fungal strains collected. To determine the virulence type of the fungal strains collected, Thrall et al. inoculated a set of standard plant genotypes for which the resistance types were known. They found that, while resistance and virulence types differ even among closely spaced populations, there was evidence that closer host populations have more similar resistance genotypes. In contrast, no evidence of spatial structure was observed for fungal strains. This study demonstrates that these 16 populations behave as expected in a metapopulation, where each population in not completely independent, but they also cannot be combined into a single population. In addition, these results indicate that the parasite is dispersing over a larger spatial scale than the host. This is not surprising considering that the fungal spores are aerially dispersed, while the host mostly selfs and has limited seed dispersion.

The fact that the interaction between *L. marginale* and *M. lini* occurs within a metapopulation, where many demes often go extinct or become recolonized (Thrall et al. 2002), raises the following questions: Is it possible for coevolution to occur? And if so, at what level does it occur? To answer these questions, a subsequent study was done in the same metapopulation using cross-inoculation experiments (Thrall et al. 2002). The Kiandra metapopulation can be subdivided into three subregions, in each of which host and fungal genotypes were collected from two demes. Thus, the results from the cross-inoculation experiments could be analyzed at the deme level, within and between subregions. It was found that the mean level of resistance (*i.e.*, number of fungal strains to which host were resistant) varied among demes, and that the number of virulence genes in a population increased proportionally to the mean number of resistance genes found in the same population, providing evidence of coevolution. Furthermore, there was found strong evidence for local adaptation of the fungus to its host, as expected when pathogens disperse more than their hosts (Gandon 2002). However, the strength of the local adaptation was stronger within subregions, indicating the importance of considering hierarchical sampling and testing of local adaptation in natural populations.

SUGGESTIONS FOR FURTHER READING

There are a number of excellent resources providing further reading on the coevolutionary genetics of hosts and parasites. Two key books cover most of the early theoretical developments in the field: Anderson and May (1992) presents detailed mathematical models that have served as the basis for many of the current models, and Fritz and Simms (1992) introduces most of the questions that are still being pursued in interactions between plants and their antagonists (parasites and herbivores). I also recommend Hamilton et al. (1990) and Parker (1994) for those interested in further reading about the possible role of host–pathogen interactions in favoring sexual reproduction.

Anderson R & R May 1992 Infectious Disease in Humans. Oxford Univ. Press.

Hamilton W, Axelrod R & R Tanese 1990 Sexual reproduction as an adaptation to resist parasites (a review). Proc. Nat. Acad. Sci. USA 87:3566–3573.

Fritz RS & EL Simms 1992 Plant Resistance to Herbivores and Pathogens. The Univ. of Chicago Press.

Parker MA 1994 Pathogens and sex in plants. Evol. Ecol. 8:560–584.

Acknowledgments I would like to thank C. Fox and J. B. Wolf for inviting me to write this chapter. S. Frank, Y. Michalakis, J. B. Wolf, and anonymous reviewers gave valuable comments on a early draft that substantially improved the text.

30

The Evolutionary Genetics of Senescence

DANIEL E.L. PROMISLOW
ANNE M. BRONIKOWSKI

Time present and time past
Are both perhaps present in time future,
And time future contained in time past.
T.S. Eliot, "Burnt Norton"

This book includes chapters on how natural selection has led to the evolution of a variety of adaptations, from social behaviors (Frank, Ch. 23 of this volume) to developmental systems (Stern, Ch. 15 of this volume; Siegal and Bergman, Ch. 16 of this volume) to pathogen resistance (Kover, Ch. 29 of this volume). We typically think of evolutionary genetics as a field that helps us to understand the genetic basis of adaptation. Here we turn our attention to senescence, a trait that is at best a product of direct selection on other traits whose side-effects lead to aging, and at worst a maladaptation, the losing outcome of a continually tossed coin, with the benefits of selection on one side countered by the costs of mutation on the other. Senescence is not an adaptation, per se. But in trying to understand how and why organisms gradually fail, we have developed fundamental insights into how and why they work. Just as we can learn a great deal about the nature of evolution from the study of the outcomes of selection, so too can we learn about evolution by studying what happens in the absence of selection.

The study of aging is the study of diverse phenomena. And perhaps more than any other topic, the study of aging has brought evolutionary biologists together with economists, demographers, physiologists, and molecular biologists. The biology of aging may serve as the touchstone by which we measure the value of interdisciplinary research in biology, with evolutionary genetics as its foundation.

Evolutionary biologists have long recognized that senescence, the intrinsic, age-related decline of physiological traits with age, is inescapable. Inspired by Haldane and Fisher, Medawar (1946) noted that a deleterious mutation that acts early in life will promptly be removed by natural selection. But if that same mutation acts late in life, individuals carrying the mutation will already have had a chance to pass it on to their offspring or grand-offspring before it has even begun to reduce the bearer's fitness (Box 30.1). In this light, he argued that over evolutionary time, late-acting mutations will accumulate at a much faster rate than early-acting mutations. These late-acting mutations will thus lead to declining viability and/or fertility as an organism ages. We can think of this simply as a special case of mutation–selection balance (Houle & Kondrashov, Ch. 3 of this volume), where the strength of selection declines with age. Medawar's "mutation accumulation" (MA) theory has since been modeled extensively and supported through numerous experimental studies.

Shortly after Medawar put forward this first evolutionary theory of senescence, George Williams suggested his "antagonistic pleiotropy" (AP) theory of aging (Williams 1957). Williams argued that selection may actually increase the frequency of

BOX **30.1.** Demography of an Age-Structured Population

The basic equation that describes the demography of an age-structured population was first discovered by Leonhard Euler in 1760, and then independently derived in 1907 by Alfred Lotka. For an individual cohort, let $l(x)$ be the probability of survival from birth to age x (survivorship), and $m(x)$ be the number of daughters produced by a

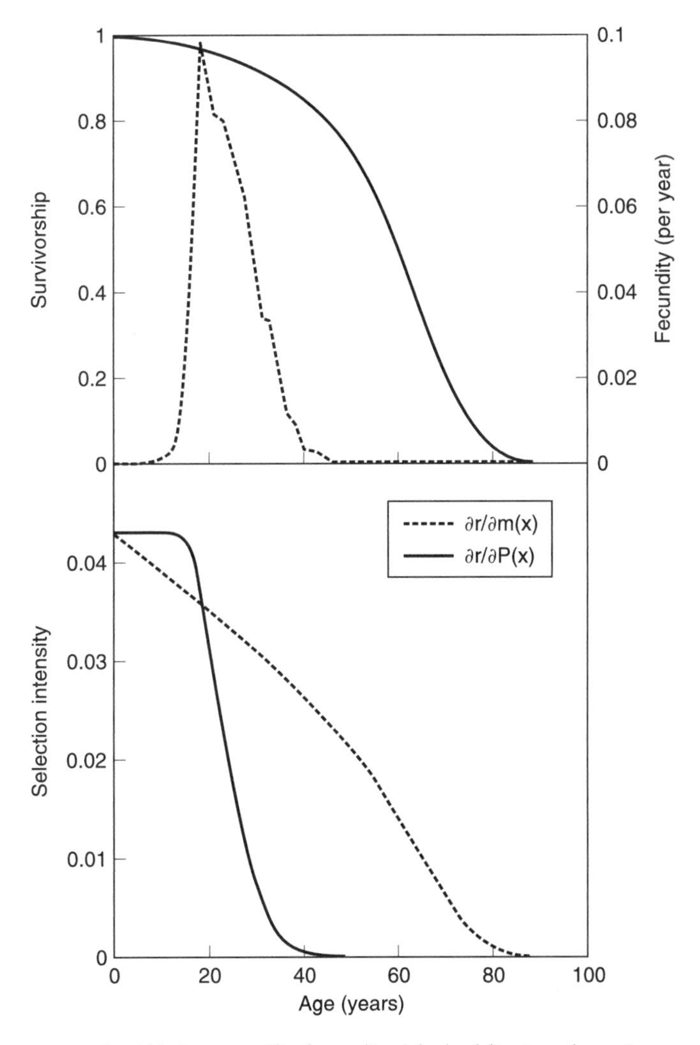

FIGURE 1. (A) Age-specific fecundity (dashed line) and survivorship (unbroken line) for a hypothetical human population. (B) Intensity of selection acting on a mutation that decreases age-specific fecundity (dashed line) or survival (dotted line) for a single age class. Note the dramatic decline in the intensity of selection on survival after age at maturity, eventually reaching zero after the last age at reproduction.

(continued)

BOX 30.1. *(cont.)*

female at age x (fecundity). These two values can then be used to determine the asymptotic growth rate in the population, r, by iteratively solving the Euler–Lotka equation:

$$1 = \sum e^{-rx} l(x)m(x). \tag{1}$$

Fisher introduced the concept that we could use r, also known as the intrinsic rate of increase, or the Malthusian parameter, as a measure of fitness.

As Medawar noted (see text), senescence evolves because the strength of selection declines with age. By "strength of selection," we really mean the sensitivity of fitness, r, to a small change in age-specific survival or fecundity. According to Medawar's thinking, the effect of this small change in age-specific survival early in life should have a much greater effect on fitness than an equally small change late in life. In a now classic paper, Hamilton used the Euler–Lotka equation to develop a formal mathematical theory for the evolution of senescence. We define age-specific survival, $P(x)$, as the probably of survival from age $x - 1$ to x, or $P(x) = l(x)/l(x - 1)$. Hamilton showed that the sensitivity of fitness to a change in $P(x)$ was given by

$$\frac{\partial r}{\partial \ln P(x)} = \frac{\sum_{y=x+1}^{\infty} e^{-ry} l(y)m(y)}{\sum x e^{-rx} l(x)m(x)} \tag{2}$$

and the sensitivity of fitness to a change in fecundity was given by

$$\frac{\partial r}{\partial m(x)} = \frac{e^{-rx} l(x)}{\sum x e^{-rx} l(x)m(x)}. \tag{3}$$

The denominator in Equations 2 and 3 is the mean age of mothers of a single cohort of newborns, commonly refered to as generation time. Note that for both equations, while the denominator is a constant value independent of x, the numerator is a declining function of x (see Figure 1). This shows that the later the age at which a mutation acts on survival or fecundity, the weaker its overall effect on fitness.

late-acting deleterious mutations if these same alleles increase fitness in juveniles or young adults. For example, a gene that decreases lifespan could be favored by selection if it increased early-age fecundity. As with mutation accumulation, theoretical and empirical studies suggest that AP is a common mechanism of senescence.

Medawar's and Williams's theories have motivated much of the experimental work on the evolution of aging for the past 25 years, standing side by side as competing models for why organisms age. And despite 25 years of research, each model continues to have its proponents (and, in some cases, detractors). In this chapter, we will discuss ways that these models have been tested, and evidence in support of each of them. But we will also argue that we are at a crossroads, where evolutionary biologists need to move beyond simple attempts to test these theories. Evolutionary studies of aging have begun to harness the power of modern molecular genomics, enabling us to test in a way that was previously impossible, evolutionary ideas about how and why organisms age. At the same time, a new generation of theoreticians has begun to expand our perspective on the relevant evolutionary forces that shape senescence.

We have divided the chapter into three sections. In the first, we explore the wealth of studies that have set out, over the past 25 years, to test models for the evolution of aging. This work has led to insights into the genetics of demography, reproductive behavior, and life history evolution. In the second section, we focus on recent ways in which biogerontologists have begun to use the power of molecular genomics to uncover fundamental evolutionary insights about the nature of senescence. Finally, we suggest avenues that we think are likely to provide rich veins for exploration in the coming years. Some of these areas are currently the focus of active research, while others have yet to be developed but may provide critical insight into the evolution of aging.

Each section is meant to be an introduction to some of the current problems that evolutionary gerontologists are working on, but is in no sense meant to be comprehensive. For additional background on the genetics and evolution of senescence, we refer the reader to the Suggestions for Further Reading at the end of the chapter.

TESTS OF EVOLUTIONARY SENESCENCE THEORY

It is obvious to all of us that mortality rates increase and fertility rates decrease with age in human populations. But there are considerable statistical challenges to obtaining accurate estimates for how these vital rates vary among age classes and between populations, and how this variation influences population dynamics and fitness. Until fairly recently, these questions were mainly the purview of demographers. Now, evolutionary gerontologists have taken on the challenge of exploring the underlying genetic basis of this variation. In the following section, we begin with a brief introduction to the demography of senescence, and then examine different ways in which genetics and demography have been combined to test evolutionary theories of aging.

Biodemography

From a demographic perspective, we can define senescence as the age-related decline in rates of survival and fecundity due to the progressive loss of physiological function (Rose 1991). Unlike most biological traits that are specific to each individual organism, we measure these vital rates in large cohorts of individuals. While aging affects both fertility and mortality, the focus of most research on the genetics of aging has focused on mortality rates. The instantaneous mortality rate at age x, $\mu(x)$, is simply a measure of the probability that an individual alive at age x dies sometime in the interval between age x and age $x + 1$. With estimates of mortality rates in hand, we can define the rate of aging as the rate at which mortality rates increase with age. Log-transformed values of mortality increase approximately linearly over much of the life course (Figure 30.1). Thus, the rate of aging is the slope of the log-transformed mortality rates versus age. This relationship is often described by the Gompertz equation,

$$\ln[\mu(x)] = \ln(A) + Bx, \qquad (30.1)$$

where A and B are the "intercept" and slope respectively of the line describing log(mortality) versus age. (Note that A is not the true y-intercept because the Gompertz equation is usually fit to the data beginning in adulthood, that is some age > 0.) For more extensive description of demographic parameters in senescence, see Tatar (2001). Numerous books provide excellent introductions to the statistical methods for analyzing demographic parameters.

The first experimental tests of senescence theory focused on survival curves, which describe the probability of surviving to a given age, x (also known as "survivorship"). In the first widely cited study, Rose used artificial selection to create strains of *Drosophila* with delayed reproduction, and found that in the late-breeding lines, survival rates were higher in older age classes, resulting in a longer lifespan (Rose & Charlesworth 1980) (later work is summarized in Rose 1991). Similarly, almost 25 years' worth of studies by Partridge and colleagues (see below) found that experimentally reducing reproductive activity at one age could increase survival at subsequent ages. More recently, our understanding of the genetics of senescence advanced considerably once we moved from the study of survivorship curves to the statistically tractable study of mortality rates. Instantaneous mortality rates depend only on the number of individuals entering and leaving an age class, and are therefore independent of events occurring in earlier age classes. By contrast, survivorship is a function of the initial cohort size (e.g., when survival equals 0.50, 50% of the initial cohort is still living), and therefore each age-specific point estimate is a function of previous events in earlier age classes.

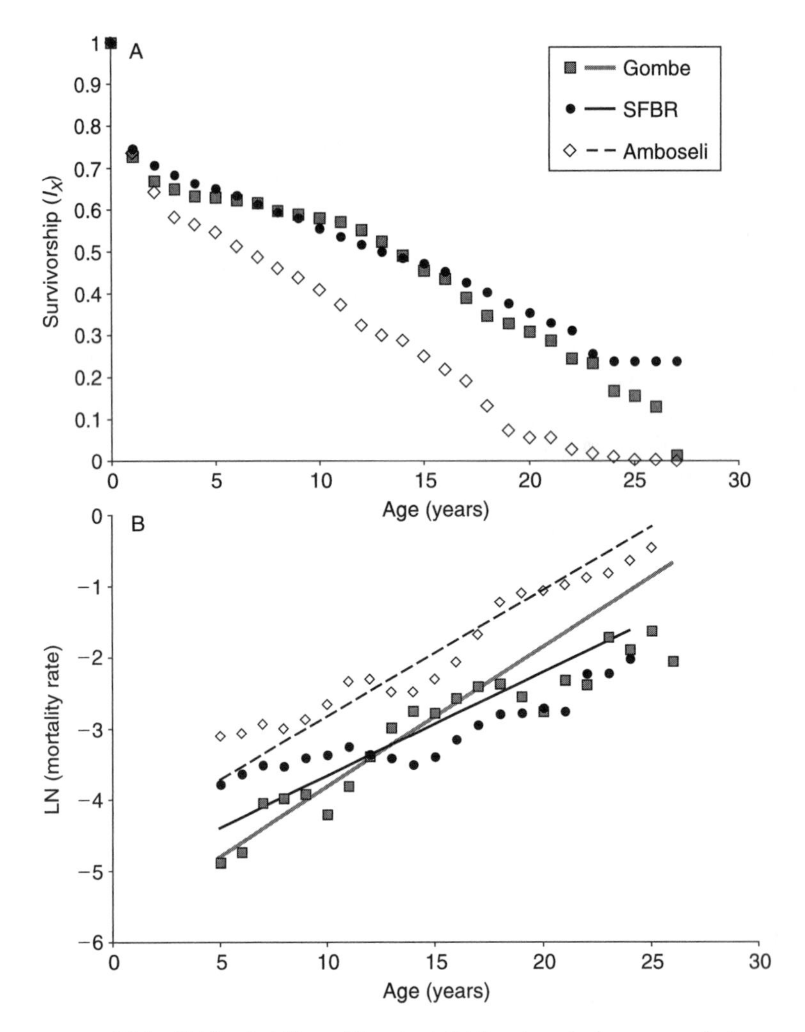

FIGURE **30.1.** (A) Probability of Survival (l_x) for three baboon populations inhabiting: Gombe National Park, Tanzania (■); Amboseli National Park, Kenya (◇); and Southwest Foundation for Biomedical Research (SFBR), San Antonio, Texas (●). (B) Natural logarithm of the instantaneous mortality rates (hazards). The lines are the Gompertz equation fitted to the hazards ($\mu(x) = Ae^{Bx}$) and are of the form ln(A) + Bx, where A is the initial adult mortality rate, B (the slope) is the rate of increasing mortality, and x is age. The Gompertz equation was modeled from the start of the adult stage (age 5 years). The line for SFBR appears skewed due to the relatively high frequency (~50%) of censored animals. From Bronikowski et al. (2002), reprinted with permission.

This new focus on mortality rates led to a variety of new and important observations. First, it became clear that senescence was a common phenomenon in natural populations of mammals (Promislow 1991), birds (Ricklefs 1998), and even insects (Bonduriansky & Brassil 2002). Second, with the analysis of ever larger cohort sizes, biodemographers discovered the Gompertz "law," which states that mortality rates increase exponentially with age, had its limits. At very late ages, mortality increases slowed and in some cases even reversed (Vaupel et al. 1998). Finally, it became apparent that we could change the age-independent component of mortality (A in the Gompertz equation) with relative ease, through environmental or genetic manipulations. However, altering the actuarial rate of senescence (B in the Gompertz equation) appears to be much more difficult.

These ideas are best illustrated with an example. Figure 30.1 is generated from survival (Figure 30.1A) and mortality (Figure 30.1B) rates computed from long-term studies of baboons (Bronikowski et al. 2002). In Figure 30.1A, we see survival curves for three populations of baboons that track the survival progress of three initial cohorts of baboons; median lifespan, defined as the 50% survival age, is lower in the Amboseli Kenya population. Without additional information, we cannot know to what this median lifespan difference should be attributed. However, by computing the age-specific mortality rates and fitting the Gompertz model to each population, we can see that the slopes of the lines do not differ, but the initial mortality rate appears to. Log-likelihood ratio tests revealed that indeed, the rates of aging (slopes) were homogeneous, whereas the initial adult mortality rates (A) were significantly different.

By focusing on mortality instead of survival, we also have greater power to examine the timing of life history events. This is well illustrated by a recent study on the effects of caloric restriction in flies. We have long known that caloric restriction can extend lifespan in rats, mice, flies, worms, and even dogs (Masoro 2000). Through a careful analysis of mortality rates, studies have shown that the effect of this treatment is only temporary (e.g., Mair et al. 2003). Flies that are started on a calorically restricted diet at eclosion have a much lower mortality rate than control flies. However, as soon as these flies are put back on a control diet, their mortality rates increase to those of the controls that had never experienced caloric restriction. More strikingly, control flies that are switched to a calorically restricted diet in middle age show immediate decreases in mortality rate comparable to those seen in flies that had been on a calorically restricted diet all along (Figure 30.2). None of these patterns would have been so clearly seen in an analysis of survival curves alone. Although these studies demonstrate a change in mortality patterns due to a phenotypic manipulation, genetic studies also suggest that it is far easier to alter the baseline mortality rate than to change the actual rate of aging (Promislow & Tatar 1998).

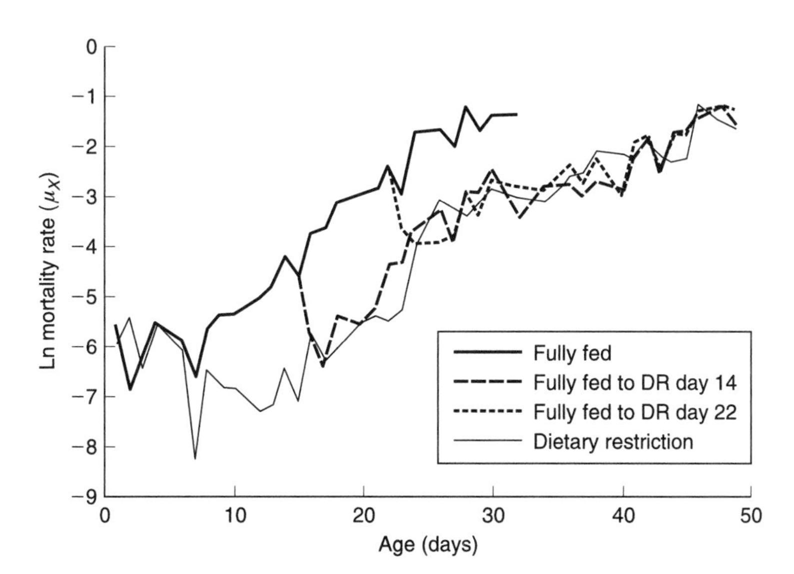

FIGURE 30.2. Age-specific mortality in female *Drosophila* subject to dietary restriction. Dietary restriction (DR) flies received a fraction of the protein and carbohydrates of fully fed flies. Analysis of mortality curves shows that DR flies have similar rates of senescence but lower baseline mortality rates. More importantly, the reduced mortality occurs as soon as the DR is initiated. Flies that experience DR from eclosion have identical mortality rates at age 40 days as flies that experience DR from age 14 or 22 days. Reprinted with permission from Mair et al. (2003) © 2003 AAAS.

The combination of demography and quantitative genetics presents particular statistical challenges. Nevertheless, as we point out in the following section, attempts to unite these two fields have led to important advances in the field.

Quantitative Genetic Tests of Evolutionary Theories

Within the realm of classical genetics, four distinct approaches have been used in evolutionary genetic studies of senescence: laboratory studies that analyze the genetic variance–covariance structure for life history traits, the use of artificial selection to extend lifespan, quantitative trait locus mapping, and studies of senescence in the wild.

Genetic Variation and Covariation

Both MA and AP theories of aging predict that one should find segregating variation for lifespan. But in addition to this prediction, two explicit quantitative genetic hypotheses arose early from MA and AP theory. First, if aging were due to MA, one would expect to see an age-related increase in additive genetic variation (V_A) for fitness traits, due to the increase in mutation–selection equilibrium with age. On the other hand, if senescence evolved due to the spread of genes with antagonistic pleiotropic effects, one would expect to see a negative genetic correlation between early-age and late-age fitness traits (Charlesworth 1994).

In the first direct test of MA theory, Rose and Charlesworth (1980) found no evidence for an age-related increase in V_A for fecundity in *Drosophila melanogaster* (Box 30.2). Almost a decade later, attempts to replicate this study using the same species found that V_A did, in fact, increase with age. Engström (1989) found a 30-fold increase in V_A for fecundity. However, both of these studies may have been confounded, in the first case by the fact that individuals vary in their intrinsic mortality rates, with high-mortality individuals dying sooner, and in the second case by the fact that fecundity is not normally distributed and falls sharply with age (see Tatar et al. 1996). In a study that shifted the focus from fecundity and longevity to age-specific mortality rates, Hughes and Charlesworth (1994) showed that V_A for age-specific mortality rates increased with age. But subsequent studies suggested that V_A may, in fact, decline at very late ages (Promislow et al. 1996), though whether this decline is a real biological phenomenon due to the genetic architecture of senescence, or an artifact of demographic heterogeneity or sampling error, is still a matter of debate. Clearly, the statistical challenges in combining large-scale demography and quantitative genetic analyses are substantial. Notwithstanding this ongoing debate, perhaps the most lasting effect of Hughes and Charlesworth's study has been to help shift the attention of geneticists away from survival curves and toward direct estimates of mortality.

As Charlesworth and Hughes (1996) later noted, the prediction that V_A for fitness traits should increase with age was problematic, as it could potentially apply to both AP and MA. Under MA, the additive variance is proportional to the strength of selection acting on a trait at a given age. Because the strength of selection declines with age, variance is predicted to increase. Under AP, if the frequency of an early-acting beneficial allele is greater than the frequency of a late-acting deleterious allele, variance is also expected to increase (see Table 2 in Charlesworth & Hughes 1996).

To come up with a means of distinguishing between these two theories, Charlesworth and Hughes (1996) were able to derive several predictions that were unique to each of the two theories. In particular, they showed that if aging were due to MA, then the deleterious effects of inbreeding would be greater for traits expressed late in life than for early-acting traits. This occurs because inbreeding load is inversely correlated with the strength of selection, or sensitivity. As sensitivity declines with age, inbreeding load should increase with age. Furthermore, the relative impact on fitness traits of genes with dominant effects relative to those with additive effects should also increase with age. In contrast, if senescence were due primarily to AP, then inbreeding load should remain constant with age as it is independent of sensitivity, and the ratio of dominance to additive genetic variance for fitness traits should decline with age. While one test of these newer predictions strongly supports MA (Charlesworth & Hughes 1996), another study suggests that both MA and AP are important (Snoke & Promislow 2003).

In a separate line of inquiry, researchers have used quantitative genetic approaches to look for trade-offs between lifespan and early-life fitness traits as a test of antagonistic pleiotropy. The results have been mixed, at best. While some studies have found negative correlations, in support of AP, others

BOX 30.2. *Drosophila* as a Model Organism in Evolutionary Biology
Jeffrey R. Powell

The first point in considering *Drosophila* as a model organism for evolutionary studies is to emphasize that this is a genus encompassing some 2000+ species. This taxonomic diversity is also reflected in the ecological diversity of the genus, which exists from the tropics to near-Arctic with larval breeding sites including rotting fruit, flowers, cacti, tree sap, soil, fungi, and crabs. Thus while most biologist think of *Drosophila* as the well-studied laboratory model, *Drosophila melanogaster,* to the evolutionary biologist it is the combination of detailed study and knowledge of this single species coupled with the evolutionary diversity of the genus that is so attractive. The phylogenetic relationships of this large group of taxa has been quite well studied, especially for species used in evolutionary studies (Powell & DeSalle 1995), so there is a rich source for analysis of evolution of traits in an historical framework. However, it must be noted that the easy culturing of *Drosophila* that has made it such a convenient laboratory animal does not hold for all species; prime examples are the 800+ endemic species comprising the magnificent adaptive radiation in Hawaii (Kaneshiro et al. 1995), most of which are cultured only with great difficulty or are impossible to culture. Nevertheless, some 250 species are routinely cultured at the National Stock Center at the University of Arizona (http://stockcenter.arl.arizona.edu) providing a good representation of the entire genus.

Perhaps the most outstanding attribute of *Drosophila* is that virtually no other organism has been so well studied on such a variety of levels: genetics, molecular biology, cell biology, development, neurobiology, behavior, ecology, and, of course, evolution. It is a prime group for integration of these diverse fields that will lead to new insights into evolution, for example the neurobiological and genetic basis of mate recognition during the evolution of new species, or the large nascent field of "evodevo." Like all model systems, detailed laboratory studies are made possible by easy culturing including large population cages with many generations per year. However, unlike many systems, field studies of natural populations and knowledge of ecology of many groups—cactophilic species in the Southwest United States being a prime example (Etges et al. 1999)—make this group suitable for integrating detailed laboratory findings with natural populations and their ecology. *Drosophila* have long provided the most detailed insights into the processes in speciation (Coyne & Orr 2004), processes that are only now being understood in any detail.

Historically, *Drosophila* have been most used to provide empirical data related to microevolution, that is, studies on the population level. One of the more surprising and fruitful demonstrations was the documentation of natural selection acting on chromosomal naturally occurring inversion polymorphisms, strong enough to be studied over tens of generations and manipulated in laboratory populations thus making natural selection a truly experimental subject (Krimbas & Powell 1992). This was possible because of another favorable attribute of *Drosophila*: the presence of giant polytene chromosomes that allow easy identification of chromosomal rearrangements. In addition, in situ hybridization to polytene chromosomes provides a rapid and convenient way to physically map genes.

Drosophila were also the first to be the subject of molecular population genetics, initially allozymes leading ultimately to DNA sequencing (Moriyama & Powell 1996). Molecular evolution studies are vast and have provided insights into the levels and patterns of molecular variation that have been proven to be widely generalizable.

(continued)

BOX 30.2. *(cont.)*

Drosophila were also one of the first two shown to have transposable elements (maize being the other), and the evolutionary consequences of transposable elements, including horizontal (interspecific) genetic transmission, has been documented in *Drosophila* (Kidwell 1993).

Evolution of karyotypes and genome size are also possible with *Drosophila*. For example, the variation in karyotypes allows for empirical studies of the evolution of sex chromosomes, including dosage compensation (Bone & Kuroda 1996). Genome size varies about 5-fold in the genus, with evidence that efficiency of removal of inessential DNA sequences (including transposable elements) may be a contributing factor in determining genome size variation (Lozovskaya et al. 1999).

The most recent exciting development in *Drosophila* evolution is the availability of whole genome sequences for a number of species. *Drosophila melanogaster* was the first complex multicellular eukaryote to have a complete genome known. Presently, three other species of *Drosophila* have been sequenced and a total of 12 are scheduled to be done in the next 1 to 2 years. One of the beauties of using this group for comparative genomics is the fact that, while it is a relatively complex organism, its genome is only 3–4% the size of a typical mammalian genome; thus the acquisition of 12 sequenced genomes requires less than half the sequencing effort of the genome of a single mammal. The 12 species to be sequenced represent varying levels of divergence, from about 2 million years to 50 million years, as well as varying ecological niches. These sequences will provide the first detailed data for in-depth analysis of entire genome evolution, including evolution of gene regulation (Rifkin et al. 2003). One species (*D. simulans*) will have several strains sequenced, providing a uniquely detailed data set for molecular population genetics.

Thus while *Drosophila* has played an important historical role in empirical evolutionary studies (for a general review see Powell 1997), the exploitation of this group for further insights into evolution is far from exhausted. Indeed, the large body of background data accumulated over nearly 100 years of study simply sets the stage for the most modern advances. We are entering into a new phase of evolutionary research and *Drosophila* is once again a leading candidate to provide exciting new insights that may well prove to be quantum leaps in understanding a variety of aspects of the evolutionary process.

have found either no correlation or positive correlations (reviewed in Tatar et al. 1996). However, many of these studies are difficult to interpret. In some cases, positive correlations between fecundity and survival could be due to gene × environment interactions (Service & Rose 1985), or the effect of inbreeding in the laboratory, which could depress fitness traits at all ages. And as Tatar et al. (1996) point out, a decrease in lifespan in response to increased reproduction could arise even if the deleterious effect on survival occurs only at early ages. In this case, the correlation between early-age reproduction and life expectancy would have little to do with senescence, per se.

At a phenotypic level, in a series of studies carried out over more than two decades, workers have shown that a decrease in reproductive effort leads to an extension of lifespan. In a series of studies on *Drosophila*, for example, Partridge and her coworkers have found that longevity is negatively correlated with numerous reproductive traits, including courtship activity, egg laying, and even exposure to male seminal fluid, and that the costs of increased mortality rates due to reproduction persist throughout life (e.g., Chapman et al. 1995; Partridge et al. 1987).

Artificial Selection

Perhaps the strongest evidence for trade-offs between life history traits, and hence the strongest evidence in support of the AP theory, comes from artificial selection experiments. The first known use of artificial selection dates back at least 10,000 years, when early human populations domesticated both plants, such as wheat and olives, and animals, including dogs, sheep, and goats. But as Darwin noted, artificial selection can also be a powerful tool in the study of evolution.

Edney and Gill (1968) first suggested that one could use artificial selection to test evolutionary theories of senescence. Although earlier studies had established a genetic basis to lifespan, and Medawar's and Williams' theories were well known at this time, no studies had directly tested these theories. Edney and Gill suggested that we could test aging theory by using artificial selection to shorten lifespan. In fact, they had considered the idea of trying to prolong lifespan but rejected it as too difficult an experiment. They surmised that a response to selection in either direction would be evidence for a genetic basis to senescence. It was not a far stretch to realize that one could test AP directly by measuring the way that correlated traits responded to a change in lifespan.

The standard approach to select for increased lifespan is to breed from individuals of ever-increasing age. The assumption is that for an individual to survival to late age, it must on average carry alleles that confer extended longevity. Thus, cohorts derived from late-age breeders will evolve longer lifespan than control lines in which individuals are forced to breed early in life (Rose & Charlesworth 1980). These studies and many others have consistently found that lines selected for increased longevity have evolved reduced early-age fecundity.

Service et al. (1988) noted that one could use this approach to test both AP and MA theories by reversing the selection process. Under MA, the effects of selection on late-acting loci should be irreversible, because their early-age effects are neutral. Selection to remove late-acting deleterious alleles should increase longevity by reducing the frequency of these alleles. But selection on early fecundity, for example, will not increase the frequency of these alleles, since they have no effect on early-age traits. (Of course, mutation accumulation will lead to the spread of new, late-acting deleterious alleles, but the process will be much slower than the effect of selection.) In contrast, under AP, traits should respond to reverse selection. In response to selection for long lifespan, an allele at locus A (call it A_1) that increases longevity at the expense of early fecundity will increase in frequency relative to an allele, A_2, with opposite effects. Reversing selection in favor of early fecundity should, in turn, reduce the frequency of A_1 and increase allele A_2. Service et al. (1988) found that some traits, such as fecundity and starvation resistance, responded to reverse selection while others, such as resistance to desiccation or ethanol, did not. This result suggests that the genetic architecture of senescence may vary among traits.

A veritable cottage industry has developed, using artificial selection in fruit flies to determine genetic correlations between rates of aging and a variety of traits, including larval development, age at reproductive maturity, stress resistance, and body size (Chippindale, Ch. 31 of this volume). These trade-offs may even turn out to be important determinants of human longevity (Westendorp & Kirkwood 1998). In Molecular Genomics Meets Evolution below we will describe how molecular gerontologists are beginning to uncover the molecular basis of these trade-offs. A more extended discussion of the power of experimental evolution is provided in Chippindale (Ch. 31 of this volume).

QTL Mapping

Whereas molecular geneticists attempt to elucidate the genetics of aging by screening entire genomes one gene at a time, the genome-wide approach of quantitative trait locus (QTL) mapping offers us a powerful tool to identify naturally segregating genes or gene regions that influence aging. Molecular methods generally screen for novel mutants that are created in the laboratory. In contrast, one can use QTL mapping to identify genes that influence lifespan and that exhibit segregating variation in natural populations.

The details of the QTL approach are laid out by Cheverud (Ch. 19 of this volume). This approach is particularly well suited to a trait such as senescence. Variation in rates of aging among genotypes is likely to occur as a result of a complex interplay of multiple genes affecting the trait, interactions among these genes (epistasis), environmental factors both internal and external to the organism, and gene × environment interactions. One of the advantages of QTL mapping is that it can shed light on such a complex genetic architecture in a relatively straightforward manner. A series of studies from Mackay's laboratory have not only found QTLs for longevity (Nuzhdin et al. 1997), but have also illustrated that the effects of these QTLs depend on whether they are found in males or females (Nuzhdin et al. 1997), as well as the environment in which flies are reared, the presence of other specific QTLs, and whether flies are virgin or mated. Equally complex patterns have been found in QTL studies of lifespan in worms and mice.

QTL mapping can identify genetic variants that influence lifespan relatively easily, but taking the step from QT locus to QT gene has proven extremely difficult. However, a recent study by De Luca et al. (2003) illustrates that QTL mapping, coupled with deletion mapping, may in fact lead us to specific genes that affect variation in lifespan in natural populations. Previous studies in the same laboratory had identified male-specific QTLs on the second chromosome in *Drosophila*. De Luca et al. used deficiency mapping to localize the QTL to a region that included DOPA decarboxylase (*Ddc*), a gene involved in the biosynthesis of neurotransmitters. They then analyzed the sequence of *Ddc* in a collection of flies caught in a local farmers' market. They were able to identify a single nucleotide polymorphism that was associated with a significant increase in mean lifespan among these wild-caught strains.

De Luca et al.'s QTL study is a success story from both an evolutionary and a molecular genetic perspective, but the approach may not be productive in other contexts. Earlier deletion mapping studies had shown that within a single QTL for longevity there may be multiple, tightly linked QTLs. Thus, the number of genes that influence senescence and that segregate in natural populations may be very large. Furthermore, if there are multiple QTLs, some of which increase a trait and some of which decrease it, and if these QTLs are kept in linkage disequilibrium due to balancing selection (Wayne & Miyamoto, Ch. 2 of this volume), their effects may cancel out and we may fail to identify them altogether.

Genetics of Senescence in the Wild

One substantial advantage of QTL mapping over other molecular approaches is its ability to study naturally occurring variation relatively easily. But when it comes to studying natural populations in their native environment, our knowledge is still fairly limited. While we may not always be able to carry out controlled crosses and other genetic experiments in the wild, there are several approaches that can inform our understanding of the genetic basis of aging in natural populations.

Historically, biologists assumed that high extrinsic mortality rates would obscure any signs of senescence in natural populations. Numerous comparative studies have shown not only that demographic senescence is widespread in natural populations of mammals and birds, but also that overall shapes of mortality curves are surprisingly consistent among diverse species, as well as among populations of the same species living in disparate habitats (Bronikowski et al. 2002). Typically, mortality rates show an initially high juvenile period, a minimum at or around the age at first reproduction, and an exponential increase thereafter. Surprisingly, we are only now beginning to see models that help to explain the complex patterns underlying these curves (Lee 2003) (see below).

One of our biggest current challenges is to move from these natural history observations to a "biodemographic genetics," to find a way to estimate natural genetic variation for senescence and, ultimately, to identify specific loci affecting senescence that segregate in natural populations. In addition to the QTL approach described above, several other approaches may help us to move toward this goal.

One approach is to compare aging among populations in different environments. Tatar et al. (1997) measured mortality rates in several grasshopper populations collected over an elevational gradient. High-elevation populations have a much shorter season, so grasshoppers should experience strong selection for early reproduction and subsequently experience little selection for extended lifespan. In accord with this expectation, in common environmental conditions, grasshoppers from the high-elevation population had a mean lifespan of 103 days whereas the population from the lowest elevation had a lifespan between 122 and 149 days (depending on rearing temperature). This increase in mean lifespan was accomplished by slower rates of senescent decline in low-elevation grasshoppers as measured by age-specific morality rates. Similarly, Bronikowski et al. (2002) measured mortality in baboon populations residing in natural and seminatural habitats. In this case, they found that the age of onset of mortality and the level of initial mortality rate differed among populations, but rate of aging remained relatively invariant (see Figure 30.1).

While these studies examined how existing environmental variation affects lifespan, others have manipulated the environment in the wild and determined the selective response. An excellent example comes from a series of studies by Reznick on populations of guppies (*Poecilia reticulata*) in Trinidad (reviewed in Reznick et al. 2001). Reznick and his colleagues changed extrinsic mortality rates by manipulating the presence or absence of predators. They found that guppy populations in high-predation environments devoted more resources to reproduction and matured at an earlier age. Of course (and fortunately for generations of graduate students), the world is not always as simple or predictable as we think. In the latest study from Reznick's laboratory (2004), guppies from high- and low-predation populations were reared in the laboratory for two generations and then measured for age-specific mortality and fecundity. High-predation populations showed higher rates of senescence but had lower early-age mortality rates, contrary to theoretical prediction. This unexpected result could have been due to several factors, including the fact that reproductive rates increase with age in guppies, or that populations with high predation have higher mortality rates but also have lower density and higher resource availability. By turning to natural populations, Reznick's studies show that aging in the wild may be far more complex than we previously imagined.

Further progress in this area may depend on finding species with relatively short lifespans but which can be followed individually in the wild throughout the life course. One promising system is that of the recently discovered antler fly (*Protopiophila litigata*), which carries out its life cycle in close proximity to discarded moose antlers. Bonduriansky and Brassil (2002) found that in spite of high extrinsic mortality (13% per day), age-specific survival and reproduction clearly declined with age. Once these behavioral and demographic studies are coupled with physiological and molecular genetic approaches, this may prove to be an ideal system in which to follow genetic variation for aging in a natural setting. For example, one might use this system to determine whether genes associated with longevity in laboratory populations also affect fitness in the wild.

It is somewhat surprising just how little is known about the genetics of lifespan in the wild. While the best evidence for natural genetic variation for life history traits comes from studies of a wide range of bird species (see Roff 2002), there are not yet any data on the heritability of avian longevity in nature. One of the few studies to measure the heritability of longevity in the wild is from the relatively long-lived bighorn sheep (*Ovis canadensis*) (Reale & Festa-Bianchet 2000). Reale and Festa-Bianchet determined heritability for longevity using parent–offspring regression in two populations of *O. canadensis*. They found that heritability for longevity varied from 32% to 46%. While these estimates are relatively high for a fitness trait, they are not far off of reported values of lifespan heritability in humans, other mammals, and invertebrates (reviewed in Martin et al. 2002). Field estimates of lifespan heritability may be inflated somewhat by common environment effects between parents and offspring. To fully correct for the common environment effect, we need manipulative experiments of the sort carried out by ornithologists, where one can swap eggs between nests before hatching.

This last point illustrates the potential weakness of field studies. In most field studies we cannot control for the potentially enormous number of confounding factors in the wild. But the laboratory environment brings with it its own set of problems. When we bring organisms into the laboratory, we may inadvertently introduce a new selective regime that can lead to genetically based changes in the life history of the organism. For example, fruit flies are commonly reared in a 2-week generation time in the laboratory. But as Promislow and Tatar (1998) point out, this rearing regime selects for high early fecundity, and entirely eliminates selection against deleterious mutations whose effects are confined to adult age 6 days and later. Both of these factors would be expected to lead to shortened lifespan. This prediction has been upheld in studies of fruit flies, as well as in a comparison of wild-caught and laboratory stocks of mice (Miller et al. 2002). Sgrò and Partridge's (2000) work suggests that the reduced longevity seen in laboratory strains is due primarily to increased early reproduction, rather than the accumulation of late-acting mutations (in support of the AP theory of senescence). It is further worth noting that screens for genes that extend lifespan are typically carried out in these relatively short-lived strains. If we were to do similar screens in long-lived stocks, we may find quite different results (e.g., Spencer et al. 2003).

MOLECULAR GENOMICS MEETS EVOLUTION

Aging results from the accumulation of deleterious mutations with late-age expression, which are either solely detrimental at late ages (MA) or that are also beneficial early in life (AP). Therefore, if we can identify the specific genes that affect aging and determine their function and regulation, we should be one step closer to understanding how senescence evolves and how it persists and is manifested in nature. Specifically, if aging results from MA, then the mutations that affect aging should differ among divergent taxa. This is because there would not be a selective advantage associated with these mutations. Alternatively, if aging results from AP, we might expect that the number of mutations would be much smaller and common across taxa. Mutations in alleles of pleiotropic genes that favorably impact such early life history traits as growth, maturation, and reproduction, while causing decline later in life, would be more likely to occur and persist in diverse taxonomic groups. Screens for mutations in model organisms have been of keen interest to evolutionary biologists because of their potential effects on other fitness traits, and because of the similarity or diversity of mechanisms that they suggest.

For the past 15 years, molecular gerontologists have been identifying single gene mutations that extend lifespan in laboratory populations of several model systems. Consider the nematode *Caenorhabditis elegans*. Mutations in genes that encode components of the insulin-like signal transduction pathway slow the rate of aging and extend lifespan relative to wild-type worms by increasing the organism's ability to respond to environmental stresses. Perhaps the best studied of these mutations are *age-1*, a catalytic subunit of a component of the insulin-like signaling pathway, and (*dauer* formation) *daf-2*, an insulin receptor that regulates the production of the transcription factor *daf16*, which is involved in insulin signaling (reviewed in Guarente & Kenyon 2000). Although the intermediate steps are not completely understood, lower insulin signaling increases lifespan by altering the phosphorylation status of a family of forkhead transcription factors (e.g., daf16), which ultimately results in an increased ability to combat oxidative stress (Guarente & Kenyon 2000).

Researchers were excited to discover that the likely cellular pathway that controls rate of aging in nematodes also controlled resistance to environmental stresses. Excitement about this insulin-signaling pathway was compounded when it was discovered that its apparent plurality of function (mediating at a minimum stress resistance and rate of aging) was shared across a diversity of organisms. In *Drosophila melanogaster*, mutations at the insulin-like receptor (*InR*) and its substrate protein chico (*chc*) also decrease the rate of insulin signaling, which results in decreased phosphorylation of these transcription factors, increased stress resistance, and increased lifespan (for references to these results and the work discussed below, see Tatar et al. (2003) and references therein). Moreover, work on *Drosophila* in Benzer's laboratory demonstrated that *methuselah*, a transmembrane receptor that extended lifespan, also increased stress resistance. In yeast (*Saccharomyces cerevisiae*), a mutation in a protein deacetylase (silent information regulator, *sir2*) acts upstream of the homologous yeast metabolic pathway and ultimately halts the expression of genes involved in glucose signaling, which has downstream effects on stress response and aging. In mice, much evidence from lines of mice with dysfunctional pituitary glands, and therefore low levels of growth hormone, supports the result that the cascade of insulin-like growth hormone signaling controls rate of aging. The laboratories of A. Bartke and R. Miller have investigated the lifespan effects of low growth hormone in the mouse lines *prop-1* and *pit-1*. They have found that production of low levels of growth hormone results in small-bodied but long-lived mice. Other work has identified a mutation in the gene for the receptor (IGF1r) of insulin-like growth factor 1, a primary ligand for this pathway, that extends lifespan by increased stress resistance in mice (Holzenberger et al. 2003). Finally, heterozygote knockouts of the forkhead transcription factor p66shc result in mice that are stress resistant and long-lived (Migliaccio et al. 1999).

The emerging picture from these life-extending mutations is that stress resistance determines longevity, particularly resistance to oxidative damage. The oxidative damage mechanism of aging posits that declining physiological functioning with age results from the accumulation of oxidative damage to the body's tissues, cells, and proteins. The damaging oxygen radicals are primarily produced as a normal by-product of aerobic respiration. Although cells mount a defense against these highly reactive oxygen metabolites (ROMs), the match between producing and destroying ROMs is imperfect and damage occurs (reviewed in Finkel & Holbrook 2000). Although all arrows point to stress resistance as the master switch of longevity, we have a long way to go toward understanding how it functions to control lifespan. Whether insulin signaling controls the allocation of resources between reproduction and stress resistance, or whether the insulin/stress/aging relationship is independent of other fitness traits, are questions that continue to motivate research into the evolutionary genetics of aging.

All of these pathways and mechanisms may seem a bit esoteric, but they actually have profound implications for the evolution of life history strategies in general, and senescence in particular. It is no surprise, then, that some of the researchers most actively involved in the search for the genes and gene pathways that influence senescence are evolutionary biologists. So what has led evolutionary biologists to think about insulin signaling pathways? The fact that the insulin signaling pathway, which enhances cellular resistance to oxidative and other stressors, is common to aging among such diverse taxa suggests a highly conserved mechanism of aging with deep phylogenetic roots. This is not to say that aging is necessarily adaptive. At the beginning of this chapter, we suggested that senescence might be the by-product of selection on something else. These molecular results give us some fairly strong clues as to what that "something else" might be. Though we still do not have an answer to the question of why the underlying cellular process, and presumed genetic architecture, of a nonadaptive trait is so highly conserved across distant taxa, the intense focus on this specific pathway should soon provide us with some excellent clues.

One interesting finding from genetic studies of aging is the relative paucity of evidence for trade-offs. Although many fertility assays of these mutant strains are preliminary and have only been measured under relatively benign laboratory environments, there are no consistent negative effects on development and fertility that have been reported. As we noted earlier, quantitative geneticists have invested heavily in the search for trade-offs between reproduction and longevity. Among life-extending mutations, why have we not found more evidence for trade-offs between longevity and reproduction? Presumably, if these mutants extended longevity without any cost, then natural selection would have driven these alternative alleles to fixation. We need to be cautious in assuming that these really are cost-free mutations, however. As Van Voorhies et al. (2003) point out, long-lived mutants need only survive in a laboratory screen for longevity, and might actually be at a selective disadvantage under natural or more stressful conditions. For example, both *age-1* in *C. elegans* and *Indy* (*I*'m *n*ot *d*ead *y*et) in *Drosophila* extend lifespan through increased stress resistance. These mutations showed no fitness cost of metabolism, development, or reproduction in standard laboratory assays. However, under conditions of nutritive stress, as might more reliably mirror natural conditions, both long-lived mutants had decreased fertility (*Indy*: Marden et al. 2003; *age-1*: Walker et al. 2000). To fully understand the consequences of novel mutants, molecular geneticists need to begin thinking about what organisms are doing in the "real world." Ultimately, if we are to develop a comprehensive understanding of the genetics of aging, we will need to determine just what constitutes an organism's natural environment and the conditions under which aging evolved.

Perhaps the solution to this two-fold challenge—that mutations may have pleiotropic effects, and that these effects may manifest themselves differently in different environments—will require a combination of field biology and new molecular techniques. While we have yet to see many molecular biologists heading into the wild with boots and binoculars in hand, we are seeing powerful molecular techniques aimed at elucidating the genetic basis of aging. For example, given that the *daf* mutations in *C. elegans* operate as transcription factors or other DNA binding proteins, the laboratory of G. Ruvkun has been successfully using RNA interference techniques to search for likely candidate genes that these transcription factors regulate. They have discovered the specific genes that are regulated by *daf16* and found that they are involved in metabolism and development as well as longevity. On another front, researchers who use artificial selection have begun to determine the myriad genes whose transcription levels have been modified by selective breeding vis-à-vis aging. Here the development of DNA microarrays to assay thousands of gene products simultaneously has enabled the identification of gene families or categories whose expression has been altered by selection. For example, Bronikowski et al. (2003) have shown that selective breeding for voluntary exercise extends lifespan in active mice, but not in sedentary mice, primarily through the induction of genes involved in immune functioning and stress response. The *pit-1* and *prop-1* mice have also been subjects of microarray analyses (e.g., Dozmorov et al. 2002). These studies have implicated the insulin-like growth factor signaling cascade in the lifespan extension seen in these lines of mice, as well as additional genes involved in metabolism and immune function. These comparative genomics approaches will undoubtedly be important in identifying the impacts of mutation and selection in longevity studies.

But microarray gene expression studies can tell us far more than just how single gene mutations affect gene expression profiles in known genetic backgrounds. By revealing the downstream effects of allelic substitution, such studies can help us to untangle the complex relationship between the underlying genotype and the expressed phenotype of an organism. They may also be able to help us understand epistatic and gene-by-environment interactions in aging. A single allele can have dramatically different effects on lifespan in different genetic backgrounds or in different environments, but the mechanisms behind these interactions have yet to be determined. Thus we will look forward to the innovative use of microarrays to take us beyond single gene mutations to the suite of their pleiotropic effects.

THE FUTURE: A NEW SET OF QUESTIONS

As evolutionary biologists and molecular geneticists have begun to work together on the problem of senescence, the advances have been dramatic. We have developed powerful tools to integrate classical and molecular genetics with large-scale demographic analyses, we have begun to identify genetic variation for senescence in natural populations, and we have identified several gene pathways associated with aging that appear in an impressively rich array of taxa, from yeast and worms to flies and mice. Despite this steady line of progress in experimental gerontology, from the earliest work by Pearl almost 100 years ago to the sophisticated molecular studies of today, there are many veins of research that have yet to be explored.

Part of the problem, we believe, lies in our continued attempts to determine which of the two classic models for the evolution of senescence are the best. In fact, we think that both likely play an important role in the origin and maintenance of genetic variation for longevity. Medawar and Williams developed their theories over half a century ago. In that time, we have had innumerable conceptual advances in our understanding of evolutionary biology, but with few exceptions MA and AP have marched along without being affected by these advances. We would argue that the time has come to incorporate into these models some of the insights we have gained from basic studies in evolutionary biology. By way of example, here we focus on two areas—behavior and pathogens—that we think may offer new insights into the evolution of aging.

Behavior and Senescence

The first area that we think will provide significant new insights into the evolution of aging in the coming years is that of behavior. Interestingly, one of the key players in bringing behavioral insights into evolution was W.D. Hamilton. This is the same Hamilton who, in 1966, developed the first formal mathematical model for the evolution of senescence (see Box 30.1). Although Hamilton continues to be cited among those working on aging, he is far better known for a pair of theoretical papers he wrote in 1964 while still a graduate student. This theoretical work on the evolution of cooperative behavior served as the cornerstone for what was to become a profound shift in evolutionary studies of behavior (Hamilton's papers are available as a collected edition: Hamilton 1996). His early behavioral work continues to exert a strong influence on evolutionary studies (see, for example, Frank, Ch. 23 of this volume).

The first indication that Hamilton's ideas might inform our understanding of aging came with attempts to explain menopause. While some argued that menopause was an inevitable, and unfavorable, consequence of aging, others suggested that it might be adaptive in light of Hamilton's theory. The kernel of Hamilton's idea was that an individual would willingly give up some of its own reproduction if that sacrifice benefited relatives. Thus, natural selection might favor females who pay the cost of menopause for the benefit of helping to rear grand-offspring (the "grandmother" hypothesis). The idea has received a mixed reception. Initial inclusive fitness models of the grandmother hypothesis suggested that the benefits of rearing offspring would not allay the costs of ceasing reproduction, and studies of wild animal populations suggested that the loss of reproductive capacity in old animals conferred no advantage to offspring or grand-offspring. On the other hand, anecdotal evidence from anthropological studies points to a concrete benefit for offspring of grandparental care, though it is not yet clear whether that benefit outweighs the costs of lost reproduction.

In what is perhaps the clearest attempt to link Hamilton's two early interests in a new theoretical framework, Lee (2003) has developed a life history model that incorporates explicitly the effects of resource transfer from parents to offspring. Lee starts with the classic renewal equation

$$1 = \int e^{-rx} l(x) m(x) dx, \qquad (30.2)$$

where $l(x)$ is the probability that a female survives to age x, $m(x)$ is the number of daughters she produces, and r is the intrinsic rate of increase (or Malthusian parameter) of the population. Onto this he maps a "balance equation," which relates population growth rate (r) to the transfer of resources between parents and offspring.

The classical model for the evolution of senescence assumed that the force of selection was high and constant until the onset of reproduction, and declined to zero after the last age at reproduction. By combining the balance equation and the renewal equation, Lee's model demonstrates that the force of selection may actually increase with age during the juvenile period (it is constant before maturity in the classic model), and will remain above zero after the last age at reproduction. This leads to a predicted mortality curve that is remarkably similar to the one we observe in most human populations, with a high initial mortality rate that drops during childhood, and then exponentially increasing mortality after the onset of reproduction. Lee's model may not be perfect (for example, it assumes clonal reproduction), but it provides a simple and general framework that relates behavior to senescence, and leads to powerful predictions.

As we turn our attention beyond the laboratory to the natural history of organisms, other fascinating behavioral questions appear. Why, for example, is female life expectancy greater than that of males in most species of mammals, while the reverse is true for birds? Recent articles (e.g., Promislow 2003) suggest that sex differences in longevity may be related to the fact that when it comes to reproductive strategies, males and females often have serious conflicts of interest. For example, studies with *Drosophila* have found that males inseminate females with accessory gland proteins (ACPs) that increase male fertility, but at a cost of reduced female lifespan and overall fitness. Furthermore, negative genetic correlations between fitness in males and females may account for differences in lifespan between the sexes, and may be an important mechanism for maintaining variation in longevity in natural populations (Fedorka & Mousseau 2004).

Eusocial organisms also present us with a fascinating puzzle. In many species of ants, workers may live for a few months but the queen can live for 20 years or more (Keller & Genoud 1997). Keller and Genoud (1997) argue that the extraordinary longevity in queens is due to the fact that she is so well protected from extrinsic mortality sources. However, the explanation may turn out to be more complicated than that. In some species, there are both single-queen (monogyne) and multiple-queen (polygyne) colonies. The polygyne colonies tend to be significantly larger, and so queens are more protected from extrinsic mortality. However, others have noted the polygyne queens have substantially shorter lifespan than the monogyne queens. It may turn out that kin selection plays a more important role than extrinsic mortality in shaping extreme lifespans in queens.

Parasites, Immunity, and the Evolution of Senescence

The previous chapter explored ways in which pathogens can influence, and be influenced by, host evolution (Kover, Ch. 29 of this volume). We suggest here that pathogens and host immune systems may turn out to be important to our understanding of the genetics of aging, from underlying molecular mechanisms to broader, population-level phenomena. Interestingly, Hamilton's name comes up again. Hamilton played a central role in leading evolutionary biologists to think that parasites may be a more potent and important force for evolutionary change than had previously been considered. He pointed out the potential role that parasites might play in the evolution of sex and sexual selection. Whereas the overlap between behavior and senescence is well established, work is just beginning on the role of parasites in the aging process.

Bell (1993) suggested that parasites may be responsible for senescence because of their rapid rate of genetic change relative to the host. Parasites and their hosts are constantly playing a molecular game of cat and mouse, with the host mounting immune responses to antigens presented by the parasite, and parasites evolving new strategies to overcome these defenses. According to Bell, this process leads inevitably to the deterioration of the host as it ages.

One problem with Bell's model is that organisms age even when held in specific-pathogen-free environments. However, rather than parasites leading directly to host deterioration, it may be that parasites have favored the evolution of costly defense mechanisms, and these defenses are what leads, ultimately, to age-related decline.

Earlier we argued that the trade-off between reproduction and lifespan may be central to the evolution of senescence, and we presented different kinds of evidence, from phenotypic trade-offs to the description of specific molecular pathways. Investment in the immune system may turn out to play a critical role in mediating life history trade-offs in general. Studies suggest that the causal arrows relating immune responses and life history strategies may go in both directions. For example, individuals exposed to pathogens have reduced stress resistance and reproductive output later in life. The cost of mounting an immune response can be seen even in the absence of pathogens. For example, injecting bees with lipopolysaccharides or glass beads reduced their survival rates dramatically. Drawing the causal arrow in the other direction, increased reproductive effort is associated with decreased immune function in a variety of species (see Zuk & Stoehr 2002 for a review).

Not surprisingly, the way in which the costs of parasitic infection or mounting an immune response influence host survival or fecundity may be complex (Figure 30.3). Not all experimental studies have found a relationship between antigen challenge and host fitness, and theoretical models suggest that the relationship between host mortality rates and costs of parasitism or immunity may be quite complex. Classic models for the evolution of virulence suggest that mortality costs of parasitism should evolve to be higher in hosts with higher background mortality rates. Williams and Day (2001) show that if host mortality and parasite virulence interact, higher host mortality can actually favor decreased virulence. Perhaps we should not be surprised, then, to learn that these same researchers have now shown the same to be true for models of senescence. In his original model of antagonistic pleiotropy, George Williams (1957) argued that increases in extrinsic mortality would favor the evolution of more rapid rates of senescence. In more recent work, Williams and Day have gone on to show that if mortality rates are condition-dependent, increased extrinsic hazards can lead to *decreased* rates of physiological decline, at least at early ages. It now remains to integrate these two models to determine just how pathogens can shape patterns of senescence.

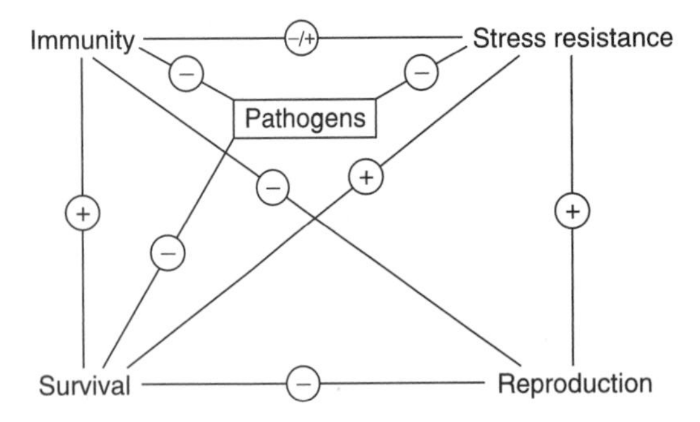

FIGURE 30.3. Pathogens may play an important role in the evolution of senescence. However, the exact correspondence between pathogens, immunity, and senescence may be difficult to disentangle, with causal arrows between various factors acting over both physiological time and evolutionary time.

CONCLUSION

We began this chapter with an apt quote from T.S. Eliot. In this last section, we have suggested that, at least within the evolutionary realm of studies into the nature of senescence, we may move forward by looking back to evolutionary ideas that have been developed over the past four decades. In this light, perhaps the following lines from Eliot are also rather appropriate.

We shall not cease from exploration
And the end of all our exploring
Will be to arrive where we started
And know the place for the first time
—T.S. Eliot, "Little Gidding"

SUGGESTIONS FOR FURTHER READING

There are several excellent book-length reviews on the evolutionary genetics of aging. Three good places to start are the books by Rose (1991) (a general introduction to the evolution of aging), Charlesworth (1994) (an excellent theoretical treatment of evolution in age-structures populations), and Finch (1990) (encyclopedic coverage of a range of topics on the genetics of aging). Tatar's (2001) chapter offers a more ecological perspective on senescence. For those interested in a more molecular perspective on the genetics of aging, several recent reviews discuss the discovery of genes within the neuroendocrine system that regulate aging (e.g., insulin-like growth factor and its homologs) throughout diverse taxa (McCarroll et al. 2004; Melov & Hubbard 2004; Tatar 2004). Finally, interested readers can find regular updates on new discoveries in aging research at SAGE-KE (http://sageke.sciencemag.org).

Charlesworth B 1994 Evolution in Age-Structured Populations. Cambridge Univ. Press.
Finch CE 1990 Longevity, Senescence and the Genome. Univ. of Chicago Press.
McCarroll SA, Murphy CT, Zou S, Pletcher SD, Chin CS, Jan YN, Kenyon C, Bargmann CI & H Li 2004 Comparing genomic expression patterns across species identifies shared transcriptional profile in aging. Nat. Genet. 36:197–204.
Melov S & A Hubbard 2004 Microarrays as a tool to investigate the biology of aging: a retrospective and a look to the future. Sci. Aging Knowledge Environ. 24:re7.
Rose MR 1991 Evolutionary Biology of Aging. Oxford Univ. Press.
Tatar M 2001 Senescence. In Fox CW, Roff DA & DJ Fairbairn, eds. Evolutionary Ecology: Concepts and Case Studies. Oxford Univ. Press.
Tatar M 2004 The neuroendocrine regulation of *Drosophila* aging. Exp. Gerontol. 39:1745–1750.

31

Experimental Evolution

ADAM K. CHIPPINDALE

For evolutionary biologists, an unfortunate feature of the past century of genetics is that much of what we know comes from broken genes. The analysis of mutants shows us what evolution by natural selection overwhelmingly discards: dysfunctional genotypes. Although our understanding of fundaments such as patterns of inheritance and gene expression patterns has been advanced extraordinarily through the mutationist program, this approach remains atomistic and focused at the individual level. While polygenic variation is the curse of the mutationist program, it is the main fuel that drives adaptation. To understand evolutionary processes, which operate at the level of populations, we are led into the more confusing and experimentally difficult world of quantitative variation for traits affecting fitness. To study variation in fitness using classical mutagenic approaches is tantamount to pulling hoses and wires under the hood of a car and then measuring its performance. Most destructive modifications will cripple or extinguish drive, telling us that the automobile is a highly integrated machine. In contrast, adaptation is predominantly a fitness *tuning* process, and studying it requires approaches that allow us to connect genetic variation to performance variation in the context of populations, and to see input from new beneficial genes as well as deleterious ones.

Historically we have used comparative methods to infer evolutionary processes. Advances in phylogenetic methods, coupled with the extraordinary boom in the availability of molecular genetic data, have made comparative evolutionary studies increasingly powerful and rigorous. At the same time, we have had many opportunities to observe evolution directly in circumstances ranging from viral adaptation in the HIV pandemic to the exquisite natural history studies of Darwin's finches on Galapagos. A third approach is to purposely engineer the experimental evolution of organisms. This last approach is the focus of the present chapter.

As will be shown, experimental evolution is one of the most powerful approaches available for the study of microevolutionary processes. No other method allows greater control, definable replication, or affords the possibility of actually doing that great *Gedankenexperiment*: playing the evolutionary tape over (or indeed, playing it backward). The approach has all of the appeal of a theoretical model: conditions are established and predictions are made to test specific hypotheses. Instead of variables in equations, populations are marshaled and then, like a computer keystroke, the simulation begins: Evolution in action. Playing Darwin. While we would like to be able to predict evolutionary outcomes over large timescales, when we do long-term experiments with evolution we often see that the genetic architecture of a population itself is malleable. Organisms in their wonderful and often intractable complexity are involved, and the simulation often produces surprises. The tape of life, it turns out, has many unexpected twists and may not play the same way fore and back. Experimental evolution is therefore an approach that informs theory at least as often as it tests it.

Practitioners of experimental evolution work on a range of problems that is at least as diverse as that studied by any other group of biologists. We study problems as broad-ranging as codon bias and courtship, enzyme efficiency and ecosystems. Yet underlying this enormous variety of intellectual problems is a set of basic operating principles and tools that makes experimental evolution identifiable as a distinct field. The aim of the present chapter is not to catalog the vast diversity of experimental evolution studies. Others, notably Bell (1997), have tackled that problem with encyclopedic thoroughness. Nor will this chapter attempt a primer of quantitative genetics, rather directing the reader to Roff (Ch. 18 of this volume). My more modest goal is to illuminate the principles of experimental design that make experimental evoluation a definable discipline, discuss the scope of this field, and highlight relatively few exemplary studies that define its bounds.

THE SCOPE AND ORIGIN OF EXPERIMENTAL EVOLUTION

Broadly defined, experimental evolution encompasses approaches as diverse as selective breeding and the introduction of new species (e.g., as in biocontrol programs). Domestic breeds provide us with a daily reminder of the power of applied directional selection. How confusing it must be for a child to be told that a Chihuahua and a Great Dane are both dogs, yet not only are these two breeds conspecific, they are the products of a mere handful of generations of artificial selection for "desirable" traits. It is currently believed that all breeds of the domestic dog descend from the East Asian grey wolf, and one or a few close domestication events. How has the gene pool that produces the relatively canalized phenotype of *Canis lupus* been morphed into the diversity of dog breeds we see today? How can we account for the behavioral and morphological gulfs between toy poodle and mastiff? Selection has taken these breeds so far beyond the confines of the wolf morphospace that they are barely recognizable as lupine. Any number of examples from domesticated species, or a stroll through the produce section at the supermarket, might be used to make the same point: Directional selection has the power to create gene combinations and effects unimaginable by variation seen in the ancestral species. It is the ability of selection to bring together the additive and nonadditive gene combinations that distinguishes it from the study of the quantitative genetics of standing genetic variation (e.g., through resemblance of relatives).

Most domestic breeds are the products of artificial selection. Artificial selection in its purest form involves evolving populations by targeting specific traits of interest and arresting all forms of natural selection acting upon the population. The entire population is measured for the trait(s) of interest and a subset of individuals is statistically selected to produce the next generation. Forestalling natural selection means regulating individual fitness so that no family contributes more or less to the next generation than any other. A pure artificial design therefore involves complete intervention by the experimenter in the choice of parents for the next generation and the regulation of their breeding success. As might be imagined, artificial selection typically steers one toward traits that lend themselves to measurement (e.g., bristle number in *Drosophila* or weaning weight in mice) and toward small population sizes.

The modern field of experimental evolution grows out of early-twentieth-century population genetic work with the fruit fly, *Drosophila*, by T.H. Morgan and others, and quantitative genetic principles developed in the middle of the century. The quantitative genetic literature was intensely concerned with the dynamics of domestic breeding, for both theoretical and applied reasons. Geneticists such as A. Robertson, F. W. Robertson, K. Mather, and D.S. Falconer, to name a few, began using laboratory populations of mice, *Drosophila*, and other easily cultured organisms in artificial selection to model processes that occur in breeding populations of larger organisms of greater economic importance. As the new field of evolutionary genetics arose, artificial selection became one of its main methodologies. By the 1970s the tools and quantitative genetic theory of artificial selection had become highly sophisticated, and the array of experiments formed a vast catalog. Then, suffering under its own weight and overshadowed by the advent of new molecular technologies such as electrophoresis, the impact of artificial selection declined. Molecular methods promised evolutionary genetics a chance to peer into selection in natural populations; experimental evolution was temporarily eclipsed.

Experimental evolution began to resurge in the 1980s with the recognition that the artificial selection approach has shortcomings as a model for the evolution of natural populations. Populations of most species are large relative to laboratory populations, emphasizing differences in two related processes: genetic drift and inbreeding. To identify drift, entire populations must be replicated in experiment. To minimize drift and its fitness manifestation, inbreeding depression, large populations must be employed. The recognition that large, replicated population designs would allow the analysis of fitness traits (e.g., life history traits) helped to spawn the approach dubbed "laboratory natural selection" (Rose et al. 1996) in which large outbred (genetically variable) populations are evolved by applying a defined selection treatment but no effort is made to forestall selection for fitness itself. In many ways, the term "laboratory natural selection" refers as much to the aims of the project as its specific methodology: the intent is to characterize evolutionary responses rather than just genetic properties underlying a specific characteristic of the study organism.

A typical laboratory natural selection experiment would involve exposing a set of populations to an environmental challenge, say elevated temperature, allowing natural selection to sort out survivors, and then allowing those survivors to reproduce. Some survivors may be sterilized by the thermal stress; they are as good as dead in the eyes of selection. Individuals surviving the stress *and* producing the most offspring will be favored by selection. Because there is no a priori trait chosen and no attempt to standardize reproductive success among survivors, evolution may proceed via any means at its disposal. If a trait even indirectly influences the condition of individuals in the breeding group then it will be subject to selection, meaning that a large fraction of the genome may contribute to the evolutionary response. Laboratory natural selection therefore sits at the other end of the spectrum of control from artificial selection. This chapter will focus on experiments at the "natural" end of the experimental continuum from this point on.

DESIGNING AN EXPERIMENTAL EVOLUTION PROJECT

A well-designed experimental evolution project allows observation of the dynamics of evolution within replicated populations under controlled conditions. There are two overarching considerations in building an evolution project: the raw material and the experimental design. The raw material for selection is defined both by the organism and by the base (starting) population structure, and presents some frequently overlooked considerations. The experimental design component consists of the selection treatment, controls, and statistical design.

The Raw Material

Although in principle we would like to conduct experiments in evolution with as many species as possible, practical problems have promoted only a handful of them to "model organism" status. The organisms most amenable to experimental evolution are those that domesticate readily and have very short generation times. The dipteran insect *Drosophila melanogaster* and the bacterium *Escherichia coli* (Box 31.1) are the undisputed champions of this literature. Fueling their precedence is a literature on nearly every aspect of their biology, a self-reinforcing community of researchers, and well-developed tools to address their genetics. Few organisms have their own meeting, let alone several international conferences attended by hundreds of scientists every year, as we see in *Drosophila*. A quick browse through Flybase (http://flybase.bio.indiana.edu/) reveals thousands of available stocks and links to thousands more, most delivered to one's laboratory with a few keystrokes. Model organisms that have similar infrastructure but have been largely passed over for experimental evolution include the zebra danio and the nematode *Caenorhabditis elegans*, favorite organisms in developmental biology. Microorganisms are increasingly popular subjects for experimental evolution. On top of the obvious advantages (rapid generations and potentially enormous population sizes) the advent of simple and cheap molecular genetic methods has supplied what microorganisms have traditionally lacked: a detailed phenotype.

Drosophila and *E. coli* reflect the single greatest divide in the selection of organism for experimental evolution: sexual versus asexual reproduction. Sexual organisms have the advantage of a built-in mechanism for generating novelty from variation present in the initial (base) population sample. Recombination has the effect of reshuffling allelic combinations to expand the variance in phenotypes and routinely exposing them to selection. Additive (consistently expressed) and nonadditive (inconsistently expressed, interacting) allelic combinations alike may change in frequency rapidly. Like the example of domestic dogs evolving from wolves, experimentally evolving sexual populations often quickly move beyond the range limits of variation segregating in the base population. For example, selection for accelerated development rate resulted in a steady increase over 125 generations with the distribution of trait values in experimental populations becoming entirely discrete from

BOX 31.1. *E. coli* as a Model Organism in Evolutionary Genetics
Richard E. Lenski

The bacterium *Escherichia coli* is a model organism in evolutionary genetics in two distinct contexts. First, *E. coli* has served as a model for examining the extent of genetic diversity in natural populations of bacteria, and elucidating the evolutionary processes that structure the patterns of diversity. Second, *E. coli* is widely used as a model for studying evolution in action in the laboratory, where researchers can test certain evolutionary hypotheses by performing replicated, controlled, and designed experiments.

Both contexts in which *E. coli* is a model organism in evolutionary genetics grew out of the fact that *E. coli* served previously as an important model for the fields of genetics, biochemistry, molecular biology, and cell physiology. Hence, there is a tremendous wealth of background information on its genomic and functional biology, as well as powerful techniques to measure and manipulate factors that are important in evolution.

E. coli as a Model for Understanding Bacterial Evolution in Nature

Bacteria are the most abundant organisms on Earth, and they play critical roles in ecosystem processes as well as in disease. For a long time, however, little attention was paid to their evolution. With the advent of molecular approaches for examining genetic diversity—including, first, protein electrophoresis and, later, DNA sequencing—it became feasible to trace the evolutionary history and study the mechanisms of bacterial evolution in nature. Not surprisingly, much of this research focused on *E. coli*, owing to the accumulated knowledge and experience working with it.

The species *E. coli* is very widespread in nature, as it lives in the colon of almost every mammal. There are probably more than 10^{20} *E. coli* alive at any moment. Most are harmless commensals, but some strains can cause severe diarrhea. Being a bacterium, *E. coli* reproduces asexually by binary fission, but it can sometimes exchange genes, even with other species, via the movement of viruses and other extrachromosomal elements between cells. Given its enormous population and deep history (it diverged from *Salmonella*, a close relative, more than 100 million years ago) it is perhaps not surprising that *E. coli* collectively harbors tremendous diversity in almost every gene. One might expect, therefore, a near-infinitude of different *E. coli* genotypes or strains, but that is not so. Instead, a relatively small number of multi-locus genotypes are found repeatedly, with minor variation, across space and over time (Whittam 1996). The reason for the wide distribution of a few strains is the asexual mode of reproduction, which maintains the statistical linkage of all the genes within the genome across generations.

This pattern of low genotypic diversity, overlying high allelic diversity, is sometimes called the "clonal paradigm." In fact, however, the population-genetic structure of *E. coli* is not strictly clonal owing to some mechanisms allowing genetic exchange, as noted above. If one constructs phylogenetic trees based on several different genes, there is compelling evidence of discordance, which indicates that occasional gene exchange among lineages is also important in structuring genetic diversity in *E. coli*. As one might expect given the tremendous diversity in the microbial world, additional research has established that different bacterial species lie at different points along the continuum between extreme clonality and substantial recombination (Maynard Smith et al. 1993).

(continued)

BOX **31.1.** *(cont.)*

Understanding the diversity and evolution of the pathogenic subset of *E. coli* strains is another facet of this research. Population-genetic analyses have shown that a group of strains that were historically given their own genus, *Shigella*, based on their pathogenic phenotype, are actually a subset of *E. coli* strains. That is, *Shigella* strains, although a well-defined group, lie well within the phylogenetic tree that comprises the diversity of *E. coli* strains (Whittam 1996). Pathogenic strains have independently evolved many times within *E. coli*. This transition typically involves the acquisition of virulence genes by genetic exchange. Strains with similar pathogenic features have sometimes emerged by the acquisition of similar virulence genes, except on distinct genetic backgrounds, providing a fascinating example of parallel evolution in the microbial world (Reid et al. 2000).

E. coli as a Model for Experimental Tests of Evolutionary Hypotheses

Many of the same features of *E. coli* that made it an excellent model organism in genetics and related fields also make it an attractive model for laboratory experiments to study evolution in action. Populations of *E. coli* are easy to grow, and key environmental factors such as resources and temperature can be readily controlled. Population sizes are typically very large, such that new mutations provide a constant supply of genetic variation. Of course, rapid generations are a key feature, especially in the context of evolutionary research, where experiments can run for hundreds or thousands of bacterial generations. And as noted before, abundant information on the genomic and functional biology of *E. coli*, and powerful molecular approaches for analysis and manipulation, provide a strong foundation for experimental studies of evolution.

Somewhat less obvious, but as important as any other feature, is the ability to freeze *E. coli* cells indefinitely and revive them for later study. A researcher can thus preserve the ancestor used to start an experiment as well as intermediate stages during evolution. In essence, the researcher is able to raise fossil organisms from the dead, and thus compare directly the properties of evolved organisms and their ancestors. It is even possible to perform head-to-head competitions between the evolved and ancestral organisms, which allow one to measure changes in fitness over time and observe the dynamics of adaptation by natural selection. But to do this, one must be able to distinguish the evolved and ancestral genotypes, which is typically done by introducing a genetic marker into one type or the other that allows them to be readily identified on an appropriate agar medium. Because the bacteria are asexual, the genetic marker remains in whichever background it is placed. Of course, when interpreting results of evolution experiments obtained using *E. coli*, it may also be important to consider how its features differ from more familiar organisms, most notably its asexual reproduction and unicellular life history.

In one long-running experiment, 12 populations of *E. coli* were started from the same ancestral genotype, and they have been evolving for 20,000 generations in identical environments (Lenski 2004). The overarching goal of this experiment is to quantify and understand the dynamics of phenotypic and genomic evolution in an asexual system. An interesting question is whether the changes are parallel across the independently evolving populations, or whether each population evolves along a different trajectory. In essence, this experiment provides a way of performing the thought-experiment of "replaying life's tape" (Gould 1989). The Figure shows the trajectories for mean fitness relative to the ancestor for all of the populations, based on competition assays. These trajectories all show a roughly similar dynamic, with the most rapid fitness improvement early in the experiment followed by decelerating gains over time.

(continued)

Another analysis of this same long-term experiment used DNA arrays to quantify the expression of more than 4000 genes in two of the evolved populations and their ancestor (Cooper et al. 2003). That study found that the global expression profiles had evolved in strikingly similar ways. By using the pattern of parallel changes in gene expression, along with the genetic information available for *E. coli*, the researchers were able to find a mutation in an underlying regulatory gene. They then moved that mutation into the ancestor, and showed that the mutation was beneficial. Interestingly, however, the parallel changes in expression in the other population they tested had a different underlying genetic basis, because that same gene had not accumulated any mutations in the other population.

In this long-running experiment, the *E. coli* populations are evolving in and adapting to a simple ecological environment, which is potentially both a strength and a limitation. Other experiments have been performed where evolving *E. coli* populations encounter diverse resources, changing thermal environments, and even predators. Certain viruses attack and kill *E. coli*, and these have been used to study hypotheses concerning the coevolution of hosts and parasites (Bohannan & Lenski 2000). In many cases, the bacteria evolve resistance to the virus. However, resistance engenders a trade-off, such that the resistant bacteria are inferior competitors for resources in comparison with their sensitive progenitors. These evolutionary changes can have important consequences for the resulting structure of microbial communities, with evolutionary outcomes often depending on subtle details of the molecular interactions between the bacteria and viruses. Also, successes in using *E. coli* as a model organism for experimental evolution have led others to perform evolution experiments with other microorganisms, including viruses and fungi as well as various bacterial species (Elena & Lenski 2003).

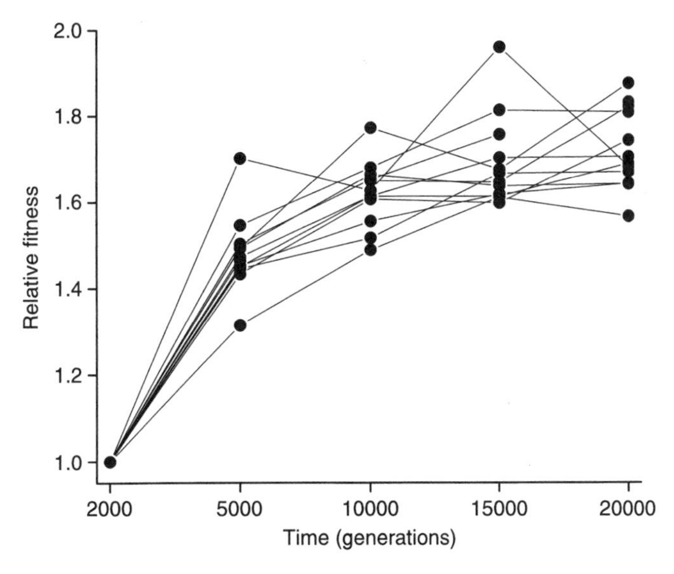

FIGURE 1. Trajectories for relative fitness from a long-term evolution experiment with *E. coli*. Lines show values for 12 independent populations. Fitness values express the ratio of growth rates for the evolved and ancestral types, obtained by competition experiments. Data from Cooper and Lenski (2000).

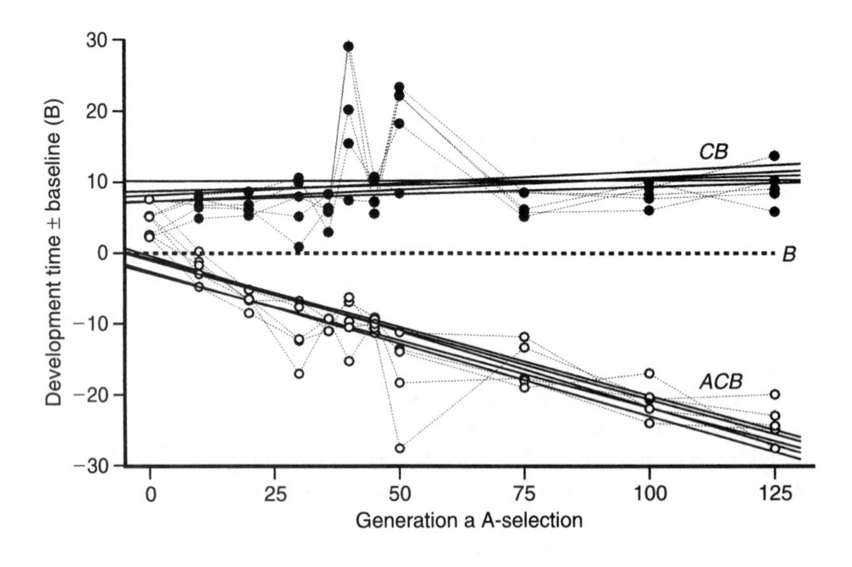

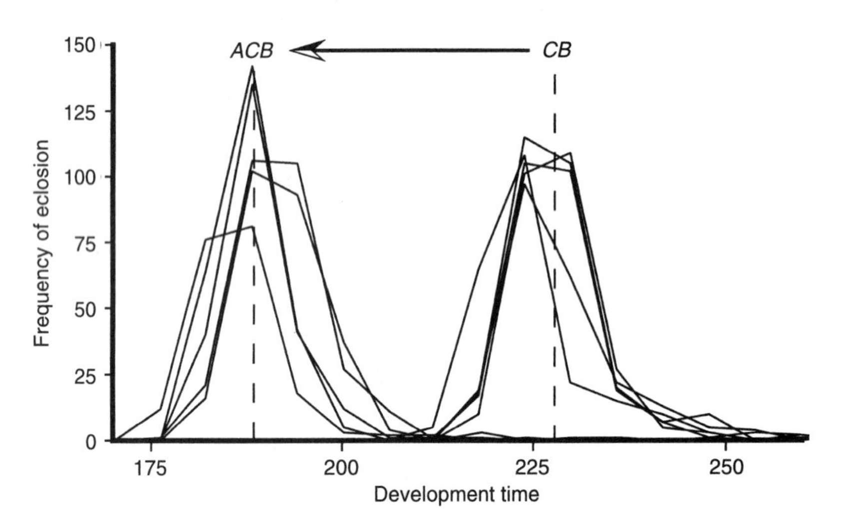

FIGURE 31.1. Evolution of *Drosophila* developmental rate in five populations selected for rapid development and extremely early fertility (ACB) relative to their matched controls (CB) and five baseline (B) populations. Selection response was unwavering for 125 generations (top panel) despite the importance of this trait in determining fitness in this species. Phenotypic variation did not decline in response to selection (bottom panel), and abundant genetic variation for this fitness character persisted throughout the experiment. A similar pattern was observed for 10 parallel (ACO and CO) populations.

controls (Chippindale et al. 1997; Figure 31.1). Even with traits such as development speed, which one might expect to already be close to its maximum in a species such as *Drosophila*, it is a common observation that variation in the target character within large populations is not eroded by even protracted selection.

Asexual organisms present a different set of dynamics under selection. Instead of pervasive rearrangement of genes, the long-term selection response of clones is largely driven by the occurrence of new mutations. The selection response of, say, *E. coli*, is typified by a stepwise progression of fitness in the selective environment. Periods of stasis are punctuated by adaptation when favorable new mutations sweep through and displace lower-fitness genotypes. This "evolution in steps" model is a simplification because a variety of factors such as frequency dependence, clonal interference, and environmental heterogeneity may sustain genetic variation, but the general point is that asexual populations depend on mutational input driving selection more than sexual populations do (Elena & Lenski 2003). The main reason asexual microbes such as *E. coli* are so appealing for experimental evolution is the rapidity of generations and potential to house large populations of genomes in small containers. Richard Lenski's 30,000 generation *E. coli* experiment (Lenski et al. 1991) chalks up generations at a rate of approximately seven per day and involves cultures in the millions of individuals per population per generation. With simple serial transfer techniques, natural selection has been allowed to screen billions of mutations arising by chance; 12 lines derived from a single, identical clone ancestor have had spectacular and different adaptive histories (Box 31.1). Moreover, these experiments approach having the perfect control because *E. coli* freeze readily and the actual ancestor can be used in competition with an evolved strain to demonstrate evolved change.

One of the more contentious questions in designing an experiment in evolution is the genetic material to start out with. Should populations be recently derived from the wild, highly domesticated, or somewhere in between? The problem is that organisms inevitably react and adapt to new environments. The laboratory is a novel environment for any free-living species and the simple act of sampling a population and bringing it into captivity becomes an evolution experiment in itself. The rapid adaptation of characters immediately important to the laboratory environment will be followed by persistent fine-tuning of less critical functions, and perhaps the loss of characters or behaviors rendered obsolete. For example, because laboratory flies on a 2-week cycle reproduce at a few days of adult age, a decline in late-life fitness and lifespan is predicted by the evolutionary theory of aging. With enforced early reproduction, the force of natural selection goes to zero at a very early age, and mutations with age-specific effects beyond a few days of adult age become selectively neutral. Similarly, any fruit flies that escape from laboratory culture, as they will during hand transfer from vial to vial, will suffer reduced fitness. Domestication therefore not only relaxes selection for migratory behavior, but actively selects for relatively sedentary flies. (Drosophilists who work with field-caught animals know this all too well.) Does laboratory adaptation strip the organism of "real" adaptations, taking populations so far from their natural genetic makeup that they become uninteresting? The answer to these questions is not straightforward, dependent upon the problem of interest, and hotly debated.

Most experimental evolutionists have opted to work with populations with long histories of laboratory cultivation—often decades under controlled environmental conditions and fixed breeding schedules. The advantages of working with laboratory populations that are at, or near to, some kind of evolutionary equilibrium are manifold. For example, it is more feasible to apply a defined selection treatment without multiple, potentially confounded sources of selection occurring. Novel environments tend to produce positive genetic correlations among performance characters (Service & Rose 1985). Genotypes that are fortuitously well-adapted to the new conditions will tend to do well for a variety of traits, while the reverse is true for poorly adapted genotypes. For example, genotypes effective at metabolizing the food resources available to them in the new environment are likely to display advantages in characters such as growth, reproduction, and survivorship even if there are trade-offs among these traits at equilibrium. Because trade-offs are such a central concern in evolutionary studies, and life history studies in particular, the conventional wisdom has been to use populations that have evolved to live under all of the general conditions of laboratory culture before applying new sources of selection.

On the other hand, laboratory populations are typically closed to immigration and are susceptible to loss of variation from two major sources: (1) selection for laboratory conditions, and (2) genetic drift and inbreeding. Because most laboratory evolution experiments depend upon the standing genetic variation in base populations, maintenance of these populations with large effective size is essential whenever they are kept for long periods. Most modern laboratories working on diploid, sexual species maintain base stocks in the range of 10^3 or greater, and population sizes of microbial systems at orders of magnitude higher. But even when inbreeding depression has been minimized, the set of responses available to a laboratory-adapted population may differ from what would happen with a natural population.

The problem of laboratory adaptation has polarized a debate over the appropriate material for experimental evolution. Harshman and Hoffmann (2000) focused on environmental stress selection experiments in *Drosophila* to point out some potential pitfalls of long-term adaptation. Among other critical points, they argued that most laboratory organisms live in "Fat City," with consistent availability of resources, and that this may predispose certain evolutionary outcomes. For example, laboratory selection for increased desiccation tolerance—likely to be experienced by wild flies, particularly in desert habitat—often results in the evolution of increased water and metabolite storage, such as lipid and glycogen (e.g., Djawdan et al. 1998). *Drosophila* selected in Plexiglas cages to resist desiccation and resource deprivation adapted by becoming obese, and further increased survival by minimizing activity until selection was stopped (personal observation). These kinds of responses are not necessarily found in desiccation-tolerant wild drosophilids where behavior (refuge-seeking, etc.) may play a major role in desiccation-avoidance. There is therefore potential for mismatch between experimental evolution and comparative studies for certain behavioral and physiological traits.

Harshman and Hoffmann (2000) advocated the use of recently sampled wild populations in addition to those already at evolutionary equilibrium laboratory conditions. They argue that more relevant responses may be observed from "fresh" material, or at least that the most meaningful responses are those observed in all circumstances. They also suggest that the major features of adaptation to the laboratory may be over within relatively few generations. If so, an ideal compromise would be use of lines which still possess abundant natural variation but have been selected to live under experimental conditions long enough to have generally adapted to them. But how much adaptation should the experimenter allow? Deciding when such a compromise is reached is an empirical matter. The transition from field to laboratory has been referred to as an "evolutionary no man's land" by Matos and colleagues, a state in which any possible quantitative genetic outcome to selection may occur—meaningful or spurious. Matos has made laboratory adaptation in *Drosophila pseudoobscura* a focus of her research program. Her studies (e.g., Matos et al. 2002) demonstrate rapid change in a number of performance traits, including larval feeding and pupation behavior, adult resource acquisition and allocation patterns, reproductive schedules, and so on. In general, separately sampled populations converged upon similar life history solutions, but did so at varying rates. Matos et al. concluded that some approach to an "adapted" state had occurred after 10 to several dozen generations, depending upon the collection. In *Drosophila* time, these results argue for a year or more of laboratory culture before such a state is reached. Nevertheless we lack a rule, and debate will continue.

To preserve genetic variation, some laboratories have set up base populations that experience variable selection regimens. For example, some laboratories maintain *Drosophila* with overlapping generations so that animals of any age can contribute to the population, and many others use self-regulation through natural cycles of boom and crash dynamics in bottles. The advantage of such systems lies in the potential for maintaining greater variation due to fluctuating selection regimens. In the case of alternately uncrowded and crowded cultures, selection presumably fluctuates in intensity and direction for characteristics such as development time, growth rate, and body size (e.g., Mueller et al. 1991; Houle & Rowe 2003). These characters appear to trade-off evolutionarily, so that a cost of rapid development is reduced body size (e.g., Nunney 1996; Chippindale et al. 1997). Similarly, if there is a negative relationship between early-life fitness and later-life fitness and survival in fruit flies, as has been widely suggested (e.g., Rose 1984; Zwaan et al. 1995), then overlapping generations in which older individuals contribute to the population may help to preserve genetic variation for later-life fitness. Work on the repeated domestication of *D. melanogaster* by Sgrò and Partridge (2000), for example, suggested that overlapping generations may forestall the convergence of life-history traits on an early schedule. However, the difficulty with such designs lies in establishing the conditions for experimentation. In a world of trade-offs, what improves fitness under one set of conditions may impair it in another. Complex base population propagation therefore presents on a miniature scale the same set of problems that natural populations do, albeit with considerably more opportunity to disentangle the sources of selection.

Still other possibilities exist for controlling laboratory adaptation and these have been little explored on a formal basis. The common practice of collecting isofemale lines in *Drosophila* (populations all descended from the same mother) may limit the potential for adaptation within each line by inbreeding. Reconstitution of mass breeding populations could be used to restore relatively normal population structure after a few rounds of recombination. We (Chippindale et al. 2001) have made use of "cytogenetic cloning" in laboratory-adapted populations. Such C-cloning makes use of the lack of molecular recombination in *Drosophila* males to transmit whole haplotypes from father to son as a single linkage unit, like a giant nonrecombining chromosome. In theory this approach could be used to "freeze" haplotypes from a population sample (wild-caught or laboratory-adapted) and then reconstitute them as a mass breeding population after any number of generations by discarding the marker chromosomes. C-cloning is a viable but cumbersome approach to storage and replication when applied to big populations. *Drosophila* biology waits anxiously for renewed interest in the cryopreservation of fly embryos to allow their long-term storage and contrasts between ancestors and evolved lines.

Note that while with sexual species we usually seek abundant genetic variation to start an experiment, in asexual organisms the diametric opposite approach is often applied: selection is initiated from a single clone. This method allows individual allelic substitutions to be detected more reliably.

Selection Treatments and Controls

The goal of most experiments in evolution is to isolate the response of populations to a well-characterized selection treatment. In a perfect design, one would alter a single factor, be it environmental or genetic, in a treatment group relative to a control, where the control is in every other way identical. This goal is rarely fully achievable. Consider the problem of selecting for greater lifespan. If the experimenter imposes selection on lifespan by postponing reproduction ever later in a set of populations (as in the *Drosophila* experiments by Rose 1984 for example), two problems immediately arise. First, selection is really on late-life fertility and not just on longevity itself. Second, if the controls are bred early in life, then they undergo more generations than the deliberately selected group. If adaptation to general conditions of laboratory culture is still occurring then it will happen more rapidly in the "controls" than the "selected" populations. Or, if effective population sizes are the same, inbreeding will occur more rapidly in the control treatment than the longevity-selected treatment.

In the fruit fly aging selection literature some elegant solutions to these confounds have been applied to isolate lifespan from age at reproduction. For example, Zwaan et al. (1995) used selection on virgin lifespan in which family members were stored at lower temperatures, and bred only after longevity of a test group was determined to decouple these characters. Stearns et al. (2000) used a manipulation of extrinsic mortality rate (risk of accidental death) to select on *Drosophila* life history without directly selecting age at reproduction. These experiments illustrate the ability of clever experimental design to overcome the intrinsic linkage among some kinds of characters. More general considerations are now addressed.

Replication

Because the population is the basic unit of evolutionary observation, it is also the unit of replication for almost all experimental evolution studies. The level of replication one chooses when establishing a set of experimental evolution populations must be done in light of the statistics to be applied later, as well as the traits measured and difficulty executing the selection protocol. A sobering thought is that the degrees of freedom used in almost any experimental evolution analysis of adaptation will be determined by the number of populations selected and not by the number of measurements made on individuals. An experiment may measure 1000 individuals in each of three experimental and three control populations yet the degrees of freedom in the analysis of variance will still be four (or d.f. = 2 in the case of a paired *t*-test) for the main effect. The measurement effort may not be wasted, insofar as it feeds into a precise estimate of the population mean, but the balance between effort expended in selecting populations and their measurement later on demands careful forethought. In general, it will pay to generate more populations when their handling is not too onerous, as is often the case in the "laboratory natural selection" design. It may be impossible to predict the range of characters ultimately analyzed once the populations have been selected because experimental results themselves generate new mechanistic hypotheses. As an example, the B and O populations created by Rose (1984) have been the focus of dozens of different analyses beyond the life history hypotheses they were originally built to examine, as well as being used to originate many new selection treatments (more on this below).

The emphasis on the population mean value further weighs into the experimental design in terms of care in handling and population size. If one sets out to test an adaptive hypothesis using selection, then minimizing nonadaptive sources of population differentiation is essential. The population size applied must be sufficient to allow realistic population processes to occur and obviate genetic drift and inbreeding depression. Typically, for sexual species, this means populations founded and maintained in the hundreds for shorter-term experiments and the thousands for longer-term ones. As noted above, it may become desirable to found new selection treatments from existing ones, once populations have been evolving for an extended period of time. This approach builds inferential power by isolating the founder populations from one another, simply through a history of isolation (as in replicate populations within a selection treatment) or by allowing evolution to start from different places (as when a new selection treatment is applied to populations with histories of different selection treatments).

In general, the design of evolution experiments should be approached as an interaction between population genetic theory and statistics from the outset. From my years of reviewing papers in the area it is an unfortunate conclusion that congenital design flaws very often compromise selection experiments that are otherwise elegant and intellectually exciting.

Using Synthetic Phylogenies

The strategy of spinning off new selection treatments from existing ones has been used to great effect by Rose and colleagues at University of California, Irvine. The synthetic phylogeny, undoubtedly the most baroque ever created, reached its peak in the 1990s with over 200 populations and dozens of distinct selection treatments. A small corner of this phylogeny is illustrated in Figure 31.2. The 55 populations shown in the figure were used to survey the relationship between development time, body size, growth rate, sexual size dimorphism, and viability by Chippindale et al. (2003). After extended laboratory evolution with varied selection treatments, the synthetic phylogeny was used (as naturally occurring phylogenetic trees are) to do comparative analyses. The knowledge of the true topology of the tree and selection history of each population allowed multiple levels of replication to be incorporated as factors: analysis by population, by selection treatment, and by ancestry. In this way, effects of the evolutionary history of populations can be factored into the analysis of selection response. Some researchers have used synthetic phylogenies to test phylogenetic methods (e.g., Hillis & Bull 1993) suggesting a powerful approach to resolving otherwise refractory issues in a field that relies on pattern to infer process.

After long-term directional selection and strong differentiation of the populations in a tree similar to that shown in Figure 31.2, Teotonio and Rose (2000) performed an "evolutionary implosion" experiment. The implosion referred to returning all populations to the same selective regime experienced by the ancestor (IV) population. Among other questions, this experiment addressed the abstract problem of the reversibility of evolution. Teotonio and Rose demonstrated the potential for altered allele frequencies, gene interactions, chance, and other phenomena to guide selection on the main and correlated characters in populations down different pathways when they are experimentally reversed to the ancestral regime. Some characters reversed readily while others showed slow, incomplete, or episodic reversion. The use of hybrid populations (crossed between replicates) by Teotonio and Rose helped to factor out some potential genetic constraints, such as unbreakable epistatic gene combinations (gene–gene interactions) or simple lack of genetic variation. Variation in the reversibility of evolution is among the questions uniquely accessible with an experimental evolution approach.

The creation of elaborate laboratory selection treatments with synthetic phylogenies leads back to one of our first concerns: the raw material for selection. The system diagrammed in Figure 31.2 all originates from a single base population. To some extent the responses of all populations derived down the line is related to the genetic variation originally captured and maintained in that sample. In this sense, subsequent experiments are unreplicated at the population level; all are explorations of the genetic architecture of a single population and its derivatives. This kind of consideration has led some researchers to build experimental systems around multiple base populations with different origins, or, increasingly, with multiple species subjected to the same kind of experimental evolution treatment. When replication is such a fundamental concern, the topologies of such designs quickly become complex.

When Good Replicates Do Bad Things

If it is the consistency of the replicate populations that one uses to define an adaptive response, variation at this level is normally bad news. The unfortunate fate of most misbehaving replica populations is to be regarded as a menace to the P value.

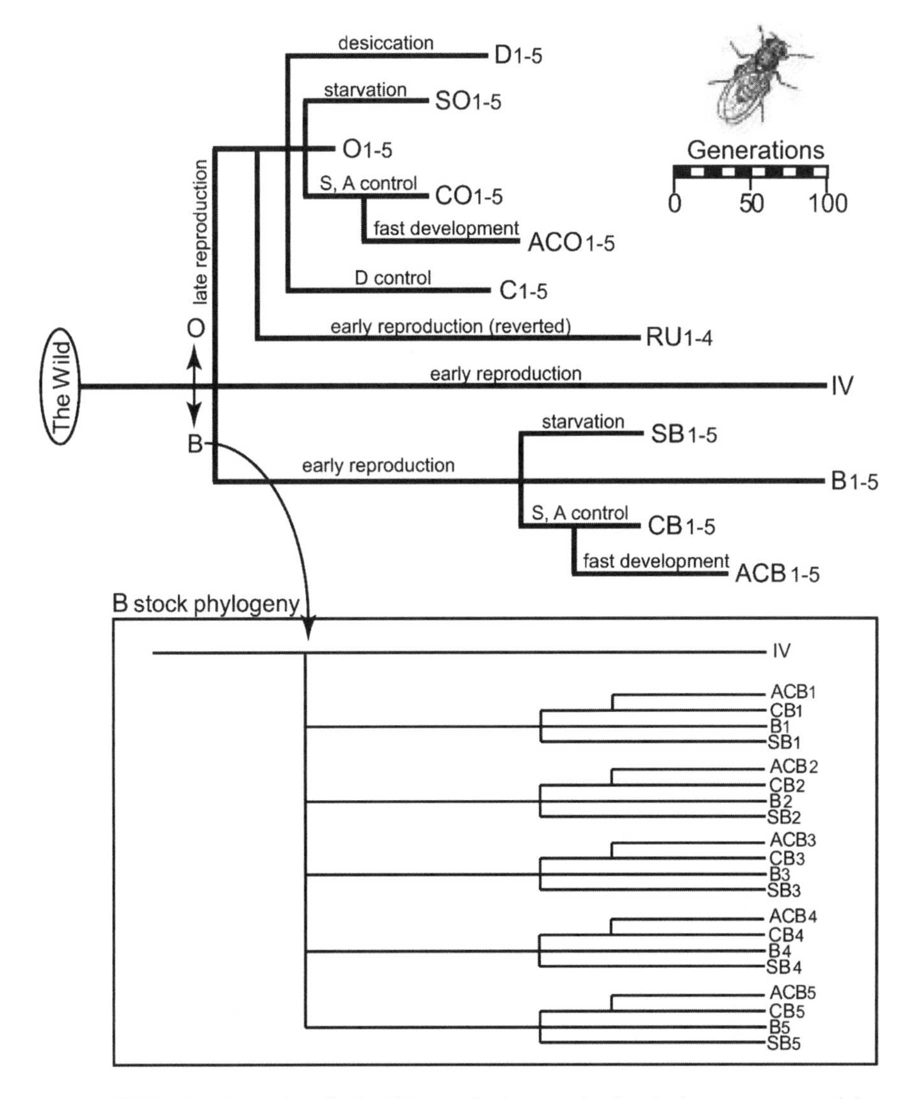

FIGURE 31.2. A schematic of the 55-population synthetic phylogeny surveyed by Chippindale et al. (2003) for developmental traits. The top panel shows the phylogeny of the selection treatments. The bottom panel shows the true phylogeny at the population level to be a "starburst" from the IV base population with subsequent radiations in parallel along each spoke. Only B-derived populations are shown at the population level because of space concerns. At its height, the laboratory system derived from IV featured over 200 populations, each with census size of over 1000 breeding adults per generation. The maintenance of such a system demands simple selection protocols and population cage culture. The ability to draw lines from pre-existing, differentiated laboratory phylogenies to comparatively test evolutionary hypotheses powerfully blends phylogenetic principles with experimental evolution.

Yet we have observed several instances in 5-fold replicated *Drosophila* experiments in which a single population does something quite different. As an example, after long-term selection for desiccation resistance, a plateau was reached and the rate of response slowed to apparent standstill. One population, however, suddenly and markedly displayed a renewed response due to an apparent mutation and posted major gains in the selected trait. Such heterogeneity should be regarded as a source of fruitful investigation rather than a hindrance to it. After all, the observation of one replica in five doing something different is a point estimate of 0.2 for that evolutionary response.

THE USES OF EXPERIMENTAL EVOLUTION: FIVE BIG EXAMPLES

Long-Term Evolution in Microbes

In 1989, Lenski and colleagues began a simple adaptation experiment with *Escherichia coli* B that has continued to yield results of fundamental importance to evolutionary genetics (Lenski et al. 1991). They founded 12 independent populations of *E. coli* (6 marked with a selectively neutral arabinose marker *Ara*$^-$, 6 *Ara*$^+$) from a single clone (i.e., with no genetic variation within or between them at the outset) in a potentially novel culture environment (glucose minimal medium). They therefore created an environment to adapt to but gave natural selection nothing to work with at the outset. Because the populations were cultivated at large sizes (roughly 10^7) and because this species undergoes rapid generations (about 2400 per year), selection could proceed based upon mutational input. This approach has the appeal of identifying explicitly genetic effects as they arise de novo. Moreover, the ability to revive the ancestor strain (or any subsequent sample) from cold storage and perform competitive fitness experiments allows a wide range of experimental possibilities. Here I will summarize only a few highlights from the first 20,000 generations.

After 2000 generations, mean fitness had increased by nearly 40% on average (Lenski et al. 1991), with the rate of adaptation decelerating subsequently and fitness increasing by about 70% after 20,000 generations (Cooper et al. 2001). Throughout their evolution, consistent changes among replicates have been observed in cell size and a number of catabolic functions (Cooper & Lenski 2000). The latter result holds particular interest from an ecological standpoint. Specialization of lines at 37 °C on glucose minimal medium led to demonstrable costs in other environments. The data suggested that antagonistic pleiotropic effects—trade-offs between environments—were preponderant, rather than the simple accumulation of mutations in less used pathways. The authors were able to take advantage of a surprise finding to further this line of analysis: some of the lines had become "mutators" over the course of the experiment, having mutation rates about 100 times greater than the ancestor or parallel-evolved lines. Sniegowski et al. (1997) had shown that three of the lines had become mutators by generation 10,000, and by generation 20,000 a fourth population had become a mutator (Cooper & Lenski 2000). Contrasts between the mutator and nonmutator lines could be used to test whether hypermutability contributed to the rate of adaptation, or to the rate of mutation accumulation in unused catabolic pathways. Evidence was suggestive but statistically equivocal on these points; however, the repeatability of the origin of mutators itself strongly supports an adaptive function.

Other evidence suggests that hypermutability can be adaptive. Giraud et al. (2001) have examined the potential for mutators to be favored in a novel environment. When germ-free mice were inoculated with both normal and mutator strains of *E. coli*, an initial advantage (during colonization) went to the mutators. Costs to high mutation rate appeared after adaptation had occurred (e.g., lower success in later colonization, reduced performance in secondary environments), suggesting a trade-off between evolutionary lability and long-term stability. The same kind of attenuation of mutation rate is often seen in epidemics, and even within single hosts in the case of HIV infection.

An evolution experiment such as the long-term adaptation of *E. coli* to a single environment embodies the design principles mapped out previously in this chapter: high levels of replication, control, and a straightforward and interpretable set of conditions for the assay of relevant characters. Indeed, even a small point like having 6 lines of each *Ara* type can facilitate analyses, since many nonparametric statistics require this level of replication for probability estimates of less than 0.05. Once established, the lines can be used to test a wide array of basic questions in evolution and ecology. Because of their ease of culture, a system like Lenski's is a phenomenal tool for building experiments that are scalable, which is important in accommodating research students at different levels.

Experimental Evolution and Ecology

Few field ecologists would accept that laboratory evolution studies adequately address the kinds of complex interactions occurring in nature. But in the same way that simple mathematical models often hold the greatest power, simplified experimental systems may provide the building blocks for understanding the major processes occurring within complicated systems. In this sense, experimental evolution is ideally poised to contribute to major questions in ecology. Here I will consider a few representative studies that have done just that.

Mueller has been one of the strongest proponents of an experimental evolution approach to ecological questions. Much of Mueller and colleagues' work has focused on the effects of population density and competition in the evolution of life histories and population stability. For example, Mueller and Ayala (1981) and Mueller et al. (1991) were among the first to experimentally test how density-dependent selection affects population growth and stability. They experimentally evolved *D. melanogaster* populations at high and low density to explicitly test the predictions of MacArthur and Wilson's powerful *r* and *K* selection hypothesis. Their work confirmed the basic predictions of that model. Subsequent exploration of their system has helped flesh out the basis of adaptation to these social environments and revealed remarkably complex ecological and evolutionary dynamics. For example, Borash et al. (1998) identified an apparently stable polymorphism that influences larval feeding behaviors. As cultures of fly larvae progress, toxic waste products (particularly ammonia) build up in the medium. Under severe crowding, Borash and colleagues discovered that larvae were playing one of two tactics: some attempt to outrun the ammonia buildup by feeding and growing rapidly while others feed slowly, processing food more efficiently. The latter group show substantially higher ammonia tolerance and cross-resistance to other environmental toxins. These experiments capture the dynamics of certain ephemeral systems, such as drying ponds or decaying food sources, or the potential impact of early versus late colonization on the fate of an individual.

Microbial evolution experiments have also addressed fundamental questions in ecology. A recent series of papers by Rainey, Travisano, Buckling, Bell and others have shown repeatable niche specialization within a very simple experimental environment in *Pseudomonas fluorescens*. Cultures of this bacterium are normally shaken during incubation and therefore represent a spatially homogeneous, well-oxygenated environment. When Rainey and Travisano (1998) allowed "experimental microcosms' (beakers) to sit, like generations before them they discovered that floating mats and scum developed in the containers. Unlike generations before them, these authors turned the "adaptive radiation" of the bacteria into an elegant experiment. The *P. fluorescens* found in the mat and those adhering to the sides of the beaker were novel mutants expressing heritable and basic changes in cell and colony morphology. The so-called fuzzy- and wrinkly-spreaders were new niche specialists along with the standard smooth morph within the (now) complex environment. Rainey and Travisano were able to demonstrate that trade-offs in competitive ability brokered by resource access and use helped to promote and maintain diversity. Later work has exploited this system to look at how adaptation limits subsequent diversification, how cooperation (adhesion into groups) might evolve in the face of cheaters, and how disturbance affects diversity. These are just a handful of studies now published on this "microcosm" of evolutionary change that reflect how simple systems may be cast in many roles to address questions in ecology and evolution.

Yoshida et al. (2003) recently showed a striking example of how selection can impinge upon a previously exclusive domain of ecology. In cyclic predator–prey systems, such as the classic lynx–hare example, peaks in predator density are seen to follow those in prey density by a quarter of a cycle. As prey increase in numbers, predators multiply and follow until they drive the prey population down again and then decline themselves from lack of food; a new cycle begins as the prey population recovers. Yoshida et al. established a rotifer (predator) and algal (prey) system in chemostats and observed out-of-phase oscillations superficially consistent with the existing predator–prey models. But the degree of phase shift (a full half-cycle) between prey and predators was impossible on the basis of existing theory. The authors had to entertain new ideas. They generated several ecological models and one in which the prey species underwent rapid evolution. In the evolution model, prey quality declined as a function of predation, becoming less nutritious as selection by rotifers reduced prey numbers. This model fit the data well because when prey densities rebound it induces a lag in predator population recovery. Yoshida et al. reasoned that this model depends upon genetic variation in the prey species undergoing selection. To limit variation, they reran their experiments with single algal clones. They then found the classic quarter-phase relationship between predator and prey. This experiment exemplifies the way in which experimental evolution can inform ecological theory. Without factoring in rapid evolutionary change in at least one of the players, the population dynamics in this system would not be explicable by conventional ecological theory.

Intersexual Coevolution and Speciation

Antagonistic coevolution between the sexes predicts a perpetual "arms race" between females and males of promiscuous species even without environmental change. Behaviors, chemical signals, or hormone analogs used by one sex to influence the other are believed to be important influences on reproductive success (e.g., Chapman et al. 1995; Rice 1996). *Drosophila melanogaster* males, for example, transfer at least 80 accessory proteins (ACPs) with their ejaculate. Those ACPs that have been studied exert an influence on female physiology that is in line with the best interests of the fertilizing male but not necessarily his mate (Chapman et al. 1995).

William Rice (1996) performed one of the most elegant evolution experiments to date to illustrate antagonistic coevolution in *Drosophila*. Rice reasoned that one way to observe the coevolutionary change was to stop evolution by one of the partners. By using special female lines, he limited entire haploid genomes to expression in males. These males were allowed to evolve to the "clone-generator" females without opportunity for the females to counter-evolve. Drawn anew from a separate population each generation, the variation in the female line was never exposed to selection. After only a few dozen generations, Rice found substantial increases in male fitness when paired with the clone-generator females. Males became more effective in remating previously mated females (females mated with a separate tester stock first) and in discouraging females from remating when they were the first mate. The "supermale" gains were coupled with increased harming effects on these females. For example, a single mating with a supermale was found to elevate mortality in the "target females" by approximately 50%. The supermale lines' increased male-benefit/female-harm effects were largely specific to the population they had coevolved with, in keeping with the idea of close coevolution happening within populations.

Holland and Rice (1999) used the reverse experimental treatment to study sexual conflict. These authors experimentally eliminated conflict by imposing strict monogamy on *D. melanogaster*, a normally promiscuous species. When all of one partner's fertility depends upon the other, mutualistic coevolution will become preponderant. In keeping with this idea, Holland and Rice observed reduced female harm in the monogamous males and increased population fitness. The females, however, became more susceptible to the harming effects of males of the polygamous treatment. These results have sustained some criticism based on an asymmetry in the design, allowing higher effective population sizes in the polygamy-treated (compared with monogamy-treated) populations. Monogamy-selected males evolved to be smaller than the control (polygamy) lines, which Holland and Rice attributed to reduced sexual selection. Add to this evidence that males made smaller by nutrient limitation and crowding are less harming to their mates than normal healthy males and some have suggested differential inbreeding or inadvertent selection on development time could explain Holland and Rice's results. These criticisms are softened somewhat because the monogamy lines posted gains in population fitness and the same number of females was used in each treatment. Last-male sperm precedence in this species is in excess of 0.8 of the offspring, limiting the disparity between population sizes. Experimental crosses between replicate populations would have provided a check on the inbreeding hypothesis: if drift fixes deleterious recessive alleles at random throughout the genome, then heterozygotes formed in the F_1 cross should recover normal performance.

Martin and Hosken (2003) performed a monogamy experiment in the yellow dung fly *Sepsis*, a species with flagrant sexual conflict (aggressive mating struggles), and showed a number of similar effects on both harm by males and susceptibility of females. In a separate study, these authors also varied the level of potential sexual conflict by varying density of cohabiting males and females, showing that high-density line females were more reluctant to mate, particularly in between-population crosses. Martin and Hosken cast their results in terms of conflict driving speciation. Higher divergence among populations was apparently correlated with the degree of potential conflict between the sexes, reinforcing the idea that sexual conflict may be a potent engine of speciation (e.g., Rice 1996).

Sexually antagonistic coevolution is expected to be ongoing in independent replicates derived from the same founding population and may take different directions in these replicates. Furthermore, the reproductive traits underlying sexual processes appear to be evolving at an accelerated rate, allowing this form of arms race to potentially take a lead role in the divergence of populations. Crosses within replicated model phylogenies may offer one of the most powerful tools available for addressing the role of intersexual coevolution in reproductive isolation.

The experiment, which so far has been done only with natural populations in a few species, typically involves a population-level version of the swinging practice of "wife-swapping." Mates are switched between geographically isolated populations and the impact of the foreign mate, relative to the local (within-population) control, is assessed. The potential difficulty with wild populations is twofold, involving first the problem of novel; "common garden" conditions used in the laboratory, which may, by chance, favor one population over others (see Figure 31.3). The second, related problem is defining what is a positive and what is a negative impact on fitness. Compared with natural populations, well-designed experimental evolution populations generate few questions about the appropriate environment in which to test for interactions between mates from different populations. Indeed, it may be possible to obtain an exact measure of the true currency of conflict—net fitness. My laboratory is currently undertaking such an experiment using the 6-fold replicated base populations from Rose's laboratory to perform the swap experiment. These populations have evolved in isolation with regulated population sizes and identical conditions of natural selection for over 600 generations. We therefore expect them to be at, or close to, evolutionary equilibrium for characters experiencing natural selection, but to have had ample opportunity for sexual coevolution. The 36-cell matrix of possible reciprocal crosses between populations has so far revealed evidence for close coevolutionary matching between the females and males of a given population.

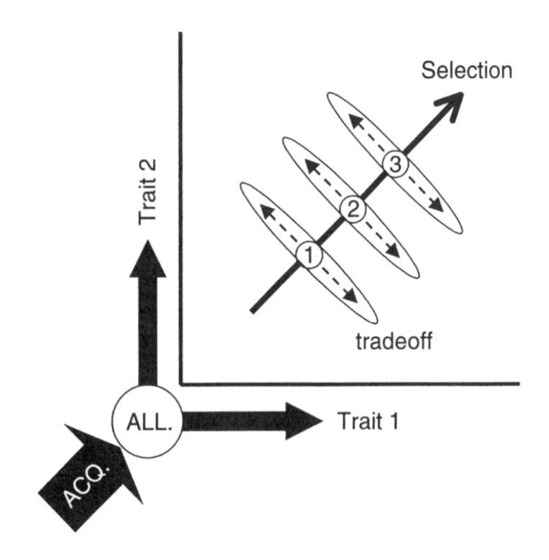

FIGURE 31.3. The problem with the "common garden" is illustrated here. Two traits may compete for allocation of a limiting resource that the organism acquires, creating a physiological trade-off. If the organism devotes energy to trait 1 then it diminishes investment in trait 2, creating a negative correlation between the traits. Genetic variation that influences *allocation* (ALL.) to trait 1 or 2 creates a negative genetic correlation (i.e., an evolutionary trade-off) between the traits. But genetic variation may also underlie the *acquisition* of energy (ACQ.). When a population is introduced to a novel environment (e.g., the laboratory or common garden) genotypes that are by chance poor resource acquirers (point 1 on the figure) have low values for both traits; genotypes that are good at getting energy (point 3) will have high values for both traits. This effect creates the appearance of a positive correlation between the traits, even when the allocation genes define a trade-off. Now, if the population evolves in the new environment, selection will favor genotype 3 over genotype 1, and both traits will be seen to improve, defining a positive *evolutionary correlation*. This problem will make the detection of trade-offs, which are expected between performance traits, difficult until the population has adapted to the new environment.

Is There a Definable "Genetic Architecture"?

The idea that organisms evolve according to a set of fixed relationships between their genes and the characters they encode leads to the idea of a "genetic architecture." This term can be interpreted several ways, but typically relates to the degree and manner in which the organism's characters are integrated. We may hope, for example, to figure out the relationships between characters by applying selection to one of the traits, thereby defining an architecture—or genetic structure—of the organism based on the correlated changes in other traits. Some characters will have strong positive associations (genetic correlations) with others, while others may be negatively genetically correlated, and some may exhibit complex interactions. Overall, the relationships or *covariances* between quantitative characters are summarized by the **G** matrix (see Phillips and McGuigan, Ch. 20 of this volume). If these relationships are fixed, then we could predict the outcome of selection on any one of the characters concerned.

Unfortunately, characters controlled by many loci (quantitative traits) often have complicated interactions and several population genetic effects predict changes in the matrix of genetic covariance that challenge our predictive abilities. This increases the importance of experimental evolution in defining how robust the relationships are. A particularly common problem with this approach is that unintended or confounding sources of selection may complicate the interpretation of an experiment to the point that the major objectives are obscured. Nowhere has this point been made more clearly than in the study of life history evolution in *Drosophila*. In the early 1990s, Leroi, myself, and colleagues found that the classic trade-off between early- and late-fertility traced by Rose's B ("young") and O ("old") populations had apparently disappeared (e.g., Leroi et al. 1994). O flies were reproductively superior at all ages, not just late in life. As new graduate students, Leroi and I were mortified that we could not reproduce our PhD advisor's well-known result. We performed a postmortem in which we tweezed apart every conceivable aspect of our experiments. It became apparent that the disappearance, or reversal, of the trade-off had occurred because of unintended adaptation to differences in their selection protocols. The conditions used to test fertility were "O-like," having low adult density and abundant dietary yeast; under "B-like" (high density and low yeast) conditions the expected differences were found. We used the phenomenon to illustrate genotype–environment interaction, adaptation, and the mutability of trade-offs. Equally, however, the story of the "B/O reversal" is a cautionary tale of a design flaw: a simple oversight in handling and testing the populations for fertility. Had we not been able to reconstruct the history of the reversal in the trade-off and trace it to its sources, that is, demonstrate that it was reversible and had evolved over a decade, then the result would stand in direct contradiction to the original finding of a trade-off. Clearly even relatively subtle environmental changes can precipitate changes in the apparent genetic correlation between traits.

The fine-tuning of the organism to its environment is underscored by a recent survey of life history correlations in *Drosophila* reciprocally tested in different laboratories by Ackerman et al. (2001). These authors found that a variety of life history traits and trait correlations were strongly influenced by the test environment, despite efforts to replicate the conditions of the original laboratories. The difficulty repeating results between laboratories with many environmental factors controlled, or in the same laboratory over time, has ominous implications for studies involving truly novel conditions. What can we say about the meaning of genetic correlations measured on wild organisms in a "common garden"?

Other examples from *Drosophila* (reviewed in Chippindale et al. 2003) have demonstrated large shifts or even reversals in character correlations during directional selection. In theory there are many reasons for quantitative and even qualitative changes in genetic correlations during selection. For examples, the fixation of alleles governing a correlation, selection on acquisition versus allocation of resources, genotype–environment interaction, and a variety of problems associated with measurement of characters can induce these changes (e.g., Figure 31.3; Houle 1991; Rose et al. 1996; Roff, Ch. 18 of this volume). The fact that the genetic architecture itself is capable of evolving means that at present we are ill-equipped to extrapolate from the standing genetic variation in a population to selection responses over even relatively short periods of evolutionary time. Investigation of how the genetic variance/covariance (**G**) matrix changes under selection through direct quantitative measurement under selection will contribute to our ability to predict evolutionary outcomes (see Phillips & McGuigan, Ch. 20 of this volume). This is one of the areas in which experimental evolution can contribute uniquely.

Experimental Evolution... In the Wild?

One of the charges repeatedly laid at the feet of experimental evolutionists is that our experiments do not capture many features of organisms in their natural habitat. One of the best examples of innovative experimental evolution design using a natural system is the work by Reznick and colleagues on Trinidadian guppies. Guppies naturally occur in a series of drainages on both slopes of the northern range of mountains in Trinidad. The streams are isolated from one another but broadly similar in character, representing the equivalent of replicates in a designed experiment. Each of the study streams is terraced, having waterfalls that create subpopulations of guppies from their lower reaches to the headwaters. And each part of a stream is relatively easy to study because of clear water and the ease with which entire populations of guppies can be netted and measured.

Beyond these general features, researchers such as Reznick, Endler, and Rodd recognized that the community of guppies and their predators typically changes with altitude. In the lower parts of the stream on the south side of the mountains, voracious predators such as the pike cichlid *Crenicichla* co-occur, eating guppies indiscriminately. These predators are usually unable to circumnavigate the waterfalls, and therefore upstream guppies have been allowed to evolve with less dangerous gape-limited omnivores such as the killifish *Rivulus*. These differences in predation translate into potentially different forces of selection acting upon how rapidly a guppy grows, when it reaches maturity, how much it reproduces and how quickly it ages and dies. In general, life history theory predicts that under the high-predation conditions, selection should favor rapid growth, early maturity, and high early investment in offspring, even if it results in reduced lifespan. Reznick and colleagues have used the natural replication and variance in predation intensity to confirm many of these predictions. However, more convincing has been a series of ingenious transplant experiments.

In one set of experiments, Reznick and colleagues (e.g., Reznick et al. 1997) moved guppies from high-predation regimes below waterfalls to low-predation sites above, and thereby theoretically instituted directional selection on their life histories. Most of the predicted changes were observed over a period of years. Interestingly, changes in male life history were more pronounced and consistent than those observed for female life history: males rapidly evolved delayed maturity at a larger size, while females changed in the predicted direction in one replicate but not the other. Guppy males are famous for their brilliant colors, whereas females are more cryptic; the same traits that make a male attractive to his prospective mates may make him a snack for *Crenicichla alta*. It is therefore perhaps not surprising that males would change most rapidly. The ease with which guppies adapt to the laboratory has enabled Reznick and coworkers to perform breeding experiments to test heritabilities of traits, and to measure a broad range of characters, from burst swimming speed to longevity. As an illuminating comparison, Reznick et al. (1997) contrasted the rates of evolution (in darwins, a proposed standard unit of morphological change) between their guppies and what is typically observed in the fossil record. They estimated that when this kind of continuous directional selection is applied, the evolution of age and size at maturity occurred at rates up to 7 orders of magnitude faster than what is typically observed in the fossil record. In other words, populations may be evolving at warp speed, but shifting selection and gene flow may homogenize these rapid shifts and give a stately appearance over geological time.

Finally, while research on the south-flowing streams incorporated most of the features of a well-controlled laboratory natural selection experiment, this research was extended to an independent, parallel set of north-flowing streams on the other slope. There, similar predation regimes were found, but the players are different. Instead of pike cichlids, gobies often play the role of top predator, and freshwater prawns accompany killifishes in the low-predation sites (Magurran 2001). The replication of selective regimes with different specific components is important in extending the generality of conclusions. With respect to guppies, feral populations are widespread and presumably experience a variety of different predation and abiotic factors that would influence their degree of showiness, behavior, and life histories, representing a kind of experiment increasingly done by accident: species invasions.

EXPERIMENTAL EVOLUTION: LIMITS AND PROSPECTS

General Considerations

We began by considering the limits of mechanistic genetics for addressing evolutionary questions and have now seen some of the problems and limits to experimental evolution as an approach. Population size, replicate population number, variety in controls and selection treatments, and the evolutionary time-course of a project are all a function of the experimenter's resources and the time available. These factors have impressed upon us a handful of convenient laboratory model systems that most experimental evolution projects exploit. Plainly this limits our view of taxonomic and ecological diversity.

But while the problem of realism deserves careful consideration, the interest value of a laboratory experiment does not usually lie in the exact simulation of adaptation in the wild. I would not assert that what is true for *E. coli* is true for elephants (only more so!). Rather laboratory evolution seeks to explore the genetic and physiological architecture of the organism and the general *form* of evolutionary change. In doing experiments in which phylogeny and environment are explicitly controlled we reduce the roles of history, phylogenetic inertia, nonindependence, and environmental interaction in patterns seen between populations. By controlling population size and structure and employing replicates, one further reduces the role of chance in the evolutionary outcome and increases the prospect of seeing the currency that evolution trades in: fitness.

Where to From Here?

As a final consideration we may ask where the field of experimental evolution is going. We have seen a number of successful applications of the approach here and overlooked hundreds. It is an approach that is growing rapidly and nothing short of a tome could capture the diversity of experimental evolution as it stands now–and that would be out of date before it hit the shelves. The application of molecular and microbial experimental evolution to problems in the evolution of pathogens (and possibly countering potential bioterrorist weapons) is developing rapidly, both directly and by analogy (e.g., Bull & Wichman 2001; Cowen et al. 2002; Taylor et al. 2002). Molecular approaches have created the possibility of experimentally evolving RNA in vitro. In this directed evolutionary approach, "sexual PCR" is used to create unnaturally high mutation rates and gene shuffling outside of the organism. The approach allows the engineering of proteins for enhanced function and the study of potential evolutionary pathways that may be constrained by genetic background, low mutation rate, and gene–gene interaction in vivo. For example, Barlow and Hall (2003) recently used the in vitro approach to predict the evolution of antibiotic resistance for cefepime. Because of the ability to experimentally generate and then compete molecules for their activity (molecular fitness), industry and academe alike are making extensive use of in vitro evolution for problems as diverse as drug resistance and recreation of ribozymes that could help explain the origins of life in self-catalytic RNA (e.g., Wilson & Szostak 1995).

The technical possibilities now available to experimental evolution from molecular genetics are tremendous. As illustrated by the success of microbial evolution experiments, experimenters increasingly have before them a quiverful of techniques designed to work in exactly the organisms favored by molecular approaches. Combining the power of natural selection to distill out variants specialized for a given treatment with modern tools of molecular biology will increasingly sharpen our understanding of subcellular and genomic processes. No other area is likely to profit more than developmental biology, where many evolutionary studies have examined the broad properties of growth and size, but the problems of pattern formation and morphogenesis have yet to be worked on in detail.

Finally, evolutionary genetics and ecology will see a broad range of basic questions addressed. Already experimental evolution has been used to probe the genetic basis of aging and the long-term advantages of sexual recombination (e.g., Rice 2002). Problems increasingly under study include mechanisms of speciation, the role of epistasis in fitness determination, gene duplication, parasitic DNA, and genomic conflict. Sexual coevolution is particularly amenable to experimental study. As discussed above, the ability to disentangle the by-products of natural selection from the targets of sexual selection will favor experimental evolution in this pursuit. Some extraordinary work (e.g., Miller & Pitnick 2002) is highlighting the interaction of gametes and reproductive systems between the sexes. Continued application of experimental evolution with the same organisms under different conditions and with new organisms will increasingly reveal what is possible, what is local, and what is fundamental in the evolution of populations.

SUGGESTIONS FOR FURTHER READING

Many classic and representative studies in selection are described by Falconer and Mackay (1996), Lynch and Walsh (1998), and Roff (1997), placing them in the context of the many tools and phenomena of quantitative genetics. Graham Bell (1997) has written a wonderful scholarly compendium of selection studies, largely focused on the experimental evolutionary approach. Bell's book argues that selection is the preponderant force of evolutionary change, making its study as an experimental force particularly important. Finally, adaptations may be defined as the products of selection. Accordingly, experimental evolution is a powerful tool for understanding adaptation as a process. This point, and an overview of laboratory selection, is made by Rose and colleagues in their chapter of the nicely conceived and prepared volume *Adaptation* (Rose and Lauder 1996).

Bell G 1997 Selection: The Mechanism of Evolution. Chapman & Hall.

Falconer DS & TFC Mackay 1996 Introduction to Quantitative Genetics, 4th Ed. Longman Group.

Lynch M & B Walsh 1998 Genetics and Analysis of Quantitative Traits. Sinauer Assoc.

Roff DA 1997 Evolutionary Quantitative Genetics. Chapman & Hall.

Rose MR & GV Lauder 1996 Adaptation. Academic Press.

32

Evolutionary Conservation Genetics

RICHARD FRANKHAM

The biodiversity of the planet is being rapidly depleted as a direct and indirect consequence of human actions (referred to as the "sixth extinction"; Leakey & Lewin 1995). An unknown but large number of species are already extinct, while many others have reduced population sizes that put them at risk. Many species now require benign human intervention to improve their management and ensure their survival.

There are four justifications for conserving biodiversity: economic value of bioresources, ecosystem services, aesthetics, and rights of living organisms to exist. Bioresources of direct economic values include food, many pharmaceutical drugs, fibers for clothing, rubber, and timber. Ecosystem services are essential functions that are provided free of charge by living organisms, including oxygen from plants, climate control by forests, nutrient recycling, natural pest control, and pollination of crop plants. In 1997, these were valued at $US33 trillion dollars annually, almost double the global national product. Humans derive aesthetic pleasure from visiting zoos and nature reserves and from ecotourism. Ethically it is difficult to justify one species exterminating many others, just as racial genocide is unacceptable.

Throughout this chapter reference will be made to threatened and endangered species, so we must define them. Threatened species are those with a high probability of extinction within a short time. The IUCN Red List categorization system is the internationally recognized means for classification threat among species (IUCN 2002), but exists besides many local systems. IUCN list species as critically endangered, endangered, or vulnerable (the combination of these three is referred to as threatened). The categorizations are based on five criteria: (i) rate of reduction in population size, (ii) extent of occurrence, (iii) adult population size and continued decline, (iv) adult population size in stable populations, and (v) projected extinction risk. For example, a critically endangered species has one or more of: 80% decline over the last 10 years or three generations, extent of occurrence in an area of less than 100 km^2, less than 250 mature individuals and continual decline, a stable population of less than 50 adults, or an estimated probability of extinction of at least 50% within 10 years or three generations, whichever is longer.

The primary initial factors causing species decline are habitat loss, introduced species, overexploitation, and pollution. These factors are caused directly or indirectly by humans, and are related to human population growth. The human-related factors reduce species to population sizes where they are susceptible to stochastic effects. These encompass environmental, demographic, and genetic (inbreeding depression, and loss of genetic diversity) stochasticity and catastrophes. Even if the original cause of population decline is removed, problems associated with small population size will still persist. Genetic concerns in conservation biology only arose in the early to-mid-1970s, largely due to the efforts of Sir Otto Frankel, an Austrian-born Australian (Frankel & Soulé 1981).

Conservation genetics deals with the genetic factors that affect extinction risk, genetic management regimes required to minimize these risks, and the use of molecular genetic methods in forensics and to determine aspects of species biology important to their conservation. There are 11 major genetic issues in conservation biology:

- Resolving taxonomic uncertainties
- Defining management units within species
- Deleterious effects on fitness that sometimes occurs as a result of outcrossing (outbreeding depression)
- The deleterious effects of inbreeding on reproduction and survival (inbreeding depression)
- Loss of genetic diversity and ability to evolve in response to environmental change
- Fragmentation of populations and reduction in gene flow
- Genetic drift overriding natural selection as the main evolutionary process
- Accumulation and loss (purging) of deleterious mutations
- Genetic adaptation to captivity and its adverse effects on reintroduction success
- Use of molecular genetic analyses in forensics
- Use of molecular genetic analyses to ascertain aspects of species biology important to conservation

Clearly, these concerns involve applied evolutionary genetics and several have been considered in previous chapters. Frankham et al. (2002) have recently reviewed all of these issues and readers are referred to that textbook for extended treatments of topics and further references. I first consider taxonomic uncertainties and management units then discuss the contentious issue of the role of genetic factors in extinctions, followed by genetic management of threatened species, and finally the contribution of molecular genetic methods to the understanding of aspects of the biology of a species important to its evolution and conservation.

RESOLVING TAXONOMIC UNCERTAINTIES AND MANAGEMENT UNITS

The first step in the conservation of a species or population is to resolve the boundaries of the taxon. Without this, undescribed threatened species may not be afforded protection, resources may be wasted on conserving populations of a common species, or distinct species may be hybridized. Legal protection is usually afforded to threatened species, subspecies, and distinct populations within species under endangered species protection laws. Trade in threatened species or their parts is prevented in countries that have signed the Convention on International Trade in Endangered Species (CITES).

The delineation of species can be aided by use of genetic information on chromosomes or molecular genetic markers. However, these efforts are seriously compromised by the plethora of definitions of species (Mallet, Species Concept box, pp. 367–373 of this volume). The Biological Species Concept has been the most influential definition of species. It and several others consider species as entities within which gene flow is possible, but restricted or impossible between species. Under this definition, the occurrence of distinct karyotypes or nonoverlapping sets of alleles for allozymes, microsatellites, etc., between sympatric populations is a clear indication that they are distinct species, as illustrated for the long-footed potoroo below. For allopatric populations, the delineation of species is more difficult. This involves comparing the extent of genetic differentiation between populations with that for known "good" species in related taxonomic groups.

Use of genetic information to resolve taxonomic uncertainties for sympatric populations is illustrated by the case of the long-footed potoroo (Seebeck & Johnston 1980; Johnston et al. 1984). Potoroos are small marsupials akin to pint-sized kangaroos. In the Gippsland area of southeastern Australia, a common potoroo was known to exist, but a few specimens of a form with larger feet were discovered, mainly from road kills. Was the long-footed form a new species? Since the two forms have sympatric distributions, they should show similar morphology, chromosomes, and molecular markers if they are the same biological species and subject to gene flow. Conversely, if the long-footed form is a new species we would expect to find differences in morphology, chromosomes, and molecular markers, due to lack of gene flow. Studies showed that the long-footed form had distinct morphological characters, especially foot length, that did not overlap with the distribution for the common potoroo. The two forms had different chromosome numbers: 12 in females and 13 in males in the common form (*Potorous tridactylus*) and 24 in the long-footed potoroo. Further, allozymes showed that the two forms had no alleles in common for five of 24 loci. Thus, there was clear evidence of lack of gene flow between the two sympatric forms and the long-footed potoroo was named as a new species (*Potorous longipes*).

The issue of defining distinct populations and subspecies is also affected by problems with definitions of distinct units. Crandall et al. (2000b) reviewed this issue and came up with a new framework for defining distinct populations that is being actively debated. They suggested that distinctiveness be based on whether populations were genetically and ecologically exchangeable on both recent and historical time frames, and developed guidelines for what degree of differentiation was required before populations were considered as sufficiently distinct to justify separate management.

GENETICS AND EXTINCTION

The main presumption underlying genetic concerns in conservation biology is that inbreeding, loss of genetic diversity, and mutational accumulation increase extinction risk. As this is a fundamental issue and is contentious, I will consider it in some detail.

Endangered species have small and/or declining populations, so inbreeding and loss of genetic diversity, are unavoidable in them, as indicated by the following equation for random mating populations:

$$H_t/H_0 = [1 - 1/(2N_e)]^t = 1 - F, \qquad (32.1)$$

where H_0 is the initial heterozygosity, H_t the heterozygosity at generation t, N_e the effective population size (Gillespie, Ch. 5 of this volume), and F the inbreeding coefficient. This equation predicts that genetic diversity will decline (and inbreeding increase) at a greater rate with generations in small than large populations. Further, it shows that the level of inbreeding is directly related to the proportionate loss of genetic diversity in random mating populations. Case Study 1 illustrates the genetic consequences of a population size bottleneck on the Mauritius kestrel.

Case Study 1. Genetic Consequences of a Population Size Bottleneck in the Mauritius Kestrel (from Box 8.1 of Frankham et al. 2002)

The decline of the Mauritius kestrel began with the destruction of native forest and the plunge toward extinction resulted from thinning of eggshells and greatly reduced hatchability following use of DDT insecticide beginning in the 1940s. In 1974, its population numbered only four individuals, with the subsequent population descending from only a single breeding pair. Under intensive management the population grew to 400–500 birds by 1997, but it experienced six generations at numbers of less than 50.

While this is a success story, the Mauritius kestrel carries genetic scars from its near extinction. It now has a very low level of genetic diversity for 12 microsatellite loci, compared with six other kestrel populations (see below). The Mauritius kestrel has 72% lower allelic diversity and 85% lower heterozygosity than the mean of the nonendangered kestrels (Table 32.1). Prior to its decline, the Mauritius kestrel had substantial genetic diversity, based on ancestral museum skins from 1829–1894, but even then its genetic diversity was lower than that of the nonendangered species. The Seychelles kestrel went through a parallel decline and recovery and also has low genetic diversity. It was rare during the 1960s and had become extinct on many outlying islands. However, it has now recovered to a size of over 400 pairs.

The reproductive fitness of the Mauritius kestrel has been adversely affected by inbreeding in the early post-bottleneck population; it has lowered fertility and productivity than in comparable falcons and higher adult mortality in captivity.

Since inbreeding reduces reproduction and survival rates, and loss of genetic diversity reduces the ability of populations to evolve to cope with environmental change, Frankel and Soulé (1981) and others suggested that genetic factors would contribute to extinction risk in threatened species.

TABLE 32.1. Genetic variation in endangered versus nonendangered kestrel species

Species	A	H_e	Sample size
Endangered			
Mauritius kestrel			
Restored	1.41	0.10	350
Ancestral	3.10	0.23	26
Seychelles kestrel	1.25	0.12	8
Nonendangered			
European kestrel	5.50	0.68	10
Canary Island kestrel	4.41	0.64	8
South African rock kestrel	5.00	0.63	10
Greater kestrel	4.50	0.59	10
Lesser kestrel	5.41	0.70	8

However, this view was challenged in the late 1980s and the contribution of genetic factors to the fate of endangered species was until recently, generally considered to be minor. A paper by Lande (1988) has been widely cited as suggesting that demographic and environmental stochasticity and catastrophes would cause extinction before genetic deterioration became a serious threat to wild populations. This controversy has persisted. However, there is now a compelling body of both theoretical and empirical evidence indicating that genetic changes in small populations are intimately involved with their fate (Spielman et al. 2004a; Frankham 2005). Specifically:

- Inbreeding causes extinctions in deliberately inbred captive populations
- Inbreeding has contributed to extinctions in some natural populations and there is circumstantial evidence to implicate it in many other cases
- Computer projections based on real-life histories, including demographic, environmental, and catastrophic factors, indicate that inbreeding will cause elevated extinction risks in realistic situations faced by natural populations
- Many surviving populations have now been shown to be genetically compromised (reduced genetic diversity and inbred)
- Loss of genetic diversity increases the susceptibility of populations to extinction under changing environmental conditions.

Inbreeding and Extinction

Populations contain a load of rare deleterious alleles from mutation–selection balance and from balancing selection. Inbreeding increases the homozygosity of these deleterious alleles and as a consequence reduces reproduction and survival (termed inbreeding depression) in essentially well-studied species (Goodnight, Ch. 6 of this volume). Consequently, it was anticipated that inbreeding would increase the risk of extinction in wild populations.

However, there was a controversy about the impacts of inbreeding on wild species, initially for captive populations in zoos and later for wild populations in natural habitats. The former issue was resolved by Ralls and Ballou (see Ralls et al. 1988) who found that inbred individuals showed higher juvenile mortality than outbred individuals in 41 of 44 captive mammal populations. On average, brother–sister mating resulted in a 33% reduction in juvenile survival. There is now also clear evidence that inbreeding adversely affects most populations in natural habitats. Crnokrak and Roff (1999) reviewed 157 valid data sets, including 34 species, for inbreeding depression in natural situations. In 90% of cases inbred individuals had poorer attributes than comparable outbreds (i.e., they showed inbreeding depression), and those that did not were based on small studies and/or paternities that were not verified using genetic markers. Results were very similar across birds, mammals, poikilotherms, and plants. Inbreeding depression is typically more severe in wild than in captive populations, as the wild is typically more stressful than captivity.

Deliberately inbred populations of laboratory and domestic animals and plants show greatly elevated extinction rates (Figure 32.1). This occurs not only with rapid inbreeding using brother–sister matings or self-fertilization, but also with slow inbreeding due to finite size where natural selection has more opportunity to remove deleterious alleles. For example, 50% of *Drosophila* populations were extinct from inbreeding at inbreeding coefficients of 0.62 for full-sib mating, 0.79 for populations with sizes of $N_e = 10$, and 0.77 for populations with $N_e = 20$ (Reed et al. 2003).

Inbreeding and loss of genetic diversity has been shown to increase the risk of extinction for two populations in nature. Inbreeding was a significant predictor of extinction risk for butterfly populations in Finland after the effects of all other ecological and demographic variables had been removed (Saccheri et al. 1998). Further, experimental populations of the *Clarkia pulchella* plant founded with a low level of genetic diversity (and high inbreeding) exhibited 69% extinction rates over three generations in the wild, while populations with low inbreeding showed only a 25% extinction rate (Newman & Pilson 1997). However, it was not clear whether these two studies were general results, or exceptions.

Computer projections incorporating factual life history information are often used to assess the combined impact of all deterministic and stochastic factors on the probability of extinction of populations. Brook et al. (2002) conducted computer projections for 20 outbreeding bird, mammal, and invertebrate threatened species that allowed for the effects of purging and encompassed population sizes typical of critically endangered, endangered, and vulnerable species. Median times to extinction

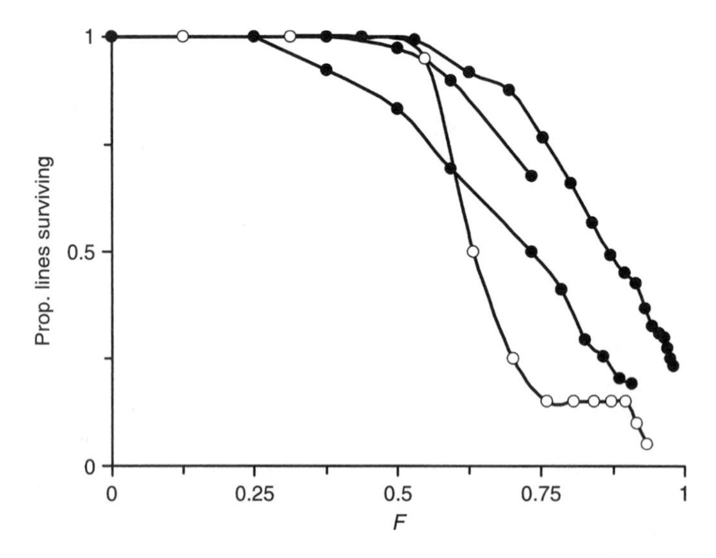

FIGURE 32.1. Inbreeding increases extinction risk in deliberately inbred populations of mice (open circles) and two species of *Drosophila* (solid circles) under laboratory conditions where other causes of extinction can be excluded. Proportion of populations surviving is plotted against the inbreeding coefficient (F). From Frankham et al. (2002), reprinted with the permission of Cambridge University Press.

were on average reduced by 25–30% when inbreeding depression was applied to juvenile survival at the level found by Ralls et al. (1988) for captive mammals, compared with cases where inbreeding depression was omitted. These results underestimate the true impact of inbreeding depression, as it affects all aspects of the life history. It is at least 4 times higher overall for populations in nature than reported for juvenile survival in captivity (Keller 1998). A related computer projection for the rare European plant *Gentiana pneumonanthe* yielded similar conclusions (Oostermeijer 2000). These computer projections indicate that the results of Saccheri et al. (1998) and Newman and Pilson (1997) are not exceptions, but are likely to apply to the majority of species.

Declines in population size or extinction in the wild have been attributed, at least in part, to inbreeding in many populations including bighorn sheep, Florida panthers, Isle Royale gray wolves, greater prairie chickens, heath hens, middle spotted woodpeckers, adders, and many island species. Further, inbreeding colonial spiders have a higher rate of colony extinction than non-inbreeding species.

Following Lande's (1988) paper, many authors suggested that species would often be driven to extinction by demographic factors before genetic factors had time to impact. While Lande (1995) has subsequently changed his views on the contribution of genetic factors to extinctions, this is due to his championing of "mutational meltdown" (described below) and not due to a retraction of his 1988 views.

If the Lande scenario is common then threatened species that are acknowledged to be at risk of extinction should show little difference in genetic diversity compared with related nonendangered species. If genetic factors were in fact impacting, then threatened species would show lower genetic diversity. We showed that the majority of threatened species do not fit the "no genetic impact" scenario. Of 170 threatened taxa, 77% had reduced genetic diversity compared with related threatened species, the median reduction being 40% (Spielman et al. 2004a). This will cause a substantial reduction in ability to evolve to cope with environmental change. Further, the 40% reduction in heterozygosity equates to the inbreeding coefficient, so most threatened species are likely to be suffering serious reductions in fitness due to inbreeding depression. Genetic diversity has been shown to be related to fitness, as expected from the relationship between genetic diversity and inbreeding in random-mating populations described in Equation 32.1 (Reed &

Frankham 2003). Consequently, most threatened species are likely to have both reduced reproductive fitness due to inbreeding depression and reduced evolutionary potential. Even those threatened species that are not currently suffering from genetic problems have time to do so. For example, vulnerable species, which form the majority of our data set, have approximately a 10% probability of extinction within 100 years.

The "no genetic impact" scenario has failed numerous tests so it must be rejected for the majority of species (Spielman et al. 2004a). Four assumptions were made by Lande (1988) that are probably incorrect based on subsequent information, involving the ratio of effective to census sizes (N_e/N), the extent of interactions among stochastic factors, the extent of inbreeding depression in the wild, and the effectiveness of purging.

Genetic impacts depend on the effective population size (N_e), so the ratio of effective to census size is critical in determining genetic impacts. Around the time of Lande's paper it was typical to talk of N_e/N ratios of 0.25–0.5, but subsequently N_e/N ratios in unmanaged populations have been found to average approximately 0.1 (Frankham 1995), so genetic factors impact sooner than Lande would have expected.

Fluctuations in population size and sex ratio and variation in family size all occur due to demographic and environmental stochasticity and catastrophes and result in reduced N_e/N ratios. Consequently, there are interactions between stochastic factors that increase genetic impacts on population persistence (van Noordwijk 1994).

Data on the full impacts of inbreeding depression for species in the wild were extremely limited in 1988. The main data available at that time were that of Ralls et al. (1988) whose estimate of the impact of inbreeding depression on juvenile survival in captive mammals was only 26% of the total inbreeding depression across the life cycle in the wild revealed in a study by Keller (1998) on the song sparrow.

Lande (1988) considered natural selection to be effective in removing (purging) deleterious alleles in small populations and markedly reducing inbreeding depression. Subsequent modeling and empirical work indicates that purging effects are typically relatively small (Byers & Waller 1999; Reed et al. 2003).

All the above points lead to greater impacts of inbreeding depression on population viability than would have been expected in 1988.

Genetic Diversity and Extinction

Natural populations face continuous assaults from environmental changes including new diseases, pests, parasites, competitors and predators, pollution, climatic fluctuations such as the El Niño–La Niña cycles, and human-induced global climate change. Species must evolve to cope with these new conditions or face extinction. Naturally outbreeding species with large populations normally possess large stores of genetic diversity that allow them to evolve in response to environmental changes.

Conversely, loss of genetic diversity in small populations is predicted to reduce the ability of populations to evolve in response to environmental change, and experimental evidence validates this prediction. Consequently, we expect a relationship between loss of genetic diversity and extinction rate due to environmental change. However, there are very few examples where extinctions of natural populations can be directly attributed to lack of genetic variation (as opposed to inbreeding), as described below.

Loss of genetic diversity at self-incompatibility loci is associated with reduced fitness and increased risk of extinction. About half of all flowering plant species have genetic systems that reduce or prevent self-fertilization (Richards 1997). Self-incompatibility (SI) is regulated by one or more loci that may have 50 or more alleles in large populations. If the same allele is present in a pollen grain and the stigma, fertilization by that pollen grain will not be successful.

SI alleles are lost by random sampling in small populations. This leads to a reduction in the number of plants that can potentially fertilize the eggs of any individual and eventually to reduced seed set and extinction. For example, the Lakeside daisy population from Illinois declined to three plants. This population did not reproduce for 15 years despite bee pollination, as it contained so few SI alleles (Demauro 1993); in other words, this population was functionally extinct. Plants did, however, produce viable seed when fertilized with pollen from large populations in Ohio or Canada. While reduced fitness due to loss of SI alleles has only been documented in a few species of plants, it is likely to be a problem, or become so, in most threatened, self-incompatible plants. The problems associated with small populations of self-incompatible plant species are elegantly demonstrated by the work of Andrew Young's group on the endangered grassland daisy in southeastern Australia (Young et al. 2000). This species exists in a series of populations with a diversity of sizes. Self-incompatibility is due to a single locus with multiple alleles. Each of the expected effects of genetic drift has been documented in these populations, namely lower genetic diversity generally in smaller populations, fewer SI alleles in smaller populations, reduced mate availability, and reduced fitness in smaller populations. Further, Young and colleagues have shown using computer simulation that smaller populations are expected to have higher extinction risks due to loss of genetic diversity at SI loci.

Populations with low genetic diversity are expected to suffer more seriously from diseases, pests, and parasites than those with high genetic diversity (Kover, Ch. 29 of this volume). Novel pathogens constitute one of the most significant challenges to all species. Associations between loss of genetic diversity and inbreeding and reduced resistance to disease and parasites have been reported in fish, Soay sheep, deer mice, California sea lions, bumblebees, and *Drosophila* (see Acedevo-Whitehouse et al. 2003; Spielman et al. 2004b).

"Mutational Meltdown"

Deleterious mutations are continually produced in all populations. In large populations, natural selection is effective in keeping them at very low frequencies. However, in small populations, mildly deleterious mutations become effectively neutral so that their fate is determined by genetic drift (Gillespie, Ch. 5 in this volume). Consequently, some go to fixation and reduce mean fitness. This process may over time drive species to extinction, a process Lynch called "mutational meltdown" (Lande 1995; Lynch et al. 1995). Mutation accumulation is generally recognized as a threat to asexual species, but its role in sexually reproducing species is controversial, in terms of theory, parameter estimates, and experimental evidence. Gilligan et al. (1997) did not find any contribution of mutation accumulation over 45–50 generations in populations of *Drosophila* with effective sizes of 25–500.

GENETIC MANAGEMENT OF THREATENED SPECIES IN CAPTIVITY

The threats faced by many species are so great that they are unlikely to survive in the wild. Captive breeding represents the only realistic salvation strategy for these species. However, individual zoos typically have limited capacity for any one species, perhaps only four individuals, and the overall capacity for threatened species over zoos is insufficient. Without active genetic management, such small populations will rapidly become inbred (as they did in the past) and suffer elevated extinction risks. Further, many individuals do not breed when brought into captivity and the contributions of the breeders differ markedly (see Case Study 2).

To overcome these difficulties, endangered species are typically managed across zoos in Species Survival Plans. The recommended genetic management regime is to minimize kinship, a procedure that chooses individuals to breed on the basis of them having the least relationship to the species as a whole (Ballou & Lacy 1995). The kinship of two individuals is the inbreeding coefficient of an offspring if they had one. When founders contribute equally, minimizing kinship is equivalent to equalizing family sizes—a procedure that doubles the effective population size. Case Study 2 describes the genetic management of the captive and wild populations of the golden lion tamarin.

To retain evolutionary potential, it is recommended that species be maintained with effective population sizes of at least 500, based on the equilibrium between the rate of loss of quantitative genetic variation by drift and its replenishment by mutation. However, captive breeding spaces are so limited that at best only about 800 of the 2000–3000 terrestrial vertebrates requiring captive breeding would be able to be conserved at this size. Consequently, the compromise has been to recommend that species be managed to conserve 90% of genetic diversity for 100 years, based on the hope that the human population will decline in this time releasing habitat and allowing reintroduction into the wild. Under ideal conditions, this corresponds to an effective size of $475/L$, where L is the generation length in years. Thus, Arabian oryx with 10 years per generation require an effective size of 48. Even with this compromise, many species are being maintained at sizes lower than required to meet this goal. About one half of captive populations of threatened mammals have sizes of less than 50 and effective sizes of less than about 15. Genetic management of captive populations is widely practised and generally well done, apart from the inadequate population sizes. Many threatened species in the wild also have population sizes that are too small to avoid genetic deterioration due to inbreeding and loss of genetic diversity.

Case Study 2. Genetic Management of Golden Lion Tamarin (after Frankham et al. 2004)

The most extensive genetic management of a species over both captive and wild population is that for the golden lion tamarin. This endangered species is found in the coastal forest of Brazil, where more than 98% of its habitat has been destroyed. Consequently, it has been subject to a captive breeding program, a reintroduction program, and wild management. Of 242 tamarins originally brought into captivity, only 48 contribute to the current captive population and two thirds of the gene pool was from only one prolific pair before active management began. The captive program was highly successful in increasing the numbers for the species, there being over 500 individuals currently in captivity.

The procedure of minimizing kinship was devised to cope with the unequal founder contributions and to equalize family sizes as much as possible, and thus maximize the effective size of the population. Genetic management by minimizing kinship has resulted in a current inbreeding coefficient of only 1.9%, in spite of the highly distorted founder contributions prior to its implementation. The species is monogamous, so it has been possible to obtain pedigree information for both the captive and the wild population.

Captive populations are spread across 140 zoos worldwide and captive management is coordinated across all these institutions. The captive population has been used to supply individuals for reintroduction into the wild. This has mainly involved founding populations in sites with suitable habitat but where the population had become extinct. This has been successful, with the reintroduced populations growing rapidly. There are currently about 600 individuals in the wild. Over 145 tamarins having been reintroduced into the wild and these have flourished, reaching 400 animals in 2002.

The small wild populations have low genetic diversity (Grativol et al. 2001) and there is considerable genetic differentiation among populations, so plans have been formulated to augment small populations with individuals from other populations to minimize the impacts of loss of genetic diversity and inbreeding. These plans are being optimized with the aid of computer simulations.

This program is very extensive and expensive, involving the Brazilian Government, Brazilian wildlife agencies, the Smithsonian National Zoo in Washington DC, and many other zoos. It involves an integrated program of ecology, behavior, genetics, management, research, and education.

USE OF MOLECULAR GENETICS IN UNDERSTANDING SPECIES BIOLOGY

Information on population size, mating system, paternity, genetic diversity, and population structure are essential to understanding the evolution of threatened species and to instituting management to recover them. Often this information is unknown or difficult to obtain by direct observation, especially on secretive, nocturnal, or fossorial species. For example, censusing may be difficult, behavioral observations on matings are notoriously inaccurate, population structure is difficult to determine, and aspects of prior population history are often unknown, but all can be determined by applying molecular genetic methods (chapter 19 of Frankham et al. 2002). Case Study 3 documents the application of molecular genetic methods to the critically endangered northern hairy-nosed wombat to determine a range of factors about its biology that provide essential background information for managing the species.

Paternities have been determined in a wide range of species using allozyme, DNA fingerprint or microsatellite markers and have often revealed a high level of extra-pair paternities, or pedigree errors. The superb fairy wren is an extreme case. Genetic analyses revealed that approximately 65% of matings were with males outside the harem, a fact that was not discovered by prior behavioral observations.

Genetic management of fragmented population depends critically upon clear knowledge of population structure and gene flow between fragments. This information can usually only be obtained from genetic analyses. Immigrants can now be detected by assignment test based on microsatellite data.

Application of coalescence theory has had a major impact on the power of genetic analyses by adding a time dimension to data analyses (Rosenberg, Ch. 12 of this volume). It has allowed approximate times to be determined for events, and inferences to be made about demographic history, population bottlenecks, geographic population structure, and selection on loci.

Case Study 3. Use of Molecular Genetic Methods with the Critically Endangered Northern Hairy-Nosed Wombat to Determine Several Aspects of Its Species Biology

The critically endangered northern hairy-nosed wombat exists in a single population of approximately 75 individuals in central Queensland, Australia. This pig-sized marsupial is difficult to study as it is nocturnal and lives in burrows. Censusing previously involved placing a large metal trap at the entrance to burrows. The process of capture was stressful and many animals became trap-shy. Taylor and colleagues (Taylor et al. 1994; Beheregaray et al. 2000; Sloane et al. 2000) developed microsatellite markers at 28 loci to study the biology of this species. Using approximately 20 loci they are able to accurately identify each individual in the colony from hair collected on adhesive tape on frames at the entrances to wombat burrows. They have estimated the population size at over 100, somewhat larger than the estimate from direct capture. Individuals have been sexed by amplifying DNA of X and Y chromosome loci. From analyses of microsatellite data Taylor et al. have determined that this species has much lower genetic diversity than the related southern hairy-nosed wombat. Analyses based on hair from museum specimens demonstrated that an extinct population at Deniliquin over 1400 km away, and nearer the location of the southern species, was the northern species and so identified a potential site for reintroduction. Genetic evidence indicates that the species is approximately random mating. Initial efforts to establish paternity based on eight or nine microsatellite loci were unsuccessful, but the subsequent development of more loci may allow this to be done. The effective size of the species has been estimated from loss of microsatellite variation and from linkage disequilibrium and shown to be very low.

CONSEQUENCES OF IGNORING GENETIC FACTORS IN THREATENED SPECIES MANAGEMENT

Overall, there is little effective genetic management of wild populations of threatened species, but a substantial need for it (chapter 16 of Frankham et al. 2002). If genetic factors are ignored, the following adverse effects may occur:

- Recovery programs may not be successful. For example, the Illinois population of greater prairie chickens declined from millions to fewer than 50 in 1993, and failed to recover following habitat restoration (Westemeier et al. 1998). It showed evidence of reduced fertility and hatchability and only recovered following outcrossing. In koalas in southeastern Australia, reintroductions using a small island population with only two or three founders have resulted in a substantial reduction in genetic diversity, a rise in inbreeding, a decrease in sperm quality, and to a marked increase in testicular aplasia (Houlden et al. 1996; Seymour et al. 2001).
- Extinction risks are likely to be underestimated
- The effects of loss of self-incompatibility alleles will not be addressed.
- Reproductive wastage from crossing between diploid and tetraploid populations of a species, resulting in sterile triploids, will not be prevented (Young et al. 2000).
- Genetic management of fragmented populations is likely to be suboptimal. Small fragmented populations with limited gene flow will lose genetic diversity and become inbred and have elevated extinction risks. The red-cockaded woodpecker in the eastern United States exhibits the genetic effects of habitat fragmentation and greatly reduced gene flow among populations. Adequate genetic management of fragmented populations is rare, and is one of the greatest unaddressed issues in conservation biology.

FUTURE DIRECTIONS

A critical issue that need to be resolved in conservation biology is agreement on a definition of species for conservation purposes, along with agreed experimental protocols for diagnosing species. There is also disagreement about the definition of management units within species that needs resolving. There is active debate on both of these issues and some progress.

Genetic management in the wild is currently in its infancy. There is a major need for implementation of rational genetic management regimes, especially for fragmented populations and for self-incompatible species.

There is limited information on genetic management of species with different breeding systems (i.e., not outbreeding diploids) and for management of groups in species where pedigrees are not available due to multi-male, multi-female breeding groups. Attention to these is needed and likely to occur.

The application of molecular genetics analyses in forensics is growing rapidly and likely to become widespread internationally. Use of molecular genetic analyses to understand aspects of species biology necessary for conservation is likely to expand markedly as both the molecular techniques and the methods of data analyses improve. Application of these to multiple nuclear DNA loci will grow and reliance on mitochondrial DNA (a single inherited unit) should decline.

Meta-analyses are an extremely important tool in medicine, evolutionary and conservation biology, as they overcome issues of low statistical power in individual studies by combining all available information on a topic (typically published information from many studies) into a single analysis. For example, they were used in both our study on the relationship between genetic diversity and fitness (Reed & Frankham 2003) and our test of the "no genetic impact" hypothesis (Spielman et al. 2004a). Their role should expand in conservation biology and evolutionary genetics.

There is a need for a better integration of genetics into conservation biology, particularly by using population viability analyses (stochastic computer projections). I expect to see increased effort in this area. It will be especially important for understanding the fate of fragmented populations, as there are so many variables affecting such populations that they are difficult to study in the wild.

Data are needed on the following topics:

- The extent of inbreeding depression in the wild for the whole life cycle for a broad range of taxa, and its role in extinctions
- Extent of inbreeding depression in polyploids compared with diploids
- Heritabilities for quantitative traits in endangered species, so that we better understand their ability to evolve
- Role of mutational accumulation in extinctions
- Relationship between inbreeding and loss of genetic diversity with disease resistance
- Extent of outbreeding depression and factors that predict its magnitude
- The magnitude of genetic adaptation to captivity in wildlife
- The role of genetic factors in reintroduction success
- Population size required for taxa to avoid genetic deterioration
- Molecular markers that track functionally important loci
- Genetic markers that predict degree of reproductive isolation between populations

CONCLUSIONS

Inbreeding and loss of genetic diversity are of conservation concern as they increase the risk of extinction. Inbreeding depression increases the risk of extinction in captive populations, and there is now strong evidence that it is one of the factors causing extinctions of wild populations. Loss of genetic diversity reduces the ability of species to evolve to cope with environmental change. Inappropriate management and allocation of resources is likely to result if genetic factors are ignored in management of threatened species.

SUGGESTIONS FOR FURTHER READING

Avise (2004) is an authoritative textbook concerned with molecular population genetics, determination of life history parameters, taxonomy, and conservation genetics. Frankham et al. (2002) is a comprehensive textbook of conservation genetics with extended treatments of the topics in this chapter, plus references. Frankham et al. (2004) is a shorter, simpler version of the above textbook. Saccheri et al. (1998) provides evidence that genetic factors contribute to extinction risk in wild populations. Young et al. (2000) is a comprehensive overview of the impacts of loss of genetic diversity for self-incompatibility alleles on fitness in a threatened Australian plant.

Avise JC 2004 Molecular Markers: Natural History, and Evolution, 2nd ed. Sinauer Assoc.

Frankham R, Ballou JD & DA Briscoe 2002 Introduction to Conservation Genetics. Cambridge Univ. Press.

Frankham R, Ballou JD & DA Briscoe 2004 A Primer of Conservation Genetics. Cambridge Univ. Press.

Saccheri I, Kuussaari M, Kankare M, Vikman P, Fortelius W & I Hanski 1998 Inbreeding and extinction in a butterfly metapopulation. Nature 392:491–494.

Young AG, Brown AHD, Murray BG, Thrall PH & CH Miller 2000 Genetic erosion, restricted mating and reduced viability in fragmented populations of the endangered grassland herb *Rutidosis leptorrhynchoides*. pp. 335–359 in Young AG & GM Clarke, eds. Genetics, Demography and Viability of Fragmented Populations. Cambridge Univ. Press.

Acknowledgments The Australian Research Council and Macquarie University have supported my research. I thank Jonathan Ballou and David Briscoe for their contributions to the material herein. This is publication number 388 of the Key Centre for Biodiversity and Bioresources.

Glossary

Note: Names in brackets after each entry indicate the author(s) who submitted the entry. Some entries include multiple definitions, and authorship is indicated for each. Entries that do not indicate authorship were written by one of the editors.

active site: the small part of an enzyme that binds substrate and contains the residues that are directly involved in catalyzing the reaction [Lovell]

ad hoc hypothesis: a hypothesis created to preserve a failed theory by explaining contradictory evidence after the fact [Egan & Crandall]

additive genetic variation: differences among individuals in a population that are governed by alleles with consistent effects on the phenotype, in relation to other alleles or the environment [Chippindale]

additivity: the situation where the effect of an allele on the phenotype is the same regardless of genotype [Houle & Kondrashov]

allopatric speciation: when species form without the opportunity to exchange genes due to geographic barriers [Johnson]

allopatry: literally means different places. See *allopatric speciation* [Johnson]

allopolyploid: a polyploid genome created via hybridization

allozymes: various alleles at a locus encoding an enzyme having different amino acid sequences. Allozymes often have different mobilities when analyzed by electrophoresis

aneuploidy: the state of having the number of chromosomes that is not an exact multiple of the haploid number (i.e., one chromosome set is incomplete) [Michalak & Noor]

antagonistic pleiotropy: effect of a single gene on two or more traits, where at least one effect increases fitness and at least one effect decreases fitness [Promislow & Bronikowski]

artificial selection: the use of selective breeding to alter traits in a population. A useful way to determine the heritability of a single trait, and the genetic correlation between traits [Promislow & Bronikowski]

autopolyploid: more than two sets of homologous chromosomes. Contrast *allopolyploid*

Bayesian inference: a method of phylogenetic analysis that incorporates prior knowledge into the posterior probability which summarizes the probability that the hypothesis is true given the data [Egan & Crandall]

benchmark set: a set of data for which you know the "right" answer. Prediction methods (e.g., to find functional sites) can then be run on this set of data to determine the accuracy [Lovell]

betweenness: a feature of nodes in interaction networks that describes how many paths the node has an effect on. The shortest path between each pair of nodes in a network can be described, and the number of such paths that a node falls on is defined as that node's betweenness [Proulx]

binding energy: the difference in free energy between two or more molecules free in solution and the same molecules when bound together. It may also be thought of as the energy "released" by molecules binding or the energy "driving" the binding of molecules [Lovell]

binding interfaces: the surface of a protein responsible for binding other molecules [Lovell]

biodemography: the study of the biological factors that influence demographic traits, such as age-specific survival and fertility [Promislow & Bronikowski]

biological species concept: two populations are separate species if they lack the potential to interbreed and are the same species if they can interbreed under natural conditions [Johnson]

bioresources: items derived from living species that are of direct economic value: for example, food, many pharmaceutical drugs, fibers for clothing, rubber, timber, etc. [Frankham]

breeders' equation: the equation describing the single-generation response of a quantitative trait to selection; $r = h^2s$ for the direct response to selection on a single trait, where r is the response to selection, s is the selection differential, and h^2 is the narrow-sense heritability

canalization: buffering of the developmental process against genetic and environmental variation [Siegal & Bergman]

cDNA: see *complementary DNA*

cell memory: the retention of a functional or structural state by cells or cell lineages [Jablonka & Lamb]

cellular inheritance (or cell heredity): the transmission of structural or functional states in cell lineages [Jablonka & Lamb]

central dogma: Crick's suggestion that protein is never used as a template to make other molecules. The term is now generally used in molecular biology to describe all flow of sequence information in the cell, but this was never the original intention [Lovell]

chromatin mark: the non-DNA part of a chromosomal locus, or a pattern of DNA methylation, that affects the nature and stability of gene expression [Jablonka & Lamb]

chromosomal inversion: when a chromosomal region is flipped such that it is now oriented in the reverse direction [Wayne & Miyamoto]

clade: a cluster of phylogenetically closely related organisms, consisting of all of the descendants of a specified common ancestor [Thornton]

cloning (of DNA): the transfer of a DNA fragment from an organism to a virus or plasmid

codon bias: unequal usage of synonymous codons within a given codon family [Chen & Stephan]

compensatory evolution: the substitution process of a pair of mutations at different loci (or nucleotide sites) that are individually deleterious but are neutral in appropriate combinations [Chen & Stephan]

complementary DNA: DNA made from an mRNA template by reverse transcription

complementation test: a mating test to determine whether two different recessive mutations on opposite chromosomes of a diploid or partial diploid will not complement each other; a test to determine whether two mutant sites are in the same functional unit or gene [Michalak & Noor]

complex (noun): two or more molecules bound together. Often refers to protein–protein complexes [Lovell]

composite interval mapping (CIM): modifies the standard interval mapping approach to include additional markers as cofactors in the regression analysis. Additional markers allow much greater power in the detection of quantitative trait loci and in estimates of position. Markers surrounding the quantitative trait locus of interest, as well as a number of linked and unlinked markers within and outside this region, are used depending on the characteristics of the neighboring area. Contrast with *multiple interval mapping* [Michalak & Noor]

consensus sequence: general DNA, RNA, or protein sequence that summarizes the common shared bases or amino acids for a particular structure and/or function, as derived from the comparison of multiple individual sequences [Wayne & Miyamoto]

consensus tree: a tree that summarizes the information in a set of trees all with the same tips [Egan & Crandall]

conspecific sperm precedence: the condition when, in the female reproductive tract, sperm from males from the same species (conspecifics) have a competitive advantage over those from males of a different species [Johnson]

contextual analysis: a statistical technique adapted by Heisler and Damuth (1987) from social sciences to analyzing selection in structured populations. Group characteristics (contextual characters) are included in an analysis of selection, where each individual has two sets of characters: individual and group. The analysis thereby separates individual and group components of selection

cost of plasticity: a decrease in the fitness of an individual with a plastic genotype even when it expresses the optimal phenotype for a given environment [Scheiner]

CpG dinucleotide pairs: CG (cytosine–guanine) couplets in which the 5′ cytosine can be methylated [Wayne & Miyamoto]

CpG islands: short sequences of DNA rich in CpG couplets (dinucleotide pairs) and thus high in CG content

deficiency: a chromosome with a deleted segment. A gene that is located within the deleted segment is uncovered in the complementation test when the deficiency is heterozygous with a chromosome that does not have the deletion. Deficiency is analogous to a loss-of-function mutation for the candidate gene and for all other genes uncovered by the deletion. See also *deficiency library* and *complementation test* [Michalak & Noor]

deficiency library: a collection of strains (currently >12,000 for *D. melanogaster*) carrying separate deletions to be used in complementation tests [Michalak & Noor]

deme: a locally interbreeding population, generally with no substructure. These are the subunits that make up a metapopulation

deterministic: indicates no role for random or probabilistic events. Given a set of starting conditions the system can be precisely and uniquely defined at some future point in time

developmental module: aspect of the phenotype that can develop all or most of its structure outside of its normal context; for example, imaginal discs of *Drosophila* [Mezey]

DNA methylation: the modification of DNA by the addition of methyl groups ($-CH_3$) to some bases; in eukaryotes the modified bases are usually cytosines [Jablonka & Lamb]

DNA microarrays: (DNA chips) arrays of DNA molecules (probes) immobilized on glass or another solid surface used to analyze gene expression of hundreds or thousands genes simultaneously [Michalak & Noor]

DNA slippage: strand slippage and mispairing of the template versus replicating strands during DNA replication [Wayne & Miyamoto]

dominance theory: an explanation for Haldane's rule based on the dominance patterns of the alleles involved in reproductive isolation [Johnson]

dominance: (1) the situation where introduction of one copy of an allele to a genotype has a greater effect on the phenotype than the introduction of a second copy [Houle & Kondrashov]; (2) interactions among alleles at a single locus leading to non-additivity

dosage (gene dosage): relative number of copies of a gene present in an individual. For example, for the sex chromosomes, heterogametic individuals have half the gene dosage of homogametic individuals. However, can also refer to multiple copies of genes on autosomes, natural or artificially altered from wild-type [Wayne & Miyamoto]

Ecdysozoans: a large clade of animals consisting of all molting invertebrates, including arthropods, nematodes, priapulids, rotifers, and several other phyla [Thornton]

ecosystem services: essential functions provided free of charge by living organisms, including oxygen from green plants, climate control by forests, nutrient recycling, natural pest control, and pollination of crop plants [Frankham]

effective population size (N_e): the number of individuals that would result in the same effects on inbreeding, loss of genetic diversity, or genetic variation among replicate populations if they reproduced in the manner of an idealized population (a conceptual random mating population with equal numbers of hermaphrodites breeding each generation, with no selection, or migration, and Poisson variation in family sizes). Often much less than the number of potential parents due to variation in sex ratio, excess variation in family sizes, and fluctuations in population size [Frankham]

endosymbiosis: a symbiotic relationship between two organisms in which one organism lives inside the other

epiallele: one of the alternative heritable forms of chromatin or DNA methylation patterns that are associated with a gene with an unchanged DNA sequence [Jablonka & Lamb]

epigenetic inheritance system: a system that enables the phenotypic expression of the information in a cell or an individual to be transmitted to the next generation [Jablonka & Lamb]

epimutation: a heritable change in phenotype that is the result of an epigenetic modification, not an altered DNA sequence [Jablonka & Lamb]

epistasis: (1) the effects of alleles at different loci are different from the sum of their individual effects [Johnson]; (2) the effect of the genotype at one locus is affected by the genotype at a second locus [Houle & Kondrashov]

eusocial: a social system characterized by cooperative brood-care, overlapping generations, and division of labor among offspring [Promislow & Bronikowski]

evolutionary capacitor: a factor that normally contributes to developmental buffering, thereby promoting the accumulation of neutral genetic variation, but occasionally fails, thereby generating potentially adaptive phenotypic variation [Siegal & Bergman]

exon: the gene segments whose transcripts are retained in the mature RNA following splicing are referred to as exons [Patthy]

exon-shuffling: the process whereby exons of genes are duplicated, deleted, or exons of different genes are joined through recombination in introns [Patthy]

fixation: an allele is said to reach fixation when it has a frequency of 1 [Houle & Kondrashov]

fluctuating asymmetry: the random component of left–right asymmetry, which can be measured as the individual deviation from the average asymmetry in the population or genotype. The asymmetry is caused by random perturbations of developmental processes, and fluctuating asymmetry has been used extensively as a measure of developmental instability [Klingenberg]

frameshift mutation: insertion or deletion of nucleotides that change the reading frame of a gene

frequency-dependence: when the fitness of an allele or phenotype depends upon its relative abundance in a population. Usually this means a selective advantage to a rare type and disadvantage to a common type [Chippindale]

functional site: a generalized term meaning any active site, regulatory site, protein–protein binding surface, or any other site in a protein to which function can be localized [Lovell]

gene family: group of genes derived from a single recent common ancestor, for example, the globin gene family as derived by gene duplication from its ancestral gene [Wayne & Miyamoto]

gene regulatory network: a set of genes that interact through transcription regulation. Each gene in such a network encodes transcription factors that alter the transcription rate of other genes in the network, or affect downstream target genes outside of the network [Proulx]

gene silencing: inactivation of a gene or gene product

genetic architecture: a description of the pattern of genetic effects underlying variation in the expression of a trait. Includes number of loci, magnitude and pattern of effects of alleles and their interactions, genomic distribution of loci, and interactions with the environment.

genetic assimilation: process through which selection for an environmentally induced phenotype leads to a canalized system that produces the phenotype in the absence of the original environmental perturbation [Jablonka & Lamb]

genetic drift: random changes in genetic composition of a population over generations due to chance sampling in finite populations. Results in random changes in allele frequencies, loss of genetic diversity, and diversification among replicate populations [Frankham]

genetic element: a feature of an individual's genome that, when altered, has an effect on fitness for some given genetic backgrounds. A classically defined gene is a genetic element, as are regulatory sequences and alternative splice points [Proulx]

genetic load: the reduction in mean fitness in a population, relative to the best possible genotype, due to deleterious alleles

genetic module: traits of the phenotype where genetic variation is statistically independent of variation at other traits; also called variational or evolutionary modules [Mezey]

genetic network: a set of genes that interact to produce a phenotype [Proulx]

genetic variation: variation in DNA base sequence across a segment of DNA, whether a coding region or one not transcribed or translated. May be detected in DNA, proteins, or phenotypes. Also referred to as genetic diversity [Frankham]

genomic imprinting: (1) a process that causes the expression of genetic information to depend on the sex of the parent from which it was inherited; also used to describe the result of this process [Jablonka & Lamb]; (2) modification of DNA such that expression differs between parental chromosomes

genomics: approach utilizing all available information from the genome simultaneously, in contrast to gene-by-gene approaches; study of the structure and function of the genome as an integrated whole [Wayne & Miyamoto]

genotype–environment interaction (GxE interaction): differences among genotypes in their amounts and forms of phenotypic plasticity [Scheiner]

genotypic value: the expected (or mean) phenotype associated with a genotype

global expression analysis: study of patterns of gene expression from effectively all genes in the genome at once (i.e., genome-wide studies of how the final production and activity of all protein and RNA products are regulated) [Wayne & Miyamoto]

guide tree: a tree that summarizes the order in which sequences are to be aligned [Egan & Crandall]

Haldane's rule: the condition where in the F_1 progeny of interspecific crosses, the heterogametic sex (the sex with heteromorphic sex chromosomes) is more adversely affected than the homogametic sex [Johnson]

heritability, broad-sense: the proportion of the total phenotypic variance due to genetic variance ($H^2 = V_G/V_P$ where V_G includes additive, dominance, and epistatic variances)

heritability, narrow-sense: the proportion of the total phenotypic variance due to additive genetic variance ($h^2 = V_A/V_P$)

heterosis: when heterozygotes have higher fitness than homozygotes, or when hybrids have higher fitness than purebreds

Hill–Robertson effect: the reduction in the efficacy of natural selection on linked loci

hitchhiking, genetic: the adaptive fixation of an advantageous mutant and the associated fixation of linked, neutral variants [Nachman]

HKA test: Hudson, Kreitman and Aguadé test; a goodness-of-fit test that evaluates the neutral hypothesis that levels of variation within species (polymorphism) and variation between species (divergence) will be correlated [Wayne & Miyamoto]

homology: similarity due to common ancestry [Egan & Crandall]

homonucleotide runs: sequential repeats of a single base, for example, AAAAAAAA [Wayne & Miyamoto]

(homo-) oligomers: "Oligo" comes from the Greek oligoi meaning "a few," and so "oligomers" is the generalization of dimers, trimers, tetramers, etc. The boundary between "oligomers" and "polymers" is as well defined as the boundary between "a few" and "lots." Homo-oligomers are those oligomers composed of identical subunits, and can be contrasted with hetero-oligomers [Lovell]

homoplasy: similarity not due to common ancestry but to convergence, parallelism, or horizontal gene transfer [Egan & Crandall]

housekeeping genes: genes coding for proteins used by nearly all cells and whose level of expression varies little or not at all among cells

hybrid speciation: process in which natural hybridization results in the production of an evolutionary lineage that is at least partially reproductively isolated from both parental lineages, and which demonstrates a distinct evolutionary and ecological trajectory [Arnold & Burke]

hybrid species: partially or completely reproductively isolated lineages arising as a result of natural hybridization. These lineages demonstrate distinct evolutionary and ecological trajectories as defined by distinguishable (and heritable) morphological, ecological, and reproductive differences relative to their progenitors [Arnold & Burke]

hybrid zone: (1) a region where two nascent species come into contact and produce at least some hybrids. The hybrids can be F_1 hybrids or later-generation hybrids [Johnson]; (2) geographical region in which natural hybridization occurs (Arnold 1997, as adapted from Harrison 1990) [Arnold & Burke]

imprint: a chromatin mark that is determined by the sex of the transmitting parent [Jablonka & Lamb]

imprinting: see *genomic imprinting*

inbreeding: (1) production of offspring from mating between related individuals, such as self-fertilization, brother–sister, or cousin matings [Frankham]; (2) at the population level, inbreeding occurs when the average relationship of mates is greater than the average relationship of randomly chosen individuals

inbreeding coefficient, F: the probability of being autozygous, typically measured as the proportional reduction in heterozygosity due to nonrandom mating, $1-H_{\text{observed}}/H_{\text{expected}}$

inbreeding depression: reduction in the mean for a quantitative trait due to inbreeding, typically found for traits associated with fitness [Frankham]

inbreeding load: the reduction in fitness in a population due to the effects of inbreeding depression [Promislow & Bronikowski]

incomplete lineage sorting: presence of polymorphism within an ancestral population transmitted through descendant lineages to the effect of misconstruing phylogenetic relationships [Egan & Crandall]

indel: insertions or deletions in a DNA sequence [Wayne & Miyamoto]

infinitesimal model: null model describing the genetic architecture of quantitative traits as an infinite number of genes, each with an infinite number of alleles with infinitesimally small, additive effects on the trait [Wayne & Miyamoto]

interaction network: a set of cellular or organismal components, and their interactions, that depend on each other to produce the phenotype [Proulx]

introgression: (1) the transfer of genomic segments from one taxon into another via hybridization followed by repeated backcrossing between hybrid and parental individuals (from Anderson & Hubricht 1938) [Arnold & Burke]; (2) replacement of a locus of one strain or species with the homologous locus of another strain or species, by serial backcrossing [Wayne & Miyamoto]

introgressive hybridization: see *introgression.*

intron: section of DNA within a gene that is transcribed but not translated

intron phase: the phase of an intron present in the translated region of a protein-coding gene refers to its position relative to the reading frame. There are three types of intron phases depending on where the intron is located: phase 0 introns lie between codons, phase 1 introns occur between the first and second bases of a codon, phase 2 introns occur between the second and third nucleotides of a codon [Patthy]

isochore: relatively long stretches of GC- or AT-rich DNA sequence. They produce chromosomal bands, with GC-rich regions associated with R bands and AT-rich regions associated with G bands

karyotype: the complement of chromosomes of a cell

knockout: (1) a gene that has been deleted from the genome through transgenic means; (2) a mutation that eliminates the function of a gene

laboratory natural selection: selection in which any attribute of the organism is allowed to evolve in response to a controlled treatment [Chippindale]

likelihood ratio test: a method for hypothesis testing in a likelihood framework. A data set's fit to a more complex model is compared with its fit to a simpler model using the likelihood ratio statistic (twice the ratio of the likelihoods of the two models). The more complex model is adopted if it increases the likelihood more than expected by chance at some critical probability. If the simpler model is a restricted version of the more complex model, the improvement in fit can be evaluated using a chi-square distribution [Thornton]

limitation of plasticity: a factor that results in a failure of a plastic genotype to match the optimal phenotype [Scheiner]

linkage disequilibrium: (1) when the inheritance of two loci is correlated. In quantitative trait locus (QTL) mapping experiments this is usually due to linkage between the loci. Loci closer to one another along the chromosome tend to be inherited together. The gene mapping approach uses the level of linkage disequilibrium between arbitrary genetic markers and genes affecting the trait of interest to locate QTLs [Cheverud]; (2) nonrandom associations among alleles at different loci. May arise from selection, small population size, or recent mutations [Frankham]

linkage equilibrium: when two loci segregate independently of each other so that their inheritance is not correlated [Cheverud]

long-branch attraction: the phenomenon within maximum parsimony in which long branches tend to group together regardless of evolutionary relationships [Egan & Crandall]

long-oligo microarrays: DNA microarrays produced by deposition of relatively long oligonucleotides (typically 50–70 nucleotides in length) on a solid substrate

massively parallel signature sequencing (MPSS): a technology that uses a DNA library embedded onto microbeads to analyze short "signature sequences" of DNA (ca. 16–20 bases in length)

maximum likelihood: a method of phylogenetic analysis that determines the best phylogeny by optimizing the likelihood of observing the data given a tree and model of nucleotide substitution [Egan & Crandall]

maximum parsimony: a method of phylogenetic analysis that determines the best tree(s) by optimizing, or rather minimizing, the number of evolutionary changes within the data [Egan & Crandall]

metapopulation: a collection of subpopulations or demes that are linked through gene flow via migration, may also include extinction and recolonization of demes

methylation (of DNA): modification of DNA by the addition of a methyl ($-CH_3$) group

microarray: a tool designed to assay levels of expression at multiple loci in the genome simultaneously, consisting of oligonucleotide or cDNA probes mounted on a substrate (glass or nylon). Test samples of nucleic acids are then allowed to hybridize to these probes via normal base pairing to track the global expression of genes for the given conditions [Wayne & Miyamoto]

microsatellite DNA: tandemly repeated short sequences of DNA, usually 2–4 bp per repeat, though the length of repeat classified as a microsatellite varies among researchers

migration: the movement of individuals among populations/demes often leading to gene flow

minimum evolution: a method of phylogenetic analysis that optimizes, or rather minimizes, the branch lengths within the tree to determine the best phylogeny [Egan & Crandall]

minisatellite DNA: tandemly repeated short sequences of DNA, with each repeat longer than a *microsatellite*, though the length of repeat classified as a minisatellite varies among researchers

missense mutation: a codon change that results in an amino acid substitution in a polypeptide chain

model of nucletoide evolution: a model that attempts to represent the underlying processes of nucleotide substitution [Egan & Crandall]

modular protein: multidomain protein containing multiple copies and/or multiple types of protein modules [Patthy]

modularity: the autonomy displayed by parts of the phenotype [Mezey]

module: a part of the phenotype that can act as a relatively independent unit; see also *genetic module* [Mezey]

module-shuffling: during evolution protein modules may be shuffled either through recombination in introns (exon-shuffling) or recombination in exons to create multidomain proteins with various domain combinations [Patthy]

monophyletic: a group that contains all of the species derived from a common ancestor

morphometrics: the quantitative study of biological forms [Klingenberg]

mosaic protein: multidomain protein containing multiple types of protein domains of independent evolutionary origin [Patthy]

multiple interval mapping (MIM): uses multiple-marker intervals simultaneously to construct multiple quantiitative trait loci (QTLs) in the model for QTL mapping. The method differs from composite interval mapping in that multiple intervals are considered rather than just additional markers [Michalak & Noor]

mutagen: an agent that increases the rate of mutation

mutation: a change in a DNA sequence that is not caused by reciprocal recombination [Houle & Kondrashov]

mutation accumulation: experimental technique designed to assess rates and properties of new mutations by "accumulating" random mutations in a previously isogenic background [Wayne & Miyamoto]

mutational meltdown: reduction in population mean fitness and downward spiral toward extinction due to chance accumulation of mildly deleterious alleles in small populations [Frankham]

natural hybrid: offspring resulting from a cross in nature between individuals from two populations, or groups of populations, that are distinguishable on the basis of one or more heritable characters (Arnold 1997, as adapted from Harrison 1990) [Arnold & Burke]

natural hybridization: successful matings in nature between individuals from two populations, or groups of populations, that are distinguishable on the basis of one or more heritable characters (Arnold 1997, as adapted from Harrison 1990) [Arnold & Burke]

node degree: the degree of a node in an interaction network is given by the number of nodes that it interacts with. Typically this is visualized as the number of neighboring nodes on the network graph [Proulx]

nonadditive genetic variation: differences among individuals in a population governed by genes with inconsistent effects due to epistasis, dominance, or environmental interaction. Usually measured as the residual of genetic effects after additive genetic variation is quantified [Chippindale]

nonsense mutation: mutation that results in a premature stop codon [Wayne & Miyamoto]

nonsynonymous mutation (substitution): nucleotide substitutions occurring in translated regions of protein-coding genes that alter the amino acid [Patthy] Contrast with *synonymous substitution*

norm of reaction: a mathematical or graphical representation of the relationship between environmental variation and phenotypic variation due to phenotypic plasticity [Scheiner]

optimality criteria: tree building methods that use a criterion to apply a score to each topology in an effort to determine which topology best represents the evolutionary relationships within the data [Egan & Crandall]

organismal complexity: a reflection of the amount of information that an organism stores in its genome about the environment in which it evolves. Organismal complexity is a function of the number of components (genes, proteins, cells, tissues, organs, etc.) of the organism and the number of their interactions [Patthy]

ortholog: genes or proteins of different species that diverged from a common ancestral gene by speciation are called orthologs. Orthologs usually fulfill essentially the same function in the different species [Patthy]. The "same" gene in more than one species. Orthologs descend from a speciation event. Contrast *paralogs* [Thornton]

outbreeding depression: a reduction in reproductive fitness that may occur in crosses between different populations (or subspecies or species). Likelihood increases with degree of genetic differentiation among populations [Frankham]

outbreeding (outcrossing): (1) mating with an individual less related than would be a randomly chosen mate; (2) at the population level, outbreeding occurs when the average relationship of mates is lower than the average relationship of randomly chosen individuals

outgroup: a taxon closely related to a group of species being studied (i.e., has a shared common ancestor with the entire group being studied) but distantly related enough that it can help differentiate ancestral from derived characters

overdominance (overdominant): the condition where heterozygotes have more extreme traits, or higher fitness, than homozygotes

P element: a transposable element that spread over *Drosophila melanogaster* populations. It was also found in species of the *D. willistoni* group, leading to speculation that *D. melanogaster* has acquired *P* via horizontal transfer from *D. willistoni* and their mites (Powell 1997, pp. 339–341). There are two broad categories of *P* elements: complete (intact) *P* elements (also called *P* factors) and defective (deleted) *P* elements. Complete *P* element is 2907 bp long and produces 87 kDa "transposase" protein essential for its mobility throughout the genome. *P* element constructs are experimentally used to tag uncharacterized genes, induce new mutations, and modify expression of genes [Michalak & Noor]

paralog: evolutionarily related genes that are produced by gene duplication [Thornton]; homologous genes or proteins derived by a duplication event. Whereas orthologs are likely to fulfill very similar functions in different species, paralogous proteins are more likely to have diversified in function [Patthy]

paramutation: the process whereby one allele in a heterozygote alters the heritable epigenetic properties of the other allele [Jablonka & Lamb]

paraphyletic: a group of species that have a common ancestor, but the group does not include all of the descendants of that common ancestor

PCR: see *polymerase chain reaction*

phenocopy: an environmentally induced phenotype that mimics a mutant phenotype [Siegal & Bergman]

phenome: sum of the phenotypes of the organism and emergent properties arising therefrom, including life history, behavior, and morphology. Includes genotype, environment, and genotype × environment interactions [Wayne & Miyamoto]

phenotypic plasticity: the expression of more than one phenotype by a single genotype [Scheiner]

phylogeny: a graphical representation of the evolutionary relationships between a group of organisms illustrating ancestor to descendant lineages [Egan & Crandall]

plasmid: an extrachromosomal sequence of circular DNA in some bacteria and some eukaryotes that replicates independently of the host chromosome. Some plasmids occur in multiple copies within a cell; some can integrate into the host genome

poly-A tract: a stretch of adenines at the 3′-end placed at messenger RNAs by the cell's RNA processing machinery [Thornton]

polymerase chain reaction: a molecular technique that uses a thermostable DNA polymerase to amplify a specific segment of DNA. By use of a cyclical program, very small quantities of DNA (i.e., a small number of copies) can be greatly amplified

polyploidy: having more than two sets of chromosomes

prion: an alternative form of a protein, differing from the usual one in structure but not in amino acid sequence, which can propagate its own structure by converting the usual protein into its own form [Jablonka & Lamb]

Procrustes superimposition: configurations of landmarks are scaled to unit size, shifted so that their centers of gravity are in the same location, and rotated so that a best overall fit to a common average configuration is achieved [Klingenberg]

promoter: a sequence of bases to which RNA polymerase binds to initiate transcription

protein domain: a structurally independent, compact spatial unit of the three-dimensional structure of a protein. Distinct structural domains of multidomain proteins usually interact less extensively with each other than do structural elements within the domains and usually fold independent of the other domains [Patthy]

protein fold: the topology of the three-dimensional structure of a protein domain. Related domains have similar protein folds [Patthy]

protein module: a structurally independent protein domain that occurs in a large variety of multidomain proteins in different architectural contexts [Patthy]

protein–protein interaction network: a network defined by pairs of proteins that are capable of physically interacting. It is often assumed that protein pairs in this network interact in vivo and that changes in their interaction affect phenotype and fitness [Proulx]

protemics: the analysis of the complete complement of proteins in a cell

proteome: the protein complement of a cell, organism, or other biological entity [Patthy]; complete complement of proteins in a cell

Protostomes: a large clade consisting of all bilaterally symmetric animals in which the mouth develops before the anus during embryogenesis. Protostomes include mollusks, annelids, worms of many sorts, and all Ecdysozoans, including mollusks and nematodes [Thornton]

pseudogene: a sequence of genomic DNA that, although derived from and similar to a normal gene, is nonfunctional [Patthy]; DNA sequence derived from a gene but which is not expressed [Houle & Kondrashov]; genes that have accumulated mutations, such as frameshift mutations or stop codons, such that they can no longer produce active proteins.

purging: reduction in frequency of deleterious alleles due to natural selection, especially in small populations where deleterious recessives are more likely to be exposed than in large populations [Frankham]

purifying selection: selection against deleterious alleles

QTL: see *quantitative trait locus*

quantitative trait locus (pl., loci): locus where allelic substitution affects a quantitative character, such as height, weight, or reproductive fitness [Frankham]

random network: although there are many "random" ways to construct networks, random networks are typically constructed by adding links at random until a fixed average node degree has been reached [Proulx]

recombinant inbred lines (RILs): a set of lineages created by the crossing of lines and subsequent inbreeding. Crossing of the lines creates recombinations between the parental genomes and inbreeding fixes each line for some recombinant haplotype

regulatory mutation: a mutation that changes timing, location, or amount of gene product but does not change the nature of the product itself [Wayne & Miyamoto]

regulatory sites: sites on a protein that bind small molecules that lead to the regulation of other functions [Lovell]

reinforcement: the evolution of increased premating isolation between nascent species due to natural or sexual selection favoring increased isolation [Johnson]

reproductive character displacement: the pattern in which premating reproductive isolation is stronger in areas where species overlap (sympatry) than in areas where they do not (allopatry). This pattern may be due to the process of reinforcement [Johnson]

reproductive isolating barriers: factors that prevent the formation of viable, fertile hybrids between nascent species. Premating barriers act prior to and during mating while postmating barriers act after mating. Prezygotic barriers act prior to the formation of zygotes while postzygotic barriers act afterward [Johnson]

retrotransposition: copying of a gene to a new location in the genome by reverse transcription of an RNA transcript expressed from the original gene [Thornton]

reverse transcriptase: an enzyme used by retroviruses (mRNA viruses) and retrotransposons for transcribing DNA from an RNA template

reverse transcription: transcription of DNA from an RNA template using reverse transcriptase

RFLP (restriction fragment length polymorphism): a collection of DNA fragments of precisely defined length that can be separated by electrophoresis (with the smaller fragments migrating farther than the larger fragments) and visualized with a probe

RNA interference: (1) a process in which small double-stranded RNA molecules silence genes with which they have sequence homology, for example, by degrading their mRNA [Jablonka & Lamb]; (2) site-directed gene expression silencing with double-stranded RNA [Michalak & Noor]

RNAi: see *RNA interference*

SAGE: see *serial analysis of gene expression*

saturated mutagenesis: used to identify (nearly) all of the genes affecting a trait or process through screening very large numbers of mutant lines until the point where all further mutant genes isolated have been isolated before [Michalak & Noor]

scale-free network: scale-free networks are characterized by node degree distributions with long tails and no modal, or characteristic, degree. Such networks have power law node degree distributions [Proulx]

secondary structure (of proteins): local interactions between amino acids giving rise to structures such as alpha helices and beta pleated sheets [Wayne & Miyamoto]

selection differential: difference between phenotype of selected parents and population mean [Wayne & Miyamoto]

selection intensity: a standardized selection differential; the selection differential divided by the phenotypic standard deviation

selective sweep: relatively rapid fixation of a favorable mutation, often associated with genetic *hitchhiking*

senescence: (1) physical deterioration with age resulting from intrinsic breakdown in physiological processes. Two evolutionary hypotheses for the evolution of senescence are the mutation accumulation hypothesis and the antagonistic pleiotropy hypothesis [Bronikowski]; (2) age-related decline in fitness components [Promislow & Bronikowski]

serial analysis of gene expression (SAGE): a technique that allows rapid, detailed analysis of thousands of transcripts. The basic concept of SAGE rests on two principles. First, a small sequence of nucleotides from the transcript, called a "tag," can effectively identify from whence the original transcript came. Second, linking these tags allows for rapid sequencing analysis of multiple transcripts. By linking the tags together, only one sequencing event is required to sequence every transcript within the cell. The counts of particular tags directly reflect respective transcript concentrations [Michalak & Noor]

sexually antagonistic genes: genes where alleles have antagonistic effects such that those that are selectively favored in one sex are disfavored in the opposite sex

silencer: regulatory sequence that reduces or prevents transcription of a gene

single nucleotide polymorphism (SNP): alteration of a single nucleotide in a stretch of DNA, found in at least 1% of the population [Promislow & Bronikowski]

SNP: see *single nucleotide polymorphism*

speciation: (1) a process whereby one species evolves into a different species (anagenesis) or one species diverges into two or more descendant species (cladogenesis) [Egan & Crandall]; (2) the process by which one species becomes two; the acquisition of reproductive isolation between nascent species [Johnson]

substitution: replacement of one allele by another over evolutionary time [Houle & Kondrashov]

supertree: a large tree constructed by combining several smaller trees or subtrees using consensus methods [Egan & Crandall]

symmetrical exon: an exon flanked by introns of the same phase at both its upstream and downstream boundaries. Since introns may be of three different phases, symmetrical exons may also be of three different types: class 0-0, class 1-1, and class 2-2. Only symmetrical exons can be deleted, duplicated, inserted into introns (of the same phase) without disrupting the reading frame of the recipient protein-coding gene and/or the shuffled exon [Patthy]

symmetrical module: a module flanked by introns of the same phase at both its upstream and downstream boundaries. Only symmetrical modules can be deleted, duplicated, or inserted into introns (of the same phase) without disrupting the reading frame of the recipient protein-coding gene and/or the shuffled module [Patthy]

sympatry: literally "in the same place"; sympatric speciation is when species form despite no initial barriers to gene flow [Johnson]

synapomorphy: a shared derived trait [Egan & Crandall]

synonymous mutation (substitution): (1) nucleotide substitutions occurring in translated regions of protein-coding genes are synonymous (or silent) if they cause no amino acid change since the original and mutant codon code for the same amino acid [Patthy]; (2) a change to a DNA base pair within a protein sequence that does not change the amino acid for which it codes [Houle & Kondrashov]; (3) a fixed change in a gene's nucleotide sequence that does not alter the protein's amino acid sequence [Thornton]. Contrast with *nonsynonymous substitution*

synteny: the presence of two or more genes on the same chromosome. It is sometimes, incorrectly, used to refer to gene loci in different organisms located on a chromosomal region of common evolutionary ancestry. This is more correctly termed "conserved synteny" [Lovell]

targeted mutations: site-directed mutations targeting specific genes. Contrast with *saturated mutagenesis* [Michalak & Noor]

tautomeric shift: a reversible change in the location of a proton that alters a molecule from one form to another

tDNA: transfer DNA; the transforming region of the Ti plasmid of *Agrobacterium tumefaciens* that is inserted into the genome of the host plant cell; a transforming vector

tertiary structure (of proteins): arrangement of secondary structure (helices, etc.) to complete assembly of a single peptide chain; for example, creation of pockets or active sites in enzymes [Wayne & Miyamoto]

tetrasomy: the state of having one or more chromosomes in four copies [Michalak & Noor]

tokogeny: the reticulate (nonhierarchical) relationships that result from sexual reproduction within a population or species

topology: branching pattern of a phylogenetic tree [Egan & Crandall]

transcription factor: a protein that binds to a *cis*-regulatory site and is involved in the regulation of transcription

transcriptome: the complete collection of mRNA transcripts in a cell

transcriptome analysis: see *transcriptome*

transduction: horizontal transmission of genes among individuals by a viral agent [Wayne & Miyamoto]

transformation: uptake and integration of extracellular DNA

transition: base pair substitution of a purine (adenine or guanine) with another purine, or of a pyrimidine (thymine or cytosine) with another pyrimidine [Houle & Kondrashov]

translocation: the product of crossing over between nonhomologous chromosomes [Wayne & Miyamoto]

transposable element: a DNA sequence capable of catalyzing the copying and reinsertion of itself elsewhere in the genome [Houle & Kondrashov]; genetic units that do not have a fixed place in the genome, but rather can move from one locus to another, sometimes by duplicating themselves and sometimes by excising themselves from the DNA [Wayne & Miyamoto]

transversion: base pair substitution of a purine (adenine or guanine) with a pyrimidine (thymine or cytosine), or vice versa [Houle & Kondrashov]

References

Note: Numbers in brackets indicate the chapters or boxes that cite that reference

Abi-Rached L, Gilles A, Shiina T, Pontarotti P & H Inoko 2002 Evidence of en bloc duplication in vertebrate genomes. Nat. Gen. 31:100–105. [10]

Abzhanov A & TC Kaufman 1999 Novel regulation of the homeotic gene Scr associated with a crustacean leg-to-maxilliped appendage transformation. Development 126:1121–1128. [15]

Acedevo-Whitehouse K, Gulland F, Greig D & W Amos 2003 Inbreeding-dependent disease susceptibility of California sea lions. Nature 422:35. [32]

Ackermann M, Bijlsma R, James AC, Partridge L, Zwaan BJ & SC Stearns 2001 Effects of assay conditions in life history experiments with *Drosophila melanogaster*. J. Evol. Biol. 14:199–209. [31]

Adams MD, Celniker SE, Holt RA, Evans CA, Gocayne JD, Amanatides PG, Scherer SE & PW Li 2000 The genome sequence of *Drosophila melanogaster*. Science 287:2185–2195. [2]

Agrawal AA 2001 Phenotypic plasticity in the interactions and evolution of species. Science 294:321–326. [21]

Agrawal AF & CM Lively 2002 Infection genetics: gene-for-gene versus matching-alleles models and all points in between. Evol. Ecol. Res. 4:79–90. [29]

Agrawal AF, Brodie ED III & LH Rieseberg 2001 Possible consequences of genes of major effect: transient changes in the G-matrix. Genetica 112–113:33–43. [20]

Akam M 1987 The molecular basis for metameric pattern in the *Drosophila* embryo. Development 101:1–22. [2.1]

Akashi H 1994 Synonymous codon usage in *Drosophila melanogaster*: natural selection and translational accuracy. Genetics 136:927–935. [9]

Akashi H 1995 Inferring weak selection from patterns of polymorphism and divergence at "silent" sites in *Drosophila* DNA. Genetics 139:1067–1076. [9]

Akashi H 2001 Gene expression and molecular evolution. Curr. Opin. Genet. Dev. 11:660–666. [9]

Akey JM, Zhang G, Zhang K, Jin L & MD Shriver 2002 Interrogating a high-density SNP map for signatures of natural selection. Genome Res. 12:1805–1814. [7]

Albert R, Jeong H & A-L Barabasi 2000 Error and attack tolerance of complex networks. Nature 406:378–382. [13.1]

Albertson RC, Streelman JT & TD Kocher 2003 Directional selection has shaped the oral jaws of Lake Malawi cichlid fishes. Proc. Natl. Acad. Sci. USA 100:5252–5257. [19.4]

Aldous DJ 2001. Stochastic models and descriptive statistics for phylogenetic trees, from Yule to today. Stat. Sci. 16:23–34. [12]

Alexander RD 1979 Darwinism and Human Affairs. Univ. of Washington Press. [23]

Alexander RD 1987 The Biology of Moral Systems. Aldine de Gruyter. [23]

Alexander RD & G Borgia 1978 Group selection, altruism, and the levels of organization of life. Annu. Rev. Ecol. Syst. 9:449–474. [23]

Alfaro ME, Zoller S & F Lutzoni 2003 Bayes or bootstrap? A simulation study comparing the performance of Bayesian Markov chain Monte Carlo sampling and bootstrapping in assessing phylogenetic confidence. Mol. Biol. Evol. 20:255–266. [28]

Anderson E 1949 Introgressive Hybridization. Wiley. [26]

Anderson E & L Hubricht 1938 Hybridization in *Tradescantia*. III. The evidence for introgressive hybridization. Am. J. Bot. 25:396–402. [26]

Anderson EC & EA Thompson 2002 A model-based method for identifying species hybrids using multilocus genetic data. Genetics 160:1217–1229. [Species Concepts box]

Anderson RM & RM May 1992 Infectious Disease of Humans: dynamics and control. Oxford Univ. Press. [29]
Andersson M 1994 Sexual Selection. Princeton Univ. Press. [22]
Andersson SG & CG Kurland 1990 Codon preferences in free-living microorganisms. Microbiol. Rev. 54:198–210. [9]
Anholt RRH, Dilda CL, Chang S, Fanara JJ, Kulkarni NH, Ganguly I, Rollmann SM, Kamdar KP & TFC Mackay 2003 The genetic architecture of odor-guided behavior in *Drosophila*: epistasis and the transcriptome. Nat. Gen. 35:180–184. [2]
Arabidopsis Genome Initiative 2000 Analysis of the genome sequence of the flowering plant *Arabidopsis thaliana*. Nature 408:796–815. [29.1]
Aravind L & G Subramanian 1999 Origin of multicellular eukaryotes: insights from proteome comparisons. Curr. Opin. Genet. Dev. 9:688–694. [14]
Aris-Brosou S & Z Yang 2002 The effects of models of rate evolution on estimation of divergence dates with a special reference to the metazoan 18S rRNA phylogeny. Syst. Biol. 51:703–714. [8]
Aris-Brosou S & Z Yang 2003 Bayesian models of episodic evolution support a late pre-Cambrian explosive diversification of the Metazoa. Mol. Biol. Evol. 20:1947–1954. [8]
Armbruster P, Bradshaw WE & CM Holzapfel 1998 Effects of postglacial range expansion on allozyme and quantitative genetic variation of the pitcher-plant mosquito, *Wyeomyia smithii*. Evolution 52:1697–1704. [27]
Arnold ML 1997 Natural Hybridization and Evolution. Oxford Univ. Press. [26]
Arnold ML 2004 Natural hybridization and the evolution of domesticated, pest, and disease organisms. Mol. Ecol. 13:997–1007. [26]
Arnold SJ 1992 Constraints on phenotypic evolution. Am. Nat. 140:S85–S107. [20]
Arnold SJ & MJ Wade 1984a On the measurement of natural and sexual selection: theory. Evolution 38:709–718. [6]
Arnold SJ & MJ Wade 1984b On the measurement of natural and sexual selection: applications. Evolution 38:720–734. [6]
Arnqvist G, Edvardsson M, Friberg U & T Nilsson 2000 Sexual conflict promotes speciation in insects. Proc. Natl. Acad. Sci. USA 97:10460–10464. [24]
Atchley WR, Wollenberg KR, Fitch WM, Terhalle W & AW Dress 2000 Correlations among amino acid sites in bHLH protein domains: an information theoretic analysis. Mol. Biol. Evol. 17:164–178. [13]
Averhoff WW & RH Richardson 1974 Pheromonal control of mating patterns in *Drosophila melanogaster*. Behav. Genet. 4:207–225. [27]
Averof M & M Akam 1995 Hox genes and the diversification of insect and crustacean body plans. Nature 376:420–423. [25]
Averof M & SM Cohen 1997 Evolutionary origin of insect wings from ancestral gills. Nature 385:627–630. [15]
Averof M & NH Patel 1997 Crustacean appendage evolution associated with changes in Hox gene expression. Nature 388:682–686. [15]
Avise JC 2000. Phylogeography: The History and Formation of Species. Harvard Univ. Press. [12]
Avise JC 2004 Molecular Markers, Natural History, and Evolution, 2nd ed. Sinauer Assoc. [32]
Avise JC & K Wollenberg 1997 Phylogenetics and the origin of species. Proc. Natl. Acad. Sci. USA 94:7748–7755. [24]
Avital E & E Jablonka 2000 Animal Traditions: Behavioural Inheritance in Evolution. Cambridge Univ. Press. [17]
Ayala FJ 1986 On the virtues and pitfalls of the molecular evolutionary clock. J. Hered. 77:226–235. [8]
Ayala FJ 1997 Vagaries of the molecular clock. Proc. Natl. Acad. Sci. USA 94:7776–7783. [8]
Badyaev AV, Hill GE, Beck ML, Dervan AA, Duckworth RA, McGraw KJ, Nolan PM & LA Whittingham 2002 Sex-biased hatching order and adaptive population divergence in a passerine bird. Science 295:316–318. [2.1]
Baines JF, Parsch J & W Stephan 2004 Pleiotropic effect of disrupting a conserved sequence involved in a long range compensatory interaction in the *Drosophila Adh* gene. Genetics 166:237–242. [2]
Bakal CJ & JE Davies 2000 No longer an exclusive club: eukaryotic signalling domains in bacteria. Trends Cell Biol. 10:32–38. [14]
Baker D & A Sali 2001 Protein structure prediction and structural genomics. Science 294:93–96. [2]
Baker RH & GS Wilkinson 2003 Phylogenetic analysis of correlation structure in stalk-eyed flies (*Diasemopsis*, Diopsidae). Evolution 57:87–103. [20]
Bakker TCM 1999 The study of intersexual selection using quantitative genetics. Behaviour 136:1237–1265. [22]
Baldwin JM 1896 A new factor in evolution. Am. Nat. 30:441–451. [21]
Baldwin JM 1902 Development and Evolution. Macmillan. [21]
Ballou J & RC Lacy 1995 Identifying genetically important individuals for management of genetic diversity in pedigreed populations. pp. 76–111 in Ballou J, Gilpin M & T Foose, eds. Population Management for Survival

and Recovery: Analytical Methods and Strategies in Small Population Conservation. Columbia Univ. Press. [32]

Bamshad M & SP Wooding 2003 Signatures of natural selection in the human genome. Nat. Rev. Genet. 4:99–111. [7]

Bányai L, Váradi A & L Patthy 1983 Common evolutionary origin of the fibrin-binding structures of fibronectin and tissue-type plasminogen activator. FEBS Lett. 163: 37–41. [14]

Barbash DA, Siino DF, Tarone AM & J Roote 2003 A rapidly evolving MYB-related protein causes species isolation in *Drosophila*. Proc. Natl. Acad. Sci. USA 100:5302–5307. [25]

Barlow M & BG Hall 2003 Experimental prediction of the natural evolution of antibiotic resistance. Genetics 163:1237–1241. [31]

Barraclough TG & AP Vogler 2000 Detecting the pattern of speciation from species-level phylogenies. Amer. Nat. 155:419–434. [Species Concepts box]

Barton NH & B Charlesworth 1984 Genetic revolutions, founder events, and speciation. Annu. Rev. Ecol. Syst. 15:133–164. [27]

Barton NH & GM Hewitt 1985 Analysis of hybrid zones. Annu. Rev. Ecol. Syst. 16:113–148. [26]

Barton NH & PD Keightley 2002 Understanding quantitative genetic variation. Nat. Rev. Genet. 3:11–21. [2]

Barton NH & M Turelli 1987 Adaptive landscapes, genetic distance, and the evolution of quantitative characters. Genet. Res. 49:157–174. [20]

Barton NH & M Turelli 1989 Evolutionary quantitative genetics: How little do we know? Annu. Rev. Genet. 23:337–370. [20]

Barton NH & M Turelli 2004 Effects of genetic drift on variance components under a general model of epistasis. Evolution 58:2111–2132. [6]

Bastolla U, Porto M, Roman HE & M Vendruscolo 2003 Connectivity of neutral networks, overdispersion, and structural conservation in protein evolution. J. Mol. Evol. 56:243–254. [8]

Bateson W 1894 Materials for the Study of Variation Treated with Special Regard to Discontinuity in the Origin of Species. Macmillan. [15]

Baum DA & KL Shaw 1995 Genealogical perspectives on the species problem. pp. 289–303 in Hoch PC & AG Stephenson, eds. Experimental and Molecular Approaches to Plant Biosystematics. Missouri Botanical Garden, St. Louis, Missouri (Monographs in Systematic Botany from the Missouri Botanical Garden; 53). [Species Concepts box]

Beatty J 1987a Dobzhansky and drift: facts, values, and chance in evolutionary biology. pp. 271–311 in Kruger L, Gigerenzer G & M Morgan, eds. The Probabilistic Revolution, vol. 2. MIT Press. [1]

Beatty J 1987b Weighing the risks: stalemate in the classical/balance controversy. J. Hist. Biol. 20:289–319. [1]

Beaumont MA & DJ Balding 2004 Identifying adaptive genetic divergence among populations from genome scans. Mol. Ecol. 13:969–980. [7]

Beaumont MA & RA Nichols 1996 Evaluating loci for use in the genetic analysis of population structure. Proc. R. Soc. Lond. B 263:1619–1626. [7]

Beaumont MA, Zhang W & DJ Balding 2002 Approximate Bayesian computation in population genetics. Genetics 162:2025–2035. [12]

Beavis W 1994 The power and deceit of QTL experiments: lessons from comparative QTL studies. pp. 250–266 in DB Wilkinson, ed. 49th Annual Corn and Sorghum Research Conference. American Seed Trade Association. [19]

Beerli P & J Felsenstein 2001 Maximum likelihood estimation of a migration matrix and effective population sizes in *n* subpopulations by using a coalescent approach. Proc. Natl. Acad. Sci. USA 98:4563–4568. [12]

Begun DJ & CF Aquadro 1991 Molecular population-genetics of the distal portion of the X-chromosome in *Drosophila:* evidence for genetic hitchhiking of the *yellow-achaete* region. Genetics 129:1147–1158. [7]

Begun DJ & CF Aquadro 1992 Levels of naturally occurring DNA polymorphism correlate with recombination rates in *D. melanogaster*. Nature 356:519–520. [7]

Begun DJ & P Whitley 2000 Reduced X-linked nucleotide polymorphism in *Drosophila simulans*. Proc. Natl. Acad. Sci. USA 97:5960–5965. [7]

Beheregaray LB, Sunnucks P, Alpers DL, Banks SC & AC Taylor 2000 A set of microsatellite loci for the hairy-nosed wombats (*Lasiorhinus krefftii* and *L. latifrons*). Conserv. Genet. 1:89–92. [32]

Beldade P, Brakefield PM & AD Long 2002 Contribution of Distal-less to quantitative variation in butterfly eyespots. Nature 415:315–318. [20]

Bell AE & MJ Burris 1973 Simultaneous selection for two correlated traits in *Tribolium*. Genet. Res. 21:29–46. [18]

Bell G 1993 Pathogen evolution within host individuals as a primary cause of senescence. Genetica 91:21–34. [30]

Bell G 1997 Selection: The Mechanism of Evolution. Chapman & Hall. [31]

Belyaev DK 1979 Destabilizing selection as a factor in domestication. J. Hered. 70:301–308. [17]

Belyaev DK, Ruvinsky AO & PM Borodin 1981a Inheritance of alternative states of the fused gene in mice. J. Hered. 72:107–112. [17]

Belyaev DK, Ruvinsky AO & LN Trut 1981b Inherited activation–inactivation of the star gene in foxes: its bearing on the problem of domestication. J. Hered. 72:267–274. [17]

Bennett MD 1971 The duration of meiosis. Proc. R. Soc. Lond. B 178:259–275. [10]

Bennett MD, Leitch IJ & L Hanson 1998 DNA amounts in two samples of angiosperm weeds. Ann. Bot. 82:121–134. [10]

Bennetzen JL & EA Kellogg 1997 Do plants have a one-way ticket to genomic obesity? Plant Cell 9:1509–1514. [10]

Bensasson D, Zhang D, Hartl DL, Hewitt GM 2001 Mitochondrial pseudogenes: evolution's misplaced witnesses. Trends Ecol. Evol. 16:314–322. [10]

Benton TG & A Grant 2000 Evolutionary fitness in ecology: comparing measures of fitness in stochastic, density-dependent environments. Evol. Ecol. Res. 2:769–789. [4.1]

Bergelson J & CB Purrington 1996 Surveying patterns in the cost of resistance in plants. Am. Nat. 148:536–558. [29]

Bergman A & ML Siegal 2003 Evolutionary capacitance as a general feature of complex gene networks. Nature 424:549–552. [2, 16]

Bergman CM & M Kreitman 2001 Analysis of conserved noncoding DNA in *Drosophila* reveals similar constraints in intergenic and intronic sequences. Genome Res. 11:1335–1345. [8]

Bergman CM, Pfeiffer BD, Rincón-Limas DE, Hoskins RA, Gnirke A, et al. 2002 Assessing the impact of comparative genomic sequence data on the functional annotation of the *Drosophila* genome. Genome Biol. 3:research 0086.1–0086.20. [8]

Bernardi G, Olofsson B, Filipski J, Zerial M, Salinas J, et al. 1985 The mosaic genome of warm-blooded vertebrates. Science 228:953–958. [9]

Berry A & M Kreitman 1993 Molecular analysis of an allozyme cline: alcohol dehydrogenase in *Drosophila melanogaster* on the east coast of North America. Genetics 134:869–893. [7]

Berry AJ, Ajioka JW & M Kreitman 1991 Lack of polymorphism on the *Drosophila* fourth chromosome resulting from selection. Genetics 129:1111–1117. [7]

Bielawski JP & Z Yang 2003 Maximum likelihood methods for detecting adaptive evolution after gene duplication. J. Struct. Funct. Genomics 3:201–212. [11]

Bierne N & A Eyre-Walker 2003 The problem of counting sites in the estimation of the synonymous and nonsynonymous substitution rates: implications for the correlation between the synonymous substitution rate and codon usage bias. Genetics 165: 1587–1597. [9]

Bierne N & A Eyre-Walker 2004 The genomic rate of adaptive amino acid substitution in *Drosophila*. Mol. Biol. Evol. 21: 1350–1360. [8]

Bird A 1995a Gene number, methylation and biological complexity [reply]. Trends Genet. 21:384. [2]

Bird A 1995b Gene number, noise-reduction and biological complexity. Trends Genet. 11:94–100. [2]

Bird A 2002 DNA methylation patterns and epigenetic memory. Genes Dev. 16: 6–21. [17]

Birky CW & JB Walsh 1988 Effects of linkage on rates of molecular evolution. Proc. Natl. Acad. Sci. USA 85:6414–6418. [7]

Blair WF 1955 Mating call and stage of speciation in the *Microhyla olivacea–M. carolinensis* complex. Evolution 9:469–480. [26]

Blows MW 2002 Interaction between natural and sexual selection during the evolution of mate recognition. Proc. R. Soc. Lond. B 269:1113–1118. [22]

Blows MW, Brooks R & PG Kraft 2003 Exploring complex fitness surfaces: multiple ornamentation and polymorphism in male guppies. Evolution 57:1622–1630. [22]

Blows MW, Chenoweth SF & E Hine 2004 Orientation of the genetic variance–covariance matrix and the fitness surface for multiple male sexually selected traits. Am. Nat. 163:329–340. [22]

Blundell TL & SP Wood 1975 Is the evolution of insulin Darwinian or due to selectively neutral mutation? Nature 257:197–203. [13]

Boag PT 1983 The heritability of external morphology in Darwin's ground finches (*Geospiza*) on Isla Daphne Major, Galapagos. Evolution 37:877–894. [18]

Boake, CRB, Arnold SJ, Breden F, Meffert LM, Ritchie MJ, Taylor B, Wolf JB & AJ Moore 2002 Genetic tools for studying adaptation and the evolution of behavior. Am. Nat. 160:S143–S159. [22]

Bogan AA & KS Thorn 1998 Anatomy of hot spots in protein interfaces. J. Mol. Biol. 280:1–9. [13]

Bohannan BJM & RE Lenski 2000 Linking genetic change to community evolution: insights from studies of bacteria and bacteriophage. Ecol. Lett. 3:362–377. [31.1]

Bohren BB, Hill WG & A Robertson 1966 Some observation on asymmetrical correlated responses to selection. Genet. Res. 7: 44–57. [27]

Bonduriansky R & CE Brassil 2002 Senescence: rapid and costly ageing in wild male flies. Nature 420:377. [30]
Bone JF & MI Kuroda 1996 Dosage compensation regulatory proteins and the evolution of sex chromosomes in *Drosophila*. Genetics 144:705–713. [30.2]
Bonner JT 1982 Evaluation and Development. Springer. [15]
Borash DJ, Gibbs AG, Joshi A & LD Mueller 1998 A genetic polymorphism maintained by natural selection in a temporally varying environment. Am. Nat. 151:148–156. [31]
Borevitz JO, Liang D, Plouffe D, Chang H-S, Zhu T, Weigel D, Berry CC, Winzeler E & J Chory 2003 Large-scale identification of single-feature polymorphisms in complex genomes. Genome Res. 13:513–523. [10]
Bork P, Schultz J & CP Ponting 1997 Cytoplasmic signalling domains: the next generation. Trends Biochem. Sci. 22:296–298. [14]
Bowler PJ 1983 The Eclipse of Darwinism: Anti-Darwinian Theories in the Decades around 1900. Johns Hopkins Univ. Press. [1]
Boyd R & PJ Richerson 1985 Culture and the Evolutionary Process. Univ. Chicago Press. [6]
Bradshaw AD 1965 Evolutionary significance of phenotypic plasticity in plants. Adv. Genet. 13:115–155. [21]
Bradshaw WE & CM Holzapfel 2001 Phenotypic evolution and the genetic architecture underlying photoperiodic time measurement. J. Insect Physiol. 47:809–820. [18]
Brakefield PM 2001 Structure of a character and the evolution of butterfly eyespot patterns. J. Exp. Zool. 291:93–104. [20]
Bremer K 1988 The limits of amino acid sequence data in angiosperm phylogenetic reconstruction. Evolution 42:795–803. [28]
Brenner SE, Chothia C & TJP Hubbard 1997 Population statistics of protein structures: lessons from structural classifications. Curr. Opin. Struct. Biol. 7:369–376. [14]
Breuker CJ & PM Brakefield 2002 Female choice depends on size but not symmetry of dorsal eyespots in the butterfly *Bicyclus anynana*. Proc. R. Soc. Lond. B 269:1233–1239. [20]
Britten RJ & EH Davidson 1969 Gene regulation for higher cells: a theory. Science 165: 349–357. [15]
Brodie ED III, Moore AJ & FJ Janzen 1995 Visualizing and quantifying natural selection. Trends Ecol. Evol. 10:313–318. [18.1, 22]
Bromham L & D Penny 2003 The modern molecular clock. Nat. Rev. Genet. 4:216–224. [8]
Bromham LD, Rambaut A, Hendy MD & D Penny 2000 The power of relative rate tests depends on the data. J. Mol. Evol. 50:296–301. [8]
Brommer JE 2000 The evolution of fitness in life history theory. Biol. Rev. 75:377–404. [4.1]
Bronikowski AM, Alberts SC, Altmann J, Packer C, Carey KD & M Tatar 2002 The aging baboon: comparative demography in a non-human primate. Proc. Natl. Acad. Sci. USA 99:9591–9595. [30]
Bronikowski AM, Carter PA, Morgan TJ, Garland T Jr, Ung N, Pugh TD, Weindruch R & TA Prolla 2003 Lifelong voluntary exercise in the mouse prevents age-related alterations in gene expression in the heart. Physiol. Genomics 12:129–138. [30]
Brook BW, Tonkyn DW, O'Grady JJ & R Frankham 2002 Contribution of inbreeding to extinction risk in threatened species. Conserv. Ecol. 6:16. [online] URL: http://www.consecol.org/vol6/iss1/art16/ [32]
Brooks DR 1988 Macroevolutionary comparisons of host and parasite phylogenies. Annu. Rev. Ecol. Syst. 19:235–259. [29]
Brooks R 2002 Variation in female mate choice within guppy populations: population divergence, multiple ornaments and the maintenance of polymorphisms. Genetica 116:343–358. [22]
Brosius J & SJ Gould 1992 On genomenclature: a comprehensive (and respectful) taxonomy for pseudogenes and other junk DNA. Proc. Natl. Acad. Sci. USA 89:10706–10710. [2]
Brown JKM 1994 Probabilities of evolutionary trees. Syst. Biol. 43:78–91. [12]
Brown JL 1987 Helping and Communal Breeding in Birds. Princeton Univ. Press. [23]
Brunet J & C Mundt 2000 Disease, frequency-dependent selection, and genetic polymorphism: experiments with stripe rust and wheat. Evolution 54:406–415. [29]
Bryant EH & LM Meffert 1988 Effect of an experimental bottleneck on morphological integration in the housefly. Evolution 42:698–707. [27]
Bryant EH & LM Meffert 1995 An analysis of selectional response in relation to a population bottleneck. Evolution 49:626–634. [27]
Bryant EH, Combs LM & SA McCommas 1986 Morphometric differentiation among experimental lines of the housefly in relation to a bottleneck. Genetics 114:1213–1223. [27]
Bucher G, Scholten J & M Klingler 2002 Parental RNAi in *Tribolium* (Coleoptera). Curr. Biol. 12:R85–R86. [15]
Bull JJ & HA Wichman 2001 Applied evolution. Annu. Rev. Ecol. Syst. 32:183–217. [31]
Bulmer MG 1985 The Mathematical Theory of Quantitative Genetics. Clarendon Press. [18]
Bulmer MG 1991 The selection–mutation–drift theory of synonymous codon usage. Genetics 129:897–907. [9]
Burdon JJ 1987 Diseases and Plant Population Biology. Cambridge Univ. Press. [29]

Burdon JJ 1994 The distribution and origin of genes for race-specific resistance to *Melampsora lini* in *Linum marginale*. Evolution 48:1564–1575. [29]

Burke JM & ML Arnold 2001 Genetics and the fitness of hybrids. Annu. Rev. Genet. 35:31–52. [26]

Burke JM, Carney SE & ML Arnold 1998 Hybrid fitness in the Louisiana irises: evidence from experimental analyses. Evolution 52:37–43. [26]

Bushman F 2002 Lateral DNA Transfer. Cold Spring Harbor Press. [12]

Buss LW 1987 The Evolution of Individuality. Princeton Univ. Press. [2.1, 23]

Byers DL & DM Waller 1999 Do plant populations purge their genetic load? Effects of population size and mating history on inbreeding depression. Annu. Rev. Ecol. Syst. 30:479–513. [32]

Cabot EL, Davis AW, Johnson NA & C-I Wu 1994 Genetics of reproductive isolation in the *Drosophila simulans* clade: complex epistasis underlying hybrid male sterility. Genetics 137:175–189. [25]

Caicedo AL, Schaal BA & BN Kunkel 1999 Diversity and molecular evolution of the RPS2 resistance gene in *Arabidopsis thaliana*. Proc. Natl. Acad. Sci. USA 96:302–306. [29]

Caicedo AL, Stinchcombe JR, Olsen KM, Schmitt J & MD Purugganan 2004 Epistatic interaction between *Arabidopsis FRI* and *FLC* flowering time genes generates a latitudinal cline in a life history trait. Proc. Natl. Acad. Sci. USA 101:15670–15675 [29.1]

Cain J 1993 Common problems and cooperative solutions: organizational activity in evolutionary studies. Isis 84:1–25. [1]

Cairns J, Overbaugh J & S Miller 1988 The origin of mutants. Nature 335:142–145. [3]

Campbell DR & NM Waser 2001 Genotype by environment interaction and the fitness of plant hybrids in the wild. Evolution 55:669–676. [26]

Carlini DB 2004 Experimental reduction of codon bias in the *Drosophila* alcohol dehydrogenase gene results in decreased ethanol tolerance of adult flies. J. Evol. Biol. 17:779–785. [9]

Carlini DB & W Stephan 2003 In vivo introduction of unpreferred synonymous codons into the *Drosophila Adh* gene results in reduced levels of ADH protein. Genetics 163:239–243. [9]

Carrière Y & DA Roff 1995 Change in genetic architecture resulting from the evolution of insecticide resistance: a theoretical and empirical analysis. Heredity 75:618–629. [18]

Carrière Y, Deland J-P, Roff DA & C Vincent 1994 Life-history costs associated with the evolution of insecticide resistance. Proc. R. Soc. Lond. B 258:35–40. [18]

Carrière Y, Roff DA & J-P Deland 1995 The joint evolution of diapause and insecticide resistance: a test of an optimality model. Ecology 76:35–40. [18]

Carroll SB, Grenier JK & SD Weatherbee 2001 From DNA to Diversity: Molecular Genetics and the Evolution of Animal Design. Blackwell Sci. [15]

Carson HL 1982 Speciation as a major reorganization of polygenic balances. pp. 411–433 in C Barigozzi, ed. Mechanisms of Speciation. Alan R. Liss. [27]

Carson HL 1990 Increased genetic variation after a bottleneck. Trends Ecol. Evol. 5:228–230. [27]

Carson HL, Hardy DE, Spieth HT & WS Stone 1970 The evolutionary history of the Hawaiian Drosphilidae. pp. 437–543 in Hecht MK & WC Steere, eds. Evolution and Genetics in Honor of Theodosius Dobzhansky. Appleton Century Crofts. [27]

Carvalho AB & AG Clark 1999 Intron size and natural selection. Nature 401:344.

Casadesús J & R D'Ari 2002 Memory in bacteria and phage. BioEssays 24:512–518. [17]

Case AL, Lacey EP & RG Hopkins 1996 Parental effects in *Plantago lanceolata* L. II. Manipulation of grandparental temperature and parental flowering time. Heredity 76:287–295. [21]

Castillo-Davis CI, Mekhedov SL, Hartl DL, Koonin EV & FA Kondrashov 2002 Selection for short introns in highly expressed genes. Nat. Genet. 31:415–418. [9]

Castle W 1905 Heredity of Coat Characters in Guinea-Pigs and Rabbits. Carnegie Institution of Washington Publication No. 23. [1]

Castresana J & D Moreira 1999 Respiratory chains in the last common ancestor of living organisms. J. Mol. Evol. 49:453–460. [8]

Caswell H 2001 Matrix Population Models, 2nd ed. Sinauer Assoc. [4.1]

Cavalier-Smith T 1978 Nuclear volume control by nucleoskeletal DNA, selection for cell volume and cell growth rate, and the solution of the DNA C-value paradox. J. Cell Sci. 34:247–278. [10]

Cavalier-Smith T 1985 The Evolution of Genome Size. Wiley. [10]

Cavalier-Smith T 2004 The membranome and membrane heredity in development and evolution. pp. 335–351 in Hirt RP & DS Horner, eds. Organelles, Genomes and Eukaryote Phylogeny: An Evolutionary Synthesis in the Age of Genomics. CRC Press. [17]

Cavalli-Sforza L 1966 Population structure and human evolution. Proc. R. Soc. Lond. B 164:362–379. [7]

Cavalli-Sforza L & M Feldman 1973 Cultural versus biological inheritance: phenotypic transmission from parents to children. Am. J. Hum. Genet. 25:618–637. [6]
Ceoighe C 2003 Turning the clock back on ancient genome duplication. Curr. Opin. Genet. Dev. 13:636–643. [10]
Chai C 1956 Analysis of quantitative inheritance of body size in mice. II. Gene action and segregation. Genetics 41:167–178. [19]
Chapman T, Arnqvist G, Bangham J & L Rowe 2003 Sexual conflict. Trends Ecol. Evol. 18:41–47. [22]
Chapman T, Liddle LF, Kalb JM, Wolfner MF & L Partridge 1995 Cost of mating in *Drosophila melanogaster* females is mediated by male accessory-gland products. Nature 373:241–244. [30, 31]
Charlesworth B 1979 Evidence against Fisher's theory of dominance. Nature 278:848–849. [2]
Charlesworth B 1990 Optimization models, quantitative genetics, and mutation. Evolution 44:520–538. [18]
Charlesworth B 1994 Evolution in Age-Structured Populations. Cambridge Univ. Press. [4.1] [30]
Charlesworth B 1998 The effect of synergistic epistasis on the inbreeding load. Genet. Res. 71:85–89. [27]
Charlesworth B 2003 The organization and evolution of the human Y chromosome. Genome Biol. 4:226. [2]
Charlesworth B & KA Hughes 1996 Age-specific inbreeding depression and components of genetic variance in relation to the evolution of senescence. Proc. Natl. Acad. Sci. USA 93:6140–6145. [30]
Charlesworth B & CH Langley 1986 The evolution of self-regulated transposition of transposable elements. Genetics 112:359–383. [2]
Charlesworth B, Coyne JA & NH Barton 1987 The relative rates of evolution of sex chromosomes and autosomes. Am. Nat. 130:113–146. [24]
Charlesworth B, Lande R & M Slatkin 1982 A neo-Darwinian commentary on macroevolution. Evolution 36:474–498. [3]
Charlesworth B, Morgan MT & D Charlesworth 1993 The effect of deleterious mutations on neutral molecular variation. Genetics 134:1289–1303. [7]
Charnov EL 1982 The Theory of Sex Allocation. Princeton Univ. Press. [23]
Chelliah V, Chen L, Blundell TL & SC Lovell 2004 Distinguishing structural and functional restraints in evolution in order to identify interaction sites. J. Mol. Biol. 342:1487–1504. [13]
Chen F-C & W-H Li 2001 Genomic divergences between humans and other hominoids and the effective population size of the common ancestor of humans and chimpanzees. Am. J. Hum. Genet. 68:444–456. [12]
Chen Y & W Stephan 2003 Compensatory evolution of a precursor messenger RNA secondary structure in the *Drosophila melanogaster Adh* gene. Proc. Natl. Acad. Sci. USA 100:11499–11504. [2, 9]
Chen Y, Carlini DB, Baines JF, Parsch J, Braverman JM, et al. 1999 RNA secondary structure and compensatory evolution. Genes Genet. Syst. 74:271–286. [9]
Cheverud JM 1984 Quantitative genetics and developmental constraints on evolution by selection. J. Theor. Biol. 110:155–171. [19, 20]
Cheverud JM & AJ Moore 1994 Quantitative genetics and the role of the environment provided by relatives in the evolution of behavior. pp. 67–100 in CRB Boake, ed. Quantitative Genetic Studies of Behavioral Evolution. Univ. of Chicago Press. [22]
Cheverud JM & EJ Routman 1995 Epistasis and its contribution to genetic variance components. Genetics 139:1455–1461. [19]
Cheverud JM & EJ Routman 1996 Epistasis as a source of increased additive genetic variance at population bottlenecks. Evolution 50:1042–1051. [6.1, 27]
Cheverud JM, Ehrich TH, Kenney JP, Pletscher LS & CF Semenkovich 2004a Genetic evidence for discordance between obesity and diabetes-related traits in the LGXSM recombinant inbred mouse strains. Diabetes 53:2700–2708. [19]
Cheverud JM, Ehrich TE, Vaughn TT, Koreishi SF, Linsey RB & LS Pletscher 2004b Pleiotropic effects on mandibular morphology. II. Differential epistasis and genetic variation in morphological integration. J. Exp. Zool. (Mol. Dev. Evol.) 302B:424–435. [19]
Cheverud JM, Routman EJ, Duarte FA, van Swinderen B, Cothran K & C Perel 1996 Quantitative trait loci for murine growth. Genetics 142:1305–1319. [19]
Cheverud JM, Vaughn TT, Pletscher LS, King-Ellison K, Bailiff J, Adams E, Erickson C & A Bonislawski 1999 Epistasis and the evolution of additive genetic variance in populations that pass through a bottleneck. Evolution 53:1009–1018. [19]
Cheverud JM, Vaughn TT, Pletscher LS, Peripato A, Adams E, Erickson C & K King-Ellison 2001 Genetic architecture of adiposity in the cross of Large (LG/J) and Small (SM/J) inbred mice. Mamm. Genome 12:3–12. [19]
Chippindale AK, Alipaz JA, Chen H-W & MR Rose 1997 Experimental evolution of accelerated development in *Drosophila*. 1. Larval development speed and survival. Evolution 51:1536–1551. [31]

Chippindale AK, Gibson JR & WR Rice 2001 Negative genetic correlation for fitness between sexes reveals ontogenetic conflict in *Drosophila*. Proc. Natl. Acad. Sci. USA 98:1671–1675. [31]

Chippindale AK, Ngo AL & MR Rose 2003 The devil in the details of life history evolution: instability and reversal of genetic correlations during selection on *Drosophila* development. J. Genet. 82:133–145. [31]

Chothia C & AM Lesk 1986 The relation between the divergence of sequence and structure in proteins. EMBO J. 5:823–826. [13]

Cianchi R, Ungaro A, Marini M & L Bullini 2003 Differential patterns of hybridization and introgression between the swallowtails *Papilio machaon* and *P. hospiton* from Sardinia and Corsica islands (Lepidoptera, Papilionidae). Mol. Ecol. 12:1461–1471. [Species Concepts box]

Clackson T & JA Wells 1995 A hot spot of binding energy in a hormone–receptor interface. Science 267:383–386. [13]

Claridge MF, Dawah HA & MR Wilson 1997 Species: The Units of Biodiversity. Chapman & Hall. [Species Concepts box]

Clark AG 2002 Sperm competition and the maintenance of polymorphism. Heredity 88:148–153. [22]

Clark AG, Glanowski S, Nielsen R, Thomas PD, A Kejariwal, et al. 2003 Inferring nonneutral evolution from human–chimp–mouse orthologous gene trios. Science 302:1876–1877. [7]

Clarke B 1970 Darwinian evolution of proteins. Science 168:1009–1011. [1]

Clay K & W van der Putten 1999 Pathogens and plant life histories. pp. 275–301 in Vuorisalo T & P Mutikainen, eds. Life History Evolution in Plants. Kluwer. [29]

Clegg SM, Degnan SM, Moritz C, Estoup A, Kikkawa J & IPF Owens 2002 Microevolution in island forms: the roles of drift and directional selection in morphological divergence of a passerine bird. Evolution 56:2090–2099. [27]

Clutton-Brock TH 1988 Reproductive Success. Studies of Individual Variation in Contrasting Breeding Systems. Univ. of Chicago Press. [4.1]

Clutton-Brock TH, Albon SD & FE Guiness 1984 Maternal dominance, breeding success and birth sex ratios in red deer. Nature 308:358–360. [2.1]

Cockerham CC 1954 An extension of the concept of partitioning heredity variance for analysis of covariance among relatives when epistasis is present. Genetics 39:859–882. [6]

Coen E 2000 The Art of Genes: How Organisms Make Themselves. Oxford Univ. Press. [15]

Colosimo PF, Peichel CL, Nereng K, Blackman BK, Shapiro MD, Schluter D & DM Kingsley 2004 The genetic architecture of parallel armor plate reduction in threespine sticklebacks. PLoS Biol. 2:635–641. [15]

Comeron JM & M Kreitman 2000 The correlation between intron length and recombination in *Drosophila*: dynamic equilibrium between mutational and selective forces. Genetics 156:1175–1190. [9]

Conant GC & A Wagner 2003 Asymmetric sequence divergence of duplicate genes. Genome Res. 13:2052–2058. [11]

Conner JK 2002 Genetic mechanisms of floral trait correlations in a natural population. Nature 420:407–410. [20]

Cooke J 1998 Reply to Toby Gibson and Jurg Spring. Trends Genet. 14:49–50. [2]

Cooke J, Nowak MA, Boerlijst M & J Maynard Smith 1997 Evolutionary origins and maintenance of redundant gene expression during metazoan development. Trends Genet. 13:360–364. [2]

Cooper TF, Rozen DE & RE Lenski 2003 Parallel changes in gene expression after 20,000 generations of evolution in *E. coli*. Proc. Natl. Acad. Sci. USA 100:1072–1077. [31.1]

Cooper VS & RE Lenski 2000 The population genetics of ecological specialization in evolving *E. coli* populations. Nature 407:736–739. [31, 31.1]

Cooper VS, Bennett AF & RE Lenski 2001 Evolution of thermal dependence of growth rate of *Escherichia coli* populations during 20,000 generations in a constant environment. Evolution 55: 889–896. [31]

Copley RR, Schultz J, Ponting CP & P Bork 1999 Protein families in multicellular organisms. Curr. Opin. Struct. Biol. 9:408–415. [14]

Corey DR & JM Abrams 2001 Morpholino antisense oligonucleotides: tools for investigating vertebrate development. Genome Biol. 2:reviews1015.1–reviews1015.3. [15]

Coulson AF & J Moult 2002 A unifold, mesofold, and superfold model of protein fold use. Proteins 46:61–71. [14]

Cowen LE, Nantel A, Whiteway MS, Thomas DY, Tessier DC, Kohn LM & JB Anderson 2002 Population genomics of drug resistance in *Candida albicans*. Proc. Natl. Acad. Sci. USA 99:9284–9289. [31]

Coyne JA 1994 Mayr and the origin of species. Evolution 48:19–30. [24]

Coyne JA & HA Orr 1989 Patterns of speciation in *Drosophila*. Evolution 43:362–381. [24, 25]

Coyne JA & HA Orr 1997 "Patterns of speciation in *Drosophila*" revisited. Evolution 51:295–303. [24, 25]

Coyne JA & HA Orr 1998 The evolutionary genetics of speciation. Phil. Trans. R. Soc. Lond. B 28:287–305. [25]

Coyne JA & HA Orr 2004 Speciation. Sinauer Assoc. [Species Concepts box, 24, 25]
Coyne JA, Barton NH & M Turelli 1997 Perspective: A critique of Sewall Wright's shifting balance theory of evolution. Evolution 51:643–671. [6]
Coyne JA, Barton NH & M Turelli 2000 Is Wright's shifting balance process important in evolution? Evolution 54:306–317. [6]
Cracraft J 1989 Speciation and its ontology: the empirical consequences of alternative species concepts for understanding patterns and processes of differentiation. pp. 28–59 in Otte D & JA Endler, eds. Speciation and its Consequences. Sinauer Assoc. [Species Concepts box]
Craig DM 1982 Group selection versus individual selection: an experimental analysis. Evolution. 36:271–282. [6]
Crandall K, Harris D & J Fetzner 2000a The monophyletic origin of freshwater crayfish estimated from nuclear and mitochondrial DNA sequences. Proc. R. Soc. Lond. B 267:1679–1686. [28]
Crandall KA, Bininda-Edmonds ORP, Mace GM & RK Wayne 2000b Considering evolutionary processes in conservation biology: an alternative to "evolutionary significant units." Trends Ecol. Evol. 15:290–295. [32]
Crawford DL, Segal JA & JL Barnett 1999 Evolutionary analysis of TATA-less proximal promoter function. Mol. Biol Evol. 16:194–207. [2]
Cresko WA, Amores A, Wilson C, Murphy J, Currey M, Phillips P, Bell MA, Kimmel CB & JH Postlethwait 2004 Parallel genetic basis for repeated evolution of armor loss in Alaskan threespine stickleback populations. Proc. Natl. Acad. Sci. USA 101:6050–6055. [15]
Crick FH 1953 The packing of alpha-helices: simple coiled coils. Acta Crystallogr. 6:689–697. [13]
Crick FH 1970 Central dogma of molecular biology. Nature 227:561–563. [13]
Crick FH 1988 What Mad Pursuit: A Personal View of Science. Basic Books. [13]
Crnokrak P & DA Roff 1999 Inbreeding depression in the wild. Heredity 83:260–270. [32]
Crouse HV 1960 The controlling element in sex chromosome behavior in *Sciara*. Genetics 45:1429–1443. [17]
Crow JF 1958 Some possibilities for measuring selection intensities in man. Hum. Biol. 30:1–13. [4]
Crow JF 1962 Population genetics: selection. pp. 53–75 in Burdett WJ, ed. Methodology in Human Genetics. Holden-Day. [4]
Crow JF 1972 Darwinian and non-Darwinian evolution. pp. 1–22 in LeCam L, et al., eds. Proceedings of the Sixth Berkeley Symposium on Mathematical Statistics. Vol. V: Darwinian, Neo-Darwinian, and Non-Darwinian Evolution. Univ. of California Press. [1]
Crow JF & M Kimura 1970 An Introduction to Population Genetics Theory. Harper & Row. [6, 18]
Cruzan MB & ML Arnold 1993 Ecological and genetic associations in an Iris hybrid zone. Evolution 47:1432–1445. [26]
Cubas P, Vincent C & E Coen 1999 An epigenetic mutation responsible for natural variation in floral symmetry. Nature 401:157–161. [17]
Cunningham EJA & AF Russell 2000 Egg investment is influenced by male attractiveness in the mallard. Nature 404:74–77. [2.1]
Curtsinger JW, Service PM & T Prout 1994 Antagonistic pleiotropy reversal of dominance and genetic polymorphism. Am. Nat. 144:210–228. [18]
Cutler DJ 2000 Understanding the overdispersed molecular clock. Genetics 154:1403–1417. [8]
Cwynar LC & GM MacDonald 1987 Geographical variation of lodgepole pine in relation to population history. Am. Nat. 129:463–469. [27]
Czesak ME & CW Fox 2003 Evolutionary ecology of egg size and number in a seed beetle: genetic trade-off differs between environments. Evolution 57:1121–1132. [21]
Daborn PJ, Yen JL, Bogwitz MR, Le Goff G, Feil E, et al. 2002 A single p450 allele associated with insecticide resistance in *Drosophila*. Science 297:2253–2256. [7]
Darbishire AD 1904 On the result of crossing Japanese waltzing with albino mice. Biometrica 3:1–51. [1]
Darlu P & G Lecointre 2002 When does the incongruence length difference test fail? Mol. Biol. Evol. 19:432–437. [28]
Darvasi A & M Soller 1995 Advanced intercross lines, an experimental population for fine genetic mapping. Genetics 141:1199–1207. [19]
Darwin CR 1859 On the Origin of Species by Means of Natural Selection or the Preservation of Favoured Races in the Struggle for Life. John Murray. [Species Concepts box, 26]
Darwin CR 1871 The Descent of Man and Selection in Relation to Sex. John Murray. [22]
David JR, Moreteau B, Gauthier JR, Pétavy G, Stockel J & A Imasheva 1994 Reaction norms of size characters in relation to growth temperature in *Drosophila melanogaster*: an isofemale lines analysis. Genet. Select Evol. 26:229–251. [21]
Davidson AR & RT Sauer 1994 Folded proteins occur frequently in libraries of random amino acid sequences. Proc. Natl. Acad. Sci. USA 91:2146–2150. [14]
Davidson CJ, Hirt RP, Lal K, Snell P, Elgar G, Tuddenham EG & JH McVey 2003a

Molecular evolution of the vertebrate blood coagulation network. Thromb. Haemost. 89:420–428. [14]
Davidson EH 2001 Genomic Regulatory Systems. Academic Press. [15, 16]
Davidson EH, McClay DR & L Hood 2003b Regulatory gene networks and the properties of the developmental process. Proc. Natl. Acad. Sci. USA 100:1475–1480. [13.1]
Day SB, Bryant EH & LM Meffert 2003 The influence of variable rates of inbreeding on fitness, environmental responsiveness, and evolutionary potential. Evolution 57:1314–1324. [27]
de Brito RA, Pletscher LL & JM Cheverud 2005 The evolution of genetic architecture. I. Diversification of genetic backgrounds by genetic drift. Evolution 59:2333–2342. [19]
de Chateau M & L Bjorck 1996 Identification of interdomain sequences promoting the intronless evolution of a bacterial protein family. Proc. Natl. Acad. Sci. USA 93:8490–8495. [14]
de Jong G 1999 Unpredictable selection in a structured population leads to local genetic differentiation in evolved reaction norms. J. Evol. Biol. 12:839–851. [21]
De Luca M, Roshina NV, Geiger-Thornsberry GL, Lyman RF, Pasyukova EG & TF Mackay 2003 Dopa decarboxylase (Ddc) affects variation in *Drosophila* longevity. Nat. Genet. 34:429–433. [30]
de Pontbriand A, Wang X-P, Cavaloc Y, Mattei M-G & F Galibert 2002 Synteny comparison between apes and human using fine-mapping of the genome. Genomics 80:395–401. [2]
de Queiroz K 1998 The general lineage concept of species, species criteria, and the process of speciation. A conceptual unification and terminological recommendations. pp. 57–75 in Howard DJ & SH Berlocher, eds. Endless Forms. Species and Speciation. Oxford Univ. Press. [Species Concepts box]
Dean AM & GB Golding 1997 Protein engineering reveals ancient adaptive replacements in isocitrate dehydrogenase. Proc. Natl. Acad. Sci. USA 94:3104–3109. [11]
Debat V & P David 2001 Mapping phenotypes: canalization, plasticity and developmental stability. Trends Ecol. Evol. 16:555–561. [16]
Degnan JH & LA Salter 2005 Gene tree distributions under the coalescent process. Evolution 59:24–37. [12]
Delneri D, Colson I, Grammenoudi S, Roberts IN, Louis EJ & SG Oliver 2003 Engineering evolution to study speciation in yeasts. Nature 422:68–72. [25]
Delon I, Chanut-Delalande H & F Payre 2003 The Ovo/Shavenbaby transcription factor specifies actin remodelling during epidermal differentiation in *Drosophila*. Mech. Dev. 120:747–758. [15]
Demas GE, Chefer V, Talan MI & RJ Nelson 1997 Metabolic costs of mounting an antigen-stimulated immune response in adult and aged C57BL/6J mice. Am. J. Physiol. 273:1631–1637. [18]
Demauro MM 1993 Relationship of breeding system to rarity in the Lakeside daisy (*Hymenoxys acaulis* var. glabra). Conserv. Biol. 7:542–550. [32]
Denver DR, Morris K, Lynch M & WK Thomas 2004 High mutation rate and predominance of insertions in the *Caenorhabditis elegans* nuclear genome. Nature 430:679–682. [3, 10]
Dermitzakis ET, Reymond A, Lyle R, Scamuffa N, et al. 2002 Numerous potentially functional but non-genic conserved sequences on human chromosome 21. Nature 420:578–582. [3]
Derrida B, Manrubia SC & DH Zanette 2000 On the genealogy of a population of biparental individuals. J. Theor. Biol. 203:303–315. [12]
DeSalle R, Giribet G & WC Wheeler 2002 Molecular Systematics and Evolution: Theory and Practice. Birkhauser. [11]
Desdevises Y, Morand S, Jousson O & P Legendre 2002 Coevolution between *Lamellodiscus* (Monogea: diplectanidae) and Sparidae (Teleostei): the study of a complex host–parasite system. Evolution 56:2459–2471. [29]
Deutsch M & M Long 1999 Intron–exon structure of eukaryotic model organisms. Nucleic. Acids Res. 27:3219–3228. [10]
Devonshire AL & LM Field 1991 Gene amplification and insecticide resistance. Annu. Rev. Entomol. 36:1–23. [14]
Devos KM, Brown JK & JL Bennetzen 2002 Genome size reduction through illegitimate recombination counteracts genome expansion in *Arabidopsis*. Genome Res. 12:1075–1079. [10]
DeWitt TJ 1998 Costs and limits of phenotypic plasticity: tests with predator-induced morphology and life history in a freshwater snail. J. Evol. Biol. 11:465–480. [21]
DeWitt TJ & SM Scheiner 2004 Phenotypic Plasticity. Functional and Conceptual Approaches. Oxford Univ. Press. [21]
DeWitt TJ, Sih A & DS Wilson 1998 Costs and limits of phenotypic plasticity. Trends Ecol. Evol. 13:77–81. [21]
Dickerson RE 1983 Hemoglobin: structure function, pathology, and evolution. Benjamin/Cummings. [2]
Dietrich MR 1994 The origins of the neutral theory of molecular evolution. J. Hist. Biol. 27:21–59. [1]
Dietrich MR 1995 Richard Goldschmidt's "heresies" and the evolutionary synthesis. J. Hist. Biol. 28:431–461. [1]

Dietrich MR 1998 Paradox and persuasion: negotiating the place of molecular evolution within evolutionary biology. J. Hist. Biol. 31:85–111. [1]
Dilda CL & TFC Mackay 2002 The genetic architecture of *Drosophila* sensory bristle number. Genetics 162:1655–1674. [2]
Dixon SA, Coyne JA & MAF Noor 2003 The evolution of conspecific sperm precedence in *Drosophila*. Mol. Ecol. 12:1179–1184. [24]
Djawdan M, Chippindale AK, Rose MR & TJ Bradley 1998 Metabolic reserves and evolved stress resistance in *Drosophila melanogaster*. Phys. Zool. 71:584–594. [31]
Dobson CM 2002 Getting out of shape: protein misfolding diseases. Nature 418:729–730. [13]
Dobzhansky T 1933 On the sterility of the interracial hybrids in *Drosophila pseudoobscura*. Proc. Natl. Acad. Sci. USA 19:397–403. [25]
Dobzhansky T 1936 Studies on hybrid sterility. II. Localization of sterility factors in *Drosophila virilis* Sturt. X *lummei* Hackman hybrids. Genetics 21:113–135. [25]
Dobzhansky T 1937 Genetics and the Origin of Species. Columbia Univ. Press. [1, 6, 24, 25, 26]
Dobzhansky T 1940 Speciation as a stage in evolutionary divergence. Am. Nat. 74:312–321. [24, 26]
Dobzhansky T 1955 A review of some fundamental concepts and problems in population genetics. Cold Spring Harbor Symp. Quant. Biol. 20:1–15. [1]
Dobzhansky T 1956 What is an adaptive trait? Am. Nat. 90:337–347. [20]
Dodd DMB & JR Powell 1985 Founder-flush speciation: an update on experimental results with *Drosophila*. Evolution 39:1388–1392. [27]
Doebley J 1992 Mapping the genes that made maize. Trends Genet. 8:302–307. [2]
Doebley J, Stec A & C Gustus 1995 Teosinte *branched1* and the origin of maize: evidence for epistasis and the evolution of dominance. Genetics 141:333–364. [6]
Doebley J, Stec A & L Hubbard 1997 The evolution of apical dominance in maize. Nature 386:485–488. [7]
Doi M, Matsuda M, Tomaru M, Matsubayashi H & Y Oguma 2001 A locus for female discrimination behavior causing sexual isolation in *Drosophila*. Proc. Natl. Acad. Sci. USA 98:6714–6719. [25]
Doiron S, Bernatchez L & PU Blier 2002 A comparative mitogenomic analysis of the potential adaptive value of Arctic Charr mtDNA introgression in Brook Charr populations (*Salvelinus fantinalis* Mitchill). Mol. Biol. Evol. 19:1902–1909. [26]
Donnelly P 1996 Interpreting genetic variability: the effects of shared evolutionary history. pp. 25–50 in Weiss K, ed. Variation in the Human Genome. Wiley. [12]
Donnelly P & S Tavaré 1997 Progress in Population Genetics and Human Evolution. Springer. [12]
Donohue K, Messiqua D, Pyle EH, Heschel MS & J Schmitt 2000 Evidence of adaptive divergence in plasticity: density- and site-dependent selection on shade-avoidance response in *Impatiens capensis*. Evolution 54:1956–1968. [21]
Doolittle RF 1995 The multiplicity of domains in proteins. Annu. Rev. Biochem. 64:287–314. [14]
Doolittle WF & C Sapienza 1980 Selfish genes, the phenotype paradigm and genome evolution. Nature 284:601–603. [10]
Douady CJ, Delsuc F, Boucher Y, Doolittle WF & EJP Douzery 2003 Comparison of Bayesian and maximum likelihood bootstrap measures of phylogenetic reliability. Mol. Biol. Evol. 20:248–254. [28]
Doulatov S, Hodes A, et al. 2004 Tropism switching in *Bordetella* bacteriophage defines a family of diversity-generating retroelements. Nature 431:476–481. [3]
Dover GA, Linares AR, Bowen T & HM Hancock 1993 Detection and quantification of concerted evolution and molecular drive. Methods Enzymol. 224:525–541. [11]
Dowling TE & BD DeMarais 1993 Evolutionary significance of introgressive hybridization in cyprinid fishes. Nature 362:444–446. [26]
Dozmorov I, Galecki A, Chang Y, Krzesicki R, Vergara M & RA Miller 2002 Gene expression profile of long-lived snell dwarf mice. J. Gerontol. A. Biol. Sci. Med. Sci. 57:B99–B108. [30]
Drake JW 1991 A constant rate of spontaneous mutation in DNA-based microbes. Proc. Natl. Acad. Sci. USA 88:7160–7164. [3]
Drake JW, Charlesworth B, Charlesworth D & JF Crow 1998 Rates of spontaneous mutation. Genetics 148:1667–1686. [2, 3]
Dryden IL & KV Mardia 1998 Statistical Shape Analysis. Wiley. [19.4]
Duboule D & AS Wilkins 1998 The evolution of "bricolage." Trends Genet. 14:54–59. [16]
Dunn KA, Bielawski JP & Z Yang 2001 Substitution rates in *Drosophila* nuclear genes: implications for translational selection. Genetics 157:295–305. [9]
Duret L 2001 Why do genes have introns? Recombination might add a new piece to the puzzle. Trends Genet. 17:172–175. [9]
Duret L & D Mouchiroud 1999 Expression pattern and, surprisingly, gene length shape codon usage in *Caenorhabditis*, *Drosophila*,

and *Arabidopsis*. Proc. Natl. Acad. Sci. USA 96:4482–4487. [9]

Durrett R 2002 Probability Models for DNA Sequence Evolution. Springer. [12]

Dyall SD, Brown MT & PJ Johnson 2004 Ancient invasions: from endosymbionts to organelles. Science 304:253–257. [2]

Dybdahl MF & C Lively 1998 Host–parasite interactions: evidence for a rare advantage and time-lagged selection in a natural population. Evolution 52:1057–1066. [29]

Dyson HJ & PE Wright 2002 Coupling of folding and binding for unstructured proteins. Curr. Opin. Struct. Biol. 12:54–60. [13]

Eanes WF, Kirchner M & J Yoon 1993 Evidence for adaptive evolution of the *G6pd* gene in the *Drosophila melanogaster* and *Drosophila simulans* lineages. Proc. Natl. Acad. Sci. USA 90:7475–7479. [7]

Eberhard W 1996 Female Control: Sexual Selection by Cryptic Female Mate Choice. Princeton Univ. Press. [22]

Ebert D & J Bull 2003 Challenging the trade-off model for the evolution of virulence: is virulence management feasible? Trends Microbiol. 11:15–20. [29]

Edney EB & RW Gill 1968 Evolution of senescence and specific longevity. Nature 220:281–282. [30]

Ehrich T, Vaughn TT, Koreishi SF, Linsey RB, Pletscher LS & JM Cheverud 2003 Pleiotropic effects on mandibular morphology. I. Developmental morphological integration and differential dominance. J. Exp. Zool. (Mol. Dev. Evol.) 296B: 58–79. [19]

Ehrlich PR & PH Raven 1969 Differentiation of populations. Science 165:1228–1232. [Species Concepts box]

Ehrman L 1969 Genetic divergence in M. Vetukhiv's experimental populations of *Drosophila pseudoobscura*. 5. A further study of rudiments of sexual isolation. Am. Midland Nat. 82:272–276. [27]

Eigen M & P Schuster 1979 The Hypercycle: A Principle of Natural Self-Organization. Springer. [23]

Eisen JA 2000 Horizontal gene transfer among microbial genomes: new insights from complete genome analysis. Curr. Opin. Genet. Dev. 10:606–611. [12]

Elena SF & RE Lenski 2003 Evolution experiments with microorganisms: the dynamics and genetic bases of adaptation. Nat. Rev. Gen. 4:457–469. [31]

Emlen DJ & HF Nijhout 1999 Hormonal control of male horn length dimorphism in the dung beetle *Onthophagus taursus* (Coleoptera: Scarabaeidae). J. Insect Physiol. 45:45–53. [21]

Emlen ST 1984 Cooperative breeding in birds and mammals. pp. 305–339 in Krebs JR & NB Davies, eds. Behavioural Ecology: An Evolutionary Approach, 2nd ed. Blackwell Scientific. [23]

Endler JA 1977 Geographic Variation, Speciation, and Clines. Princeton Univ. Press. [26]

Endo T, Ikeo K & T Gojobori 1996 Large-scale search for genes on which positive selection may operate. Mol. Biol. Evol. 13:685–690. [29]

Engström G, Liljedahl L-E, Rasmuson M & T Björklund 1989 Expression of genetic and environmental variation during ageing. I. Estimation of variance components for number of adult offspring in *Drosophila melanogaster*. Theor. Appl. Genet. 77:119–122. [30]

Eshel I & C Matessi 1998 Canalization, genetic assimilation and preadaptation: a quantitative genetic model. Genetics 149:2119–2133. [16]

Etges WJ, Johnson WR, Duncan GA, Huckins G & WB Heed 1999 Ecological genetics of cactophilic *Drosophila*. pp. 164–214 in RH Robichaux ed. Ecology of Sonoran Desert Plants and Plant Communities. Univ. of Arizona Press. [30.2]

Ewald P 1983 Host–parasite relations, vectors, and the evolution of disease severity. Annu. Rev. Ecol. Syst. 14:465–485. [29]

Ewald P 1994 The Evolution of Infectious Diseases. Oxford Univ. Press. [29]

Ewens WJ 2004 Mathematical Population Genetics. I. Theoretical Introduction. Springer. [5, 12]

Eyre-Walker AC 1991 An analysis of codon usage in mammals: selection or mutation bias? J. Mol. Evol. 33:442–449. [9]

Fairbairn DJ & J Reeve 2001 Natural selection. pp. 29–43 in Fox CW, Roff DA & DJ Fairbairn, eds. Evolutionary Ecology: Concepts and Case Studies. Oxford Univ. Press. Oxford. [4.1, 18]

Fairbairn DJ & DE Yadlowski 1997 Coevolution of traits determining a migratory tendency: correlated response to a critical enzyme, juvenile hormone esterase, to selection on wing morphology. J. Evol. Biol. 10:495–513. [18]

Falconer DS & TFC Mackay 1996 Introduction to Quantitative Genetics. Longman. [2, 4, 19, 27]

Fang S, Takahashi A & C-I Wu 2002 A mutation in the promoter of desaturase 2 is correlated with sexual isolation between *Drosophila* behavioral races. Genetics 162:781–784. [25]

Farber GK & GA Petsko 1990 The evolution of α/β barrel enzymes. Trends Biochem. Sci. 15:228–234. [13]

Farris JS, Kallersjo M, Kluge AG & C Bult 1994 Testing significance of incongruence. Cladistics 10:315–319. [28]

Farris JS, Kallersjo M, Kluge AG & C Bult 1995 Constructing a significance test for incongruence. Syst. Biol. 44:570–572. [11]

Fay JC & CI Wu 2000 Hitchhiking under positive Darwinian selection. Genetics 155:1405–1413. [7]

Fay JC, Wycoff, GJ & C-I Wu 2002 Testing the neutral theory of molecular evolution with genomic data from *Drosophila*. Nature 415:1024–1026. [8]

Feder JH 1998 The apple maggot fly, *Rhagoletis pomonella*: flies in the face of conventional wisdom about speciation? pp. 130–144 in Howard DJ & SH Berlocher, eds. Endless Forms: Species and Speciation. Oxford Univ. Press. [24]

Fedorka KM & TA Mousseau 2004 Female mating bias results in conflicting sex-specific offspring fitness. Nature 429:65–67. [2.1, 30]

Fedoroff N & D Botstein 1992 The Dynamic Genome: Barbara McClintock's Ideas in the Century of Genetics. Cold Spring Harbor Laboratory Press. [17]

Feller W 1968 An Introduction to Probability Theory and Its Applications. Wiley. [5]

Felsenstein J 1976 The theoretical population genetics of variable selection and migration. Annu. Rev. Genet. 10:253–280. [5]

Felsenstein J 1985 Confidence limits on phylogenies: an approach using the bootstrap. Evolution 39:783–791. [28]

Felsenstein J 1988 Phylogenies from molecular sequences: inference and reliability. Annu. Rev. Genet. 22:521–565. [28]

Felsenstein J 2001 The troubled growth of statistical phylogenetics. Syst. Biol. 50:465–467. [28]

Felsenstein J 2004 Inferring Phylogenies. Sinauer Assoc. [11, 12, 28]

Ferea TL, Botstein D, Brown PO & RF Rosenzweig 1999 Systematic changes in gene expression patterns following adaptive evolution in yeast. Proc. Natl. Acad. Sci. USA 96:9721–9726. [2]

Ferguson HM, Mackinnon MJ, Chan BH & AF Read 2003 Mosquito mortality and the evolution of malaria virulence. Evolution 57:2792–2804. [29]

Fernández Iriarte P, Céspedes W & M Santos 2003 Quantitative-genetic analysis of wing form and bilateral asymmetry in isochromosomal lines of *Drosophila subobscura* using Procrustes methods. J. Genet. 82:95–113. [19.4]

Finch CE 1990 Longevity, Senescence and the Genome. Univ. of Chicago Press. [30]

Finkel T & NJ Holbrook 2000 Oxidants, oxidative stress and the biology of ageing. Nature 408: 239–247. [30]

Fisher RA 1922 On the dominance ratio. Proc. R. Soc. Edinb. 42:321–341. [5]

Fisher RA 1930 The Genetical Theory of Natural Selection. Clarendon Press. [1, 4, 6, 19, 22]

Fisher RA 1958 The Genetical Theory of Natural Selection. Dover. [5]

Fishman L, Kelly AJ & JH Willis 2002 Minor quantitative trait loci underlie floral traits associated with mating system divergence in *Mimulus*. Evolution 56:2138–2155. [26]

Fitch WM 1970 Distinguishing homologous from analogous proteins. Syst. Zool. 19:99–113. [11]

Fitch WM 1971a Rate of change of concomitantly variable codons. J. Mol. Evol. 1:84–96. [8]

Fitch WM 1971b Toward defining the course of evolution: minimal change for a specific tree topology. Syst . Zool. 20:406–416. [11]

Fitzpatrick MJ 2004 Pleiotropy and the genomic location of sexually-selected genes. Am. Nat. 163:800–808. [22]

Flor HH 1947 Host–parasite interactions in flax rust: its genetics and other implications. Phytopathology 45:680–685. [29]

Florin AB & A Ödeen 2002 Laboratory environments are not conducive for allopatric speciation. J. Evol. Biol. 15:10–23. [27]

Force A, Lynch M, Pickett FB, Amores A, Yan Y-L & JH Postlethwaite 1996 Preservation of duplicate genes by complementary degenerative mutations. Genetics 151:1531–1545. [11]

Ford EB 1975 Ecological Genetics. Chapman & Hall. [4]

Ford MJ 2002 Applications of selective neutrality tests to molecular ecology. Mol. Ecol. 11:1245–1262. [7]

Foster PL 2000 Adaptive mutation: implications for evolution. BioEssays 22:1067–1074. [3, 16]

Fox CW & TA Mousseau 1996 Larval host plant affects fitness consequences of egg size variation in the seed beetle *Stator limbatus*. Oecologia 107:541–548. [21]

Fox CW & UM Savalli 2000 Maternal effects mediate diet expansion in a seed-feeding beetle. Ecology 81:3–7. [21]

Fox CW, Czesak ME, Mousseau TA & DA Roff 1999 The evolutionary genetics of an adaptive maternal effect: egg size plasticity in a seed beetle. Evolution 53:552–560. [21]

Fox CW, Thakar MS & TA Mousseau 1997 Egg size plasticity in a seed beetle: an adaptive effect. Am. Nat. 149:149–163. [21]

Fox CW, Waddell KJ & TA Mousseau 1994 Host-associated fitness variation in a seed beetle (Coleoptera: Bruchidae): evidence for local adaptation to a poor quality host. Oecologia 99:329–336. [21]

Franck P, Garnery L, Celebrano G, Solignac M & J-M Cornuet 2000 Hybrid origins of honeybees from Italy (*Apis mellifera ligustica*) and Sicily (*A. m. sicula*). Mol. Ecol. 9:907–921. [26]

Frank SA 1986 Hierarchical selection theory and sex ratios. I. General solutions for structured populations. Theor. Pop. Biol. 29:312–342. [23]

Frank SA 1993 Coevolutionary genetics of plants and pathogens. Evol. Ecol. 7: 45–75. [29]
Frank SA 1994a Coevolutionary genetics of hosts and parasites with quantitative inheritance. Evol. Ecol. 8:74–94. [29]
Frank SA 1994b Genetics of mutualism: the evolution of altruism between species. J. Theor. Biol. 170:393–400. [23]
Frank SA 1994c Kin selection and virulence in the evolution of protocells and parasites. Proc. R. Soc. Lond. B 258:153–161. [23]
Frank SA 1995a George Price's contribution to evolutionary genetics. J. Theor. Biol. 175:373–388. [4]
Frank SA 1995b Mutual policing and repression of competition in the evolution of cooperative groups. Nature 377:520–522. [23]
Frank SA 1995c The origin of synergistic symbiosis. J. Theor. Biol. 176:403–410. [23]
Frank SA 1996 Models of parasite virulence. Q. Rev. Biol. 71:37–78. [23]
Frank SA 1997a Models of symbiosis. Am. Nat. 150:S80–S99. [23]
Frank SA 1997b The Price equation, Fisher's fundamental theorem, kin selection, and causal analysis. Evolution 51:1712–1729. [23]
Frank SA 1997c Models of symbiosis. Am. Nat. 150:S80–S99. [23]
Frank SA 1998 Foundations of Social Evolution. Princeton Univ. Press. [23]
Frank SA 2003 Repression of competition and the evolution of cooperation. Evolution 57:693–705. [23]
Frankel OH & ME Soulé 1981 Conservation and Evolution. Cambridge Univ. Press. [32]
Frankham R 1995 Effective population size/adult population size ratios in wildlife: a review. Genet. Res. 66:95–107. [32]
Frankham R 2005 Genetics and extinction. Biol. Cons. 126:131–140. [32]
Frankham R, Ballou JD & DA Briscoe 2002 Introduction to Conservation Genetics. Cambridge Univ. Press. [32]
Frankham R, Ballou JD & DA Briscoe 2004 A Primer of Conservation Genetics. Cambridge Univ. Press. [32]
Frazer KA, Chen X, Hinds DA, Pant PV, Patil N & DR Cox 2003 Genomic DNA insertions and deletions occur frequently between humans and nonhuman primates. Genome Res. 13:341–346. [10]
Freeman S & JC Herron 2001 Evolutionary Analysis, 2nd ed. Prentice Hall. [4.1]
Friedman R & AL Hughes 2001 Pattern and timing of gene duplication in animal genomes. Genet. Res. 11:1842–1847. [10]
Fritz RS & EL Simms 1992 Plant resistance to herbivores and pathogens. Univ. of Chicago Press. [29]
Fu H, Park W, Yan X, Zheng Z, Shen B & HK Dooner 2001 The highly recombinogenic *bz* locus lies in an unusually gene-rich region of the maize genome. Proc. Natl. Acad. Sci. USA 98:8903–8908. [10]
Fu Y-X 1997 Statistical tests of neutrality of mutations against population growth, hitchhiking and background selection. Genetics 147:915–925. [7]
Fu Y-X & W-H Li 1993 Statistical tests of neutrality of mutations. Genetics 133:693–709. [7, 12]
Fujinaga M, Cherney MM, Oyama H, Oda K & MNG James 2004 The molecular structure and catalytic mechanism of a novel carboxyl peptidase from *Scytalidium lignicolum*. Proc. Natl. Acad. Sci. USA 101:3364–3369. [13]
Furlong RF & PW Holland 2002 Were vertebrates octoploid? Phil. Trans. R. Soc. Lond. B 357:531–544. [11]
Futuyma DJ 1998 Evolutionary Biology, 3rd. ed. Sinauer Assoc. [4.1, 6, 11]
Galant R & SB Carroll 2002 Evolution of a transcriptional repression domain in an insect Hox protein. Nature 415:910–913. [15]
Galiana A, Moya A & FJ Ayala 1993 Founder-flush speciation in *Drosophila pseudoobscura*: a large-scale experiment. Evolution 47:432–444. [27]
Gallardo MH, Bickham JW, Kausel G, Kohler N & RL Honeycutt 2003 Gradual and quantum genome size shifts in the hystricognath rodents. J. Evol. Biol. 16:163–169. [10]
Galton F 1889 Natural Inheritance. Macmillan. [1]
Gandon S 2002 Local adaptation and the geometry of host–parasite coevolution. Ecol. Lett. 5:246–256. [29]
Gaucher EA, Xu G, Miyamoto MM & SA Benner 2002 Predicting functional divergence in protein evolution by site-specific rate shifts. Trends Biochem. Sci. 27:315–321. [2, 11]
Gaut BS & JF Doebley 1997 DNA sequence evidence for the segmental allotetraploid origin of maize. Proc. Natl. Acad. Sci. USA 94:6809–6814. [10]
Gavrilets S 1997 Coevolutionary chase in exploiter–victim systems with polygenic characters. J. Theor. Biol. 186:527–534. [29]
Gavrilets S & SM Scheiner 1993 The genetics of phenotypic plasticity. V. Evolution of reaction norm shape. J. Evol. Biol. 6:31–48. [21]
Gayon J & M Veuille 2001 The genetics of experimental populations: L'Heritier and Teissier's population cages. pp. 77–102 in Singh R, Krimbas CB, Paul D & J Beatty, eds. Thinking about Evolution: Historical, Philosophical, and Political Perspectives. Cambridge Univ. Press. [1]
Gehring WJ & K Ikeo 1999 Pax 6:mastering eye morphogenesis and eye evolution. Trends Genet. 15:371–377. [15]

Gerhart J & M Kirschner 1997 Cells, Embryos, and Evolution. Blackwell Sci. [15, 16]
Gerlt JA & PC Babbitt 1998 Mechanistically diverse enzyme superfamilies: the importance of chemistry in the evolution of catalysis. Curr. Opin. Chem. Biol. 2:607–612. [13]
Gerstein M 1997 A structural census of genomes: comparing bacterial, eukaryotic, and archaeal genomes in terms of protein structure. J. Mol. Biol. 274:562–576. [14]
Giaever G, Chu AM, Ni L, et al. 2002 Functional profiling of the *Saccharomyces cervesiae* genome. Nature 418:387–391. [3]
Gibert P, Moreteau B, David JR & SM Scheiner 1998 Describing the evolution of reaction norm shape: body pigmentation in *Drosophila*. Evolution 52:1501–1506. [21]
Gibson G 2002 A genetic attack on the defense complex. BioEssays 24:487–489. [20]
Gibson G & DS Hogness. 1996. Effect of polymorphism in the *Drosophila* regulatory gene *Ultrabithorax* on homeotic stability. Science 271:200–203. [16]
Gibson G & G Wagner 2000 Canalization in evolutionary genetics: a stabilizing theory? BioEssays 22:372–380. [16]
Gibson JR, Chippindale AK & WR Rice 2001 The X chromosome is a hot spot for sexually antagonistic fitness variation. Proc. R. Soc. Lond. B 269:499–505. [22]
Giddings LV, Kaneshiro KY & A Moya 1989 Genetics, Speciation and the Founder Principle. Oxford Univ. Press. [27]
Gilbert SF 1991 Induction and the origins of developmental genetics. pp. 181–206 in SF Gilbert, ed. A Conceptual History of Modern Embryology. Plenum Press. [16]
Gilbert W 1978 Why genes in pieces? Nature 271:501. [11, 14]
Gilchrist AS & L Partridge 2001 The contrasting genetic architecture of wing size and shape in *Drosophila melanogaster*. Heredity 86:144–152. [20]
Gillespie JH 1984 The molecular clock may be an episodic clock. Proc. Natl. Acad. Sci. USA 81:8009–8013. [1]
Gillespie JH 1989 Lineage effects and the index of dispersion of molecular evolution. Mol. Biol. Evol. 6:636–648. [8]
Gillespie JH 1991 The Causes of Molecular Evolution. Oxford Univ. Press. [5]
Gillespie JH 1993 Substitution processes in molecular evolution. I. Uniform and clustered substitutions in a haploid model. Genetics 134:971–981. [8]
Gillespie JH 2000 Genetic drift in an infinite population: the pseudohitchhiking model. Genetics 155:909–919. [5]
Gillespie JH 2001 Is the population size of a species relevant to its evolution? Evolution 55:2161–2169. [5]
Gillespie JH 2004 Why $k = 4Nus$ is silly. pp. 178–192 in Singh RS & MK Uyenoyama, eds. The Evolution of Population Biology. Cambridge Univ. Press. [5]
Gilligan DM, Woodworth LM, Montgomery ME, Briscoe DA & R Frankham 1997 Is mutation accumulation a threat to the survival of endangered populations? Conserv. Biol. 11:1235–1241. [32]
Giraud A, Matic I, Tenaillon O, Clara A, Radman M, Fons M & F Taddei 2001 Costs and benefits of high mutation rates: adaptive evolution of bacteria in the mouse gut. Science 291:2606–2608. [31]
Gleason JM, Nuzhdin SV & MG Ritchie 2002 Quantitative trait loci affecting a courtship signal in *Drosophila melanogaster*. Heredity 89:1–6. [22]
Godfray HCJ & JH Werren 1996 Recent developments in sex ratio studies. Trends Ecol. Evol. 11:59–63. [23]
Goldman N 1994 Variance to mean ratio, $R(t)$, for Poisson processes on phylogenetic trees. Mol. Phylogenet. Evol. 3:230–239. [8]
Goldman N, Anderson JP & AG Rodrigo 2000 Likelihood-based tests of topologies in phylogenetics. Syst. Biol. 49:652–670. [28]
Goloboff PA 1999 Analyzing large data sets in reasonable times: solutions for composite optima. Cladistics 15:415–428. [28]
Gompel N & SB Carroll 2003 Genetic mechanisms and constraints governing the evolution of correlated traits in drosophilid flies. Nature 424:931–935. [25]
Goodale H 1938 A study of the inheritance of body weight in the albino mouse by selection. J. Hered. 29:101–112. [19]
Goodman M, Czelusniak J, Moore GW & G Matsuda 1979 Fitting the gene lineage into its species lineage: a parsimony strategy illustrated by cladograms constructed from globin sequences. Syst. Zool. 28: 132–163. [11]
Goodnight CJ 1985 The influence of environmental variation on group and individual selection in a cress. Evolution 39:545–558. [6]
Goodnight CJ 1987 On the effect of founder events on the epistatic genetic variance. Evolution 41:80–91. [6]
Goodnight CJ 1988 Epistasis and the effect of founder events on the additive genetic variance. Evolution 42:441–454. [6.1, 27]
Goodnight CJ 1990a Experimental studies of community evolution. I. The response to selection at the community level. Evolution 44:1614–1624. [6]
Goodnight CJ 1990b Experimental studies of community evolution. II. The ecological basis of the response to community selection. Evolution 44:1625–1636. [6]

Goodnight CJ 2000a Quantitative trait loci and gene interaction: the quantitative genetics of metapopulations. Heredity 84:587–598. [6, 19]

Goodnight CJ 2000b Modeling gene interaction in structured populations. pp. 129–145 in Wolf JB, Brodie ED III & MJ Wade, eds., Epistasis and the Evolutionary Process. Oxford Univ. Press. [6]

Goodnight CJ & L Stevens 1997 Experimental studies of group selection: what do they tell us about group selection in nature? Am. Nat. 150:S59–S79. [6]

Goodnight CJ & MJ Wade 2000 The ongoing synthesis: a reply to Coyne, Barton and Turelli. Evolution 54:317–324. [6]

Goodnight CJ, Schwartz JM & L Stevens 1992 Contextual analysis of models of group selection, soft selection, hard selection and the evolution of altruism. Am. Nat. 140:743–761. [6]

Goodrich JA, Cutler G & R Tijan 1996 Contacts in context: promoter specificity and macromolecular interactions in transcription. Cell 84:825–830. [2]

Gould SJ 1977 Ontogeny and Phylogeny. Harvard Univ. Press. [15]

Gould SJ 1983 The Hardening of the Modern Synthesis. pp. 71–93 in M Greene, ed., Dimensions of Darwinism. Cambridge Univ. Press. [1]

Gould SJ 1989 Wonderful Life: The Burgess Shale and the Nature of History. Norton. [31.1]

Gould SJ & N Eldredge 1993 Punctuated equilibrium comes of age. Nature 366:223–227. [3]

Gould SJ & RC Lewontin 1979 The spandrels of San Marco and the Panglossian paradigm: a critique of the adaptationist programme. Proc. R. Soc. Lond. B 205:581–598. [3]

Grant BR & PR Grant 1989 Evolutionary Dynamics of a Natural Population. Univ. of Chicago Press. [18]

Grant PR & BR Grant 1995 Predicting microevolutionary responses to directional selection on heritable variation. Evolution 49:241–251. [18]

Grant PR & BR Grant 2002 Unpredictable evolution in a 30-year study of Darwin's finches. Science 296:707–711. [2, 7, 26]

Grant V 1981 Plant Speciation. Columbia Univ. Press. [26]

Grantham R, Gautier C, Gouy M, Mercier R & A Pave 1980 Codon catalog usage and the genome hypothesis. Nucleic Acids Res. 8:r49–r62. [9]

Grativol AD, Ballou JD & RC Fleischer 2001 Microsatellite variation within and among recently fragmented populations of the golden lion tamarin (*Leontopithecus rosalia*). Conserv. Genet. 2:1–9. [32]

Graur D & W-H Li 2000 Fundamentals of Molecular Evolution. Sinauer Assoc. [2]

Graybeal A 1994 Evaluating the phylogenetic utility of genes: a search for genes informative about deep divergences among vertebrates. Syst. Biol. 43:174–193. [8]

Greenberg AJ, Moran GR, Coyne JA & C-I Wu 2003 Ecological adaptation during incipient speciation revealed by precise gene replacement. Science 302:1754–1757. [25]

Greene E 1989 A diet-induced developmental polymorphism in a caterpillar. Science 243:643–646. [21]

Greenfield MD, Tourtellot MK, Tillberg C, Bell WJ & N Prins 2002 Acoustic orientation via sequential comparison in an ultrasonic moth. Naturwissenschaften 89:376–380. [22]

Gregory TR 2001a Animal genome size database. http://www.genomesize.com [10]

Gregory TR 2001b Coincidence, coevolution, or causation? DNA content, cell size, and the C-value enigma. Biol. Rev. Camb. Phil. Soc. 76:65–101. [10]

Gregory TR & PD Hebert 1999 The modulation of DNA content: proximate causes and ultimate consequences. Genome Res. 9:317–324. [10]

Greig D, Borts RH, Louis EJ & M Travisano 2002a Epistasis and hybrid sterility in *Saccharomyces*. Proc. R. Soc. Lond. B 269:1167–1171. [25]

Greig D, Louis EJ, Borts RH & M Travisano 2002b Hybrid speciation in experimental populations of yeast. Science 28:1773–1775. [25]

Gribaldo S, Casane D, Lopez P & H Philippe 2003 Functional divergence prediction from evolutionary analysis: a case study of vertebrate hemoglobin. Mol. Biol. Evol. 20:1754–1759. [11]

Grimes GW & KJ Aufderheide 1991 Cellular Aspects of Pattern Formation: The Problem of Assembly (Monographs in Developmental Biology, vol. 22). Karger. [17]

Grishin NV 2001 Fold change in evolution of protein structures. J. Struct. Biol. 134:167–185. [13]

Groeters FR & H Dingle 1987 Genetic and maternal influences on life history plasticity in response to photoperiod by milkweed bugs (*Oncopeltus fasciatus*). Am. Nat. 129:332–346. [21]

Gromko MH 1995 Unpredictability of correlated response to selection: pleiotropy and sampling interact. Evolution 49:685–693. [19, 20]

Grosberg RK & MW Hart 2000 Mate selection and the evolution of highly polymorphic self/nonself recognition genes. Science 289:2111–2114. [29]

Gu X 2003 Functional divergence in protein family sequence evolution. Genetica 118:133–141. [11]
Gu Z, Steinmetz LM, Gu X, Scharfe C, Davis RW & W-H Li 2003 Role of duplicate genes in genetic robustness against null mutations. Nature 421:63–66. [16]
Guarente L & C Kenyon 2000 Genetic pathways that regulate ageing in model organisms. Nature 408:255–262. [30]
Gustafsson Å 1979 Linnaeus' Peloria: the history of a monster. Theor. Appl. Genet. 54:241–248. [17]
Hafner MS & RDM Page 1995 Molecular phylogenies and host–parasite cospeciation: gophers and lice as a model system. Phil. Trans. R. Soc. Lond. B 349:77–83. [29]
Haig D 2000 The kinship theory of genomic imprinting. Annu. Rev. Ecol. Syst. 31:9–32. [2]
Haig D 2002 Genomic Imprinting and Kinship. Rutgers Univ. Press. [17]
Haldane JBS 1922 Sex ratio and unisexual sterility in hybrid animals. J. Genet. 12:101–109. [25]
Haldane JBS 1932 The Causes of Evolution. Longman, Green, & Co. [1]
Haldane JBS 1938 The nature of interspecific differences. pp. 19–94 in G de Beer, ed. Evolution. Clarendon Press. [25]
Haldane JBS 1949 Disease and Evolution. La Ric. Sci. Suppl. 19:68–76. [29]
Haley CS & SA Knott 1992 A simple regression method for mapping quantitative trait loci in line crosses using flanking markers. Heredity 69:315–324. [19]
Haley CS, Knott SA & JM Elsen 1994 Mapping quantitative trait loci in crosses between outbred lines using least squares. Genetics 136:1195–1207. [19]
Hall BG 2004 Phylogenetic Trees Made Easy: A How To Manual. Sinauer Assoc. [28]
Hall BK 1992 Waddington's legacy in development and evolution. Am. Zool. 32:113–122. [16]
Hamilton W, Axelrod R & R Tanese 1990 Sexual reproduction as an adaptation to resist parasites (a review). Proc. Natl. Acad. Sci. USA 87:3566–3573. [29]
Hamilton WD 1964a The genetical evolution of social behaviour: I. J. Theor. Biol. 7:1–16. [6, 23]
Hamilton WD 1964b The genetical evolution of social behaviour: II. J. Theor. Biol. 7:17–52. [6, 23]
Hamilton WD 1967 Extraordinary sex ratios. Science 156:477–488. [23]
Hamilton WD 1970 Selfish and spiteful behaviour in an evolutionary model. Nature 228:1218–1220. [23]
Hamilton WD 1972 Altruism and related phenomena, mainly in social insects. Annu. Rev. Ecol. Syst. 3:193–232. [23]
Hamilton WD 1980 Sex vs non-sex vs parasites. Oikos 35:282–290. [29]
Hamilton WD 1996 Narrow Roads of Gene Land: The Collected Papers of W. D. Hamilton. WH Freeman. [30]
Hammer MF, Blackmer F, Garrigan D, Nachman MW & JA Wilder 2003 Human population structure and its effects on sampling Y chromosome sequence variation. Genetics 164:1495–1509. [7]
Hansen M & JC Koella 2003 Evolution of tolerance: the genetic basis of a host's resistance against parasite manipulation. Oikos 102:309–317. [29]
Hansen TF & D Houle 2004 Evolvability, stabilizing selection, and the problem of stasis. pp. 130–154 in Pigliucci M & K Preston, eds. The Evolutionary Biology of Complex Phenotypes. Oxford Univ. Press. [3]
Hanski IA & OE Gaggiotti 2004 Ecology, Genetics, and Evolution of Metapopulations. Elsevier. [5]
Hard JJ, Bradshaw WE & CM Holzapfel 1993 The genetic basis of photoperiodism and its evolutionary divergence among populations of the pitcher-plant mosquito, *Wyeomia smithii*. Am. Nat. 142:457–453. [27]
Hardin G 1993 Living within Limits: Ecology, Economics, and Population Taboos. Oxford Univ. Press. [23]
Harper JL 1977 Population Biology of Plants. Academic Press. [6]
Harrison RG 1990 Hybrid zones: windows on evolutionary process. Oxf. Surv. Evol. Biol. 7:69–128. [26]
Harrison S & A Hastings 1996 Genetic and evolutionary consequences of metapopulation structure. Trends Ecol. Evol. 11:180–183. [6]
Harshman LG & AA Hoffmann 2000 Laboratory selection experiments using *Drosophila*: what do they really tell us? Trends Ecol. Evol. 15:32–36. [31]
Hart RW & RB Setlow 1974 Correlation between deoxyribonucleic acid excision-repair and life span in a number of mammalian species. Proc. Natl. Acad. Sci. USA 71:2169–2173. [8]
Hartl DL 1980 Principles of Population Genetics. Sinauer.
Hartl DL, Dykhuizen DE & AM Dean 1985 Limits of adaptation: the evolution of selective neutrality. Genetics 111: 655–674. [9]
Hawkins MB, Thornton JW, Crews D, Skipper JK, Dotte A & P Thomas 2000 Identification of a third distinct estrogen receptor and reclassification of estrogen receptors in teleosts. Proc. Natl. Acad. Sci. USA 97:10751–10756. [11]

Hay DA 1976 The behavioral phenotype and mating behavior of two inbred strains of *Drosophila melanogaster*. Behav. Genet. 6:161–170. [27]

Hedges SB, Bogart JP & LR Maxson 1992 Ancestry of unisexual salamanders. Nature 356:708–710. [26]

Hedrick PW 2005 Genetics of Populations. Jones and Bartlett. [2, 4, 6, 27]

Hein JJ 1994 TreeAlign. pp. 349–364 in Griffin AM & HG Griffin, eds. Computer Analysis of Sequence Data. Humana Press. [28]

Heisler L & JD Damuth 1987 A method for analyzing selection in hierarchically structured populations. Am. Nat. 130:582–602. [6]

Hellman I, Ebersberger I, Ptak SE, Paabo S & M Przeworski 2003 A neutral explanation for the correlation of diversity with recombination rate in humans. Am. J. Hum. Genet. 72:1527–1535. [7]

Hendry AP, Vamosi SM, Latham SJ, Heilbuth JC & T Day 2000 Questioning species realities. Conserv. Genet 1:67–76. [Species Concepts box]

Hendy MD & D Penny 1982 Branch and bound algorithms to determine minimal evolutionary trees. Math. Biosci. 59:277–290. [28]

Henikoff S, Furuyama T & K Ahmad 2004 Histone variants, nucleosome assembly and epigenetic inheritance. Trends Genet. 20:320–326. [17]

Hennig W 1968 Elementos de una Sistemática Filogenética (Translation of *Grundzüge einer Theorie der phylogenetischen Systematik*), 2nd ed. Editorial Univ. de Buenos Aires. [Species Concepts box]

Hey J 2001 Genes, Categories, and Species. The Evolutionary and Cognitive Causes of the Species Problem. Oxford Univ. Press. [Species Concepts box]

Hey J & CA Machado 2003 The study of structured populations: new hope for a difficult and divided science. Nat. Rev. Genet. 4:535–543. [12]

Hill RE & ND Hastie 1987 Accelerated evolution in the reactive center regions of serine protease inhibitors. Nature 326:96–99. [7]

Hill WG & A Robertson 1968 Linkage disequilibrium in finite populations. Theor. Appl. Genet. 38:226–231. [4]

Hillis DM & JJ Bull 1993 An empirical test of bootstrapping as a method for assessing confidence in phylogenetic analysis. Syst. Biol. 42:182–192. [28, 31]

Hillis DM, Huelsenbeck JP & CW Cunningham 1994 Application and accuracy of molecular phylogenies. Science 264:671–677. [28]

Hine EM, Higgie S, Lachish S & MW Blows 2002 Positive genetic correlation between female preference and offspring fitness. Proc. R. Soc. Lond. B 269:2215–2219. [22]

Hoekstra HE & MW Nachman 2003 Different genes underlie adaptive melanism in different populations of rock pocket mice. Mol. Ecol. 12:1185–1194. [15]

Hoekstra HE, Drumm KE & MW Nachman 2004 Ecological genetics of adaptive color polymorphism in pocket mice: geographic variation in selected and neutral genes. Evolution 58:1329–1341. [7]

Hoffmann AA 1994 Genetic analysis of territoriality in *Drosophila melanogaster*. pp. 188–205 in CRB Boake, ed. Quantitative Genetic Studies of Behavioral Evolution. Univ. of Chicago Press. [22]

Holbrook GL & C Schal 2004 Maternal investment affects offspring phenotypic plasticity in a viviparous cockroach. Proc. Natl. Acad. Sci. USA 101:5595–5597 [2.1]

Holland B & WR Rice 1998 Perspective: Chase-away sexual selection: antagonistic seduction versus resistance. Evolution 52:1–7. [24]

Holland B & WR Rice 1999 Experimental removal of sexual selection reverses intersexual antagonistic coevolution and removes a reproductive load. Proc. Natl. Acad. Sci. USA 96:5083–5088. [31]

Holland JJ 1990 Defective viral genomes. pp. 151–165 in BN Fields, ed. Virology, 2nd ed. Raven Press. [23]

Holland PWH 1999 Gene duplication: past, present and future. Semin. Cell Dev. Biol. 10:541–547. [15]

Holliday R 1996 DNA methylation in eukaryotes: 20 years on. pp. 5–27 in Russo VEA, Martienssen RA & AD Riggs, eds. Epigenetic Mechanisms of Gene Regulation. Cold Spring Harbor Laboratory Press. [17]

Hollocher H, Hatcher JL & EG Dyreson 2000 Genetic and developmental analysis of abdominal pigmentation differences across species in the *Drosophila dunni* subgroup. Evolution 54:2057–2071. [25]

Hollocher H, Ting C-T, Wu ML & C-I Wu 1997 Incipient speciation by sexual isolation in *Drosophila melanogaster*: extensive genetic divergence without reinforcement. Genetics 147:1191–1201. [25]

Holzenberger M, Dupont J, Ducos B, Leneuve P, Geloen A, Even PC, Cervera P & Y Le Bouc 2003 IGF-1 receptor regulates lifespan and resistance to oxidative stress in mice. Nature 421:182–187. [30]

Horn C & EA Wimmer 2000 A versatile vector set for animal transgenesis. Dev. Genes Evol. 210:630–637. [15]

Horne TJ & H Ylonen 1998 Heritabilities of dominance-related traits in male bank voles (*Clethrionomys glareolus*). Evolution 52:894–899. [22]

Horowitz NH 1945 On the evolution of biochemical synthesis. Proc. Natl. Acad. Sci. USA 31:153–157. [13]
Horowitz NH 1965 The evolution of biochemical synthesis: retrospect and prospect. pp. 15–23 in Bryson V & HJ Vogel, eds. Evolving Genes and Proteins. Academic Press. [13]
Houlden BA, England PR, Taylor AC, Greville WD & WB Sherwin 1996 Low genetic variability of the koala *Phascolarctos cinereus* in southeastern Australia. Mol. Ecol. 5:269–281. [32]
Houle D 1991 Genetic covariance of fitness correlates: what genetic correlations are made of and why it matters. Evolution 45:630–648. [31]
Houle D 1992 Comparing evolvability and variability of quantitative traits. Genetics 130:195–204. [19]
Houle D 1998 How should we explain variance in the genetic variance of traits? Genetica 102/103:241–253. [3]
Houle D & L Rowe 2003 Natural selection in a bottle. Am. Nat. 161:50–67. [31]
Houle D, Hughes KA, Hoffmaster DK, Ihara J, Assimacopoulos S, Canada D & B Charlesworth 1994 The effects of spontaneous mutation on quantitative traits. I. Variance and covariance of life history traits. Genetics 138:773–785. [3]
Houle D, Morikawa B & M Lynch 1996 Comparing mutational variabilities. Genetics 143:1467–1483. [3]
Howard DJ 1986 A zone of overlap and hybridization between two ground cricket species. Evolution 40:34–43. [26]
Howard DJ 1993 Reinforcement: origin, dynamics, and fate of an evolutionary hypothesis. pp. 46–69 in RG Harrison, ed. Hybrid Zones and the Evolutionary Process. Oxford Univ. Press. [24, 26]
Howard DJ 1999 Conspecific sperm and pollen precedence and speciation. Annu. Rev. Ecol. Syst. 30:109–132. [24]
Howard DJ & SH Berlocher 1998 Endless Forms: Species and Speciation. Oxford Univ. Press. [Species Concepts box, 24]
Howard DJ, Reece M, Gregory PG, Chu J & ML Cain 1998 The evolution of barriers to fertilization between closely related organisms. pp. 279–288 in Howard DJ & SH Berlocher, eds. Endless Forms: Species and Speciation. Oxford Univ. Press. [24]
Hsu TC 1975 A possible function of constitutive heterochromatin: the bodyguard hypothesis. Genetics 79 (Suppl):137–150. [10]
Hu Z, Ma B, Wolfson H & R Nussinov 2000 Conservation of polar residues as hot spots at protein interfaces. Proteins: Struct. Funct. Genet. 39:331–342. [13]
Huang XP, Kagami N, Inoue H, Kojima M, Kimura T, Makabe O, Suzuki K & K Takahashi 2000 Identification of a glutamic acid and an aspartic acid residue essential for catalytic activity of aspergillopepsin II, a non-pepsin type acid proteinase. J. Biol. Chem. 275:26607–26614. [13]
Hubbs CL 1955 Hybridization between fish species in nature. Syst. Zool. 4:1–20. [26]
Hubby JL & RC Lewontin 1966 A molecular approach to the study of genic heterozygosity in natural populations. I. The number of alleles at different loci in *Drosophila pseudoobscura*. Genetics 54:546–595. [1]
Hudson RR 1983 Properties of a neutral allele model with intragenic recombination. Theor. Pop. Biol. 23:183–201. [12]
Hudson RR 1990 Gene genealogies and the coalescent process. Oxford Surv. Evol. Biol. 7:1–44. [12]
Hudson RR 2001 Two-locus sampling distributions and their application. Genetics 159:1805–1817. [12]
Hudson RR & JA Coyne 2002 Mathematical consequences of the genealogical species concept. Evolution 56:1557–1565. [Species Concepts box]
Hudson RR, Kreitman M & M Aguadé 1987 A test of neutral molecular evolution based on nucleotide data. Genetics 116:153–159. [2, 7]
Huelsenbeck JP & JJ Bull 1996 A likelihood ratio test to detect conflicting phylogenetic signal. Syst. Biol. 45:92–98. [11, 28]
Huelsenbeck JP & KA Crandall 1997 Phylogeny estimation and hypothesis testing using maximum likelihood. Annu. Rev. Ecol. Syst. 28:437–466. [8, 28]
Huelsenbeck JP, Bull JJ & CW Cunningham 1996 Combining data in phylogenetic analysis. Trends Ecol. Evol. 11:152–158. [28]
Huelsenbeck JP, Larget PD & DL Swofford 2000 A compound Poisson process for relaxing the molecular clock. Genetics 154: 1879–1892. [8]
Huelsenbeck JP, Ronquist F, Nielsen R & JP Bollback 2001 Bayesian inference of phylogeny and its impact on evolutionary biology. Science 294:2310–2314. [8, 28]
Hughes AL & R Friedman 2003 2R or not 2R: testing hypotheses of genome duplication in early vertebrates. J. Struct. Funct. Genomics 3:85–93. [11]
Hughes AL & M Nei 1988 Pattern of nucleotide substitution at major histocompatibility complex class-I loci reveals overdominant selection. Nature 335:167–170. [7]
Hughes AL & M Nei 1989 Nucleotide substitution at major histocompatibility complex class II loci: evidence for overdominant selection. Proc. Natl. Acad. Sci. USA 86:958–962. [3]

Hughes AL, Friedman R, Ekollu V & JR Rose 2003 Non-random association of transposable elements with duplicated genomic blocks in *Arabidopsis thaliana*. Mol. Phylogenet. Evol. 29:410–416. [11]

Hughes KA & B Charlesworth 1994 A genetic analysis of senescence in *Drosophila*. Nature 367:64–66. [30]

Hurst LD 1997 Evolutionary theories of genomic imprinting. pp. 211–237 in Reik W & A Surani, eds. Genomic Imprinting. IRL Press. [17]

Hurst LD, Atlan A & BO Bengtsson 1996 Genetic conflicts. Q. Rev. Biol. 71:317–364. [23]

Huson DS, Nettles S & T Warnow 1999 Disk-covering, a fast converging method for phylogenetic tree reconstruction. J. Comp. Biol. 6:369–383. [28]

Hutter H, Vogel BE, Plenefisch JD, Norris CR, Proenca RB, et al. 2000 Conservation and novelty in the evolution of cell adhesion and extracellular matrix genes. Science 287:989–994. [14]

Huttley GA, Smith MW, Carrington M & SJ O'Brien 1999 A scan for linkage disequilibrium across the human genome. Genetics 152:1711–1722. [7]

Huxley JS 1938 Darwin's theory of sexual selection and the data subsumed by it, in the light of recent research. Am. Nat. 72:416–433. [22]

Huxley JS 1942 Evolution: The Modern Synthesis. Allen & Unwin. [1]

Hynes RO & Q Zhao 2000. The evolution of cell adhesion. J. Cell Biol. 150:F89–F96. [14]

Ikemura T 1985 Codon usage and tRNA content in unicellular and multicellular organisms. Mol. Biol. Evol. 2:13–34. [2]

Ilic K, SanMiguel PJ & JL Bennetzen 2003 A complex history of rearrangement in an orthologous region of the maize, sorghum, and rice genomes. Proc. Natl. Acad. Sci. USA 100:12265–12270. [10]

Imhof M & C Schlotterer 2001 Fitness effects of advantageous mutations in evolving *Escherichia coli* populations. Proc. Natl. Acad. Sci. USA 98:1113–1117. [3]

Innan H & W Stephan 2001 Selection intensity against deleterious mutations in RNA secondary structures and rate of compensatory nucleotide substitutions. Genetics 159:389–399. [9]

Innes RW 1995 Plant–parasite interactions: has the gene-for-gene model become outdated? Trends Microbiol. 3:483–485. [29]

Isaac NJB, Mallet J & GM Mace 2004 Taxonomic inflation: its influence on macroecology and conservation. Trends Ecol. Evol. 19:464–469. [Species Concepts box]

IUCN 2002. IUCN Red List of Threatened Species. http://www.redlist.org/ [32]

Jablonka E & MJ Lamb 1989 The inheritance of acquired epigenetic variations. J. Theor. Biol. 139:69–83. [17]

Jablonka E & MJ Lamb 1995 Epigenetic Inheritance and Evolution: The Lamarckian Dimension. Oxford Univ. Press. [17]

Jablonka E & MJ Lamb 1998 Epigenetic inheritance in evolution. J. Evol. Biol. 11:159–183. [17]

Jablonka E & MJ Lamb 2002 The changing concept of epigenetics. Ann. N. Y. Acad. Sci. 981:82–96. [17]

Jablonka E & MJ Lamb 2005. Evolution in Four Dimensions: Genetic, Epigenetic, Behavioral, and Symbolic Variations in the History of Life. MIT Press. [17]

Jablonka E, Lamb MJ & E Avital 1998 "Lamarckian" mechanisms in darwinian evolution. Trends Ecol. Evol. 13:206–210. [17]

Jacob F 1977 Evolution and tinkering. Science 196:1161–1166. [15]

Jang Y & MD Greenfield 2000 Quantitative genetics of female choice in an ultrasonic pyralid moth, *Achroia grisella*: variation and evolvability of preference along multiple dimensions of the male advertisement signal. Heredity 84:73–80. [22]

Janoušek B, Široký J & B Vyskot 1996 Epigenetic control of sexual phenotype in a dioecious plant, *Melandrium album*. Mol. Gen. Genet. 250:483–490. [17]

Jasny BR 2000 The universe of *Drosophila* genes. Science 287:2181. [14]

Jayakar SD 1970. A mathematical model for interaction of gene frequencies in a parasite and its host. Theor. Pop. Biol. 1:140–164. [29]

Jeanmougin F, Thompson JD, Gouy M, Higgins DG & TJ Gibson 1998 Multiple sequence alignment with ClustalX. Trends Biochem. Sci. 23:403–405. [8]

Jensen MA, Charlesworth B & M Kreitman 2002 Patterns of genetic variation at a chromosome 4 locus of *Drosophila melanogaster* and *D. simulans*. Genetics 160:493–507. [7]

Jia F-Y, Greenfield MD & RD Collins 2000 Genetic variance of sexually selected traits in waxmoths: maintenance by genotype × environment interactions. Evolution 54:953–967. [22]

Jiang Y & RF Doolittle 2003 The evolution of vertebrate blood coagulation as viewed from a comparison of puffer fish and sea squirt genomes. Proc. Natl. Acad. Sci. USA 100:7527–7532. [14]

John B & GLG Miklos 1988 The Eukaryotic Genome in Development and Evolution. Allen & Unwin. [10]

Johnson NA 2000 Gene interaction and the origin of species. pp. 197–212 in Wolf JB, Brodie ED III & MJ Wade, eds. Epistasis

and the Evolutionary Process. Oxford Univ. Press. [24]
Johnson NA 2002 Sixty years after "Isolating mechanisms, evolution, and temperature": Muller's legacy. Genetics 161:939–944. [24]
Johnson NA & AH Porter 2000 Rapid speciation via parallel, directional selection on regulatory genetic pathways. J. Theor. Biol. 205:527–542. [24, 25]
Johnson T 1999 Beneficial mutations, hitchhiking and the evolution of mutation rates in sexual populations. Genetics 151:1621–1631. [3]
Johnston JA, Wesselingh RA, Bouck AC, Donovan LA & ML Arnold 2001 Intimately linked or hardly speaking? The relationship between genotypic variation and environmental gradients in a Louisiana Iris hybrid population. Mol. Ecol. 10:673–681. [26]
Johnston PG, Davey RJ & JH Seebeck 1984 Chromosome homologies in *Potorous tridactylus* and *P. longipes* (Marsupialia: Macropodidae) based on G-banding patterns. Aust. J. Zool. 32:319–324. [32]
Jones AG, Arnold SJ & R Burger 2003 Stability of the G-matrix in a population experiencing pleiotropic mutation, stabilizing selection, and genetic drift. Evolution 57:1747–1760. [20]
Jones PA & SB Baylin 2002 The fundamental role of epigenetic events in cancer. Nat. Rev. Genet. 3:415–428. [17]
Jordan IK, Rogozin IB, Glazko GV & EV Koonin 2003 Origin of a substantial fraction of human regulatory sequences from transposable elements. Trends Genet. 19:68–72. [2]
Kacser H & JA Burns 1981 The molecular basis of dominance. Genetics 97:639–666. [2]
Kaltz O & Shykoff JA 1998 Local adaptation in host–parasite systems. Heredity 81:361–374. [29]
Kaneshiro KY 1980 Sexual selection, speciation, and the direction of evolution. Evolution 34:437–444. [27]
Kaneshiro KY 1988 Speciation in the Hawaiian *Drosophila*. BioScience 38:258–263. [27]
Kaneshiro KY, Gillespie RG & HL Carson 1995 Chromosomes and male genitalia of Hawaiian *Drosophila*: tools for interpreting phylogeny and geography. pp. 57–71 in Wagner WL & VA Funk, eds. Hawaiian Biogeography: Evolution on a Hot Spot Archipelago. Smithsonian Press. [30.2]
Kapitonov VV & J Jurka 2003 Molecular paleontology of transposable elements in the *Drosophila melanogaster* genome. Proc. Natl. Acad. Sci. USA 100:6569–6574. [10]
Kaplan NL, Hudson RR & CH Langley 1989 The "hitchhiking effect" revisited. Genetics 123:887–899. [5, 7]
Kardong KV 2005 An Introduction to Biological Evolution. McGraw-Hill. [4.1]
Kareiva P 2000 Coevolutionary arms races: is victory possible? Proc. Natl. Acad. Sci. USA 96:8–10. [29]
Karn RC & MW Nachman 1999 Reduced nucleotide variability at the salivary androgen-binding locus in house-mice: evidence for positive natural selection. Mol. Biol. Evol. 16:1192–1197. [7]
Kassen R 2002 The experimental evolution of specialists, generalists, and the maintenance of diversity. J. Evol. Biol. 15:173–190. [21]
Katz LA 1998 Changing perspectives on the origin of eukaryotes. Trends Ecol. Evol. 13:493–497. [2]
Kayser M, Brauer S & M Stoneking 2003 A genome scan to detect candidate regions influenced by local natural selection in human populations. Mol. Biol. Evol. 20:893–900. [7]
Kazazian HH 2004 Mobile elements: drivers of genome evolution. Science 303:1626–1632. [2]
Keightley PD & A Eyre-Walker 2000 Deleterious mutations and the evolution of sex. Science 290:331–333. [3]
Keightley PD & M Lynch 2003 Toward a realistic model of mutations affecting fitness. Evolution 57:683–685. [2]
Keller LF 1998 Inbreeding and its fitness effects in an insular population of song sparrows (*Melospiza melodia*). Evolution 52: 240–250. [32]
Keller LF & M Genoud 1997 Extraordinary lifespans in ants: a test of evolutionary theories of ageing. Nature 389:958–960. [30]
Kellis M, Birren BW & ES Lander 2004 Proof and evolutionary analysis of ancient genome duplication in the yeast *Saccharomyces cerevisiae*. Nature 428:617–624. [2, 11]
Kellis M, Patterson N, Endrizzi M, Birren B & ES Lander 2003 Sequencing and comparison of yeast species to identify genes and regulatory elements. Nature 423:241–254. [25]
Kelly JK 1997 A test of neutrality based on interlocus associations. Genetics 146:1197–1206. [7]
Kent WJ, Baertsch R, Hinrichs A, Miller W & D Haussler 2003 Evolution's cauldron: duplication, deletion, and rearrangement in the mouse and human genomes. Proc. Natl. Acad. Sci. USA 100:11484–11489. [11]
Kentner EK & MR Mesler 2000 Evidence for natural selection in a fern hybrid zone. Am. J. Bot. 87:1168–1174. [26]
Kidwell MG 1993 Lateral transfer in natural populations of eukaryotes. Annu. Rev. Genet. 27:235–256. [2, 30.2]
Kim CH, Oh Y & TH Lee 1997 Codon optimization for high-level expression of human

erythropoietin (EPO) in mammalian cells. Gene 199:293–301. [9]
Kim K 1994 Explaining Scientific Consensus: The Case of Mendelian Genetics. Guilford Press. [1]
Kim S-C & LH Rieseberg 1999 Genetic architecture of species differences in annual sunflowers: implications for adaptive trait introgression. Genetics 153:965–977. [26]
Kim Y & W Stephan 2000 Joint effects of genetic hitchhiking and background selection on neutral variation. Genetics 155:1415–1427. [7]
Kimura M 1968 Evolutionary rate at the molecular level. Nature 217:624–626. [1, 7, 8, 13]
Kimura M 1969a The number of heterozygous nucleotide sites maintained in a finite population due to steady flux of mutations. Genetics 61:893–903. [7]
Kimura M 1969b The rate of molecular evolution considered from the standpoint of population genetics. Proc. Natl. Acad. Sci. USA 63:1181–1188. [1]
Kimura M 1970 The length of time required for a selectively neutral mutant to reach fixation through random frequency drift in a finite population. Genet. Res. 15:1131–1133. [1]
Kimura M 1981 Possibility of extensive neutral evolution under stabilizing selection with special reference to nonrandom usage of synonymous codons. Proc. Natl. Acad. Sci. USA 78:5773–5777. [9]
Kimura M 1983 The Neutral Theory of Molecular Evolution. Cambridge Univ. Press. [1, 7, 8, 9]
Kimura M 1985 The role of compensatory neutral mutations in molecular evolution. J. Genet. 64:7–19. [9]
Kimura M & JF Crow 1964 Number of alleles that can be maintained in a finite population. Genetics 49:725–738. [7]
Kimura M & T Ohta 1969 The average number of generations until fixation of a mutant gene in a finite population. Genetics 61:763–771. [8]
Kimura M & T Ohta 1971a Protein polymorphism as a phase in molecular evolution. Nature 229:467–469. [1]
Kimura M & T Ohta. 1971b On the rate of molecular evolution. J. Mol. Evol. 1:1–17. [8]
King J & T Jukes 1969 Non-Darwinian evolution: random fixation of selectively neutral mutations. Science 164:788–798. [1, 7, 13]
King M-C & AC Wilson 1975 Evolution at two levels in humans and chimpanzees. Science 188:107–116. [15]
Kingman JFC 1982 On the genealogy of large populations. J. Appl. Prob. 19A: 27–43. [12]
Kingsolver JG & DC Wiernasz 1987 Dissecting correlated characters: adaptive aspects of phenotypic covariation in melanization pattern of *Pieris* butterflies. Evolution 41:491–503. [20]
Kingsolver JG & DC Wiernasz 1991 Development, function and the quantitative genetics of wing melanin pattern in *Pieris* butterflies. Evolution 45:1480–1492. [20]
Kingsolver JG, Hoekstra HE, Hoekstra JM, Berrigan D, Vignieri SN, Hill CE, Hoang A, Gilbert P & P Beerli 2001 The strength of phenotypic selection in natural populations. Am. Nat. 157:245–261. [18, 18.1]
Kirby DA, Muse SV & W Stephan 1995 Maintenance of pre-mRNA secondary structure by epistatic selection. Proc. Natl. Acad. Sci. USA 92:9047–9051. [9]
Kirik A, Salomon S & H Puchta 2000 Species-specific double-strand break repair and genome evolution in plants. EMBO J. 2000:5562–5566. [10]
Kirkpatrick M 1996 Genes and adaptation: a pocket guide to theory. pp. 125–146 in Rose MJ & G Lauder, eds. Evolutionary Biology and Adaptation. Sinauer Assoc. [4]
Kirkpatrick M & N Heckman 1989 A quantitative genetic model for growth, shape, reaction norms, and other infinite-dimensional characters. J. Math. Biol. 27:429–450. [21]
Kirkpatrick M & R Lande 1989 The evolution of maternal characters. Evolution 43:485–503. [2.1, 18]
Kirkpatrick M, Johnson T & N Barton 2002 General models of multilocus evolution. Genetics 161:1727–1750. [20]
Kishino H & M Hasegawa 1989 Evaluation of the maximum likelihood estimate of the evolutionary tree topologies from DNA sequence data, and the branching order in Hominoidea. J. Mol. Evol. 29:170–179. [28]
Kishino H & M Hasegawa 1990 Converting distance to time: application to human evolution. Methods Enzymol. 183:550–570. [8]
Klingenberg CP 2003 Developmental instability as a research tool: using patterns of fluctuating asymmetry to infer the developmental origins of morphological integration. pp. 427–42 in M Polak ed. Developmental Instability: Causes and Consequences. Oxford Univ. Press. [19.4]
Klingenberg CP & LJ Leamy 2001 Quantitative genetics of geometric shape in the mouse mandible. Evolution 55:2342–2352. [19.4]
Klingenberg CP & SD Zaklan 2000 Morphological intergration between development compartments in the *Drosophila* wing. Evolution 54:1273–1285. [20]
Klingenberg CP, Barluenga M & A Meyer 2002 Shape analysis of symmetric structures: quantifying

variation among individuals and asymmetry. Evolution 56:1909–1920. [19.4]

Klingenberg CP, Leamy LJ & JM Cheverud 2004 Integration and modularity of quantitative trait locus effects on genometric shape in the mouse mandible. Genetics 166:1909–1921. [19.4, 19.5]

Klingenberg CP, Leamy LJ, Routman EJ & JM Cheverud 2001 Genetic architecture of mandible shape in mice: effects of quantitative trait loci analyzed by geometric morphometrics. Genetics 157:785–802. [19.4]

Klingler M, Soong J, Butler B & JP Gergen 1996 Disperse versus compact elements for the regulation of runt stripes in *Drosophila*. Dev. Biol. 177:73–84. [15]

Knight CA & DD Ackerly 2002 Variation in nuclear DNA content across environmental gradients: a quantile regression analysis. Ecol. Lett. 5:66–76. [10]

Knight CA, Molinari N & DA Petrov 2005 The large genome constraint hypothesis: evolution, ecology, and phenotype. Ann. Bot. 95:177–190. [10]

Knowles LL & WP Maddison 2002 Statistical phylogeography. Mol. Ecol. 11:2623–2635. [12]

Kohler RE 1991 *Drosophila* and evolutionary genetics: the moral economy of scientific practice. History Sci. 29:335–375. [1]

Kokko H, Brooks R, Jennions MD & J Morley 2003 The evolution of mate choice and mating biases. Proc. R. Soc. Lond. B 270:653–664. [22]

Kokko H, Brooks R, McNamara JM & AI Houston 2002 The sexual selection continuum. Proc. R. Soc. Lond. B 269:1331–1340. [22]

Kolaczkowski B & JW Thornton 2004 Performance of maximum parsimony and likelihood phylogenetics when evolution is heterogeneous. Nature 431: 980–984. [11]

Kondrashov AS 1995 Modifiers of mutation–selection balance: general approach and the evolution of mutation rates. Genet. Res. 66:53–69. [3]

Kondrashov AS 1998 Measuring spontaneous deleterious mutation process. Genetica 103:183–197. [3]

Kondrashov AS 2002 Direct estimates of human per nucleotide mutation rates at 20 human loci causing Mendelian diseases. Hum. Mut. 21:12–27. [3]

Kondrashov AS & IB Rogozin 2004 Context of deletions and insertions in human coding sequences. Hum. Mut. 23:177–185. [3]

Koonin EV 2003 Horizontal gene transfer: the path to maturity. Mol. Microbiol. 50:725–727. [12]

Koonin EV, Wolf YI & GP Karev 2002 The structure of the protein universe and genome evolution. Nature 420:218–223. [14]

Koornneef M, Alonso-Blanco C & D Vreugdenhil 2004 Naturally occurring genetic variation in *Arabidopsis thaliana*. Annu. Rev. Plant Biol. 55:141–172. [29.1]

Kopp A, Duncan I & SB Carroll 2000 Genetic control and evolution of sexually dimorphic characters in *Drosophila*. Nature 408:553–559. [15, 25]

Kover PX & AL Caicedo 2001 The genetic architecture of resistance and the role of parasites in maintaining sexual recombination. Mol. Ecol. 10:1–17. [29]

Kover PX & BA Schaal 2002 Genetic variation for disease resistance and tolerance among *A. thaliana* accessions. Proc. Natl. Acad. Sci. USA 99:11270–11274. [29]

Kraaijveld A & H Godfray 1997 Trade-off between parasitoid resistance and larval competitive ability in *Drosophila melanogaster*. Nature 389:278–280. [29]

Kreitman M 2000 Methods to detect selection in populations with applications to the human. Annu. Rev. Genomics Hum. Genet. 1:539–559. [1, 7]

Kreitman M & H Akashi 1995 Molecular evidence for natural selection. Annu. Rev. Ecol. Syst. 26:403–422. [7, 8]

Kreitman M & RR Hudson 1991 Inferring the evolutionary histories of the ADH and ADH-DUP loci in *Drosophila melanogaster* from patterns of polymorphism and divergence. Genetics 127:565–582. [7]

Kretsinger RH, Ison RE & S Hovmoller 2004 Prediction of protein structure. Methods Enzym. 383:1–27. [2]

Krimbas CB & JR Powell 1992 Drosophila Inversion Polymorphism. CRC Press. [30.2]

Krishna SS & NV Grishin 2004 Structurally analogous proteins do exist! Structure 12:1125–1127. [13]

Kroymann J, Donnerhacke S, Schnabelrauch D & T Mitchell-Olds 2003 Evolutionary dynamics of an *Arabidopsis* insect resistance quantitative trait locus. Proc. Natl. Acad. Sci. USA 100:14587–14592. [29.1]

Kumar S & SB Hedges 1998 A molecular timescale for vertebrate evolution. Nature 392:917–920. [8]

Kuris AM 1974 Trophic interactions: similarity of parasitic castrators to parasitoids. Q. Rev. Biol. 49:129–148. [29]

Kyriacou CP 2002 Single gene mutations in *Drosophila*: what can they tell us about the evolution of sexual behaviour? Genetica 116:197–203. [22]

Lachmann M & E Jablonka 1996 The inheritance of phenotypes: an adaptation to fluctuating environments. J. Theor. Biol. 181:1–9. [17]

Lahn BT & DC Page 1999 Four evolutionary strata on the human X chromosome. Science 286:964–967. [11]
LaMunyon CW & S Ward 2002 Evolution of larger sperm in response to experimentally increased sperm competition in *Caenorhabditis elegans*. Proc. R. Soc. Lond. B 269:1125–1128. [22]
Lande R 1979 Quantitative genetic analysis of multivariate evolution, applied to brain:body allometry. Evolution 33:402–416. [20]
Lande R 1980a Sexual dimorphism, sexual selection, and adaptation in polygenic characters. Evolution 34:292–305. [19]
Lande R 1980b The genetic covariance between characters maintained by pleiotropic mutations. Genetics 94:203–215. [20]
Lande R 1988 Genetics and demography in biological conservation. Science 241:1455–1460. [32]
Lande R 1995 Mutation and conservation. Conserv. Biol. 9:782–791. [32]
Lande R & SJ Arnold 1983 The measurement of selection on correlated characters. Evolution 37:1210–1226. [18.1, 20]
Lande R & GS Wilkinson 1999 Models of sex-ratio meiotic drive and sexual selection in stalk-eyed flies. Genet. Res. 74:245–253. [22]
Lander ES & D Botstein 1989 Mapping Mendelian factors underlying quantitative traits using RFLP linkage maps. Genetics 121:185–199. [19]
Lander ES, Linton LM, Birren B, et al. 2001 Initial sequencing and analysis of the human genome. Nature 409:860–921. [10, 11, 13,14]
Langley C & W Fitch 1974 An examination of the constancy of the rate of molecular evolution. J. Mol. Evol. 3:161–177. [1]
Lawrence JG & H Ochman 2002 Reconciling the many faces of lateral gene transfer. Trends Microbiol. 10:1–4. [2]
Leakey R & R Lewin 1995 The Sixth Extinction: Biodiversity and its Survival. Phoenix. [32]
Leamy LJ, Routman EJ & JM Cheverud 1998 Quantitative trait loci for fluctuating asymmetry of quasi-continuous skeletal characters in mice. Heredity 80:509–518. [19]
Leamy LJ, Routman EJ & JM Cheverud 1999 Quantitative trait loci for early and late developing skull characters in mice: a test of the genetic independence model of morphological integration. Am. Nat. 153:201–214. [19]
Leamy LJ, Routman EJ & JM Cheverud 2002 An epistatic genetic basis for fluctuating asymmetry of mandible size in mice. Evolution 56:642–653. [19]
Lee RD 2003 Rethinking the evolutionary theory of aging: transfers, not births, shape senescence in social species. Proc. Natl. Acad. Sci. USA 100:9637–9642. [30]
Leigh EG Jr 1971 Adaptation and Diversity. Freeman. [23]
Leigh EG Jr 1977 How does selection reconcile individual advantage with the good of the group? Proc. Natl. Acad. Sci. USA 74:4542–4546. [23]
Lemmon AR & MC Milinkovitch 2002 The metapopulation genetic algorithm: an efficient solution for the problem of large phylogeny estimation. Proc. Natl. Acad. Sci. USA 99:10516–10521. [28]
Lenski RE 2004 Phenotypic and genomic evolution during a 20,000-generation experiment with the bacterium *Escherichia coli*. Plant Breed. Rev. 24:225–265. [31.1]
Lenski RE, Rose MR, Simpson SC & SC Tadler 1991 Long-term experimental evolution in *Escherichia coli*. 1. Adaptation and divergence during 2000 generations. Am. Nat. 138:1315–1341. [31]
Lerner IM 1954 Genetic Homeostasis. Oliver & Boyd. [21]
Leroi AM, Chippindale AK & MR Rose 1994 Long-term laboratory evolution of a genetic life-history trade-off in *Drosophila melanogaster*. 1. The role of genotype-by-environment interaction. Evolution 48:1244–1257. [31]
Lesk AM, Branden C-I & C Chothia 1989 Structural principles of α/β barrel proteins: the packing of the interior of the sheet. Proteins: Struct. Funct. Genet. 5:139–148. [13]
Letunic I, Copley RR, Schmidt S, Ciccarelli FD, Doerks T, Schultz J, Ponting CP & P Bork 2004 SMART 4.0: towards genomic data integration. Nucleic Acids Res. 32(Database issue): D142–D144. [14]
Levin B, Bull J & F Stewart 1996 The intrinsic rate of increase of HIV/AIDS: epidemiological and evolutionary implications. Math. Biosci. 132:69–96. [29]
Levin BR & RE Lenski 1983 Coevolution in bacteria and their viruses and plasmids. pp. 99–127 in Futuyma DJ & M Slatkin, eds. Coevolution. Sinauer Assoc. [23]
Levin DA 1963 Natural hybridization between *Phlox maculata* and *Phlox glaberrima* and its evolutionary significance. Evolution 50:714–720. [26]
Lewin B 2004 Genes VIII. Pearson Education. [2, 3]
Lewis EB 1978 A gene complex controlling segmentation in *Drosophila*. Nature 276:565–570. [15]
Lewis P 1998 A genetic algorithm for maximum-likelihood phylogeny inference using nucleotide sequence data. Mol. Biol. Evol. 15:277–283. [28]
Lewontin RC 1970 The units of selection. Annu. Rev. Ecol. Syst. 1:1–18. [4, 20]
Lewontin RC 1974 The Genetic Basis of Evolutionary Change. Columbia Univ. Press. [1, 5]

Lewontin RC 1981 Introduction: the scientific work of Theodosius Dobzhansky. pp. 93–115 in Lewontin RC, Moore JA, Provine WB & B Wallace, eds. Dobzhansky's Genetics of Natural Populations I–XLIII. Columbia Univ. Press. [1]
Lewontin RC 1991 Perspectives: 25 years ago in genetics. electrophoresis in the development of evolutionary genetics: milestone or millstone? Genetics 128:657–662. [1]
Lewontin RC & LC Birch 1966 Hybridization as a source of variation for adaptation to new environments. Evolution 20:315–336. [26]
Lewontin RC & JL Hubby 1966 Molecular approach to the study of genic heterozygosity in natural populations. II. Amount of variation and degree of heterozygosity in natural populations of *Drosophila pseudoobscura*. Genetics 54:595–609. [1]
Lewontin RC & J Krakauer 1973 Distribution of gene frequency as a test of the theory of the selective neutrality of polymorphisms. Genetics 74:175–195. [7]
Li N & M Stephens 2003 Modeling linkage disequilibrium and identifying recombination hotspots using single-nucleotide polymorphism data. Genetics 165:2213–2233. [12]
Li W-H 1978 Maintenance of genetic variability under the joint effect of mutation, selection, and random drift. Genetics 85:331–337. [2]
Li W-H 1987 Models of nearly neutral mutations with particular implications for nonrandom usage of synonymous codons. J. Mol. Evol. 24:337–345. [9]
Li W-H & Y-X Fu 1999 Coalescent theory and its applications in population genetics. pp. 45–79 in Halloran ME & S Geisser, eds. Statistics in Genetics, Springer. [12]
Li W-H, Ellsworth DL, Krushkal J, Chang BH-J & D Hewett-Emmett 1996 Rates of nucleotide substitution in primates and rodents and the generation-time effect hypothesis. Mol. Phylogenet. Evol. 5:182–187. [8]
Li W-H, Gu Z, Wang H & A Nekrutenko 2001 Evolutionary analyses of the human genome. Nature 409:847–849. [14]
Li W-H, Wu C-I & CC Luo 1985 A new method for estimating synonymous and nonsynonymous rates of nucleotide substitution considering the relative likelihood of nucleotide and codon changes. Mol. Biol. Evol. 2:150–174. [7]
Liao D 1999 Concerted evolution: molecular mechanism and biological implications. Am. J. Hum. Genet. 64:24–30. [11]
Lichtarge O, Bourne HR & FE Cohen 1996 The evolutionary trace method defines the binding surfaces common to a protein family. J. Mol. Biol. 257:342–358. [13]
Lijtmaer DA, Mahler B & PL Tubaro 2003 Hybridization and post-zygotic isolation patterns in pigeons and doves. Evolution 57:1411–1418. [24]
Lindholm A & F Breden 2002 Sex chromosomes and sexual selection in poecilid fishes. Am. Nat. 160:S214–S224. [22]
Lints FA & M Bourgois 1982 A test of the genetic revolution hypothesis of speciation. pp. 157–180 in S Lakovaara, ed. Advances in Genetics, Development and Evolution of *Drosophila*. Plenum Press. [27]
Lipman DJ, Altschul SF & JD Kececioglu 1989 A tool for multiple sequence alignment. Proc. Natl. Acad. Sci. USA 86:4412–4415. [28]
Lipsitch M & ER Moxon 1997 Virulence and transmissibility of pathogens: what is the relationship? Trends Microbiol. 5:31–37. [29]
Little JW, Shepley DP & DW Wert 1999 Robustness of a gene regulatory circuit. EMBO J. 18:4299–4307. [16]
Liu J, Mercer JM, Stam LF, Gibson GC, Zeng Z-B & CC Laurie 1996 Genetic analysis of a morphological shape difference in the male genitalia of *Drosophila simulans* and *D. mauritiana*. Genetics 142:1129–1145. [25]
Lively CM & MF Dybdahl 2000 Parasite adaptation to locally common host genotypes. Nature 405:679–681. [29]
Llopart A, Elwyn S, Lachaise D & JA Coyne 2002 Genetics of a difference in pigmentation between *Drosophila yakuba* and *Drosophila santomea*. Evolution 56:2262–2277. [25]
Long AD, Lyman RF, Langley CH & TF Mackay 1998 Two sites in the Delta gene region contribute to naturally occurring variation in bristle number in *Drosophila melanogaster*. Genetics 149:999–1017. [2]
Long M & K Thornton 2001 Gene duplication and evolution. Science 293:1551. [11]
Long M, Betran E, Thornton K & W Wang 2003 The origin of new genes: glimpses from the young and old. Nat. Rev. Genet. 4:865–875. [11]
López-Fanjul C & A Villaverde 1989 Inbreeding increases genetic variance for viability in *Drosophila melanogaster*. Evolution 43:1800–1804. [27]
López-Fanjul C, Fernández A & MA Toro 2000 Epistasis and the conversion of non-additive to additive genetic variance at population bottlenecks. Theor. Pop. Biol. 58:49–59. [6.1]
López-Fanjul C, Fernández A & MA Toro 2004 Epistasis and the temporal change in the additive variance–covariance matrix induced by drift. Evolution 58:1655–1663. [20]
Lotsy JP 1931 On the species of the taxonomist in its relation to evolution. Genetica 13:1–16. [26]
Lovell SC 2003 Are non-functional, unfolded proteins ("junk proteins") common in the genome? FEBS Lett 554:237–239. [13]

Lovell SC, Word JM, Richardson JS & DC Richardson 1999 Asparagine and glutamine rotamers: B-factor cutoff and correction of amide flips yield distinct clustering. Proc. Natl. Acad. Sci. USA 96:400–405. [13]
Lovell SC, Word JM, Richardson JS & DC Richardson 2000 The penultimate rotamer library. Proteins: Struct. Funct. Genet. 40:389–408. [13]
Löytynoja A & MC Milinkovitch 2001 SOAP, cleaning multiple alignments from unstable blocks. Bioinformatics 17:573–574. [8]
Lozovskaya ER, Nurminsky D, Petrov DA & DL Hartl 1999 Genome size as a mutation–selection–drift process. Genes Genet. Syst. 74:201–207. [30.2]
Ludwig MZ 2002 Functional evolution of noncoding DNA. Curr. Opin. Genet. Dev. 12:634–639. [9]
Ludwig MZ, Bergman CM, Patel NH & M Kreitman 2000 Evidence for stabilizing selection in a eukaryotic enhancer element. Nature 403:564–567. [8, 9, 15]
Ludwig MZ, Patel NH & M Kreitman 1998 Functional analysis of eve stripe 2 enhancer evolution in *Drosophila*: rules governing conservation and change. Development 125:949–958. [15]
Luikart G, England PR, Tallmon D, Jordan S & P Taberlet 2003 The power and promise of population genomics: from genotyping to genome typing. Nat. Rev. Genet. 4:981–994. [7]
Luning ET & HH Kazazian 2000 Mobile elements and the human genome. Nat. Rev. Genet. 1:134–144. [2]
Lupas AN, Ponting CP & RB Russell 2001 On the evolution of protein folds: are similar motifs in different protein folds the result of convergence, insertion, or relics of an ancient peptide world? J. Struct. Biol. 134:191–203. [13]
Lyman RF & TF Mackay 1998 Candidate quantitative trait loci and naturally occurring phenotypic variation for bristle number in *Drosophila melanogaster*: the Delta-Hairless gene region. Genetics 149:983–998. [2]
Lynch M 1990 The rate of morphological evolution in mammals from the standpoint of the neutral expectation. Am. Nat. 136:727–741. [3]
Lynch M 2002 Intron evolution as a population-genetic process. Proc. Natl. Acad. Sci. USA 99:6118–6123. [9]
Lynch M & JS Conery 2000 The evolutionary fate and consequences of duplicate genes. Science 290:1151–1155. [2, 3, 11]
Lynch M & JS Conery 2003 The origins of genome complexity. Science 302:1401–1404. [10]
Lynch M & AG Force 2000 The origin of interspecific genomic incompatibility via gene duplication. Am. Nat. 156:590–605. [2]
Lynch M & WG Hill 1986 Phenotypic evolution by neutral mutation. Evolution 40:915–935. [20]
Lynch M & B Walsh 1998 Genetics and Analysis of Quantitative Traits. Sinauer Assoc. [2, 18, 19, 19.4, 19.5]
Lynch M, Blanchard J, Houle D, Kibota T, Schultz S, Vassilieva L & J Willis 1999 Perspective: spontaneous deleterious mutation. Evolution 53:645–663. [3]
Lynch M, Conery J & R Bürger 1995 Mutational meltdowns in sexual populations. Evolution 49:1067–1080. [32]
Lynch M, O'Hely M, Walsh B & A Force 2001 The probability of preservation of a newly arisen gene duplicate. Genetics 159:1789–1804. [10]
Lyon MF 1999 Imprinting and X-chromosome inactivation. pp. 73–90 in R Ohlsson, ed. Genomic Imprinting: An Interdisciplinary Approach. Springer. [17]
Lyttle TW 1991 Segregation distorters. Annu. Rev. Genet. 25:511–557. [23]
Lyytinen A, Brakefield PM & J Mappes 2003 Significance of butterfly eyespots as an anti-predator device in ground-based and aerial attacks. Oikos 100:373–379. [20]
Ma RZ, Black WC & JC Reese 1992 Genome size and organization in an aphid (*Schizaphis graminum*). J. Insect Physiol. 38:161–165. [10]
MacArthur J 1944 Genetics of body size and related characters. I. Selection of small and large races of the laboratory mouse. Am. Nat. 78:142–157. [19]
Macdonald PM 1990 *bicoid* mRNA localization signal: phylogenetic conservation of function and RNA secondary structure. Development 110:161–171. [9]
Machado CA & J Hey 2003 The causes of phylogenetic conflict in a classic *Drosophila* species group. Proc. R. Soc. Lond. B 270:1193–1202. [12]
Mackay TFC 2004 The genetic architecture of quantitative traits: lessons from *Drospohila*. Curr. Opin. Genet. Dev. 14:253–257. [19]
Maddison DR 1991 The discovery and importance of multiple islands of most-parsimonious trees. Syst. Zool. 40:315–328. [28]
Maddison DR & WP Maddison 2000 MacClade 4:Analysis of Phylogeny and Character Evolution, version 4.0. Sinauer Assoc. [28]
Maddison WP 1997 Gene trees in species trees. Syst. Biol. 46:523–536.
Magurran AE 2001 Sexual conflict and evolution in Trinidadian guppies. Genetica 112/113:463–474. [31]
Mair W, Goymer P, Pletcher SD & L Partridge 2003 Demography of dietary restriction and death in *Drosophila*. Science 301:1731–1733. [30]
Mallet J 1995 A species definition for the Modern Synthesis. Trends Ecol. Evol. 10:294–299. [Species Concepts box]

Mallet J 2001 Species, concepts of. pp. 427–440 in SA Levin, ed. Encyclopedia of Biodiversity, vol. 5. Academic Press. [Species Concepts box]
Mallet J 2004 Poulton, Wallace and Jordan: how discoveries in *Papilio* butterflies initiated a new species concept 100 years ago. Syst. Biodiv. 1:441–452. [Species Concepts box]
Malmos KB, Sullivan BK & T Lamb 2001 Calling behavior and directional hybridization between two toads (*Bufo microscaphus* × *B. woodhousii*) in Arizona. Evolution 55:626–630. [26]
Marden JH, Rogina B, Montooth KL & SL Helfand 2003 Conditional tradeoffs between aging and organismal performance of Indy long-lived mutant flies. Proc. Natl. Acad. Sci. USA 100:3369–3373. [30]
Margulis L 1993 Symbiosis in Cell Evolution. W. H. Freeman. [2]
Martin AP 2000 Choosing among alternative trees of multigene families. Mol. Phylogenet. Evol. 16:430–439. [11]
Martin LJ, Mahaney MC, Bronikowski AM, Dee Carey K, Dyke B & AG Comuzzie 2002 Lifespan in captive baboons is heritable. Mech. Ageing Dev. 123:1461–1467. [30]
Martin OY & DJ Hosken 2003 Costs and benefits of evolving under experimentally enforced polyandry or monogamy. Evolution 57:2765–2772. [31]
Masel J & A Bergman 2003 The evolution of the evolvability properties of the yeast prion [PSI+]. Evolution 57:1498–1512. [16, 17]
Masoro EJ 2000 Caloric restriction and aging: an update. Exp. Gerontol. 35:299–305. [30]
Matos M, Avelar T & MR Rose 2002 Variation in the rate of convergent evolution: adaptation to a laboratory environment in *Drosophila subobscura*. J. Evol. Biol. 15:673–682. [31]
Matthews REF 1991 Plant Virology, 3rd ed. Academic Press. [23]
May RM & MA Nowak 1995 Coinfection and the evolution of parasite virulence. Proc. R. Soc. Lond. B 261:209–215. [29]
May RM, Endler JA & RE McMurtrie 1975 Gene frequency clines in the presence of selection opposed by gene flow. Am. Nat. 109:659–676. [26]
Maynard Smith J 1964 Group selection and kin selection. Nature 201:1145–1147. [6]
Maynard Smith J 1976 Group selection. Q. Rev. Biol. 51:277–283. [6]
Maynard Smith J 1978 Optimization theory in evolution. Annu. Rev. Ecol. Syst. 9:31–56. [4]
Maynard Smith J 1988 Evolutionary progress and levels of selection. pp. 219–230 in MH Nitecki ed. Evolutionary Progress. Univ. of Chicago Press. [23]
Maynard Smith J 1992 Age and the unisexual lineage. Nature 356:661–662. [26]
Maynard Smith J & J Haigh 1974 The hitch-hiking effect of a favorable gene. Genet. Res. 23:23–35. [5, 7]
Maynard Smith J & E Szathmáry 1993 The origin of chromosomes. I. Selection for linkage. J. Theor. Biol. 164:437–446. [23]
Maynard Smith J & E Szathmáry 1995 The Major Transitions in Evolution. Freeman. [23]
Maynard Smith J, Smith NH, O'Rourke M & BG Spratt 1993 How clonal are bacteria? Proc. Natl. Acad. Sci. USA 90:4384–4388. [31.1]
Mayr E 1942 Systematics and the Origin of Species. Columbia Univ. Press. [1]
Mayr E 1944 Chromosomes and phylogeny. Science 100:11–12. [1]
Mayr E 1954 Change in the genetic environment and evolution. pp. 157–180 in J Huxley, ed. Evolution as a Process. Allen & Unwin. [27]
Mayr E 1963 Animal Species and Evolution. Harvard Univ. Press. [6, 19, 24, 26]
Mayr E 1970 Populations, Species, and Evolution. Harvard Univ. Press. [Species Concepts box]
Mayr E 1982 The Growth of Biological Thought. Harvard Univ. Press. [17]
McAdam AG & S Boutin 2004 Maternal effects and the response to selection in red squirrels. Proc. R. Soc. Lond. B 271:75–79. [2.1]
McCarroll SA, Murphy CT, Zou S, Pletcher SD, Chin CS, Jan YN, Kenyon C, Bargmann CI & H Li 2004 Comparing genomic expression patterns across species identifies shared transcriptional profile in aging. Nat. Genet. 36:197–204. [30]
McDonald JF 1999 Genomic imprinting as a coopted evolutionary character. Trends Ecol. Evol. 14:359. [2]
McDonald JF 2000 Transposable Elements and Genome Evolution. Kluwer Academic. [2]
McDonald JH & M Kreitman 1991 Adaptive protein evolution at the *Adh* locus in *Drosophila*. Nature 351:652–654. [7, 8]
McGinnis W, Garber RL, Wirz J, Kuroiwa A & WJ Gehring 1984a A homologous protein-coding sequence in *Drosophila* homeotic genes and its conservation in other metazoans. Cell 37:403–408. [15]
McGinnis W, Levine MS, Hafen E, Kuroiwa A & WJ Gehring 1984b A conserved DNA sequence in homoeotic genes of the *Drosophila Antennapedia* and *bithorax* complexes. Nature 308: 428–433. [15]
McGuigan KL, Chenoweth SF & MW Blows 2004 Phenotypic divergence along lines of genetic variance. Am. Nat. 165:32–43. [20]
McKenzie A & M Steel 2000 Distributions of cherries for two models of trees. Math. Biosci. 164:81–92. [12]
McLysaght A, Hokamp K & KH Wolfe 2002 Extensive genomic duplication during early

chordate evolution. Nat. Genet. 31:200–204. [2, 13]
McNamara JM, Houston AI, dos Santos MM, Kokko H & R Brooks 2003 Quantifying male attractiveness. Proc. R. Soc. Lond. B 270:1925–1932. [22]
McVean GA & J Vieira 2001 Inferring parameters of mutation, selection and demography from patterns of synonymous site evolution in *Drosophila*. Genetics 157:245–257. [9]
Mead LS & SJ Arnold 2004 Quantitative genetic models of sexual selection. Trends Ecol. Evol. 19:264–271. [22]
Medawar PB 1946 Old age and natural death. Modern Quarterly. 2:30–49. [30]
Meffert LM 1995 Bottleneck effects on genetic variance for courtship repertoire. Genetics 139:365–374. [27]
Meffert LM 1999 How speciation experiments relate to conservation biology. BioScience 49:701–715. [27]
Meffert LM 2000 The evolutionary potential of morphology and mating behavior: the role of epistasis in bottlenecked populations. pp. 177–193 in Wolf JB, Brodie ED III & MJ Wade, eds. Epistasis and the Evolutionary Process. Oxford Univ. Press. [27]
Meffert LM & EH Bryant 1991 Mating propensity and courtship behavior in serially bottlenecked lines of the housefly. Evolution 45:293–306. [27]
Meffert LM & JL Regan 2002 A test of speciation via sexual selection on female preferences. Anim. Behav. 64:955–965. [27]
Meffert LM, Hicks SK & JL Regan 2002 Nonadditive genetic effects in animal behavior. Am. Nat. 160S:S198–S213. [27]
Meffert LM, Regan JL & BW Brown 1999 Convergent evolution of the mating behaviour of founder-flush populations of the housefly. J. Evol. Biol. 12:859–868. [27]
Meiklejohn CD & DL Hartl 2002 A single mode of canalization. Trends Ecol. Evol. 17:468–473. [16]
Mello CC & D Conte Jr 2004 Revealing the world of RNA interference. Nature 431:338–342. [17]
Melov S & A Hubbard 2004 Microarrays as a tool to investigate the biology of aging: a retrospective and a look to the future. Sci. Aging Knowl. Environ. 2004:re7. [30]
Mendelson TC 2003 Sexual isolation evolves faster than hybrid inviability in a diverse and sexually dimorphic genus of fish (Percidae: Etheostoma). Evolution 57:317–327. [24]
Meyer K 1998 Estimating covariance functions for longitudinal data using a random regression model. Genet. Select Evol. 30:221–240. [21]
Mezey JG & D Houle 2003 Comparing G matrices: are common principal components informative? Genetics 165:411–425. [19.5, 20]
Mezey JG, Cheverud JM & GP Wagner 2000. Is the genotype–phenotype map modular? A statistical approach using mouse quantitative trait loci data. Genetics 156:305–311. [19.5]
Michalak P & MAF Noor 2003 Genome-wide patterns of expression in *Drosophila* pure-species and hybrid males. Mol. Biol. Evol. 20:1070–1076. [24, 25]
Michod RE & D Roze 2001 Cooperation and conflict in the evolution of multicellularity. Heredity 86:1–7. [23]
Migliaccio E, Giorgio M, Mele S, Pelicci G, Reboldi P, Pandolfi PP, Lanfrancone L & PG Pelicci 1999 The p66shc adaptor protein controls oxidative stress response and life span in mammals. Nature 402:309–313. [30]
Miller GT & S Pitnick 2002 Sperm–female coevolution in *Drosophila*. Science 298:1230–1233. [31]
Miller RA, Harper JM, Dysko RC, Durkee SJ & SN Austad 2002 Longer life spans and delayed maturation in wild-derived mice. Exp. Biol. Med. 227:500–508. [30]
Mirsky AE & H Ris 1951 The DNA content of animal cells and its evolutionary significance. J. Gen. Physiol. 34:451–462. [10]
Mishler BD 1999 Getting rid of species? pp. 307–315 in R. Wilson, ed. Species: New Interdisciplinary Essays. MIT Press. [Species Concepts box]
Mitchell GA, Labuda D, Fontaine G, Saudubray JM, Bonnefont JP, Lyonnet S, Brody LC, Steel G, Obie C & D Valle 1991 Splice-mediated insertion of an Alu sequence inactivates ornithine-aminotransferase: a role for Alu elements in human mutation. Proc. Natl. Acad. Sci. USA 88:815–819. [2]
Mitchell-Olds T 2001 *Arabidopsis thaliana* and its wild relatives: a model system for ecology and evolution. Trends Ecol. Evol. 16:693–700. [29.1]
Mitra S, Landel H & S Preutt Jones 1996 Species richness covaries with mating system in birds. Auk 113:544–551. [24]
Miyata T & N Suga 2001 Divergence pattern of animal gene families and relationship with the Cambrian explosion. BioEssays 23:1018–1027. [14]
Miyata T & T Yasunaga 1980 Molecular evolution of mRNA: a method for estimating evolutionary rates of synonymous and amino acid substitutions from homologous nucleotide sequences and its application. J. Mol. Evol. 16:23–36. [7]
Möhle M 2000 Ancestral processes in population genetics: the coalescent. J. Theor. Biol. 204:629–638. [12]
Møller AP & H Tegelstrom 1997 Extra-pair paternity and tail ornamentation in the barn

swallow *Hirundo rustica*. Behav. Ecol. Sociobiol. 41:353–360. [18]

Monteiro A, Brakefield PM & V French 1994 The evolutionary genetics and developmental basis of wing pattern variation in the butterfly *Bicyclus anynana*. Evolution 48:1147–1157. [20]

Monteiro A, Prijs J, Bax M, Hakkaart T & PM Brakefield 2003 Mutants highlight the modular control of butterfly eyespot patterns. Evol. Dev. 5:180–187. [20]

Montrose VT, Harris WE & PJ Moore 2004 Sexual conflict and cooperation under naturally occurring male enforced monogamy. J. Evol. Biol. 17:443–452. [22]

Moore AJ & PJ Moore 1999 Balancing sexual selection through opposing mate choice and male competition. Proc. R. Soc. Lond. B 266:711–716. [22]

Moore AJ & T Pizzari 2005. Quantitative genetic models of sexual conflict based on interacting phenotypes. Am. Nat. 165:S88–S97. [22]

Moore AJ, Brodie ED III & JB Wolf 1997 Interacting phenotypes and the evolutionary process. I. Direct and indirect genetic effects of social interactions. Evolution 51:1352–1362. [22]

Moore AJ, Haynes KF, Preziosi RF & PJ Moore 2002 The evolution of interacting phenotypes: genetics and evolution of social dominance. Am. Nat. 160:S186–S197. [22]

Moore AJ, Wolf JB & ED Brodie III 1998 The influence of direct and indirect genetic effects on the evolution of behavior: social and sexual selection meet maternal effects. pp. 22–41 in Mousseau TA & CW Fox, eds. Maternal Effects as Adaptations. Oxford Univ. Press. [22]

Moore PJ & AJ Moore 2001 Reproductive aging and mating: the ticking of the biological clock in female cockroaches. Proc. Natl. Acad. Sci. USA 98:9171–9176. [22]

Moore PJ, Harris WE, Montrose VT, Levin D & AJ Moore 2004 Constraints on evolution and post-copulatory sexual selection: trade-offs among ejaculate characteristics. Evolution 58:1773–1780 [22]

Moore WS & JT Price 1993 Nature of selection in the northern flicker hybrid zone and its implications for speciation theory. pp. 196–225 in RG Harrison, ed. Hybrid Zones and the Evolutionary Process. Oxford Univ. Press. [26]

Morgan GJ 1998 Emile Zuckerkandl, Linus Pauling, and the molecular evolutionary clock, 1959–1965. J. Hist. Biol. 31:155–178. [1]

Morgan HD, Sutherland HGE, Martin DIK & E Whitelaw 1999. Epigenetic inheritance at the agouti locus in the mouse. Nat. Genet. 23:314–318. [17]

Morgan TH 1932 The Scientific Basis of Evolution. Norton. [2]

Moritz C, Donnellan S, Adams M & PR Baverstock 1989 The origin and evolution of parthenogenesis in *Heteronotia binoei* (Gekkonidae): extensive genotypic diversity among parthenogens. Evolution 43:994–1003. [26]

Moriyama EN & JR Powell 1996 Intraspecific nuclear DNA variation in *Drosophila*. Mol. Biol. Evol. 13:261–277. [30.2]

Morrell PL, Lundy KE & MT Clegg 2003 Distinct geographic patterns of genetic diversity are maintained in wild barley (*Hordeum vulgare* ssp. *spontaneum*) despite migration. Proc. Natl. Acad. Sci. USA 100:10812–10817. [12]

Morris RF 1971 Observed and simulated changes in genetic quality in natural populations of *Hyphantria cunea*. Can. Entomol. 103:893–906. [18]

Morris RF & W Fulton 1970 Heritability of diapause intensity in *Hyphantria cunea* and correlated fitness responses. Can. Entomol. 102:927–938. [18]

Morrow J, Scott L, Congdon B, Yeates D, Frommer M & J Sved 2000 Close genetic similarity between two sympatric species of Tephritid fruit fly reproductively isolated by mating time. Evolution 54:899–910. [26]

Mousseau TA & CW Fox 1998 Maternal Effects as Adaptations. Oxford Univ. Press. [2.1]

Mousseau T & D Roff 1987 Natural selection and the heritability of fitness components. Heredity 59:181–197. [19]

Moya A, Galiana A & F Ayala 1995 Founder-effect speciation theory: failure of experimental corroboration. Proc. Natl. Acad. Sci. USA 92:3983–3986. [27]

Moyle LC, Olson MS & P Tiffin 2004 Patterns of reproductive isolation in three angiosperm genera. Evolution 58:1195–1208. [24]

Mueller LD & FJ Ayala 1981 Trade-off between *r*-selection and *K*-selection in *Drosophila* populations. Proc. Natl. Acad. Sci. USA 78:1303–1305. [31]

Mueller LD, Guo PZ & FJ Ayala 1991 Density-dependent natural selection and trade-offs in life history traits. Science 253:433–435. [31]

Mukai T, Chigusa SI, Mettler LE & JF Crow 1972 Mutation rate and dominance of genes affecting viability in *Drosophila melanogaster*. Genetics 72:335–355. [3]

Müller F, Bladre P & U Strähle 2002 Search for enhancers: teleost models in comparative genomic and transgenic analysis of *cis*-regulatory elements. BioEssays 24: 564–572. [8]

Muller HJ 1939 Reversibility in evolution considered from the standpoint of genetics. Biol. Rev. Camb. Phil. Soc. 14:261–280. [6]

Muller HJ 1940 Bearing of the *Drosophila* work on systematics. pp. 185–268 in J Huxley ed. The New Systematics. Clarendon Press. [25]

Muller HJ 1942 Isolating mechanisms, evolution and temperature. Biol. Symp. 6:71–125. [24]

Muller HJ 1950 Our load of mutations. Am. J. Hum. Genet. 2:111–176. [1]

Muller HJ & G Pontecorvo 1940 Artificial mixing of incompatible germ-plasms in *Drosophila*. Science 92:476. [25]

Mundy NI, Badcock NS, Hart T, Scribner K, Janssen K & NJ Nadeau 2004 Conserved genetic basis of a quantitative plumage trait involved in mate choice. Science 303:1870–1873. [15]

Murphy SK & RL Jirtle 2003 Imprinting evolution and the price of silence. BioEssays 25:577–588. [2]

Muse SV 1995 Evolutionary analyses of DNA sequences subject to constraints of secondary structure. Genetics 139:1429–1439. [9]

Mylius SD & O Diekmann 1995 On evolutionary stable life histories, optimization and the need to be specific about density-dependence. Oikos 74:218–224. [4.1]

Nachman MW 2001 Single nucleotide polymorphisms and recombination rate in humans. Trends Genet. 17:481–485. [7]

Nachman MW & SL Crowell 2000 Estimate of the mutation rate per nucleotide in humans. Genetics 156:297–304. [3]

Nachman MW, Bauer VL, Crowell SL & CF Aquadro 1998 DNA variability and recombination rates at X-linked loci in humans. Genetics 150:1133–1141. [7]

Nachman MW, Hoekstra HE & SL D'Agostino 2003 The genetic basis of adaptive melanism in pocket mice. Proc. Natl. Acad. Sci. USA 100:5268–5273. [7, 15]

Nagano N, Orengo CA & JM Thornton 2002 One fold with many functions: the evolutionary relationships between TIM barrel families based on their sequences, structures and functions. J. Mol. Biol. 321:741–765. [13]

Nakamura Y, Gojobori T & T Ikemura 2000 Codon usage tabulated from international DNA sequence databases: status for the year 2000. Nucleic Acids Res. 28:292. [9]

Nanney DL 1960 Microbiology, developmental genetics and evolution. Am. Nat. 94:167–179. [17]

Nath HB & RC Griffiths 1993 The coalescent in two colonies with symmetric migration. J. Math. Biol. 31:841–852. [12]

Needleman SB & CD Wunsch 1970 A general method applicable to the search for similarities in the amino acid sequence of two proteins. J. Mol. Biol. 48:443–453. [28]

Nei M 1987 Molecular Evolutionary Genetics. Columbia Univ. Press. [28]

Nei M & T Maruyama 1975 Lewontin–Krakauer test for neutral genes. Genetics 80:395. [7]

Newman D & D Pilson 1997 Increased probability of extinction due to decreased genetic effective population size: experimental populations of *Clarkia pulchella*. Evolution 51:354–362. [32]

Newman MEJ 2003 The structure and function of complex networks. SIAM Rev. 45:5167–5256. [13.1]

Nielsen R & J Wakeley 2001 Distinguishing migration from isolation: a Markov chain Monte Carlo approach. Genetics 158:885–896. [12]

Nijhout HF 1991 The Development and Evolution of Butterfly Wing Patterns. Smithsonian Inst. Press. [20]

Nijhout HF 2002. The nature of robustness in development. BioEssays 24:553–563. [16]

Nijhout HF, Berg AM & WT Gibson 2003 A mechanistic study of evolvability using the mitogen-activated protein kinase cascade. Evol. Dev. 5:281–294. [16]

Nixon KC 1999 The parsimony ratchet, a new method for rapid parsimony analysis. Cladistics 15:407–414. [28]

Noller HF & CR Woese 1981 Secondary structure of 16S ribosomal RNA. Science 212:403–411. [9]

Noor MAF 1999 Reinforcement and other consequences of sympatry. Heredity 83:503–508. [24]

Noor MAF, Grams KL, Bertucci LA, Almendarez Y, Reiland J & KR Smith 2001 The genetics of reproductive isolation and the potential for gene exchange between *Drosophila pseudoobscura* and *D. persimilis* via backcross hybrid males. Evolution 55:512–521. [24]

Nordborg M 2001 Coalescent theory. pp 179–212 in Balding DJ, Bishop M & C Cannings, eds. Handbook of Statistical Genetics. Wiley. [12]

Nordborg M & SM Krone 2002 Separation of time scales and convergence to the coalescent in structured populations. pp. 194–232 in Slatkin M & M Veuille, eds. Modern Developments in Theoretical Population Genetics. Oxford Univ. Press. [12]

Nordborg M, Borevitz JO, Bergelson J, Berry CC, Chory J, et al. 2002 The extent of linkage disequilibrium in *Arabidopsis thaliana*. Nat. Genet. 30:190–193. [29.1]

Nordskog AW 1977 Success and failure of quantitative genetic theory in poultry. pp. 47–51 in Pollack E, Kempthorne, O & TB Baily Jr, eds. Proceedings of the International Conference on Quantitative Genetics. Iowa State Univ. Press. [18]

Nunney L 1996 The response to selection for fast larval development in *Drosophila* and its effect on adult weight: an example of a fitness tradeoff. Evolution 50:1193–1204. [31]

Nuzhdin SV, Pasyukova EG, Dilda CL, Zeng Z-B & TFC Mackay 1997 Sex-specific quantitative trait loci affecting longevity in *Drosophila melanogaster*. Proc. Natl. Acad. Sci. USA 94:9734–9739. [30]

Ny T, Elgh F & B Lund 1984 The structure of human tissue-type plasminogen activator gene: correlation of intron and exon structures to functional and structural domains. Proc. Natl. Acad. Sci. USA 81:5355–5359. [14]

Ohno S 1970 Evolution by Gene Duplication. Springer. [11, 14]

Ohno S 1972 So much "junk" in our genomes. pp. 366–370 in HH Smith, ed. Evolution of Genetic Systems. Brookhaven Symp. Biol. [10]

Ohno S 1985 Dispensable genes. Trends Genet. 1:160–164. [2]

Ohta T 1972 Evolutionary rate of cistrons and DNA divergence. J. Mol. Evol. 1:150–157. [8]

Ohta T 1973 Slightly deleterious mutant substitutions in evolution. Nature 246:96–97. [13]

Ohta T 1992 The nearly neutral theory of molecular evolution. Annu. Rev. Ecol. Syst. 23:263–286. [2, 8]

Ohta T 1995 Synonymous and nonsynonymous substitutions in mammalian genes and the nearly neutral theory. J. Mol. Evol. 40:56–63. [8]

Ohta T & JH Gillespie 1996 Development of neutral and nearly neutral theories. Theor. Pop. Biol. 49:128–142. [9]

Ohta T & M Kimura 1971 On the constancy of the evolutionary rate of cistrons. J. Mol. Evol. 1:18–25. [1, 8]

Ohta T & M Kimura 1973 A model of mutation appropriate to estimate the number of electrophoretically detectable alleles in a finite population. Genet. Res. 22:201–204. [7]

Olby R 1979 Mendel no Mendelian? History Sci. 17:53–72. [1]

Oleksiak MF, Churchill GA & DL Crawford 2002 Variation in gene expression within and among natural populations. Nat. Genet. 32:261–266. [2, 25]

O'Neill RJ, O'Neill MJ & JAM Graves 1998 Undermethylation associated with retroelement activation and chromosome remodelling in an interspecific mammalian hybrid. Nature 393:68–72. [17]

Online Mendelian Inheritance in Man, OMIM (TM) 2000 McKusick-Nathans Institute for Genetic Medicine, Johns Hopkins Univ. (Baltimore, MD) and National Center for Biotechnology Information, National Library of Medicine, Bethesda, MD. World Wide Web URL: http://www.ncbi.nlm.nih.gov/omim/ [2]

Ono K, Suga H, Iwabe N, Kuma K & T Miyata 1999 Multiple protein tyrosine phosphatases in sponges and explosive gene duplication in the early evolution of animals before the parazoan–eumetazoan split. J. Mol. Evol. 48:654–662. [14]

Oostermeijer JGB 2000 Population viability analysis of the rare *Gentiana pneumonanthe*: the importance of genetics, demography and reproductive biology. pp. 313–334 in Young AG & GM Clarke, eds. Genetics, Demography and Viability of Fragmented Populations. Cambridge Univ. Press. [32]

Orel N & H Puchta 2003 Differences in the processing of DNA ends in *Arabidopsis thaliana* and tobacco: possible implications for genome evolution. Plant Mol. Biol. 51:523–531. [10]

Orgel LE & Crick FHC 1980 Selfish DNA: the ultimate parasite. Nature 284:604–607. [10]

Orr HA 1995 The population genetics of speciation: the evolution of hybrid incompatibilities. Genetics 139:1805–1813. [6, 24]

Orr HA 1996 Dobzhansky, Bateson and the genetics of speciation. Genetics 144:1331–1335. [24]

Orr HA 1997 Haldane's Rule. Annu. Rev. Ecol. Syst. 28:195–218. [24, 25]

Orr HA 2001 The genetics of species differences. Trends Ecol. Evol. 16:343–350. [25]

Orr HA 2002 The population genetics of adaptation: the adaptation of DNA sequences. Evolution 56:1317–1330. [5]

Orr HA & JA Coyne 1992 The genetics of adaptation revisited. Am. Nat. 140:725–742. [25]

Orr HA & DC Presgraves 2000 Speciation by postzygotic isolation: forces, genes and molecules. BioEssays 22:1085–1094. [25]

Otto SP & CD Jones 2000 Detecting the undetected: estimating the total number of loci underlying a quantitative trait. Genetics 156:2093–2107. [2]

Otto SP & T Lenormand 2002. Resolving the paradox of sex and recombination. Nat. Genet. 3:252–261. [2]

Ouellette M, Hettema E, Wüst D, Fase-Fowler F & P Borst 1991 Direct and inverted DNA repeats associated with P-glycoprotein gene amplification in drug resistant *Leishmania*. EMBO J. 10:1009–1016. [14]

Overington J, Donnelly D, Johnson MS, Sali A & TL Blundell 1992 Environment-specific amino acid substitution tables: tertiary templates and prediction of protein folds. Protein Sci. 1:216–226. [13]

Page RDM 1993a Genes, organisms, and areas: the problem of multiple lineages. Syst. Biol. 42:77–84. [11]

Page RDM 1993b On islands of trees and the efficacy of different methods of branch swapping in finding most-parsimonious trees. Syst. Biol. 42:200–210. [28]

Page RDM & Homes EC 1998 Molecular Evolution: A Phylogenetic Approach. Blackwell Sci. [13]
Palopoli MF & C-I Wu 1994 Genetics of hybrid male sterility between *Drosophila* sibling species: a complex web of epistasis is revealed in interspecific studies. Genetics 138:329–341. [25]
Pamilo P & M Nei 1988 Relationships between gene trees and species trees. Mol. Biol. Evol. 5:568–583. [12]
Parejko K & SI Dodson 1991 The evolutionary ecology of an antipredator reaction norm: *Daphnia pulex* and *Chaoborus americanus*. Evolution 45:1665–1674. [21]
Park T 1948 Experimental studies of interspcies competition. I. Competition between populations of the flour beetles, *Tribolium confusum* Duval and *Tribolium castaneum* Herbst. Ecol. Monog. 18:265–308. [6]
Park T 1954 Experimental studies of interspecies competition. II. Temperature, humidity, and competition in two species of *Tribolium*. Physiol. Zoöl. 27:177–238. [6]
Parker GA 1979 Sexual selection and sexual conflict. pp. 123–166 in Blum MS & NA Blum, eds. Sexual Selection and Reproductive Competition in Insects. Academic Press. [22]
Parker MA 1994 Pathogens and sex in plants. Evol. Ecol. 8:560–584. [29]
Parsch J, Braverman JM & W Stephan 2000 Comparative sequence analysis and patterns of covariation in RNA secondary structures. Genetics 154:909–921. [9]
Parsch J, Tanda S & W Stephan 1997 Site-directed mutations reveal long-range compensatory interactions in the *Adh* gene of *Drosophila melanogaster*. Proc. Natl. Acad. Sci. USA 94:928–933. [9]
Partridge L & NH Barton 2000 Evolving evolvability. Nature 407:457–458. [16]
Partridge L, Green A & K Fowler 1987 Effects of egg-production and of exposure to males on female survival in *Drosophila melanogaster*. J. Insect Physiol. 33:745–749. [30]
Paterson HEH 1985 The recognition concept of species. pp. 21–29 in ES Vrba, ed. Species and Speciation. Transvaal Museum, Pretoria (Transvaal Museum Monograph; 4). [Species Concepts box]
Patthy L 1985 Evolution of the proteases of blood coagulation and fibrinolysis by assembly from modules. Cell 41:657–663. [14]
Patthy L 1987 Intron-dependent evolution: preferred types of exons and introns. FEBS Lett. 214:1–7. [14]
Patthy L 1991 Modular exchange principles in proteins. Curr. Opin. Struct. Biol. 1:351–361. [14]
Patthy L 1994 Exons and introns. Curr. Opin. Struct. Biol. 4:383–392. [14]
Patthy L 1995 Protein Evolution by Exon-Shuffling. Molecular Biology Intelligence Unit. R.G. Landes/Springer. [14]
Patthy L 1999a Genome evolution and the evolution of exon-shuffling: a review. Gene 238:103–114. [11, 14]
Patthy L 1999b Protein Evolution. Blackwell Sci. [14]
Patthy L 2003 Modular assembly of genes and the evolution of new functions. Genetica 118:217–231. [14]
Paulsen SM 1996 Quantitative genetics of the wing color pattern in the buckeye butterfly (*Precis coenia* and *Precis evarete*): evidence against the constancy of **G**. Evolution 50:1585–1597. [20]
Payre F, Vincent A & S Carreno 1999 *ovo*/*svb* integrates Wingless and DER pathways to control epidermis differentiation. Nature 400:271–275. [15]
Payseur BA & MW Nachman 2000 Microsatellite variation and recombination rate in the human genome. Genetics 156:1285–1298. [7]
Payseur BA, Cutter AD & MW Nachman 2002 Searching for evidence of positive selection in the human genome using patterns of microsatellite variability. Mol. Biol. Evol. 19:1143–1153. [7]
Perez DE & C-I Wu 1995 Further characterization of the *Odysseus* locus of hybrid sterility in *Drosophila*: one gene is not enough. Genetics 140:201–206. [25]
Perez DE, Wu C-I, Johnson NA & ML Wu 1993 Genetics of reproductive isolation in the *Drosophila simulans* clade: DNA marker-assisted mapping and characterization of a hybrid-male sterility gene, *Odysseus* (*Ods*). Genetics 134:261–275. [25]
Peripato AC, de Brito RA, Matioli SR, Pletscher LS, Vaughn TT & JM Cheverud 2004 Epistasis among quantitative trait loci affecting litter size in mice. J. Evol. Biol. 17:593–602. [19]
Peripato AC, de Brito RA, Vaughn TT, Pletscher LS, Matioli SR & JM Cheverud 2002 Quantitative trait loci for maternal performance for offspring survival in mice. Genetics 162:1341–1353. [19]
Pétavy G, Morin JP, Moreteau B & JR David 1997 Growth temperature and phenotypic plasticity in two *Drosophila* sibling species: probable adaptive changes in flight capacities. J. Evol. Biol. 10:875–887. [21]
Peters AD & CM Lively 2000 The Red Queen and fluctuating epistasis: a population genetic analysis of antagonistic coevolution. Am. Nat. 154:393–405. [29]
Petrov DA 2001 Evolution of genome size: new approaches to an old problem. Trends Genet. 17:23–28. [10]

Petrov DA 2002a DNA loss and evolution of genome size in *Drosophila*. Genetica 115:81–91. [10]

Petrov DA 2002b Mutational equilibrium model of genome size evolution. Theor. Pop. Biol. 61:531–544. [10]

Petrov DA & DL Hartl 1998 High rate of DNA loss in the *Drosophila melanogaster* and *Drosophila virilis* species groups. Mol. Biol. Evol. 15:293–302. [10]

Petrov DA & DL Hartl 2000 Pseudogene evolution and natural selection for a compact genome. J. Hered. 91:221–227. [10]

Petrov DA, Aminetzach YT, Davis JC, Bensasson D & AE Hirsh 2003 Size matters: non-LTR retrotransposable elements and ectopic recombination in *Drosophila*. Mol. Biol. Evol. 20:880–892. [10]

Petrov DA, Lozovskaya ER & DL Hartl 1996 High intrinsic rate of DNA loss in *Drosophila*. Nature 384:346–349. [9, 10]

Petrov DA, Schutzman JL, Hartl DL & ER Lozovskaya 1995 Diverse transposable elements are mobilized in hybrid dysgenesis in *Drosophila virilis*. Proc. Natl. Acad. Sci. USA 92:8050–8054. [10]

Phillips A, Janies D & W Wheeler 2000 Multiple sequence alignment in phylogenetic analysis. Mol. Phylogenet. Evol. 16:317–330. [28]

Phillips PC 1998 The language of gene interaction. Genetics 149:1167–1171. [6.1]

Phillips PC & SJ Arnold 1989 Visualizing mutlivariate selection. Evolution 43:1209–1222. [18.1, 20]

Phillips PC & SJ Arnold 1999 Hierarchical comparison of genetic variance–covariance matrices. I. Using the Flury hierarchy. Evolution 53:1506–1515. [20]

Phillips PC, Whitlock MC & K Fowler 2001 Inbreeding changes the shape of the genetic covariance matrix in *Drosophila melanogaster*. Genetics 158:1137–1145. [20]

Phinchongsakuldit J, MacArthur S & JFY Brookfield 2004 Evolution of developmental genes: molecular microevolution of enhancer sequences at the *Ubx* locus in *Drosophila* and its impact on developmental phenotypes. Mol. Biol. Evol. 21:348–363. [2]

Piatigorsky J & G Wistow 1991 The recruitment of crystallins: new functions precede gene duplication. Science 252:1078–1079. [11]

Pigliucci M 2001 Phenotypic Plasticity. Beyond Nature and Nurture. Johns Hopkins Univ. Press. [21]

Pigliucci M & CJ Murren 2003 Genetic assimilation and a possible evolutionary paradox: can macroevolution sometimes be so fast as to pass us by? Evolution 57:1455–1464. [21]

Pikaard CS 2001 Genomic change and gene silencing in polyploids. Trends Genet. 17:675–677. [17]

Pizzari T & TR Birkhead 2002 The sexually-selected sperm hypothesis: sex-biased inheritance and sexual antagonism. Biol. Rev. 77:183–209. [22]

Pizzari T & RR Snook 2003 Sexual conflict and sexual selection: chasing away the paradigm shifts. Evolution 57:1223–1236. [22]

Plowman GD, Sudarsanam S, Bingham J, Whyte D & T Hunter 1999 The protein kinases of *Caenorhabditis elegans*: a model for signal transduction in multicellular organisms. Proc. Natl. Acad. Sci. USA 96:13603–13610. [14]

Pluzhnikov A & P Donnelly 1996 Optimal sequencing strategies for surveying molecular genetic diversity. Genetics 144:1247–1262. [12]

Porter AH & NA Johnson 2002 Speciation despite gene flow when developmental pathways evolve. Evolution 56:2103–2111. [24, 25]

Posada D & KA Crandall 1998 Modeltest: testing the model of DNA substitution. Bioinformatics 14:817–818. [8, 28]

Posada D & KA Crandall. 2001 Selecting the best-fit model of nucleotide substitution. Syst. Biol. 50:580–601. [28]

Poulton EB 1904 What is a species? Proc. Entomol. Soc. Lond. 1903:lxxvii–cxvi. [Species Concepts box]

Poulton J & AA Winn 2002 Costs of canalization and plasticity in response to neighbors in *Brassica rapa*. Plant Spec. Biol. 17:109–118. [21]

Powell JR 1978 The founder-flush speciation theory: an experimental approach. Evolution 32:465–474. [27]

Powell JR 1994 Molecular techniques in population genetics: a brief history. pp. 31–156 in Schierwater B, Streit B, Wagner G & R DeSalle, eds. Molecular Ecology and Evolution: Approaches and Applications. Birkhauser. [1]

Powell JR 1997 Progress and Prospects in Evolutionary Biology: The *Drosophila* Model. Oxford Univ. Press. [2, 25, 30.2]

Powell JR & R DeSalle 1995 *Drosophila* molecular phylogenies and their uses. Evol. Biol. 28:87–138. [30.2]

Powell JR & L Morton 1979 Inbreeding and mating patterns in *Drosophila pseudoobscura*. Behav. Gene. 9:425–429. [27]

Prager E & A Wilson 1988 Ancient origin of lactalbumin from lysozyme: analysis of DNA and amino acid sequences. J. Mol. Evol. 27:326–335. [28]

Presgraves DC 2002 Patterns of post-zygotic isolation in Lepidoptera. Evolution 56:1168–1183. [24]

Presgraves DC, Balagopalan L, Abmayr SM & HA Orr 2003 Adaptive evolution drives divergence of a hybrid inviability gene between two species of *Drosophila*. Nature 423:715–719. [25]

Preziosi RF & DJ Fairbairn 2000 Lifetime selection on adult body size and components of body size in a water strider: opposing selection and maintenance of sexual size dimorphism. Evolution 54:558–566. [4.1]

Price CSC, Kim CH, Gronlund CJ & JA Coyne 2001 Cryptic reproductive isolation in the *Drosophila simulans* clade. Evolution 55:81–92. [24]

Price CSC, Kim CH, Poslusky J & JA Coyne 2000 Mechanisms of conspecific sperm precedence in *Drosophila*. Evolution 54:2028–2037. [24]

Price G 1970 Selection and covariance. Nature 227:520–521. [4]

Price G 1972 Extension of covariance selection mathematics. Ann. Hum. Genet. 35:485–490. [4]

Price PW 1996 Biological Evolution. Saunders College Publ. [4.1]

Price TD & A Bontrager 2001 Evolutionary genetics: the evolution of plumage patterns. Curr. Biol. 11:R405–R408. [15]

Price TD & MD Bouvier 2002 The evolution of F1 post-zygotic incompatibilities in birds. Evolution 56:2083–2089. [24]

Prince VE & FB Pickett 2002 Splitting pairs: the diverging fates of duplicated genes. Nat. Rev. Genet. 3:827–837. [11]

Pritchard JK 2001 Are rare variants responsible for susceptibility to complex diseases? Am. J. Hum. Genet. 69:124–137. [12]

Pritchard JK, Stephens M & P Donnelly 2000 Inference of population structure using multilocus genotype data. Genetics 155:945–959. [Species Concepts box]

Promislow DEL 1991 Senescence in natural populations of mammals: a comparative study. Evolution 45:1869–1887. [30]

Promislow DEL 2003 Mate choice, sexual conflict, and evolution of senescence. Behav. Genet. 33:191–201. [30]

Promislow DEL 2004 Protein networks, pleiotropy and the evolution of senescence. Proc. R. Soc. Lond. B 271:1225–1234. [13.1]

Promislow DEL & M Tatar 1998 Mutation and senescence: where genetics and demography meet. Genetica 102/103:299–314. [30]

Promislow DEL, Tatar M, Khazaeli A & JW Curtsinger 1996 Age-specific patterns of genetic variance in *Drosophila melanogaster*. I. Mortality. Genetics 143:839–848. [30]

Proulx SR & PC Phillips 2005 The opportunity for canalization and the evolution of genetic networks. Am. Nat. 165:147–162 [13.1]

Provine WB 1971 The Origins of Theoretical Population Genetics. Univ. of Chicago Press. [1]

Provine WB 1981 Origins of the genetics of natural populations series. pp. 1–76 in Lewontin RC, Moore JA, Provine WB & B Wallace, eds. Dobzhansky's Genetics of Natural Populations I–XLIII. Columbia Univ. Press. [1]

Provine WB 1986 Sewall Wright and Evolutionary Biology. Univ. of Chicago Press. [1]

Provine WB 1988 Progress in evolution and the meaning of life. pp. 49–74 in M Nitecki, ed. Evolutionary Progress. Univ. of Chicago Press. [1]

Provine WB 1989 Founder effects and genetic revolutions in microevolution and speciation: a historical perspective. pp. 43–78 in Giddings LV, Kaneshiro K & W Anderson, eds. Genetics, Speciation and the Founder Principle. Oxford Univ. Press. [1]

Provine WB 1990 The neutral theory of molecular evolution in historical perspective. pp. 17–31 in Takahata N & J Crow, eds. Population Biology of Genes and Molecules. Baifukan. [1]

Provine WB 1992 The R. A. Fisher–Sewall Wright controversy. pp. 201–229 in S Sarker, ed. The Founders of Evolutionary Genetics. Kluwer. [1]

Prowell DP 1998 Sex linkage and speciation. pp. 309–319 in Howard DJ & SH Berlocher, eds. Endless Forms: Species and Speciation. Oxford Univ. Press. [24]

Prusiner SB 1998 Prions. Proc. Natl. Acad. Sci. USA 95:13363–13383. [17]

Ptacek MB, Gerhardt HC & RD Sage 1994 Speciation by polyploidy in treefrogs: multiple origins of the tetraploid, *Hyla versicolor*. Evolution 48:898–908. [10]

Ptak SE & DA Petrov 2002 How intron splicing affects the deletion and insertion profile in *Drosophila melanogaster*. Genetics 162:1233–1244. [10]

Ptashne M & A Gann 2002 Genes & Signals. Cold Spring Harbor Lab. Press. [15]

Queitsch C, Sangster TA & S Lindquist 2002. Hsp90 as a capacitor of phenotypic variation. Nature 417:618–624. [16]

Queller DC 1992 A general model for kin selection. Evolution 46:376–380. [23]

Raff RA 1996 The Shape of Life. Univ. of Chicago Press. [19.5]

Raff RA & TC Kaufman 1983 Embryos, Genes, and Evolution. Indiana Univ. Press. [15]

Rainey PB & M Travisano 1998 Adaptive radiation in a heterogeneous environment. Nature 394:69–72. [31]

Rakyan VK, Chong S, Champ ME, Cuthbert PC, Morgan HD, Luu KVK & E Whitelaw 2003 Transgenerational inheritance of epigenetic states at the murine *Axin*^{Fu} allele occurs after maternal and paternal transmission. Proc. Natl. Acad. Sci. USA 100:2538–2543. [17]

Ralls K, Ballou JD & A Templeton 1988 Estimates of lethal equivalents and the cost of inbreeding in mammals. Conserv. Biol. 2:185–193. [32]

Rambaut A 2002 Se-Al: Sequence alignment editor, version 2.0. http://evolve.zoo.ox.ac.uk [28]
Rambaut A & L Bromham 1998 Estimating divergence dates from molecular sequences. Mol. Biol. Evol. 15:442–448. [8]
Ramos-Onsins SE & J Rozas 2002 Statistical properties of new neutrality tests against population growth. Mol. Biol. Evol. 19:2092–2100. [12]
Randolph LF, Nelson IS & RL Plaisted 1967 Negative evidence of introgression affecting the stability of Louisiana Iris species. Cornell Univ. Ag. Exp. Station Mem. 398:1–56. [26]
Rannala B & Y Michalakis 2002 Population genetics and cospeciation: from process to pattern. pp. 120–143 in R Page, ed. Tangled Trees: Phylogeny, Coespeciation and Coevolution. Univ. of Chicago Press. [29]
Ranz JM, Namgyal K, Gibson G & DL Hartl 2004 Anomalies in the expression profile of interspecific hybrids of *Drosophila melanogaster* and *Drosophila simulans*. Genome Res. 14:373–379. [25]
Reale D & M Festa-Bianchet 2000 Quantitative genetics of life-history traits in a long-lived wild mammal. Heredity 85:593–603. [30]
Reddy BVB, Nagarajaram HA & TL Blundell 1999 Analysis of interactive packing of secondary structural elements in alpha/beta units in proteins. Protein Sci. 8:573–586. [13]
Reed DH & R Frankham 2003 Population fitness is correlated with genetic diversity. Conserv. Biol. 17:230–237. [32]
Reed DH, Lowe E, Briscoe DA & R Frankham 2003 Inbreeding and extinction: effects of rate of inbreeding. Conserv. Genet. 4:405–410. [32]
Reeve HK & DW Pfennig 2002 Genetic biases for showy males: are some genetic systems especially conducive to sexual selection? Proc. Natl. Acad. Sci. USA 100:1089–1094. [22]
Reeve JP 2000 Predicting long-term response to selection. Genet. Res. 75:83–94. [20]
Regan JL, Meffert LM & EH Bryant 2003 A direct experimental test of founder-flush effects on the evolutionary potential for assortative mating. J. Evol. Biol. 16:302–312. [27]
Reid SD, Herbelin CJ, Bumbaugh AC, Selander RK & TS Whittam 2000 Parallel evolution of virulence in pathogenic *Escherichia coli*. Nature 406:64–67. [31.1]
Riedl R 1978 Order in Living Organisms. Wiley. [19]
Reinhold K 1998 Sex linkage among genes controlling sexually selected traits. Behav. Ecol. Sociobiol. 44:1–7. [22]
Relyea RA 2002 Costs of phenotypic plasticity. Am. Nat. 159:272–282. [21]
Rendel JM 1967 Canalisation and Gene Control. Logos Press.
Reznick DN, Bryant MJ, Roff D, Ghalambor CK & DE Ghalambor 2004 Effect of extrinsic mortality on the evolution of senescence in guppies. Nature 431:1095–1099. [30]
Reznick D, Buckwalter G, Groff J & D Elder 2001 The evolution of senescence in natural populations of guppies (*Poecilia reticulata*): a comparative approach. Exp. Gerontol. 36:791–812. [30]
Reznick DN, Shaw FH, Rodd FH & RG Shaw 1997 Evaluation of the rate of evolution in natural populations of guppies (*Poecilia reticulata*). Science 235:1934–1937. [31]
Rhymer JM & D Simberloff 1996 Extinction by hybridization and introgression. Annu. Rev. Ecol. Syst. 27:83–109. [26]
Rice S 2004 Developmental associations between traits: covariance and beyond. Genetics 166:513–526. [19]
Rice WR 1996 Sexually antagonistic male adaptation triggered by experimental arrest of female evolution. Nature 381:232–234. [31]
Rice WR 2002 Experimental tests of the adaptive significance of sexual recombination. Nat. Rev. Genet. 3:241–251 [31]
Rice WR & EE Hostert 1993 Laboratory experiments on speciation: what have we learned in 40 years? Evolution 44:1140–1152. [27]
Richards AJ 1997 Plant Breeding Systems. Chapman & Hall. [32]
Richmond R 1970 Non-Darwinian evolution: a critique. Nature 225:1025–1028. [1]
Ricklefs RE 1998 Evolutionary theories of aging: confirmation of a fundamental prediction, with implications for the genetic basis and evolution of life span. Am. Nat. 152:24–44. [30]
Ridley M 1989 The cladistic solution to the species problem. Biol. Philos. 4:1–16. [Species Concepts box]
Ridley M 2004 Evolution, 3rd ed. Blackwell Sci. [4.1]
Riedl R 1978 Order in Living Organisms. Wiley. [19]
Rieseberg LH 1997 Hybrid origins of plant species. Annu. Rev. Ecol. Syst. 28:359–389. [26]
Rieseberg LH, Raymond O, Rosenthal DM, Lai Z, Livingstone K, Nakazato T, Durphy JL, Schwarzbach AE, Donovan LA & C Lexe 2003 Major ecological transitions in wild sunflowers facilitated by hybridization. Science 301:1211–1216. [2, 26]
Rieseberg LH, Van Fossen C & AM Desrochers 1995 Hybrid speciation accompanied by genomic reorganization in wild sunflowers. Nature 375:313–316. [26]
Rifkin SA, Kim J & KP White 2003 Evolution of gene expression in the *Drosophila melanogaster* subgroup. Nat. Genet. 33:138–144. [25, 30.2]
Ringo J, Barton K & H Dowse 1986 The effect of genetic drift on mating propensity, courtship behaviour, and postmating fitness

in *Drosophila simulans*. Behaviour 97:226–233. [27]

Riska B, Rutledge JJ & WR Atchley 1984 A genetic analysis of targeted growth in mice. Genetics 107:79–101. [19]

Ritchie MG 1996 The shape of female mating preferences. Proc. Natl. Acad. Sci. USA 93:14628–14631. [22]

Ritchie MG 2001 The inheritance of female preference functions in a mate recognition system. Proc. R. Soc. Lond. B 267:327–332. [22]

Ritchie MG, Saarikettu M, Livingstone S & A Hoikkala 2001 Characterization of female preference functions for *Drosophila montana* courtship song and a test of the temperature coupling hypothesis. Evolution 55:721–727. [22]

Ritland K, Newton C & HD Marshall 2001 Inheritance and population structure of the white-phased "Kermode" black bear. Curr. Biol. 11:1468–1472. [15]

Rivera MC & JA Lake 2004 The ring of life provides evidence for a genome fusion origin of eukaryotes. Nature 431:152–155 [2]

Roach DA & RD Wulff 1987 Maternal effects in plants. Annu. Rev. Ecol. Syst. 18:209–235. [2.1]

Robertson A 1966 A mathematical model of the culling process in dairy cattle. Anim. Prod. 8:95–108. [4]

Robertson A 1975 Remarks on the Lewontin–Krakauer test. Genetics 80:396. [7]

Robinson T, Johnson NA & MJ Wade 1994 Postcopulatory, pre-zygotic isolation: intraspecific and interspecific sperm precedence in *Tribolium* spp., flour beetles. Heredity 73:155–159. [24]

Rodin A & W Li 2000 A rapid heuristic algorithm for finding minimum evolution trees. Mol. Phylogenet. Evol. 16:173–179. [28]

Rodriguez RL & MD Greenfield 2003 Genetic variance and phenotypic plasticity in a component of female mate choice in an ultrasonic moth. Evolution 57:1304–1313. [22]

Rodríguez-Trelles F, Tarrío R & FJ Ayala 1999 Switch in codon bias and increased rates of amino acid substitution in the *Drosophila saltans* species group. Genetics 153:339–350. [8]

Rodríguez-Trelles F, Tarrío R & FJ Ayala 2000 Evidence for a high ancestral GC content in *Drosophila*. Mol. Biol. Evol. 17:1710–1717. [8]

Rodríguez-Trelles F, Tarrío R & FJ Ayala 2001 Erratic overdispersion of three molecular clocks: GPDH, SOD, and XDH. Proc. Natl. Acad. Sci. USA 98:11405–11410. [8]

Rodríguez-Trelles F, Tarrío R & FJ Ayala 2002 A methodological bias toward overestimation of molecular evolutionary time scales. Proc. Natl. Acad. Sci. USA 99:8112–8115. [8]

Rodríguez-Trelles F, Tarrío R & FJ Ayala 2003 Evolution of *cis*-regulatory regions versus codifying regions. Int. J. Dev. Biol. 47:665–673. [8]

Roff DA 1986 The genetic basis of wing dimorphism in the sand cricket, *Gryllus firmus* and its relevance to the evolution of wing dimorphisms in insects. Heredity 57:221–231. [18]

Roff DA 1992 The Evolution of Life Histories: Theory and Analysis. Chapman & Hall. [4.1]

Roff DA 1997 Evolutionary Quantitative Genetics. Chapman & Hall. [18]

Roff DA 2000 The evolution of the **G** matrix: selection or drift? Heredity 84:135–142. [18]

Roff DA 2002 Life History Evolution. Sinauer Assoc. [18, 30]

Roff DA & DJ Fairbairn 1999 Predicting correlated responses in natural populations: changes in JHE activity in the Bermuda population of the sand cricket, *Gryllus firmus*. Heredity 83:440–450. [18]

Roff DA & TA Mousseau 1999 Does natural selection alter genetic architecture? An evaluation of quantitative genetic variation among populations of *Allonemobius socius* and *A. fasciatus*. J. Evol. Biol. 12:361–369. [18]

Roff DA, Mousseau TA, Møller AP, Lope FD & N Saino 2004 Geographic variation in the **G** matrices of wild populations of the barn swallow. Heredity 93:8–14. [18]

Rohde DLT, Olson S & JT Chang 2004 Modelling the recent common ancestry of all living humans. Nature 431:562–566. [12]

Roldan ERS & M Gomendio 1999 The Y chromosome as a battle ground for sexual selection. Trends Ecol. Evol. 14:58–62. [22]

Ronquist F & JP Huelsenbeck 2003 MrBayes 3: Bayesian phylogenetic inference under mixed models. Bioinformatics 19:1572–1574. [28]

Ronshaugen M, McGinnis N & W McGinnis 2002 Hox protein mutation and macroevolution of the insect body plan. Nature 415:914–917 [15]

Rose LE, Bittner-Eddy PD, Langley CH, Holub EB, Michelmore RW & JL Beynon 2004 The maintenance of extreme amino acid diversity at the disease resistance gene, RPP13, in *Arabidopsis thaliana*. Genetics 166:1517–1527. [29]

Rose MR 1984 Laboratory evolution of postponed senescence in *Drosophila melanogaster*. Evolution 38:1004–1010. [31]

Rose MR 1991 Evolutionary Biology of Aging. Oxford Univ. Press. [30]

Rose MR & B Charlesworth 1980 A test of evolutionary theories of senescence. Nature 287:141–142. [30]

Rose MR & GV Lauder 1996 Adaptation. Academic Press. [31]

Rose MR, Nusbaum TJ & AK Chippindale 1996 Laboratory selection: the experimental wonderland and the Cheshire Cat Syndrome. pp. 221–242 in Rose MR & GV Lauder, eds. Adaptation. Academic Press. [31]
Rosenberg NA 2002 The probability of topological concordance of gene trees and species trees. Theor. Pop. Biol. 61:225–247. [12]
Rosenberg NA 2003 The shapes of neutral gene genealogies in two species: probabilities of monophyly, paraphyly, and polyphyly in a coalescent model. Evolution 57:1465–1477. [12]
Rosenberg NA & MW Feldman 2002 The relationship between coalescence times and population divergence times. pp. 130–164 in Slatkin M & M Veuille, eds. Modern Developments in Theoretical Population Genetics. Oxford Univ. Press. [12]
Rosenberg NA & AE Hirsh 2003 On the use of star-shaped genealogies in inference of coalescence times. Genetics 164:1677–1682. [12]
Rosenberg NA & M Nordborg 2002 Genealogical trees, coalescent theory and the analysis of genetic polymorphisms. Nat. Rev. Genet. 3:380–390. [12]
Rousset F, Pelandakis M & M Solignac 1991 Evolution of compensatory substitutions through G-U intermediate state in *Drosophila* rRNA. Proc. Natl. Acad. Sci. USA 88:10032–10036. [9]
Routman EJ & JM Cheverud 1997 Gene effects on a quantitative trait: two-locus epistatic effects measured at microsatellite markers and at estimated QTL. Evolution 51:1654–1662. [19]
Rowe L & D Houle 1996 The lek paradox and the capture of genetic variance by condition dependent traits. Proc. R. Soc. Lond. B 263:1415–1421. [22]
Roy BA & JW Kirchner 2000 Evolutionary dynamics of pathogen resistance and tolerance. Evolution 54:51–63. [29]
Rubin GM, Yandell MD, Wortman JR, Gabor Miklos GL, Nelson CR, et al. 2000 Comparative genomics of the eukaryotes. Science 287:2204–2215. [11, 14]
Rundle HD, Mooers AØ & MC Whitlock 1998 Single founder-flush events and the evolution of reproductive isolation. Evolution 52:1850–1855. [27]
Russell AF, Brotherton PNM, McIlrath GM, Sharpe LL & TH Clutton-Brock 2003 Breeding success in cooperative meerkats: effects of helper number and maternal state. Behav. Ecol. 14:486–492. [2.1]
Russell LB & WL Russell 1996 Spontaneous mutations recovered as mosaics in the mouse specific-locus test. Proc. Natl. Acad. Sci. USA 93:13072–13077. [3]
Russell RB, Saqi MA, Sayle RA, Bates PA & MJ Sternberg 1997 Recognition of analogous and homologous protein folds: analysis of sequence and structure conservation. J. Mol. Biol. 269:423–439. [13]
Russell RB, Sasieni PD & MJE Sternberg 1998 Supersites within superfolds: binding site similarity in the absence of homology. J. Mol. Biol. 282:903–918. [13]
Rutherford SL 2000 From genotype to phenotype: buffering mechanisms and the storage of genetic information. BioEssays 22:1095–1105. [16]
Rutherford SL & S Lindquist 1998 Hsp90 as a capacitor for morphological evolution. Nature 396:336–342. [16, 17]
Rutledge JJ, Eisen EJ & JE Legates 1973 An experimental evalution of genetic correlation. Genetics 75:709–726. [18]
Sabeti PC, Reich DE, Higgins JM, Levine HZ, Richter, et al. 2002 Detecting recent positive selection in the human genome from haplotype structure. Nature 419:832–837. [7]
Saccheri I, Kuussaari M, Kankare M, Vikman P, Fortelius W & I Hanski 1998 Inbreeding and extinction in a butterfly metapopulation. Nature 392:491–494. [32]
Saitou N & M Nei 1987 The neighbor-joining method: a new method for reconstructing phylogenetic trees. Mol. Biol. Evol. 4:406–425. [28]
Salter LA & DK Pearl 2001 Stochastic search strategy for estimation of maximum likelihood phylogenetic trees. Syst. Biol. 50:7–17. [28]
Sanderson MJ 1995 Objections to bootstrapping phylogenies: a critique. Syst. Biol. 44:299–320. [28]
Sanderson MJ 1997 A nonparametric approach to estimating divergence times in the absence of rate constancy. J. Mol. Evol. 14:1218–1231. [8]
Sangster TA, Lindquist S & C Queitsch 2004 Under cover: causes, effects and implications of Hsp90-mediated genetic capacitance. BioEssays 26:348–362. [16, 17]
SanMiguel P, Tikhonov A, Jin YK, et al. 1996 Nested retrotransposons in the intergenic regions of the maize genome. Science 274:765–768. [10]
Sarkar IN, Thornton JW, Planet PJ, Figurski DH, Schierwater B & R DeSalle 2002 An automated phylogenetic key for classifying homeoboxes. Mol. Phylogenet. Evol. 24:388–399. [11]
Sasa MM, Chippindale PT & NA Johnson 1998 Patterns of post-zygotic isolation in frogs. Evolution 52:1811–1820. [24]
Saunders IW, Tavaré S & GA Watterson 1984 On the genealogy of nested subsamples from a haploid population. Adv. Appl. Prob. 16:471–491. [12]

Saunders MA, Hammer MF & MW Nachman 2002 Nucleotide variability at *G6pd* and the signature of malarial selection in humans. Genetics 162:1849–1861. [7]

Schadt EE, Monks SA, Drake TA, Lusis AJ, Che N, Colinayo V, et al. 2003 Genetics of gene expression surveyed in maize, mouse, and man. Nature 422:297–302. [2]

Schalet A 1960 A study of spontaneous visible mutations in *Drosophila melanogaster*. PhD dissertation, Indiana Univ. [3]

Scharloo W 1991 Canalization: genetic and developmental aspects. Annu. Rev. Ecol. Syst. 22:65–93. [16]

Schartl M 1995 Platyfish and swordtails: a genetic system for the analysis of molecular mechanisms in tumor formation. Trends Genet. 11:185–189. [25]

Scheiner SM 1993a Genetics and evolution of phenotypic plasticity. Annu. Rev. Ecol. Syst. 24:35–68. [21]

Scheiner SM 1993b Plasticity as a selectable trait: reply to Via. Am. Nat. 142:372–374. [21]

Scheiner SM 1998 The genetics of phenotypic plasticity. VII. Evolution in a spatially structured environment. J. Evol. Biol. 11:303–320. [21]

Scheiner SM 2002 Selection experiments and the study of phenotypic plasticity. J. Evol. Biol. 15:889–898. [21]

Scheiner SM & D Berrigan 1998 The genetics of phenotypic plasticity. VIII. The cost of plasticity in *Daphnia pulex*. Evolution 52:368–378. [21]

Scheiner SM & RF Lyman 1989 The genetics of phenotypic plasticity. I. Heritability. J. Evol. Biol. 2:95–107. [21]

Scheiner SM & RF Lyman 1991 The genetics of phenotypic plasticity. II. Response to selection. J. Evol. Biol. 4:23–50. [21]

Schlenke TA & DJ Begun 2004 Strong selective sweep associated with a transposon insertion in *Drosophila simulans*. Proc. Natl. Acad. Sci. USA 101:1626–1631. [7]

Schlichting CD 2004. The role of phenotypic plasticity in diversification. pp. 191–200 in DeWitt TJ & SM Scheiner, eds. Phenotypic Plasticity: Functional and Conceptual Approaches. Oxford Univ. Press. [21]

Schlichting CD & M Pigliucci 1993 Control of phenotypic plasticity via regulatory genes. Am. Nat. 142:366–370. [21]

Schlosser GS & GP Wagner 2004 The modularity concept in developmental and evolutionary biology. pp. 1–11 in Schlosser GS & GP Wagner, eds. Modularity in Development and Evolution. Univ. of Chicago Press. [19.5]

Schlötterer C 2002 A microsatellite-based multilocus screen for the identification of local selective sweeps. Genetics 160:753–763. [7]

Schlotterer C, Vogl C & D Tautz 1997 Polymorphism and locus-specific effects on polymorphism at microsatellite loci in natural *Drosophila melanogaster* populations. Genetics 146:309–320. [7]

Schluter D 1996 Adaptive radiation along genetic lines of least resistance. Evolution 50:1766–1774. [20]

Schluter D 2000 The Ecology of Adaptive Radiation. Oxford Univ. Press. [Species Concepts box]

Schmalhausen II 1949 Factors of Evolution. Blakiston. [16]

Schmitt J, Dudley SA & M Pigliucci 1999 Manipulation approaches to testing adaptive plasticity: phytochrome-mediated shade-avoidance responses in plants. Am. Nat. 154:S43–S54. [21]

Schmitt J, Nils J & RD Wulff 1992 Norms of reaction of seed traits to maternal environments in *Plantago lanceolata*. Am. Nat. 139:451–466. [21]

Schug MD, Mackay TFC & CF Aquadro 1997 Low mutation rates of microsatellite loci in *Drosophila melanogaster*. Nat. Genet. 15:99–102. [3]

Schweitzer JA, Martinsen GD & TG Whitham 2002 Cottonwood hybrids gain fitness traits of both parents: a mechanism for their long-term persistence? Am. J. Bot. 89:981–990. [26]

Seebeck JH & PG Johnston 1980 *Potorous longipes* (Marsupialia: Macropidae): a new species from Eastern Victoria. Aust. J. Zool. 28:119–134. [32]

Semple C & M Steel 2003 Phylogenetics. Oxford Univ. Press. [28]

Semple C & KH Wolfe 1999 Gene duplication and gene conversion in the *Caenorhabditis elegans* genome. J. Mol. Evol. 48:555–564. [11]

Servedio MR & MAF Noor 2003 The role of reinforcement in speciation: theory and data. Annu. Rev. Ecol. Syst. 34:339–364. [2, 25]

Service PM & MR Rose 1985 Genetic covariation among life-history components: the effects of novel environments. Evolution 39:943–945. [30, 31]

Service PM, Hutchinson EW & MR Rose 1988 Multiple genetic mechanisms for the evolution of senescence in *Drosophila melanogaster*. Evolution 42:708–716. [30]

Seymour AM, Montgomery ME, Costello BH, Ihle S, Johnsson G, St John B, Taggart D & BA Houlden 2001 High effective inbreeding coefficients correlate with morphological abnormalities in populations of South Australian koalas (*Phascolarctos cinereus*). Anim. Conserv. 4:211–219. [32]

Sgrò CM & L Partridge 2000 Evolutionary responses of the life history of wild-caught *Drosophila melanogaster* to two standard methods of laboratory culture. Am. Nat. 156:341–353. [30, 31]
Shabalina SA, Ogurtsov AY, Kondrashov VA & AS Kondrashov 2001 Selective constraint in intergenic regions of human and mouse genomes. Trends Genet. 17:373–376. [3]
Shabalina SA, Yampolsky LY & AS Kondrashov 1997 Rapid decline of fitness in panmictic populations of *Drosophila melanogaster*. Proc. Natl. Acad. Sci. USA 94:13034–13039. [3]
Shapiro MD, Marks ME, Peichel CL, Blackman BK, Nereng KS, Jonsson B, Schluter D & DM Kingsley 2004 Genetic and developmental basis of evolutionary pelvic reduction in threespine sticklebacks. Nature 428:717–723. [2, 15]
Shaw FH, Shaw RG, Wilkinson GS & M Turelli 1995 Changes in genetic variances and covariances: G whiz! Evolution 49:1260–1267. [20]
Shaw KL & YM Parsons 2002 Divergence of mate recognition behavior and its consequences for genetic architectures of speciation. Am. Nat. 159:S61–S75. [22]
Shields DC, Sharp PM, Higgins DG & F Wright 1988 "Silent" sites in *Drosophila* genes are not neutral: evidence of selection among synonymous codons. Mol. Biol. Evol. 5:704–716. [9]
Shimizu KK, Cork JM, Caicedo AL, Mays CA, Moore RC, Olsen KM, Ruzsa S, Coop G, Bustamante CD, Awadalla P & Purugganan MD 2004 Darwinian selection on a selfing locus. Science 306:2081–2084 [29.1]
Shimodaira H 2002 An approximately unbiased test of phylogenetic tree selection. Syst. Biol. 51:492–508. [28]
Shimodaira H & M Hasegawa 1999 Multiple comparisons of log-likelihoods with applications to phylogenetic inference. Mol. Biol. Evol. 16:1114–1116. [28]
Shimodaira H & M Hasegawa 2001 CONSEL: for assessing the confidence of phylogenetic tree selection. Bioinformatics 17:1246–1247. [28]
Shine R 1989 Ecological causes for the evolution of sexual dimorphism: a review of the evidence. Q. Rev. Biol. 64:419–461. [22]
Shoemaker JS, Painter IS & BS Weir 1999 Bayesian statistics in genetics: a guide for the uninitiated. Trends Genet. 9:354–358. [8]
Shrager J 2003 The fiction of function. Bioinformatics 19:1934–1936. [13]
Shuster SM & MJ Wade 2003. Mating Systems and Mating Strategies. Princeton Univ. Press. [4, 22]
Siegal ML & A Bergman 2002 Waddington's canalization revisited: developmental stability and evolution. Proc. Natl. Acad. Sci. USA 99:10528–10532. [16]
Simcox AA & JH Sang 1983 When does determination occur in *Drosophila* embryos? Dev. Biol. 97:212–221. [19.5]
Simillion C, Vandepoele K, Van Montagu MC, Zabeau M & Y Van de Peer 2002 The hidden duplication past of *Arabidopsis thaliana*. Proc. Natl. Acad. Sci. USA 99:13627–13632. [11]
Simmons LW 2003 The evolution of polyandry: patterns of genotypic variation in female mating frequency, male fertilisation success and a test of the sexy-sperm hypothesis. J. Evol. Biol. 16:624–634. [22]
Simmons LW & JS Kotiaho 2002 Evolution of ejaculates: patterns of phenotypic and genotypic variation and condition dependence in sperm competition traits. Evolution 56:1622–1631. [22]
Simonsen KL, Churchill GA & CF Aquadro 1995 Properties of statistical tests of neutrality for DNA polymorphism data. Genetics 141:413–429. [7]
Simpson GG 1944 Tempo and Mode in Evolution. Columbia Univ. Press. [1]
Simpson GG 1951 The species concept. Evolution 5:285–298. [Species Concepts box]
Simpson GG 1953 The Baldwin effect. Evolution 7:110–117. [21]
Singer SS, Mannel DN, Hehlgans T, Brosius J & J Schmitz 2004 From "junk" to gene: curriculum vitae of a primate receptor isoform gene. J. Mol. Biol. 341:883–886. [13]
Singh ND & DA Petrov 2004 Dramatic sequence turnover at an intergenic locus in *Drosophila*. Mol. Biol. Evol. 21:670–680. [10]
Sjödin P, Kaj I, Krone S, Lascoux M & M Nordborg 2005 On the meaning and existence of an effective population size. Genetics 169:1061–1070. [12]
Skipper R 2002 The persistence of the R. A. Fisher–Sewall Wright controversy. Biol. Philos. 17:341–367. [1]
Slack JM 2002 Conrad Hal Waddington: the last Renaissance biologist? Nat. Rev. Genet. 3:889–895. [16]
Slatkin M 1996 In defense of founder-flush theories of speciation. Am. Nat. 147:493–505. [27]
Slatkin M & RR Hudson 1991 Pairwise comparisons of mitochondrial DNA sequences in stable and exponentially growing populations. Genetics 129:555–562. [12]
Slatkin M & M Veuille. 2002 Modern Developments in Theoretical Population Genetics. Oxford Univ. Press. [12]
Sloane MA, Sunnucks P, Alpers D, Beheregaray LB & AC Taylor 2000 Highly

reliable genetic identification of individual northern hairy-nosed wombats from single remotely collected hairs: a feasible censusing method. Mol. Ecol. 9:1233–1240. [32]
Slowinski JB & C Guyer 1989 Testing the stochasticity of patterns of organismal diversity: an improved null model. Am. Nat. 134:907–921. [12]
Smirle MJ, Vincent C, Zurowski CL & B Rancourt 1998 Azinphosmethyl resistance in the oblique-banded leafroller, *Choristoneura rosaceana*: reversion in the absence of selection and relationship to detoxication enzyme activity. Pesticide Biochem. Physiol. 61:183–189. [18]
Smith NGC & A Eyre-Walker 2002 Adaptive protein evolution in *Drosophila*. Nature 415:1022–1024. [3, 8]
Smith NGC & A Eyre-Walker 2003 Partitioning the variation in mammalian substitution rates. Mol. Biol. Evol. 20:10–17. [8]
Smocovitis VB 1996 Unifying Biology: The Evolutionary Synthesis and Evolutionary Biology. Princeton Univ. Press. [1]
Sniegowski PD, Gerrish PJ & RE Lenski 1997 Evolution of high mutation rates in experimental populations of *E. coli*. Nature 387:703–705. [31]
Sniegowski PD, Gerrish PJ, Johnson T & A Shaver 2000 The evolution of mutation rates: separating causes from consequences. BioEssays 22:1057–1066. [3]
Snoke MS & DEL Promislow 2003 Quantitative genetic tests of recent senescence theory: age-specific mortality and male fertility in *Drosophila melanogaster*. Heredity 91:546–556. [30]
Snyder EE, Walts B, Pérusse L, Chagnon YC, Weisnagel SJ, Rankinen T & C Bouchard 2004 The human obesity gene map: the 2003 update. Obes. Res. 12:369–439. [19]
Sokal RR & TJ Crovello 1970 The biological species concept: a critical evaluation. Am. Nat. 104:107–123. [Species Concepts box]
Sollars V, Lu X, Xiao L, Wang X, Garfinkel MD & DM Ruden 2003 Evidence for an epigenetic mechanism by which Hsp90 acts as a capacitor for morphological evolution. Nat. Genet. 33:70–74. [17]
Soltis DE & PS Soltis 1999 Polyploidy: recurrent formation and genome evolution. Trends Ecol. Evol. 14:348–352. [2]
Soltis DE, Soltis PS & JA Tate 2004 Advances in the study of polyploidy since Plant Speciation. New Phytol. 161:173–191. [26]
Spencer CC, Howell CE, Wright AR & DEL Promislow 2003 Testing an "aging gene" in long-lived *Drosophila* strains: increased longevity depends on sex and genetic background. Aging Cell 2:123–130. [30]
Spencer HG, Clark AG & MW Feldman 1999 Reply from H.G. Spencer, A.G. Clark & M.W. Feldman. Trends Ecol. Evol. 14:359. [2]
Spielman D, Brook BW & R Frankham 2004a Most species are not driven to extinction before genetic factors can impact. Proc. Natl. Acad. Sci. USA 101:15261–15264. [32]
Spielman D, Brook BW, Briscoe DA & R Frankham 2004b Does inbreeding and loss of genetic diversity reduce disease resistance? Conserv. Genet. 5:439–448. [32]
Spring J 1997 Vertebrate evolution by interspecific hybridisation: are we polyploid? FEBS Lett. 400:2–8. [2]
Stahl EA, Dwyer G, Mauricio R, Kreitman M & J Bergelson 1999 Dynamics of disease resistance polymorphism at the *Rpm1* locus of *Arabidopsis*. Nature 400:667–671. [29.1]
Staskawicz BJ 2001. Genetics of plant–pathogen interactions specifying plant disease resistance. Plant Physiol. 125:73–76. [29]
Stearns SC & TJ Kawecki 1994 Fitness sensitivity and the canalization of life-history traits. Evolution 48:1438–1450. [16]
Stearns SC & P Magwene 2003 The naturalist in a world of genomics. Am. Nat. 161:171–180. [21]
Stearns SC, Ackermann M, Doebeli M & M Kaiser 2000 Experimental evolution of aging, growth, and reproduction in fruitflies. Proc. Natl. Acad. Sci. USA 97:3309–3313. [31]
Stearns SC, Kaiser M & TJ Kawecki 1995 The differential genetic and environmental canalization of fitness components in *Drosophila melanogaster*. J. Evol. Biol. 8:539–557. [16]
Stebbins GL 1950 Variation and Evolution in Plants. Columbia Univ. Press. [1]
Stebbins GL 1959 The role of hybridization in evolution. Proc. Am. Philos. Soc. 103:231–251. [26]
Stebbins GL & RC Lewontin 1972 Comparative evolution at the level of molecules, organisms, and populations. pp. 23–42 in LeCam L et al., eds. Proceedings of the Sixth Berkeley Symposium on Mathematical Statistics. Vol. V: Darwinian, Neo-Darwinian, and Non-Darwinian Evolution. Univ. of California Press. [1]
Steel M & A McKenzie 2001 Properties of phylogenetic trees generated by Yule-type speciation models. Math. Biosci. 170:91–112. [12]
Steel M, Huson DS & PJ Lockhart 2000 Invariable sites models and their use in phylogeny reconstruction. Syst. Biol. 49:225–232. [28]
Stenberg P, Lundmark M, Knutelski S & A Saura 2003 Evolution of clonality and polyploidy in a weevil system. Mol. Biol. Evol. 20:1626–1632. [10]

Stephan W & DA Kirby 1993 RNA folding in *Drosophila* shows a distance effect for compensatory fitness interactions. Genetics 135:97–103. [9]

Stephan W, Xing L, Kirby DA & JM Braverman 1998 A test of the background selection hypothesis based on nucleotide data from *Drosophila ananassae*. Proc. Natl. Acad. Sci. USA 95:5649–5654. [7]

Stephens M 2001 Inference under the coalescent. pp. 213–238 in Balding DJ, Bishop M & C Cannings, eds. Handbook of Statistical Genetics. Wiley. [12]

Steppan SJ 1997 Phylogenetic analysis of phenotypic covariance structure. II. Reconstructing matrix evolution. Evolution 51:587–594. [20]

Steppan SJ, Phillips PC & D Houle 2002 Comparative quantitative genetics: evolution of the G matrix. Trends Ecol. Evol. 17:320–327. [19.5, 20]

Stern C 1944 A study of race. J. Hered. 35:314–316. [1]

Stern DL 1998 A role of *Ultrabithorax* in morphological differences between *Drosophila* species. Nature 396:463–466. [15]

Stern DL 2000 Perspective: Evolutionary developmental biology and the problem of variation. Evolution 54:1079–1091. [15, 20]

Stevens L, Goodnight CJ & S Kalisz 1995 Multilevel selection in natural populations of jewelweed, *Impatiens capensis*. Am. Nat. 145:513–526. [6]

Storz JF, Payseur BA & MW Nachman 2004 Multilocus scans of microsatellite variability in humans reveal evidence for selective sweeps outside of Africa. Mol. Biol. Evol. 21:1800–1811. [7]

Strauss SY & AA Agrawal 1999 The ecology and evolution of plant tolerance to herbivory. Trends Ecol. Evol. 4:179–185. [29]

Strimmer K & A von Haeseler 1996 Quartet puzzling: a quartet maximum likelihood method for reconstructing tree topologies. Mol. Biol. Evol. 13:964–969. [28]

Sturtevant AH 1920 Genetic studies on *Drosophila simulans*. I. Introduction. Hybrids with *Drosophila melanogaster*. Genetics 5:488–500. [25]

Sucena E & DL Stern 2000 Divergence of larval morphology between *Drosophila sechellia* and its sibling species caused by *cis*-regulatory evolution of *ovo/shaven-baby*. Proc. Natl. Acad. Sci. USA 97:4530–4534. [15]

Sucena E, Delon I, Jones I, Payre F & DL Stern 2003 Regulatory evolution of *shavenbaby/ovo* underlies multiple cases of morphological parallelism. Nature 424:935–938. [15, 25]

Suga H, Koyanagi M, Hoshiyama D, Ono K, Iwabe N, et al. 1999 Extensive gene duplication in the early evolution of animals before the parazoan–eumetazoan split demonstrated by G proteins and protein tyrosine kinases from sponge and hydra. J. Mol. Evol. 48:646–653. [14]

Sultan SE & HG Spencer 2002 Metapopulation structure favors plasticity over local adaptation. Am. Nat. 160:271–283. [21]

Sun S, Ting C-T & C-I Wu 2004 The normal function of a speciation gene, *Odysseus*, and its hybrid sterility effect. Science 305:81–83. [25]

Suzuki Y & T Gojobori 1999 A method for detecting positive selection at single amino acid sites. Mol. Biol. Evol. 16:1315–1328. [11]

Suzuki Y & M Nei 2002 Simulation study of the reliability and robustness of the statistical methods for detecting positive selection at single amino acid sites. Mol. Biol. Evol. 19:1865–1869. [8, 11]

Sweigart AL & JH Willis 2003 Patterns of nucleotide diversity in two species of *Mimulus* are affected by mating system and asymmetric introgression. Evolution 57:2490–2506. [26]

Swofford DL 2002 PAUP*: Phylogenetic Analysis Using Parsimony (* and other methods). Sinauer Assoc. [28]

Syvanen M & CI Kado 2002. Horizontal Gene Transfer. Academic Press. [2]

Tachida H & CC Cockerham 1989 A building block model for quantitative genetics. Genetics 121:839–844. [6]

Tajima F 1983 Evolutionary relationship of DNA sequences in finite populations. Genetics 105:437–460. [12]

Tajima F 1989 Statistical method for testing the neutral mutation hypothesis by DNA polymorphism. Genetics 123:585–595. [7, 12]

Takahashi A, Tsaur SC, Coyne JA & C-I Wu 2001 The nucleotide changes governing cuticular hydrocarbon variation and their evolution in *Drosophila melanogaster*. Proc. Natl. Acad. Sci. USA 98:3920–3925. [25]

Takahata N 1987 On the overdispersed molecular clock. Genetics 116:169–179. [8]

Takahata N & M Nei 1985 Gene genealogy and variance of interpopulational nucleotide differences. Genetics 110:325–344. [12]

Takahata N & M Slatkin 1990 Genealogy of neutral genes in two partially isolated populations. Theor. Pop. Biol. 38:331–350. [12]

Takezaki N, Rzhetsky A & M Nei 1995 Phylogenetic test of the molecular clock and linearized trees. Mol. Biol. Evol. 12:823–833. [8]

Tao Y & DL Hartl 2003 Genetic dissection of hybrid incompatibilities between *Drosophila simulans* and D. *mauritiana*. III.

Heterogeneous accumulation of hybrid incompatibilities, degree of dominance and implications for Haldane's rule. Evolution 57:2580–2598. [24]

Tao Y, Chen S, Hartl DL & CC Laurie 2003 Genetic dissection of hybrid incompatibilities between *Drosophila simulans* and *D. mauritiana*. I. Differential accumulation of hybrid male sterility effects on the X and autosomes. Genetics 164:1383–1397. [24]

Tarrío R, Rodríguez-Trelles F & FJ Ayala 2001 Shared nucleotide composition biases among species and their impact on phylogenetic reconstructions of the Drosophilidae. Mol. Biol. Evol. 18:1464–1473. [8]

Tatar M 2001 Senescence. pp. 128–141 in Fox CW, Roff DA & DJ Fairbairn, eds. Evolutionary Ecology: Concepts and Case Studies. Oxford Univ. Press. [30]

Tatar M 2004 The neuroendocrine regulation of *Drosophila* aging. Exp. Gerontol. 39:1745–1750. [30]

Tatar M, Bartke A & A Antebi 2003 The endocrine regulation of aging by insulin-like signals. Science 299:1346–1351. [30]

Tatar M, Gray DW & JR Carey 1997 Altitudinal variation for senescence in *Melanoplus* grasshoppers. Oecologia 111:357–364. [30]

Tatar M, Promislow DEL, Khazaeli AA & JW Curtsinger 1996 Age-specific patterns of genetic variance in *Drosophila melanogaster*. II. Fecundity and its genetic correlation with age-specific mortality. Genetics 143:849–858. [30]

Tavaré S, Balding DJ, Griffiths RC & P Donnelly 1997 Inferring coalescence times from DNA sequence data. Genetics 145:505–518. [12]

Taylor AC, Sherwin WB & RK Wayne 1994 Genetic variation of microsatellite loci in a bottlenecked species: the northern hairy-nosed wombat *Lasiorhinus krefftii*. Mol. Ecol. 3:277–290. [32]

Taylor DR, Zeyl C & E Cooke 2002 Conflicting levels of selection in the accumulation of mitochondrial defects in *Saccharomyces cerevisiae*. Proc. Natl. Acad. Sci. USA 99:3690–3694. [31]

Taylor JS, Braasch I, Frickey T, Meyer A & YV de Peer 2003 Genome duplication, a trait shared by 22,000 species of ray-finned fish. Genome Res. 13:382–390. [2]

Taylor PD & SA Frank 1996 How to make a kin selection model. J. Theor. Biol. 180:27–37. [23]

Templeton AR 1979 The unit of selection in *Drosophila mercatorum*. II. Genetic revolution and the origin of coadapted genomes in parthenogenetic strains. Genetics 92:1265–1282. [27]

Templeton AR 1980 The theory of speciation via the founder principle. Genetics 94:1101–1038. [27]

Templeton AR 1983 Phylogenetic inference from restriction endonuclease cleavage site maps with particular reference to the evolution of humans and apes. Evolution 37:221–244. [28]

Templeton AR 1996 Experimental evidence for the genetic-transilience model of speciation. Evolution 50:909–915. [27]

Templeton AR 1998 Species and speciation: geography, population structure, ecology, and gene trees. pp. 32–43 in Howard DJ & SH Berlocher, eds. Endless Forms: Species and Speciation. Oxford Univ. Press. [Species Concepts box]

Templeton AR, Maxwell T, Posada D, Stengård JH, Boerwinkle E & CF Sing 2005 A method for using haplotype trees in phenotype/genotype association studies. Genetics 169:441–453. [Preface]

Teotonio H & MR Rose 2000 Variation in the reversibility of evolution. Nature 408:416–417. [31]

Teshima KM & F Tajima 2002 The effect of migration during the divergence. Theor. Pop. Biol. 62:81–95. [12]

The Gene Ontology Consortium 2000 Gene ontology: tool for the unification of biology. Nat. Genet. 25:25–29. [13]

Theron E, Hawkins K, Bermingham E, Ricklefs RE & NI Mundy 2001 The molecular basis of an avian plumage polymorphism in the wild: a melanocortin-1-receptor point mutation is perfectly associated with the melanic plumage morph of the bananaquit, *Coereba flaveola*. Curr Biol 11:550–557. [15, 25]

Thieffry D & L Sánchez 2002 Alternative epigenetic states understood in terms of specific regulatory structures. Ann. N. Y. Acad. Sci. 981:135–153. [17]

Thomas CA 1971 The genetic organization of chromosomes. Annu. Rev. Genet. 5:237–256. [10]

Thompson JD, Gibson TJ, Plewniak F, Jeanmougin F & DG Higgins 1997 The ClustalX Windows interface: flexible strategies for multiple sequence alignment aided by quality analysis tools. Nucleic Acids Res. 24:4876–4882. [28]

Thompson JN 1999. Specific hypothesis on the geographical mosaic of coevolution. Am. Nat 153:S1-S14. [29]

Thorne JL, Kishino H & IS Painter 1998 Estimating the rate of evolution of the rate of molecular evolution. Mol. Biol. Evol. 15:1647–1657. [8]

Thornton JW 2001 Evolution of vertebrate steroid receptors from an ancestral estrogen receptor by ligand exploitation and serial genome expansions. Proc. Natl. Acad. Sci. USA 98:5671–5676. [11]

Thornton JW 2004 Resurrecting ancient genes: experimental analysis of extinct molecules. Nat. Rev. Genet. 5:366–375. [11]

Thornton JW & R DeSalle 2000a Gene family evolution and homology: genomics meets phylogenetics. Annu. Rev. Genomics Hum. Genet. 1:41–73. [11]

Thornton JW & R DeSalle 2000b A new method to localize and test the significance of incongruence: detecting domain shuffling in the nuclear receptor superfamily. Syst. Biol. 49:183–201. [11]

Thornton JW & DB Kelley 1998. Evolution of the androgen receptor: structure–function implications. BioEssays 20:860–869. [11]

Thornton JW, Need E & D Crews 2003 Resurrecting the ancestral steroid receptor: ancient origin of estrogen signaling. Science 301:1714–1717. [11]

Thrall PH & JJ Burdon 2002 Evolution of gene-for-gene systems in metapopulations: the effect of spatial scale of host and pathogen dispersal. Plant Pathol. 51:169–184. [29]

Thrall PH & JJ Burdon 2003 Evolution of virulence in a plant host–pathogen metapopulation. Science 299:1735–1737. [29]

Thrall PH, Burdon JJ & JD Bever 2002 Local adaptation in the *Linum marginale–Melampsora lini* host–pathogen interaction. Evolution 56:1340–1351. [29]

Thrall PH, Burdon JJ & A Young 2001 Variation in resistance and virulence among demes of a plant host–pathogen metapopulation. J. Ecol. 89:736–748. [29]

Tian D, Araki H, Stahl E, Bergelson J & M Kreitman 2002 Signature of balancing selection in *Arabidopsis*. Proc. Natl. Acad. Sci. USA 99:11525–11530. [29]

Tian D, Traw MB, Chen JQ, Kreitman M & J Bergelson 2003 Fitness costs of R-gene-mediated resistance in *Arabidopsis thaliana*. Nature 423:74–77. [29, 29.1]

Ting C-T, Takahashi A & C-I Wu 2001 Incipient speciation by sexual isolation in *Drosophila*: concurrent evolution at multiple loci. Proc. Natl. Acad. Sci. USA 98:6709–6713. [22, 25]

Ting C-T, Tsaur S-C, Wu M-L & Wu C-I 1998 A rapidly evolving homeobox at the site of a hybrid sterility gene. Science 282:1501–1504. [25, 26]

Todd AE, Orengo CA & JM Thornton 2002 Plasticity of enzyme active sites. Trends Biochem. Sci. 27:419–426. [13]

Toomajian C & M Kreitman 2002 Sequence variation and haplotype structure at the human *HFE* locus. Genetics 161:1609–1623. [7]

Torgerson DG, Kulathinal RJ & RS Singh 2002 Mammalian sperm proteins are rapidly evolving: evidence of positive selection in functionally diverse genes. Mol. Biol. Evol. 19:1973–1980. [24]

Trivers R 1971 The evolution of reciprocal altruism. Q. Rev. Biol. 46:35–57. [23]

True HL & SL Lindquist 2000 A yeast prion provides a mechanism for genetic variation and phenotypic diversity. Nature 407:477–483. [16, 17]

True HL, Berlin I & SL Lindquist 2004 Epigenetic regulation of translation reveals hidden genetic variation to produce complex traits. Nature 431:184–187. [17]

Trut LN 1999 Early canid domestication: the farm-fox experiment. Am. Sci. 87: 160–169. [17]

Turelli M 1988a Phenotypic evolution, constant covariances, and the maintenance of additive variance. Evolution 42:1342–1347. [20]

Turelli M 1988b Population genetic models for polygenetic variation and evolution. pp. 601–618 in Weir BS, Eisen EJ, Goodman MM & G Namkoong, eds. Proceedings of the Second International Conference on Quantitative Genetics. Sinauer Assoc. [19]

Turelli M & NH Barton 1990 Dynamics of polygenic characters under selection. Theor. Pop. Biol. 38:1–57. [18]

Turelli M & NH Barton 2004 Polygenic variation maintained by balancing selection: pleiotropy, sex-dependent allelic effects and G×E interactions. Genetics 166:1053–1079. [22]

Turelli M & DJ Begun 1997 Haldane's rule and X-chromosome size in *Drosophila*. Genetics 147:1799–1815. [24]

Turelli M & HA Orr 1995 The dominance theory of Haldane's rule. Genetics 154:1663–1679. [24]

Turelli M, Barton NH & JA Coyne 2001 Theory and speciation. Trends Ecol. Evol. 16:330–343. [27]

Uetz P, Giot L, Cagney G, Manseld TA, Judson RS, et al. 2000 A comprehensive analysis of protein–protein interactions. Nature 403:623–627. [13.1]

Urabe K, Aroca P & VJ Hearing 1993 From gene to protein: determination of melanin synthesis. Pigment Cell Res. 6:186–192. [25]

Urnov FD & AP Wolffe 2001 Above and within the genome: epigenetics past and present. J. Mammary Gland Biol. Neoplasia 6:153–167. [17]

Uyenoyama MK 1997 Genealogical structure among alleles regulating self-incompatibility in natural populations of flowering plants. Genetics 147:1389–1400. [12]

Uzzell T & KW Corbin 1971 Fitting discrete probability distributions to evolutionary events. Science 172:1089–1096. [8]

Vallender EJ & BT Lahn 2004 Positive selection on the human genome. Hum. Mol. Genet. 13:R245–R254. [7]

Van de Peer Y, Taylor JS & A Meyer 2003 Are all fishes ancient polyploids? J. Struct. Funct. Genomics 3:65–73. [11]
van Noordwijk AJ 1994 The interaction of inbreeding depression and environmental stochasticity in the risk of extinction of small populations. pp. 131–146 in Loeschcke V, Tomiuk J & SK Jain, eds. Conservation Genetics. Birkhäuser. [32]
Van Speybroeck L, Van de Vijver G & D De Waele 2002 From epigenesis to epigenetics: the genome in context. Ann. N.Y. Acad. Sci. 981. [17]
Van Valen L 1976 Ecological species, multispecies, and oaks. Taxon 25:233–239. [Species Concepts box]
Van Voorhies WA, Khazaeli AA & JW Curtsinger 2003 Long-lived *Drosophila melanogaster* lines exhibit normal metabolic rates. J. Appl. Physiol. 95:2605–2613. [30]
Vaughn TT, Pletscher LS, Peripato A, King-Ellison K, Adams E, Erikson C & JM Cheverud 1999 Mapping quantitative trait loci for murine growth: a closer look at genetic architecture. Genet. Res. 74:313–322. [19]
Vaupel JW, Carey JR, Christensen K, Johnson TE, et al. 1998 Biodemographic trajectories of longevity. Science 280:855–860. [30]
Veiga JP, Salvador A, Merino S & M Puerta 1998 Reproductive effort affects immune response and parasite infection in a lizard: a phenotypic manipulation using testosterone. Oikos 82:313–318. [18]
Venter JC, Adams MD, Myers EW, Li PW, Mural RJ, et al. 2001 The sequence of the human genome. Science 291:1304–1351. [14]
Vermaak D, Ahmad K & S Henikoff 2003 Maintenance of chromatin states: an open-and-shut case. Curr. Opin. Cell Biol. 15:266–274. [2]
Via S 1993 Adaptive phenotypic plasticity: target or byproduct of selection in a variable environment. Am. Nat. 142:352–365. [21]
Via S, Gomulkiewicz R, de Jong G, Scheiner SM, Schlichting CD & PH Vantienderen 1995 Adaptive phenotypic plasticity: consensus and controvessy. Trends Ecol. Evol. 10:212–217. [21]
Vilà C, Savolainen P, Maldonado JE, Amorimi IR, Rice JE, et al. 1997 Multiple and ancient origins of the domestic dog. Science 276:1687–1689. [26]
Vinogradov AE 1998 Buffering: a possible passive-homeostasis role for redundant DNA. J. Theor. Biol. 193:197–199. [10]
Vinogradov AE 1999 Intron-genome size relationship on a large evolutionary scale. J. Mol. Evol. 49:376–384. [10]
Vinogradov AE 2003 Selfish DNA is maladaptive: evidence from the plant Red List. Trends Genet. 19:609–614. [10]
Vision TJ, Brown DG & SD Tanksley 2000 The origins of genomic duplications in *Arabidopsis*. Science 290:2114–2117. [10]
Vos R 2003 Accelerated likelihood surface exploration: the likelihood ratchet. Syst. Biol. 52:368–373. [28]
Vrana PB, Guan X-J, Ingram RS & SM Tilghman 1998 Genomic imprinting is disrupted in interspecific *Peromyscus* hybrids. Nat. Genet. 20:362–365. [17]
Waddington CH 1940 Organisers and Genes. Cambridge Univ. Press. [16]
Waddington CH 1942 Canalization of development and the inheritance of acquired characters. Nature 150:563–565. [16]
Waddington CH 1952 Selection of the genetic basis for an acquired character. Nature 169:278. [16]
Waddington CH 1953 Genetic assimilation of an acquired character. Evolution 7:118–126. [21]
Waddington CH 1957 The Strategy of the Genes. Allen & Unwin. [16]
Waddington CH 1961 Genetic assimilation. Adv. Genet. 10:257–293. [16, 17]
Waddington CH 1975 The Evolution of an Evolutionist. Edinburgh Univ. Press. [17]
Wade MJ 1977 An experimental study of group selection. Evolution. 31:134–153. [6]
Wade MJ 1978 A critical review of the models of group selection. Q. Rev. Biol. 53:101–114. [6]
Wade MJ 1979 Sexual selection and variance in reproductive success. Am. Nat. 114:742–747. [4]
Wade MJ 1982a Group selection: migration and the differentiation of small populations. Evolution 36:945–961. [6]
Wade MJ 1982b The evolution of interference competition by individual, family, and group selection. Proc. Natl. Acad. Sci. USA 79:3575–3578. [4]
Wade MJ 1991 Genetic variance for rate of population increase in natural populations of flour beetles, *Tribolium* spp. Evolution 45:1574–1584. [27]
Wade MJ 1995 Mean crowding and sexual selection in resource polygynous mating systems. Evol. Ecol. 9:118–124. [4]
Wade MJ 1996 Adaptation in subdivided populations: kin selection and interdemic selection. pp. 381–405 in Rose MR & G Lauder, eds. Evolutionary Biology and Adaptation. Sinauer Assoc. [4]
Wade MJ & SJ Arnold. 1980. The intensity of sexual selection in relation to male sexual behavior, female choice, and sperm precedence. Anim. Behav. 28: 446–461. [4]

Wade MJ & CJ Goodnight 1991 Wright's shifting balance theory: an experimental study. Science 253:1015–1018. [6]
Wade MJ & CJ Goodnight 1998 Genetics and adaptation in metapopulations: when nature does many small experiments. Evolution 52:1537–1553. [6]
Wade MJ & S Kalisz 1989 The additive partitioning of selection gradients. Evolution 43:1567–1569. [4]
Wade MJ & S Kalisz 1990 The causes of natural selection. Evolution 44:1947–1955. [4]
Wade MJ & DE McCauley 1988 The effects of extinction and colonization on the genetic differentiation of populations. Evolution 42:995–1005. [4]
Wade MJ & SM Shuster 2004 Estimating the strength of sexual selection from Y chromosome and mitochondrial DNA diversity. Evolution 58:1613–1616. [4]
Wade MJ, Patterson H, Chang NW & NA Johnson 1994 Postcopulatory, pre-zygotic isolation in flour beetles. Heredity 72:163–167. [24]
Wagner A 1996 Does evolutionary plasticity evolve? Evolution 50:1008–1023. [16]
Wagner A 2000 Robustness against mutations in genetic networks of yeast. Nat. Genet. 24:355–361. [16]
Wagner A 2001 The yeast protein interaction network evolves rapidly and contains few redundant duplicate genes. Mol. Biol. Evol. 18:1283–1292. [13.1]
Wagner A 2003 Does selection mold molecular networks? Science's STKE 202:pe41. [13.1]
Wagner GP & L Altenberg 1996 Perspective: complex adaptations and the evolution of evolvability. Evolution 50:967–976. [3, 19, 19.5]
Wagner GP & JG Mezey 2004 The role of genetic architecture constraints in the origin of variational modules. pp. 338–358 in Schlosser GS & GP Wagner, eds. Modularity in Development and Evolution. Univ. of Chicago Press. [19.5]
Wagner GP, Booth G & H Bagheri-Chaichian 1997 A population genetic theory of canalization. Evolution 51:329–347. [16]
Wakeley J 1998 Segregating sites in Wright's island model. Theor. Pop. Biol. 53: 166–174. [12]
Wakeley J 2000 The effects of subdivision on the genetic divergence of populations and species. Evolution 54:1092–1101. [12]
Walbot V & DA Petrov 2001 Gene galaxies in the maize genome. Proc. Natl. Acad. Sci. USA 98:8163–8164. [10]
Walker DW, McColl G, Jenkins NL, Harris J & GJ Lithgow 2000 Evolution of lifespan in *C. elegans*. Nature 405:296–297. [30]
Wallace AR 1889 Darwinism, 2nd ed. Macmillan. [22]
Wallace B & JC King 1951 Genetic changes in populations under irradiation. Am. Nat. 85:209–222. [1]
Wallis M 1994 Variable evolutionary rates in the molecular evolution of mammalian growth hormones. J. Mol. Evol. 38:619–627. [8]
Walsh B 2003 Population genetic models of the fates of duplicate genes. Genetica 118:279–294. [2, 11]
Wang RL, Stec A, Hey J, Lukens L & J Doebley 1999 The limits of selection during maize domestication. Nature 398:236–239. [7]
Warren RW, Nagy L, Selegue J, Gates J & S Carroll 1994 Evolution of homeotic gene regulation and function in flies and butterflies. Nature 372:458–461. [15, 25]
Waterland RA & RL Jirtle 2003 Transposable elements: targets for early nutritional effects on epigenetic gene regulation. Mol. Cell. Biol. 23:5293–5300. [17]
Waterston RH, et al. 2002 Initial sequencing and comparative analysis of the mouse genome. Nature 420:520–562. [13]
Watterson GA 1978 The homozygosity test of neutrality. Genetics 88:405–417. [7]
Wayne ML & K Simonsen 1998 Statistical tests of neutrality in the age of weak selection. Trends Ecol. Evol. 13:236–240. [8]
Weatherbee SD, Halder G, Kim J, Hudson A & S Carroll 1998 *Ultrabithorax* regulates genes at several levels of the wing-patterning hierarchy to shape the development of the *Drosophila* haltere. Genes Dev. 12:1474–1482. [15]
Weatherhead PJ & RH Robertson 1979 Offspring quality and the polygyny threshold: the "sexy son" hypothesis. Am. Nat. 113:201–208. [22]
Weber JL & C Wang 1993 Mutation of human short tandem repeats. Hum. Mol. Genet. 2:1123–1128. [3]
Weigensberg I & DA Roff 1996 Natural heritabilities: can they be reliably estimated in the laboratory? Evolution 50:2149–2157. [18]
Weinig C, Dorn LA, Kane NC, German ZM, Halldorsdottir SS, Ungerer MC, Toyonaga Y, Mackay TF, Purugganan MD & J Schmitt 2003 Heterogeneous selection at specific loci in natural environments in *Arabidopsis thaliana*. Genetics 165:321–329. [29.1]
Weissmann F & F Lyko 2003 Cooperative interactions between epigenetic modifications and their function in the regulation of chromosome architecture. BioEssays 25:792–797. [17]
Wendel JF 2000 Genome evolution in polyploids. Plant Mol. Biol. 42:225–249. [10]
Wendel JF & RC Cronn 2003 Polyploidy and the evolutionary history of cotton. Adv. Agron. 78:139–186. [10]

Wendel JF, Cronn RC, Alvarez I, Liu B, Small RL & DS Senchina 2002a Intron size and genome size in plants. Mol. Biol. Evol. 19:2346–2352. [10]

Wendel JF, Cronn RC, Johnston JS & HJ Price 2002b Feast and famine in plant genomes. Genetica 115:37–47. [10]

Werner T 2003 The state of the art of mammalian promoter recognition. Brief Bioinf. 4:22–30. [2]

Werren JH, Nur U & C-I Wu 1988 Selfish genetic elements. Trends Ecol. Evol. 3:297–302. [23]

Wesselingh RA & ML Arnold 2000 Pollinator behaviour and the evolution of Louisiana Iris hybrid zones. J. Evol. Biol. 13:171–180. [26]

West-Eberhard MJ 2003 Developmental Plasticity and Evolution. Oxford Univ. Press. [17, 21]

Westemeier RL, Brawn JD, Simpson SA, Esker TL, Jansen RW, Walk JW, Kershner EL, Bouzat JL & KN Paige 1998 Tracking the long-term decline and recovery of an isolated population. Science 282:1695–1698. [32]

Westendorp RG & TB Kirkwood 1998 Human longevity at the cost of reproductive success. Nature 396:743–746. [30]

Westerman M, Barton NH & GM Hewitt 1987 Differences in DNA content between two chromosomal races of the grasshopper *Podisma pedestris*. Heredity 58:221–228. [10]

Wheeler NC & RP Guries 1987 A quantitative measure of introgression between Lodgepole and Jack pines. Can. J. Bot. 65:1876–1885. [26]

Wheeler QD & R Meier 1999 Species Concepts and Phylogenetic Theory: A Debate. Columbia Univ. Press. [Species Concepts box]

Wheeler WC 1996 Optimization alignment: the end of multiple sequence alignment in phylogenetics? Cladistics 12:1–9. [28]

Wheeler WC 2003 Implied alignment: a synapomorphy-based multiple-sequence alignment method and its use in cladogram search. Cladistics 19:261–268. [28]

Wheeler WC & DS Gladstein 1994 Malign: a multiple sequence alignment program. J. Hered. 85:417–418. [28]

Wheeler WC & DS Gladstein 2000 POY: the optimization of alignment characters. Program and documentation available at ftp: //ftp.amnh.org/pub/molecular/poy/. American Museum of Natural History. [28]

White MJD 1978 Modes of Speciation. W. H. Freeman. [27]

Whitham TG 1989 Plant hybrid zones as sinks for pests. Science 244:1490–1493. [26]

Whitlock MC, Phillips PC & K Fowler 2002 Persistence of changes in the genetic covariance matrix after a bottleneck. Evolution 56:1968–1975. [20, 27]

Whitlock MC, Phillips PC, Moore FB-G & SJ Tonsor 1995 Multiple fitness peaks and epistasis. Annu. Rev. Ecol. Syst. 26:601–629. [18, 20]

Whitlock MC, Phillips PC & MJ Wade 1993 Gene interaction affects the additive genetic variance in subdivided populations with migration and extinction. Evolution 47:1758–1769. [6]

Whittam TS 1996 Genetic variation and evolutionary processes in natural populations of *Escherichia coli*. pp. 2708–2720 in Neidhardt FC, et al., eds. *Escherichia coli* and *Salmonella*: Cellular and Molecular Biology. ASM Press. [31.1]

Wilkins AS 1997 Canalization: a molecular genetic perspective. BioEssays 19: 257–262. [16]

Wilkins AS 2002 The Evolution of Developmental Pathways. Sinauer Assoc. [15, 16]

Wilkinson-Herbots HM 1998 Genealogy and subpopulation differentiation under various models of population structure. J. Math. Biol. 37:535–585. [12]

Williams GC 1957 Pleiotropy, natural selection, and the evolution of senescence. Evolution 11:398–411. [30]

Williams PD & T Day 2001 Interactions between sources of mortality and the evolution of parasite virulence. Proc. R. Soc. Lond. B 268:2331–2337. [30]

Willis JH & HA Orr 1993 Increased heritable variation following population bottlenecks: the role of dominance. Evolution 47:949–956. [27]

Wilson AC, Carlson SS & TJ White 1977 Biochemical evolution. Annu. Rev. Biochem. 46:573–639. [1, 2]

Wilson C & JW Szostak 1995 In-vitro evolution of a self-alkylating ribozyme. Nature 374:777–782. [31]

Wilson DS 1980 The Natural Selection of Populations and Communities. Benjamin/Cummings. [23]

Wilson IJ, Weale ME & DJ Balding 2003 Inferences from DNA data: population histories, evolutionary processes and forensic match probabilities. J. R. Stat. Soc. Ser. A 166:155–187. [12]

Wittkopp PJ, Carroll SB & A Kopp 2003a Evolution in black and white: genetic control of pigment patterns in *Drosophila*. Trends Genet. 19:495–504. [25]

Wittkopp PJ, Vaccaro K & SB Carroll 2002 Evolution of *yellow* gene regulation and pigmentation in *Drosophila*. Curr. Biol. 12:1547–1556. [15]

Wittkopp PJ, Williams BL, Selegue JE & SB Carroll 2003b *Drosophila* pigmentation evolution: divergent genotypes underlying convergent phenotypes. Proc. Natl. Acad. Sci. USA 100:1808–1813. [15, 25]

Wolf JB & ED Brodie III 1998 Coadaptation of parental and offspring characters. Evolution 52:535–544. [2.1]
Wolf JB & MJ Wade 2001 On the assignment of fitness to parents and offspring: whose fitness is it and when does it matter? J. Evol. Biol. 14:347–358. [22]
Wolf JB, Brodie ED III & AJ Moore 1999 The role of maternal and paternal effects in the evolution of parental quality by sexual selection. J. Evol. Biol. 12:1157–1167. [22]
Wolf JB, Brodie ED III & MJ Wade 2000 Epistasis and the Evolutionary Process. Oxford Univ. Press. [6.1, 18]
Wolf JB, Brodie ED III & MJ Wade 2004 The genotype–environment interaction and evolution when the environment contains genes. pp. 173–190 in DeWitt TJ & SM Scheiner, eds. Phenotypic Plasticity: Functional and Conceptual Approaches. Oxford Univ. Press. [21]
Wolf JB, Moore AJ & ED Brodie III 1997 The evolution of indicator traits for parental quality: the role of maternal and paternal effects. Am. Nat. 150:639–649. [22]
Wolf JB, Vaughn TT, Pletscher LS & JM Cheverud 2002 Contribution of maternal effect QTL to the genetic architecture of early growth in mice. Heredity 89:300–310. [19]
Wolf YI, Brenner SE, Bash PA & EV Koonin 1999 Distribution of protein folds in the three superkingdoms of life. Genome Res. 9:17–26. [14]
Wolfe KH & DC Shields 1997 Molecular evidence for an ancient duplication of the entire yeast genome. Nature 387:708–713. [2, 10]
Wolfenbarger LL & GS Wilkinson 2001 Sex-linked expression of a sexually selected trait in the stalk-eyed fly, *Cyrtodiopsis dalmanni*. Evolution 55:103–110. [22]
Wollenberg K & JC Avise 1998 Sampling properties of genealogical pathways underlying population pedigrees. Evolution 52:957–966. [12]
Wollenberg KR & WR Atchley 2000 Separation of phylogenetic and functional associations in biological sequences by using the parametric bootstrap. Proc. Natl. Acad. Sci. USA 97: 3288–3291. [13]
Woltereck R 1909 Weitere experimentelle Untersüchungen über Artveränderung, speziell über das Wesen quantitativer Artunterschiede bei Daphniden. Verh. Zool. Ges. 19:110–172. [21]
Wooding S, Kim UK, Bamshad MJ, Larsen L, Jorde LB & D Drayna 2004 Natural selection and molecular evolution in PTC, a bitter-taste receptor gene. Am. J. Hum. Genet. 74:637–646. [7]
Woodruff RC, Slatko BE & JN Thompson Jr 1983 Factors affecting mutation rates in natural populations. pp. 37–124 in Ashburner M, Carson HL & JN Thompson Jr, eds. The Genetics and Biology of *Drosophila*, vol. 3c. Academic Press. [3]
Wray GA, Hahn MW, Abouheif E, Balhoff JP, Pizer M, Rockman MV & LA Romano 2003 The evolution of transcriptional regulation in eukaryotes. Mol. Biol. Evol. 20:1377–1419. [8, 15, 16]
Wright S 1931 Evolution in Mendelian populations. Genetics 16:97–159. [1, 6, 21]
Wright S 1955 Classification of the factors of evolution. Cold Spring Harbor Symp. Quant. Biol. 20:16–24. [5]
Wright S 1969 Evolution and the Genetics of Populations. Vol II: The Theory of Gene Frequencies. Univ. of Chicago Press. [6]
Wright S 1977 Evolution and Genetics of Populations. Vol III. Experimental Results and Evolutionary Deductions. Univ. of Chicago Press. [6, 27]
Wright S 1980 Genic and organismic evolution. Evolution 34:825–843. [19]
Wu C-I 2001 The genic view of the process of speciation. J. Evol. Biol. 14:851–865. [25]
Wu C-I & AW Davis 1993 Evolution of postmating isolation: the composite nature of Haldane's rule and its genetic bases. Am. Nat. 142:187–212. [24]
Wu C-I & MF Palopoli 1994. Genetics of postmating reproductive isolation in animals. Annu. Rev. Genet. 28:83–308. [25]
Wu C-I, Johnson NA & MF Palopoli 1996 Haldane's rule and its legacy: why are there so many sterile males? Trends Ecol. Evol. 11:281–284. [24]
Wu KH, Tobias ML, Thornton JW & DB Kelley 2003 Estrogen receptors in *Xenopus*: duplicate genes, splice variants, and tissue-specific expression. Gen. Comp. Endocrinol. 133:38–49. [11]
Yamamoto Y, Stock DW & WR Jeffery 2004 *Hedgehog* signalling controls eye degeneration in blind cavefish. Nature 431:844–847. [15]
Yang AS, Jones PA & A Shibata 1996 The mutational burden of 5-methylcytosine. pp. 77–94 in Russo VEA, Martienssen RA & AD Riggs, eds. Epigenetic Mechanisms of Gene Regulation. Cold Spring Harbor Laboratory Press. [17]
Yang Z 1994 Maximum likelihood phylogenetic estimation from DNA sequences with variable rates over sites: approximate methods. J. Mol. Evol. 39:306–314. [28]
Yang Z 1996a Maximum likelihood models for combined analyses of multiple sequence data. J. Mol. Evol. 42:587–596. [8]

Yang Z 1996b Among-site rate variation and its impact on phylogenetic analyses. Trends Ecol. Evol. 11:367–372. [8]

Yang Z 1997 PAML: a program package for phylogenetic analysis by maximum likelihood. CABIOS 13:555–556. [8]

Yang Z 2002 Inference of selection from multiple species alignments. Curr. Opin. Genet. Dev. 12:688–694. [11]

Yang Z & B Bielawski 2000 Statistical methods for detecting molecular adaptation. Trends Ecol. Evol. 15:496–503. [7, 8]

Yang Z & R Nielsen 2002 Codon-substitution models for detecting molecular adaptation at individual sites along specific lineages. Mol. Biol. Evol. 19:908–917. [7]

Yang Z & WJ Swanson 2002 Codon-substitution models to detect adaptive evolution that account for heterogeneous selective pressures among site classes. Mol. Biol. Evol. 19:49–57. [8]

Yang Z, Kumar S & M Nei 1995 A new method of inference of ancestral nucleotide and amino acid sequences. Genetics 141:1641–1650. [11]

Yoder AD & Z Yang 2000 Estimation of primate speciation dates using local molecular clocks. Mol. Biol. Evol. 17:1081–1090. [8]

Yoon HS & DA Baum 2004 Transgenic study of parallelism in plant morphological evolution. Proc. Natl. Acad. Sci. USA 101:6524–6529. [15]

Yoshida T, Jones JE, Ellner SP, Fussmann GF & NG Hairston 2003 Rapid evolution drives ecological dynamics in a predator–prey system. Nature 424:303–306. [31]

Young AG, Brown AHD, Murray BG, Thrall PH & CH Miller 2000 Genetic erosion, restricted mating and reduced viability in fragmented populations of the endangered grassland herb *Rutidosis leptorrhynchoides*. pp. 335–359 in Young AG & GM Clarke, eds. Genetics, Demography and Viability of Fragmented Populations. Cambridge Univ. Press. [32]

Yuh, C-H, Bolouri H & EH Davidson 1998 Genomic *cis*-regulatory logic: experimental and computational analysis of a sea urchin gene. Science 279:1896–1902. [15]

Zeh JA & DW Zeh 2003 Toward a new sexual selection paradigm: polyandry, conflict and incompatibility. Ethology 109:929–950. [22]

Zeng Z-B, Liu J, Stam LF, Kao C-K, Mercer JM & CC Laurie 2000 Genetic architecture of a morphological shape difference between two *Drosophila* species. Genetics 154:299–310. [25]

Zhan J, Mundt C, Hoeffer ME & BA McDonald 2002 Local adaptation and effect of host genotype on the rate of pathogen evolution: an experimental test in a plant pathosystem. J. Evol. Biol. 15:634–647. [29]

Zhang J 2004 Frequent false detection of positive selection by the likelihood method with branch-site models. Mol. Biol. Evol. 21:1332–1339. [8]

Zhang J & X Gu 1998 Correlation between the substitution rate and rate variation among sites in protein evolution. Genetics 149:1615–1625. [8]

Zhang J & M Nei 1997 Accuracies of ancestral amino acid sequences inferred by the parsimony, likelihood, and distance methods. J. Mol. Evol. 44 (Suppl 1):S139–S146. [11]

Zhang L, Gaut BS & TJ Vision 2001 Gene duplication and evolution. Science 293:1551. [11]

Zheng Q 2001 On the dispersion index of a Markovian molecular clock. Math. Biosci. 172:115–128. [8]

Zimmerman E, Palsson A & G Gibson 2000 Quantitative trait loci affecting components of wing shape in *Drosophila melanogaster*. Genetics 155:671–683. [19.4, 20]

Zuckerkandl E 2001 Intrinsically driven changes in gene interaction complexity. I. Growth of regulatory complexes and increase in number of genes. J. Mol. Evol. 53:539–554. [16]

Zuckerkandl E 2002 Why so many noncoding nucleotides? The eukaryote genome as an epigenetic machine. Genetica 115:105–129. [10]

Zuckerkandl E & L Pauling 1965 Evolutionary divergence and convergence in proteins. pp. 97–166 in Bryson V & H Vogel, eds. Evolving Genes and Proteins. Academic Press. [1, 8]

Zuk M & AM Stoehr 2002 Immune defense and host life history. Am. Nat. 160:S9–S22. [30]

Zuker M, Jaeger JA & DH Turner 1991 A comparison of optimal and suboptimal RNA secondary structures predicted by free energy minimization with structures determined by phylogenetic comparison. Nucl. Acids Res. 19:2707–2714. [9]

Zwaan B, Bijlsma R & RF Hoekstra 1995 Direct selection on life-span in *Drosophila melanogaster*. Evolution 49:649–659. [31]

Index

Page numbers in **bold** refer to glossary entries

A

A4 family (see *aspartic proteinases*)
Acacia greggii, 326, 334
accessory gland proteins (ACP), 347, 377, 472, 479, 496
Achroia grisella (lesser wax moth), 345
adaptive evolution, 49–51, 54, 56, 64, 77, 78, 94–95, 276, 307
  of coloration, 116–117
  epigenetics, 255, 259–261
  molecular, 69–70, 116–117, 129–131, 166
  population bottlenecks, 414–416, 419–420
adaptive fixation, 106, 116
adaptive introgression, 400
adaptive landscape, 238, 270, 316, 317
  peaks, 94–95, 312
adaptive mutation(s), 43, 231
adaptive peaks, 94–95, 312
  local, 95, 312
adaptive radiation, 373, 471, 495
adaptive topography, 94–95
adaptive value, 8
additive genetic covariance (correlation) (see also *G matrix*), 59–60, 273, 305, 315, 316
additive genetic effect, 50, 87–89, 277, 290
additive genetic variance, V_A (variation) (see also *G matrix*), 21, 22–24, 50, 85, 86–90, 95–97, 268, 272–273, 275–279, 287, 290, 297, 302–303, 308, 310, 316, 414, 415, 417–421, 423, 424, 470, **513**
additivity (see also *additive genetic variance*), 288, 289, 290, 417, **513**
age-structured population, 465–466, 480
aging (see *senescence*)
agouti gene, 231, 258
Akaike information statistics, 120
alleles
  average effect, 86, 89–91, 95, 96, 98, 290, 297
  deleterious, 8, 55, 108, 422, 424, 470, 473, 505, 507
  local average effect, 86, 89–90, 95, 98
  mildly deleterious, 108
  rare, 65, 66, 67, 78, 108–109, 348, 417
Allonemobius, 377, 402
allopatric/allopatry, 373, 374, 376, 378, 380, 383, 384, 385, 400, 404, 407, 458–459, 503, **513**
allopolyploid (see *polyploidy*)
allozyme(s), 104, 114, 378, 381, 409, 471, 503, 509, **513**
altruism, 350, 351, 355, 356
  reciprocal, 350
Alu elements, 151, 207, 210
amino acids
  asparagine (Asp), 196
  aspartate (Asn), 196, 197, 198, 199
  aspartic proteinases, 197, 204
  codon usage/redundancy (see also *codon bias*), 16, 134, 195
  CRESCENDO, 197, 205
  cysteine (Cys), 25
  difference among taxa, 127, 129, 166
  effect of mutations, 15–16, 25, 35, 111, 119, 134, 195
  effect on protein structure/function, 25, 166, 196–197, 198, 199, 209, 210, 211, 227, 231, 389
  glutamate (Glu), 25, 196, 199
  homeobox/homeodomain, 224, 394
  poly-alanine tract, 227
  proline (Pro), 25
  proteases, 212
  rate of replacements, 10–11, 30, 41, 121–123, 129–131, 164, 166, 170, 196–197, 198, 211, 394
  selection, 26, 40, 41, 111, 114, 122, 130–131, 133, 164, 195, 460
  sequence alignment, 428, 429
  sequence differences/divergence, 10, 120–121
  serine proteinases, 213
ancestral gene resurrection, 166–167
androgen receptor, 168, 169–170
androgens, 20, 167–169
aneuploidy, 18, 34, 396, **513**
angiosperms, 17, 145, 380
antagonistic pleiotropy (AP)
  maintenance of genetic variation, 276
  tests of, 470
  theories of aging, 203, 464, 470, 480, 494, **513**
anterior–posterior axis, 224, 227
Apis mellifera, 410

Arabidopsis
ecotypes, 453
lyrata, 453
as a model organism, 453
R gene(s), 452, 458, 459, 460
recombinant inbred lines, 454
RPM1 (disease resistance gene), 452, 454, 455
RPS4 gene, 460
self-incompatibility gene (*SCR*), 453, 455
suecica, 454
thaliana, 135, 144, 146, 217, 242, 452, 453–455, 460
Archaea, 28, 145, 214, 219
artificial selection, 238, 241, 250, 251, 273, 274, 293, 320, 336–337, 346, 421, 424, 467, 472–473, 477, 483, **513**
aspartic proteinases, 197, 204
association mapping (see *association studies*)
association studies, 117, 228, 231, 306, 454
assortative mating, 81–82, 315–316, 317, 323, 417, 420, 422, 423, 449,
autotetraploid (see *polyploidy*)
average effect 96, 290, 297
local, 86, 89–90, 95, 98

B
Bactrocera (Dacus) spp., 406, 407, 408
balancer chromosomes (see *chromosomes*)
Baldwin effect (genetic assimilation), 331
basement membranes, 218
Bateson, William, 4–5, 224, 376
Bayesian inference/methods, 121, 122, 131, 157–158, 434, 436, 438, 439, **513**
Bayesian Markov Chain Monte Carlo (BMCMC), 157, 186, 436
Beavis effect, 23, 297
benchmark set, 204, **513**
Bicyclus anynana, 320
biodemography (see *demography*)
biodiversity, 49, 157, 369, 373, 502
biometricians, 3, 4, 12
biometrics, 4
birds, 159, 349
chicken, 274
Galapagos/Darwin's finches, 281–282, 283, 400, 410, 412, 482
greater prairie chickens, 506, 510
kin selection, 351, 355, 356
maternal effects, 20, 21, 277
Mauritius kestrel, 504–505
melanism, 231, 389
reproductive isolation and hybridization, 378, 381, 382, 385, 386, 404
senescence, 468, 474, 475, 479
species richness, 385
bootstrapping, in phylogenetic estimation, 439
Bordetella, 42
bottleneck(s), 275, Ch. 27, 135
effects on genetic load, 415, 419, 422, 423
effects on genetic variation/variance, 87–89, Ch. 27, 275, 276, 302, 309, 322–324
bottleneck *(Continued)*
effects on speciation, Ch. 27
founder-flush, Ch. 27
Mauritius kestrel, 504–505
quantitative genetic models, 417
boundary process, 65, 69, 76, 78
branch length(s), 163, 164, 165, 169, 174, 187, 426, 427, 429, 432–436, 440
Brassica rapa, 327
breeders' equation, 23, 268, 310, 312, **514**
maternal effects, 277
multivariate, 273, 277, 282, 283, 287, 295, 310

C
C-value (chromosome complement), 145, 155, 407
angiosperms, 155
paradox, 145, 147
CAAT box, 133
Caenorhabditis elegans
age-1 gene, 476, 477
aging (genetics of), 477
codon bias, 135
daf16 gene (dauer formation), 476, 477
daf-2 gene (dauer formation), 476
deletions, 152
duplications, 160
extracellular domains, 217
genome size, 40, 146
intron length, 139
multidomain proteins, 217, 219
mutation rates, 38
number of genes, 146
sperm competition, 346
canalization, 204, Ch. 16, 253, **514**
developmental, 260
environmental, 235, 239, 247, 248–249
genetic, 236, 239, 246–249
Castle, William, 5
catalytic conundrum, 198
cDNA (complementary DNA), 167, 168, 171, **514**
cellular competition, 359–360
central dogma, 25, 193, 195, 210, **514**
central limit theorem, 268
Cepaea, 6
Chaetodipus intermedius (pocket mice), 117
chimeric gene, 160, 162, 230
chimpanzee(s), 37, 44, 144, 186
chordates, 17, 218–219
Choristoneura rosaceana, 278–279
chromatin, 259
marks/marking, 254, 255, 259, 263, **514**
structure, 258, 260, 261
chromosome(s)
acrocentric, 18
autosomes, 112, 300, 301, 348, 382, 385, 390, 393, 394
balancer, 45, 47, 392, 394
complement (see *C-value*)
fusion, 34
inversions, 3, 7, 8, 18, 34, 161, 410, 471, **514**

chromosome *(Continued)*
metacentric, 18
monosomy, 18
number (see also *ploidy*), 14, 18, 34, 379, 401, 407
paracentric, 18
pericentric, 18
rearrangements, 7, 14, 30, 145, 161, 410, 471
sex, 27, 301, 348, 385, 472
hemizygous, 385
heterogametic/heteromorphic, 27, 347, 378, 381, 382, 385
and speciation, 375, 382–383
W, 348, 382
X, 109, 112, 187, 230, 253, 262, 347, 348, 381–383, 385, 390, 393, 394, 395
Y, 27, 63, 174, 186, 347, 348, 358, 383, 395
Z, 382
structure, 7, 17–18, 263, 402, 410
submetacentric, 18
translocations, 18, 396, 410, **525**
trisomy, 18
chromosome complement (see *C-value*)
cis-regulation (*cis*-acting), 27, 131, 136, 138, 223, 226, 229–233, 234, 243, 245, 249, 315, 390
cladistics, 434
Clarkia pulchella, 505
classical–balance controversy, 8–9, 10, 11
CNG triplets (see also *CpG couplets*), 254
co-adapted gene complexes, 18, 95
coalescence (coalescent), Ch. 12, 370, 509
distribution, 177, 180, 182–183, 186
effective size, 180, 182–183
sequence, 177, 184, 185
simulations, 108, 110
time, 105, 106, 174–178, 180, 183, 185
coancestry(ies), 82–84, 85
codominance (see *dominance*)
codon bias (see also *codons*), 16, 110, 133, 134–135, 140–141, **514**
Drosophila, 139–141
codons (see also *codon bias*; *mutation: substitutions, missense, non-synonymous, synonymous*)
Adh gene, 140–141
nonoptimal vs optimal (see also *codon bias*), 135–136, 140–141
rates of substitutions, 11, 111
redundancy, 16, 134–136, 195
stop codons, 16, 25, 27, 163, 206, 242, 260
transposable elements, 27
coefficients of relatedness, 352–354
coevolution
antagonistic, 496–497
arms race, 448, 451, 455, 496–497
empirical evidence, 458–461
Farhenholz rule, 460
gene-for-gene models (GFG), 450–452
host–parasite coevolution, Ch. 29
intersexual, 340, 496–497, 500
matching-alleles models (MAM), 449–450
mate choice, 340
coevolution *(Continued)*
metapopulations, 461
multiple traits within species, 302, 340
phylogenetic congruence, 460–461
quantitative genetic models, 452, 457–458
red queen hypothesis, 450, 451, 458, 459, 460
comparative phylogenetic method, 142
compensatory evolution, **514**
Adh gene, 141–142
RNA secondary structure, 136–137, 141, 142
compensatory mutations, 26, 137, 141, 142, 230
complex traits (see *genetic architecture*; *quantitative trait loci; senescence, quantitative genetics; sexual selection, quantitative genetics*)
concerted evolution, 163
conjugation, 28, 182
conservation genetics, Ch. 32
captive management of threatened species, 508
effects of inbreeding, 504
genetic diversity and extinction, 507
management units, 503
conspecific sperm precedence (see *sperm competition*)
constraints, 43, 152, 164, 165, 270, 415
adaptational, 326
developmental, 236, 275, 236
evolutionary, 194, 196–197, 199, 204, 205, 275, 317, 424
functional, 122, 125, 126, 138, 198, 274, 326
genetic, 295, 337, 339, 492
historical, 326
hypothesis, 43
physical, 208
selective, 111, 125, 129, 133, 134, 385
structural, 197, 211
contextual analysis, 93–94, 99–100, **515**
convergent evolution, 208, 231, 232, 321, 389, 433, 490
cooperation, 350–351, 354–355, 358, 360–362, 478, 495
cospeciation, 460–461
cotton (see *Gossypium hirsutum*)
covariance, genetic (see *additive genetic covariance*; ***G*** *matrix*)
CpG couplets, 16, 34, 254, **515**
mutation rate, 16, 34, 38, 44
CpG islands, 135, **515**
crayfish, freshwater, 441–442
Crick, Francis, 9, 193–194, 195
crickets (see *Allonemobius*; *Gryllus*)
Crow, James, 10, 11, 64
crustaceans, 225, 226, 227, 441–442
crystallins, 162
crystallography, 25, 210
cytogenic cloning (C-cloning), 490–491

D
daisy, lakeside, 507
Darwin, Charles, 4, 14, 65, 262, 281, 339, 367, 374, 401, 455, 473
DDT, 115, 504
Cyp6g1 gene and resistance to DDT, 115
dead on arrival (DOA) elements, 152

deficiency mapping, 395, 474
deleterious mutations (see *mutations*)
deletions, 16, 17, 18, 25, 33, 34, 36, 37, 38, 44, 119, 139, 151–152, 154, 155, 249, 362
demes, 50, Ch. 6
  differentiation of, 90–91, 92–93, 94
  variance among, 50, 80, 82, 83, 85, 86, 89–91, 94, 97
  variance within, 50, 80, 82, 83, 84–86, 90, 95
demography (biodemography), 467–470, **514**
  age-structured populations, 465–466
density dependence (dependent), 53, 457, 495
developmental buffering (see also *canalization*), 235–241, 247
diapause, evolution of, 20, 279–280
differential dominance, 299, 308, 309
differentiation (see also *divergence*)
  of average effects of alleles, 90–91, 98
  effect of G X E on, 331
  geographic, 414–417, 419, 420, 423, 424
  in molecular evolution, 110–111, 112, 117
  morphological, 388–389
  population (of demes), 90–94, 421, 422, 423, 424, 491, 492, 503–504, 509
  of species, 369, 387, 388, 389, 392, 407
disease resistance (see also *selective sweep, parasite resistance*), 448, 449, 450, 454, 511
  genetic architecture, 452, 461
dispersal (see also *gene flow*)
  cotton, 148, 149
dispersion, index of, 123, 124
divergence
  behavioral, 418, 420
  effects of G X E, 335
  epigenetic, 253, 262–263
  of genes, 104, Ch. 11, 375, 408, 427, 460
  of genomes, 148–150
  and hybridization, 399–402
  morphological, 222, 230, 262, 320, 369, 420
  of populations, 91, 92, Ch. 12, 302, 313, 317–318, 322, 374, 385, 497
  of proteins, 211–213, 219
  sequence, 104–107, 109–110, 114–115, 117, 120–123, 126, 130–131, 136–138, 148, 199, 208–209, 230, 378, 429, 434–435
  of species (interspecific), 35–36, 37, 39, 44–45, Ch. 12, 321, 324, 369–371, 373, 378, 392–394, 416–417, 427–428, 472, 518
DNA
  junk, 147, 151
    Drosophila vs mammals, 152
    rate of evolution, 124
  mitochondrial, 37, 116, 117, 127, 152, 370, 407, 410, 436, 441, 511
  noncoding, 16, 38, 109, 115, 123, 131, Ch. 9, 147, 152, 164, 186, 194, 196, 207, 209, 428
  nongenic, 144, 147, 156
  repair, 33, 42–43, 127–128, 151, 152, 154–156, 259, 397
  "selfish", 147, 193
  sequence gaps, 119, 120, 131, 428, 430
DNA binding domains (DBD), 169–171
Dobzhansky, Theodosius, 3, 6–9, 13, 376, 383, 390, 396, 400
Dobzhansky–Muller model (see *speciation*)
domain shuffling, 161–162, 212
domestication, 261, 483, 489, 490
  corn, 98, 115–117
dominance, genetic, 22, 35, 268, 274, 298
  bottlenecks, 415, 417–419
  codominance, 298, 299
  conversion to additive genetic variance, 419
  developmental buffering, 236
  differential, 299, 308
  dominance epistasis, 86, 90, 91, 277, 300, 302, 417–419
  genetic variance, 86, 268, 274–276, 290, 417–419
  genotypic values, 289
  inbreeding, 86, 276
  mechanism 23, 24
  overdominance/underdominance, 91, 92, 298, 299, 308, 336
  phenotypic plasticity, 336
  QTLs, 298–299
  speciation, 91
  structured populations, Ch. 6
  Wright & Fisher debate, 22
dominance, social, 343, 345, 359
dominance theory (see *Haldane's rule*)
dosage compensation, 262, 472
  gene silencing, 18
  X-inactivation, 253, 262
Draba, 409
drift (see *genetic drift*)
Drosophila
  abd-a (*abdominal-A*) locus/protein, 226, 389
  abd-b (*abdominal-B*) locus, 389, 390
  accessory gland proteins (ACP), 479, 496
  adh gene/protein, 12, 25, 113–114, 140–142
  affinis, 390
  americana, 230, 390
  ananassae, 389, 393
  antennapedia complex, 224
  antagonistic coevolution, 496
  apterous gene– 225
  arawakana, 390
  balancer chromosomes, 45
  bicoid (bcd) gene/protein/gradient, 136, 137, 227, 228
  bithorax complex, 224
  bric-a-brac (*bab*) gene, 226, 389–390
  chc gene (chico), 476
  codon usage/bias, 134, 135, 139–141
  cyp6g1 gene, 114–115
  delta gene, 23
  desat2 gene, 393
  DOPA decarboxylase (*Ddc*), 474
  dunni subgroup, 390
  duplications, 160
    fixation rate, 39
  ebony locus [N-,-alanyldopamine (NBAD) synthetase], 390

Drosophila (Continued)
epigenetic inheritance, 258–259
epistasis, 30
even skipped (*eve*) gene, 131, 138, 230
experimental evolution, 483, 484, 488, 489, 490, 491, 493, 498
founder effect/bottlenecks, 415–417, 421, 422
gene genealogies, 187
generation time, 40
genetic assimilation, 331
genome size, 40, 146, 472
GPDH (glycerol-3-phosphate dehydrogenase) locus, 128
haldane's rule, 381, 382
haltere(s), 226
Hawaiian, 142, 414, 415, 416–417
heat shock, 238, 240–242, 260
heterozygosity/polymorphism, 9, 12, 77, 113, 131
hitchhiking, 106
hox genes, 17, 224
hybrid dysfunction/sterility, 376, 377, 378, 385, Ch. 25
hybrid make rescue (*Hmr*) gene/protein, 393, 394, 396
imaginal discs, 304–305
inbreeding depression, 505–506, 508
Indy gene (I'm not dead yet), 477
insularis, 127
insulin-like receptor (*InR*), 476
introns, 138
kikkawai, 390
Krüppel gene, 258–259
maternal effects, 20, 47
mauritiana, 377, 382, 385, 388, 393, 394, 397
mercatorum, 418, 423
as a model organism, 471
montium, 389, 390
mutations, 8, 38, 41, 44, 106, 114, 140, 378
genetic draft, 77
indels, 17, 38, 151, 152
mutation accumulation, 44–47, 126, 128, 470, 508
rate, 38, 46, 126, 128
nigrodunni, 390
novamexicana, 230, 390
number of genes, 30, 146, 388
obscura, 127
obscura group, 126, 127, 142
odysseus (ods) gene, 394
omb locus, 390
pallidosa, 393
pdm/nubbin gene, 225
P-elements, 28
persimilis, 376
phenotypic plasticity, 336
pigmentation, 226, 228, 230, 389–391, 396
abdominal-A (*abd-A*) gene, 226, 389–391
abdominal-B (*abd-B*) gene, 389–391
bric-a-brac (*bab*) gene, 226, 228, 389–391, 392
doublesex (dsx) gene, 389
yellow gene, 228–229, 230, 391
proteins
multidomain, 217
pseudoobscura, 9, 11, 230, 376, 377, 418, 490

Drosophila (Continued)
radiation exposure, 8
recombination, 106, 113, 138
regulatory elements, 138, 230
RNA secondary structure, 136, 141
saltans, 126
santomea, 390
sechellia, 230, 377, 389, 393
segmentation, 230
selection
codon usage, 136
molecular, 113
positive selection, 110, 131
senescence, 467, 469–470, 472, 474, 476, 477, 491, 498
sequence divergence, 131
serrata, 345, 390
sex-limited gene expression, 474
sex-linkage, 348
sexual dimorphism, 389–390, 492
sexual selection, 345, 346, 347, 377, 384, 422, 496
shavenbaby/ovo (*svb*) gene/transcription factor, 231, 232, 396
simulans, 77, 114–115, 136, 138, 227, 230, 377, 382, 385, 388, 393–395, 397–398, 418, 421, 472
sophophora subgroup, 126
speciation, Chs. 24, 25, 422, 423, 496
sperm storage, 347
subobscura, 228, 390
transgenesis, 228
transposable elements, 18, 151, 152
ultrabithorax (*Ubx*) gene, 226, 227
virilis, 228, 390
virilis group, 152
white locus, 17, 22
willistoni, 11, 127, 390
willistoni group, 126–127, 521
wing development/shape, 225, 226, 238, 304–305, 322, 328, 331
xdh (xanthine dehydrogenase) locus, 126
yakuba, 390, 394, 397
yellow locus, 228–229, 230, 390
duplications (see also *mutation*), 17–18, 29–30, 147–151
gene, 17–18, 29–30, 130–131, Ch. 11, 211–212
dating, 157–159, 166–167
duplication, degeneration, and complementation (DDC) model, 162–163, 165, 169
evolutionary dynamics, 162–165
gene loss, 159–160
mechanisms, 160–162
more-of-the-same model, 162, 163
neofunctionalization, 30, 162, 164, 169
new functions, 165–166
nonfunctionalization, 163
steroid hormone receptors (see *hormones and receptors*)
subfunctionalization, 30, 163, 165, 169
genome (see *polyploidy*)
mutational hot spots, 38
tandem, 17, 34, 160

E

Ebenopsis ebano, 326, 335
effective population size (N_e), 29, 63, 133, 317, 507, **516**
  in captivity, 508
  effect on genetic drift, 29, 76
  effect of selection, 29, 63
  heterozygosity, 104
  males vs. females, 63
  maternal inheritance, 63
  number of mutations, 121
electrophoresis (history), 9
embryogenesis, 227, 261
endosymbiosis, 28, **516**
environmental variance (see *variance*)
epiallele(s), 254, 256, 259, 260, **516**
epigenesis, 253
epigenetic inheritance, 18, Ch. 17, **516**
  chromatin marking (see also *methylation of DNA*), 254
  Drosophila, 258–257, 260
  effect on evolution, 259
  effect on genetic variation, 260
  evolution of, 18, 255
  genomic imprinting, 18 , 22, 253, 256, 348, **517**
  history 252–253
  Linaria vulgaris, 256–257
  Mus, 257–258
  self-sustaining feedback loops, 254
  sexual selection, 348
  single-celled organisms, 255
  speciation, 262–263
  structural inheritance, 254, 255
epigenetic landscape, 235, 236, 237, 238, 239, 251, 253
epigenetics (see *epigenetic inheritance*)
epistasis, (see under specific topics)
epistatic genetic variance (see *epistasis*)
equilibrium
  evolutionary, 56, 489, 490, 497
  gene-expression, 244–247, 249
  Hardy–Weinberg (–Castle), 51, 81, 83, 89, 290, 369
  mutation–drift, 104, 105, 317
  mutation–selection, 419, 470
  selection–mutation–drift, 136
Escherichia coli, 485
  adaptation, rate of, 250, 487
  experimental evolution, 484, 486, 488, 494
  gene expression, 487
  genome size, 40
  genotypic diversity, 485
  horizontal inheritance, 28
  mutation rate, 41, 43
  natural history, 485
  selection, 135, 136
  Shigella strains, 486
estrogens (and estrogen receptors), 167–171
Etheostoma, 380, 385
Eubacteria, 28, 214, 219
Euler–Lotka equation, 466
evo-devo (see *evolutionary developmental biology*)
evolution in steps model, 488
evolutionary capacitance, 249–250
evolutionary developmental biology (EDB), Ch.15
  comparative, 223–228
  genetical, 228–231
evolutionary novelty, 17, 25
  gene duplications, 29, 166, 169
  hybridization, 17, 28, 402, 404, 406, 410, 412
  recombination, 484
evolvability
  gene networks, 249–250
  Hsp90, 241–242
  selection for, 42–43
exon shuffling, 160, 207, 219–221, **516**
experimental evolution, Ch. 31
  adaptation to the lab, 489–490
  asexual organisms, 488
  in *Drosophila*, 490, 493, 498
  and ecology, 495
  in *E. coli*, 486–487, 494–495
  evolution of **G**, 321
  experimental design, 484
  synthetic phylogenies, 492
  in the wild, 499
expression domains (see also *subfunctionalization*), 165, 169

F

F-statistics, 63, 81
  F_{IS}, 84
  F_{ST}, 84, 110
fast evolving genes, 126
Faunis menado, 320
fecundity, age-specific [m(x) or m_x], 52, 466, 479
Fisher, Ronald (R. A.), 3, 5–6, 12, 23, 64, 72, 74, 80, 276, 301, 464, 466
Fisher's fundamental theorem, 23
Fisher's geometric model, 69
Fisher's large population theory, 6
fitness (see also *selection*), Ch. 4
  absolute, 53
  age-specific, 466, 490
  covariance with phenotypes (see *selection*)
  definition of, 52, 54, 244
  density-dependence (see *density dependence*; *selection*), 53
  effect of generation time, 52–53
  effect of mutation, 27, Ch. 3, 40–41, 47, Chs. 7–8, 466
  effect of radiation, 8
  effect of recombination, 27
  Euler–Lotka/characteristic equation, 53, 466
  evolution of, *E. coli*, 487, 494
  Fisher's fundamental theorem, 23
  frequency-dependence (see *frequency dependence*; *selection*), 53
  heritability of, 296
  inclusive fitness, Ch. 23
  intrinsic rate of increase, 53, 466
  life history traits, 41
  lifetime reproductive success/net reproductive rate, 52–53
  measurement of, 52, 342–343
    partitioning, 50, 60–62

fitness *(Continued)*
overdominance, 299
parental vs offspring fitness, 52
relative, 53–54, 55
surface/landscape, 269–272, 312
surrogate measures, 54
flour beetle(s) (see *Tribolium*)
fluctuating asymmetry, 295, 299, 301, **516**
fluctuating neutral space model, 129
founder event/effect (see *bottleneck*)
founder-flush models, Ch. 27
free energy minimization, 136
frequency dependence (dependent), 489, **516**
estimates of fitness, 53
Fu and Li's test, 108
functional annotation, 206, 453
functional constraint (see *constraints*)

G
G matrix (see also *additive genetic covariance*; *additive genetic variance*), 203, 273, 277–278, 287, 305–306, Ch. 20, 344, 345, 349, 498
Gaussian distribution (see *normal distribution*)
gene conversion, 34, 147, 163
gene expression, 26, 30, 250, 327, 389, 455, 478, 515, 517, 523
and codon use, 135, 140
and epigenetics (imprinting), 18, 253, 261, 348
equilibrium, 244–247, 249
patterns, 222–223, 229, 250, 322, 397
regulation of, 131, 141, 142, 232, 233, 386, 397, 487
and reproductive isolation (speciation), 386, 388, 397–398
senescence, 477–478
gene families, 130, Ch. 11, 455, 477, **517**
and domain shuffling, 161–162
and duplication, 160–161
phylogenies, 157–159, 161
superfamilies, 157, 198, 208
trees, 157–159
gene flow, 335, 343, 409, 499
barriers to, 91, 387, 392, 524
and coevolution, 449, 461–462
in conservation, 503, 509–510
between demes, 95, 110, 520
in hybrid zones, 407, 409, 412–413
and speciation, 368, 373, 386, 387, 392
gene genealogy(ies), (see *genealogies*)
gene interaction(s) (see *epistasis* under specific topics)
gene knockout(s), 17, 40, 202, 247–249, 250, 476, **519**
gene loss, 148, 150, 159–160, 206, 224
gene ontology (GO) system, 206
gene regulation, 118, 136, 142, 232–233, 325, 385–386, 388–389, 397, 472
gene regulatory networks, 201, 243–245, 250, **517**
gene rescue, 229
gene silencing, **517**
duplicate genes, 29–30, 408
epigenetic inheritance, 255, 259, 263
methylation, 18, 263
gene silencing *(Continued)*
mutations, 30
RNA-mediated (see also *RNA, interference*), 254, 255
silencer, 26, **524**
transposable elements, 18
gene superfamilies, (see *gene families*)
gene tree(s), (see *phylogeny*)
genealogies
of alleles, 104
C-equivalent, 179
collapsed, 179, 180
C-type, 179, 180
D-equivalent, 179
D-type, 179, 180, 184
gene, 105, 106, 110, Ch. 12, 368, 369, 370, 460
H-equivalent, 177
M-equivalent, 178, 179
M-type, 178–179, 180, 184
P-equivalent, 179
P-type of, 179, 180
phyletic status, 178–179. 184
random, 173, 177, 180, 183–184, 185, 187–188
species, 185–186, 187
star like, 184
T-equivalent, 178
generation time effect, 123, 133
genetic architecture (see *quantitative trait loci*; *quantitative traits*)
genetic assimilation, 236–237, 238–240, 242, 251, 260, 331, 400, **517**
genetic correlations, 285, 302, 345, 346, 489
antagonistic, 276, 278, 470, 497
as constraints, 274, 317
contribution of pleiotropy, 278, 302–303, 307, 323
cross-environment, 329
estimating, 281, 320, 473, 498, 513
linkage (disequilibrium), 315, 323
in response to selection, 273, 274, 282–283, 307, 320, 415, 473, 498
between sexes (intersex), 301, 340, 343, 344, 347, 479
genetic distance, 293, 378–381, 382–384, 385
genetic draft (see *hitchhiking*)
genetic drift (see also *neutral theory*), 29, 36, 49–50, 63, Ch. 5, Ch. 6, 104–105, **517**
effect on genetic variance/G matrix, 89, 90, 275, 302, 317–318
wing shape, 322–323
mutation–drift equilibrium, 104, 135
selection–mutation–drift, 136, 141
Wright vs. Fisher, 6, 8
genetic load (see *mutations*)
purging, 419, 422, 423, 503, 505, 507, **523**
genetic variance, (see *variance*)
genic values, 290
genome duplication (see *polyploidy*)
genome sequencing (projects), 47, 111, 118, 143, 144, 145–146, 157, 171, 210, 233, 251, 453, 454, 472
genome sizes, 40, Ch. 10, 146
Drosophila, variation among species, 472
effect on mutation rate, 38, 41, 42

genome sizes *(Continued)*
evolution of, 145–147, 148, 149, 151, 472
rate of evolution, 153–154
functional significance, 154
insertions/deletions, 151–153, 154–156
phage, 40
transposable elements, 148, 151
genomic imprinting, 18, 22, 253, 256, 348, **517**
genotype-×-environment (G X E) interaction, Ch. 21, **517**
effect on evolution, 331
genotypic values, 49, 94, 288, **517**
additive, 96, 276, 289–290, 292, 297, 298, 300, 302
dominance, 96, 289–290, 292, 298, 299, 300, 302
epistasis (epistatic), 87, 89, 96, 289–290, 300, 302
estimating, 292, 296
and variance components, 86, 89, 95, 96–97, 288, 309
Gentiana pneumonanthe, 506
Geospiza spp., 281, 412
Gila spp., 409
Gillespie, John, 12
glucocorticoids, 167
Glycine spp., 379, 380
Goldschmidt, Richard, 7
Gompertz equation, 467–469
good-sperm hypothesis, 347
gorilla(s), 186
Gossypium hirsutum, 147–149, 153
GPDH (glycerol-3-phosphate dehydrogenase), 128–129
green fluorescent protein (GFP), 229
Gryllus firmus, 282, 284, 286

H

Hagedoorn effect, 72
hairpin structure(s), 141–142, 208
Haldane, J. B. S., 5–6, 381
Haldane's rule, 377, 381, **517**
dominance theory/model, 381, 382, 397, **515**
faster male theory, 381
faster X theory, 382
Hamilton, W. D., 94, 351, 357, 450, 466, 478, 479
Hamilton's rule, 94, 351, 354, 355, 357
Hardy–Weinberg equilibrium (also Hardy–Weinberg–Castle), 51, 81
genetic variation at, 83, 89, 290
heat shock proteins (see *Hsp90*)
Helianthus spp., 400, 402, 410, 412
linkage map, 411
Hennig, Willi, 370, 434
herbivory, 332
herbivory genes, 455
heritability (h^2) (see also *additive genetic variance*; *breeders' equation*; **G** *matrix*), 23, 57, 268, 330
broad sense, 268, **517**
environmental effects on, 330–331
estimates
critical photoperiod, 287
diapause, 280
egg size, 330
lifespan, 475
heritability *(Continued)*
phenotypic plasticity, 336
social dominance, 345
estimation, 268, 281, 337
evolution of, 272, 273
Hamilton's rule, 94
maternal effects on, 21, 277
narrow sense, 23, 50, 268, **518**
threshold trait, 282
variation among trait types, 296
heterochromatin (heterochromatic), 145, 151
microchromosomes, 261
Heteronotia binoei, 409
heterosis, 8, 104, **518**
heterozygosity, 9, 73–75, 77, 78, 104, 106, 108, 110, 112–113, 115, 133, 302, 336, 415, 504, 506, 508
Hill–Robertson effect, 139, **518**
histone modification, 254, 259
histones, 109, 124, 126
hitchhiking, 72, 74–75, 77, 79, 106–107, 112, **518**
HKA test (see *Hudson–Kreitman–Aguadé test*)
HLA, 40
homeobox/homeodomain, 224, 394
homologous sequences, 33, 34, 119, 121, 197
homology
genome assembly, 210
identification/testing, 208, 223, 225
insect wings, 225
positional, 119, 122
protein domains, 215
sequence alignment, 428, 429
homonucleotide runs, 17, **518**
homoploidy, 401, 409–410
horizontal inheritance/transfer (see also *endosymbiosis*; *hybridization*; *plasmids*; *transduction*; *transformation*)
among cells within an individual, 255
E. coli, 28
interspecific, 27–29, 181, 207, 362, 428, 456, 472
hormones and receptors
activation, 197
antagonistic coevolution, 496
growth hormone, 128, 476
insulin-signaling and aging, 476–477
juvenile hormone esterase (JHE), 282, 284, 330
maternal effects, 20
melanocortin-1-receptor (*Mc1r*) in *Mus,* 231
steroid hormones, 167
evolution of, 169–171
mechanism of action, 168
receptor physiology, 167, 204
host–parasite interactions, Ch. 29
host–pathogen interactions (see *host–parasite interactions*)
hox genes, 17, 223, 224, 227–228
Scr gene/protein, 227, 453, 455
Hsp90
Arabidopsis, 242
canalization/buffering –237, 240–242, 249, 260
function/pleiotropic effects, 240, 241, 258

Hsp90 *(Continued)*
genetic assimilation/evolutionary capacitor, 237, 240–242, 249–250, 260
inhibition, 258
Hubby, Jack L., 9
Hudson–Kreitman–Aguadé (HKA) test, 12, 26, 105, 109–110, 112, 115, **518**
Adh, 114
human(s)
chromosome numbers, 18
chromosome size, 145
common ancestry, 174, 186
gene/genome duplications, 144, 160, 206–207
genome size, 40, 145
human pancreatic lipase, 199
linkage disequilibrium, 111
mutations/rates, 27, 37–40, 44, 126, 131
noncoding DNA, 164
protein domains, 217
PTC locus, 109
rare alleles, 109
RNA genes, 194
sickle-cell anemia, 25
social evolution, 359
steroid hormones/receptors, 167–168, 170
transposable elements, 27, 151, 152, 160
Huxley, Thomas (T. H.), 4
Hyalodaphnia cucullata, 329
hybrid(s)
breakdown, 30, 276
dysfunction, 376–377, 381, 383, 387–388, 390, 393, 397
fitness, 376, 378, 396, 397, 402, 412, 413
Hybrid male rescue (Hmr), 393
incompatibility(ies), 187, 263, 276, 376, 378, 380–383, 385, 390, 396–397
inviability, 376, 377, 378, 380, 382, 385, 388, 390, 393–394, 396, 417
lethality, 393–395
sink, 28, 400
speciation (see also *polyploidy*), Ch. 26, **518**
sterility, 187, 263, 376, 377, 378, 380–382, 385, 388, 390, 393–394, 396–397
zone(s), 276, 369, 399, 402, 403, 405, 407, 412–413, **518**
evolutionary novelty, 404, 406, 410–412
mosaic , 377, 403–404
selection in, 403
stable, 399, 403, 413
tension zone (TZ) (dynamic equilibrium), 403–404, 410
hybridization, 28, 147–148, 160, 173, 371, 378, 381, 385, 428
history, 3, 399–402
interspecific, 17, 28
introgressive, 404, 409–410, **519**
natural, Ch. 26, 518, **521**
and ploidy, 17, 34, 148, 401, 402, 407–410, 513
hypermutability, 43, 494
Hyphantria cunea, 279

I

identical by descent (IBD), 63, 81, 83, 276
immunity (see also *host–parasite interactions*)
evolution of, 278, 456
and senescence, 477, 479–480
immunoglobulin, 213, 216, 218, 220
immunohistochemistry, 10, 226
Impatiens capensis, 99
imprinting (see *genomic imprinting*)
inbreeding, 81–82, **518**
coefficient (Wright's), 81, 83–84
depression, 276–277, 379, 505–507, 510
age-specific, 470
bottlenecked populations, 415, 419, 421, 422, 424
effect of epistasis, 423
Mauritius kestrel, 504
effect on experimental evolution studies, Ch. 31
effect on genetic variation
among demes, 90–91, 96–97
within demes, 86, 504, 505
population extinction, 504, 505–507
purging deleterious alleles, 423
incomplete lineage sorting, 427, 428, **518**
indirect genetic effects, 50, 343
infinite alleles model, 10, 104, 108, 297
infinite sites model, 104, 108
infinitesimal model, 23, 519
intercross design, 291–292, 293
intersexual conflict (see *sexual selection*)
intragenomic conflict, 18
introgression, adaptive, 400
introns, 16, 44, 130, 133, 160, 194, 206, 219–221, 257, 394, 516, **519**, 520, 524
compensatory evolution in, 26, 141–142
homologous, 40
length (size), 133, 134, 138–139, 153–154
orthologous, 153
splicing, 133, 136, 138–139
inversions, chromosomal (see *chromosomes*)
Iris spp. 404, 406
irradiation (see *radiation*)
isochore, 135, **519**

J

jackknifing (see *phylogenetic estimation*)
Jukes, Thomas, 10–11, 103, 195
jumping genes (see *transposable elements*)
Junonia spp.– 320–321

K

K_a, rate of nonsynonymous substitution, 111, 164, 211
K_a/K_s ratio, 111–112, 164, 166, 211
K_s, rate of synonymous substitution, 111, 164, 211
kestrel, Mauritius, conservation genetics, 504
Kimura, Moto (see also *molecular clock*; *neutral theory*), 10–12, 103, 121, 123, 135
kin selection, 62, 94–95, 351, 354–358
coefficient of relatedness, 352–354
Hamilton's rule, 94, 351, 354, 355, 357
mortality, ants, 479

King, Jack C., 8–9, 10–11, 103, 195
knockouts, (see *gene knockouts*)
Kölreuter, Joseph, 401
Kreitman, Martin, 12
Kringle domain, 213, 216

L

labeled history, 177–180, 183, 184
  collapsed, 179–180
labeled topologies, 176–180, 181, 184, 186–187
  collapsed, 179–180, 181–182, 184, 187
Lamarckian inheritance (Lamarckism), 237, 252, 255
Lcyc, 256–257
Lepidoptera
  Choristoneura rosaceana, 278–279
  Haldane's rule, 381
  hybrid sterility, 378, 383
  Hyphantria cunea, 279
  Nemoria arizonaria, 328
  reproductive isolation, 378, 382
  X-chromosome, 382–383
Lethal hybrid rescue (Lhr), 394
Lewontin, Richard, 9–12
Lewontin's paradox, 77
lifetime reproductive success (LRS), 21, 52, 54
ligand binding domain (LBD), 167, 169–171, 216
likelihood ratio test, 120, **519**
  duplication rates, 164
  molecular clock, 124, 125, 126, 128
  phylogenetics, 164, 431
  RNA secondary structure, 136
Linaria vulgaris, 256–257
lineage splitting, 157, 370, 371
linearized tree method, 128
line-cross analysis, 276–277, 288, 420
linkage
  disequilibrium (LD), 85, 278, 369, 403, 452, 453, 510, **519**
    as a covariance, 59
    and genetic covariance, 316, 317, 323
    and maternal effects, 21
    in QTL mapping, 292, 293, 308, 454, 474
    and selection, 111, 137, 360–361, 403, 474
  equilibrium, 273, 290, 293, 323, **519**
Linnaeus, Carl, 206, 256, 401
Linnean rank (see *species rank*)
Linum spp., 462
local adaptation, 110, 458–459, 462
local average effect (of an allele), 86, 89–90, 95, 98
LOD score, 292

M

maize (see also *teosinte*)
  cross with teosinte, 95–98
  genome duplication & transposable elements, 147–148, 150, 151, 153, 472
  selection on teosinte branched1 (TB1), 115–117
maladaptation, 458–459, 464
Malthusian parameter (r), 53, 466, 479
Markov chain model,
  genetic variances, 84–86
  phylogenetics and genealogies, 157, 186, 436
matching allele mechanism (see *coevolution*)
maternal effects, 19–21
  effect on response to selection, 21, 277
mating success, 54, 339, 340, 376, 380, 392–393, 421
mating system(s), 59, 62, 341, 349, 385, 447, 509
maxillipeds, 226–227
maximum likelihood, **519**
  nucleotide substitutions, 111, 120–121, 122, 130, 135
  phylogenetic analyses, 159, 166, 169–170, 206, 431–442
maximum parsimony, 157, 166, 433–435
Mayr, Ernst, 6, 7, 252, 302
McDonald–Kreitman (MK) test, 12, 105, 114, 110, 118
  Adh, 114
meiosis, 27, 38, 256, 261, 396
  errors in, 17, 45, 160, 263, 407
  "fair", 350, 358, 359–360
meiotic drive, 51, 358, 362
Melampsora lini, 462
melanism (see *pigmentation*)
Mendel, Gregor, 4, 14, 22
Mendelian traits, 267–268
Mendelians, 3–5, 12
meningitis, 456
menopause, 478
metabolic pathways, 22, 24, 199–200, 476
metapopulations, 4, 50, Ch. 6, 276, 334, 438, 452, 461–462, 515, **520**
Metazoa(ns), 131, 167, 214, 216–218, 220–221, 224, 233, 359–360
methylation of DNA, 16, 18, 254–255, 255, **515**
  Axin^{Fu} in *Mus,* 257
  gene silencing, 18, 254
  Lcyc gene in *Linaria,* 256–258
  mutation rates, 38, 259
  transposable elements, 256, 259, 263
mice (see *Mus*)
microarray(s), 24, 26, 228, 385, 388, 397–398, 454, 477–478, 515, 519, **520**
microcosm experiments, 495
Microphallus, 459
microsatellite DNA, 17, 34, 37, 104, 291, 293, 503, 504, 509–510, **520**
migration (see also *gene flow*), 11, 332, **520**
  and genetic variation, 6, 11, 28, 63, 80, 81, 83, 93, 103, 108, 173, 186, 302, 312, 319, 403, 414
  and host–parasite evolution, 459, 461
  island model, 184
  in metapopulations, 80, 82, 83, 95, 520
  rate, 6, 11, 63–64, 65, 92–93, 184, 186
  sex-specific, 63–64
minisatellite DNA, 33, **520**
mitochondrial DNA (see *DNA*)
mitochondrial genes, 124, 126–127, 129, 408, 441
mitochondrial genome, 29, 63, 124, 152, 174, 362, 410
MK test (see *McDonald–Kreitman test*)
modularity
  genetic modules, 304, **520**
  regulatory, 233

molecular clock, Ch. 8, 377
calibration, 37, 122, 128, 380, 382
constancy, 10–11, 121–123, 124, 125
generation time effect, 122–123, 133
history of, 10–12, 121
index of dispersion, 123, 124, 132
likelihood ratio test, 120, 124, 125, 126, 128
lineage effects, 123, 124, 126–128,
local or local models, 127–128, 131
locus effect(s), 124, 126, 128
multiplicity, 121
overdispersed, 123–124, 128, 132
reference substitution rate, 122
relative rate test, 124, 125, 126, 128
tests of (testing), 125, 128
timing events with, 121, 122, 126, 128, 131, 159, 427, 432–433
monophyletic (monophyly), 178–180, 184, 186–187, 368–370, 427, 434, 436, 439–442, **520**
reciprocal, 178, 184, 370
status, 178
Moran model, 181
Morgan, Thomas Hunt (T. H.), 7, 8, 483
morpholinos, 229
morphometrics, Box 19.4, 420, **520**
mortality, age-specific, 469, 470, 474–475
most recent common ancestor MRCA, 174–176
time to the most common ancestor (T_{MRCA}), 174, 176
mouse *(see Mus)*
Muller, Hermann Joseph, 8–9, 22, 91, 376
Dobzhansky–Muller model, 30, 376, 380–381, 390, 392
Mus, 257, 293
aging, 476, 477
Axin gene (*Axin*^{FU}), 257
body size, genetics, 293, 297–299, 303, 308
duplications, 160, 207
Fused gene, 257, 261
genome size, 40, 146
IGF1r gene (insulin-like growth factor 1), 476
inbreeding depression, 506
lifespan, 476
mandible morphology, 295, 299, 306
mutation rate, 32, 38, 111, 126
p66shc trascription factor, 476
pigmentation, 4–5, 115–117, 231, 256, 258
pit-1 gene, 476, 477
prop-1 gene, 476, 477
tail morphology, 257–258, 274
Musca domestica, 420
mutagenesis, 22, 171, 229
Bicyclus anynana, 321
saturated, **523**
site-directed/targeted, 166, 171, **525**
mutagens, 16, 42
EMS (ethylmethane sulfonate), 16
mutation(s), 15–18, Ch. 3
accumulation (MA), 35–37, **520**
balancer chromosomes, 45
Drosophila, 44–47, 470, 508
fitness effect, 40
inversion effect, 18
mutation *(Continued)*
mutation rates, 38–39, 46–47
mutational meltdown, 508, **520**
phenotypic effects, 41, 47
senescence, 47, 464, 470, 473
adaptive, 43, 231
alcohol dehydrogenase (Adh), 141–142
compensatory, 26, 136–138, 141, 142, 230, **514**
complementation test, 395, **514**
copy-error, 127
correlated mutations, 198–200
CpG dinucleotide pairs, 16, 34, 38, 44
deficiency, **515**
libraries, 388, **515**
mapping, 395, 474
directed, 43, 44
duplications, 17–18, 29–30, 147–151
gene, 17–18, 29–30, 130–131, Ch. 11, 211–212
dating, 157–159, 166–167
duplication, degeneration, and complementation (DDC) model, 162–163, 165, 169
evolutionary dynamics, 162–165
gene loss, 159–160
mechanisms, 160–162
more-of-the-same model, 162, 163
neofunctionalization, 30, 162, 164, 169
new functions, 165–166
nonfunctionalization, 163
steroid hormone receptors (see *hormones and receptors*)
subfunctionalization, 30, 163, 165, 169
genome (see *polyploidy*)
mutational hot spots, 38
tandem, 17, 34, 160
effect on G matrix, 278, 314, 316, 317, 321, 323, 324
effect sizes, 23–24, 27, 40–41, 195–196, 378, 508
advantageous/beneficial, 35
deleterious, 8, 35, 104, 123
effect of genetic background, Ch. 16
indels, 108
epimutation (see *epigenetic inheritance*)
fixation probability (see also *genetic drift*; *molecular clock*; *selection*), 66–77, 124–130
advantageous, 66–67
deleterious, 133, 134
hitchhiking, 72, 74–75, 77, 79, 106–107, 112, **518**
infinite alleles model, 104, 108
neutral mutation, 29, 36, 66, 68, 122
frameshift, 15, 25, 35, 393, **516**
gain-of-function, 22, 162
genetic/mutational load, 8, 135, 415, **517**
purging, 419, 422, 423, 503, 505, 507, 523
history of ideas, 6–7, 8–9, 10–11
homeotic transformations (see *hox genes*)
hypermutability, 43, 494
indels (insertions, deletions), 16–17, 25, 33–34, 139, 151
DNA slippage, 16, 17, 151, 160, 209, **515**
genome size, 152–154
pseudogenes, 152
rate, 151–153, 394

mutation *(Continued)*
sequence alignment, 119, 428, 430
test for neutrality, 108
variance among organisms, 152
intron length, 138–139
length changes, 147
lethal mutations, 35, 37, 41, 249
Hsp90, 241
loss of function (see also *mutation, deficiency*), 22, 162, 249, 250, 394
mRNA secondary structure, 137–138, 141–142
mutational variance (CV_M), 41, 46
nearly neutral, 27, 103–104
nearly neutral theory, 123–124, 127, 133, 195
neutral, 10, 11, 35, Ch. 8
tests of neutrality, 107–111
point mutation (see *mutation, substitutions*)
promoter, 162
rates, 32, 37–43, 105
adaptive increases in, 43
Drosophila, 38, 44–47, 112–115
E. coli, 41, 43
evolution of, 41–43, 494
gain-of-function, 162
humans, 38, 44
indels, 44, 151–153
lethals, 46, 248
phage, 39
promoters, 131
pseudogenes, 44, 152
sex differences, 38
theory, 69, 71–72
transitions, 44
transversions, 44
regulatory, 25–26, 131, 134, 138
structural, 24–25
substitution(s), 33, 119–121
missense, 16, 25, 30, **520**
nonsense mutations, 16
nonsynonymous, 15–16, 125, 195
synonymous (see also *codon bias*), 11, 15–16, 134–136, 140, 195
transition(s) –16, 33
transversion(s), 16, 33
tautomeric shifts, 16, **525**
transposable elements, 16–17, 27, 34, 147–151, 160
mutation–drift balance, 104, 105, 135, 136, 317
mutation–selection balance, 136, 139, 141, 278, 419, 422, 464, 470
mutational bias, 37, 134, 152, 154, 317
mutational landscape model, 70–71
mutational meltdown, 508, **520**
mutator genes, 242, 250, 494

N

natural selection (see *selection*)
Nauphoeta cinerea (cockroach), 342, 346
near-isogenic lines (NILs), 454
nearly neutral theory, 123–124, 127, 133, 195
neighbor-joining (see *phylogenetic estimation*)
Nemoria arizonaria, 328
neo-Darwinian (Darwinism), 6, 7, 11
neofunctionalization, 30, 162, 164, 169
networks, gene interaction, 200–204
betweenness, 201, **513**
clustering, 201
coefficient, 202
degree distribution, 201–203
node degree, 201, **521**
scale-free, 201, **524**
neutral distribution, 104, 109
neutral mutations (see *mutations*)
neutral theory, 36, 103–104, 105, 133, 371
history of, 11–12
and the molecular clock, 121, 123, 131, 133
tests of neutrality, 107–111, 133
neutralist–selectionist controversy (see *neutralists*)
neutralists, 3, 10–12, 79, 129, 195, 197, 210
New Synthesis, 223
Nicotiana, 152, 155
noncoding DNA, (see *DNA*)
noncoding gene features, (see *DNA*)
nondeterministic polynomial, 429
nonfunctionalization, 162–164
nongenic DNA, (see *DNA*)
nonhomologous recombination, (see *recombination*)
non-LTR (long terminal repeat) retrotransposable elements, 152
nonmonophyly, 186
norm of reaction, 328, **521**
normal distribution, 4, 268
novelty, evolutionary origins, 17, 25, 27, 28, 29, 119, 169, 402, 404, 406, 410–412, 484
number of genes, 146, 206–207, 288, 344, 387, 389, 396, 474
numts, 152

O

Occam's razor (see also *maximum parsimony*), 433
Ohta, Tomoko, 10, 11, 123, 133
Onthophagus taurus (dung beetle), 330, 346
operational taxonomic units (OTUs), 426–427
orthologs/orthologous (see also *paralogs*), 44, 104, 158, 211, 219, **521**
bicoid orthologs, 228
estrogen receptors, 168
introns, 153
mammalian blood clotting factors, 219
Ods gene, 394 (see also *hox genes*)
outgroup(s), 125, 165, 169, 427, 441, 453, **521**
overdominance (see *dominance*)
Ovis canadensis (bighorn sheep), 475

P

Panaxia, 6
paralogs/paralogous (see also *orthologs*), 158, 160, 162, 163, 164, 211, 236, 250, **522**
sister, 160
yeast, 207
paraphyly/paraphyletic, 178, 179, 370–371, **522**

parent–offspring regression, 268, 475
parental care
  and the evolution of senescence, 478
  fitness estimates, 52
  genetics of, 299, 301, 308
  maternal effects, 20, 277, 281
Parkinsonia florida, 326, 334
parsimony (see *maximum parsimony*)
parthenogenesis
  production of polyploids, 401
  speciation, 417
paternity analyses, 509, 510
Pauling, Linus, 10, 121
Pax-6 gene, 233
peak shift(s) (see *adaptive landscape*)
pedigrees (see also *gene genealogies*), Ch. 12
Peromyscus, 263
pesticide resistance, evolution of, 278–279
phage(s) (see also *viruses*)
  genome sizes, 40
  mutation rates, 39
phenetics, 434
phenocopy(ies), 237–238
phenology, 327, 375–376
phenotypic plasticity, 21, Ch. 21, 239, 253, 330, 345, **515**, 517, 519, 521, **522**
phylogenetic congruence, 460–461
phylogenic estimation, Chs.11 and 28
  Bayesian Markov Chain Monte Carlo methods (BMCMC), 157, 186, 436
  beam-bootstrap algorithm (BB), 437
  bootstrap methods, 439
  Bremer support, 439
  cladistics, 370, 434
  combining data, 441
  confidence assessment, 438–440
  direct optimization (DO), 430
  disc covering method, 438
  distance methods, 120, 432–433
  Felsenstein 1981 model (F81), 431
  general time reversible model (GTR), 431
  genetic algorithms, 438
  Hasegawa, Kishino and Yano model (HKY85), 431
  hypothesis testing, 440–441
  jackknife methods, 439
  Jukes–Cantor model (JC69), 431, 432
  Kimura 2-parameter model (K2P), 431
  Kishino–Hasegawa test (KH), 440
  likelihood ratio test, 164, 431
  MacClade, 429
  MALIGN, 429
  maximum likelihood (ML), 159, 166, 169–170, 206, 431–442
  maximum parsimony, 157, 166, 433–435
  METAPIGA, 438
  metapopulation genetic algorithm (metaGA), 438
  model optimization, 431–432
  ModelTest, 432, 441
  MrBayes, 436
  multiple sequence alignment (MSA), 429–430
phylogenic estimation *(Continued)*
  nearest-neighbor interchanges (NNI), 437
  Needleman and Wunsch algorithm, 429
  neighbor-joining method/algorithm (NJ), 186, 429, 432, 433, 439
  non-parametric techniques, 157, 440
  optimality criteria, 433–436
  parametric techniques, 157, 440
  parsimony ratchet, 437
  phenetics, 434
  quartet puzzling, 437–438
  reconciled tree, 159
  searching tree space, 436–438
  sequence alignment, 119, 197, 428–430
    multiple, 429–430
    simultaneous, 430
  Shimodaira and Hasegawa test (SH), 440
  simulated annealing algorithm, 438
  subtree pruning and regrafting (SPR), 437
  Swofford–Olsen–Waddell–Hillis test (SOWH), 440
  Templeton test, 440, 441
  TreeAlign, 430
  tree bisection and reconnection (TBR), 437
  tree drifting (DFT), 437
  tree fusing (TF), 437
  unweighted pair-group method of arithmetic means (UPGMA), 432
  weighted Shimodaira test (WSH), 440
  winning sites test, 440
phylogenies, inferring (see *phylogenetic estimation*)
phylogeny (see also *phylogenetic estimation*), 426, 321, 367–368, 460, **522**
  gene family, 157–159, 161, 165, 166–168, 171
  gene trees, 158–159, 168, 427–428, 460
  population, 184
  species phylogenies/trees, 158–159, 179–180, 182, 187, 427–428, 371–372, 391, Ch. 28
  synthetic, 492–493, 500
  tree space, 436–437, 438
Physa, 327
Pieris occidentalis, 321
pigmentation
  dominance, 22
  Drosophila, 226, 228, 230, 389–391, 396
    abdominal-A (*abd-A*) gene, 226, 389–391
    abdominal-B (*abd-B*) gene, 389–391
    bric-a-brac (*bab*) gene, 226, 228, 389–391, 392
    doublesex (dsx) gene, 389
    yellow gene, 228, 230, 391
  melanism, 228, 231, 389
    butterfly wings, 321
    mice, 117, 231
  transposons, 258
plasmid(s) (see also *horizontal inheritance*), 28, 155, 182, 362, **522**
plasticity (see *phenotypic plasticity*)
pleiotropy (see also *genetic correlations*), 94, 205, 273, 301, 302–303, 307, 313–315, 318
  antagonistic, 203, 276, 464, 480, **513**
  distinguishing from linkage, 323
  estimation, 324

pleiotropy *(Continued)*
hsp90, 249
modularity, 233, 305, 307, 309
ploidy (see *aneuploidyl; polyploidy*)
Poecilia reticulata (guppies)
experimental evolution, 475, 499
senescence, 475
polymorphism (see also *heterozygosity*; *variation*), 104, 144, 276
allozyme, 9, 11, 12, 113–115, 131
balanced, 8
chromosome structure, 18
color, 231
host–parasite interactions, 447, 448, 450, 452, 458, 461
inversions, 471
length, 17, 291, 523
maintenance of, 77, 78, 105, 108, 276, 448, 458, 461, 495
nucleotide (see also *single nucleotide polymorphism*), 16, 24, 77, 79, 110, 130, 136, 291, 394, 474
phylogenetics (see also *incomplete lineage sorting*), 427, 428
regulatory regions, 116, 138
sequence length/repeats, 17, 108, 291, 293
tests (see *F-statistics*; *Hudson–Kreitman–Aguadé (HKA) test*)
polyphyly/polyphyletic, 178, 184, 371
protein domains, 209
polyploidy, **522**
allopolyploidy, 17, 28, 34, 148, 401, 402, 407–409, **513**
autopolyploidy, 17, 34, **513**
cotton, 147–151
homoploidy, 409, 150
maize, 147–151
via parthenogenesis, 401
plants vs animals, 17, 22, 34, 149, 160, 379
polyploidization, 15, 149, 151, 153, 154, 160, 169, 206
speciation, 263, 379, 407–410, 454
polytypic species, 368, 369, 373
population
contraction, 108, 109
differentiation, (see *differentiation*)
expansion, 108–109, 453
structure, 6, Ch. 6, 110, 184, 484, 490, 509, 515
subdivided (subdivision), 6, 50, 84, 91, 92, 108, 184, 369, 462
Populus
angustifolia, 410
fremontii, 410
posterior probability, 436, 439
postmating isolating mechanisms (barriers), 91, 380, 385
post-transcriptional
modification, 223
regulation, 227, 231, 244
Potamopyrgus, 459
Potorous spp. (potoroos), 503
predator–prey dynamics, 495–496
principal components analysis (PCA), 311, 322
prion(s), 197, 254, **522**
PSI^{+}, 242, 260
Procrustes superimposition, 294–295, **522**
progesterone receptor, 170,171
promoters, 26, 30, 131, 133, 162, 167–168, 194, 206, 232, 397, 516, **522**
protein(s)
active site(s), 197, 198–199, 204, 208, **513**, 517, 525
basic helix–loop–helix (bHLH) transcription factors, 198
binding domain, 167, 169, 170
binding energy, 204, **514**
binding interface, 197, 205, **514**
binding residues, 204–205
binding sites, 131, 138, 194, 197, 198, 208, 230, 232, 233
collagens, 214, 218–219
complexes, 200, 514
crystal structure, 25, 165, 170, 171, 197, 209, 212
domain architecture(s), 213, 214, 217, 219, 221
domain joining, 214, 219
domain shuffling, 161, 212
domains, 34, 161, 207, 208, 209, 212–213, 214, 217, 221, 520, **522**
evolutionary constrain(s), 194, 196–197, 199, 204, 205
evolutionary trace method, 197
extracellular, 214, 215, 216–221
folds or folding, 129, 137, 193, 196–197, 198, 200, 206–209, 211, 212, 214, 221, 240, **522**
function, Ch. 13, 211–214, 217, 218, 221
functional constraints, 122, 126, 135, 198
functional sites, 197, 198–200, 204, 513
heat shock (see also *Hsp90*), 240, 258
hot spot(s), 204
hydrogen bonds, 196, 197–198
integrins, 218
interaction network, 193, Box 13.1, 513, 514, 519, **522**
interaction surfaces, 200, 204–205
intracellular, 167, 214–216, 218, 221
multidomain, 30, Ch. 14, 514. 520, 522
paralogous, 213, 522
secondary structure, 25, **524**, 525
structural conservation, 209
structural constraints, 129, 197
structural domain, 212, 522
structure, 25, 194, 196, 198, 209, 210, 212, 224
three-dimensional structure, 129, 165, 193, 194, 195, 196, 197, 199, 207, 259, 522
TIM barrel(s), 208
transmembrane, 213, 216, 218, 220, 221, 231, 476
unstructured, 209–210
protein complexes, 200, 514
protein domains, 161, 207–209, 212–214, 217, 221, 520, **522**
proteome(s), 214, **522**
Arabidopsis, 217
C. elegans, 218
Protopiophila litigata (antler fly), senescence, 475
pseudogene(s), 152, 162–164, 212, **523**
rate of evolution, 44, 152–153
selection on, 36

Pseudomonas
experimental evolution, 495
host–pathogen interactions, 455, 460
pyruvate kinase, 205–206

Q

Q matrix, 120
QTL (see *quantitative trait loci*)
quantitative trait locus/loci (QTL), Ch. 19, **523**
Arabidopsis, 454
Beavis effect, 23, 297
body size, mouse, 293, 297–299,
candidate genes, 117, 228, 230, 291, 393, 477
differential dominance, 299, 308, 309
effect sizes, distribution, 23, 296–298
methods, 291–292, 293, 296
composite interval mapping, 388, **515**
limitations, 297–298, 307–308, 324, 388
LOD score, 292
multiple interval mapping, 388, **520**
pigmentation, *Drosophila*, 230, 390
senescence genes, 473–474
species differences, *Drosophila*, 388
teosinte, 95–98
quantitative traits, 22–23, 31, Ch. 18, 288, 325, 419, 498, 511, 514, 518, 519

R

radiation, adaptive, 373, 495
Hawaiian *Drosophila*, 414, 416, 471
radiation (irradiation) experiments, 8–9
receptors (see *hormones and receptors*)
recombinant inbred lines (RILs), **523**
Arabidopsis, 454
QTLs, 293
recombination (see also *linkage*), 27, 33
bottleneck effects, 416, 423
compensatory evolution, rate of, 137
effect on **G** matrix, 316, 317
effect on hitchhiking, 75–76, 77, 106–107
effect on intron length, 139
effect on mutation, 106–107
effect on nucleotide variability, 112–114
host–parasite coevolution, 450, 452
linkage mapping (see *quantitative trait loci*)
nonhomologous, 38
pseudogenes, 36
suppression, 18
balancer chromosomes, 45
cytogenetic cloning, 490
unequal crossing over, 16, 17, 33, 34, 160, 161, 207, 209
Red-Queen theory/hypothesis, 450, 451, 458–460
regulatory genetic pathways (see *gene regulation*)
regulatory region(s), 114, 118, 134, 138, 230, 259, 317
cis, 131, 223, 226, 229, 230–231, 232–233
in *Tb1*, 116–117
regulatory sequence(s), 17, 118, 131, 133–134, 243, 259, 516, 517, 524
reinforcement, 383–384, 400, **523**
reproductive character displacement (RCD), 383, **523**
reproductive isolating barriers (RIBs) (see also *hybridization*; *speciation*; *species, definition of*), Ch. 24, Ch. 25
resistance to parasites (see *host–parasite interactions*)
restriction fragment length polymorphism (RFLP), 291, **523**
retrotransposable elements, 148, 151, 152, 153
retrotransposition, 160, **523**
RNA
functional, 194–195
interference (RNAi), 229, 255, 388, 477, **523**
messenger (mRNA), 25, 26, 44, 133–134, 135, 136, 138, 142, 194, 226, 227, 229, 244, 260, 327, 514, 523, 525
micro-RNAs, 36
precursor messenger (pre-mRNA), 133, 134, 137, 138, 141
ribosomal, 163, 195
RNA–RNA long range interactions, 142
secondary structure, 26, 133–134, 136–138, 139, 141–142, 143, 195
small nuclear (snRNA), 195
small nucleolar, 195
transfer (tRNA), 16, 135, 136
Woese–Noller criterion, 136
RNA viruses, 229, 362
ROMs (highly reactive oxygen metabolites), 476
root node, 175

S

Saccharomyces cerevisiae
aging genetics, 476
codon bias, 16, 135, 136
developmental buffering, 250
extracellular domains (proteins), 217
genome duplication, 17, 149, 164, 207
genome size, 40, 145, 146
hybridization, 396
intracellular modeules (proteins), 215
mutation effect sizes, 40, 250
number of genes, 146
proteins, 202, 203, 213
PSI$^+$ protein (prion), 242, 260
reproductive isolation, 388, 394, 396
sir2 gene (silent information regulator), 476
subfunctionalization and neofunctionalization, 30
Sup35 gene, 242
translocations, 396
sampled lineages, 174–176, 178–179, 180, 183
saturation theory, 140
secondary structure, 429
DNA, 136
free energy minimization, 136
protein, 25, 196, 198, **524**, 525
RNA (mRNA), 26, 133–134, 136–138, 139, 141–142, 143, 195
Woese–Noller criterion, 136
segregation distortion (see *meiotic drive*)
selection
artificial (see *artificial selection*)
background, 106, 107, 112–113

selection *(Continued)*
balancing, 8, 55, 59, 78, 104, 106, 108, 109, 110, 114, 299, 308, 309, 341, 455, 460, 474, 505
canalizing, 236, 239
coefficient, 29, 67, 69, 70, 77, 79, 103, 123, 133, 135–136, 137, 138, 141, 143, 188, 282
contextual analysis of, 93–94, 99–100, **515**
correlated response, 282–287, 304, 305, 307, 317
correlational, 271–272, 312, 314, 316, 317, 321, 323
density-dependent, 495
differential, 23, 268, 273, 277, 282, 342, 514, **524**
directional, 50, 58–59, 64, 67, 78, 92, 95, 104, 106, 108, 109, 111, 114, 271, 281, 282, 297, 299, 304–305, 308, 316, 317, 336, 345, 414, 455, 459, 483, 492, 498, 499
disruptive, 58, 59, 271, 276
frequency-dependent, 448, 450, 458, 459, 460
gametic, 49, 51, 57
gradient, 271–272, 273, 277, 281–282, 304, 343, 403
group, 50, 51 60, 62, 92–95, 99
hard, 93
indirect, 59, 347
intensity (*i*), 268, 273, 282, **524**
inter-demic, 49, 51
kin, 62, 94, 351, 352, 354, 358, 360, 362, 363, 479
laboratory (natural) selection, 484, 490, 491, 492, 499, **519**
multilevel, 92–94, 99–100, 333
multivariate, 269–272, 287, 344
negative, 29, 55, 104, 109, 112, 113
positive, 29, 40–41, 47, 104, 106–107, 109, 110, 111, 115, 126, 130, 131, 147, 162, 164, 207, 211, 212, 327, 385, 394, 396, 460
purifying, 30, 104, 107, 111, 121, 130, 138, 162–163, 164, 171, 195, **523**
regression analysis of, 99–100, 271–272,
sexual (see *sexual selection*)
soft, 93
stabilizing, 41, 43, 58, 59, 135, 163, 169, 229–230, 236, 237, 239, 246–248, 251, 297, 321, 368
strength of, 56–57, 117, 141, 164, 245, 269, 466, 470
strong, 6, 25, 66–67, 115, 133, 137, 139, 194, 196, 202, 239, 245–246, 272, 273, 339, 448, 450, 474
weak, 111, Ch. 9, 230
selectionist(s), (see also *neutralist*), 3, 10–12, 27, 79, 129, 195, 197, 210
selective sweep(s), 75–77, 106, 107, 108, 109, 111, 112, 117, 136, **524**
parasite resistance, 450, 451, 459–460, 461
self-incompatability, 453, 455, 507–508, 510, 511
self-sustaining feedback loops, 254
selfish DNA, 147, 193,
selfishness (selfish behavior), 351, 354, 355, 356, 357
seminal fluid (see *accessory gland proteins*)
senescence, Ch.30, **524**
antagonistic pleiotropy (AP), 203, 464, 466, 470, 472–473, 475–476, 478 480, 494, **513**
artificial selection, 467, 470, 472–473, 477
and behavior, 478
deleterious alleles/mutations, 32, 464, 466, 470, 473, 475
senescence *(Continued)*
Drosophila, 467, 469, 470, 472, 474, 476, 477, 479
eusocial insects, 479
gene expression studies, 477–478
and immunity, 479
insulin signaling pathways, 476–477
menopause, 478
mutation accumulation (MA) theory, 47, 464, 466, 470, 473, 475–476, 478
in nature (natural populations), 468, 474, 475, 478, 479
number of genes, 474
population comparisons, 474, 475
quantitative genetics, 470, 472, 477
QTL mapping, 473–474
stress resistance, 473, 476–477, 480
Sepsis (yellow dung fly), 496
sequence alignment, 119, 198, 428–430
direct optimization (DO) (optimization alignment), 430
gaps, 119, 120, 131, 428, 430
MALIGN, 429
multiple, 197, 198, 429–430
Se-Al, 429
sequence hypothesis, 193
sex, consequences of, 42, 450, 479
sex chromosomes
decay, 27
epistasis, 300–301
evolutionary rates, 382
reproductive isolation (see also *Haldane's rule*), 378, 381–383, 385
sex linked genes, 347, 348, 397
sex determination, 20, 226, 262, 300, 347, 389
sex hormones (see *hormones and receptors*)
sex-limited gene expression, 21, 301, 347
sex ratio
distortion, 358, 362, 382
effect on fitness, 60, 63
effective population size (N_e), 507
evolution, 351, 353, 357, 358
maternal effects, 21
sex-specific gene expression, 301, 348
sexual conflict, 18, 22, 339–340, 341–342, 348, 496–497
rate of evolution, 382, 385
sexual dimorphism
Etheostoma spp., 385
gene expression, 301, 390
genetic correlation between the sexes, 301
methylation pattern, 18
sex-specific gene expression, 301, 348
sexual selection, Ch. 22
cryptic mate choice, 341, 346
fitness, 54, 342
genetic architecture, 315, 344
good-sperm hypothesis, 347
intersexual coevolution, 496–497
mate choice, 340
male–male competition, 341
opportunity for selection, 60, 63
parasites, 479
partitioning selection, 60, 63, 342
quantitative genetics, 343

sexual selection *(Continued)*
reinforcement, 383–384, 388, 400, **523**
sex linkage, 347
sexual conflict, 341
sexy-sperm hypothesis, 347
speciation, 383, 384–385
sexy-sperm hypothesis (see *sexual selection*)
shifting balance (see *Wright's shifting balance*)
sickle cell, 25
signal transduction pathways, 171, 476
Silene, 379, 380
Simpson, George Gaylord (G.G.), 7
single nucleotide polymorphisms (SNPs), 16, 26, 291, **524**
slippage (DNA slippage), 16, 17, 151, 160, 209, **515**
social selection, Ch. 23
Society for the Study of Evolution, 7
speciation
allopatric, 373, 374, 378, 380, 383, 384, 385, 407, 503, **513**
clock, 375, 377–378
Dobzhansky–Muller model, 30, 376, 380, 381, 390, 392
effects of population bottlenecks, Ch. 27
epigenetic inheritance, 262–263
founder induced, 414, 416–417, 422, 424
founder-flush, 416–419, 420–422, 423
genes, 186–187, 385, 388, 394, 396
hybrid, 400, 401, 404, 407, 409–410, 412, **518**
parapatric, 373
recombinational, 410
sympatric, 383, 384, 503, 524
speciation clock, 375, 377–378
species, definition of (see *species concepts*)
species concepts, 367
biological (BSC), 7, 91, 369, 374, 503, **514**
cladistic, 367, 370
cohesion concept, 369
consequences for conservation, 510
diagnostic, 368, 370–371
ecological, 368
evolutionary, 368, 370
Farhenholz rule, 460
genealogical, 368, 370
general lineage, 368, 372
morphological, 368
phenetic, 368, 369
phylogenetic, 368, 370–371
polytypic, 368, 369, 371, 373
recognition, 368, 369
species differences, genetics of, Ch. 25
species identification in conservation biology, 503
species rank, 370, 371, 373
species survival plans, 508
species tree(s), 158, 159, 427–428
sperm competition, 341, 346–348, **515**
Allonemobius, 377
C. elegans, 346
Drosophila, 347, 377, 496
good-sperm hypothesis, 347
reproductive isolation, 375–377
sexy-sperm hypothesis, 347
Tribolium, 376
sperm displacement (see *sperm competition*)
sperm precedence (see *sperm competition*)
sperm proteins, 385
spermatogenesis
acceleration of hybrid male sterility, 382
genes on the Y chromosome, 348
male sterility, 396
splicing, 133, 138, 139
Adh, 141–142
Stator limbatus, 20, 326, 329, 330, 331, 334, 335, 336
Stebbins, G. Ledyard, 6, 7, 11, 402
stepwise mutation model, 104
Stichophthalma camadeva, 320
Streptanthus, 379, 380
structural inheritance, 254, 255
structurally constrained neutrality (SCN) model, 129
structured populations, Ch. 6, 338, 357, 515
subfunctionalization, 30, 163, 165, 169
subgenealogy, 176
substitutions (see *mutations*)
superoxide dismutase (SOD), 128
survivorship, age-specific [$l(x)$ or l_x], 465, 467
synergistic symbiosis, 360–361
synonymous sites, 12, 38, 40, 111, 133, 134, 135, 136, 166
synteny, **525**
conserved, 207
systems biology, 193, 195

T

Taenaris macropus, 320
Tajima's D statistic, 108–109
tamarin, golden lion, captive management, 508, 509
tandem duplication (tandem repeats) (see also *Hox genes*; *microsatellite DNA*; *minisatellite DNA*), 17, 34, 160, 161, 163, 208
TATA box, 133
taxonomic rank (see *species rank*)
teosinte (see also *maize*), 95–98, 115–117
teosinte branched1 (*Tb1*), 115–117
Tokogenetic relationships, 370, 372
trade-offs (see also *antagonistic pleiotropy*), 497
across environments, 490, 494
early vs late fertility, 498
fitness, 54
genetic, 276
lability vs stability, 494
maintenance of diversity, 495
reproduction vs survival, 470, 472–473, 477, 480, 489
resistance to parasites, 279, 455, 487
selfing vs outcrossing, 62
virulence vs transmission in parasites, 452, 456, 457
tragedy of the commons, 353, 356, 362
transcription factors, 26–27, 163, 167, 171, 198, 224, 226, 227, 229, 231, 232–233, 242, 243, 245–246, 247, 315, 389, 393, 476, 477, **525**
binding sites, 131, 194, 230, 232–233
expression, 234
maternally derived, 20
networks, 200–201, 247, 517

GOSHEN COLLEGE - GOOD LIBRARY
3 9310 01049343 3

QH
390
.E94
2006

transcriptional regulation, 231, 232–233, 234
  post-, 227, 244
transduction, 28, 182, **525**
transformation, 28, 140, 155, 182, 452, **525**
  reciprocal, 229
  stable, 229
  transient, 229
transgenesis, 223, 228
translocations, 18, **525**
  Helianthus, 410
  yeast, 396
transposable element(s) (TEs), 16, 34, **525**
  Alu, 151
  Axin gene, 257
  C. elegans, 139
  distribution, 148
  domain shuffling, 161
  Drosophila, 151, 472
    Accord, 115
    Doc, 115
    P element, 28, 393, **521**
  euchromatin, 151
  genome size, 147, 153
  indel variation, 16–17, 34
  junk DNA, 147
  maize, 147–150
  mammals, 257
    humans, 151, 152
  methylation, 257–258, 259
  mutation rates, 18
  retrotransposons, 151, 152, 153
  silencing, 18
  transformation, 229
transposition (see also *transposable elements*), 34, 160
  retrotransposition, 42, 147, 160
transposon (see *transposable element*)
Tribolium, 421
  artificial selection, 274
  castaneum, 92, 274, 376
  confusum, 92, 376
  founder-flush speciation, 418
  freemani, 376group selection, 92
  population size, 93
  reproductive isolation, 376
  sperm competition, 377
trichomes, *Drosophila*, 227, 230–231, 232, 390
two-population divergence model, 184

U
urochordate(s), 219
UTR (untranslated regions), 16, 26, 134, 136, 137, 142, 160, 229

V
variance (variation)
  additive (genetic), V_A, 22–24, 85, 86, 87–89, 90, 95–97, 268, 272, 273, 275–276, 277–279, 287, 290, 297, 302, 308, 310, 316, Ch. 27, 470, 518
variance *(Continued)*
  among-deme, 90
  conversion, 86, 87–89, 97, 272, 275, 276, 419, 420
  dominance, 22, 86, 268, 274–276, 275, 290,
  epistatic, V_I, 24, 88–89, 268, 275, 276–277, 290, 301, 419
  nonadditive, 22, 24, 93, 94, 97, 268, 272, 274–277, 417, 419, 420, 422, **521**
  partitioning, 50, 60, 62, 82, 84–85, 89, 95, 96–97, 267–268, 275, 330
  within-deme, 80, 82, 83, 84, 95
virulence (see also *host–parasite interactions*), Ch. 29
  altruism/cooperation, 350, 351, 356
  cost of, 452, 480
  E. coli, 486
  evolution of, 456–457
  gene-for-gene coevolution, 450–452
  host–parasite coevolution, 449–452, 457–458
  Linum marginale and *Melampsora lini*, 462
  matching-alleles models, 449–450
  multiple parasites, 461
  and senescence, 480
  trade-off hypothesis, 452, 456
viruses, 38, 130, 145, 193, 206, 207, 229, 362, 447, 482, 485, 487, 523

W
Waddington, Conrad, 235–240, 242, 246, 250, 251, 252–253, 260, 331
wallaby, 263
Watson, James, 9, 193
whole genome duplications (see *polyploidy*)
whole genome sequence(s), (see *genome sequencing*)
Williams, George, 239, 464, 480
Woese–Noller criterion, 136
wombat, hairy-nosed, conservation genetics, 509, 510
Wright, Sewall, 3, 5–8, 10, 22, 72, 81, 84, 94–95, 238, 276, 301, 302, 331
Wright–Fisher model, 180
Wright's shifting balance, 5, 6, 7, 94–95
Wyeomyia smithii, 420

X
Xdh (xanthine dehydrogenase) locus/protein, 126, 128
  amino acid replacements, 129
Xenopus laevis, 169
Xiphophorus hybrids (swordfish), 397

Y
yeast (see *Saccharomyces*)
Yule, G. Udny, 5
Yule distribution, 177, 180, 183, 184, 186

Z
zebrafish, 131
Zuckerkandl, Emile, 10, 121